HD
9519
.S68
L49
1994

Lewis, W. David
(Walter David),
1931-

Sloss Furnaces and
the rise of the
Birmingham

$39.95

DATE			

BAKER & TAYLOR

SLOSS FURNACES AND THE RISE
OF THE BIRMINGHAM DISTRICT

History of American Science and Technology Series

General Editor, LESTER D. STEPHENS

SLOSS FURNACES AND THE RISE OF THE BIRMINGHAM DISTRICT
An Industrial Epic

W. DAVID LEWIS

The University of Alabama Press

Tuscaloosa and London

The paper on which this book is printed meets the minimum
requirements of American National Standard for Information
Science-Permanence of Paper for Printed Library Materials,
ANSI Z39.48-1984.

Library of Congress Cataloging-in-Publication Data

Lewis, W. David (Walter David), 1931–
    Sloss Furnaces and the rise of the Birmingham district : an
industrial epic / W. David Lewis.
        p.   cm.—(History of American science and technology series)
    Includes bibliographical references and index.
    ISBN 0-8173-0708-7 (alk. paper)
    1. Sloss Furnace Company—History.   2. Iron industry and trade
—Alabama—Birmingham—History.   3. Iron foundries—Alabama
—Birmingham—History.   4. Iron-founding—Alabama—Birmingham
—History.   5. Iron—Metallurgy.   6. Birmingham (Ala.)—Industries
—History.   7. Birmingham (Ala.)—Economic conditions.   8. Birmingham
(Ala.)—Social conditions.   9. Sloss Furnaces National Historic
Landmark—History.   I. Title.   II. Series.
HD9519.S68L49   1994
338.7'6722'09761781—dc20   93-48178

British Library Cataloguing-in-Publication Data available

for Mel Kranzberg,

and in loving memory of Randy Lawrence

Within the last half century the South has gone through a series of political, social, and industrial upheavals and readjustments; and yet the South of today is the historical product of the South of old, with much the same conditions and problems.

ULRICH BONNELL PHILLIPS
South Atlantic Quarterly, January 1904

Contents

Preface

From the outset, this book has had a mind of its own. The process of gestation through which it has gone, spanning more than eight years since its inception, has yielded results far different from what I had expected when I began working on it. Along the way, an undertaking that I had initially viewed as a temporary diversion from other interests became an adventure that I pursued with mounting excitement as unanticipated discoveries continued to unfold. Each new piece of information that I uncovered merely increased my zest to find more; something about a massive array of dormant furnace stacks, skip hoists, blowing engines, hot blast stoves, and casting sheds that had once roared and pulsated with life gripped my imagination with an intensity that I had never felt in previous research projects. In time, what started as a short work about a single industrial site became an extended history of the city in which it is located and an effort to reach a better understanding of southern industrialization in general. Ending an unrelenting quest for knowledge that has meant more to me than words can express leaves me hoping that I will find my future expeditions into the past equally rewarding, because in tracing the origins and development of Sloss Furnaces and the Birmingham District I have felt a sense of fulfillment and joy in being a historian that I had earlier encountered only in the classroom as a teacher and storyteller who revels in celebrating the human experience.

I am first and foremost a historian of technology. As far as I am aware, this book is the first major reinterpretation of southern industrialization to be published by a scholar in my discipline. But I have not conceived the volume with only a professional audience in mind and have done my best to make it accessible to general readers. At many points the story that I tell reflects my zest for observing how things work and explaining the operation of complex devices to students who take my courses in Technology and Civilization at Auburn University. Because I had to be selective in choosing illustrations and could not rely on the large number of visual aids that I use in my classes, I have tried to verbalize about industrial processes and mechanical inventions as clearly as possible. I have also made a special effort to supply basic information about ferrous metals and other scientific and technical subjects in such a way that readers who have little or no familiarity with this type of material will be able to follow the narrative without difficulty. Only by understanding the way in which intractable problems presented by regional raw materials led to technological choices that were governed to a large extent by distinctively southern economic needs and cultural traditions can one begin to appreciate the forces that shaped the Birmingham District's dramatic rise and ultimately frustrated its lofty ambitions.

Subtle differences in the metallurgical composition of three industrial commodities—pig iron, cast iron, and steel—had an enormous impact on Birmingham's tortuous development. All three of these products are alloys composed mostly of iron, but the uses for which they were suited greatly affected the markets that they could command and the profits that could be earned in making them. Because the Birmingham District is geologically unique in possessing all of the raw materials needed for iron manufacture in one place, its prospects for wealth appeared to be boundless. But nature is fickle, and attempts to capitalize on the economic potential of Jones Valley, where Birmingham is located, yielded problematical results. For three decades the district could make only pig iron, a semi-finished product that resulted in limited financial rewards. Searching for the riches that seemed to be the area's birthright, promoters and boosters pinned their hopes on steel, an eagerly coveted commodity for which the mineral resources and social institutions of north-central Alabama were unfortunately ill-fitted. Instead, Jones Valley's raw materials and cultural traditions were more appropriate for the fabrication of goods made from steel's relatively humble but nonetheless very useful metallurgical cousin, cast iron. The history of Sloss Furnaces is a case study in the implications of this fortuitous circumstance.

In 1899, with much difficulty, the art of steelmaking with local coal and ore was finally mastered by Birmingham's biggest enterprise, the Tennessee Coal, Iron, and Railroad Company (TCI). This achievement led boosters to predict that Jones Valley would soon supplant the Pittsburgh District as the ruling center of the steel age. Three years later, the same boosters were dismayed when the owners of Sloss Furnaces decided not to emulate TCI and chose instead to concentrate on supplying high-quality pig iron to foundries for conversion into cast iron that would be molded in turn into a great variety of products including soil and pressure pipe, engine components, stove plates, steam radiators, and ceramic-coated kitchen and bathroom fixtures. Because the phosphorus content of local ores enhanced moldability, firms that made these and other goods became heavily concentrated in the South. In time, Sloss Furnaces ranked among the world's largest manufacturers of foundry pig iron, catering to enterprises whose owners preferred a sand-cast type of metal that could be produced in relatively small batches by low-paid black labor.

Meanwhile, in 1907, TCI was acquired by the recently formed United States Steel Corporation in one of the most famous mergers in American history. Birmingham's boosters exulted, thinking that this development would assure the city's future industrial preeminence. Instead, the new owners, who were deeply committed to stability, used discriminatory pricing formulas to prevent TCI from competing vigorously with steel mills in other parts of the country. Unlike the production of foundry iron, steelmaking also required skills that were scarce in a region whose segregated black workers had no economic or social motivation to strive for efficiency and productivity. By the time the Great Depression hit Birmingham with the force of a tidal wave, it was obvious that the city's faith in steel as a magic key that would unlock future riches had been badly misplaced. After World War II, Birmingham stagnated while its near neighbor, Atlanta, burgeoned.

Part of what makes the history of technology such an exciting discipline to pursue is the satisfaction that comes from discovering connections between seemingly unrelated developments. The disillusionment resulting from the collapse of Birmingham's hopes of becoming the world's greatest steelmaking center played no small role in generating the violent response to social change that took place when Bull Connor ordered fire hoses and police dogs to be turned on black children marching for racial justice in 1963. Realizing the existence of this linkage was one of the turning points in the writing of this book, changing it from a narrowly focused study of iron and steel manufacturing into something that was much broader and promised far more interesting results. What made the process increasingly compelling was my growing sense that I had only begun to fathom the connections between Birmingham's virulent resistance to racial integration and strategies of industrialization that had been present since the time of the city's founding.

A growing mass of evidence that I encountered as I pursued these connections forced me to change the way I had looked at post–Civil War southern industrial development at the time I began studying the history of Sloss Furnaces. Although I had already written extensively about the evolution of the American iron and steel industry, I had approached that subject only from a northern perspective. From that vantage point, I believed that the rise of such ironmaking centers as Birmingham had resulted from efforts to transplant a northern model of economic growth to the South in a process that was based mainly on outside capital, business leadership, and technical expertise. Manufacturing and urbanization, I thought, were contrary to the spirit of traditional southern institutions. One of the chief road maps that I used as I began my journey into unfamiliar territory was Jonathan Wiener's *Social Origins of the New South* (1978), which led me to believe that downstate Alabama planters were hostile to Birmingham and regarded industrialization as an affront to agrarian values.

Early drafts of my book reflected these assumptions, and some traces of their influence still remain. I continue to believe that the South could not have achieved the industrial growth that took place within the region in the late nineteenth century without a great deal of outside capital, credit, and technical assistance. But I am impressed by the extent to which southerners drew upon their own human and financial resources and the degree to which, at least before the watershed of the 1890s, they industrialized in their own way, on their own terms. I have learned to admire southern entrepreneurship and to appreciate the skill with which investors and managers who were native to the region compensated for a lack of capital by choosing congenial north-ern and foreign associates, blending labor-intensive methods with selective mechanization, and securing outside technical help as they happened to need it. I will argue that the South pursued its own indigenous approach to indus-trial growth, implementing strategies already in existence before 1861. Al-though Birmingham did not exist until the 1870s, it was the brainchild of wealthy slaveholding planters who promoted internal improvements, urban planning, and manufacturing in the antebellum period. These agribusiness-men and their descendants were the driving force behind the development of Birmingham as Radical Reconstruction came to an end.

Birmingham was rooted in human bondage and would have been a center of industrial slavery had the South won the Civil War. The growth of the city was based on a labor-intensive approach to manufacturing that had been widely practiced in the South before 1861. The coal mines and blast furnaces that sprang up in Jones Valley after the war used ill-paid, severely disadvan-taged black workers. Many of these persons were convicts, perpetuating tradi-tions of involuntary servitude inherited from antebellum times. When necessary, mine and furnace owners used machinery that was as up-to-date as most northern equipment and imported outside technical help to keep it in good running order. But mechanization was used only selectively, and day-to-day supervision of operations was kept largely in southern hands.

The southern strategy of industrial development that underlay the rise of Birmingham demanded a social order in which white supremacy was rigidly enforced. Upward mobility on the part of African-Americans was kept within strict bounds. Congregating large numbers of black males in one place to mine coal, smelt iron, and perform many other tasks resulted in an unyield-ing determination to maintain stringent forms of racial discrimination that were based partly on fear for the safety of white women in a fiercely exploita-tive environment. Preserving antebellum custom, the owners, managers, and their families occupied large residences in choice enclaves while African-Americans lived in squalid areas resembling earlier slave quarters. Birming-ham was an iron plantation in an urban setting.

It is paradoxical that Birmingham and enterprises like Sloss Furnaces came to symbolize the New South. Readers may find it strange that this well-known term, made famous by Henry W. Grady and repeated like a mantra by his-torians since he used it in a speech to northern investors in 1886, seldom appears in this volume, dealing as it does with the origins and growth of a city that supposedly epitomized his artful metaphor. But the omission is deliber-ate. I talked about the New South frequently in my early drafts and even included the expression for a time in the title of this book. But I finally decided that it would be better to jettison words that, in my opinion, do more to obscure what happened in the post–Civil War period than to illuminate it. To some degree, it would not have been inappropriate for me to use Grady's familiar phrase, because the new order that emerged in places like Jones Valley was indeed different from the way things were in antebellum times when small landholders and other plain folk dominated Alabama's state government and prevented wealthy planters from achieving much of their prewar agenda for economic growth. But I do not believe that this divergence is what most people think about when they hear a time-worn turn of speech that Grady used mainly as a rhetorical device to convince Yankee businessmen that a slaveholding region had somehow become a "perfect democracy."

Throughout this book there is much emphasis on continuity, as opposed to change. This too is deliberate. But change does occur, and today's Birmingham is much the better for it. As I write the last of millions of words that have appeared in various drafts of my manuscript, I think of sitting alone at night in a room on the eleventh floor of The University of Alabama at Birmingham's Denman Hall in the summer of 1986, not long after I had begun studying the history of Sloss Furnaces. Looking out my window and watching the WZZK sign flashing on and off atop the Bank for Savings Building, I realized how much this transplanted northerner had already come to care for the place that spread up and down Jones Valley and how earnestly he wanted to do justice to its long and storied past. My feelings are still the same, only deeper. Just as the acknowledgments that follow end with a tribute to Alabama's pioneer industrial historian, Ethel Armes, this preface closes with a salute to Birmingham, which, at least in my estimation, truly deserves to be known as the Magic City.

Acknowledgments

When Randall G. Lawrence commissioned this study in 1985, I had no idea what an enormous task I was undertaking when I began work on what was originally conceived as a modest guidebook. As director of Sloss Furnaces National Historic Landmark, a large outdoor industrial museum in Birmingham, Alabama, Lawrence wanted to interpret the site to as broad a public as possible, ranging from casual visitors to specialists in scholarly disciplines. He pointed out that many questions remained unanswered about the evolution of the dormant ironmaking facility, which had been shut down in 1970 amid economic and technological changes that were already having an adverse impact on Rust Belt communities across America. Neither he nor I, however, knew about the wealth of information that was waiting to be discovered about Sloss Furnaces in archives, libraries, and other research institutions. It is a tribute to Randy's greatness as a museum director, and the depth of his commitment to our mutual quest for knowledge about the site and its heritage, that he steadfastly supported an increasingly time-consuming project whose final outcome he never lived to see. Without his patience and faith, this book could not have been written. Dedicating it to his memory is the best way I know to acknowledge how much I owe to him and how deeply I cherished his friendship.

Robert Casey, who was curator at Sloss Furnaces National Historic Landmark when I began my work, shared with me his unusual combination of talents as a metallurgical engineer and historian of technology and placed at my disposal valuable source materials that he had collected in efforts to fill gaps in what was then known about the development of the site. Casey's strong sense of the property's significance had much to do with the steadily growing scope of my research, which he continued to encourage after he left Alabama to take other curatorial posts. Among American cities, he believed, only Birmingham had an industrial installation that so fully epitomized its history.

Casey also emphasized the regional and national significance of the furnaces, pointing out that their early development and eventual closing paralleled the rise of the United States as the world's foremost industrial power and the shift from producing goods to providing services that is now taking place all across the country.

Soon after I began my project, the Birmingham Public Library's Department of Archives and Manuscripts acquired a substantial collection of business records kept by firms that had manufactured pig iron on the furnace site. The most notable of these entities was the Sloss-Sheffield Steel and Iron Company, which had once been the second-largest industrial enterprise in the Deep South. (Despite its name, the firm never made steel, creating a paradox that would become a major theme as my research progressed.) Marvin Y. Whiting, Archivist and Curator of Manuscripts at the Library, secured the records from the Jim Walter Corporation, which had gained possession of them when it absorbed the United States Pipe and Foundry Company and its Sloss-Sheffield division in 1969. Whiting and his staff extended me many courtesies as I spent two summers studying the collection. Whiting also read a draft of my book, which benefited greatly from his rich store of knowledge about every aspect of Birmingham's history.

The frequency with which I saw the names of Virginians in the stock ledgers that I examined led me to visit Richmond, where I found an excellent source of previously untapped information about Sloss Furnaces at the Virginia Historical Society. Among the Society's most important holdings is a series of letterbooks kept by Joseph Bryan, a wealthy industrialist who invested heavily in Sloss and was for many years one of Birmingham's most devoted champions. Sara B. Bearss, Nelson D. Lankford, Frances Pollard, Joseph Robertson, Janet Schwarz, and other staff members were helpful to me as I made repeated trips to Richmond to study Bryan's correspondence. E. Lee Shepard, senior archivist, permitted me to quote numerous passages from Bryan's letters in this book, and Annmarie Price supplied photographs. I also appreciate aid given me by Brent Tarter and other officials at the Virginia State Library and Archives and am grateful to Conley L. Edwards for authorizing me to quote from letters in a collection of Bryan Family Papers at that institution. Gregg D. Kimball and Teresa Roane of Richmond's Valentine Museum also helped me locate photographs of Bryan and other Virginians whose names appear frequently in my narrative.

Hundreds of letters written by Bryan, an aristocrat who had been raised on a tobacco plantation, forced me to reevaluate my early assumption that southern industrialization had been an imitative process based on northern models of economic development and antithetical in spirit to the traditions of the antebellum planter class. Even before my first trip to Richmond, however, I had begun to change my views as a result of discussions with Robert J. Norrell, who had gathered impressive evidence that wealthy Alabama planters were heavily involved in town planning, railroad promotion, and industrial activities. As I became increasingly convinced that the rise of Birmingham was rooted in plantation agribusiness, Norrell shared advance copies of forthcoming publications with me and invited me to present a paper in a session that he helped organize for a meeting of the Southern Historical Association at Fort Worth, Texas. Taking part in the session gave me an opportunity to receive valuable commentary from David L. Carlton and other scholars. An eminent historian, Bertram Wyatt-Brown, also read a draft of my book as a result of hearing my presentation. I appreciated his advice and encouragement, which came at a time when I was still struggling to assimilate insights I had gained from the Bryan letterbooks.

As my work progressed, I submitted a preliminary manuscript of my book to The University of Alabama Press. I am grateful to the staff at the Press and to its director, Malcolm M. MacDonald, for many things but especially for MacDonald's willingness to grant me time for more research and a major revision of my manuscript even though referees reviewed it enthusiastically and the Press had accepted it for publication essentially as it stood. I also appreciate the professionalism and courtesy shown by Suzette Griffith, Project Editor, after I completed a final draft that satisfied me much better in the way in which it attempted to demonstrate that both Birmingham and Sloss Furnaces were rooted in prewar southern agribusiness and industrial slavery. I owe special thanks to Jane Powers Weldon, who displayed rare skills in copy-editing an extremely long manuscript that went through many last-minute changes.

Numerous persons read and criticized drafts of my work. I am particularly indebted to Edwin T. Layton, who demonstrated the depth of his friendship for me by commenting at great length on a preliminary version of this book and writing a detailed report recommending its publication by The University of Alabama Press. Layton also invited me to give a colloquium presentation at the University of Minnesota, giving me an opportunity to discuss my findings with graduate students in that institution's program in the history of science and technology. Layton's expertise and advice benefited nearly every page of the massive text that I sent him. He also suggested the subtitle of this book by stating that I had produced an "industrial epic."

Robert J. Norrell read my manuscript when I first submitted it to The University of Alabama Press and offered constructive criticisms that improved its final form. Gary B. Kulik also read the same version of the manuscript and made many useful comments. I was particularly glad to have Kulik's advice because he had written a research report for the Historic American Engineering Record in 1976 that, up to now, has been the best secondary source on the origins and development of Sloss Furnaces. I admire that pioneering document and a thoughtful essay that Kulik published on connections between black labor and technological change in the Birmingham District. André B. Millard and Merritt Roe Smith also made valuable suggestions about parts of my book and helped me in other much-appreciated ways.

Leah Rawls Atkins read repeated drafts of my work and gave me the benefit of her great familiarity with the history of Birmingham, about which she has written books and articles that have been very helpful to me. Together with Marvin Y. Whiting, Atkins helped me reexamine difficult factual and interpretive questions involving the determination of Birmingham's location by rival groups of railroad promoters during the Reconstruction Era.

The chemical peculiarities of Alabama's raw materials, and the problems that they posed for generations of industrialists and engineers, are repeatedly analyzed in the pages that follow. As I indicate, many of the insights that I present are derived from the work of Jack R. Bergstresser, Sr., who wrote a doctoral dissertation under my direction that searchingly examines the relationship between raw material constraints and the development of coal mining and ironmaking in the Birmingham District. It has been a real privilege to enjoy the interchange of ideas that has taken place in my discussions with Bergstresser, who possesses a rare talent for industrial archaeology.

Maury Klein made important contributions to my understanding of southern railroad development and sagely advised me not to forget that I was writing about human beings. Klein's masterful study, *The Great Richmond Terminal* (1970), was conceptually valuable and served as an indispensable guide to the endless maneuvering that took place between Joseph Bryan and other southern railroad leaders. Without such aid, I could not have made sense of many references to people and situations in Bryan's letters that would otherwise have been unintelligible.

No one probing the history of Jones Valley can fail to be impressed by its remarkable geological features. I derived much knowledge about Alabama's physical evolution from interviews with Thomas J. Carrington of the Auburn University Geology Department, Charles Copeland of the Alabama Geological Survey, and Lewis S. Dean of the Alabama Petroleum Board. Dean, a historian in his own right, read my first chapter and made many helpful suggestions stemming not only from his scientific expertise but from his published research on the life and career of Michael Tuomey, Alabama's first state geologist.

Clarence E. Mason, who was the last superintendent of Sloss Furnaces, reviewed parts of my manuscript and granted me a series of valuable interviews. I benefited greatly from his intimate knowledge of ferrous metals technology and his recollections of events that occurred during and after World War II. Most of all, Mason taught me that southern blast furnace practice was a world unto itself and helped me avoid making false comparisons with northern conditions that were irrelevant to Alabama ironmaking.

Edward R. Uehling made a major contribution to this book by permitting me to use an unpublished autobiography and several letters written by his grandfather, Edward A. Uehling, an outstanding engineer and inventor who was furnace superintendent of the Sloss Iron and Steel Company at a critical stage in its development. But for these materials I would not have known about several developments that were of fundamental importance to the narrative that follows. Jane G. Hartye of the S. C. Williams Library at Stevens Institute of Technology provided supplementary materials that helped clarify events in which Uehling and his mentor, Henry Morton, figured significantly. Hartye also supplied a photograph of Uehling that is included in this book. Dudley Dovel Shearburn, whose grandfather, James Pickering Dovel, brought southern merchant blast furnace design to a climax in the early twentieth century and was responsible for most features of Sloss Furnaces as the installation exists today, granted me a helpful interview and supplied a photograph of Dovel that also appears in this volume.

It is a pleasure to express thanks for funding from various sources without which this book could not have been written. The City of Birmingham and the Sloss Furnace Association provided a series of grants that supported my early research. Later, Paul F. Parks, then Vice President for Research at Auburn University, where he is now Provost and Vice President for Academic Affairs, helped finance many of my trips to Richmond. The Virginia Historical Society awarded me a series of Mellon Fellowships, and the Hagley Museum and Library gave me a grant-in-aid that made possible two weeks of research on connections between railroads and the development of Sloss Furnaces. The National Endowment for the Humanities supported my work with a Summer Stipend, No. FT-33831, in 1990. The Auburn University Humanities Development Fund provided funding for travel and supplied a subvention that aided publication of my manuscript.

Portions of this book are based on papers that I gave at professional meetings and publications that resulted from these presentations. I appreciate the help of persons who were responsible for these opportunities. Bernard Carlson invited me to give a presentation on the significance of the foundry trade at a conference honoring the 150th anniversary of the College of Engineering at The University of Alabama, and Howard L. Hartman edited my paper for publication in a book of proceedings. A paper that I gave at a meeting of the Alabama Historical Association in Auburn, Alabama, dealing with Joseph Bryan's role in the industrial development of north-central Alabama, resulted in an article published in the *Virginia Magazine of History and Biography*. I appreciate the help given me by the editor of that journal, Nelson D. Lankford. Alexandre Herlea invited me to speak about the contrast between Sloss-Sheffield and its great rival, TCI, at a meeting of the International Committee for the History of Technology at Paris, France, and edited my paper for publication in the resulting volume of proceedings. Graham Hollister-Short, editor of a British journal, *History of Technology*, published a paper that I gave about Edward A. Uehling at the 1992 annual meeting of the Society for the History of Technology at Uppsala, Sweden. Edwin Perkins, editor of *Essays in Business and Economic History*, published an article by me about competition between southern furnace operators and northeastern iron producers based on a paper that I gave at a meeting of the Economic and Business History Society at Toronto, Canada. David O. Whitten, who invited me to give a paper at a conference at Auburn University honoring the 200th anniversary of Eli Whitney's cotton gin, later arranged to have it published in *Agricultural History*.

I visited many archives, libraries, and museums and am deeply indebted to the officers and staff members of these institutions, beginning with Sloss Furnaces National Historic Landmark. I have already mentioned the roles that Randall G. Lawrence and Robert Casey played from the outset of my project, but I also wish to express appreciation to James A. Burnham, who succeeded Lawrence as director of the Landmark and kindly arranged for publication in pamphlet form of a speech, "Sloss Furnaces: The Heritage and the Future," that I gave honoring his appointment to that post; Paige Wainwright, assistant director, who took the helm in the difficult transition period following Lawrence's death; Margo M. Hays, who answered many research-related questions and located illustrations for this book; and Sherrie M. Jones, who rendered many acts of kindness. Ronnie Bates, George Brown, Joy L. Griner, Matt Landers, Charles Lanier, Barbara Nunn, and John Scott were also helpful to me. I owe special thanks to Brown, a veteran Sloss-Sheffield employee, for an enlightening impromptu presentation that he gave on an unforgettable evening.

Edwin Bridges, Norwood Kerr, Debbie Pendleton, Willie Maryland, Franzine Taylor, and Tanya Zanish aided my research at the Alabama Department of History and Archives, and Barbara Taylor invited me to share my findings in a public presentation at that institution. Joyce Lamont and the staff of the W. S. Hoole Special Collections Library of The University of Alabama provided access to materials on James Bowron and John T. Milner.

The Birmingham Historical Society was a source of help as my research progressed. I am grateful to its director, Marjorie Longenecker White, and to other staff members including Brenda Howell and William D. Jones. I also appreciate aid given me by Linda Cohen, Yvonne Crumpler, Deborah Dahlin, Jane E. Keeton, Anne Knight, William A. Tharpe, and Don M. Veasey in my work at the Birmingham Public Library.

I began my career at the Hagley Museum and Library and have benefited on numerous occasions from return visits to that institution. I wish to thank Charles Foote, Glenn D. Porter, and Jon M. Williams for their help and hospitality during my latest trip to Hagley and for assistance in providing illustrations for this book.

I owe a particular debt of gratitude to Auburn University's Ralph Brown Draughon Library and its director, William C. Highfill. Glenn Anderson, Barbara Bishop, Harley Brooks, Boyd Childress, Dwayne Cox, Grady E. Geiger, Joyce Hicks, Claudine Jenda, Yvonne Kozlowski, David Rosenblatt, Harmon Straiton, Linda L. Thornton, Lorna Wiggins, and many other persons rendered help without which this book would not have been possible. I would also like to acknowledge assistance by the late Felix Pretsch, who aided me early in my research.

Among colleagues and friends at Auburn University whom I have not yet mentioned, I am particularly indebted to Gordon C. Bond, who was my department head at the beginning of this project and later became Dean of Liberal Arts. Bond arranged a leave of absence at a critical stage in my writing and was a constant source of encouragement. Wayne Flynt read material from several chapters before they were published as journal articles and made helpful suggestions. Larry G. Gerber and Lawrence F. Owsley, Jr., made books available to me from their personal libraries. Thomas A. Belser, Jr., Anthony G. Carey, Hines H. Hall, Joseph Harrison, and James L. McDonough shared valuable insights. I am also thankful for the intellectual stimulation and encouragement from my fellow historians of technology at Auburn: Guy V. Beckwith, Lindy Biggs, James R. Hansen, and William F. Trimble. I owe a special debt of gratitude to graduate students for their invaluable assistance. James W. Sledge made an important contribution to one of my chapters by sharing with me his collection of newspaper clippings and other materials pertaining to the citizen effort to preserve Sloss Furnaces after the installation went out of operation in 1970. Sledge also advised me on specific research-related questions, as did Roy F. Houchin and Melvin C. Smith.

Among historians at other institutions, I owe special gratitude to Melvin Kranzberg, who has been a constant source of inspiration throughout my career. Without his personal magnetism and his role in making the Society for the History of Technology a true community of scholars, my professional development would have been much less rewarding. As a result, I have dedicated this book to him and to Lawrence. I also wish to acknowledge the help and encouragement of my dear friend, Hans-Joachim Braun, a distinguished European scholar who has visited Sloss Furnaces and understands how I feel about the installation.

Donna M. Adams, Merle L. Coalglazier, Barbara C. Connell, Tom D. Crouch, Claire-Louise Datnow, Alison W. Franks, Marvin Guilfoyle, John G. Heilman, Curtis T. Henson, David R. Hiley, Jay Lamar, Gene T. and Bernice S. Lawrence, J. L. Lee, Margaret W. Manley, Patricia L. Manos, Eric L. Mundell, Nancy J. Nowicki, Mary Pavlovich, Jennifer V. Pennington, William Warren Rogers, Lester D. Stephens, Serlester Williams, and Robert L. Yuill rendered gracious help to me at various stages of my project, for which I wish to express my thanks.

I could not have completed this project without the benefit of word processing technology. I am therefore grateful to Minnie Bryant, Edgar R. Mallory, Stephen McFarland, and William F. Trimble for teaching me whatever I know about the techniques involved in using the two computers that have served me well. I particularly appreciated the word processing talents of Sandra Rose, who handled numerous drafts of my work and of publications that stemmed from it, gave me literary advice, and was patient with my tendency never to be satisfied with anything I gave her until I had revised it repeatedly.

The deepest debt of all I owe to Pat, who is at one and the same time my wife, my best friend, and a superb editor to whom this book owes far more than I can express. She shared every moment of my project, and I could not have completed it without her steadfast love and support.

I close with a final salute. On the road along which my research took me, I became increasingly aware that I was retracing the footsteps of an extraordinary pathfinder whose remarkable book, *The Story of Coal and Iron in Alabama*, originally published in 1910 and repeatedly reprinted since that time, has been for more than eighty years the main point of entry into the uncharted regions that it opened for investigation. One could no more undertake an expedition like the one I have conducted without using the map drawn by Ethel Armes than one could try to retrace the Persian Wars without reading Herodotus. My book therefore ends with a thinly-veiled tribute that will be easily recognized by anyone familiar with the works of a pioneering precursor who blazed the trail for my own journey across the wondrous terrain that she was the first person to explore. Long may her memory live!

SLOSS FURNACES AND THE RISE
OF THE BIRMINGHAM DISTRICT

Statue of Vulcan by Giuseppe Moretti, Overlooking Birmingham, Alabama, with Sloss Furnaces Barely Visible in the Far Right. The colossus, symbolizing the spirit of industry, is the world's largest cast-iron statue. (Jet Lowe, photographer; courtesy of Historic American Engineering Record, Washington, D.C.)

The Inheritance

From a lofty eminence atop Red Mountain in north-central Alabama, a huge statue of Vulcan, Roman god of fire and the forge, looks across one of the world's most remarkable geological settings: Jones Valley. Here, the three raw materials needed to make iron—coal, ore, and limestone—are closer together than at any other place on earth. The colossus symbolizes the spirit of industry. Towering 56 feet high and standing on a 124-foot base, it is the largest cast-iron statue ever erected.[1]

Below Vulcan's feet lies Birmingham, Alabama's biggest city. Standing by an anvil and clutching a hammer in his left hand, the fire god holds aloft a torch pointing toward a long-dormant industrial complex on the valley floor. Much of the metal that forms his body came from this once-vibrant facility.[2] Motorists entering the city from the south on the Red Mountain Expressway catch a momentary glimpse of it, neatly framed between two high-rise apartment buildings, but may be in too great a hurry to wonder about its significance. Just as the lofty spires, pointed arches, and flying buttresses of Gothic cathedrals symbolize the unquestioning religious faith that once prevailed in such places as Bayeux or Chartres, the massive stacks, dome-topped stoves, and slanting skip hoists of Sloss Furnaces are mute reminders of the industrial heritage of the Birmingham District. Now officially designated a National Historic Landmark, the complex has become an outdoor museum that attracts up to 100,000 visitors a year. On summer nights, musical groups ranging from the Alabama Symphony Orchestra to rock bands give concerts in one of its cavernous casting sheds. Throughout most of its history, however, it played a dramatically different role. As one of the district's economic mainstays, it epitomized the dynamic spirit of southern businessmen who believed that Birmingham was destined to become the world's greatest center for the manufacture of iron and steel.

Sloss Furnaces derive their name from Col. James Withers Sloss, one of

Birmingham's greatest pioneers, who founded the Sloss Furnace Company in 1881. His creation was at least partly demonic from the beginning; like Vulcan himself and his Greek cousin Hephaestus before him, the installation had a sinister aspect. Due to not only the fires that metalsmiths tended but also the crippling effects of arsenic, mercury, and other substances with which they came into daily contact, many of the ancient gods identified with metallurgy were linked with destruction and depicted as lame and misshapen.[3] Sloss Furnaces lived up to such mythological associations. People who lived in Birmingham when the complex was still active recall riding at night along the viaduct running past the site and seeing giant "bull ladles," each capable of holding 125 tons of white-hot pig iron, pouring incandescent casts of molten metal into an endless chain of molds, lighting the sky with an eerie glow. Others drove to the top of Red Mountain to get a different, but equally spectacular, view. It was a fascinating and terrifying sight, from which those who witnessed it formed lasting ideas of what Hell might be like. Belching soot and smoke, the furnaces blanketed the city with a pall of dust that blackened buildings, clogged window screens, soiled clothing, and helped raise Birmingham's abnormally high incidence of lung disease.[4]

Like Vulcan and other mythic figures that embodied tension between opposites, however, Sloss Furnaces also played a creative role. Their acrid fumes imparted the breath of economic life to the aggressive, dynamic community they helped spawn. Along with other ironmaking enterprises that sprang up in Jones Valley during the 1880s, the stacks erected by Colonel Sloss undergirded Birmingham's transformation from an empty field into a thriving urban center with such swiftness that it became known as the Magic City. Giving substance to the visions of such prophets as Richard Hathaway Edmonds and Henry W. Grady, the sudden metamorphosis typified what Paul M. Gaston has called the "New South Creed," premised on a belief that industrialization was the key to fulfilling the region's hopes for the future.[5]

But the status that Birmingham attained as a leading symbol of the New South was paradoxical. The city was in fact deeply rooted in the old plantation order. It had been conceived before the Civil War by a group of wealthy Alabama landowners who invested in railroads, banks, ironworks, and textile factories in the belief that such ventures were vital to the diversification and growth of a predominantly agricultural state whose prosperity was based on its role as an exporter of a single cash crop, cotton, to other parts of a vast international business system. Rejecting a northern style of industrialization based on free labor, these planter-entrepreneurs favored a southern model of manufacturing built upon the institution of slavery. Deeply involved in the development of towns and cities, they promoted the growth of Montgomery and helped it replace Tuscaloosa as Alabama's state capital. During the 1850s, amid a nationwide surge of economic expansion, they became fascinated by the prospect of exploiting the rich mineral resources of Jones Valley by creating an ironmaking city at the base of Red Mountain and using slaves to operate its blast furnaces and rolling mills. Realization of their plans was delayed by opposition from communities opposed to Montgomery's interests, the outbreak of the Civil War, and the hardships of the arduous period immediately following the Confederate surrender at Appomattox. Not until the 1870s, as Radical Reconstruction came to an end and white supremacists took control of Alabama's state government, did surviving members of the group finally succeed in founding the city about which they had dreamed, brashly naming it Birmingham after its famous British industrial counterpart.

Much had changed in two decades since Birmingham had first been envisioned, and the onset of a severe business depression in 1873 nearly killed the infant city. But there were still many reasons to believe that the industrial exploitation of Jones Valley would ultimately yield large profits. Slavery had become illegal, but furnaces and other installations could be operated by low-paid, severely disadvantaged black workers. Some of these laborers were convicts, leased at cheap rates from state and county governments. By capitalizing on low labor costs and abundant raw materials located in close proximity to one another, Birmingham's promoters expected to make iron at much less expense than it could be produced anywhere else. Using labor-intensive methods would help compensate for one of the South's greatest problems, a scarcity of capital. Levels of mechanization that were already becoming common in the North could be avoided by adopting new and advanced equipment only selectively, as required for operations that could not be performed in traditional ways. Skilled workers or engineers who were needed to provide expertise unobtainable within the region could be imported from other places and paid for out of the savings that would result from Birmingham's unique locational advantages. All that remained to complete the picture was a way to carry iron from Jones Valley to the outside world. Railroads and oceangoing steamships would supply that crucial missing element.

All of these developmental strategies had already proved successful to a limited degree in the South's prewar experience. In antebellum Alabama, they had been stubbornly resisted by plain folk who realized that the resulting profits would enrich a small planter aristocracy. Now, after Reconstruction, it was precisely such privileged people and their descendants who ran the state. Birmingham's founders did not expect that wealth derived from shipping iron to the outside world would be more widely distributed than the profits that large landowners had earlier gained from the cotton export trade. From the outset, the fiercely aggressive young city that emerged in Jones Valley was a distinctly southern enterprise. But it also had much in common with new communities that were springing up all over the country in the late nineteenth century, for whatever was unique about the South was far overshadowed by the ways in which the region was similar to other parts of a recently reunited nation. Putting sectional animosities aside, Americans returned to the relentless drive for material acquisition that had been so much in evidence in the 1850s. Birmingham's headlong growth was not merely a sectional phenomenon but part of a frenzied scramble for wealth that would make the United States the world's greatest industrial power within only a few decades.[6]

I

Like the economic transformation of the nation as a whole, Birmingham's burgeoning development was based on seemingly limitless natural resources inherited from the remote past. The foundations of the city's sudden rise and of Alabama's equally rapid emergence as the South's leading industrial state rested on events that began in the Cambrian period of geologic time, 570 to 500 million years ago, when a shallow inland sea extended like an enormous crescent between what later became the states of New York and Oklahoma.[7] As eons passed, sediments flowed into the sea from a low-lying area to the northwest, and volcanic debris fell upon it from earthquake-prone islands to the southeast. As these materials accumulated, the floor of the sea sank deeper and deeper under their weight. During Ordovician times, 500 to 440 million years ago, deposits of calcium, magnesium, and silica were transformed into limestone and dolomite. Known as fluxes, such materials promote fluidity; in time, they helped remove impurities from iron ore in the roaring bellies of Alabama blast furnaces.[8]

Geologists differ about how iron ore came into being in the inland sea, agreeing only that deposits began to be formed during the Silurian age, 440 to 400 million years ago. Iron is one of the commonest elements in the earth's crust, but it is not clear how large amounts of it became united with oxygen and other elements to form vast ore deposits that later proved commercially valuable. According to one theory, grains of sand, rolled back and forth across the floor of the inland sea by the motion of water, were slowly surrounded by layers of rusty scum excreted by tiny organisms. It is also possible that a process similar to petrification occurred as iron-rich waters invaded porous limestone. Iron-fixing bacteria, inhabiting restricted arms of the inland sea, may also have produced ore formation. One way or another, enormous amounts of iron ore were laid down.

Two types of iron ore evolved in what later became north-central Alabama. One, containing variable amounts of iron and water in irregular surface masses embedded in clay, was limonite, better known as brown ore. Well suited to the needs of charcoal-fired furnaces built before the Civil War, vast quantities of it underlay at least twenty Alabama counties. Not utilized until later, but extremely important after the Civil War, was a second type of ore: red hematite. Containing no water, it took its name from the Greek word *haima*, "blood," because of its color when exposed to air. Red Mountain, stretching ninety miles across north-central Alabama, contained huge amounts of this substance. Some seams were thirty-five feet thick.[9]

In mining and metallurgical terminology, Alabama's red hematite deposits were "lean," containing only modest amounts of iron proportionate to other materials. Much ingenuity was required to make their commercial exploitation profitable. Early accounts boasting Jones Valley's advantages produced false impressions about its industrial potential; prospectors were misled by the richness of weathered deposits near the surface, where nature had leached out impurities. Once surface ores had been mined, remaining deposits produced much lower yields; it took about three tons of red ore to make one ton of iron. A high silica content resulted in large quantities of thick, lava-like slag, more than double the amount yielded by northern furnaces. Pittsburghers contemptuously called Red Mountain ore "ferruginous sandstone."[10]

Because of its leanness, red ore required abnormally high temperatures for smelting. "The Southern operator blows as much wind to produce 500 tons of iron per day as the Northern operator does in producing 700 tons," stated one expert. Making a single ton of pig iron required about 2,500 pounds of coke, giving the district the highest rate of fuel consumption in the entire nation. Red ore's refractory qualities made it highly abrasive, and its distinctive chemical composition produced an above-average incidence of "scaffolding," a rapid buildup of deposits on the inner walls of blast furnaces. This necessitated frequent relining, impeded the flow of heat, and, by damaging furnace walls, led to periodic "breakouts," in which incandescent raw materials burst through weak spots with explosive force and sometimes lethal results.

High concentrations of phosphorus and silica made red ore unsuitable for steel production by the Bessemer process that was dominant in the United States in the late nineteenth century. By frustrating boosters who wanted to make Birmingham the world's greatest steelmaking center, raw material constraints cast a long shadow over Alabama's future. Local ore seams were severely faulted by vertical displacement of strata, impairing accessibility; they also lay far below ground level, slanting deep under Red Mountain before leveling off beneath Shades Valley to the southeast. This formation prevented open-pit mining and required costly underground operations; by the early twentieth century, some subterranean workings were fifteen miles long. Such conditions lessened the potential advantage of having vast deposits of ore close to huge amounts of coal and flux, forcing industrialists to find ways of maximizing the benefits of proximity while compensating for the district's less fortunate peculiarities.[11]

Red ore did have some valuable characteristics. The lime carbonate content in deep deposits, for example, made them virtually self-fluxing. The chief industrial advantage inherent in the district's red hematite, however, was its sheer abundance. Avoided by ironmakers prior to the Civil War because it was hard to smelt in charcoal-fired furnaces, red ore came into its own in the postwar era when coke, made from bituminous coal, won preference over charcoal due to its greater heat potential and mechanical strength. Proximity to vast deposits of red ore, coking coal, and flux, local boosters believed, would enable iron to be made at less cost in Birmingham than anywhere else.

Alabama's enormous coal reserves were as basic to this vision as were the

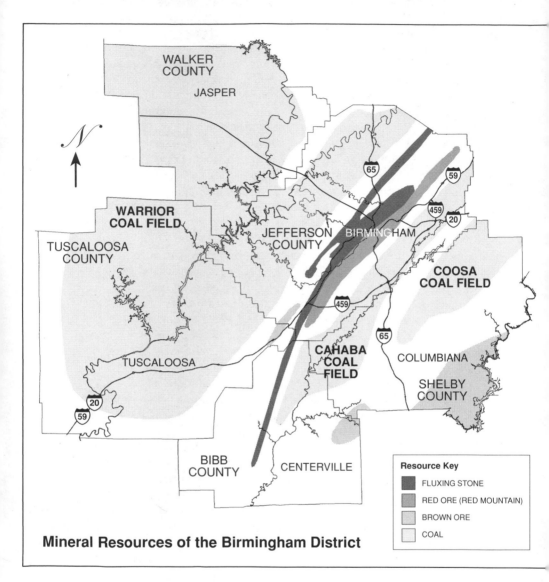

Mineral Resources of the Birmingham District

Mineral Resources of the Birmingham District. This map shows deposits of ore, coal, and fluxing stone in close proximity, which made the area geologically unique as a natural center for iron manufacturing. Opportunities and problems inherent in the location and chemical characteristics of the district's raw materials did much to shape a distinctively southern style of ferrous metals production and helped make Alabama the heart of the American foundry trade. (Courtesy of Birmingham Historical Society and icon graphics, based on research by Jack Roland Bergstresser, Sr.)

district's abundant ore deposits. Brown and red ore are both oxides, and their oxygen content had to be removed before anything useful could be done with the iron they contained. Removal required bringing the ore into contact with carbon monoxide, a strong reducing agent, at high temperatures. Charcoal, an impure form of carbon resulting from incomplete combustion of wood in the absence of air, met this need prior to the Civil War. Coke, however, was preferable, not only because it had greater latent heat but also because its superior mechanical strength enabled it to be mixed with a heavier burden of ore and flux without crumbling and thus impeding air circulation in a blast furnace. In coke, postwar ironmasters found a fuel suitable for larger, more productive furnaces than any that had previously existed in Alabama.[12]

Massive deposits of coal, from which coke was later made, were formed during the Pennsylvanian era, 320 to 280 million years ago. By then the surface of the inland sea covering north-central Alabama abounded with swamps that teemed with plant and animal life. The decaying remains of these marshy areas were slowly incorporated into layers of carbon-rich materials; heat and pressure caused by the accumulating weight ultimately produced enormous concentrations of peat, lignite, and bituminous coal. Originally, all of Alabama's coal deposits lay in one vast formation, stretching all the way from the Coosa River on the southeast to the Sipsey on the northwest. This, however, became separated into three distinct fields. Two of these, the Cahaba and the Coosa, were less significant than their far more richly endowed neighbor, the Warrior. Seventy miles long by sixty-five miles wide at its broadest point, it covers about 3,500 square miles and encompasses ten counties, of which Jefferson, Tuscaloosa, and Walker are the most well endowed. Unlike coal from the Coosa and Cahaba beds, best suited for steam engines, the Warrior variety was good for coking.[13]

Like Red Mountain ore, Warrior coal contained impurities that complicated its industrial use. The most important of these was sulfur, which made iron "hot short," or excessively brittle. "Sulphurous coal," stated a leading nineteenth-century metallurgist, "will produce sulphurous coke, and consequently sulphurous metal, which, in all subsequent manipulations, will be injurious, troublesome, and expensive."[14] Removing sulfur from Warrior coal required heavy investments in washing apparatus, forcing Birmingham's industrialists to become pioneers in developing such machinery. Partly because the seams in which it was embedded contained large amounts of slate and shale, Warrior coal also yielded a high residual ash content when it was coked. In addition, seams in the Warrior Field were often too narrow for easy mining. Methane, a colorless and odorless gas that is potentially explosive in the presence of air, was also prevalent, resulting in mine disasters. Though superior to charcoal in mechanical strength, coke made from Warrior coal was not as high in this property as that from such places as Connellsville in western Pennsylvania; this quality limited the height of furnaces in the Birmingham District and the blast pressures under which they could operate. Such problems intensified the challenge of profiting from the district's close access to coal, ore, and flux while minimizing the difficulties stemming from its complex resource endowments.

Jones Valley became Alabama's most heavily populated site because it gave ironmakers closer access to a broad range of natural resources than did any other place in the north-central part of the state. No place on earth, other than the Birmingham District, contained within a thirty-mile radius all three raw materials required for iron production. This fortuitous condition resulted from cataclysmic geological events at the end of the Paleozoic era, about 285 million years ago. Far below ground level, tectonic plates on which the North American and African continents rested moved steadily closer together until they collided. The terrific grinding and thrusting heaved up deposits that had accumulated in the inland seabed, causing them to fold and buckle in spasms of mountain-building. Fierce lateral pressures caused masses of rock to break into slanting configurations that geologists would later call anticlines and synclines, forming a mountain chain that once towered three miles high.

Jones Valley occupies what was once the crest of an anticline. Sedimentary deposits were thrust upward to form an inverted **V**, the fractured top of which reflected the violence of the process that created it. Entering cracks in the broken surface, rain washed away large amounts of soft rock, forming a valley flanked by ridges that resisted erosion. On the south was Red Mountain, overlooking rich deposits of limestone and dolomite; to the northwest were the huge Warrior coal beds.[15]

The geological events that made the Birmingham District so rich in coal, ore, and limestone, however, left it short of another industrial asset, water. Occupying a narrow plateau between two river basins, drained by the Cahaba to the southeast and the Black Warrior to the northwest, it had only one notable stream, Village Creek, which pursued a meandering forty-mile course from near present-day Huffman to a confluence with the Black Warrior River. Like Turkey and Five Mile creeks to the north and Valley Creek to the south, it was fed only by runoff water and limestone springs and thus proved insufficient to cool blast furnaces and simultaneously meet the drinking, cooking, and sanitary needs of a large population. Meanwhile, the lack of navigable outlets to other Alabama river systems blocked access to the Tennessee Valley to the north or the Gulf of Mexico to the south. Even the nearby Cahaba River was too treacherous for anything but the most primitive forms of water-borne traffic, and access to places below Centreville was impossible after 1849 when a bridge was built at that point. Surrounded by wilderness, Jones Valley remained largely undeveloped until railroads entered it after the Civil War.[16]

Because of its favorable and unfavorable characteristics, Jones Valley afforded postwar industrialists a unique mix of opportunities and drawbacks. Lured by its abundant raw materials in close proximity, entrepreneurs had to make difficult strategic choices to compensate for its geological peculiarities. In 1902, at a key turning point in the history of Sloss Furnaces, owners chose to capitalize on the special fitness of local raw materials for one product that the Birmingham District was possibly the best place on earth to manufacture: foundry pig iron. Only a few years before, another group of businessmen had staked the future of a competing enterprise upon another product, steel, for which the district was much less well suited. These contrasting decisions helped shape Birmingham's destiny for generations.

II

Natural resources have no value unless human knowledge makes it possible to exploit them. For this reason, Jones Valley's economic potential was long unrealized. Despite the ingenuity with which they made weapons, tools, and containers from stone and clay, the Native Americans who began entering Alabama about 9,000 years ago had no metallurgical knowledge and never made iron. Although they came in contact with Red Mountain ore, they used it only for decorative purposes, smearing it on their bodies as war paint or staining their clothes with its dark-hued pigment.[17] Alabama's first European settlers also made little use of its mineral wealth, but they did possess a cultural background rich in knowledge about ferrous metals and practice in working with them. This heritage stretched back to the dawn of the Iron Age, 1,500 years before the birth of Christ.[18]

For centuries, iron was made by what became known as the direct method. Early ironworks, called bloomeries, could not melt iron, a process that required temperatures exceeding 1,500 degrees Centigrade. Such intense heat, however, was not necessary for making a valuable product, wrought iron. This type of metal could be made by combining iron ore and carbon in a charcoal fire that was hot enough—about 800 degrees Centigrade—to overcome the chemical affinity between iron and oxygen and remove the latter from the ore. Such temperatures could be achieved with the aid of simple bellows. Because the iron did not melt, the process produced lumps of iron—called blooms—mixed with carbon and traces of other chemical elements. After the blooms were taken out of the fire, oxidized impurities could be removed by repeatedly heating and hammering the red-hot metal, giving it a tough, fibrous structure. Sometimes temperatures did actually rise to a point at which melting occured, but this outcome was avoided because the resulting product, known as cast iron, contained too much carbon and was therefore brittle, unsuitable for tools or weapons. For this reason, ironmakers regarded cast iron as worthless. Eventually, however, it became a prized commodity, one on which the owners of Sloss Furnaces staked the future of their enterprise in 1902.

The main drawback of bloomeries was low output for a large input of labor. To obtain higher yields with less human effort, European ironmakers developed larger furnaces with water-powered bellows that pumped more air into a bigger volume of space. Such installations were known as blast furnaces, high structures within which iron liquefied because of the large amounts of charcoal that could be fed into the stack and the high temperatures that could be attained. The product of a blast furnace became known as pig iron because the liquefied metal ran into channels that led in turn to oblong molds whose layout reminded ironworkers of a litter of piglets suckling at the belly of a mother sow.

Pig iron was a form of cast iron, the waste material previously shunned by ironworkers. It had a high carbon content, resulting from protracted exposure to burning charcoal, making it brittle and hence unsuitable for implements that had to sustain rough use. This drawback, however, could be removed by reheating and hammering the metal at a separate installation known as a finery, producing a tough, fibrous product that was essentially the same as wrought iron and was therefore called by the same name. To secure greater output, ironworkers had substituted an elaborate two-stage method for the earlier bloomery process; the dramatically enhanced productivity per unit of labor more than made up for the added expense caused by the scale and complexity of the equipment needed, yielding what would later be called economies of scale.

By the time the blast furnace had been developed, moreover, it was not necessary to refine all the pig iron that could be made in such an installation, because cast iron was no longer a waste product. Gunpowder had spread from Asia to Europe, resulting in the development of artillery. Crude cannons were made at first of wrought iron; better ones were then made of bronze or brass; still later, cast-iron cannons became popular because they were less expensive. Cannonballs, earlier made of stone, were now made of molded cast iron. What had once been an unwanted material had now become a military necessity.[19]

Cast iron was also used for fire backs, stove plates, andirons, and hollowware. The high carbon content that made unrefined pig iron unsuitable for tools and implements was no drawback in these products; on the contrary, it actually enhanced them, because cast iron is excellent for radiating heat. Demand for cast iron therefore remained brisk until the late twentieth century. This produced a new type of installation that ultimately created substantial markets for Sloss Furnaces. For centuries, cast-iron products were molded at the mouth of a blast furnace. It became more efficient, however, to take pig iron to a separate facility, known as a foundry, where it was remelted in a new type of furnace, known as a cupola. The foundry trade became a key component of the iron industry.[20] For a time, the firm that owned Sloss Furnaces became the world's largest maker of foundry pig iron, and Alabama was the chief center of the American foundry trade.

In colonial America, cast-iron products were still made at the furnace mouth; separate foundries did not become common in the United States until the early nineteenth century. Bloomeries, blast furnaces, and fineries, however, were numerous before the Revolution, particularly in Pennsylvania, which seized an early lead as the heart of the American iron industry.[21] Blast furnaces were built in the shape of flattened pyramids with hand-hewn stone

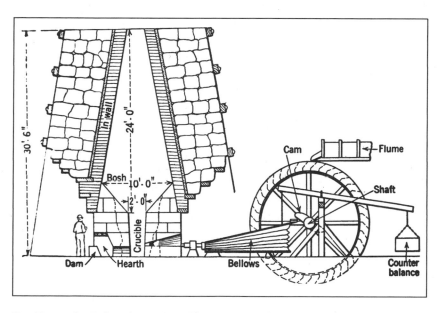

Two Views of a Colonial American Blast Furnace, Showing Working Parts and Other Structural Details. Such installations epitomized a plantation style of operation that survived in the South longer than in any other part of the United States, contributing to the distinctiveness of the region's industrial development. (Courtesy of Hagley Museum and Library, Wilmington, Delaware.)

blocks and were about thirty-five feet high. The stack had a hollow interior, lined with fire-resistant brick; this widened from the top until it reached the bosh, about two-thirds of the way down. After this, the interior walls sloped inward and downward to support the charcoal, limestone, and ore with which the furnace was charged. The slanting walls below the bosh directed these materials toward the crucible, a chamber at the base of the furnace that received the melted iron and liquefied waste running down from the fiery mass above. Water-powered bellows (two of these were sometimes operated in tandem to maintain a continuous blast) pumped air into the crucible through a nozzle called the tuyére, adding oxygen to raise the temperature. In front of the crucible was an open area, hollowed out of the side of the furnace, known as the hearth. Here workers periodically released clay plugs with pronged tools to allow slag and molten iron to burst from the blazing interior. In front of the hearth, in a roofed enclosure known as the casting shed, were a series of channels laid out in complex patterns in layers of heat-resistant sand.

Charged from the top through an opening called the tunnel head, a blast furnace was built by a hill, the brow of which was connected to the stack by a trestle across which raw materials were trundled in barrows and dumped into the inferno below. Before being used, such materials were kept in a stock house near the trestle. Nearby, a fast-flowing stream diverted water through a channel, called a race, into a wooden conduit, known as a flume, to power the waterwheel that drove the bellows. Early Alabama blast furnaces had all these features.

As iron ore and limestone moved downward through a furnace together with burning charcoal, the ore was gradually reduced to a molten liquid that gathered in the crucible. Its high porosity, which permitted the use of a relatively weak air blast, and also its lack of sulfur and other contaminants, made charcoal an ideal fuel for small blast furnaces in early America, because modest levels of productivity were acceptable.[22] Because of its chemical affinity for carbon, oxygen in the ore combined with the charcoal to produce carbon monoxide and carbon dioxide. The resultant waste formed a mixture known as slag. Because slag was lighter than iron, it collected in the upper part of the crucible and was periodically removed by opening a hole known as the cinder notch. After spurting out, it was allowed to cool and taken to a dump. Molten iron, which sank to the bottom of the crucible because of its greater weight, was tapped at intervals and gushed out into the channels and molds covering the floor of the casting shed. After the white-hot metal had cooled and solidified, individual billets of pig iron were broken off from the sows by workers wielding sledgehammers.

Pig iron that was intended for implements requiring ductility or ability to withstand abrasion was next taken to a finery for conversion into bars of wrought iron. These could be taken in turn to a rolling mill and flattened into sheets, which might then be slit into rods from which blacksmiths could make nails. Blacksmith shops also made tools and horseshoes. Alabama's earliest ironworks were little more than primitive smitheries; John Hanby, a machinist from Virginia who had fought under Andrew Jackson in the War of 1812, operated such an installation in the hills near Blount Springs and Mount Pinson, using brown ore he had found near Oneonta. One of his grandsons later became a mine superintendent for Sloss Furnaces.[23]

As previously indicated, some of the molten metal from a colonial blast furnace never became pig iron. Instead, it could be channeled directly into molds at the furnace mouth to be cast into fire backs, stove plates, cannonballs, pots, and other products that did not require the elimination of carbon. Stove plates were elaborately decorated with artistic motifs and later became prized as collector's items. Pennsylvania, where Benjamin Franklin invented an early cast-iron stove, was famous for these. Stove iron later became a staple commodity of the southern foundry trade, supplying lucrative markets for Sloss Furnaces.

A major reason why the owners of Sloss Furnaces decided in 1902 to specialize in foundry pig iron was that Alabama's natural resources were not well suited for a type of ferrous metal that was highly prized for its ability to stand abrasion and endure repeated flexing. This was steel, an alloy of iron and a small but critical percentage of carbon. Used for swords, springs, tools, and other highly specialized items, it was made by baking strips of wrought iron for long periods in charcoal dust at extremely high temperatures in a process known as cementation. As enough carbon from the dust—about 1 percent—entered the iron, it became steel. Because it was so hard to make, steel was prohibitively expensive for most uses and remained so until the late nineteenth century, when new methods of production led to a technological revolution. Steel then became the glamour metal of American industry, protected by tariff barriers that ensured high profits. The difficulty of converting southern raw materials into steel, caused by the presence of phosphorus and silica in regional ores and of sulfur in Warrior coal, became a constant source of frustration to Alabama furnace operators, who experimented with a succession of new methods in their efforts to solve the problems involved. Swimming against the tide, the owners of Sloss Furnaces chose in 1902 to stick with pig iron and concentrate on supplying the needs of the foundry trade, which was moving into the southeastern United States because the phosphorus content of local ore enhanced fluidity and moldability. This decision was unpopular in Birmingham at the time, but it turned out to be a sensible move, epitomizing what would later be called "appropriate technology."

By foregoing steel production, the businessmen who ran Sloss Furnaces were taking a conservative approach rooted in the values and traditions of an earlier, agricultural era. Because of the need for charcoal, colonial American blast furnaces had to be located on large, heavily forested tracts known as "iron plantations." These rural locations furnished all the raw materials needed for making iron, including ore and limestone. Workers rented small dwellings on the grounds and bought goods on credit at stores kept by the owners, who lived in large, handsomely furnished houses appropriate to their superior status. The system, which enabled owners to minimize wages and control costs, was common not only in such places as Pennsylvania but also in the South. In Virginia, for example, slaves operated blast furnaces and fineries in a setting similar to that of tobacco plantations. This way of conducting operations was followed at Alabama ironworks throughout the antebellum period and remained a model after 1865, using cheap, servile black labor.

Even into the twentieth century, Alabama's blast furnaces were run much like colonial iron plantations; company stores and company-owned housing were maintained by the owners of Sloss Furnaces until the 1950s. Further conforming to the plantation model, some of the managers who directed Birmingham's blast furnaces came from agricultural backgrounds. Thomas Seddon, who ran a Louisiana sugar plantation before coming to Birmingham to take charge of Sloss Furnaces in 1887, was a case in point. Disinclination to depart from this style of management and reluctance to adopt mechanization as a substitute for manual labor had much to do with the choice that was made in 1902 to concentrate on the foundry trade and avoid entering the strange new world of high-volume steel production. The continuity with plantation-bred traditions that was manifested in this critical decision will constitute one of the major themes of this book.

III

By the eve of independence, colonial America was already producing about one-seventh of all the iron that was made in the western world, but Alabama was not yet part of the picture. The earliest settlements around Mobile Bay and in the Tombigbee Valley were remote from the mineral-rich interior. Settlers who flocked into the territory as "Alabama Fever" raged after the War of 1812 tended to avoid the mountainous north-central region, in which Indian claims had barely been extinguished, in favor of agricultural lands in the Tennessee Valley or the Black Belt, a crescent-shaped area of rich, dark limestone-based soil stretching across the south-central region between the mountains and the coastal plain. The attainment of statehood in 1819 did nothing to change this situation. Agriculture and mercantile enterprises continued to predominate, centering around cotton, which reigned supreme in Alabama more than anywhere else. By 1850, 564,429 bales, nearly one-quarter of the nation's cotton crop, were raised in this state alone.[24]

Contrary to images of the Old South that have more to do with stereotypes that later became common than with economic realities prevailing before the Civil War, Alabama was neither poor nor backward in antebellum times. This was particularly true in the 1850s, when cotton prices were high. One of the South's greatest industrial prophets, John T. Milner, accurately stated that Alabama had been "at the height of her glory" immediately prior to secession. "In 1860," he said, "Alabama was a great and rich State—the seventh in aggregate of wealth of the Union of States, and exceeded in the aggregate production of agricultural values by only Illinois, Pennsylvania, New York and Mississippi. . . . Agricultural labor was better rewarded here than in any other State, except in Louisiana, Mississippi and California. . . . At that time the value of her agricultural products, per capita of her farm population, was double that of any of the free States . . . except California, Illinois and Iowa." Milner buttressed his claims with reams of statistics and tables, stressing particularly the self-sufficiency of Alabama's antebellum farms and plantations, which produced great yields of corn, peas, beans, buckwheat, barley, oats, and hay and possessed large numbers of hogs and dairy cows in addition to leading the nation in cotton production. The findings of such historians as Stanley Engerman, Robert Fogel, Eugene Genovese, and Gavin Wright are congruent with Milner's conclusions, presenting a picture strikingly at variance with previous assumptions about southern economic growth prior to 1861. As Milner acknowledged, not all parts of Alabama were equally prosperous. The southeastern Wiregrass area was sparsely populated prior to 1860, and farming was much more remunerative in areas with good soils and access to water transportation than it was in less favored locations. Nevertheless, one of the main reasons for the failure of the state to industrialize was simply that agriculture was so rewarding.[25]

Despite the primacy of agriculture, interest in manufacturing and urban development was not absent in Alabama, even prior to its admission to statehood. Much of the impetus for such activity came from members of the Broad River group, wealthy planters who migrated from Virginia to Georgia in the 1780s, began buying land in the Tennessee Valley in 1809, and acquired large holdings in the Black Belt a decade later. Broad River people played an important role in early Alabama politics and were largely responsible for drafting a liberal state constitution that was adopted in 1819. After this period, they lost power to small farmers and other citizens of more modest means who marched under the Jacksonian banner and dominated the state government during most of the antebellum period. The loss of power did not stop the Broad River elite from attaining many of its goals; as J. Mills Thornton has pointed out, it played a key role in founding five communities—Cahaba, Florence, Huntsville, Montgomery, and, after 1870, Birmingham. But the unpopularity of the Broad River agenda among a large number of voters, together with the superior profitability of agriculture in many parts of the state, helps explain Alabama's failure to pursue industrial growth more vigorously before the Civil War.

Along with their ownership of large estates and many slaves, the strong interest of Broad River people in promoting internal improvements and manufacturing disproves the idea that wealthy planters, as a class, opposed industrial development. What prevented them from implementing their visions was a lack of political power. The state's largest planters and cotton factors were generally Whigs and were in a minority as opposed to Democrats. As Thornton indicates, the Broad River program had little chance of winning support among Jacksonian legislators who were constantly on guard against threats to "individual autonomy"; shared a common "opposition to the use of power in institutions"; harbored "a profound hostility to corporations"; and imposed taxes as much as possible on "excess capital," luxuries, slaveholdings, and various forms of "wealth derived from nonagricultural sources." Not large plantation owners but the less affluent followers of Old Hickory tried to block the chartering of banks, turnpikes, canals, and railroads that ran counter to visions of a simple, rural, egalitarian society free from aristocratic pretensions and special privilege. "Throughout most of the antebellum period," Thornton writes, "Alabama's rulers strove vigorously to limit the influence of wealth—and particularly of corporate wealth—in the life of the state."[26]

Much of the manufacturing that took place in Alabama was so scattered that it is hard for historians to detect. Farmers and planters, pursuing self-sufficiency, used slave labor to provide many of their own needs by performing such tasks as blacksmithing, coopering, and making clothes. Assessments of the degree of manufacturing that actually prevailed are also skewed by neglecting to include cotton ginning and flour milling on farms and plantations. In view of economic and political circumstances, moreover, it is perhaps remarkable that Alabama developed as many industrial establishments as it did. What is not surprising, given the nature of the state's economy, is the extent to which the manufacturing that existed was rooted in agriculture. Cotton, not iron, spawned Alabama's first important industrial enterprises. Shortly after 1815, a few textile plants appeared in the Tennessee Valley. Among them were the Bell Factory, located on the Flint River near Huntsville, and the Globe Company's cotton mill at Florence, both of which were thriving by the outbreak of the Civil War. The Bell Factory—so-called because workers were summoned to work by a bell instead of a whistle—burned to the ground in 1841 but was quickly rebuilt and had a capital investment of $400,000 in 1858. Using slave labor, it produced material that included cheap grades of osnaburg cloth for plantation hands. That same year, the Globe works had 23,000 spindles and employed 800 people. Wealthy planters invested surplus capital in such enterprises.[27]

Alabama's most famous antebellum manufacturer, Daniel Pratt, combined textile production with other industrial pursuits, again related mainly to agriculture. A native of New Hampshire who migrated to Georgia in 1821, he came to Alabama in 1833 and built what became the world's largest cotton gin factory at Prattville in Autauga County. In 1847, he started a cotton mill, hiring poor whites from nearby pine forests. The new installation had about 3,000 spindles and 100 looms and employed 160 workers. Using water power, Pratt also built a sash, door, and blind factory, a foundry, a horse-powered grist mill, and machine and blacksmith shops. In 1857, his business receipts exceeded $500,000.

Pratt was Alabama's best-known industrial propagandist. He urged the creation of small manufacturing communities that would inculcate respect for manual labor among poor whites, free the state from dependence on northern and foreign goods, and develop technical skills among its people. Despite amassing a fortune, he failed to achieve his goals. As the antebellum period came to a close, his northern origins and fidelity to the Union aroused increasing suspicion among Jacksonian secessionists about his fidelity to southern institutions. Meanwhile, his belief in the dignity of work was scorned by many of his white workers, who looked on industrial labor as only a temporary way to gain the means to buy land and slaves and thus pursue upward mobility. In both cases, Pratt's aims were thwarted not by wealthy planters but by small landowners and poor whites. Because it was hard to find enough white operatives, Pratt had to staff his factories with increasing numbers of slaves. [28]

Other Alabama industrialists shared Pratt's experience. Founded by a small cadre of entrepreneurs, a few textile mills sprang up along the fall lines of fast-rushing streams at such places as Autaugaville, Scottsville, and Tallassee. A good example was David Scott's factory at Scottsville in Bibb County, which had 25,000 spindles and 50 looms by 1858 and shipped cloth as far away as Boston. Tuscaloosa, for a time the state capital, was also a center of industrial enthusiasm; referring to the famous textile mills along the Merrimack River in Massachusetts, one of the town's business leaders predicted that it would someday become the "Lowell of the South." Despite such visions, factories remained only a small part of Alabama's prewar economy. By 1860, the state had fourteen cotton mills; heavily dependent on slave labor that could be spared from agricultural production, these establishments made goods valued at $1,040,147. Though higher than average for the South, the extent of the industry was modest by northern standards. [29]

Despite being supportive of manufacturing, wealthy planters did not want Alabama to emulate too closely what was taking place along the banks of the Merrimack or other northern rivers. Genovese has made a useful distinction between sympathy among leading slaveholders for a limited number of factories, which was widespread, and opposition to the degree of industrialization that took place in many parts of the North during the 1840s and 1850s. In any case, there is no convincing evidence that planters as a class believed that manufacturing threatened basic southern institutions. A careful study by three economic historians has shown a considerable degree of linkage between agriculture and manufacturing in the South prior to the Civil War, even though industrial investment fell short of what was required for large-scale growth. As Genovese points out, wealthy planters were the best customers of factory operators. As long as the latter did not aspire to social and political dominance, the two groups were partners in a common alliance. [30]

That this alliance would continue in the postwar period is not surprising. Broad River associates and other wealthy southern landowners were not simply farmers but agribusinessmen who were deeply involved in a web of international trade in basic commodities like cotton and tobacco. As Walter L. Fleming pointed out in analyzing Alabama's industrialization after 1865, antebellum planters were highly experienced in running large-scale organizations that depended on specialization of labor and skilled management. Ulrich B. Phillips made similar observations in an early essay, calling plantation slavery an "essentially capitalistic industry to concentrate wealth" and comparing it to the factory system in the way it subdivided labor for the performance of regular and repetitive tasks. Viewed in this perspective, the role of John T. Milner and other Broad River partners in conceiving such future industrial cities as Birmingham was no aberration.[31]

Nor was the presence of large numbers of black workers at the factories operated by men like Daniel Pratt or David Scott without significance for Alabama's later industrial development. The most obvious link between the prewar Broad River associates and the postwar members of the same group who built Birmingham was their common commitment to using a predominantly black labor force, enslaved before the war and kept in subjection by low wages and various forms of legal oppression thereafter. As Robert S. Starobin has shown, industrial slavery was widespread in the South prior to the Civil War, both in rural districts and in what relatively few cities the region possessed; between 160,000 and 200,000 slaves worked in textile mills, iron works, coal mines, tobacco factories, sugar refineries, saltworks, and a host of other enterprises. Southern factory owners preferred slave to free white labor, which they regarded as undependable and prone to strike. As Starobin's research indicates, slave-operated industrial enterprises were generally remunerative and in some cases highly profitable. Furthermore, as both Genovese and Starobin have indicated, leasing slaves for industrial employment provided a lucrative outlet for surplus field hands prior to 1861. The low-wage strategy for competing with northern manufacturers that set the South apart from the rest of the nation until just before World War II, analyzed closely by Gavin Wright, was a logical extension of the earlier system. The use of convict labor by Sloss Furnaces and other enterprises in the Birmingham District in the late nineteenth and early twentieth centuries provides further evidence of a distinctively southern style of industrialization that was deeply rooted in antebellum conditions.[32]

Part of what was distinctive about the style of industrialization that emerged in Jones Valley in the late nineteenth century, also clearly foreshadowed before the Civil War, was a heavy reliance on a labor-intensive strategy that contrasted sharply with the policy of mechanization followed in northern centers of iron and steel production. As Gavin Wright has indicated, antebellum plantation owners were much slower to mechanize agricultural operations than were their northern counterparts, and they also provided their slaves with fewer agricultural implements than did farmers in the free states.[33] The development of Sloss Furnaces would later epitomize a style of industrialization that mechanized only selectively, in specific cases in which the need for innovative equipment was imperative, preferring instead to employ large numbers of ill-paid, servile, and inefficient workers using manual techniques that became increasingly anachronistic by northern standards. Not until the eve of the Great Depression would the owners of Sloss Furnaces embrace mechanical methods of charging blast furnaces and casting pig iron that had already been standard for decades in such places as western Pennsylvania.[34]

IV

None of the ironworks that appeared in Alabama prior to the Civil War rivaled the Prattville complex in scope. Many were merely primitive bloomeries, of which the state had at least eight by 1849; their proprietors, bearing such names as Adams, Brantley, Claubaugh, Gray, and Scott, were among Alabama's earliest industrialists. Costing $3,000 or less, the installations they built had few workers; a typical force consisted of two men to tend the fire, one or two hammermen, one person to roast ore, two colliers, four woodcutters, and three teamsters. Some plants produced as little as forty-five tons of iron per year. A loup of wrought iron weighing from 100 to 135 pounds could be made every three hours; four loups were a good day's output. However small the yield, owners of such works took pride in them; the Riddle brothers, from Pennsylvania, stamped iron from Maria Forge in Talladega County with the symbol of a boar's head and advertised that it was equal to the best Swedish product.[35]

Slowly, larger installations began to appear. About 1815, a Pennsylvanian, Joseph Heslip, built Alabama's first blast furnace on Cedar Creek near Russellville in Franklin County, where the owners of Sloss Furnaces would later strip-mine brown ore. Heslip's facility, using charcoal to smelt the same raw material, did not flourish and went out of blast by no later than 1837. During its brief existence, however, it found buyers as far away as England, reaching the outside world through New Orleans by way of the Tennessee and Mississippi rivers. This export strategy, related to the limited nature of markets in the immediate vicinity, would remain a characteristic feature of Alabama's iron industry throughout the nineteenth century. Heslip did, however, ship cast-iron kettles to Louisiana sugar planters. He was also typical of southern furnace operators in using slave labor.[36]

By the time of the Civil War, six more blast furnaces had been built in north Alabama. All used charcoal, typically made from cedar, and all but one used brown ore. Usually they were built by settlers with northern roots, particularly members of the Stroup family from Pennsylvania. Its members helped operate three Alabama furnaces and also built pioneer ironworks in Georgia and South Carolina. Built and staffed mostly by slaves, Alabama furnaces and forges made various cast- and wrought-iron products. In 1855, Cane Creek Furnace, built by Jacob Stroup at Polkville, near present-day Anniston, advertised iron columns, lintels, rods, pinions, gudgeons, and gearing for mills. Its daily production was about three tons, of which one ton was made into hollowware and other castings, the second into pig iron, and the third into wrought-iron bars. It used about 600 bushels of charcoal per day and got its ore from nearby deposits, some of which were located within a few hundred yards of the stack. Its output went down the Coosa in flatboats to such places as Wetumpka, Montgomery, and Mobile, indicating that it had to look far afield for markets despite the small amount of iron it produced.

An unusually innovative Alabama ironworks was Shelby Furnace, which went into blast near Columbiana between 1846 and 1849. It was built by Horace Ware, a native of Massachusetts who came south with his parents, settling in North Carolina before coming to Alabama. Learning ironmaking from his father, Ware produced charcoal iron for markets as far away as Sheffield, England, again showing the importance of the export trade. Closer to home, however, were such customers as Pratt's cotton gin factory, indicating that the modest degree of manufacturing underway in the state by the late 1840s was having a favorable impact that may help explain Ware's interest in using relatively new techniques. An irascible but highly ingenious person, Ware was Alabama's most creative early ironmaster; he used a steamboat engine instead of a waterwheel for blowing power and experimented with a heated air blast in 1855. Although it was already common in the North, this was the first time that this method had been tried in Alabama. In 1858, Ware also began building Alabama's first rolling mill, which started making merchant bar iron in 1860. His installation, the most progressive in the state, also included a cupola, foundry, and machine shop.[37]

Another important furnace complex emerged in Roupes Valley, in Tuscaloosa County. In 1830, Daniel Hillman, from New Jersey, built a forge that made horseshoes, plowshares, and other agricultural products. Ralph McGehee, a member of the Broad River group, was among the promoters who funded the venture. Following Hillman's death in 1832, the property was bought by a cotton planter, Ninian Tannehill, whose name became permanently associated with the site. Moses Stroup started building a blast furnace there in 1859; two additional stacks were soon added; and other facilities were built during the Civil War with help from the Confederate government. Known as Old Tannehill, the installation fell into decay after the war but was restored in the 1950s by faculty members at the University of Alabama, with aid from the state government.[38]

In a perceptive account of how Alabama ironmakers utilized problematical raw materials, Jack R. Bergstresser has emphasized that some of the state's early blast furnaces were not technologically backward compared to charcoal-fired northern counterparts. One reason for this situation was the ease with which ironmasters and technical information moved from north to south. In 1825, the Oliphant brothers, owners of an innovative furnace near Connellsville, Pennsylvania, advised the managers of the Cedar Creek installation in northwest Alabama how to place a boiler atop the stack for increased efficiency in utilizing heat. Thus, as Bergstresser states, "an isolated blast furnace in Alabama was . . . among the nation's first to experiment successfully with waste furnace gases." Continuation of technological transfer through such persons as the Stroups, and the later prevalence of European and northern experts in Birmingham, which was easily accessible to Pittsburgh and other leading centers of ferrous metal manufacture by rail, must be taken into account in assessing the degree to which southern industrial growth suffered from the lack of a "strong indigenous technological tradition," as Gavin Wright has argued. On the other hand, the small number of Alabama's antebellum furnaces and its lack of progress in using coal for smelting show how far ironmaking in the state lagged behind northern standards. As previously noted, the Cedar Creek installation did not prosper for very long and went permanently out of blast before 1840. Nor should the fact that practically all of Alabama's early ironmasters came from outside the South be overlooked. In Dixie, young men generally aspired to vocations in such fields as agriculture, the military, preaching, politics, and law and thought industrial work best suited for slaves. Although engineers could always be imported, Sloss Furnaces and other Alabama enterprises would later experience hardship due to a lack of skilled technicians who were familiar with southern conditions and resources.[39]

Despite Jefferson County's preeminence in ironmaking after 1865, none of Alabama's antebellum blast furnaces were located there. Named for John ("Devil") Jones, a Tennessean who came to Alabama in 1815, Jones Valley remained a rural arcadia. An early inhabitant, Baylis Grace, was reputedly the first person to prove that its red hematite was actually iron ore; he did this by sending a shipment to a Bibb County forge in the 1840s and getting back some wrought-iron bars. Shortly afterward, Francis M. (Frank) Gilmer, a wealthy planter who belonged to Montgomery's Broad River group, came through the area on his way south from Tennessee and was impressed by the dark red rock that was crushed into powder by the pounding hooves of his horse. Filling his pockets and saddlebags with the substance, he took it to Montgomery and found that it was "solid iron ore and that he had come over a mountain of iron, named Red Mountain." From then on, Gilmer's imagination would be captivated by the district's industrial potential.

For more than a decade, nothing much was done to exploit Jones Valley's unique endowments. Riding through it in 1858, John T. Milner, who two years earlier had become chief engineer of a railroad that was projected to run north from Montgomery to Decatur through the heart of the Mineral District, described it as "one vast garden as far as the eye could reach." Born in 1826 in Pike County, Georgia, Milner was already on his way to becoming one of Alabama's greatest industrial pioneers. The son of a slaveholding planter who combined farming with railroading and gold mining, he had begun working in his father's mine when he was only ten years old. With his sharp eye for metallic deposits, he could not fail to notice Jones Valley's potential advantages.

Milner's career exemplified the way in which native southerners became engineers in the absence of educational institutions offering formal curricula in technical subjects. After studying at the University of Georgia and learning engineering through on-the-job training in building the Macon and Western Railroad, he went to California in 1849 in an ox-drawn wagon to take part in the gold rush. While there, he became a surveyor for the city of San Jose. After returning home in 1853, he moved to Alabama and became an engineer for the Montgomery and West Point line. As it neared completion to Pensacola toward the end of the decade, he accepted a commission from governor Andrew B. Moore to run a survey for a railroad, chartered in 1854, that Montgomery's Broad River elite wanted to build into the mineral-rich wilderness north of the Alabama capital. While conducting this mission, Milner first saw Jones Valley. Like Gilmer, he became obsessed by the dream of creating an industrial city on the banks of Village Creek.[40]

V

Coal mining in antebellum Alabama was even more underdeveloped than ironmaking. In 1855, northern interests organized the Alabama Coal Mining Company and opened a deep-shaft mine in the Cahaba field near Montevallo, the first in the state to use steam to hoist coal to the surface. The engine was purchased from a firm in Pennsylvania, and two English craftsmen supplied technical expertise. In every respect, the enterprise was exceptional by Alabama standards. Most early operations in the state, dating to the late 1820s when Levi Reid and James Grindle of Jefferson County settled along the Locust Fork of the Warrior River and sent small amounts of coal by flatboat to Mobile, exploited surface deposits, and it was not until 1839 that the first shaft mine, a small operation with a hand windlass, was sunk in Walker County. Legends that upstate coal had to be given away in Mobile for trial use and that one operator, David Hanby, had to send a slave to teach citizens of the city how to burn it are probably apocryphal because local residents were accustomed to importing it from England, but this reliance on outside supply in itself emphasizes the neglect of native mineral resources.

As late as 1850, Alabama's coal trade employed only about 200 persons, many of whom derived most of their living from farming and dug coal when time allowed. Hanby's business was bigger than most; he sent up to a dozen boatloads of coal to Mobile each year and earned modest profits despite shoals that capsized some of his boats. A Mobile gasworks was his main customer. Coal mining also lured a few operators in Tuscaloosa and Walker counties, but profits were low because of the lack of downstate markets and the perils of reaching them. Especially feared were Squaw Shoals, a dangerous stretch on the Black Warrior River near Tuscaloosa, and the Devil's Staircase, twisting whitewater rapids on the Coosa between Talladega and Wetumpka.[41]

Coal was often dug from outcroppings in stream beds. While making a pioneering survey of the state's mineral resources in 1849, University of Alabama geologist Michael Tuomey found that a novel technique, diving for coal, was widely used. Mooring a flatboat near an underwater outcropping, miners drove crowbars into the seam with mauls. After prying loose a mass of coal, a few men would plunge into the stream, "lift the coal bodily to the surface, and place it in the boat." Miners sometimes used cranes, raising blocks of coal with chains. The work was seasonal, done when water was low and agricultural needs were not pressing. Flatboats used in the trade held 200 bushels at most and cost about $70; on reaching Mobile, they were sold for lumber. Most operators were uninterested in underground deposits that cost too much to mine. "I did not meet with two persons," Tuomey said, "who were acquainted with the character or thickness of the beds at which they were at work." Underground labor, he was told, was "inimical to the free hunter habits of our working population."[42]

The experience of William Phineas Browne, a Vermonter who came to Alabama in 1831, indicates many of the problems that the state's early mine operators encountered when they tried to advance beyond rudimentary techniques. For almost two decades, Browne pursued varied business ventures, alternately gaining and losing large sums of money, until he finally located three seams of coal near Montevallo and went into partnership to exploit them. From the start, he was plagued by the difficulty of getting coal to market, and he spent much time and effort trying to create a spur line to the nearest railhead, ten miles away. He also had trouble raising capital to screen his coal, causing customers to complain about its dirty condition. As would be true in the future, impurities in Alabama coal were already plaguing the state's pioneer mine operators. Using slave labor exclusively after concluding that "no reliance whatever" could be placed on free operatives, Browne found it hard to lease blacks in Alabama because plantation owners worried about permitting them to work underground. As a result, he had to look as far away as Louisiana, where a cousin ran a sugar plantation, and Virginia, where he bought four slaves in 1860. "I must have a negro force or give up my business," he stated in 1859.[43]

VI

Alabama's mineral deposits attracted attention from geologists despite cultural, economic, and technological barriers that retarded their exploitation. Most of the impetus came from the University of Alabama, which opened its doors at Tuscaloosa in 1831. Eight years later, Richard T. Brumby, who became professor of natural history in 1834, published the first discussion of the state's subterranean wealth in *Barnard's Almanac*. When British geologist Sir Charles Lyell visited Alabama in 1846, Brumby took him to see the Warrior coal field. In an account of his travels, published in 1849, Lyell praised Brumby's work but stated that prospects for exploiting the area were dim, citing the "unthriftiness of slave labor."[44]

After Brumby left Tuscaloosa in 1848, his work was carried on by Michael Tuomey, an accomplished Irish-born scientist with an engineering degree from Rensselaer Polytechnic Institute. Prior to coming to Alabama, Tuomey had been state geologist of South Carolina. A bright-eyed man with muttonchop whiskers, whose face resembled that of a leprechaun, he looked like a person well suited to find buried treasure. Commissioned by the university's trustees to spend up to four months a year examining Alabama's mineral resources, he was appointed state geologist by the legislature in 1848 but received no appropriations to go along with his title. He did, however, receive university funds that enabled him to buy a wagon and hire an assistant. Two Tuscaloosa coal operators, James Hogan and Benjamin Whitfield, sometimes accompanied him in his travels.

Ignorant of scientific methods, residents of the Alabama hill country wondered why Tuomey did not use a divining rod. "When I exhibited my geological hammer," he wrote in a Tuscaloosa newspaper, "I could plainly see that the idea of hammering information out of rocks was either too simple or too preposterous to be entertained for a moment." Driving himself relentlessly and enduring privations that undermined his health, he found ten separate coal beds, large deposits of iron ore, and vast quantities of limestone and dolomite. In 1850, he published a detailed report describing these and other mineral resources. Four years later, the legislature gave him $10,000 to support his survey and provided a salary of $2,500 so that he could devote full time to the project. Tuomey made a second report to the legislature in 1855, but before it was published his appropriation had been used up and he left the university. He returned in 1856, but his labors had worn him out and he died the next year. After numerous delays due to lack of funds, a former colleague, J. W. Mallet, got the second report printed in 1858. Both reports would later be valuable when serious development of Alabama's mineral wealth finally began; meanwhile, taxpayers were disappointed that Tuomey had not found commercially valuable gold and silver deposits.[45]

VII

At one point, Browne hired Tuomey as a consultant. Like other advocates of Alabama's industrial development, both men were railroad enthusiasts, as were members of the Broad River group. During the 1850s, meetings of railroad promoters often took place in Alabama, usually at Prattville. Daniel Pratt urged state officials to support the building of rail lines that would bring outside capital into Alabama, enhance real estate values, break down rural isolation, yield fresh tax revenue, and lessen the cost of bringing agricultural and industrial goods to market. Jacksonian lawmakers, however, resisted his pleas.[46]

The railroad boom in England and America began in 1829 after British inventor George Stephenson's steam locomotive, the Rocket, won a competition that became famous as the Rainhill Trials. Within a few years, steam locomotives became widely used in the United States, which, because of its vast size, lack of internal political barriers, and commitment to aggressive economic growth, soon became the world's chief railroading nation. By 1840, it already had about 3,000 miles of railroads; by contrast, continental Europe as a whole had only 1,818.[47]

It is virtually impossible to exaggerate the importance of steam locomotion to American industrialization. Because of its speed, year-round operation, and ability to haul freight over any type of terrain, railroading made it possible to bring raw materials to factories and manufactured goods to market faster than ever before. Freed from carrying large inventories that turned over only slowly, merchants could now order goods overnight and charge lower mark-ups because of increased velocity of trade. Railroads also sped the movement of passengers. Along with seagoing steamships, they figuratively shrank the world.[48]

Revisionist historians have emphasized that it is unfair to judge southern progress in railroading simply by contrasting it with what happened in the North. By 1861, the South had much more railroad mileage than most other parts of the world. Nevertheless, it is undeniable that northeastern and midwestern states adopted the railroad much more avidly than those below the Ohio and Potomac rivers. Although the Charleston and Hamburg in South Carolina was for a time the longest line in North America, southern states soon fell behind their northern counterparts in using the new mode of transport. In 1860, the eleven states that later seceded to form the Confederacy had 9,001 miles of railroads; the national total was 30,636. That the North had more than twice the railroad mileage of the South had much to do with the fact that the Confederacy lost the Civil War. It also helps explain why Alabama, with only 743 miles of railroads in 1860, was relatively slow to develop its industrial potential.[49]

Wealthy planters were in the vanguard of what little railroad development took place in Alabama prior to the Civil War. Starting with the Decatur and Tuscumbia in 1830, built to haul cotton around the rapids of the Tennessee River at Muscle Shoals, most of the state's early railroads served agricultural needs. The Decatur and Tuscumbia was typical in being built by slave labor, which southern railroad builders preferred to free workers. Plentifully endowed with navigable rivers, the state depended on steamboats plying such streams as the Alabama, Black Warrior, Chattahoochee, Coosa, Tallapoosa, Tennessee, and Tombigbee to bring cotton and other agricultural commodities to Mobile, New Orleans, and Pensacola for export to outside markets. Enthusiasm for railroads, however, was widespread, and at least twenty-nine lines were chartered before the Panic of 1837 halted the drive. One of these, the Mobile and Cedar Point, was planned to take cotton from docks along Mobile's shallow harbor to deep water farther down the bay, thus eliminating the need for lighters to carry it to ocean-going ships. Other projected rail lines, such as the Selma and Tennessee, were intended to link major rivers. A lack of trained southern engineers to survey and build such lines caused promoters to hire talent from outside the region; David Deshler, an engineer from Pennsylvania, surveyed the Decatur and Tuscumbia. The desire to train skilled technicians closer to home prompted the establishment of an engineering school at the University of Alabama, where several trustees were active in promoting railroads.[50]

The Panic of 1837 killed the early drive for railroads in Alabama. More than a decade later, the Decatur and Tuscumbia was still the only completed rail line in the state, and its trackage was in such bad condition that steam locomotives had to be replaced by horses. Construction of the Montgomery and West Point, aimed at making connections with Georgia's rail system, stopped after only thirty-two miles had been finished. Lack of funds, scandals affecting the Alabama banking system, and voter indifference among the largely Jacksonian electorate strangled most projected lines. The University of Alabama's engineering program withered and was temporarily abandoned in 1846. Five years later, the Montgomery and West Point reached the Georgia line, but by 1852 Alabama still had only 165 miles of track.[51]

During the 1850s the winds began to shift, producing what Thornton has described as an "economic miracle." Overcoming its opposition to banks, of which the state had none from 1842 to 1850, the legislature had a change of heart because even small farmers suffered from lack of an adequate circulating medium and business suffered from an influx of "shinplasters," small notes issued outside the state. A free banking law was passed in 1850, and liberal chartering ensued; ten years later, eight banks were doing business, with $4.5 million of loans outstanding. Even more remarkable, in view of Thornton's statement that "before 1860 few Alabamians would have trusted a bank to hold their money," was the fact that deposits totaled $4.8 million. Manufacturing took on new life, and the value of industrial output doubled to $10.5 million by the end of the decade. Most of this growth took place in such small establishments as grist and saw mills, but twenty foundries and machine shops were making about $800,000 worth of products in 1860, and textile output doubled in the preceding ten years. The support given to Tuomey's geological explorations, however limited in extent, was a sign of the times. It was within this context that Horace Ware began experimenting at Shelby Furnace with apparatus already becoming common in the North.[52]

Amid these conditions, a flurry of railroad-building got underway. Construction of the Mobile and Ohio, conceived by M. J. D. Baldwyn and other Mobile businessmen to permit that city to compete with New Orleans for trade between the Gulf of Mexico and the upper Mississippi Valley, began early in 1851. Initially intended to connect Mobile and St. Louis, it shifted its sights toward Chicago with support from U.S. senator Stephen A. Douglas and the Illinois Central and benefited from passage of a pioneering land-grant bill that Douglas and Alabama senator William R. King pushed through Congress in 1850. By the end of the decade, the Mobile and Ohio, running mainly through eastern Mississippi, was carrying over 100,000 passengers a year. It also brought large amounts of raw cotton to Mobile for shipment to New York City and England. In March 1861, the driving of a silver spike at Corinth, Mississippi, completed rail connections from Mobile to the Ohio River at Columbus, Kentucky. From there, steamboats plied the remaining twenty miles to Cairo, the Illinois Central's southernmost terminus. Still, the Mobile and Ohio encountered stiff competition when the Crescent City forged an even better midwestern link, the New Orleans, Jackson, and Great Northern. [53]

Meanwhile, Montgomery, a bastion of Broad River strength and a strong rival of Mobile, launched efforts to lay track in three directions: south to Pensacola, west to Meridian, and north to Decatur, where linkage was to be effected with lines coming south from Nashville. The main force behind the drive was the head of the Montgomery and West Point, Charles T. Pollard, a Broad River associate who looked on railroads as ways not simply of connecting navigable waterways but also of forging independent systems optimizing the unique advantages of steam locomotion. By 1861, however, only the Alabama and Florida, the line to Pensacola, had been completed.

Because of the role that it would play in the future, Montgomery's most potentially important railroad venture was the South and North Alabama line, which was projected to penetrate the wilderness to the north of the capital city and open up the Mineral District that Tuomey had explored. Under a legislative grant of $10,000, John T. Milner conducted a survey of the area beginning in 1858. He recommended that a start be made toward forging two complementary systems, one coming south through the Tennessee Valley and Decatur and the other moving north from Montgomery through Wetumpka toward Limekiln Station, a place on the Alabama and Tennessee Rivers Railroad that later became better known as Calera. From that point, the road would proceed northward through Jones Valley. Milner's report, which became a mine of information for industrialists, was presented to state officials in the summer of 1859 but was derided by legislators, one of whom declared that much of the area through which the South and North would pass was "so poor that a buzzard, flying over it, would have to carry provisions on its back or starve to death." Nevertheless, the legislature adopted the report the following year and authorized $663,135 worth of financial incentives to begin work. Because of the need to overcome opposition from Selma, which was extremely jealous of Montgomery, efforts to penetrate the area immediately north of the capital city were deferred. In 1860, however, trackage began to be laid from Limekiln Station toward Jones Valley under the aegis of the Mountain Railroad Construction Company, chartered in February of that year. [54]

Amid conflict between rival groups of urban boosters and opposition from Jacksonian politicians, an alliance of planters, merchants, and railroad promoters tried to bind new transportation ventures to agricultural development. Against opposition from Montgomery, which already had its own steamboat line to New Orleans, Mobile attempted to forge a rail link with the Chattahoochee Valley before the outbreak of the Civil War. Coming from both directions, it featured two components, the Mobile Great Northern, heading east, and the Mobile and Girard, going southwest from Georgia's western entrepot, Columbus. Continuing an alliance with Selma to outflank the

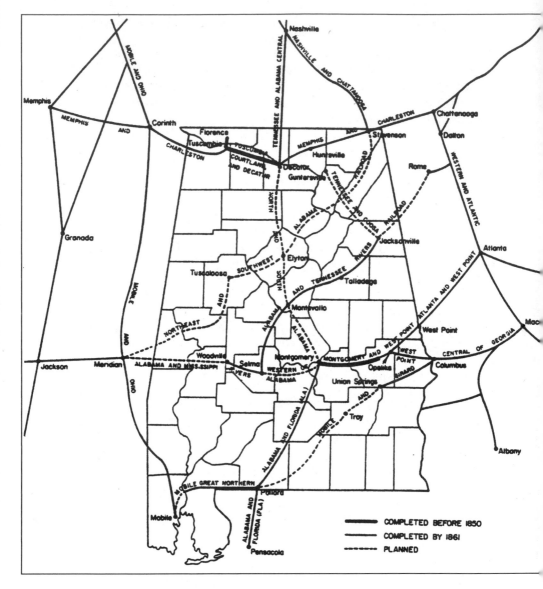

Location of Alabama Railroads in 1861, Showing Completed and Projected Lines
and Connections with Major Destinations Outside the State. The fact that lines
projected to penetrate the Mineral District of north-central Alabama remained
unbuilt at this time underscores the unreadiness of the state to make the type of
contribution to the Confederate cause that its natural resources might otherwise have
permitted. (Reprinted by permission of Louisiana State University Press from *Politics
and Power in a Slave Society: Alabama 1800–1860* by John Mills Thornton III,
copyright © 1978.)

northward thrust of Montgomery's Broad River group, Mobile also backed the Alabama and Tennessee Rivers Railroad, as the abortive Selma and Tennessee was now named. Starting at Selma, where enthusiasts bought a Philadelphia-built locomotive and christened it the Alabama, the projected artery had a northeastern trajectory toward Montevallo, Talladega, and Jacksonville, where it was to meet the Tennessee and Coosa Railroad coming south from Guntersville. The fact that the line would intersect with the South and North at Limekiln Station resulted in Selma's lending limited support to that Montgomery-backed enterprise, permitting Broad River associates and up-state allies to start construction from that point northward toward Jones Valley and Decatur in 1860. Meanwhile, Tuscaloosa promoted an ambitious project, the Northeast and Southwest Alabama Railroad, cutting diagonally across the state from Meridian toward Chattanooga, where it would connect with the Memphis and Charleston, coming eastward through the Tennessee Valley. In the absence of native-born workers to build such roads, technicians were imported from as far away as England. Instruction in engineering was resumed at the University of Alabama in 1851 to remedy the lack of local talent.[55]

Members of the Broad River group were not alone among large slave-holders backing the railroad drive. Noah B. Cloud, whose *American Cotton Planter* promoted crop diversification, soil conservation, scientific methods of cultivation, and educational reform, was among the important landowners active in the crusade.[56] Opponents, however, fought against the spread of the Iron Horse. Jacksonians, remembering the failure of the state bank in 1842, remained hostile to corporations, were afraid of urbanization, and believed that speculation would only invite disaster when the bubble finally burst. Secessionists, alarmed by abolitionism, also opposed rail connections to parts of the country that might soon become enemy territory.[57]

Opponents of transportation and manufacturing flocked to the banner of John A. Winston, a tight-fisted native of the Tennessee Valley who became governor of Alabama in 1853 and vetoed no less than thirty-three bills aimed at aiding railroad projects. Although many of these measures were passed over his opposition, requiring only a simple majority in the state legislature, Winston found ways to prevent allocation of authorized funds and block needed land transfers. Not until 1859, when he was out of office, was a general railroad aid law passed over protests that it would make the state subservient to corporations. Rivalry between Alabama cities also continued to hinder progress; Mobile, Montgomery, Selma, and Tuscaloosa fought stubbornly to advance their respective commercial interests. Unable to raise sufficient capital, a host of railroads that were launched in the early 1850s, including the Alabama and Mississippi, Mobile Great Northern, Mobile and Girard, Northeast and Southwest Alabama, South and North Alabama, Tennessee and Coosa, and Western of Alabama, remained unfinished by 1861. Backers of one of the few to be completed, the Alabama and Tennessee

Rivers, learned the hard way that its northeast terminus, Jacksonville, was little more than a dead end.[58]

VIII

During the 1850s, Alabama experienced an unprecedented burst of urban development. Towns and cities grew twice as fast as the population at large, and the number of citizens employed directly in agriculture declined from 75 to about 70 percent. Urban institutions, including libraries, increased dramatically; public schools received much greater support; colleges grew from five to seventeen; newspapers became more numerous, reaching nearly 100; and illiteracy among the white population per 1,000 population fell from eighty to seventy-two.[59] Such gains, however, were impressive chiefly because of the low base numbers from which they started. At the end of the decade, Alabama society continued to epitomize what Carl Degler has aptly called the "rurality of the South."[60]

Urban economies were tied tightly to agricultural interests. Mobile, with a population of 29,258 in 1860, was by far Alabama's largest city; its sixth ward alone had more people than any other place in the entire state. Despite boasting such firms as the Dog River Manufacturing Company, which made cotton cloth, and a foundry run by two Pennsylvanians, Mobile was almost wholly absorbed in exporting raw cotton. A visitor described it as a place "where people live in cotton houses and ride in cotton carriages. They buy cotton, sell cotton, think cotton, eat cotton, drink cotton, and dream cotton."[61]

Montgomery, chief center of the Black Belt and the state's second-largest city with 8,843 residents, was also oriented toward agriculture. The aggressiveness of its Broad River elite was epitomized by the success of the community in wresting the statehouse from Tuscaloosa. Aside from reaping the economic benefits of being Alabama's capital after 1847, Montgomery prospered mainly from selling cotton and slaves and from various professional pursuits, particularly law and medicine. Tuscaloosa ranked third with a population of only 3,989; although disappointed after losing out to Montgomery as state capital, it retained a modest trade in coal, iron, and textiles and avoided collapse mostly because of agricultural distribution, the continued presence of the university, and the decision of the state to build an insane asylum in the city. Despite widespread zeal for railroading within its business community, Tuscaloosa had no rail connections to the outside world and depended on steamboats plying the Black Warrior River to carry its goods to market.

Huntsville, which was virtually isolated from the southern parts of the state, had 3,634 people in 1860. It made modest amounts of textiles but was mainly an entrepot for agricultural goods, sending these commodities to market via the Tennessee and Mississippi rivers. Selma, the chief center of the western Black Belt, was a wealthy community in the state's largest cotton-producing county. With a population of 3,177, it continued to promote railroads, had a thriving steamboat trade, and tried with limited success to develop an iron foundry. Essentially, however, it was an agricultural distribution center despite the industrial role that it would later play in the Civil War. Alabama, with no other places having 2,000 or more people, offered relatively few models for the promoters of such later industrial cities as Birmingham.[62]

Cultural similarities between the antebellum and postwar eras, however, are evident. Like other parts of the early republic, in which a scramble for wealth was replacing other ways of determining status, pre–Civil War Alabama bequeathed a spirit of hard-driving enterprise to the industrial cities that emerged after 1865. In his *Flush Times of Alabama and Mississippi*, Joseph Baldwin described a social order "where the stranger of yesterday is the man of mark today. . . . where amidst a host of competitors in an open field of rivalry, every man . . . enters the course with a race-horse emulation, to win the prize which is glittering within sight of the rivals." Writing of "feverish speculation" among Alabama settlers, Virginia Hamilton later commented that "men were rewarded in the Black Belt, not on the basis of ancestry, but in proportion to their boldness, enterprise, or skill at gambling, swindling, or cheating." In such a society, as Baldwin said, "he who does not go ahead is run over and trodden down."[63]

Antebellum planters and industrialists who later spearheaded the growth of Birmingham shared these attitudes as deeply as other Alabamians and transmitted them to the city that would rise in Jones Valley. Perhaps the most enduring prewar legacy that was handed on to Birmingham was the antebellum attitude that black people were members of an inferior race, best suited for menial tasks in a society where white men reigned supreme. Historians, who until recently wrote mostly from a white perspective, differ about how badly slaves were treated, whether working in agricultural or industrial capacities. But it is clear that many practices later used by firms that employed black convicts in the Birmingham District, including flogging and punitive confinement in "nigger boxes" that were "barely large enough for a prisoner to stand erect and lighted only by air holes to prevent suffocation," were also used by slaveholders before the Civil War. Another carryover from prewar days was the conviction, both in Alabama and in other southern states, that cheap, servile black labor would enable the region to compete with northern manufacturers who had to cope with strikes and other problems that southerners believed they could avoid.[64]

IX

Despite the success achieved by small landholders and other Jacksonians in delaying Alabama's drive toward industrialization, the "economic miracle" of the 1850s figured critically in the later development of Jones Valley by bringing to the forefront two key figures who were destined to become Birmingham's greatest founding fathers. One of these was James Withers Sloss, who emerged in the 1850s as the leader in efforts to link the Tennessee Valley with Nashville on the north and the Mineral District to the south; the other was John T. Milner, whose surveys for the South and North Railroad opened his eyes to Jones Valley's vast industrial potential. Milner's move from Georgia to Alabama in 1853 to help his father build a railroad between Montgomery and Pensacola brought into the state one of the South's greatest prophets of economic growth. Having only modest financial means—he had been swindled out of a small fortune in California by unscrupulous prospectors with far more experience than he possessed at the time—Milner allied himself with Frank Gilmer and other wealthy Broad River associates, commencing a lifelong mission aimed at promoting Alabama's industrial development through forging links between agriculture, commerce, transportation, manufacturing, and urban growth.

Aptly characterized by Dorothea Orr Warren as a "practical dreamer," Milner possessed a fluent writing style, making it all the more unfortunate that his main published works are now available only in rare book collections. As a propagandist for southern economic development, he had few equals; worthy to be ranked with J. D. B. DeBow, William Gregg, and Daniel Pratt among antebellum protagonists, he was no less important in the postwar era than his more renowned contemporaries, Richard Hathaway Edmonds and Henry W. Grady.

Milner's steadfast conviction that Alabama was a "great and rich State" was rooted in its abundant natural resources; one cannot read his descriptions of the areas through which his railroad surveys took him without realizing that he had a poetic imagination matching his shrewd business sense and technological talents. Like other members of the Broad River group, he was deeply devoted to cotton, which he regarded as the agricultural equivalent of the gold that he and his father had spent so much of their careers in mining. On the other hand, he was passionately convinced that cotton must be raised in conjunction with diversified food crops so that his adopted state would not have to import these commodities from the outside. In his postwar writings he never tired of preaching that Alabama had enjoyed this type of agricultural diversification and self-sufficiency prior to 1861, only to become impoverished after Appomattox by concentrating too much on cotton cultivation.[65]

Milner typified Broad River attitudes by believing that railroads were vital to agricultural prosperity. In his reports to the state government and the owners of the South and North, written in 1859, he expressed amazement at the "holy horror . . . of all railroad corporations" that prevailed so widely among Alabama's Jacksonians and set forth reams of statistics to show how the Iron Horse had benefited such states as Georgia and South Carolina. Inveighing against the "mad fanaticism" that led such persons as John A. Winston to impede the progress of railroads, he declared that "the great State of Alabama is furrowed all over with half graded Railways and her bosom stuck full of engineer's stakes, yet the means are wanting to complete her improvements.

She gives to the people the richest and most abundant crops from her exuberant soil; while entombed in her mountains, lie millions of ores. Yet amid all this greatness, this wealth and native generosity, the *iron rail* is wanting to render her people prosperous and independent, and place her in the rank she ought to occupy among the States of the Union."[66]

Of all Alabama's cities, Milner declared, Montgomery was in the best position to benefit from railroads; indeed, it occupied "the most favored position for an unexampled improvement of any city in the South." Scorning arguments of "do-nothing men" that it could safely rely on its "incomparable river," Milner stated that such myopia could only play into the hands of the capital city's arch-rivals, Columbus, Eufaula, and Selma. "Men who have lived to green old age tell us that the battle of life is a perpetual struggle to maintain our position among our fellows," he said. "So it is with nations, and States, and even cities." By forging a rail link through the Mineral District to Decatur, intersected by branch lines that would exact tribute from would-be rivals, Montgomery would amass an immense trade not only in cotton but in groceries, which it would then funnel toward the Gulf. Foreshadowing what would ultimately happen late in the century, Milner urged that instead of continuing to wage economic warfare with Mobile, Montgomery had everything to gain by seeking a community of interest with that port to make it a "commercial emporium" for the entire state.[67]

Mining and manufacturing, far from being antithetical to agricultural interests, were in Milner's view vital components of a complex pattern of economic development from which Montgomery stood to derive great benefit; his report to the president and directors of the South and North bristled with statistics showing how fostering the coal and iron trade could aggrandize the city's wealth. He looked at Jefferson County not simply as a source of pig iron but as a place where iron could be rolled into rails. It was in this connection that he most clearly articulated the conviction, common among the planters with whom he was allied, that mining and manufacturing were entirely consistent with the interests of large landholders and slave owners and that the use of slave labor in industrial installations constituted the South's best weapon in competing with northern interests.[68]

Milner saw blacks as fitted only for servile occupations. "The negro is a peculiar being, and differs widely from all other races of men—in that they have no ambition, no aspiration, no care for to-morrow," he once wrote, adding that white men would "always look upon and treat the negro as an inferior being."[69] On the other hand, he believed, blacks were well suited for work not only as field hands but also as industrial operatives, particularly so long as the tasks they performed did not call for the exercise of great skill or fine judgment. In a prophetic passage in an 1859 report to the officers of the South and North, Milner declared,

I am clearly of opinion, that negro labor can be made exceedingly profitable in manufacturing iron, and in rolling mills. The want of skillful labor, has hitherto been the grand objection to the manufacture of rails in the South. This is compensated for by the freedom of the Southern States, from that curse of Northern works, "strikes among their workmen." Making Railroad bars is a monotonous process. Each bar is the facsimile of the other, and the great labor consists in the heavy lifting, and managing the heated masses in the machines or rolls.—It requires no great mechanical skill, even, for every part is done by machinery, which simply requires to be fed. A negro who can set a saw, or run a grist mill, or work in a blacksmith shop, can do work as cheaply in a rolling mill, even now, as white men do in the North, provided he has an *overseer—a southern man*, who knows how to manage negroes. I have long since learned that negro labor is more reliable and cheaper for any business connected with the construction of a railroad, than white.[70]

Here, in the words of one of Birmingham's founders, the nature of the city's future labor force and the attitudes underlying its exploitation were foreordained. Had the Civil War and the Thirteenth Amendment not intervened between 1859 and Birmingham's creation, the city's mines and furnaces would have been manned by slaves. From the perspective of the Old South, no other outcome was imaginable.

X

Because of the opposition of Jacksonians who wanted to restrain large-scale enterprise, and also because superior financial rewards from agriculture limited investment in transportation and manufacturing by planters otherwise sympathetic to industrial development, Alabama entered the Civil War unready to make the contribution to the Confederate cause that its natural resources might have permitted. Nevertheless, the Confederate government mounted a vigorous program of industrialization within the state during the war. Recognizing the dependence of the Southern cause on railroads, particularly because of the Northern blockade of sea routes, the Confederate congress spearheaded the completion of such earlier projects as connecting Selma and Meridian by rail. Although that goal was attained, failure to bridge the Tombigbee required a steamboat portage between Demopolis and McDowell's Bluff, and river boats were still needed to carry goods between Selma and Montgomery. With government help, work also proceeded on the projected rail line from Selma to Chattanooga via Blue Mountain—later Anniston—but this too remained unfinished by the end of the war.[71]

More successful than the drive to expand the rail network were efforts by Confederate officials to expand iron production in Alabama. No fewer than thirteen blast furnaces were built or completed within the state during the war years. Typical was the role played by Confederate authorities in building a furnace to produce naval iron in Bibb County after purchasing an ironworks built by Caswell C. Huckabee and other promoters in 1862. In some cases, as in the construction and operation of Oxford Furnace in Calhoun County, skilled manpower was provided by the army. Confederate troops also helped to build Cornwall Furnace in Cherokee County, while hundreds of slaves dug a canal and drove a tunnel through Dirtseller Mountain to provide water power for the bellows. The stack was erected by members of the Noble family, Cornish emigrants who operated a foundry at nearby Rome, Georgia, that made ordnance for the Southern cause. Attempts were also made with government support to use bituminous coal at Shelby Furnace and smelt red hematite at Cornwall Furnace.[72]

Alabama's blast furnaces made important contributions to the Confederate war effort. Iron from Cane Creek furnace helped transform the captured Union warship *Merrimack* into the ironclad *Virginia*, which defeated a federal naval squadron before it fought its famous encounter with the *Monitor* off Hampton Roads in March 1862. It was difficult, however, to develop Alabama's industrial potential under emergency conditions. The experience of the Shelby Iron Works is typical. Shortly after war broke out, Horace Ware sold most of his stake in the installation to six investors who reorganized it as the Shelby Iron Company. Throughout the war, the new management squabbled continually with the Nitre and Mining Bureau, which had charge of mobilizing production for military purposes, and with its head, Josiah Gorgas, a Pennsylvanian who had embraced the Southern cause. In addition to suffering from the steady depreciation of Confederate currency, which increased the price of pig iron to about $500 per ton by 1865, Shelby's managers had constant trouble finding and retaining skilled workers, many of whom were lost to the armed services at the worst possible times. Despite such problems, the Shelby company produced iron throughout the war. It furnished much of the armor plate, for example, for the C.S.S. *Tennessee*, which became famous for its ability to withstand repeated Northern broadsides in the Battle of Mobile Bay.[73]

Significantly, in light of Birmingham's later development, the first two blast furnaces in Jefferson County were built during the war, becoming the first fruits of the visions that Milner and the Broad River group had entertained of creating a great ironmaking center in the heart of the Mineral District. During the war, the Mountain Railroad Construction Company pushed the unfinished trackage of the South and North closer and closer to Jones Valley from Limekiln Station; by 1865, work crews had reached Brock's Gap. According to family tradition, Milner, with Confederate support, scavenged scrap iron to build a flimsy, twisting railway, full of sharp curves, steep grades, and sudden descents, across Shades Mountain to meet the South and North at Brock's Gap and thus enable pig iron to be taken from the new furnaces to Selma via the connection with the Alabama and Tennessee Rivers Railroad at Limekiln Station. Partly because all traces of Milner's line later disappeared, this story is hard to substantiate; instead, pack animals may have been used to haul iron to the South and North's railhead.[74]

Because it ultimately became the first installation in Alabama to smelt iron with coke, one of the Jefferson County stacks built during the war was particularly significant. Later known as Oxmoor Furnace, it was constructed in

1863 at Grace's Gap by the Red Mountain Iron and Coal Company, in which both Daniel Pratt and such Broad River associates as Frank Gilmer were active. Erected by Moses Stroup, the modestly proportioned stack was only thirty-two feet high and could not produce more than five or six tons of iron per day from the red hematite with which it was charged.

Jefferson County's other Civil War blast furnace was built at Irondale by Wallace S. McElwain, a native of Massachusetts who had moved to Mississippi in 1859. Making a cupola from the shell of a locomotive boiler that he fished from the Tallahatchie River, he helped establish a foundry at Holly Springs, Mississippi, not far from the Tennessee line. Before the war, it shipped ornamental iron to such places as the French Quarter of New Orleans; after 1861, it was enlarged and became the first installation to make small arms for the Confederacy under contract. Facing its capture by Union forces late in 1862, McElwain took its equipment to Jefferson County and built a furnace in Shades Valley. There, in 1864, he made a brief but encouraging experiment using coke for smelting. This fuel had been virtually unknown in Alabama before the war.[75]

Confederate authorities also promoted the expansion of the Alabama coal industry. Mines were opened in St. Clair County, but boats could reach downstate areas only "when the rivers were swollen, and sometimes would wait for months for enough water to float them over the shoals." Because of such problems, coal commanded prices in Montgomery ranging as high as $125 per ton. The Alabama Coal Mining Company sold much of its output to the Shelby works at Columbiana, but shipments were often held up for lack of rail cars, and skilled workers were in short supply. Coal mines that would figure importantly in the early history of Birmingham were also opened at another Shelby County site, Helena, in 1863.[76]

Much of the coal and iron that was produced in Alabama during the war years went to a large naval yard, arsenal, and ordnance works at Selma. That city was chosen for this important complex because of its location on the Alabama and Tennessee Rivers Railroad, its water connections with various places via the Alabama River, and its remoteness from the front lines. Skilled workers were recruited from all over the Confederacy, and Selma became a hive of industry, famed in later years for the construction and arming of the *Tennessee* and for production of the Confederacy's famous rifled Brooke guns, which were originally developed at the Tredegar Iron Works in Richmond. In the absence of rail links to most parts of the state, pig iron often had to be brought to Selma by wagons or river boats.[77]

Although not as large as the Tredegar works, which remained the Confederacy's main source of munitions and ordnance, the facilities at Selma, sprawling over fifty acres, contributed importantly to the war effort in 1864 and 1865 as prospects for the Southern cause became increasingly grim. About 10,000 persons, some of whom were only nine or ten years old, worked there under such managers as Colin J. McRae, a politician, railroad promoter, and Mobile cotton factor who established the arsenal as a privately owned venture in 1862 before becoming a Confederate financial agent in Europe; Catesby ap R. Jones, who commanded the *Virginia* in its encounter with the *Monitor* before taking charge of the Selma plant in June 1863; and George Peacock, an English-born ironmaster who devised new core-making techniques for producing shells and spearheaded an effort to find deposits of good coking coal within Alabama. "In addition to the government plants," states one historian, "there was a host of private firms producing rifles, shovels, oil, whiskey, cotton cards, canteens, and a variety of other goods." After 1865, skilled workers from Selma fanned out over the South, contributing greatly to its industrial development by performing tasks for which black workers were regarded as unsuitable.

During the war, however, officials at Selma endured numerous difficulties showing the inability of Alabama and the Confederacy alike to assume the industrial demands of a major war. In 1862, lack of funds forced McRae to use barter to obtain labor and raw materials. Skilled workers, particularly molders, were hard to find and retain. Poor rail connections hampered deliveries of munitions to distant military theaters. In its frustration, the Confederate government took over the complex in February 1863, but even this action failed to solve the problems plaguing the facility's operation; one exasperated official wrote that "its cost has been too great to put it aside, and like the old man of Sinbad's it will continue to cling tenaciously around the neck of the Ordnance Department notwithstanding all its unavailing struggles." Deliveries of cannon did not begin until January 1864, almost twenty-two months after the facility was created. As Edwin T. Layton has stated, it is remarkable, given such circumstances, that the Selma works was able to play even "a modest, but measurable role in the defense of the Confederacy." Harassed officials experienced a similar outcome at Tallassee, to which a carbine factory was moved from Richmond, Virginia, in 1864 and reestablished on the site of textile mills at the falls of the Tallapoosa River. Because of many of the same problems that affected Selma, the Tallassee plant made only about 600 cavalry arms before the war ended.[78]

Alabama's wartime industrialization terminated abruptly in March and April 1865, when 14,000 Northern troops under Gen. James H. Wilson swept through the state in what one historian has aptly called a "Yankee Blitzkrieg," systematically destroying or damaging every ironworks they could find. Among others, both Jefferson County stacks, the Tannehill installation, and the Bibb, Oxford, and Shelby furnaces were put out of commission. After defeating Confederate general Nathan Bedford Forrest's forces on April 2, the invaders captured Selma, leveling what remained of the arsenal and naval foundry. Before abandoning the city, the defenders dumped much of its industrial equipment into the Alabama River to prevent it from falling into enemy hands. Tuscaloosa suffered a similar fate as Union forces under Gen. John T. Croxton burned most of the buildings at the University of Alabama.[79]

By the time Wilson's raid was over, the fire-blackened ruins of foundries and furnaces were all that remained of Alabama's already limited capacity to produce ferrous metals. Still, the war had helped demonstrate the state's industrial potential. Within a few decades, Jones Valley would be transformed, and an aggressive young city would stand where nothing but "sagebrush, woodland, and cotton fields" had previously existed.[80] This change required a furious struggle between carpetbaggers, scalawags, and members of the Broad River group. In the end, planter-dominated forces, preaching white supremacy and bent on carrying out Milner's mandate to build an industrial metropolis manned by cheap, servile black labor, would triumph. Jacksonian Democrats would no longer control the state government, and ex-Whigs with strong ties to railroads and other corporations would be ascendant. Amid the strife for which the Reconstruction Era became proverbial, the South's greatest center of heavy industry was soon to be born.

James W. Sloss and the Birth of Birmingham

Like every other southern state, Alabama had a hard time during Reconstruction. Harsh realities had to be overcome to make even limited gains. Despite all odds, it was an era of achievement. Amid recovery from a devastating war, Alabama became the South's leading producer of two basic industrial commodities, coal and iron. Fundamental to this accomplishment was the fulfillment of Milner's prewar dream that a manufacturing center might rise under the shadow of Red Mountain, but this took place only after a bitter struggle that left enduring marks on what Ethel Armes would later call a "bold, mean, little town."[1] Without intense, unyielding determination, Birmingham could not have survived infancy.

Birmingham's seedtime was an era notable for two significant changes in ferrous metal production, both of which had a lasting impact on Alabama's Mineral District. The first of these transformations, the emergence of bituminous coal as the iron industry's main blast furnace fuel, greatly benefited Jones Valley. The other change, however, had problematical results. By the time the Sloss Furnace Company was founded in 1881, new methods had been brought to the United States from Europe that made it possible to produce steel much more cheaply than ever before. For decades, learning how to make this profitable alloy from intractable southern resources would lure Alabama's ironmasters like a magnet, only to recede like a mirage.

Above all, it was a ruthless age, that of the "Robber Barons." Business leaders enriched themselves by linking a constantly growing network of towns and cities with far-flung rail networks, creating an impersonal, ruthlessly competitive market system that relentlessly destroyed old ways of conducting trade.[2] No place in America epitomized the competitive impulses of the era better than did Birmingham, a city whose emergence was full of intrigue and double-dealing. And no person was more responsible for Birmingham's birth than James W. Sloss, who forged an enduring link between Jones Valley's locational advantages and the interests of a great railroad, the Louisville and Nashville.

I

After Lee surrendered at Appomattox, the defeated South lay in ruins. The times demanded an almost desperate type of tenacity, rooted in the belief that things must inevitably get better because it was hard to see how they could get much worse. Such industrial capacity as the defeated Confederacy possessed had been either destroyed by invading Northern armies or used up in the wear and tear of the four-year conflict. The southern rail network had been dismantled by federal troops who burned bridges and trestles, demolished depots, and ripped up trackage. Heating rails in bonfires and twisting them around trees or telegraph poles, they left the landscape festooned with grotesque relics that were sardonically called "Sherman's hairpins." Confederate bonds and currency were worthless, credit institutions virtually nonexistent, and money scarce. Agriculture recovered at a snail's pace; not until 1878 did cotton production regain 1860 levels. In 1880, southern per capita wealth stood at $375; the corresponding figure for other parts of the nation was $1,086. Poverty was endemic; even plantation owners were often shabbily dressed.[3]

Ruined cities were forlorn re..iinders of the Lost Cause. Columbus, Georgia, was typical. "The destruction of the city's economic system was so thorough and the disruption of its social organization so complete that simple survival became the paramount concern of the people during the months immediately following Wilson's raid," states one account. "Many people in the city were left destitute by the war, and many others were homeless as well." Some places, like Memphis and Nashville, were relatively unscathed because they had been occupied by Northern troops during most of the conflict; others, like Atlanta and Richmond, recovered briskly despite having been gutted in its final stages. Far different, however, were conditions in coastal cities that had dominated the antebellum cotton trade. Charleston's burned-out business district still lay in ruins fifteen years after Appomattox. New Orleans, which had once exported half the nation's crop, long remained deeply depressed.[4]

Alabama did not escape the prevailing distress. In 1866, its governor, Robert M. Patton, calculated the state's losses since 1861 at roughly half a billion dollars. Cotton production had shrunk by more than 50 percent, from 990,000 to 430,000 bales; real estate, valued at $500 million in 1860, was worth less than half that much ten years later. Over the same period, manufacturing capital plummeted from $9 million to $5.7 million. Even by 1880, the state's cotton crop was less than 700,000 bales, and the yield per acre was still much less than it had been two decades before.[5]

What few cities the state had possessed prior to 1861 lay prostrate as the

Reconstruction Era began. Mobile, a center of blockade-running throughout much of the war, had somehow averted destruction despite Farragut's success in closing its harbor in August 1864. Soon after the war ended, however, an explosion of unknown origin devastated much of the city, which remained demoralized and mired in debt for decades. Much of its prewar cotton trade disappeared, and its population shrank from 32,034 to 31,076 between 1870 and 1890. Montgomery, which had become a thriving manufacturing center during the war, lost many of its industrial facilities at the hands of Wilson's raiders but was luckily spared destruction. Its recovery was therefore faster than that of Mobile, but nonetheless difficult. After the war, the capital city forged rail links with Selma and Eufaula, but completion of a line northward to Decatur was delayed by strife between carpetbaggers and local investors. Partly because it could not absorb an influx of destitute freedmen, Montgomery was also hard hit by the nationwide depression that began in 1873. Tuscaloosa also floundered. In 1866, torrential rains tore a ravine, known as the "Big Gully," through the heart of town, but money was so scarce that the damage was not repaired for two decades.[6]

Industry grew at a snail's pace in Alabama during the fifteen years after 1865. In Jefferson County, Irondale furnace was returned to blast in 1866 by Wallace McElwain, with help from investors in Cincinnati. Raising the height of the stack to forty-six feet, he used a hot-air blast and a steam-powered blowing engine but gave up in the mid-1870s after seven years of heartache. Later, the installation was dismantled. Bibb Furnace, now known as the Briarfield Iron Works, was returned to life in 1866 by Josiah Gorgas, but his efforts were ultimately fruitless. He had a hard time recruiting workers and was constantly short of cash. Chronic squabbling with the owners of the Alabama and Tennessee Rivers Railroad weighed him down; eventually, bitter and disillusioned, he decided to sell out. Taking a post as headmaster of the preparatory school at Sewanee, Tennessee, in 1869, he later became president of the University of Alabama. The Briarfield Works stayed in operation under new owners, but were shut down by the depression that began in 1873 and remained dormant for nearly a decade. Another installation that had escaped wartime destruction due to its remote location, Hale and Murdock Furnace in Lamar County, was temporarily active after the war but could not overcome the handicap of being twenty-five miles from the nearest railhead. It shut down in 1870.[7]

An important episode in Alabama's reconstruction began in 1867 when the Noble family, which continued to operate a foundry in Rome, Georgia, began reconstructing Cornwall Furnace in Cherokee County with help from northern investors. The stack went into blast in 1869, but after it was destroyed by fire the next year, the Nobles sold out. Undaunted, they moved to nearby Calhoun County and, once more aided by northern financiers, acquired what remained of Oxford Furnace, which had been left in ruins by Union raiders at the end of the Civil War. In 1872 a fateful meeting occurred between Samuel Noble, who by this time had become the family's dominant member, and Daniel Tyler, a Connecticut capitalist. Impressed by Noble's ambition and drive, Tyler provided funding to organize a firm that built a new charcoal-fired furnace near the old Oxford stack. The venture was named the Woodstock Iron Company, after a village near Oxford, England. A carefully planned model village was established near the new furnace; called Anniston in honor of Annie Tyler, Daniel Tyler's daughter-in-law, it ultimately became a showcase of postwar southern industrialization. Progress, however, was difficult. Launching a northern sales campaign, Samuel Noble secured a contract to supply iron to the Springfield Armory in Massachusetts, and the Woodstock Company earned about $87,000 in its first year. But the Panic of 1873 hit the enterprise hard, and profits plummeted to less than $26,000 in the next twelve months.[8]

The Shelby Iron Company, which resumed production in 1869, was among a few firms that came through the Reconstruction period relatively unscathed. Obtaining capital from businessmen in Connecticut and New York, who also supplied trained managers, it benefited from a location that gave it access to two railroads: the Selma, Rome, and Dalton, as the former Alabama and Tennessee Rivers Railroad was now known after having been linked with the Georgia rail system, and the Louisville and Nashville. Developing profitable northern markets for charcoal car wheel iron, Shelby's owners added a new stack in 1875 and survived the ongoing depression with relative ease. Another successful enterprise, the Tecumseh Iron Company in Cherokee County, was organized by a northerner, Willard Warner, who had served with Sherman's army and shrewdly gave the enterprise his old commander's middle name to attract Yankee funding. Aided by good management and northern money, the firm also benefited from the fact that, like the Shelby Iron Company, it could ship iron to market via the Selma, Rome, and Dalton.[9] The Shelby and Tecumseh ventures, however, were atypical. Six of Alabama's wartime furnaces never resumed operations, and most of the ones that did fared poorly. In 1875, Alabama produced only 25,108 tons of pig iron. Compared with northern output, this figure was tiny. Pennsylvania alone produced 960,894 tons of pig iron that same year.[10]

II

Alabama's poor showing was not measurable merely numerically. Its blast furnaces, like those of other southern states, were obsolete. Every ounce of Alabama's minuscule production in 1875 came from furnaces that still used charcoal. Above the Mason-Dixon Line, conditions were much different. Before the war, furnaces in northeastern Pennsylvania had begun smelting iron with anthracite, named after *anthrax*, a Greek word for coal. This mineral substance was sometimes called "stone coal" due to its glossy finish and hard texture. Lacking hydrogen, it burned with a virtual absence of flame. A hot-blast technique imported from Great Britain, beginning in

1839, made its utilization possible after early efforts to smelt iron with anthracite had failed. By 1861, more than 100 furnaces, mostly in eastern Pennsylvania, were using this fuel, which contained more latent heat than charcoal and was also cheaper, selling for prices as low as $3 per ton in the early 1850s. Anthracite was superior to charcoal in its ability to sustain a greater weight of ore without crumbling; furnaces that used it could therefore be taller than charcoal-burning stacks, resulting in greater efficiency and lower costs. [11]

As Alfred D. Chandler has pointed out, the use of anthracite led to the rapid development of coal-based industries in the United States. Because of its low cost and easy availability after canals and railroads had penetrated northeastern Pennsylvania in the 1830s and 1840s, older methods of refining wrought iron were superseded by "puddling," which had been developed in Great Britain by Henry Cort. Through use of this method, pig iron was decarbonized by being heated in coal-fired reverberatory furnaces before passing through rolls, emerging as wrought-iron bars suitable for conversion into boiler plate, sheets, and rails. Factories, now able to adopt all-metal machinery because iron was less expensive, multiplied throughout the Northeast; steam engines replaced waterwheels; and productivity increased rapidly in a wide range of industries that were dependent upon heat. By the time of the Civil War, anthracite was the nation's leading blast furnace fuel. [12]

Even more significant was the increasing use by northern and midwestern ironmakers of bituminous coal, particularly after the war. This switch resulted from converting coal into coke, which rivaled anthracite in mechanical strength but burned more fiercely because of its greater porosity. Because railroads did not cross the Appalachian barrier until the 1850s, the only deposits of bituminous coal readily accessible to eastern manufacturers before that time were those near Richmond, Virginia. Once trackage had scaled the Allegheny Mountains, however, vast bituminous coal fields were opened in western Pennsylvania, Kentucky, Ohio, and West Virginia, and transportation costs plummeted. In 1875, a watershed was reached when, for the first time, more pig iron was smelted in the United States with bituminous coal and coke—947,545 as against 908,046 tons—than with anthracite. By this time, only 410,990 of the national output of 2,266,581 tons of pig iron was still made with charcoal. This decline was not because charcoal iron was inferior in quality; it remained the material of choice for such specialized products as railroad car wheels and was also suitable for agricultural implements. Coke, however, was cheaper and superior for high-volume production. [13]

Alabama and other southern states were technologically handicapped because they produced virtually no steel at a time when this much-prized alloy was beginning to be made in large quantities in the North and marketed at high profits under tariff protection. While the war was still raging, two groups of businessmen, one in the Northeast and the other in the Midwest, began to make steel in a revolutionary new way. During the 1850s, a British inventor, Henry Bessemer, had discovered how to eliminate carbon by blowing air through molten pig iron in a pear-shaped converter. As oxygen in the air united with carbon in the iron, the carbon burned away. Because of the heat that was liberated by the combustion of carbon, no other fuel was needed. By stopping the air blast at the right time, Bessemer achieved the small but critical percentage of carbon needed to produce steel.[14]

After securing British patents, Bessemer sought similar protection in the United States but was challenged by an American inventor, William Kelly, who claimed that he had discovered the same principle earlier while running a Kentucky ironworks. Although this claim was doubtful, Kelly won an American patent covering the air-blowing concept. The U.S. Patent Office, however, upheld the originality of Bessemer's converter. Things became even more complicated when another British inventor, Robert Mushet, found that Bessemer's method would work better if the air blast was continued until carbon had been completely removed from the iron, after which enough carbon for steel could be reintroduced through the addition of spiegeleisen, a compound of iron, manganese, and carbon. Mushet won British and American patents for his process.

For the first time, large amounts of steel could now be made at low cost. Producing it legally in America, however, required access to the combined Bessemer, Kelly, and Mushet patents, which proved impossible for any one manufacturer to obtain. Despite this obstacle, two groups of promoters used the new technology illegally because they needed steel to replace iron rails that were constantly wearing out under the abrasive pounding of larger locomotives and heavier rolling stock. After obtaining rights from Bessemer, owners of the New York Central began making steel at a plant in Troy, New York, in 1865. Meanwhile, winning authorization from Kelly and Mushet, another syndicate made steel at Wyandotte, Michigan, near Detroit. Infringing on each other's rights, the two groups battled in court until 1866, when they created the Pneumatic Steel Company to pool their patents and license steel production, on a royalty basis, to firms wishing to use Bessemer technology. By 1876, eleven additional northern plants had secured licenses. Most flourished, and American output of Bessemer steel soared from 2,000 tons in 1867 to 470,000 tons in 1876.

Late in the antebellum era, northern progress in iron and steel production was enhanced by the discovery of large ore deposits in the Lake Superior region. Because that area lacked coal deposits, it was necessary to ship vast quantities of ore to blast furnaces in western Pennsylvania and eastern Ohio via the Great Lakes. Completion in 1855 of the Sault Ste. Marie ("Soo") Canal between Lake Superior and Lake Huron, which became the world's most heavily traveled artificial waterway, stimulated the subsequent growth of iron and steel production in places like Pittsburgh.[15]

Fortunately for northern interests, and for the Pittsburgh area in particular, Lake Superior ore and bituminous coal from such places as Connellsville, Pennsylvania, made an ideal combination for Bessemer steel. Southern conditions, however, were different. Tests at Troy, New York, in 1867 showed that

Alabama pig iron, made from brown ore in a charcoal furnace, could be converted into an acceptable grade of steel, but this discovery was not particularly helpful. What was needed in Jones Valley was a way of producing commercially significant quantities of steel in blast furnaces that used red hematite and coke made from Warrior coal. Such a combination resulted in pig iron that had too much phosphorus, silica, and sulfur for the Bessemer process to utilize.[16]

Because much European iron ore was also high in phosphorus, inventors in Great Britain and France experimented with ways of dealing with problems caused by this troublesome element. One method, developed in England by Sidney G. Thomas and Percy C. Gilchrist, replaced the acid lining of silica brick in a Bessemer converter with dolomite, a basic material that combined easily with phosphorus. They also added limestone to the charge of molten iron as a flux. But few ores proved initially suitable, and Bessemer steel manufacture in the United States remained almost totally committed to the acid process.

Throughout the late nineteenth century, southern industrialists hoped to find economical ways to make basic iron instead of the acid variety and also to deal with the high sulfur content of southern coal, so that steel could be produced in the region. The transfer to America in 1868 of the open-hearth technique of making steel encouraged such dreams. Developed in Great Britain and France, it used a reverberatory furnace in which heat transferred by hot gases was radiated downward from the ceiling of a chamber in which iron ore and other materials were converted into steel. Air and coal gas were preheated in two regenerative chambers containing stacks of brick checkerwork, and efficiency was maximized by alternately switching intake and exhaust. Because the fuel that heated the bricks did not come in contact with the metallic charge, cheap coal could be used, and problems posed by impurities were minimized. By adopting the open-hearth process and lining the reverberatory chamber with dolomite or other materials that combined readily with phosphorus, southern managers hoped to make steel from regional raw materials. This goal, however, was hard to implement and failed to address the problems caused by silica and sulfur. As a result, steelmaking did not begin on a significant scale in the South until the end of the nineteenth century.

Fortunately for the region, the growth of the southern iron industry did not require success in making steel. Had this been true, Birmingham would never have existed. For many years, most American steel went into producing rails, and cast and wrought iron remained useful for a host of commodities, including agricultural implements, animal traps, boiler plate, firearms, machinery, nails, locomotive and steam engine components, stove grates, tools, and wire. Major cities abounded in foundries and rolling mills that used large quantities of pig iron for conversion into either cast or wrought iron, depending on whether or not tensile strength was needed. Many potential outlets for southern blast furnaces therefore existed, but deriving profit from these required something more up to date than the charcoal-burning stacks that were scattered throughout the region in the mid-1870s. Gaining access to outside markets also required completion of rail connections with major northern and midwestern cities.[17]

III

Tennessee was the first southern state in which ironmasters responded effectively to the need for change. Long before the Civil War, blast furnaces were numerous in its mountainous eastern counties, which abounded in iron ore and limestone. Partly because of the progress that the Volunteer State made in substituting coke for charcoal, it temporarily surpassed Alabama in iron production. A significant breakthrough leading to Tennessee's ability to use coke in ironmaking took place in 1850, as the Nashville, Chattanooga, and St. Louis Railroad pushed through the Cumberland Plateau. An Irish prospector, Leslie Kennedy, discovered an outcropping of coal near Montgomery's Gap. Aided by Nashville lawyer William M. Bilbo, he persuaded a group of financiers in New York, headed by Samuel F. Tracy, to establish the Sewanee Mining Company. After incurring heavy costs to build a rail spur to reach its deposits at Tracy City, this enterprise was reorganized in 1860 as the Tennessee Coal and Railroad Company. During the Civil War, its mines and railways were exploited alternately by Confederate and Union forces and suffered heavy damage. After 1865, however, the enterprise recovered under the leadership of another Nashville attorney, Arthur S. Colyar. By 1873, it was producing almost half the coal mined in Tennessee. Partly by using convict labor, but also by proving the suitability of its coal for coke production, it survived the Panic of 1873, using a crude experimental furnace known as the Fiery Gizzard to make a run of pig iron with this fuel. Although this primitive pilot plant was short lived, the company capitalized on its temporary success by selling coke to nearby ironmaking installations. Through its leadership, 10,300 of the 28,311 tons of pig iron made in Tennessee in 1875 were smelted with bituminous coal or coke.[18]

The Nashville area, which was already an important ironmaking district before the war, remained so after 1865. The chief postwar center of Tennessee's iron industry, however, was the Chattanooga District, just east of Tracy City. During the war, John T. Wilder, a Union cavalry officer, discovered ore deposits in Roane County. Aided by backers in Indiana and Ohio, he returned afterward to establish a highly progressive firm, the Roane Iron Company, at Rockwood. Beginning in 1878, using ore from the nearby Cranberry deposits in North Carolina and adopting the open-hearth acid process, Roane became the first enterprise in the Deep South to make steel. By that time, fourteen ironworks and machine companies were operating in the Chattanooga area. Roane employed about 600 men and was capitalized at $1 million.

Foreign capital also boosted iron production in east Tennessee. Through the influence of Wilder and that of Sir Lowthian Bell, a British metallurgist who predicted that the southern region would "prove a match for any part of the world in the production of cheap iron," a syndicate from England acquired large coal and ore properties in the Sequatchie Valley near Chattanooga. Establishing the Southern States Coal, Iron, and Land Company, it built furnaces at South Pittsburg, on the Tennessee River just north of the Alabama line. James Bowron, Sr., a Quaker metallurgist and industrialist from Stockton-on-Tees, was chief executive officer of the group, and Thomas Whitwell, a prominent British inventor of hot-blast stoves, was affiliated with it. English engineers and operatives came to Tennessee to open coal mines, erect furnaces and coke ovens, and smelt pig iron. When Bowron died late in 1877, his son, James Bowron, Jr., became general manager of the venture, which was floundering because of a lack of investment capital and British unfamiliarity with American conditions. Within a few years, it was acquired by the ever-growing mining and ironmaking enterprise centered around Sewanee and Tracy City. This aggressive entity was now known as the Tennessee Coal, Iron, and Railroad Company, or more simply as TCI. Alabama would soon become extremely familiar with those three initials.[19]

IV

Despite Tennessee's progress, its iron production fell behind that of Alabama by 1879. That year, Alabama made 49,841 tons of pig iron, while the Volunteer State made only 41,475. Key to the turnaround was the fact that 17,850 tons of Alabama's output now came from the use of bituminous coal or coke. Within only a few years, Alabama had become the leading southern ironmaking state.[20]

This dramatic reversal was traceable to events involving Jones Valley. On December 20, 1870, a group of promoters, "knowing the fact that there were immense deposits of coal and iron ore in Jefferson County," met in Montgomery and organized the Elyton Land Company. Named for what was then Jefferson County's leading town, this enterprise was created to acquire real estate and build a new city in the heart of the Mineral District. Josiah Morris, a prominent Montgomery banker, was the leading stockholder, with 437 shares; ultimately, his investment made him the wealthiest person in Alabama. At the time the Elyton venture was established, Morris had just purchased 4,150 acres of land in the Jones Valley for himself and nine associates, many of whom were members or descendants of the prewar Broad River group. Like them, Morris had Georgia antecedents, having lived in the Peach State for sixteen years and married into a prominent Columbus family. Most members of the group were heavily involved in the South and North Railroad. They now aimed to fulfill the dreams of Frank Gilmer, John T. Milner, and other prewar Broad River associates by founding a great ironmaking center at the base of Red Mountain.[21]

Two members of the Elyton syndicate rivaled Morris in the size of their stockholdings. One, James R. Powell, was a tall, square-jawed man who had migrated to Alabama in 1833 from Virginia, where his father had owned a plantation in Brunswick County before losing his fortune in land speculation. Arriving in Montgomery with only twenty dollars and a few personal belongings, Powell was befriended by Abner McGehee and John B. Scott, both prominent members of the Broad River group, and became involved in their transportation ventures. Launching a highly successful career, he became a mail contractor, stagecoach operator, cotton planter, and state legislator. During the war, he bought heavily in Mississippi agricultural lands; afterward, he pursued a colonization venture in the Yazoo region, hoping to establish "a cotton world where he might rule like a shah," as Ethel Armes later stated.[22]

Powell's subscription in the Elyton venture, 360 shares, was matched by that of Samuel Tate, who had figured prominently during the 1850s in the construction of the Memphis and Charleston Railroad along the old route of the Decatur and Tuscumbia. After the war, Tate became president of the Memphis and Chattanooga before signing a contract in 1869 to complete the South and North between Montgomery and Decatur. One of his sons, Thomas S. Tate, later became Birmingham's second mayor in 1873 when the first incumbent, Robert Henley, died of tuberculosis soon after taking office.

Four Broad River associates subscribed 120 shares each in the Elyton venture. One of them, Henry M. Caldwell, was a native of Greenville, Alabama, who had earned a medical degree from the University of Pennsylvania. Epitomizing the kinship ties connecting Broad River people, he was married to a sister of Willis J. Milner, who was John T. Milner's brother. Another Elyton subscriber, James N. Gilmer, belonged to one of the Broad River group's most prominent families. Born at Montgomery in 1839, he had graduated from Georgia Military Institute at Marietta in 1858 and served with distinction in the Confederate army before resuming a business career in Alabama after the war. Among its other properties, his family owned a plantation, Ellerslie, in Autauga County, again epitomizing the close connections between large-scale agriculture and postwar industrial growth. Bolling Hall, another Broad River associate whose Elyton holdings matched those of Caldwell and Gilmer, had established a law practice in Montgomery after graduating from the University of Georgia in 1831. As one account of his life stated, however, he preferred "the independent life of the planter" and "devoted himself largely to agriculture." According to family tradition, he had recouped his fortunes after the Civil War by raising strawberries. Along with Morris and Powell, Hall served on the South and North's board of directors. Campbell Wallace, a Georgian like so many other Broad River affiliates, was a former Confederate officer who held another block of 120 shares in the Elyton company. Like Tate, he was a veteran railroad builder and promoter who by 1870 was one of the South and North's principal contractors.

Three members of the Elyton syndicate came from Jefferson County. The most prominent, owning 180 shares, was William S. Mudd, a native of Kentucky whose parents migrated to Madison County, Alabama. After graduating from St. Joseph's College in Bardstown, Kentucky, Mudd practiced law at Elyton before becoming an assemblyman and circuit judge. After the war, he resumed his judicial career. One of Mudd's friends, William F. Nabers, was a key member of the Elyton company because he had owned much of the tract on which the city of Birmingham would soon be built. He held a block of 180 shares. "His life has been principally devoted to agriculture," stated a brief sketch of his career. Nabers's father-in-law, Benjamin P. Worthington, who held 133 shares, was also involved in the Elyton venture. [23]

Collectively, the Elyton group exemplified strong connections between prewar plantation agriculture, transportation, and urban planning and postwar southern industrialization. All native southerners, they also typified the extent to which the South developed its own mining and ironmaking enterprises after Appomattox. From the start, the city that the group founded was a brainchild of southern entrepreneurship, rooted in Old South traditions.

At a meeting held at Montgomery on January 26, 1871, Powell was elected president of the Elyton company. At his suggestion, the city that was about to be created was named "Birmingham" in honor of the famous British industrial center, which he had visited during a recent trip to Europe. Powell proceeded at once to the site selected for the new community, setting up his office in a two-room dwelling that had originally been built as a section house for the Alabama and Chattanooga Railroad. Soon, surveyors headed by a civil engineer, William P. Barker, were laying out a gridwork of streets stretching from Flint Ridge to Red Mountain on a north-south axis. Most of it had a checkerboard pattern, but a more irregular pattern of streets was plotted where the South Highlands would later materialize. [24] Powell arranged with a Montgomery firm to make bricks for new buildings on the spot at cost, agreeing to pay for them "as fast as they will burn."

On June 1, 1871, the first lot, on the corner of First Avenue and Nineteenth Street, was sold for $100. About six months later, on December 19, Birmingham received its charter of incorporation from the state legislature. Its initial boundaries stretched 3,000 feet on either side of the Alabama and Chattanooga Railroad tracks running through the area that Morris had purchased a year earlier. From the start, the founders had no doubt that the city they were creating was destined for greatness. Because of proximity to raw materials, cheap black labor, newly forged rail connections to the outside world, and a benign climate, they were confident that iron could be made more cheaply in Jones Valley than anywhere else on earth. Powell captured the aspirations of Birmingham's pioneers in a speech delivered at the Elyton Land Company's annual meeting in February 1873, in which he coined the community's enduring nickname by referring to "this magic little city of ours." [25]

V

Behind the formation of the Elyton Land Company was a story of intrigue and cunning that showed how effectively southerners could play the game of business competition, even against well-seasoned northern rivals. Like almost every other aspect of American economic life at the time, the founding of Birmingham reflected the politics of railroading, resulting from a dramatic struggle between two groups of promoters pursuing radically different goals.

During the first decade after the war, the southern rail network grew by only 4,187 miles (46 percent) while that of the United States as a whole increased by 39,529 miles (113 percent). Among all the former Confederate states east of the Mississippi, Alabama stood out by adding 927 miles of track to the 805 it had possessed in 1865, registering an increase of 115 percent. Support for railroads in the state legislature, which was seeking fresh tax revenue now that slavery no longer existed, lay behind this achievement. Ten railroads received help in the late 1860s under laws enabling city and county governments to authorize bond issues that facilitated their development.[26]

The growth of Alabama's railroads contributed greatly to the state's emergence as an important producer of coal and iron, both heavy, bulky commodities requiring rail transport for fast, effective distribution. But Alabama's growing rail system was built at great cost by promoters who took advantage of legislative liberality by misappropriating the proceeds of bond issues for their own benefit. Among them were two carpetbaggers from Massachusetts, John C. and Daniel N. Stanton, who were allied with financial interests in Chattanooga. The Stantons became key figures in the postwar development of the Alabama and Chattanooga Railroad, which they formed by combining two antebellum ventures, the Northeast and Southwest Alabama and the Wills Valley Railroad. Their line ran from a junction with the Mobile and Ohio at Meridian, Mississippi, to the Tennessee border along a projected 295-mile route crossing Jones Valley. By bribing members of the Alabama legislature and enlisting support from Republican governor William H. Smith, who endorsed the bonds of the Alabama and Chattanooga well beyond limits authorized by state law, the Stantons obtained at least $4,700,000 in bond endorsements and loans, plus 600,000 acres of land grants. Like former Confederate general Nathan Bedford Forrest, who was building the Selma, Marion and Memphis Railroad during the same period, the Stantons hired Chinese coolies, some of whom encamped near Elyton, to lay track. This idea proved a fiasco when the Stantons fell six months behind in paying wages, leading to the seizure of their trains by irate workers. Nevertheless, they completed their line in 1871, providing express service to communities along the route and greatly shortening the time required for trains to run from New York to New Orleans.[27]

As the Stantons pushed ahead, they clashed with Gilmer, Milner, and other Broad River associates who had never abandoned their hopes of developing an industrial city in Jones Valley. This struggle later came to be portrayed as a battle between good and evil by historians who abhorred Radical Reconstruction and took a heroic view of Birmingham's emergence, but it was more complex than the legends surrounding it might lead one to believe. Despite being condemned by such scholars as Walter L. Fleming and Albert B. Moore for typifying Radical Republicanism at its worst, the Stantons were neither more nor less villainous than other promoters at a time when such railroad scandals as the Credit Mobilier were rife. Charges by Fleming that the Stantons squandered public funds on a hotel and opera house in Chattanooga, for example, are exaggerated. As John R. Scudder, Jr., has indicated, some sections of the Alabama and Chattanooga were well built, the line served a valuable public purpose, and completion of the road might have been long delayed had the New Englanders not undertaken the task. In any case, there was plenty of conniving and double-dealing on both sides.

Some of the issues involved in the struggle between the Stantons and their opponents were rooted in prewar rivalries between Montgomery and Mobile. The fact that the Stantons were financially interested in Chattanooga, and therefore wished to block developments contrary to the interests of that city, made them natural allies of downstate planters and merchants who had traditionally favored Mobile over Montgomery. Mobile wanted to see the coal and iron ore of the Mineral District used mainly as raw materials to be exported to the outside world via the Gulf of Mexico instead of being converted into pig iron and shipped to northern and midwestern markets by rail. As always, boosters of Selma and Tuscaloosa had no reason to favor aggrandizing Montgomery. It was also unclear to some of Montgomery's leaders, wedded as that city was to the cotton trade, that the Broad River group was wise in trying to create a potential rival in Jones Valley.

As Horace Mann Bond has noted, the battle between the Stantons and their opponents had a larger national context. On one side were carpetbaggers and scalawags allied with northern investors like Henry Clews, Jay Cooke, and William D. ("Pig Iron") Kelley; on the other were Alabamians tied to Wall Street financiers including J. P. Morgan. But even more fundamental political, economic, and social matters were at stake in the struggle attending Birmingham's birth. At issue, as Eric Foner has indicated in a searching analysis of Reconstruction, was the extent to which ex-slaves would really be liberated, whatever the Thirteenth Amendment might stipulate about involuntary servitude. Believing that black people would not work unless they were kept in a servile state, advocates of white rule fought for control of the southern labor force and freedom from northern interference in racial problems. The successful outcome of this battle, resulting in state governments whose leaders were called Redeemers because of their role in saving the South from Radical Republicanism, was epitomized in Alabama by George S. Houston's election as governor in 1874. Houston, a white supremacist from the Tennessee Valley, was a plantation owner and ex-slaveholder who had been allied with the Broad River group since the 1850s. A former Whig, he was also closely connected with Luke Pryor, James W. Sloss, and other residents of north Alabama who championed railroads and manufacturing. Houston's election, marred by violence, was a key factor in the chain of events that resulted in the emergence of Birmingham, ensuring that severely disadvantaged black workers would operate the city's industrial plants in fulfillment of Milner's prewar visions. [28]

The details of the conflict between the Stantons and Birmingham's pioneers have been repeatedly discussed, but most scholars have taken them largely as they are given in one book, that of Ethel Armes. Few topics in Alabama history merit closer reexamination. Full of drama and suspense, the episode was noteworthy not only because it led to Birmingham's emergence but also because it laid the basis for the chain of developments in which James W. Sloss and his furnace company later figured prominently. [29]

After Alabama's first postwar legislature enacted a law designed to encourage development of the state's mineral endowments, Gilmer and his Broad River associates revived their dream of building an industrial city in Jones Valley and secured an appropriation to complete the South and North Railroad, which had been moribund for many years. Funds now available would permit the line to proceed as far north as Calera, as the place formerly called Limekiln Station was now known. As chief engineer, John T. Milner prepared to resume laying track. At this point, the furnace that Jacob Stroup had built in Jefferson County during the war was in ruins, and the rickety railway that Milner had constructed to carry its pig iron to Brock's Gap, if indeed it ever had existed, had disappeared.

Trouble arose in 1868 when a Republican legislature came to power in Montgomery under Reconstruction Acts recently passed in Washington. At this point the Stantons began efforts to thwart the Broad River group by having Gilmer ousted as president of the South and North and replaced by John Whiting, a Montgomery cotton dealer who was allied with south Alabama planters and New York capitalists friendly to the Alabama and Chattanooga. Playing on Whiting's identification with the state's chief crop, the Stantons urged that the cotton trade would be best served by abandoning the idea of making Decatur the ultimate terminus of the South and North and stopping that line at Elyton, from which point the road the Stantons were building would take passengers and freight directly to Chattanooga.

Seeing that fulfillment of the Stantons' plans would mean death for the projected industrial city in Jones Valley, Milner, who remained chief engineer of the South and North despite Gilmer's dismissal, hastened to Montgomery. There, he tried to convince Whiting that aggrandizing Chattanooga would be "an irreparable loss to Alabama" and managed through adroit lobbying to save the South and North's appropriation. Meanwhile, Gilmer scoured the Alabama countryside to secure proxies from widely scattered stockholders so that he might regain control of the line. At a critical juncture, he returned, haggard but triumphant, with the needed votes. Whiting was now dead, and

Gilmer was reinstated as president. Temporarily, the Stantons had been foiled.

By the end of the year, friends of the South and North had rushed through legislation permitting its construction all the way to Decatur.[30] In April 1869, Sam Tate became chief contractor and started laying track north from Montgomery to Calera. To stay within his budget, he was ordered by the directors of the line to use the utmost economy. The result was a jerry-built road, full of detours, twisting curves, poor grading, rickety trestles—"anything, everything, to save money," as Armes later stated. Late in 1870, Tate and his crews reached Calera, sixty-three miles from Montgomery and thirty-three miles from Elyton. Beyond them lay the stretch that the Mountain Company had already developed during the war and the still-missing link between Brock's Gap and Jones Valley.

As the South and North proceeded toward Decatur, it was obvious that it would have to cross the Alabama and Chattanooga line somewhere in Jones Valley. Much depended, however, on where the crossing might be and who owned the site involved. Knowing that the Stantons had strained their financial resources and thinking that he might be able to work out a deal with them, Gilmer made an agreement with John Stanton under which the two parties secured options on a 7,000-acre tract along Village Creek. This was favored by Milner because it contained a large number of springs, maximizing the potential water supply for the city that he and his partners wanted to build. Had this arrangement been carried out, Birmingham would have been located northwest of the site ultimately chosen.

As Leah Rawls Atkins has pointed out, it is not clear why Gilmer and his associates thought that they could deal successfully with Stanton, given his enmity for their plans and his reputation for untrustworthiness. Predictably, Stanton, who did not want to create a rival to Chattanooga, soon reneged. Changing the projected route that the Alabama and Chattanooga was to follow through Jones Valley, he negotiated sixty-day options on another 4,150-acre tract closer to Elyton, payable at Josiah Morris's bank in Montgomery. Keeping a level head, Milner refused to panic. According to Dorothea Warren, his biographer, he drew on acumen that he had gained in being outsmarted by swindlers in California. Atkins has also speculated that Milner knew about kinship ties among Alabamians of which Stanton was unaware, dooming the New Englander's plans from the start. In any case, Milner conceived an artful ruse. He saw that Stanton could not take up his new options so long as he did not know exactly where the South and North would lay its track through Jones Valley. He was also cognizant that Stanton doubted the ability of the owners of the South and North to raise enough money to acquire alternate tracts of land on their own. The Chattanooga and Alabama was running short of cash, leading to a bankruptcy judgment that would be rendered against it by June 1871; because of this, Milner probably surmised that Stanton lacked the means to take up his new options and was bluffing. This game was one that two could play. Pretending to search up and down the valley for a suitable crossing site, Milner publicly wavered between one location after another, keeping Stanton guessing about his true intentions. As the cat-and-mouse game continued, the owners of the land on which Stanton had taken out his option became worried about what would happen and placed their deeds in the hands of a local lawyer, Alburto Martin. He in turn put them in escrow at Morris's bank.

The denouement was swift and stunning. By the time the sixty-day period expired, Stanton had not exercised his options. The law governing such cases provided for a three-day grace period. Just as it elapsed, Morris himself doled out "one hundred thousand dollars in cold cash," as Ethel Armes later put it, to buy up the deeds that Martin had placed in escrow. Morris thus became temporary owner of the entire 4,150-acre tract. Within hours, he turned over part of his holdings to various members of the Broad River group and a party of Jefferson County landowners. While Stanton had waited to see what would happen, Morris had secretly become Milner's ally after the latter had promised to form a syndicate to pay back part of his investment and cut him in as the largest stockholder in a bold new enterprise based upon Jones Valley's potential for coal mining and ironmaking. It was in this manner that the Elyton Land Company had been formed.

The outcome was full of paradox; as it ultimately turned out, the site on which Birmingham was built had not been originally selected by the victorious South and North group, but was instead co-opted by them through Milner's adroit maneuvering. The person who had chosen the place where the new city would rise, by taking out options that he never exercised, was John C. Stanton, who had used every resource at his command to keep it from being born.[31]

Furious that he had been outsmarted, Stanton had one last chance for revenge. Because they had strained their resources in building the South and North and organizing the Elyton Land Company, Gilmer and his allies could not pay interest on state-backed bonds when an installment fell due in April 1871. By this time, construction of the South and North had reached Jones Valley, increasing Stanton's determination to stop the line once and for all. Seeing a final opportunity to defeat his financially embarrassed enemies, Stanton forged an alliance with a group of capitalists who had acquired most of the South and North's first mortgage bonds, including Boston financier Russell Sage and Vernon K. Stevenson, president of the Nashville and Chattanooga Railroad. As the interest deadline neared, Stanton, Sage, and Stevenson notified Gilmer that default would cause control of the South and North to be transferred to Stevenson's line. Stevenson would then halt further construction between Jones Valley and Decatur, where the South and North planned to make connections with rail lines already built between that city and Nashville.

After making a fruitless search for northern financial aid, Gilmer seemed ready to surrender. Then, suddenly, the tide of battle turned, and the Stantons

were finally defeated. James W. Sloss, a close friend of Morris and an ally of
the Broad River group, had gone on a secret mission that thwarted the Stantons and their associates. In one bold stroke, Sloss saved the projected industrial city, still only a dream, from oblivion.

VI

Sloss was born at Mooresville, Limestone County, Alabama, on 7 April 1820.
He was of Scotch-Irish descent, his father having emigrated from County
Derry to Virginia in 1803 and fought in the War of 1812. Receiving little
formal education, Sloss became an apprentice bookkeeper for a butcher at age
fifteen, serving the traditional seven years. At the end of his term, he married
Mary Bigger, whose parents had come to America from Belfast. Using his
savings to buy a store at Athens, Alabama, he launched a career as a merchant
and plantation owner that made him one of the wealthiest men in the state.
By the time of the Civil War he was well known and respected for his Irish wit
and business acumen.[32]

James W. Sloss. A railroad promoter and founder of the Sloss Furnace Company,
James Withers Sloss played a crucial role in a long-term alliance between upstate
landowners and members of Montgomery's Broad River group that resulted in the
founding of Birmingham. Because of his success in bringing the L&N Railroad into
Jones Valley in 1871 and his many other contributions to the early growth of the
Birmingham District, Sloss was Alabama's most important post–Civil War industrialist despite the greater fame of his more colorful contemporary, Henry F
DeBardeleben. Sloss's face, as shown in a photograph of undetermined date, reveals
much about the inner essence of a person who was both a visionary and a man of
action. (Courtesy of Sloss Furnaces National Historic Landmark, Birmingham,
Alabama.)

Sloss became involved in railroads in the 1850s. As president of the Tennessee and Alabama Central, chartered by the Alabama legislature in December 1853, he completed construction on that line between Alabama's northern boundary and Decatur by 1860. Meanwhile, he worked closely with Gilmer and other Broad River associates as they spearheaded development of the South and North after its chartering in 1854. Not until 1859, after Milner had recommended a route and the Alabama legislature had chartered the Mountain Company, could work begin on even part of that line, and wartime exigencies blocked its completion. After 1865, Sloss watched in frustration as his Broad River friends in Montgomery tried to resume progress. Above the Alabama line, however, things went better; two connecting railroads, the Tennessee and Alabama and the Central Southern, were completed to Nashville, giving that city connections all the way south to Decatur. In 1867, all the lines between Nashville and the Tennessee River were consolidated into one company, the Nashville and Decatur, with Sloss as president.[33] Partly because of the lack of connections in Montgomery's direction, however, profits were meager.

Sloss was one of the chief proponents of Alabama's postwar industrial development. As Horace Mann Bond later stated, "His influence, on close inspection, will be found connected with every important industrial and commercial enterprise in the State during the latter half of the nineteenth century."[34] Besides continuing his prewar ties with the Broad River group, Sloss also became allied with an enterprise that had emerged by 1865 as the strongest railroad in the South: the Louisville and Nashville, better known as the L&N. As Stanton, Stevenson, and Sage prepared to foil the Elyton Land Company's plans in the spring of 1871, Sloss was well aware that his own interests were in jeopardy. Because the Nashville and Decatur was already a satellite of the L&N, it was equally obvious where he would turn for help when the time was ripe.

At this time, the L&N had reached a critical stage at which it needed to find an outlet through Alabama to the Gulf of Mexico. Its early history had been a saga of relentless expansion. Organized in 1850 by Louisville businessmen to compete for southern trade with that city's arch rival, Cincinnati, it had completed its main line from Louisville to Nashville by 1860 and reached Memphis soon after the outbreak of the Civil War. Its operations were particularly prosperous after mid-1863, when the territory through which it ran was no longer a combat zone, and it came out of the war with a good credit rating. Building new trackage after Appomattox, it formed a pool of five railroads to provide fast freight service to southern merchants who ran general stores on which farmers depended for credit. Further growth, however, required improving its access to areas south of Nashville.[35]

Much of the L&N's success was due to the leadership of its German-born general superintendent, Albert Fink. Known as the "Teutonic Giant" because of his massive build, Fink was a pioneer in devising rational railroad rate structures. He was also a master strategist deeply committed to territorial expansion as a way of maximizing long-distance traffic. The fact that the L&N chronically had a larger volume of goods moving in a north-south direction than shipments going the other way worried him because this imbalance meant lost revenue on northbound trains. Fink also wanted access to new markets. As a result, he was eager to establish connections with Montgomery and Mobile. [36]

Well aware of Fink's interests, Sloss seized the opportunity to forge an alliance between the L&N and the beleaguered Elyton Land Company as the deadline for the South and North's interest payment loomed. Just as the Stantons seemed poised to achieve final victory in their battle with Gilmer and his forces, Sloss hurried to Louisville and presented Fink with a glowing picture of the Mineral District and the future rail traffic it was capable of generating. He offered to lease the Nashville and Decatur to the L&N if that line would assume the South and North's debts, pay the interest due on its bonds, and complete work on the gap between Decatur and Birmingham. After this, he promised, the L&N could run the combined system, all the way from Nashville to Montgomery, as a single unit. [37]

Fink was receptive. Along with his other worries, he was concerned about a threat from the Cincinnati and Southern Railroad to move southward through Kentucky and Tennessee, outflanking the L&N and putting itself in position to reach the Gulf of Mexico. Fink therefore quickly convened the directors of the L&N to consider Sloss's plan. Because it was a risky proposition that would add greatly to the L&N's debts, and also because of the unbroken wilderness between Decatur and Birmingham, the directors were skeptical. Without waiting for them to make up their minds, Fink went to Montgomery, told the Stantons and their associates that the L&N was likely to redeem the South and North's obligations before that line's bond interest fell due, and returned to Louisville with an Alabama delegation including Francis Gilmer, John T. Milner, and Sam Tate. A stormy meeting held at the Galt House to discuss the proposals that Sloss had advanced nearly erupted in fisticuffs when Tate brazenly demanded $100,000 for relinquishing his existing rights as a contractor for the South and North, but Fink broke up the impending melee by ordering a round of whisky. The next day, with Sloss playing a mediating role, the directors of the L&N decided by a narrow vote to endorse his plan.

Birmingham was thus saved. Interest on the South and North's bonds was paid in time to prevent the takeover threatened by the Stantons and their allies. Soon the Stantons themselves were deep in trouble. The Alabama and Chattanooga went bankrupt and fell under state control in July 1871. As political tides shifted and Alabama prepared to shed Republican rule, the Stantons and their northern associates lost influence in Montgomery. Election to the governorship in 1874 of Sloss's friend, Houston, solidified the victory that Birmingham's founders had won. By this time the Alabama and Chattanooga was in receivership. In 1877, it fell into the hands of European investors headed by Emile Erlanger and Company of London and was reorganized as the Alabama Great Southern Railroad. Under its new owners, it became one of Birmingham's main links to the outside world.[38]

Later events showed how crucial was the corporate marriage that Sloss had arranged between the South and North and the L&N. In the words of Jean Keith, entry of the L&N into Jones Valley in 1871 "determined the site of the future city of Birmingham as a focal point for southern industrial enterprise." The Panic of 1873 jarred the owners of L&N into realizing that it would sink or swim on the basis of the gamble that it had made in rescuing the Elyton Land Company. Through investments, loans, advances, and freight rates that fluctuated upward or downward with the price of pig iron, the line carried out a massive colonization program in north Alabama. In return, firms that benefited from the L&N's largesse signed contracts guaranteeing it lucrative freight privileges. Ultimately reaching the Gulf of Mexico, the L&N invested more than $30 million in furnaces, mines, wharves, rail spurs, steamship lines, and other Alabama operations. By 1888, it was hauling annual tonnages of iron, coal, and other mineral products outweighing the nation's entire cotton crop.[39]

VII

The decision made at the Galt House to ratify Sloss's proposal transformed Birmingham from a squalid jumble of tents, shanties, and boxcars into a bustling community. Prior to 1870, the site of what became Alabama's largest city was "just an old cornfield, all overgrown with weeds and briers," as Milner's son, Henry, later recalled. Compared with the Black Belt to the south and the Tennessee Valley to the north, Jefferson County had never amounted to much. A spoil of the draconian peace settlement imposed by Andrew Jackson on a broken Creek Nation after his victory at Horseshoe Bend, it had first been settled by whites in 1815 and was split off from Blount County four years later. By 1860, it had only 11,727 people, of whom 2,649 were slaves. The small black population in itself indicated that the chief staple of the area was not cotton but corn, and most farmers who did own slaves worked side by side with them in the fields. Some prominent landowners in 1860 were still illiterate, signing deeds with marks when registering them at the courthouse. Williamson Hawkins, the leading planter, owned 107 slaves and farmed 2,000 acres in a hilly area where Pennsylvania ironmaster Samuel Thomas would erect two blast furnaces only a few decades later. Judge Mudd owned an eight-room house, The Grove, later known as Arlington. It had mahogany furniture, crystal chandeliers, and a rosewood piano. Most people, however, eked out a bare living, dwelling in nothing more pretentious than one-story dogtrot houses, so called because of their central passageways flanked by two rooms on each side.[40]

The county seat was Elyton, originally known as Frog Level. It had been named for William H. Ely, a New Englander representing the Connecticut Asylum for the Deaf and Dumb, which held land preferred by settlers because

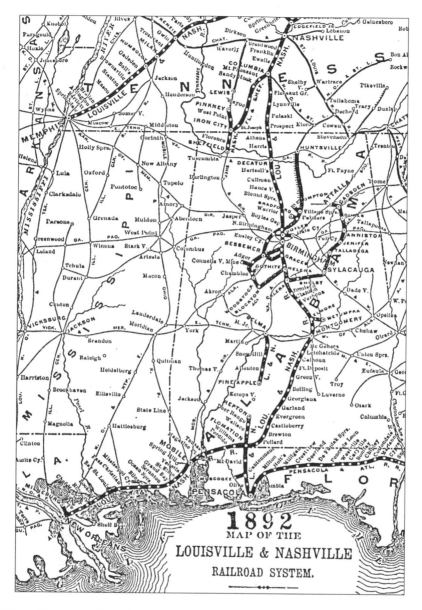

Map of the Louisville and Nashville Railroad System in 1892, Showing its Crucial Importance as an Outlet for the Birmingham District. The success of James W. Sloss in attracting the L&N to expand its operations into Alabama in 1871 did more than any other single development to make Birmingham's dramatic emergence possible. (Courtesy of Birmingham Historical Society, Birmingham, Alabama.)

titles to other tracts were defective or had been cornered by speculators. Loathing the county, which did not have a single pane of window glass in 1820, Ely went home the following year after writing his wife that the district's houses looked "more like the haunts of wild & savage man, than residences of civilized members of Society."[41] Ely did, however, donate lots for a court-house and jail. As a result, Elyton became the county seat, replacing Car-rollsville. Elyton also benefited from having a bountiful spring, assuring residents an ample water supply. By 1870, it had about 1,000 people, a corner store known as the Exchange, a two-story red brick courthouse and tavern, churches maintained by the Baptists, Episcopalians, and Methodists, and a newspaper, *The Jefferson Independent*. It lay at the junction of three stage-coach and express lines and was also situated on the route planned for the Alabama and Chattanooga Railroad.[42] Residents wanted it to be chosen as the place where that line would intersect with the South and North, but, as a result of the deal worked out between Milner and Morris, the crossing took place slightly to the east, consigning Elyton to swift oblivion. Ultimately it became part of Birmingham.

Elyton's fate was sealed in 1872 when Birmingham launched a campaign to replace it as county seat. Under existing state law, ballots could be cast at any place in the county regardless of where voters actually lived, and lax practices made it easy for recently enfranchised freedmen to cast multiple ballots. Taking advantage of the situation, Powell arranged to have black voters brought free of charge to Birmingham by rail and treated to a lavish barbecue. "About noon on the day of the election," as one account later recalled, "trains packed almost to suffocation with a dark mass of perspiring, hungry hu-manity, rolled into Birmingham." Powell, mounted on a pony and believed by many of the passengers to be Ulysses S. Grant, met them at the depot. Brandishing a sword, he escorted them to the scene of the canvass, where tables were "groaning beneath the load of savory meats just from the smoking pits."[43] The balloting resulted in an overwhelming triumph for Birmingham, and the old court house in Elyton was abandoned. Before being torn down, it was used as a paint factory.

Birmingham had 4,000 persons and approximately 500 houses by 1873. A thirty-room frame structure on Nineteenth Street, known as the Relay House, was the largest of four hotels; in addition, the community had approximately 125 stores, six churches, two grist mills, a national bank, various manufactur-ing facilities, "and all the other concomitants which go to making up a thriving and prosperous city," as Powell said in one of numerous publicity releases that he wrote to attract settlers. Emphasizing the importance of the Iron Horse to Birmingham's development was a wide strip of land running through the middle of town, reserved "for railroad purposes only." Excursion trains brought a stream of visitors from all points of the compass in response to Powell's predictions, boasting of "mountains groaning to be delivered of their wealth" and calling Birmingham "the Eldorado of the ironmasters." Powell and other boosters were confident that the city would soon replace Montgomery as Alabama's capital. An area at the head of Twentieth Street was set aside as "Capitol Park" in expectation that this change would occur.[44]

Knowledgeable outsiders echoed Powell's rhetoric. Abram S. Hewitt, who introduced the open-hearth method to the United States, stated that Jones Valley would be "a region of coke made iron on a scale grander than has ever been witnessed on the habitable globe." British expert Sir Lowthian Bell pronounced Alabama's resources to be on a par "with the most favored of those which I have examined in Europe." Calculations by E. M. Norton, a prominent Kentucky ironmaster, indicated that Jones Valley pig iron cost only $8.51 per ton as against $21.50 in Pittsburgh, $19.70 in Kentucky, and $16.00 in eastern Ohio.[45] Such pronouncements and projections nurtured a visionary boosterism in Birmingham that endured into the twentieth century.

VIII

Birmingham's leaders worked hard to make their dreams come true. With their encouragement, Daniel Pratt performed the last of many services on behalf of Alabama's industrial progress in 1872 by helping establish a new enterprise in the district, the Eureka Mining and Transportation Company. Its aim was to rebuild the old furnace at Oxmoor, as the site where Stroup had built the stack in 1863 was now called. Since being destroyed by Wilson's raiders at the end of the war, it had been only a heap of rubble. Raising the height of the original stack from thirty-two to sixty feet and widening the bosh, the new Eureka ownership also built a second furnace on the site, spending about $200,000. Early in 1873, the furnaces went into blast. As workers waited for the first runs of iron, dogs flushed a deer that ran headlong through the pig beds; wild turkeys swarmed throughout the nearby woods, which teemed with game. Supplied by a narrow-gauge railroad leading to ore mines a few miles away, the furnaces smelted red ore with a mixture of charcoal and coke. Because of inexperience on the part of both the management and the workers, however, output was meager.[46]

Despite the low productivity of the furnaces, the Eureka Company's venture at Oxmoor made a significant contribution to the development of Birmingham by bringing into the area Henry Fairchild DeBardeleben, characterized by Ethel Armes as "the most picturesque and dramatic character in the coal and iron history of the South." Descended from a Hessian mercenary who had fought for Great Britain in the American Revolution, DeBardeleben had been born in Autauga County in 1841. Like many other natives of the region who later became industrialists, he came from an agricultural background; his father was a cotton planter who had migrated to Alabama from South Carolina.

Orphaned at ten, DeBardeleben became Daniel Pratt's legal ward. Working for the latter, he advanced from head of the teamsters to boss of the lumber yard and superintendent of the cotton gin factory. After serving in the Confederate army and fighting in the Shiloh campaign, he went home to take charge of a bobbin factory owned by Pratt and married his daughter, Ellen. As Wayne Flynt has indicated, the union epitomized the alliance between planters and manufacturers that would ultimately result in the founding of Birmingham. Though wild and impulsive, DeBardeleben gained his father-in-law's confidence. In 1872, Pratt put him in charge at Oxmoor, where his inexperience—DeBardeleben freely acknowledged at the time that he knew nothing about making iron—helped bring about the project's failure. His spirit, however, aptly described by Armes as "savagely energetic, restless, impatient," epitomized the zest for exploiting the natural environment that was fundamental to Birmingham's formative style.

DeBardeleben compared himself with the "piney rooter," a skinny hog that was well known to local residents and survived by using its snout to dig for food. Black-haired, with an aquiline nose, bushy mustache, and piercing eyes, "quick as a bird's," he had a ruddy, hawk-like face matching his fierce determination and hard-bitten attitude. "There's nothing like taking a wild piece of land, all rock and woods, ground not fit to feed a goat on, and turning it into a settlement of men and women, making pay rolls, bringing the railroads in, and starting things going," he liked to say. "There's nothing like boring a hillside through and turning over a mountain." Boiling with energy, "he seemed to have one foot always in the stirrup and to be itching to mount and ride away," an admirer recalled. "I had rather be out in the woods on the back of a fox-trotting mule with a good seam of coal under my feet than to be president of the United States," he once said. "I never get lonely in the woods . . . and the rocks and the forests are the only books I read."[47]

Birmingham needed people of DeBardeleben's spirit. In 1873, it was devastated by a cholera epidemic that decimated its population. While Mortimer H. Jordan, Mudd's son-in-law, headed a cadre of physicians who worked around the clock to care for the sick and dying, madams opened houses of ill fame to serve as makeshift hospitals, and prostitutes volunteered to serve as nurses. The Panic of 1873 added to the misery, sending the price of iron plummeting and forcing businesses to shut down. Settlers fled the area, and the value of stock in the Elyton Land Company declined precipitously. Traffic on Alabama railroads that were now operated by the L&N took a nosedive. At one point during the debacle, Fink met Milner in Montgomery and turned on him bitterly. "You have ruined me, you fool," he exclaimed. "Where are those coal mines and those iron mines you talked so much about. . . . Where are they? I look, but I see nothing! All lies!—lies!" Soon, Fink resigned from the L&N, feeling disgraced by the failure of events in Alabama to turn out as he had hoped. Even Powell, who had become the city's first elected mayor in 1873, lost faith. Resigning as president of the Elyton Land Company in March 1875, he was succeeded by Henry Caldwell. Returning to Mississippi, where he founded a town on the Yazoo River, Powell resurfaced in Birmingham in 1878 to make an unsuccessful run for mayor. His career ended in 1883 when he was shot to death in Mississippi by a former employee whom he had recently discharged. Morris and other local leaders tried to have him buried in Birmingham, but he was instead interred in his family's cemetery plot at Montgomery.[48]

Despite their discouragement, most of Birmingham's leaders did not give up. Stung by Fink's remarks, Milner organized the Newcastle Coal and Iron Company, which soon produced seventy tons of coal per day. The most important step taken to revive local industry, however, was at Oxmoor, where the Eureka Company was reorganized by a syndicate headed by Daniel S. Troy, a Montgomery lawyer and friend of Francis Gilmer who had served the Confederacy with valor in the Civil War. Securing a legislative grant of limited liability and virtual exemption from taxes, the group hired Levin S. Goodrich, an experienced Tennessee ironmaster, to head a fresh campaign at Oxmoor. Using charcoal, he put one of the furnaces in blast, increasing its output and reducing fuel consumption. Still unsatisfied, he sent specimens of pig iron to Pittsburgh for chemical analysis. The results were encouraging, but Goodrich realized that the future of the Alabama iron industry hinged upon switching from charcoal to coke.[49] No other solution was compatible with high-volume production.

Oxmoor's financial backers were skeptical about abandoning charcoal. A few earlier experiments using coke on a small scale had taken place, but it was not clear that coal from the Warrior Field was suitable for coking. Making iron with coke might also require skills that local workers did not possess. Having already failed once, most of Birmingham's promoters were not inclined to throw good money after bad. Switching to coke, however, was the only way out. Already deeply in debt, and well aware of the added outlays that Goodrich's plan would require, the Oxmoor syndicate offered to turn the furnace over to any entrepreneurs who were willing to gamble on it. At this juncture, Milner called a meeting at the offices of the Elyton Land Company and urged those who attended to form a new enterprise. His leadership resulted in a venture known as the Cooperative Experimental Coke and Iron Company. It was headed by a three-man board of managers including James W. Sloss, who once again demonstrated his faith in the district by risking his fortune in its development.

Under Goodrich's direction, five coke ovens were built. Samples of coal submitted by local mines were tested to determine which would be most suitable for smelting Red Mountain ore, and a variety from the "Browne seam" was selected. Named after pioneer coal operator William Phineas Browne, it had been discovered by a local prospector, "Uncle Billy" Gould, who had emigrated from Scotland to Pennsylvania before coming to Alabama.[50] The ovens, some of which were of a new Belgian design, were highly innovative, being equipped with oscillating bottoms for easier loading and lateral flues for the controlled removal of gases that would otherwise have gone to waste. As Jack R. Bergstresser has stated, this installation contradicts impressions that early industrial technology in the Birmingham District was backward by comparison with that in other parts of the country.[51]

Meanwhile, the newer of the two furnaces at Oxmoor was remodeled for use of a hot-air blast. Innovative equipment contrived by Goodrich included a cone-shaped charging apparatus designed to retain furnace gases in the stack. Combustible gas recovered from the coke ovens was used to heat the blast and fire the boilers that provided steam to the blowing engine. With such apparatus, modern by all prevailing standards, a crucial experiment took place to smelt iron with coke at the revamped Oxmoor furnace. The event later became the subject of so much folklore that it is hard to determine exactly what took place. A fictional account published eighty years later spoke of a crowd whose members "pressed as close as they dared while the fiery soup rushed down the main sow out into the dozens of pig beds" and compared the scene to "the Fourth of July in hell. . . . It had a brutal magnificence that gave it dignity and an awful kind of beauty, dwarfing its human masters. Even the sky had a hectic flush."[52] Much of this description, however, is sheer imagination. Even the date later reported by Ethel Armes and repeated in numerous accounts including the novel just cited—28 February 1876—is wrong; the first blow actually occurred on 11 March of that year, yielding thirteen tons of "silver grey iron." A brief report made to a local newspaper by the secretary of the Eureka Company three days later indicated that output had been raised to about twenty-five tons, using various mixtures of coke and red hematite. "This is an important epoch in the industrial history of Alabama, upon which we congratulate her people as well as the Eureka Company," the newspaper stated. "It is no longer a question as to whether our coals will make good iron from our ores." The coke had "proved sufficiently strong to sustain the burden required and the product of the furnace is all that was hoped for."[53]

Almost three weeks later, the news from Red Mountain was reported in New York City by a leading trade journal, *Iron Age*, which indicated that output for the first full week of operations at Oxmoor had been 175 tons. "This is large work for a 60 × 12 furnace, but whether it is good for the furnace is questionable," stated a notice that gave the event scant attention. "The best practice is not to push the furnace at first, but to let it come up to its work very gradually."[54] It was not the first time, or the last, that eastern experts would take a condescending attitude toward the upstart industry that was slowly beginning to appear in Alabama.

IX

Despite the derision with which the Oxmoor experiment was greeted on the eastern seaboard, Birmingham had reached a critical turning point in its emergence as an industrial center. That coke made from Warrior coal could be successfully used to smelt pig iron became the salvation of the district. In modern terms, however, what had happened at Oxmoor was merely a "pilot project." It did not prove that pig iron could be made consistently on a large scale in the Birmingham area and marketed profitably outside the region. It did, however, encourage further investment.

The ensuing search for outside capital led to new problems. Large sums of money, far more than existed in Alabama itself, were needed to fulfill the prospects that the results of the Oxmoor experiment had created. Pratt was now dead, and his heir, DeBardeleben, held title to the Oxmoor property. His interests, plus those of local businessmen who had backed the Cooperative Experimental Coke and Iron Company, had to be combined with outside funds under a new financial arrangement. Because of the role that the L&N had played in developing the area, its stake in Birmingham's future, and its function as chief distributor of the region's products to the outside world, investors in Louisville were primed to help. Their bitter rival, Cincinnati, was also a natural market for Birmingham pig iron because of its geographical location and its thriving foundries and rolling mills. By this time, Ohio's Queen City had been connected to the L&N by rail; despite their dislike of Cincinnati, financial leaders in Louisville had approved the new line because it used a gauge different from that of the L&N, thus enabling Louisville to profit from being the place where freight between Cincinnati and the South had to be reloaded. Both cities therefore had a strong community of interest with Birmingham.[55]

Soon after the Oxmoor experiment, two of Birmingham's most prominent business leaders volunteered to make arrangements between local investors and their counterparts in Louisville and Cincinnati. James Thomas, a northern emigrant who belonged to the Pennsylvania ironmaking family that had introduced hot-blast technology to the United States, served as an intermediary with Cincinnati entrepreneurs. Meanwhile, Sloss once again helped preserve Birmingham's momentum by agreeing to represent the community in meetings with financiers from Louisville. After negotiations had been completed, the L&N over a two-year period subscribed $60,000 in the stock of the Oxmoor venture, built a spur line to serve Birmingham coal and iron companies, and made other commitments that brought its total stake in the district to a reported $125,000.[56] Cincinnati businessmen also risked heavily in the area. As the two cities vied to gain the upper hand, local leaders, particularly DeBardeleben, played one off against the other in an attempt to protect local interests.

With the infusion of fresh capital, both of the Oxmoor furnaces were rebuilt for expanded production, using coke. Although their sixty-foot height made them relatively small by northern standards, the new stacks had iron shells, which were atypical of Alabama furnaces at this time and reflected continued adoption of up-to-date equipment. Batteries of coke ovens, some of which were equipped with reversible bottoms recently developed in Belgium, were built at Helena and Oxmoor. Despite such progress, the chemical peculiarities of Red Mountain ore and the inexperience of the crews at Oxmoor resulted in output well below northern standards. Goodrich soon returned to Tennessee to rebuild an ironworks in Nashville, and quarrels between rival

Cincinnati and Louisville backers continued to plague the Oxmoor enterprise.[57]

With Powell gone, Milner assumed his mantle as the district's chief promoter. In 1876, he published a book, *Alabama As It Was, As It Is, and As It Will Be,* to expound its advantages. He took a broad view, beginning with an analysis of the depths into which agriculture had fallen in the state since the Civil War. This decline he blamed upon two things: the overweening concentration on cotton cultivation to the detriment of foodstuffs, requiring imports from other states, and the emancipation of blacks, who, he asserted, were inherently incapable of becoming productive as free citizens. The remedy, he said, was to encourage white Anglo-Saxon immigrants from Europe to learn American methods of farming and fill the inexhaustibly fertile agricultural lands along the corridor that the South and North Railroad was now operating as a subsidiary of the L&N. As an illustration of what could be done, he pointed to the accomplishments of German immigrants now settling in Cullman County. Buttressing his argument was a dismal estimate of agricultural prospects in the western states, based partly on his experiences during the Gold Rush. Capitalizing on the still-prevalent belief that a "Great American Desert" lay west of the one-hundredth parallel, he painted a harsh picture of arid lands, scorching winds, and other adverse climatic conditions, all aimed at convincing readers that settling in Alabama's potential agricultural paradise was preferable to life even in such supposedly favored places as California. Quoting from eyewitness accounts, he claimed that irrigation was no remedy for such desperate circumstances; even including hay and wild grass, he declared, the whole state of Nevada could produce only one-fifth the crop yield of Montgomery County in 1860.

Milner closed his book with a glowing vision of the prospects of Alabama's Mineral District, pointing out its advantages over the Chattanooga region. No place on earth, he claimed, could compete with Alabama pig iron, given the proximity and quality of Jones Valley's raw materials and the low cost of black labor. As Alabama attracted more and more Caucasian immigrants, he asserted, its black population would decline and wither away. Meanwhile, blacks, who were better suited for industrial tasks under white supervision than for tilling fields, could make a useful contribution by mining coal and tending blast furnaces.[58]

The district's switch to coke set off a wave of speculation in the Warrior coal field. Among the many new arrivals swarming into the area was Truman H. Aldrich, an engineer from New York who had moved to Alabama after graduating from Rensselaer Polytechnic Institute in 1869. After starting a bank in Selma, he had acquired and modernized coal mines at Montevallo before coming to Jefferson County in 1877. Prospecting in the Warrior field, he discovered the rich Jefferson and Black Creek seams. Sloss, admiring his engineering expertise, went into partnership with him. Soon, DeBardeleben also joined. In January 1878, after acquiring 30,000 acres in the Warrior field, Aldrich, DeBardeleben, and Sloss formed the Pratt Coal and Coke

Company, named in memory of Daniel Pratt. With help from Joseph Squire, a British engineer who advised DeBardeleben on the location of coal deposits, it soon became the largest mining enterprise in the district. The Browne seam was now renamed the Pratt seam, also in honor of DeBardeleben's deceased father-in-law. Not long after its establishment, the Pratt enterprise built a spur line into Birmingham and began shipping coal and coke in February 1879. The opening of this artery played a vital role in reanimating the community, whose population had shrunk to only 1,200 people.[59]

Without pausing, DeBardeleben joined other investors in creating Alice Furnace, named for his oldest daughter. Building was begun in September 1879 on a twenty-acre tract at the western edge of the railroad reservation, donated by the Elyton Land Company, and the stack went into operation on 23 November 1880. The largest blast furnace erected in Alabama to that date, it was also the first to be designed from the start to use coke. It had the distinction of making the first coke-fired pig iron from the area to win favor in the North and sell there at competitive prices.[60] Sixty-three feet high and fifteen feet wide at the bosh, it achieved an average daily production of fifty-three tons of pig iron under the supervision of Thomas T. Hillman, a veteran ironmaster whose grandfather had built the Roupes Valley forge in 1830. Known as "Little Alice" or "Alice No. 1" to distinguish it from a second and larger installation, "Big Alice," built in 1883, it set off a mania of furnace construction that Robert H. McKenzie later called "The Great Birmingham Iron Boom."[61]

Attempts were made to fulfill Milner's dream of building rolling mills in the district. DeBardeleben offered investors from Cincinnati and Louisville cheap land along the spur line leading to the Pratt mines and agreed to provide them with coking coal at low prices. At his invitation, capitalists from Louisville visited Birmingham; the group included A. Bidermann du Pont, who belonged to a branch of the famous Delaware family that had moved to Kentucky. Samples of red ore were shipped to Louisville and tested at a rolling mill operated by du Pont; when the results were successful, du Pont and other Kentucky businessmen established the Birmingham Rolling Mill on a tract at Twelfth Street and Avenue B where the L&N crossed the Alabama Great Southern. The first bar iron ever rolled in Birmingham was produced there in October 1880. Saloons soon sprang up across the street so that puddlers seeking refuge from the searing heat of the plant could quaff large quantities of beer.

The mill, which had twelve puddling furnaces, was staffed by skilled workers drawn to Jones Valley from various places, particularly Great Britain. One employee, John H. Adams, was an Englishman who had worked in Pittsburgh and Chattanooga; later, he would supervise ore mines owned by Sloss Furnaces. The manager of the new plant, Arthur J. Moxham, had been born in Wales, migrated to America as a teenager, and learned the craft of ironmaking at A. Bidermann du Pont's Louisville plant, where he invented puddling and rolling techniques using hydraulic pressure to make wrought iron of great strength at low cost. Related to du Pont by marriage, he moved into a small cottage near the rolling mill with his wife and four-year-old son. Adams and Moxham typified an influx of new arrivals bringing advanced technological skills to Alabama's industrial frontier.[62]

As newcomers poured into the district, Birmingham was transformed from an industrial campsite into a settled community. Substantial buildings of brick and stone now housed business firms; more than a century later, preservationists would restore such architectural treasures as the three-story Harris Building, erected in 1883, with its richly ornamented mouldings and cornices and its rows of pilasters framing long, narrow windows. It was originally occupied by George Harris, the young city's leading dealer in paint and wallpaper. Other structures still surviving in the late twentieth century included the McAdory Building, with an ornate terra-cotta facade and elaborately varied fenestration, and the Iron Age Building, featuring an exquisitely detailed cast-iron front. Both were erected in 1888; the latter served as headquarters for one of the city's earliest newspapers.[63]

Churches multiplied in the 1880s as places of worship were erected by Baptists, Episcopalians, Methodists, Presbyterians, and Roman Catholics. A Baptist church, whose members had previously met in private homes, was completed in 1886, providing seats for more than 1,000 worshippers. Having had no place to immerse converts, its congregation had earlier dammed a local stream. Jewish residents were also "erecting a handsome synagogue." Blacks, who formed about 40 percent of the population, had their own churches; most of these structures were "wooden chapels," but two, built by Baptist and Methodist congregations, were made of brick. A Young Men's Christian Association and a Women's Christian Union Temperance Association were active. Promotional publications boasted that houses of worship were "filled to overflow every Sunday with well dressed and attentive audiences."

Industrial leaders were conspicuous among Birmingham's congregations. Sloss was superintendent of the Sunday school at the First Methodist Episcopal Church; promotional literature described him as "an active, and more than ordinarily intelligent, sympathizer with all social reforms." Echoing a recurrent theme in the development of American culture, widespread fears were entertained of potential threats to stability and order, resulting in an emphasis on the need for social control. Civic leaders worried about the impact of dime novels and other "poisonous literature" upon impressionable young minds and lamented that many homes had only one book, a largely ornamental Bible displayed on a parlor table. To help rectify the situation, a public library, where "magazines, newspapers and numerous valuable books are kept in good order," was established. Furnaces necessarily ran seven days a week, but virtually everything else in the city came to a standstill on the Sabbath, which was rigidly observed. "The streets are quiet and orderly on that day," a local publication asserted; city ordinances forbade business to be conducted, and bars and barber shops were closed. Birmingham already had a professional baseball team, which tried to play on Sunday, but "the public soon frowned them down."

Cultural aspirations were much in evidence. By the mid-1880s there were enough people of Teutonic origin in the city to form a German Society. Frank O'Brien, an Irish immigrant who had been an officer in the "Montgomery Blues" during the war, built a four-story opera house that was also used for services by congregations that had not yet finished constructing their own churches. Sloss contributed to cultural uplift by heading the Alabama Chautauqua Association. Three "elegant grammar school buildings" and a high school were built for whites, and two segregated structures were set aside for black children. Smith & Montgomery's book store provided reading fare that its owners carefully selected each fall on trips to the East Coast. J. H. Shepherd, Birmingham's leading photographer, specialized in "Crayon Work, in which he has no superior in the State." The Herma Vista and Alabama clubs, composed of eligible young men, held decorous parties for eligible females, and well-dressed people strolled on weekends in Lakeview Park, developed by the Elyton Land Company.

Birmingham had come a long way in a short time. Looking back at the devastation of the Reconstruction Era and the troubled early days of the community, Mary Gordon Duffee of nearby Blount Springs saw nothing incongruous about upholding the "stately dignity and gracious manners" of the prewar planter aristocracy at the same time that she praised the work of industrialists who had erected blast furnaces and rolling mills in Jones Valley:

> Thro' the night there arose a mighty voice
> Of manhood calling to the strength of men,
> And morning dawned upon old, red field
> Made glorious by a City's sudden birth. . . .

> The little babe, that lay within the arms
> Of that poor, weary, weakened mother-state
> Hath grown to such imperial stature now
> Its arms uphold her in her golden pride
> And she forgets the past, when hope had died.[64]

During the 1880s, as pig iron production in Alabama soared from 68,995 to 706,629 gross tons, no fewer than nineteen blast furnaces would be built in Jefferson County alone. The second of these stacks, built by Colonel Sloss, launched the series of enterprises with which the rest of this book is concerned. Sloss had not remained long with the Pratt Coal and Coke Company, leaving it to manage the technologically and financially troubled Oxmoor installation. Partly because of its poor performance, and also because of constant bickering between Cincinnati and Louisville investors, he was not happy there. Leaving Oxmoor in 1881, he started his own furnace company.[65] For almost a century, this enterprise and the firms that it spawned would exploit the opportunities unlocked by Sloss's successful visit to Louisville in 1871.

3 The Sloss Furnace Company

Undaunted by the turmoil of the 1870s, Birmingham plunged into the next decade on a wave of optimism. Boosters hailed the birth of the Sloss Furnace Company in 1881 as one of many signs of progress. The fledgling community, however, faced enormous problems. Some of these were typical of any recently founded American industrial city, particularly one so dependent on a small number of basic commodities for which demand gyrated because of sudden upswings and downturns in the business cycle. Other difficulties, however, were more distinctively local or regional, stemming from the peculiarities of Jones Valley's raw materials, problems dealing with an unmotivated black labor force, and a lack of foundries and rolling mills capable of absorbing anything but a tiny fraction of the district's pig iron.

The early history of Sloss Furnaces was a case study of the headaches that resulted from such problems. In 1886, during a brief but powerful economic upsurge, Colonel Sloss, weary from wrestling with a succession of financial and technological difficulties, decided that it was time to sell out. But his spirit remained as indomitable as ever, and he relinquished control of his enterprise as part of a larger vision, aimed at making steel. If his plan succeeded, he believed, it would bring his remarkable career to a climax, guaranteeing future industrial supremacy to the district in which he had invested so many hopes and dreams.

I

After taking charge at Oxmoor in 1879, Sloss became more and more frustrated by strife between the rival Louisville and Cincinnati factions that controlled the installation, the uneven quality of the pig iron produced by the stacks, and the venture's inability to earn consistent profits. The success of the Little Alice furnace and an upswing in markets for iron that began near the end of the decade caused orders to pour into Birmingham faster than they could be filled. Unable to keep up with demand, DeBardeleben urged Sloss to build his own furnace.[1]

DeBardeleben agreed to supply Sloss with coking coal from the Pratt mines at cost plus 10 percent for five years. Mark W. Potter, who owned ore deposits on Red Mountain, made a similar pledge. After both men had put their commitments in writing, Sloss took the resulting contracts to Louisville and won financial backing from E. D. Standiford, president of the L&N. Upon Sloss's return to Birmingham, he and his sons, Maclin and Frederick, filed papers at the Jefferson County Court House on 21 March 1881 to incorporate the Sloss Furnace Company. Probate judge John Calhoun Morrow recorded the petition the next day.[2]

The Sloss Furnace Company was an ambitious enterprise, capitalized at $300,000. More than half this amount, $160,000, was subscribed on 28 March 1881, the day Morrow specified for opening its books. Pledges for the rest were in hand by May 3, when a board of directors was named. It included B. F. Guthrie, a wholesale grocer from Louisville who had been active in the L&N for many years, and Thomas J. Martin, who had headed the rail line in the mid-1870s. Guthrie had also been a partner in the Oxmoor venture. All told, the L&N invested $80,000 in the venture. Colonel Sloss was named president, Guthrie vice president, and Frederick Sloss secretary and treasurer. On May 4, after 20 percent of the stock had been paid in, Morrow issued a certificate of incorporation.[3]

The Sloss Furnace Company was born at a time when the doldrums of the postwar era had ended and the South was feeling a measure of confidence for the first time since the opening years of the Civil War. Politically, it had been in charge of its own institutions since Rutherford B. Hayes removed the last remaining Northern troops in the wake of the disputed 1876 presidential election. Alabama itself had achieved home rule in 1874 as a result of the victory won by Sloss's friend, George S. Houston, in the latter's campaign for governor against Republican David P. Lewis. Redeemers like Houston now ran state governments throughout Dixie. Economically, the long depression of the 1870s was finally over, and demand for industrial products was picking up throughout the nation.[4]

The time was ripe for a resurgence of moves toward southern industrialization, continuing efforts already underway in the 1850s and during the Civil War. Agriculture was still the basis of the region's economy, but it was not the source of prosperity that it had once been. Even after prewar levels of cotton production were regained in the South as a whole, profits shrank. Demand for cotton on the world market, which had been high during the early phase of the Industrial Revolution in Great Britain, declined in the late nineteenth century, depressing prices. But for demand for cottonseed, which had no value in antebellum times but had become a fertilizer and source of cash income by 1890, the situation would have been much worse. Meanwhile, the

South became locked in on cotton production as creditors demanded that tenants and sharecroppers concentrate on raising it and charged ruinous interest rates that prevented them from breaking free of what became a self-perpetuating cycle of grinding rural poverty. In the process, the high degree of self-sufficiency that the South had enjoyed with regard to foodstuffs before the Civil War was lost, and sustenance for both people and animals had to be bought from the outside. Once the basis of southern well-being, agriculture was now a source of chronic concern. [5]

Alabama was worse off in this respect than some other southern states. Partly because of poor yields per acre, cotton production was still less in 1890, at 915,210 bales, than it had been in 1860, at 989,955. Over the same period, yield per acre shrank from 175 pounds of lint to 165 and was at times as low as 120. Three of the state's closest neighbors, Arkansas, Georgia, and Mississippi, had all done appreciably better. Meanwhile, production of foodstuffs in Alabama had plummeted. As early as 1876, Milner had expressed alarm about precipitously declining output in Alabama of such commodities as corn, oats, potatoes, pork, and wheat, warning that the state could not "produce cotton properly and well, unless she makes her grain and meat on the farm." The failure of such admonitions to be heeded, as tenants and sharecroppers continued to concentrate on raising relatively poor yields of cotton, resulted in increasingly desperate conditions in Alabama's agricultural districts. As a result, there was an exodus of ragged, malnourished people to towns and cities where employment was available at mines, mills, and blast furnaces. [6]

Editorials in what one historian has misleadingly called the "planter press" sometimes praised the superior virtue of agrarian life, criticized the social ambience of places like Birmingham, and urged that Alabama's mineral resources be regarded as export commodities rather than a basis for industrial growth at home. Claims that these views were preponderant, however, are as questionable as the idea that planters as a class opposed industrialization. Agriculture was too self-evidently in trouble throughout the state, and rural districts too obviously overpopulated by surplus labor, for such arguments to be taken seriously by perceptive observers. Much of the animosity that was expressed toward Birmingham, such as the statement of a Selma newspaper that "Jefferson County would be better off if all its furnaces were torn down, its mines filled with dirt, and its inhabitants put to planting cotton," reflected residual Jacksonianism and the continuation of prewar urban rivalries, in this case Selma's long-standing resentment toward Montgomery's Broad River group. Such rhetoric had little or no adverse impact on the growth of Jones Valley and other industrial centers. It took only the return of political stability after 1874 and the first stirrings of economic recovery later in the decade to unleash a mania for manufacturing. [7]

Amid rising expectations, Alabamians—particularly those living in the mineral-rich upstate areas—vied eagerly with their recent northern enemies in embracing industrialization. Despite irregular cycles of boom and bust that made the 1880s an economically treacherous decade, they responded avidly to the New South Creed articulated by spokesmen including Richard H. Edmonds, editor of the *Manufacturers' Record*, and Henry W. Grady, the fiery Atlanta journalist whose speeches and editorials preached that northern and southern entrepreneurs could join hands in exploiting the South's abundant natural resources without endangering established regional customs and folkways. [8]

Feverish urban development, particularly in the Appalachian piedmont, was characteristic of the times. Boom towns—Justin Fuller has identified twenty-four in Alabama alone—sprouted throughout the 1880s as promoters took options on favorably situated tracts of land, established corporations capitalized at millions of dollars, staked out lots and street plans, issued brochures, and trumpeted the "astounding" and "inexhaustible" nature of local coal, limestone, and ore deposits. The sudden rise and equally rapid decline of Fort Payne, Alabama, into which New England capitalists poured millions of dollars in the hope that it would become a future industrial metropolis, typified the false hopes and unethical tactics inherent in much of the boosterism that permeated the period, repelling fastidious onlookers. Every aspiring community adopted a title epitomizing its dreams: Attalla was "The Model City of the South," Bridgeport the "Key of the Sequachee Valley," and Nottingham "Alabama's Young Napoleonic City." As James C. Cobb has indicated, speculators often spoke with forked tongues about the resources of the areas from which they hoped to enrich themselves, bidding "astronomical but fictitious sums for choice lots in order to trigger a 'bandwagon' effect." After visiting Anniston, one critic lampooned the promotional rhetoric abounding there: "Every bunch of sassafras is a mighty forest, every frog pond a sylvan lake, every waterfall a second Tallulah, every ridge of rocks a coal mine . . . and every man a liar." Few heeded such croakers as the boom intensified. [9]

Everyone knew that railroads were the key to the future. It is impossible to read Alabama newspapers during the 1880s without recognizing the avidity with which editors and readers greeted every scrap of information heralding the prospects and progress of this or that line as trackage spread through the wilderness toward such communities as Anniston, Coketon, Goodwater, Jasper, and Tuscumbia; the awe with which northern and European capitalists who backed the ventures were regarded; the ardor with which reporters interviewed promoters, surveyors, and engineers representing faraway investors;. and the vigor with which fledgling towns fought to be located on routes over which the Iron Horse would make its way to Atlanta, Memphis, New Orleans, St. Louis, or San Francisco. Knowing that the outcome of their struggles meant economic life or death, civic leaders battled ruthlessly against counterparts in other communities. [10]

Under these circumstances, industrialists, railroad magnates, and town planners received a degree of adulation previously reserved for military heroes. Typical of the regard in which they were held was the praise heaped upon James W. Sloss by the Birmingham press as his new furnace began to take shape. In November 1881, he was boomed for the governorship of Alabama by the Birmingham *Weekly Independent*. Sloss stood "without a peer for this office," the journal declared. "His excellent business qualifications, brilliant intellect, splendid character and fine executive ability, all combined, make him the grandest man in Alabama today for our chief executive. He is the very personification of Christian manhood and integrity, possessing the qualifications of head and heart which we should emulate."[11] Impelled by such rhetoric, Alabama eagerly embraced the gospel of industrialism.

II

Construction of Sloss's new furnace began in June 1881, when ground was broken on a fifty-acre site that had been donated by the Elyton Land Company on 3 May "provided said Sloss Furnace Company will erect thereon and put into operation within about twelve months from this date, a blast furnace, not less than sixteen by sixty-five feet, for the manufacture of pig iron." The plot, which had access to both the L&N and Alabama Great Southern, lay between Twenty-eighth and Thirty-second streets along Birmingham's northeast edge, at the opposite end of the railroad reservation from DeBardeleben's Alice furnaces. Work proceeded briskly. By early July the foundations of the stack had been laid, and a "handsome brick office with four rooms" was nearly completed.[12]

Harry Hargreaves, a European-born engineer, was in charge of construction. He had grown up in Liverpool and studied at Harrow before becoming a pupil and protégé of Thomas Whitwell, a British inventor who had designed the stoves that would supply the hot-air blast for the new furnace. Whitwell had transformed early inefficient arrays of pipes that used coal or waste furnace gases to heat such a blast. His methods were based on regenerative methods initially conceived by a German emigrant, William Siemens; the latter had designed heating chambers that were equipped with firebricks stacked in such a way that gas could circulate through passageways between them and absorb heat en route to furnaces used for making crucible steel. Because of the high pressures needed for blast furnace operation, Whitwell had to modify Siemens's original design to resist deformation. This resulted in tall, air-tight, cylindrical structures with iron shells and dome-shaped tops.[13] Such stoves, as distinctive in form as the crenellated towers of medieval castles, became familiar sights in southern cities like Birmingham.

In the early 1880s, Great Britain was still the world's leading manufacturer of iron, and British furnace practice remained state-of-the-art despite strides that were being made by such Americans as Andrew Carnegie to overtake it. Representing a continuous line of technical evolution since James B. Nielson had introduced the hot blast in 1828, British stove design set the pace for the ferrous metals industry, and Whitwell was its acknowledged master. As the latter's technical agent, Hargreaves had represented his mentor at such events as the Philadelphia Centennial Exposition in 1876. He was no newcomer to the South, having installed four Whitwell stoves at Rising Fawn, Georgia, and built two furnaces embodying the same principles at South Pittsburg, Tennessee.[14]

Sixty feet high and eighteen feet in diameter, Colonel Sloss's new Whitwell stoves were the first of their type ever built in Birmingham. After being cleaned, waste gases from the furnace would pass upward and downward, circulating through pipes enclosed in a checkerwork of hot bricks before reentering the furnace through the tuyéres. Noted for their power to maintain a steady, even temperature, the domed structures were modern in every respect, comparable to similar equipment used in the North. Stoves with exactly the same dimensions had recently been installed with good results at McCormick and Company's Paxton Iron Works near Harrisburg, Pennsylvania. Others of comparable size and design were being used at such places as Chicago. Like other features of Sloss's new installation, they supported his later claim that he and his partners had "spared neither money nor pains" to put the South "upon an equality with any other section of the United States."[15]

The identity of the principal contractor for the project, J. P. Witherow and Company of New Castle, Pennsylvania, provided yet another indication of Colonel Sloss's determination to use the best available technology. This firm had built blast furnaces all over the North, including Carnegie's Lucy No. 2. Along with another prominent northern firm, Gordon, Stroebel, and Laureau, from Philadelphia, it was extremely active in serving the needs of southern ironmakers. During the 1880s, these two Pennsylvania enterprises built numerous furnaces in Alabama; Witherow alone built seven such installations in the state prior to 1886. As Jack R. Bergstresser has observed, Alabama was part of a technological community embracing every major iron and steel producing region in the country and was not backward by comparison with any of them.[16]

Local observers were proud that much of the machinery used by Sloss's new furnace would be of southern manufacture. It included two blowing engines to force hot air into the furnace through the tuyéres at a rate of 10,000 cubic feet per minute, more than eight times as much as a charcoal furnace required. Such machinery was used in "hard driving," a practice pioneered in western Pennsylvania during the 1870s. Driven by steam, Sloss's blowing apparatus was made at one of Birmingham's foremost enterprises, the Linn Iron Works; it was owned by Charles Linn, a Finnish-born entrepreneur who came from Montgomery to Jefferson County in 1871 to establish a bank that was derided by croakers as "Linn's Folly" because of the large size of the building in which it was housed. It prospered as the city grew. Linn typified the resourcefulness of local industrialists; some of the machinery used in his plant had been scavenged from McElwain's old furnace at Irondale.[17]

The blowing engines that the Linn works made for Sloss were worthy of the latter's optimism and penchant for bigness. Noting that they weighed eighty-five tons each and that their cylinders were eighty-four inches in diameter, a local reporter described them as "the largest engines ever made south of Pittsburg [*sic*]," showing "what can be done by the sons of Alabama in the development of her unapproached and inexhaustible resources." Ten boilers, each thirty feet long and forty-six inches in diameter, were too complex to be built in Birmingham at this time. They too, however, came from the South, being made by Walton & Company in Louisville.

The furnace stack was sixty feet high by sixteen feet wide at the bosh. It had a capacity of eighty tons of pig iron per day, about 25,000 tons per year. Although design drawings have disappeared, it is logical to assume from Hargreaves's role in its construction that it embodied a principle, also developed by Whitwell, of configuring the charging bell so that the raw materials distribution method would save wear and tear on the inner furnace walls. In addition to the scouring action of the pressurized air blast itself, coke is extremely abrasive, and iron is a highly active chemical element; Whitwell's bell was intended to minimize the need for frequent relining. It had been adopted only four years before by Carnegie himself and was being used to good effect in his competition with rivals in western Pennsylvania.

If not outstandingly large by standards of the time, neither was Sloss's stack small when compared with furnaces elsewhere. The Pittsburgh District's Lucy furnaces, owned by Carnegie and his partners, were exceptional in being seventy-five feet high and twenty feet wide at the bosh. Since 1872, they had been pitted against the Isabella furnaces, also of the same size, which were owned by a western Pennsylvania syndicate that Carnegie ultimately absorbed. In 1881, because of their innovative use of hard driving, the Lucy and Isabella installations were smashing previous production records with outputs ranging from 1,000 to more than 1,400 tons a week. This output, however, was atypically high for the industry as a whole; many recently built northern stacks were no larger than the one that Sloss was erecting. It was still conventional wisdom, as a new edition of Frederick Overman's *Treatise on Metallurgy* stated in 1882, that "beyond a certain limit, there is no advantage in increasing the height of a furnace. . . . A lower furnace works easier, makes better iron for the forge, and, when well arranged—that is, of sufficient capacity—does not consume more coal in proportion to the iron smelted, than a high furnace. It is, therefore, desirable to operate with the least height."[18]

Like up-to-date counterparts all over the country, Sloss's new furnace looked much different from earlier truncated stone pyramids that had been built in the colonial and antebellum periods. Unlike such furnaces, which had been placed near the side of a hill, it was a free-standing structure; raw materials and workers would be carried up the side by an eighty-foot vertical steam-powered hoist to a platform encircling a single-hopper charging bell,

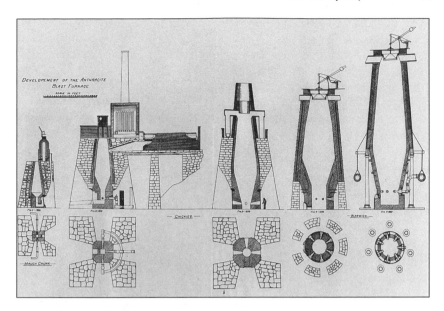

Hot Blast Furnaces. These diagrams show key stages in the development of the hot-blast furnace from 1838 to 1892 and the significant increases in scale that occurred within less than six decades. By the 1870s the stoves that supplied the hot blast had become too large to be perched atop the stacks and were built as independent structures located near the furnaces. (Courtesy of Hagley Museum and Library, Wilmington, Delaware.)

into which coke, ore, and flux would be dumped from hand-operated wheelbarrows. Round brick walls, resting on a foundation of cast-iron columns, were enclosed by shells made of riveted iron plates held together by flat, horizontal iron bands. The bosh had to be built with special care, because the outward and downward thrust of the burden was especially heavy at this point, the heat was particularly intense, and the upward thrust of the blast was highly abrasive. This part of the furnace was jacketed with cast-iron cooling plates that were interlaced with pipes through which water could circulate.

The entire structure, like the stoves supplying it with hot air, had to be gas tight and pressure resistant. The hearth necessitated separate foundations because of the extreme weight of the accumulated slag and molten iron that it would be required to bear, plus the pressure of the blast itself; the slightest weakness in the base or jacket might lead to a dreaded occurrence known as a breakout, in which liquefied iron would force an opening, swiftly enlarge it, and flow outward "with a rapidity which is almost beyond belief except by those unfortunates who have had experience with this action." One expert called such an accident "too harrowing to discuss." Because such incidents were still looked upon as inevitable, the furnace was surrounded by a brick-lined ditch that would also carry off water used to cool the bosh and tuyéres.[19]

Among the other facilities that were taking shape under the direction of Hargreaves was a casting shed, 128 long by 50 feet wide, directly in front of the furnace. Molten metal would periodically flow from the mouth of the furnace into molds of heat-resistant sand on the floor of the shed to solidify into pig iron. The crucible would have a capacity of 125 tons per heat. There would also be an engine house, 48 by 28 feet; a stock house, 250 by 52 feet, capable of accommodating 10,000 tons of ore; and twenty-eight dwellings for workers. The entire plant, praised by local reporters as "modern in all respects," would cost approximately $180,000 and employ about 250 men.

Although Colonel Sloss did not own his own coal mines, the new installation would include 120 coke ovens, each twelve feet in diameter. Embodying the same design used in the southwestern Pennsylvania community of Connellsville, the nation's foremost coking center, they were built under the supervision of a recent arrival from that place whose name was variously reported in the local press as Dutillas or Gutilius. He typified an influx of skilled northern technicians now swarming into Alabama's new industrial frontier. The ovens were of the beehive variety, which was standard throughout the industry. They had domed ceilings, were built close together to maximize the benefits of their collective heat, and were designed to be charged through a hole in the roof with coal that would gradually be reduced to coke from the top of the oven downward in a process that might take as long as three days. After this was completed, the coke would be removed through a door at the bottom by workers who had a particularly unenviable job because of the searing heat they had to endure; coke ovens had to be discharged as soon as possible to make sure the next charge would ignite at once. The coking procedure was wasteful, in that all of the gaseous and liquid constituents of the coal, representing approximately 40 percent of its potential heat content, were lost to the open air. Nevertheless, the ovens could produce excellent coke, depending on the quality of the coal with which they were charged.[20]

As construction proceeded, Sloss enjoyed escorting visitors around the site. E. W. Bailey, a visiting clergyman from England, came away convinced that the new facility was "undoubtedly the finest furnace in America . . . a model of beauty and perfection." Asking Dutillas about the prospects of the southern iron and coal region, he was told that "the north and west cannot compete with it" and was assured that local coke was "equal in every respect" to the best Connellsville product. Full of enthusiasm, Sloss told Bailey that he planned to build another furnace without delay. Preparations for groundbreaking got underway in March 1882, a month before the initial furnace was slated to go into production. The second stack would be even larger than the first, measuring seventy feet high and seventeen feet in diameter at the bosh—almost as large as the Lucy and Isabella furnaces. Its yearly capacity would be 30,000 tons. Reflecting the financial demands that would result from this expansionist move, the firm's capital stock was soon increased to $500,000.[21]

III

On 12 April 1882, the first furnace went into blast. Apparently not wanting to be embarrassed if things went badly, Colonel Sloss kept the event relatively private, but one of the few invited guests who witnessed it reported his impressions in a local newspaper. Taken by the vertical hoist to the top of the stack, the visitors gazed from the "dizzy height" at the forests still dominating the surrounding landscape and then descended to watch Sloss lead his five-year-old daughter, Rosa, toward the hearth. "Shouts and hurrahs" resounded as she applied a torch to the fuel, starting operations at what was proudly described as "the largest furnace ever erected in the Southern States."[22]

Despite high expectations, the company was having trouble with southern raw materials within a few months. Toward the end of August, a coal crusher was installed in an effort to improve the density of the coke produced by the new ovens. Satisfaction, however, was expressed with the Whitwell stoves, which were delivering air to the tuyéres at an average temperature of 1,600 degrees Fahrenheit. By the end of the year, pig iron from the furnace, made at a cost of $12 to $14 per ton, was being sold in Pittsburgh, 800 miles away, at what was termed a "liberal profit." After its first year of operations, the furnace had sold 24,000 tons of iron, only 1,000 tons below its rated capacity. At the 1883 Louisville Exposition, the company won a bronze medal for "best pig iron."[23]

Construction of the second furnace proceeded smoothly after groundbreaking took place on 1 May 1882. Within a few months, most of the ironwork for the stack was in place, the blast ovens had been nearly finished, the casting house was nearly all under roof, and work on the boilers and blowing machines was going smoothly. Completed by the following May and put into blast on a trial basis later that month, the furnace received ore from a stock house 500 feet long and 100 feet wide, with a capacity of 200,000 tons. At the far end of this structure stood the new coal crusher, fed from a storehouse with a capacity of 10,000 tons. An endless chain of buckets dumped coal into the crusher at a rate of 250 buckets per minute, again showing the company's willingness to mechanize selectively in cases of need. Crushed coal was taken to the furnace by 5-ton buggies running up an inclined track.

Ogden Street, a recent arrival who had worked at one of the Lucy furnaces in western Pennsylvania, was now in charge of operations. "He is like a furnace, itself," a local newspaper stated, "always heated up and making an output." The crowd attending the startup of the second furnace marveled at the extent of the facility. Even the waste material it would generate caused celebration; observers predicted that slag would "soon cover many feet high the 50 acres of ground adjoining the plants and belonging to the company" and noted that it would ultimately fill the entire V-shaped intersection of the Alabama Great Southern and L&N. Fifty "neat, comfortable tenement houses," now standing elsewhere, would then be raised and put atop the slag bed, from which railroads could obtain ballast.[24]

Sloss Furnaces in 1885, Showing the Furnaces, Vertical Hoists, Engine House, Hot-blast Stoves, Stock House, Casting Sheds, Coke Ovens, Office Building, and the Rail Lines Serving the Facility. Although a few northern plants were larger at the time, the Sloss Furnace Company's installation was modern by all prevailing standards and epitomized the confidence in Alabama in the 1880s that the state's coke-fired stacks could compete successfully with rival facilities in other parts of the country. (Courtesy of Birmingham Public Library, Department of Archives and Manuscripts, Birmingham, Alabama.)

Birmingham was now a mushrooming community of at least 10,000 people. "It strikes the visitor as a new Western town rather than as a Southern community," stated one account, "or rather it would so strike him but for the presence of so many negroes and the unmistakable Southern dialect of so many of the whites." Under the leadership of Henry Caldwell, the Elyton Land Company's new president, steps were being taken to improve the city's inadequate water supply by boring artesian wells. Pending completion of the project, municipal officials shuddered at the prospect of a major fire that they would not be able to extinguish. The city's only two hotels were described by one visitor as "really nothing more than wooden shanties," and the unpaved streets were a sea of mud when it rained. Amid the bustle and squalor, industrial operations never ceased. At night, the sky was brilliantly illuminated by the fiery glow of the furnaces. As the shrill sound of locomotive whistles and the rumble of railroad cars resounded through the air, hundreds of workers tended "the red hot blasts where molten iron is poured forth in solid streams."[25]

The environment was not only noisy but also dangerous. Late in November 1882, a crowd gathered in the eerie gleam of the No. 1 furnace to gape at the remains of two black laborers, Aleck King and Bob Mayfield, who had died horribly after being lowered on a platform into the interior of the banked installation to dislodge a scaffold of unconsumed ore and coke that was adhering to the inner walls. As they went about the job with picks, some of the congealed materials suddenly "gave way and fell with a crash to the bottom," and a dense cloud of smoke and gas ascended from still-smoldering materials in the hearth. The fumes suffocated the men, who fell to their death.[26]

That same week, an equally terrible incident took place at the Alice furnaces. Unlike the tragedy that had killed King and Mayfield, it was intentional, revealing the anomie prevalent among transient workers who were crowding into Jones Valley. Samuel Cunningham, a depressed rolling mill employee who had just arrived from Kentucky, committed suicide by gaining unauthorized access to the grounds at Alice No. 1, ascending the hoist to the top of the stack, and plunging headlong into the roaring furnace interior while horrified workers looked on.[27]

Such events continued to occur from time to time and were for the most part quickly forgotten. One, however, led to an enduring part of the city's industrial folklore. In September 1887, Richard Jowers, an assistant foundryman at Alice No. 1, fell to a fiery death while helping repair the charging apparatus. After the incident, workers at the furnace began reporting eerie encounters with a spectral presence, believed to be that of Jowers. Ultimately, the luckless employee "became something of a folk hero, a prototype of the thousands of men who dug the ore and mined the coal and tended the furnaces." Apparitions of his presence sometimes appeared in the searing heat when molten iron was released from Alice No. 1. These continued until the furnace was torn down in 1905, after which the same phenomenon began to take place at Alice No. 2 until it, too, went out of blast for the final time in 1927. From then on, it was said, Jowers's shade transferred its appearances across town and became "The Ghost in Sloss Furnaces," making its presence felt as white-hot iron flowed into the casting shed. Some local residents believe that it continues to brood over the idle facility and will do so as long as the stacks exist. [28]

IV

Jowers, who worked in a supervisory capacity, was white. That King and Mayfield were not typified one of the most distinctive features of the southern iron industry: its continuing use of an almost exclusively black labor force. From hundreds of rural hamlets and once-prosperous plantations, swarms of former slaves flocked to the Birmingham District to work at its mines, furnaces, and rolling mills. However bad conditions might be at such installations, blacks saw them as preferable to peonage in the poverty-stricken countryside. They were also lured by the prospect of earning cash on a regular basis instead of having to wait for the uncertain prospect of a pittance at the end of a long, hot growing season. Colonel Sloss and other owners were happy to hire them, believing that low wages, proximity to raw materials, a benign climate that permitted cheap, flimsy housing, and freedom from municipal taxes on land situated just outside the city limits made an unbeatable strategy for competing with northern firms.

In 1886, a book published by a group of Birmingham's leading citizens to advertise the city's advantages stated that about 90 percent of all workers at blast furnaces in the area were black. "Besides the manufacturing element of negro population," it said, "almost all the drays here are driven by negroes; the merchants employ them largely as porters in confidential positions; the hacks are all driven by negroes; the livery stables employ them; the domestic service of the town is practically monopolized by them." Birmingham's black population, starting with 2,506 in 1870, doubled by 1880, multiplied six times by 1890, and kept growing steadily after the turn of the twentieth century. Comprising 43 percent of the city's residents in 1890, and still forming 39 percent in the early 1900s, it gave Birmingham more black people per 100,000 population than any other place in America. [29]

Competition between industrial cities in an age of falling prices encouraged Birmingham's reliance on low-paid black labor. Like their counterparts in other parts of the country, owners and managers in Jones Valley regarded workers as what Maury Klein and Harvey A. Kantor have called "one more resource to exploit, an input to production that must be obtained as cheaply as possible." The implacable demands of the market system and the high cost of equipment made such an attitude inevitable. In such an atmosphere, as Gavin Wright has shown in a searching analysis of industrialization below the Mason-Dixon Line, black work forces became vital components of a separate "regional labor market" based upon low wages as a primary weapon to overcome the South's disadvantages in competing with the North, just as Milner had advocated in 1859. The use of black workers by southern furnace operators as an essential element of this strategy linked it to racist attitudes that were deeply imbedded in southern culture, producing a particularly severe form of exploitation and social problems of enormous long-range consequence.[30] Only in South Africa, where black workers burrowed for gold in the Main Reef, Johannesburg's equivalent of Red Mountain and the Warrior coal beds, did a similar situation come about. Pictures taken in the 1880s of Birmingham and the faraway city on the Witwatersrand have an uncanny likeness.[31]

Because of the crucial role that reliance on black labor played in Birmingham's development, the antebellum background out of which this practice evolved bears further discussion here, emphasizing the city's Old South roots. As has already been shown, the use of an almost exclusively black labor force at mines and furnaces in the Birmingham District was not merely a response to the competitive position of the South vis-à-vis the North in the late nineteenth century, but a continuation of policies prevailing long before the war, fulfilling plans outlined by Milner in his 1859 report to the officers of the South and North Railroad. Partly because many southern whites disdained manual labor, and also because white workers imported from outside the region proved unruly and undependable, the industrial employment of black workers in transportation and industry was practically universal throughout the region before the Civil War. Only a few factory owners, most notably William Gregg, favored the exclusive use of white operatives. The use of blacks in skilled occupations was just as frequent as their employment at tasks requiring little or no training. Census figures indicate that the South had only 20,000 white tradesmen in 1865. By contrast, approximately 100,000 slaves were employed as blacksmiths, gunsmiths, cabinet makers, plasterers, shipbuilders, stone masons, and bricklayers. Blacks were also steamboat pilots and engineers, displaying skills comparable to those of white workers. Horace King, a black architect and bridge builder, did such outstanding work as a bondsman that he was emancipated under special legislation passed by the state legislature.[32]

Because of a prevailing belief that blacks were particularly well suited for hot and strenuous occupations, their use as ironworkers was virtually universal in the South during antebellum times, as previously indicated. In Virginia, for decades the leading southern iron-manufacturing state, slaves tended charcoal-fired blast furnaces up and down the Blue Ridge. They were also employed as miners in the Midlothian coal field, boatmen on the James River and Kanawha Canal, and construction and maintenance hands on railroads.[33] Virginia's capital city, which became a role model for Birmingham because of the heavy involvement of such Richmond industrialists as Joseph Bryan in the development of north-central Alabama, was well-known for its use of black labor in factories prior to the war. Slaves did almost every type of operation, skilled and unskilled, in tobacco plants; watching them and listening to their work songs as they separated, rolled, pressed, and cut leaves into plugs was a popular tourist attraction. As was common throughout the South, they also worked in Richmond's textile mills.[34]

Richmond's most prominent antebellum industrialist, Joseph R. Anderson, was the state's leading urban employer of slaves, using many of them in iron manufacturing at his Tredegar and Armory works. His use of black workers in skilled operations at the Armory mill, after he forced whites to sign contracts to teach them such trades, precipitated a strike that was crushed by Anderson with help from the municipal government. Slaves, whom he either owned outright or leased from their masters, helped him cut costs, offer better wages to skilled whites, compete more effectively for overseas markets, and plan ahead with greater freedom from cyclical fluctuations. Prior to 1861, he used blacks chiefly in his rolling mills and blacksmith shops. After the outbreak of hostilities, however, his use of slaves in other capacities escalated sharply, partly because whites were pressed into military service. By 1864, Anderson was employing 550 blacks at his blast furnaces. Hundreds more worked for him as coal miners and canal workers.[35]

Slaves predominated among the workers staffing the relatively few factories that existed in Alabama before the war. Under questioning by members of a committee from the United States Senate that investigated the condition of southern labor in 1883, a witness recalled that, of the several hundred workers who made tickings and other textile products at the Bell factory, only two had been white. As earlier indicated, even Daniel Pratt, who had originally intended to employ only poor whites, had to use slaves in operating his mills, partly because many white workers would not accept his ideas about the dignity of manual labor.[36]

The use of slaves as ironworkers was as universal in antebellum Alabama as it was in the Old Dominion. Furnace owners usually leased them from planters on an annual basis, assuming the cost of keeping them housed, fed, clothed, and guarded. Foreshadowing the fate of King and Mayfield, the first known fatality of an ironworker in Alabama occurred when a slave was killed by falling rocks in an ore pit at the Cane Creek Furnace. One of its owners, John M. Moore, supplied approximately fifty slaves as part of his investment in the enterprise. Black labor was also used at Shelby Furnace, where the cost of leasing a slave at the beginning of the Civil War was about $175 per year, and at Tannehill, where slaves built the furnaces and quarried the stones out of which they were made. During the Civil War, slave labor continued to be indispensable at Alabama ironworks; slaveowners who lived near advancing federal armies in such places as Tennessee and Virginia frequently leased field hands to furnace operators, welcoming the chance to place their bondsmen in relative safety at a time when Alabama was not a major theater of combat.[37]

Despite their prewar use in textile manufacturing, blacks were almost totally excluded from working in the cotton mills that sprang up throughout the South after the Civil War. This rejection occurred mainly because of the nature of the work itself, the state of the labor market, and southern convictions involving racial mixing in the workplace. Although textile production was physically and mentally tiring and was also somewhat dangerous because of the prevalence of cotton dust and the possibility of getting caught in the machinery, it was not as hot, arduous, or potentially unhealthful as ironworking or coal mining. Because of the postwar climate of enthusiasm for industry, the existence of large numbers of potential textile workers among the region's poor whites, and the depressed state of agriculture, southern industrialists readily embraced Daniel Pratt's earlier vision of neat, orderly mill villages where the daily lives of white operatives who were drawn from the poverty-stricken uplands would be tightly controlled by a paternalistic management. Such entrepreneurs avoided racial mixing because they believed that the resulting discord would jeopardize discipline and the stability of their mill villages. The widespread use of white females as textile workers also made the hiring of black males anathema to those employers, who were haunted by fears of rape, miscegenation, and racial intermarriage.[38]

The considerations just noted were not relevant to the operation of blast furnaces and mines, where conditions of extreme arduousness, danger, heat, and poor ventilation were chronic. Black males were regarded as ideal for such an environment, in which it was widely assumed that brute force, not finesse, was needed. Congregating hordes of black workers in one place, however, intensified off-the-job tensions because they exacerbated deep-seated southern fears about the safety of white women. In such an environment, what W. J. Cash later called the region's "rape complex," based on concern about preserving the inviolability of white females, promoted extreme forms of social hostility toward black males. Much of Birmingham's legacy of violence against blacks resulted from this unintentional by-product of the southern low-wage labor strategy.[39]

Although Birmingham did not exist before 1871, the fact that it had been conceived in the 1850s as a slave-operated industrial city by John T. Milner and his Broad River associates raises questions about the degree to which urbanization was incompatible with slavery, as some historians, including Richard C. Wade, have asserted.[40] A careful study by Claudia Dale Goldin indicates that, to the contrary, demand for urban slaves was actually growing in many parts of the South in the last decade of the antebellum era, even though slave populations were declining in some cities at that time. This increase was at least partly because heavy foreign demand for cotton in the 1850s made it temporarily profitable to divert relatively unskilled slaves from cities to rural areas. Goldin has suggested that, if slavery still had been legal after 1870, declining prices for cotton on the world market would have caused an increasing use of slave labor in southern cities.[41] Except for the fact that the workers were former bondsmen and their offspring instead of actual slaves, this situation is exactly what took place in Birmingham.

Goldin's research further indicates that the decline in absolute numbers of slaves working in southern cities during the 1850s had little or nothing to do with fear on the part of southern managers that such workers were any more difficult to control in an urban setting than would have been the case in rural surroundings. Indeed, as she points out, antebellum slave insurrections that occurred after the 1822 Vesey uprising near Charleston were largely rural, not urban.[42] Milner's prewar visions of black operatives manning mines and ironworks in the industrial city about which he dreamed gave no hint that he feared control problems; indeed, he believed that the skill of southern bosses in managing black labor would be a prime weapon favoring the South in competition with northern industries. Robert Starobin's research on industrial slavery in the Old South also indicates that the region was confident of its ability to keep bondsmen under effective control in occupations involving manufacturing and transportation. As Goldin has shown, the prewar industrial use of male slaves in cities such as Richmond also provided a convenient source of domestic help because their wives could be employed as maids and housekeepers in private homes.[43] The same pattern was subsequently followed with regard to black workers and their spouses in Birmingham.

Southern entrepreneurs were well aware that the existence of black workers in urban settings could create problems involving social control. Concerns about the use of alcohol among urban slaves in the antebellum era, for example, resulting in efforts to prohibit the sale of intoxicating beverages to them or to require written permission from owners before such drinks could be purchased, were analogous to anxieties in postwar Birmingham about the effects of drunkenness among black operatives, contributing to the strength of prohibitionist sentiment in the Magic City. Above all, there was nothing in the antebellum era, even in such industrial cities as Richmond, comparable to the massive numbers of black workers and their families who thronged into the Birmingham District in the late nineteenth and early twentieth centuries, leading to mounting concerns about the safety of white females in such a situation and a corresponding determination to protect them at any cost. One should not, however, take this to mean that those who founded Birmingham appreciated the magnitude of the fears and pressures that might result from such a situation. As already indicated, Milner believed that the black population of the South would wither away due to the collapse of slavery but that in the meantime the region should do its best to capitalize on the advantages of cheap black labor. All available evidence indicates that Milner and others were confident that any problems resulting from the creation of an industrial city using a servile black labor force could be successfully managed.[44]

The work required of operatives at blast furnaces like those erected by Colonel Sloss typified the labor that Milner and other southern entrepreneurs thought particularly well suited for blacks. Hand-loading heavy raw materials onto the vertical hoists to be carried to the top of the stacks and filling the furnaces with them were characteristic of the arduous and dangerous processes involved. Manual filling required much skill to make sure that burdens would reduce evenly but had to be done under great time pressure, increasing the risk that workers might accidentally plunge into the flaming interior.

Fillers were also constantly exposed to poisonous fumes escaping from the stack.[45]

Work in the searing heat of casting sheds was equally grueling and perilous. As the seething brew within one of Birmingham's blast furnaces reached the point at which it was ready to be tapped, sweating, heavily muscled black workers prepared for operations that would test to the limit their endurance and ability to withstand danger and stress. Besides requiring quick reactions and agility to avoid injury, such tasks also demanded a great deal of judgment, giving the lie to stereotypes that furnace work was best suited for brainless brutes. Using hand drills, sledgehammers, and crowbars, workers periodically released plugs at the base of the furnaces to allow white-hot slag or molten iron to gush out. This process required heavy exertion by six to eight men for at least ten minutes, and it might require up to one hour. Because of the intensity of the heat, no worker could endure opening the notch for more than a few minutes at a time, requiring the men to work in spells. Under the downward pressure of the materials coming through the stack, the iron burst from the furnace with explosive force in a turbulent flow that spattered and burned the flesh of operatives who did not quickly get out of the way. After the torrent had subsided, crew members closed the notches, using clay balls, rams, and stopping hooks. Though not as arduous as opening the notch, this process might take up to half an hour.

Showering the shed with sparks, the iron ran into a trench, called the runner, sloping toward the opposite end of the cast house. Other channels, called sows, branched off at right angles from the runner toward the walls; the sows led in turn to the pigs, in which individual billets of iron were formed. At the place where each sow joined the runner, cast-iron stops, called shutters, were opened and closed with hand-held implements to regulate the flow of the molten iron into the sows and pigs. This job was not only extremely hot, but it called for a high degree of skill to ensure that each cluster of pigs filled evenly. Danger was ever present; even a small amount of moisture in the molds could produce an explosion that would send molten metal flying in all directions. Even so, the work had to be done as quickly as possible, because the casting floor had to be cleared before the furnace could be tapped again.

Gradually solidifying, but still extremely hot, the iron in the pig molds was covered with sand. Putting on shoes with heavy wooden soles, workers walked along the beds, separating the pigs from the sows with sledgehammers. It was because of such tasks, resulting in burned and blistered feet, that so many mythical metallurgical gods were depicted as being lame. "This was as hot a job as a man can stand, but it had to be done," said Edward A. Uehling, an engineer who served for a time as superintendent at Sloss Furnaces. "If the iron were allowed to get cold, it would not only be more difficult to break the pigs off the sow, and the sow into pieces, but the iron could not be carried away in time to mould up for the next cast." After being broken apart, the pigs, weighing up to 125 pounds each, were cooled by being sprayed with water. Workers then loaded them onto mule-drawn tramcars that carried the

Sloss Casting Floor. This photograph of the maze of channels in the sand beds at the mouth of one of the Sloss City Furnaces in 1906 shows the long central runner that branched off into sows and individual pig beds; cast-iron stops that were manually opened and closed to make sure that clusters of pigs filled evenly; hand-held tools that were used in this hot and dangerous environment; hardened pigs ready to be moved to the furnace yard; and a few black workers and white supervisors preparing for the next melt. (Courtesy of Henry Ford Museum and Greenfield Village, Dearborn, Michigan.)

metal to the furnace yard to await shipment. Only one man in ten was strong enough for this type of work, and even such persons could not endure it for more than a few years.[46]

This type of work, reserved mostly for immigrant workers in the North, became the daily lot of the black labor crews upon which the industrial order that emerged in Birmingham in the 1880s was built. As Milner had forecast in 1859, whites were to be found only in supervisory posts or positions requiring skills that the owners did not think black men possessed. Testifying to senate investigators at Birmingham in 1883, Sloss stated that "at our place nearly all our labor is colored . . . except our skilled men; they are all white. . . . The balance of the iron men, the coke men, the yard men, the furnace men, and some of the helpers and stock men are all colored."[47]

Sloss admitted that using free black workers created difficulties that antebellum slaveholders had not fully anticipated. The problems he was encountering revealed a great deal, not simply about his company, or merely about the development of southern ironmaking, but about what managers were experiencing throughout the United States as they tried to break workers to the demands of an industrial regime. Herbert G. Gutman has observed that the working class that was formed during the Gilded Age was made up of people "who brought into industrial society ways of work and other habits and values not associated with industrial necessities and the industrial ethos."[48] This drawback was particularly true for former Alabama field hands, who had to make difficult psychological adjustments in moving to such places as Birmingham from rural plantations.

Before the war—especially after Nat Turner's rebellion in 1832—slaves had been rigidly bound to their workplaces. In Birmingham, however, in spite of vagrancy laws and every other expedient that industrialists and civic officials could devise to keep freedmen immobilized, this control was impossible. Like other southern industrialists, Sloss complained about the transiency of free black workers. His operatives came from all over the South, drawn to Birmingham by the lure of urban life. "They are a moving, restless, migratory class, quite different from the farm or plantation negroes," he declared, ignoring that many of them came from rural environments. "They come down here, they work on railroads, they work on the streets, they work in the coal mines to some extent, and largely in the ore mines and about the furnaces and rolling mills, and it is no terror to them to be discharged, none in the world."

Subjecting such persons to industrial discipline, Sloss declared, was extremely hard. "If there is a show in town, or an excursion on the Fourth of July, or a burial, it makes no difference what the excitement may be, they will just drop their work and go off." Welding blacks into an effective work force was also complicated because some of them had two or more identities, making it easy to flit from one job to another. "A man may be with me as John Brown," Sloss said, "and with somebody else at another time as Dick Jones."

As Starobin has pointed out, wandering off the job, deliberate slowdowns, and other types of soldiering were widespread occurrences among industrial slaves prior to the war.[49] Now that blacks had been emancipated and could no longer be subjected easily to various forms of coercion and social control that had been used in antebellum times, the challenge of securing proper levels of discipline among them was even more formidable. Because of high labor turnover and chronic absenteeism, Sloss testified, he had to employ 569 men to assure his company of the 269 workers that it actually needed. On any given Monday morning, half the persons who were needed to run coke ovens or perform other tasks might not show up. Agents were then sent into downtown Birmingham to places where idle blacks, called "floaters," were known to congregate and told to round up as many substitutes as they could find. Meanwhile, workers who had actually reported for duty would be shifted from one part of the plant to another, depending on what tasks had to be done. It might take two or more days before things got back to normal.

One way of combatting such instability was to hire as many workers as possible who had wives and families. Sloss did so. Furnishing houses to workers at reasonable rents also helped promote stability; he charged $4 per month for a one-room dwelling, $5 for two rooms, and $6 for three, asserting that this was about half of what similar accommodations cost in the city. According to him, wage incentives were not much help, but he still used them, paying what he claimed to be the highest rates in the district. Workers at the coke ovens averaged about $1.47 per day. Those at the furnaces got $1.36, while operatives in the iron yard, where the lifting was particularly strenuous, earned $1.58.

Sloss said that a worker and his wife, living in a two-room house, could get by on $10 to $12 per month, leaving $15 to $20 for savings. He lamented, however, that operatives often squandered their pay despite repeated admonitions to plan ahead: "They work and get their money, but they have nothing to invest it in, and while it is in their pockets it burns, and they are never satisfied until it is gone." Saying that there were "very few means of safe investment in this country," he urged that state and national governments issue bonds in small denominations to encourage thrift among workers.

Because of the lack of dependability that Sloss complained about, southern employers used what was known in other places as the "padrone" system, dividing their work forces into gangs and contracting them to bosses, who were usually black. These recruiters took responsibility for seeing that specific operations were adequately staffed. Sloss followed this practice at the Potter ore mines, which he had bought by the mid-1880s. Although he said nothing about using the same policy at the furnaces, it undoubtedly prevailed there as well. Another common custom required newly hired workers to stay on the job for two weeks or more before they could collect their pay; money forfeited by failure to meet such requirements was carried on the books as "unclaimed wages." Sloss did not mention this tactic, but it was used systematically by new owners who took control of the company in 1887, and he probably used it, too. [50]

Sloss admitted that he did not know how to get "continuous service" from his black operatives. Even the promotional literature issued by local boosters confessed that their use created problems. "It is not to be denied," stated a book trumpeting the advantages of the Birmingham District, "that as a laborer, the negro here is below the old slave standard of efficiency in his race, and is here below the standard of white labor engaged elsewhere in the United States in like industries." [51] Nevertheless, Sloss defended using black workers, saying they were best suited for his needs. Pressed to justify his stand, he gave no economic arguments except to repeat a familiar antebellum theme by stating that blacks were less likely than whites to strike. Otherwise, his answers seemed vague and dismissive. Stressing his affection for blacks, a common stance among both prewar plantation owners and postwar industrialists, he said that "the colored man likes the furnace business; he has a fondness for it." Asked whether blacks "can stand that kind of work better than the whites," he replied, "Yes, sir. The white man sees the furnace belching out and it suggests to him his future home." After laughter had subsided, he ventured that more dependable workers might reduce labor costs by at least 10 percent but said that the problem "has got to work its own cure."

What Sloss did not tell the investigators, but what everybody knew, was that he and other operators in the district clung to the almost exclusive use of black workers at least partly because they feared the social problems that would result from trying to mix them on an equal basis with whites who might be willing to accept the onerous labor required. As Anderson's experience in Richmond had shown, and as Sterling Spero and Abram Harris confirmed through later historical research, white workers on both sides of the Mason-Dixon Line objected in antebellum times to being forced to compete with blacks.[52] Propagandists for Birmingham made a virtue of its labor policy despite the lack of industrial discipline that resulted from it. "The manifest result of the presence of the negro labor here is that we have a more intelligent and orderly white laboring population than otherwise might be anticipated," stated one account. "The negro of Birmingham fills the industrial position, which elsewhere in great manufacturing towns is filled by a low class of whites. The negro here is satisfied and contented; the low whites elsewhere are dissatisfied and turbulent. The white laboring classes here are separated from the negroes . . . by an innate consciousness of race superiority. This sentiment dignifies the character of white labor."[53]

It is clear from such statements that the rationale for relying on a predominantly black labor force in Birmingham's primary industries was as much social as economic. Sloss's testimony concerning the undependability of transient and poorly disciplined workers casts doubt on the extent to which the use of poorly paid black workers actually enhanced the ability of Birmingham furnace owners to compete with ironmasters in other parts of the country. His failure to present effective arguments in support of the system indicates that, instead of being an asset, it may have been a drawback, now that the legal constraints associated with slavery no longer prevailed. If true, this situation would support Gavin Wright's argument that the South's dependence on a restricted regional labor market cost it heavily in the long run.[54] In any case, manufacturers in the Birmingham District did their best to devise ways of managing black workers that would preserve as much as possible of the old slave system, and they continued a massive commitment to servile, low-paid labor as a key part of their strategy for controlling costs.

V

As already noted, Birmingham's other primary weapon against northern competition was the close proximity that Jones Valley afforded to all of the raw materials needed for ironmaking. This was the logistical equivalent of having short supply lines in wartime, something that former Confederate officers who supplied much of the South's industrial leadership after the war could readily understand. Birmingham's situation contrasted dramatically with that of western Pennsylvania and eastern Ohio, which had access to excellent coking coal but were located far from the rich ore deposits along Lake Superior. Combining these two widely separated raw materials required massive investments in transportation systems and handling facilities, resulting in huge fleets of ore-carrying vessels, large batteries of highly sophisticated unloading devices, and rail systems connecting such places as Pittsburgh to ports along Lake Erie. By contrast, Birmingham's blast furnaces were only a few miles away from ore, fuel, and flux. [55]

Birmingham's advantage in this respect, however, was not as great as might be assumed. The cost per ton of transporting ore on the Great Lakes, low to begin with, was constantly reduced through streamlining operations. Furthermore, ore from the upper Michigan peninsula, northern Wisconsin, and northeastern Minnesota had a much higher percentage of metallic ore than Alabama ore did. Entrepreneurs in western Pennsylvania and eastern Ohio also enjoyed convenient access, both by rail and water, to the best metallurgical coke in the country and, unlike ironmasters in Jefferson County, employed workers who were highly experienced in coke-making. It would be hard to find a problem that frustrated iron manufacturers in Alabama more consistently than that of obtaining adequate high-quality coke, partly because of the chemical makeup of Warrior coal but also because they had to rely on inexperienced and badly trained operatives. Further complicated by the social imperatives that required using unmotivated blacks to staff the district's coke ovens, this situation proved to be more of a drawback than did a scarcity of engineers or technical supervisors, who could readily be imported from the North or from abroad. To this extent, perhaps, Wright's arguments about the region's lack of technical personnel require further refinement. As Starobin has pointed out, slaveholding southern industrialists had already profitably used in the antebellum era a deliberate strategy of combining cheap slave labor with the hiring of relatively high-priced northern or foreign technicians and engineers. [56] During the late nineteenth century, however, the lack of middle-level technicians became more and more of a problem.

Because northern entrepreneurs had access to enough capital to organize the massive infrastructure that was required to bring high-quality ore where it was needed, the distance between blast furnaces in Pittsburgh or the Mahoning Valley and the ore beds of upper Michigan was less of a disadvantage to them than Birmingham's boosters recognized. Once Lake Superior ore reached its destination in northeastern Ohio or southwestern Pennsylvania, its favorable chemical characteristics paid off. Unlike southern producers who had to use lean ores containing high concentrations of phosphorus and silica, northern operators could capitalize on the easy convertability of pig iron made from Lake Superior ore into Bessemer steel, the most profitable investment open to the ferrous metals industry at the time. Because steel rails were in high demand due to the constant growth of the country's railroad network, pig iron could be produced in Pittsburgh on a scale that Birmingham could only dream about. This resulted in low unit costs of production and even greater profits. Carnegie discovered this fortunate combination of circumstances as soon as he opened his Edgar Thomson Steel Works in 1875; from then on, he used every resource at his command to pare costs as ruthlessly as possible while maximizing output. Wanting to play the same game, Birmingham's boosters longed to overcome the technical problems that prevented them from making steel. [57]

Northern furnaces had another strategic advantage over southern installations by virtue of being located close to large numbers of processing plants that shaped metal into various forms—foundries, rolling mills, wireworks, and the like—and to a host of industries, ranging from axe factories to shipyards, that consumed vast quantities of iron and steel. Northern furnaces were also close to the nation's largest and most prosperous railroads, which had a continuing need for rails, boiler plate, car wheels, and numerous other products that were made from ferrous metals. Because the North was both more populous and more prosperous than the South, it also had the benefit of larger tertiary markets for products made of ferrous metals. Throughout its history, even after it began to make steel and to draw foundries southward because of the peculiar suitability of its phosphorus-rich ore for molding purposes, the Birmingham District was hampered by having an inadequate infrastructure of home markets for the output of its furnaces and mills. This was true with a vengeance in the 1880s. A spokesman for the iron trade in Tennessee noted in 1883 that southern pig iron production was "far beyond the capacity of local mills and foundries to consume." Four years later, *Iron Age* stated that 90 percent of all iron made in the South was shipped to other parts of the country. "Export or Die" would have been an appropriate motto for Birmingham's furnace operators, continuing a state of affairs already familiar to Joseph Heslip and other Alabama ironmakers in the antebellum period. [58]

In order to understand the market opportunities that Birmingham furnacemen enjoyed in the late nineteenth century despite such problems, it is necessary to recognize that pig iron that was made for conversion into steel, on the one hand, and pig iron that was meant to be converted into wrought or cast iron, on the other, formed two separate market categories. Normally, these two classes of metal competed only when downswings in railroad building forced steel producers who ordinarily made pig iron for their own rail mills to find other outlets. As already indicated, most of the steel that was made in America at this time went only into rail production. This vast market was closed to Birmingham because its raw materials were poorly suited for the acid Bessemer process. Birmingham's pig iron was also unsuitable for a few specialized products, such as railroad car wheels, which were still made from charcoal iron. But there remained a multitude of uses that Jones Valley entrepreneurs could supply in an economy that still used large amounts of iron for agricultural implements, animal traps, hollowware, pipes, stove plates, and other products made by rolling mills and foundries.

Because of distance alone, the availability of southern pig iron did not usually pose a threat to furnace operators in eastern Ohio, West Virginia, and western Pennsylvania. It could adversely affect such producers, however, when strikes in those areas shut down their facilities and forced customers to turn temporarily to southern installations. This type of situation arose in 1882, when a walkout by the Amalgamated Association of Iron and Steel Workers shut down large numbers of plants in Pittsburgh, Wheeling, the Mahoning Valley, and other places. Birmingham and Chattanooga briefly enjoyed surging demand because of this fortuitous circumstance, but the effect was short-lived.[59]

But for the decisive advantage that southern pig iron enjoyed in the lower Ohio Valley, cities like Birmingham and Chattanooga could not have survived. Not only were Cincinnati and Louisville close to Alabama and Tennessee, but, because influential members of their business communities had large investments in southern blast furnaces, they established low rates on such railroads as the L&N to enhance the competitiveness of these installations. Furnace operators in Birmingham and Chattanooga could therefore count on at least two major outlets when all else failed.

Demand for pig iron in Cincinnati and Louisville was heavy. Cincinnati had been an important center of rolling mills and foundries since the early nineteenth century. It had an abundance of skilled ironworkers, many of whom were German immigrants. Louisville was also a thriving city whose population grew from about 100,000 in 1870 to double that by 1887; during the same period, the value of its manufactured goods increased from $18 million to $66 million, and the number of its industrial workers rose from 10,365 to 39,125. Gas and water pipe were among its leading products. Despite an influx of northerners after the war, it remained strongly southern in its orientation. "Towards the South, Louisville occupies the attitude held by Chicago towards the Northwest," stated the *Manufacturers' Record*. "Situated in a State where slavery was once acknowledged, the people of Louisville are naturally more in sympathy with the people of the South generally than any other large city supplying Southern demand."[60]

In the early years of the Birmingham District, Cincinnati and Louisville together absorbed more than 90 percent of its pig iron. Commission merchants specializing in ferrous metals, particularly Matthew Addy in Cincinnati and George H. Hull in Louisville, handled most of the trade. This method of distribution, well suited to the diffuse nature of local markets and the inability of furnace operators to identify specific customers, had become well established throughout the northeast and midwest by the time of the Civil War. Because of their proximity and accessibility by rail and water, Evansville, Nashville, and St. Louis also provided convenient targets for the Sloss Furnace Company and other Birmingham producers in the 1880s and 1890s. So, too, did Chicago, Cleveland, Detroit, and Indianapolis, because of connections with railroads that led southward into the L&N. Colonel Sloss told senate investigators in 1883 that he sold pig iron in most of these places.[61]

Commission merchants who handled Birmingham pig iron assumed a number of functions that were later played by the furnace companies themselves. Such vendors were particularly frustrated by the lack of attention that furnace operators paid to grading their output. Pig iron was marketed in a bewildering variety of up to fifteen grades, ranging from "No. 1 Extra Selected" at the top to "White" on the bottom. Each grade fetched a different price. Sample pigs were broken apart with sledgehammers before they were shipped from the furnace and analyzed by "fracture." Contrary to arguments that would later be advanced by advertising executives when Sloss Furnaces switched to the laboratory testing of pig iron in the late 1920s, fracture testing was actually a good empirical method that had been used with satisfactory results by ferrous metal manufacturers since its introduction by the French metallurgist, Reaumur, in the eighteenth century; the basic problem was that Birmingham furnace crews often failed to perform this operation, leaving it up to commission merchants. One of the latter subsequently recalled that "his daily task was to go down to the Cincinnati railroad yards and go over the cars and, after a fashion, separate the pigs into various and sundry piles according to fracture." Another problem that vexed commission merchants was haggling with the owners of rolling mills and foundries that converted pig iron into finished products. Manufacturers debated endlessly with merchants about such questions as whether a given lot of iron was "No. 3 Foundry, No. 4 Foundry, or Foundry Forge." Ultimately, Sloss Furnaces and other pig iron producers circumvented the entire process by dealing directly with customers through their own sales representatives, but throughout most of the late nineteenth century the old system still prevailed.[62]

In addition to the lower Ohio Valley, there was yet another area in which southern ironmasters developed large markets in the 1880s. Despite its distance from the South, the pig iron industry in the Middle Atlantic states was vulnerable to competition from below the Mason-Dixon Line because it was accessible to steamships operating out of such ports as Savannah. Furnacemen in the Adirondack region of New York and the mountains and pine barrens of New Jersey were plagued by antiquated equipment and dwindling supplies of high-quality ore. Meanwhile, the iron industry in New England was virtually dead except for a few furnaces that made high-quality charcoal pig iron for the railroad car-wheel trade. Commission merchants in Boston, New York City, and Philadelphia who had earlier marketed the output of nearby charcoal furnaces to foundries and rolling mills were therefore likely customers for southern pig iron.[63]

Even more important was the declining vigor of the anthracite pig iron industry, centered in eastern Pennsylvania. Peter Temin and other historians have searchingly analyzed its problems. Dating back to the late 1830s, much of its equipment was old and obsolescent. In addition, its raw material costs were high. More anthracite was required than coke to produce an equivalent amount of heat; it was estimated that an anthracite furnace needed twice as much fuel to operate as a coke-fired furnace of the same size. Cost differentials based on such considerations effectively eliminated anthracite from trans-Appalachian markets after the Connellsville and other western coal fields were opened. Occurring in irregular deposits, anthracite was more costly to mine than bituminous coal, partly because of the skills demanded of the workers who extracted it. Furnacemen in the Lehigh and Schuylkill valleys railed against the steep prices charged by the coal companies upon which they depended and also complained about the freight rates charged by lines including the Reading Railroad. Because of high costs, averaging $18.30 per ton in 1884, anthracite furnaces were highly vulnerable to outside competition; many were forced to shut down whenever the price of pig iron dropped. Whereas 161 were in blast at the beginning of 1883, only 80 remained active by mid-1885.[64]

Distress among furnace operators in the Middle Atlantic states spelled opportunity for ironmasters in Birmingham and Chattanooga, where pig iron could be produced for $10 or $11 per ton. Southern railroad and steamship companies established rates that brought the price of shipping iron to such ports as Boston, New York, and Philadelphia to as little as $3.75 per ton. Taking advantage of those conditions, furnaces in Alabama and Tennessee mounted a determined invasion of Middle Atlantic markets in 1884, when approximately 20,000 tons of southern iron were sold in Philadelphia alone. The onslaught continued throughout most of 1885, tapered off toward the end of that year, and resumed in 1886 when southern furnaces sold at least 140,000 tons of pig iron in northern markets. Virginia, benefiting from proximity to northern ports, shipped 80,000 tons; furnaces in Alabama and Tennessee sent 47,000. Throughout the closing years of the decade, whenever the business cycle turned downward and demand in the Ohio Valley shrank, Alabama furnacemen focused on northeastern markets as if impelled by a conditioned reflex.[65]

VI

Taking maximum advantage of the market opportunities open to southeastern ironmasters in the 1880s, however, required good timing. DeBardeleben's first Alice furnace was built during an upswing in demand between 1879 and 1881 and went into blast while the price of good foundry-grade pig iron was still high, at times reaching $28 to $30 per ton. Colonel Sloss was less fortunate; when his first furnace went into blast in 1882, the market was already declining, causing prices to drop. The situation got worse in 1883, and markets continued to stagnate throughout the middle of the decade. By the end of 1885, less than half the country's 591 blast furnaces were still active. American pig iron production had shrunk to 4,529,869 tons, compared to 5,178,122 in 1882.[66]

Because of their proximity to raw materials and their new furnaces, operators in the Birmingham District fared better in this period than did their counterparts in most of the country's other ironmaking regions. Despite sagging prices, pig iron production in Alabama rose from 100,683 gross tons in 1882 to 253,445 in 1886.[67] Making profits, however, demanded maximizing Birmingham's locational benefits by means of what came to be called "straight line production" or "vertical integration," involving careful coordination of raw materials and transportation so as to yield the lowest possible costs. In this respect, the Sloss Furnace Company was not as successful as were some other firms in the district.

One firm that succeeded dramatically in vertical integration was the Woodward Iron Company. It was established late in 1881, not long after Colonel Sloss founded his own enterprise. In 1867, S. H. Woodward, a West Virginia ironmaster, learned about Alabama's mineral resources from northern veterans who had fought there. Visiting the state in 1869, he bought large acreages of mineral lands in Jefferson and Tuscaloosa counties but wisely refrained from exploiting them until the area's prospects became more evident. While the Oxmoor experiments were taking place in 1876, he sent a son, Joseph H. Woodward, to observe them, but he remained unconvinced that the time was ripe to move south. Not until another son, William H. Woodward, reported favorably on the operation of DeBardeleben's first Alice furnace, after it had gone into blast in 1880, did the patriarch make up his mind to capitalize on Jones Valley's potential for making pig iron with coke.[68]

By the time the Woodward Iron Company was organized, S. H. Woodward was dead, but his sons carried on in his place, building a blast furnace on a site twelve miles southwest of Birmingham. It was located in the best possible place, in the midsection of Red Mountain where the ore was the highest average quality. The Woodward furnace was also higher than either of the ones built by Colonel Sloss, permitting greater economies of scale. Seventy-five feet high and seventeen feet wide at the bosh, it went into blast for the first time in August 1883, using coke from a battery of 150 beehive ovens. John H. McCune, a Union veteran who had fought at Gettysburg and supervised the construction of furnaces in Pennsylvania, Ohio, and West Virginia, directed the project.

From the beginning, the new installation demonstrated the fruits of the careful planning that had gone into it. By practicing vertical integration to the highest possible degree, the Woodwards emulated tactics that were also being introduced at the same time by Carnegie and other northern producers. The Woodward Iron Company controlled its own coal and ore mines, which were only eight miles apart, giving them interior supply lines that would have been the envy of a battlefield commander. The Woodwards linked these natural resources to their blast furnaces with their own railroad line, thus avoiding freight charges. They also acquired their own limestone quarries. Because of the extremely short distances over which they had to convey raw materials, they had the most cost-effective operation in the Birmingham District, and quite possibly in the entire world. Within three years, they had liquidated construction costs. Then, and only then, did they erect a second blast furnace, with dimensions identical to those of the first; build 215 new coke ovens; and open up a second coal mine. In early 1887 the cost of their ore and coal, delivered at the furnace, was sixty-eight and ninety cents per ton respectively. Hailing their achievement, *Iron Age* stated that the Woodward company made the only truly "cheap iron" in Alabama.

The disappearance of the Sloss Furnace Company's business records makes it hard to establish close comparisons between its performance and that of the Woodward enterprise. All available evidence, however, shows that Colonel Sloss and his partners achieved nothing comparable to the Woodwards' success. The key to the disparity between the two firms was the way in which they obtained and marshaled raw materials.[69]

As previously indicated, Colonel Sloss never controlled his coal supply. Instead, he secured it under contract from DeBardeleben's Pratt Coal and Coke Company at rates ranging from $1.25 to $1.50 per ton. This was far higher than the $.90 figure posted by the Woodwards in 1887. In addition to paying more for his coal, Sloss could not depend on regular deliveries from the Pratt Company, which was often plagued by strikes, fires, and other problems. Even though his second furnace was ready for operation in May 1883, it produced only a test run and did not begin full-scale production until October because it could not get enough coal to make coke for both furnaces.[70]

By the time Colonel Sloss testified to senate investigators in 1883, he had purchased the Potter mine from which he had originally secured red ore under contract. The quality of the red hematite that he obtained from this source, however, was inferior to that of the ore available to the Woodwards, and the distance from the Potter mine to his furnaces was much greater than that between the mines and furnaces operated by the West Virginians. Another drawback inherent in Sloss's operations was that he lacked his own railroad to deliver ore to Birmingham; instead, he had to rely upon the Alabama Great Southern, near whose tracks the Potter mine was located. Here again, the contrast between his operation and that of the Woodwards was evident; in 1883, Sloss estimated that his ore cost at least $1.60 per ton. This was much higher than the $.68 figure posted in 1887 by the Woodwards.[71]

The penalty that Colonel Sloss paid for inadequate control over raw materials and the means by which they were transported to his furnaces was compounded by undependable performance from his equipment. Unlike the Woodwards, he did not wait until his first furnace had proved itself before building a second; instead, he impetuously erected a new stack before the first one went into blast. As a result, because of the difficulties that he encountered in securing coke, he had more equipment than he could use when the second furnace was ready for operation in April 1883. His predicament became even worse in 1884, when both his furnaces required major repairs. Reporting on this overhaul in mid-June of that year, *Iron Age* did not go into specific details but indicated that, however well designed Colonel Sloss's stacks had been by northern standards, they were ill suited to the specific characteristics of southern raw materials. "Constructing Southern furnaces without regard to the difference in fuel and ores from those worked by the Northern furnaces, whose lines have been copied, has been, it seems, the great error of Southern iron-makers," the article stated, "and accounts for the irregularity in many cases of output, both in quantity and quality, as well as the short life of the furnaces in this and other Southern territory."[72]

This statement provides another key to many of the difficulties experienced by Colonel Sloss and other southern furnace operators. It was not that these entrepreneurs were technologically backward or lacked access to the best engineering advice that the United States and Great Britain could afford, but that they were only beginning to adapt to the deficiencies and peculiarities of local raw materials. This problem, plus their use of unmotivated and ill-disciplined workers, would plague them for decades. Several weeks after *Iron Age* commented on the difficulty of obtaining good results by attempting to use northern or British furnace designs in combination with southern natural resources, the *Manufacturers' Record* sounded the same theme. "Companies have learned, at great cost, that a Pittsburg [sic] blast furnace cannot be planted in Birmingham and the expected results secured," said Bradstreet's Cincinnati correspondent in discussing southern pig iron manufacture. "The special local conditions of ore and fuel must govern largely the manner of building and operating the furnaces. The neglect of this and other similar considerations, born of too limited experience, has added hundreds of thousands of dollars to the cost of Alabama and Tennessee pig iron not counted into the original estimates, and hence we have the anomaly of bonded debts growing and some absolute failures in the case of companies that can show you any day a comfortable profit on sales. It seems a pretty strong statement, but I venture to say that if every leading furnace company in Alabama and Tennessee had at the outset charged to the cost of iron for five years $2 per ton for scaffolds, burning out of linings, breaks, accidents and the various vicissitudes that actually occur, the result would not have been far from right."[73]

Despite his previous experience at Oxmoor, Colonel Sloss had not anticipated that the silica content of ore from the Potter mine, running as high as 15 percent, and the large residues of ash that resulted from the Pratt coal upon which he depended for coke, would result in frequent scaffolding of the type that had cost King and Mayfield their lives soon after his first furnace began production. By the time the constant accumulation of deposits on the inner walls of his stacks made extensive repairs necessary, he had two furnaces to overhaul instead of one, greatly increasing the expense of the modifications he was forced to make.

Even after repairs to his furnaces had been completed in June 1884, Sloss had more difficulty than most local ironmasters making a profit under the conditions that were depressing markets for pig iron. His was the only furnace company in the district that expressed interest later that summer in a nation-wide scheme, conceived by Louisville commission merchant George H.

Hull, to restrict production until iron prices went up. Among the Birmingham producers that responded negatively to the proposal was the Woodward Iron Company, which stated, "cannot see it to our interest." Only a few large producers in the entire country became engaged in the scheme; most participants were small and antiquated operations. The plan soon collapsed.[74]

Instead of cutting back production, it made more sense to such operators as DeBardeleben and the Woodwards to capitalize on midwestern and northeastern distress by increasing output. DeBardeleben negotiated a contract in 1884 to supply large quantities of pig iron to a Nashville foundry at a price rumored to be only $12.50 per ton.[75] After the collapse of Hull's scheme, Colonel Sloss joined the onslaught on northeastern markets; in February 1885 his company was said to be making 160 tons of foundry and forge iron per day and shipping it "largely to Eastern points." The following month, it sold 5,000 tons of pig iron in Philadelphia. Later that year, however, both the Sloss furnaces again went out of blast for further repairs and did not resume production until November.[76]

By May 1886 things were once more going well. Colonel Sloss's two stacks produced 5,365 tons of pig iron that month, only slightly less than a record output of 5,670 tons recently made by the Alice Company. In June, however, an unexplained stoppage in Sloss's ore supply forced him to blow out one of his furnaces. *Iron Age* announced that his son, Frederick Sloss, would remodel it before production was resumed. Later that year, commenting on the impending start of operations at the Woodward Iron Company's second furnace, the same journal reported that both of the ones operated by Sloss had been "cold for some time."[77]

Such enforced idleness came at a bad point, because the long recession of the mid-1880s had finally ended and business had been booming in Birmingham for many months. Reports coming out of the city in 1886 teemed with news about a steady influx of capital, burgeoning prices for real estate, and orders for pig iron from all points of the compass, including Pittsburgh itself, that were going begging for lack of capacity to fill them. In October, a Cincinnati trade correspondent, calling southern ironmasters "evidently the autocrats of the market for Pig Iron," stated that "there is not a Southern furnace of any note which has not sold its entire capacity up to the first of next year." All Colonel Sloss had been able to do, however, was sit on the sidelines and watch.[78] Long into the future, the firm that he had founded would struggle with the consequences of his failure to secure convenient access to raw materials and the poor performance of the stacks he had erected.

VII

Birmingham seethed with excitement as 1886 came to an end. Even in the worst of times, the district's furnaces had been able to compete successfully with northern and midwestern pig iron producers; now that prosperity had returned, the Magic City's prospects seemed limitless.[79] The Sloss Furnace Company, however, had missed most of the boom. By the time it was ready to resume production in October, it had gone through three cycles of renovation and repair within the same number of years, enough to discourage even its most optimistic backers.

Colonel Sloss seems to have been a good financier and promoter but a poor manager. Nothing in his background as a merchant and railroad executive had given him experience in running a manufacturing firm that depended upon its ability to marshal equipment and raw materials. While promoters capitalize on big ideas, managers must make many incremental decisions that collectively spell the difference between success and failure. Different temperaments are required for such functions. By 1886, Sloss was depending more and more on his sons to make managerial decisions, but they apparently were not up to the task.

Colonel Sloss's failure to run a successful industrial enterprise was poignant. Clearly, he had acted on the best available advice. Everything about Hargreaves's background, including his close association with Whitwell, suggested his capacity to superintend furnace construction. Similarly, the northern experience of Witherow's Company augured well for its ability to design ironmaking installations. Sloss spared no effort to get up-to-date equipment. Despite the experience that he had gained by installing Whitwell stoves and furnaces in Georgia and Tennessee, however, Hargreaves lacked familiarity with southern raw materials. For the same reason, as both the *Manufacturers' Record* and *Iron Age* indicated, the northern background of such designers as Witherow was likely to be more of a liability than an asset. Such were the risks of technological transfer on a new industrial frontier.

Sloss was now almost seventy years old. Seeking to explain why he decided at this time to sell the enterprise he had created five years earlier, Ethel Armes later wrote that "the strain of incessant toil began to tell on him."[80] Actually, the flame of the aging entrepreneur's creative energy still burned as brightly as ever; what really prompted his decision to sell was his desire to climax his remarkable career with one final achievement that would equal his success in bringing the L&N railroad to Jones Valley in 1871. Organizing a syndicate that included some of Birmingham's leading businessmen, he spearheaded a drive to manufacture the wonder alloy that boosters were confident would ensure the city's dominance as an industrial center once and for all: steel.

Sloss was well aware that an achievement of this magnitude would require a great deal of outside capital, just as the emergence of Birmingham itself had required financial aid from Cincinnati and Louisville. The Magic City could at best supply seed money with which to get his project launched. Sloss also realized that any group of investors who could plausibly finance a steel mill would have the power to take over the Sloss Furnace Company in the process. But he had always been beholden to outside interests and was more than willing to sacrifice control of his brainchild to attain the goal to which he was now committed. The fact that the new owners who assumed control were southerners, coming from Virginia, probably made the outcome all the more satisfying to him. Certainly the purchase price of $2 million that he and his associates negotiated was a handsome reward for the owners of a company capitalized at one-fourth that amount.

A new era in the history of Sloss Furnaces, which would last for three decades, was about to begin. As had been true from the beginning, railroad men would continue to call the tune. Now, however, they came not from Louisville and Cincinnati but from a city that had once been the greatest industrial center of the antebellum South before becoming capital of the Confederacy. Heading a syndicate from Richmond, Joseph Bryan would now become the dominant figure in the enterprise that Colonel Sloss had founded in 1881.

Joseph Bryan and the Virginia Connection

4

Virginia suffered more terribly during the Civil War than any other state, northern or southern.[1] Partly because of this ordeal, but also because of pride in a political tradition unmatched by any other part of the Confederacy, it was natural for the Old Dominion, home of Robert E. Lee and mother of presidents, to aspire to continued leadership in regional affairs. Nowhere was this attitude more true than in Richmond, one of a relatively few American cities ever to have been the capital of a nation. One aspect of Richmond's postwar activity, long unrecognized, was its role in nurturing Alabama's industrial development. Sloss Furnaces became a prime example of its economic outreach.

I

In November 1886, John W. Johnston secured an option, acting on behalf of himself and a group of associates, to buy "all the property of the Sloss Furnace Company, and all rights of every character owned by it . . . for the sum of Two Million Dollars."[2] Johnston was president of the Georgia Pacific Railway, which in 1883 had completed the first trackage to connect Atlanta and Birmingham. He was also a Virginian, as were most of the people he represented. His action was part of an ongoing process of investment in north Alabama by residents of the Old Dominion, mostly from Richmond, who for several decades played a major role in Birmingham's development.

The importance of foreign and northern capital in southern economic growth after the Civil War is well known.[3] Less familiar, however, is the way in which intraregional investment, flowing from such places as Kentucky and Virginia into the Deep South, supplemented money from the outside. Writing to the *Manufacturers' Record* in 1884, I. W. Avery of Augusta, Georgia, voiced frustration that Yankees were getting too much credit for their role in funding southern enterprises. Visiting a new cotton mill at Pacolet, South Carolina, he had found that more than two-thirds of the $300,000 required to build it had been subscribed by southerners. All of the capital behind the largest textile plant in the region, the Eagle & Phenix mill in Columbus, Georgia, had come from Dixie, as had the cost of erecting a new knitting mill in Atlanta. In Bainbridge, Georgia, most of the money invested in the yellow pine industry had come from "Southern men with Southern capital." Quarrying operations at Stone Mountain near Atlanta had been similarly financed. Without wishing to deny the role being played by the North in postwar southern growth, Avery wanted to make sure that investors from below the Mason-Dixon Line also got due credit.[4]

The vital part played by Virginia capital in Alabama's industrial development during the late nineteenth and early twentieth centuries supports Avery's argument. It helps demonstrate the truth of Gavin Wright's claim, made with specific reference to textiles but equally pertinent in other contexts, that the South industrialized "with mostly southern money and some northern help."[5] It also reveals the key contribution made to Birmingham's growth by a long-ignored figure, Richmond investor Joseph Bryan, whom Alabama historians have neglected until now because of the lack of attention paid to him by the pioneer chronicler of the state's industrial development, Ethel Armes.

Despite a devastating fire that swept Richmond during its evacuation in April 1865, economic recovery was well underway in the city by 1870. Having been one of the South's most dynamic centers of commerce and manufacturing in antebellum times, it still had a solid core of aggressive business leaders, including Joseph R. Anderson, whose Tredegar Iron Works had escaped destruction. Within five years after Appomattox, the Tredegar was flourishing, and a number of foundries, including the Old Dominion Iron and Nail Works on Belle Isle, were also busy. Tobacco production had reached 80 percent of prewar levels, and the city's three largest flour mills—the Gallego, Haxall, and Dunlop—were back in operation after having been destroyed or damaged in the war. All told, 543 industrial firms were making goods valued at more than $7 million. Five railroads led in and out of Richmond, and the same number of steamship lines linked it to seaports up and down the coast. In Douglas Southall Freeman's words, a person visiting this city of 51,000 people in 1870 "would have had to look carefully before he could have found evidence that he was walking the streets of a Southern Troy."[6]

The dynamism that Richmond displayed, and the way in which it channeled funds into Alabama, provide keys to a better understanding of postwar southern industrial growth. One such key was the existence of scattered pockets of southern wealth, like Richmond, that somehow survived the war. Another factor, also evident in the city on the James, was a substantial pool of southern managerial and entrepreneurial skills. Part of this came from military sources; both in Virginia and elsewhere, there were many former Confederate officers whose experience in commanding armies and meeting their logistical needs was potentially useful in running large enterprises. By gravitating into commerce and manufacturing after 1865, such veterans heeded the advice of Virginia's most venerated citizen, Robert E. Lee, who as president of Washington College (later Washington and Lee) during the last five years of his life strongly backed southern industrial development and emphasized commercial and technological subjects. In tracing what happened to almost 600 former Confederate leaders after Appomattox, William B. Hesseltine found that a large number of ex-generals subsequently became executives of banks, insurance companies, railroads, and manufacturing firms.[7]

Another key to postwar southern industrialization, one that formed part of the background for the way in which capital ultimately flowed into a group of Alabama enterprises including Sloss Furnaces, was an antebellum heritage of close business relationships between southern planters and wealthy merchants in northern and eastern seaports. Baltimore had particularly strong traditional affinities with the South and would be an important source of support for Richmonders who later became involved in the development of Birmingham. Virginians also had strong ties with bankers, cotton brokers, and railroad promoters in Philadelphia.

The most important of the South's prewar connections was with New York City, which dominated most of the region's export trade with Great Britain and continental Europe in such commodities as cotton and tobacco. New York was well known for the solicitude with which its leading merchants regarded southern interests prior to 1861. Abolitionists complained about the "ten thousand cords" that linked the nation's largest metropolis to slaveholders. Admiration for southern aristocrats was widespread among rich New Yorkers who regarded their political power and genteel traditions as counterbalancing forces against leveling tendencies in northern society. Marriages between mercantile and planter families were common, and Wall Street bankers vied with one another in organizing relief drives when yellow fever swept southern communities. New York capitalists invested heavily in southern mines and railroads, winked at the continuation of slave trading through such places as Cuba, and supported filibustering expeditions in Central America. Ties with the South were also strong among transplanted British agents who lived in New York and served as middlemen in the cotton and tobacco trades.

New York City's pro-Southern leanings had encouraged rebel leaders to believe that the city would follow them into secession after Lincoln's election to the presidency precipitated the worst political crisis in American history during the winter of 1860–61. Although financial and municipal leaders ultimately rejected Mayor Fernando Wood's call to withdraw from the Union and become a "free port" allied with the Confederacy, Gotham remained throughout the war what historian Basil Lee has called "the strongest center of anti-administration feeling, bitterness, and even disloyalty." Continuing support for the Southern cause helped bring about such episodes as the draft riots that swept New York City in July 1863. During the 1864 presidential campaign, the metropolis was a hotbed of support for George McClellan, who, it was believed, would sue for peace if elected.[8]

After the war, leading merchants in New York City and other northern seaports adopted an attitude of "forgive and forget" toward the conquered South. It did not take long for close business ties, based on mutual affection, trust, and respect, to be restored between the two sections. Hesseltine's references to "a New South which acquiesced in its subservience to Northern countinghouses" should not be taken to indicate that southerners acted abjectly toward Yankee capitalists, however devastated their region had become as a result of the war.[9] It would have been unnatural for Confederate veterans, who had met defeat only because of overwhelming Union superiority in numbers and war materiel, to come to Wall Street and other financial centers crawling on their knees. Nor did they have to do so; northern investors who wanted to make money below the Mason-Dixon Line needed the help of southern allies just as much as the converse was true. Earning profits in the South demanded ability to handle its distinctive labor force, familiarity with southern folkways, and a knowledge of the land itself that few outsiders possessed. Northern financiers therefore treated southern businessmen with respect.

Despite the speed with which they moved to reestablish earlier ties with the South, northern capitalists were limited in the amount of funding that they could supply for the region's postwar industrial development. As Paul H. Buck has stated in his classic study *The Road to Reunion*, the North had to devote most of its financial resources to its own explosive economic growth. Moreover, the West, with its transcontinental railroads, cattle drives, and mining bonanzas, offered more lucrative profits than did the South, where cotton production languished long after the war and sharecroppers now tilled once-thriving plantations. Texas, with its potential to furnish vast quantities of meat to northeastern cities, was the only former Confederate state that offered much excitement to Yankee investors in the early postwar era.

For at least fifteen years after 1865, therefore, most northern funds that gravitated southward went either into railroads or state bond issues, the second of which usually turned out to be poor risks. It was partly for this reason that the South turned in many cases to Great Britain for industrial funding, as in securing the capital that built such manufacturing facilities as the ironworks at Rising Fawn, Georgia, and South Pittsburg, Tennessee. Mostly, however, the region was forced to depend on its own resources to finance factories, mills, and furnaces. "No lesson was more valuable or more thoroughly learned," Buck stated, "than that the section's redemption would have to be achieved through the efforts of Southern people."[10]

Despite the large amounts of northern aid that they absorbed, even postwar southern railroads were to a significant degree the region's own creation. Often, they were administered by natives of the region including ex-Confederate general Edward P. Alexander, a plantation owner's son who pursued a distinguished postwar career as a railroad executive with such lines as the Charlotte, Columbia and Augusta, Savannah and Memphis, Western Railroad of Alabama, Louisville and Nashville, and Central of Georgia. Nathan Bedford Forrest and Edmund W. Rucker, who spearheaded the building of the Selma, Marion & Memphis Railroad in west Alabama, were other cases in point.[11]

In many instances, southern railroads also received substantial amounts of cash from natives of the region who worked closely with trusted northern allies. In these cases, rival groups whose members were drawn from both sides of the Mason-Dixon Line sometimes fought bitterly with one another, displaying radically different entrepreneurial styles. Some southern investors favored sound, conservative policies that resulted in solid economic growth, but others watered stock as freely as their northern associates did and resorted to predatory tactics that seriously harmed the region's best interests. The contrasting roles played by Virginians in the railroad system that ultimately became involved in the development of Sloss Furnaces epitomized the complexity of this multifaceted process.

II

Sloss Furnaces ultimately became part of an industrial empire that was related to an offshoot of Richmond's main rail connection to the Deep South: the Richmond and Danville Railroad, which had originally been chartered in 1847. Shifting relationships between local and northern investors marked the line's progress during Reconstruction. Aided by northern financiers, particularly Thomas A. Scott of the Pennsylvania Railroad, it expanded all the way to Atlanta by 1873. Although Richmond was hard hit by the financial panic that struck that year, the Danville, led by former Confederate general Algernon S. Buford, got through the ensuing depression with the help of the Southern Railway Security Company (SRSC), a group of Pennsylvania Railroad stockholders dominated by Scott.[12]

In 1880, however, the Pennsy disposed of its southern interests, and the Danville became controlled by a syndicate that formed a holding corporation, the Richmond and West Point Terminal Railway and Warehouse Company, to manage its properties. Necessitated by Virginia laws that prevented the Danville itself from owning rail lines with which it was not contiguous, this arrangement was part of a plan aimed at further expansion through the acquisition and development of new and existing railroads.

Under the new plan, the power that Virginians had previously enjoyed in the Danville was diluted. Nevertheless, Richmond businessmen tried to retain as much control as possible over further expansion by maintaining a balance of power between two groups of northern investors. One of these consisted of former participants in the SRSC; the other was led by Thomas and William Clyde of Philadelphia, owners of northern steamship lines. The

main tactic used by Virginians to keep their remaining power was the formation in 1881, when securities of the Danville and Terminal companies were
first listed on the New York Stock Exchange, of an "ironclad pool" of syndicate members who controlled a majority of Danville stock. More than half
those involved, who also held most of the seats on the pool's executive committee, came from Richmond. [13]

Pools, however, had no standing in American law and were therefore
inherently unstable. Like so many other arrangements of this type, the plan
soon broke down. With stock now publicly listed, large shareholdings fell into
unfamiliar hands, and a bewildering sequence of plots and counterplots ensued between groups of old and new investors.

Discussing the troubled history that followed, Maury Klein has argued
persuasively that the growth of the Danville system after the formation of the
Terminal Company involved conflict between two business strategies, one
"developmental" and the other "opportunistic." Developmental strategy was
aimed at making gradual profits through careful management of resources,
prudent reinvestment of income, and close attention to promoting the economic growth of areas served by the rail system. By contrast, opportunistic
strategy was based on pursuing quick returns through over-rapid expansion,
frequent reorganizations, stock manipulation, and other predatory tactics at
the expense of sound financial management, effective maintenance of equipment and facilities, efficient service to the public, and patient development of
newly entered areas. [14]

The acquisition of the Sloss Furnace Company by Virginians in the winter
of 1886–87 resulted from a strong commitment by one group of Richmond
financiers to the developmental approach. During the infighting of the early
1880s, much of this group's activity became concentrated in one part of the
Terminal Company's ever-growing system, stretching westward from Atlanta
to the Mississippi River. Increasing amounts of Richmond capital were invested in this area, at the heart of which lay Alabama's massive coal and iron
deposits. Tied together not only by common values but also by comradeship
stemming from their participation in the Confederate cause, entrepreneurs
from the Old Dominion created a new economic empire and tried to control
its development, both for their own financial benefit and also as an expression
of their shared belief that the South could industrialize without compromis-
ing its distinctive cultural traditions. Here, as in many other respects, the New
South was an extension of the Old.

The colonization effort mounted by the Virginians centered around an
enterprise that began in 1854 as the Georgia Western, a railroad projected to
run from Atlanta to the Alabama state line. Little or no progress was made in
its construction, and it ultimately went bankrupt in the mid-1870s. In 1881,
however, one of Georgia's greatest military heroes, ex-Confederate general
John B. Gordon, obtained its charter rights from the L&N Railroad, which
had earlier planned to use them to gain entry to Atlanta. By coupling the
Georgia Western with narrow-gauge lines already partly built in north-central

Mississippi and adjacent parts of Alabama—including the Greenville, Columbus and Birmingham Railroad, incorporated in 1873—Gordon hoped to lay the basis for a major transcontinental route from Atlanta to the Pacific Coast, linking with the Texas & Pacific system at Texarkana.[15]

Gordon's visions were the latest manifestation of an old southern dream. As far back as the early 1850s, when South Carolina railroad promoter James Gadsden negotiated American purchase of the Gila Valley from Mexico, regional promoters had wanted to develop a transcontinental line to California, following a southern route. Gordon, however, quickly realized that he and his friends lacked sufficient funds to carry out his aims. He therefore persuaded the Terminal Company to acquire a controlling interest in his brainchild, resulting in a reconstituted enterprise known as the Georgia Pacific Railway Company. Its creation brought Virginia capital for the first time into the development of north-central Alabama. To connect Atlanta with Greenville on the Mississippi River, Terminal officials organized the Richmond and Danville Extension Company, chartered in May 1881, with a capitalization of $5 million. It quickly began laying track westward from Atlanta toward the Alabama line, using steel rails and adopting a five-foot gauge. Gordon remained temporarily active in the Georgia Pacific as president of the enterprise but had no effective control over its affairs. Ultimately he sold his interest in the venture at a reputed $200,000 profit.[16]

Whatever hopes may have been entertained by the Terminal Company of fulfilling Gordon's transcontinental dreams, the immediate aim of the Georgia Pacific was to create outlets for Alabama's coal and iron. Enthusiasts for the line, which would intersect several railroads and four rivers between Atlanta and Columbus, Mississippi, predicted that it would command "the best local business of any road in the South" and become one of the most profitable railroads in the country.[17]

John W. Johnston, who became general manager of the Extension Company, was put in charge of surveys to determine the exact location of the Georgia Pacific's right-of-way. He was also responsible for deploying work crews, laying track, and constructing depots and other facilities required for operating the line. Johnston's background was typical of Virginians who became involved in Alabama's industrial development. Born in 1839 in Botetourt County, he had served gallantly as an artillery officer in the Confederate army, rising by 1865 to the rank of major. Wounded on three occasions, he fought in a succession of engagements including First Manassas, the siege of Vicksburg, and the battle of Resaca in north Georgia, where he was badly wounded when a minié ball lodged in one of his thighs. After the war he became a lawyer and was elected to the Virginia state senate. Turning to business, he administered the James River and Kanawha Canal Company and the Buchanan and Clifton Forge Railway. Involvement in the Danville system led to his appointment in 1881 to take charge of the Extension Company.[18]

As construction proceeded west of Atlanta and trains began to move over completed portions of the Georgia Pacific, Johnston secured choice properties along its projected right-of-way, both for himself and as a representative of other Virginia investors. Among his chief clients were two brothers, John and Daniel Kerr Stewart, who had come to Virginia from Scotland in the 1820s to join their uncle, Norman Stewart, a tobacco merchant. As the business grew, the Stewarts amassed wealth. The way in which they protected their fortune after 1861 shows how some blocks of southern capital survived the Civil War. Sensing from the start of hostilities that the Confederate cause was hopeless, John Stewart stayed in Richmond, supported the war effort, and shared his house with Gen. and Mrs. Robert E. Lee. Daniel, however, took the bulk of the family fortune to New York City and stayed there for the duration. Taking advantage of the fact that he was still a British subject, he speculated in land, invested in railroads, and traded in agricultural commodities. Returning home after the war, he lived with John at the latter's Henrico County estate, Brook Hill. Thereafter the two brothers invested heavily in various southern enterprises, including the Danville and Terminal systems. The formation of the Georgia Pacific led them to invest in Alabama mineral lands, guided by Johnston's advice.[19]

III

During the 1870s and 1880s, the Stewarts' financial affairs became increasingly managed by John Stewart's son-in-law, Joseph Bryan. Born in 1845 at the Eagle Point tobacco plantation of his father, John Randolph Bryan, in Gloucester County, he was steeped from birth in the traditions of Virginia's planter aristocracy. Too young for military service at the start of the war, he enlisted upon reaching his eighteenth birthday and fought bravely, at times recklessly, for the Confederate cause. Serving first with the Richmond Howitzers, he spent the rest of the war with Col. John S. Mosby's guerilla force, known as Mosby's Raiders. Fiercely committed to a code of chivalry that sometimes led him to feats of almost suicidal bravado, he charged single-handedly against a squadron of Sheridan's cavalry in one encounter and was twice wounded in other engagements. In the words of his son, John Stewart Bryan, his wartime experiences "always remained the most deeply etched of all his remembrances."[20]

After Appomattox, Bryan studied law at the University of Virginia and started practicing his profession at Palmyra, in Fluvanna County. This vocational choice was fortunate; his career epitomized the rewards available to lawyers who thrived from the growing need of business firms for expert guidance in coping with the increasingly intricate legal structures within which they had to operate.[21] Bryan, however, became much more than a mere corporation lawyer; instead, he built an extremely successful career as a lawyer-entrepreneur who used his knowledge of business law to advance the fortunes of enterprises in which his status was that of a principal owner, not simply that of an advisor on legal matters.

The Civil War devastated the Bryan family's fortunes, and Joseph's economic outlook was temporarily bleak. Soon, however, his prospects improved. Meeting Isobel Lamont (Belle) Stewart, one of John Stewart's seven daughters, at the plantation home of a client, he married her at Brook Hill in 1871. Establishing residence there, he moved his practice to Richmond and won a growing reputation by successfully representing canal companies, railroads, and other corporations. Within a decade, his abilities, his connections with the Stewart family, and his ceaseless devotion to work had put him firmly on the road to wealth and power.

Convinced that agriculture in the Old Dominion was past its prime, Bryan became a leading champion of southern industrialization. He was a conspicuous exception to Jonathan Wiener's unproved thesis that exponents of the New South ideology usually "had their social origins outside the mainstream of planter society" or "came from families that were not enthusiastic about slavery or secession."[22] Patrician to his fingertips, Bryan epitomized the aristocratic heritage of the Old South.

Although James M. Lindgren has characterized Bryan as a "southern economic nationalist," he might better be described as an unrepentant ex-Confederate who willingly followed the road to reunion and saw clearly what the South had to gain from being part of a united nation that was rapidly becoming the world's foremost industrial giant. Like his planter ancestors, he was well aware of the need for outside capital and credit if the South were to develop its natural and human resources. Nevertheless, he clung fervently to a belief that his native region could control its economic destiny. Along with other Virginians who allied themselves with trusted northerners, he saw industrialization not simply as a source of profit but as a peacetime equivalent of the struggle for self-determination that the South had lost in battle.[23]

Bryan was an intense, prodigiously energetic man with a piercing gaze and a short, neatly trimmed Vandyke beard. "When his gray eyes look at you," it was said of him, "they seem to be sighting you along a pistol barrel."[24] His strongly held convictions reflected his aristocratic roots; his rigid conservatism was grounded in what Lindgren has aptly characterized as a tradition of "economic orthodoxy, elite rule, lower-class deference, and black subservience." After home rule had been restored in Virginia, his belief in the sanctity of contracts led him to oppose efforts made by people like William Mahone, whom he detested, to readjust state debts to provide money for such purposes as public education. The same views ultimately made him an unyielding supporter of the gold standard in the 1890s. His strong sense of honor and moral rectitude was closely connected to deeply held religious beliefs. Like the Stewarts, he was an Episcopalian who tried diligently to apply Christian principles, as he understood them, in conducting his business affairs. Throughout his career he gave liberally of his time and money to charitable causes.

Bryan's commitment to industrialism led him into a host of entrepreneurial activities. Together with Richmond's mayor, Lewis Ginter, he took over the

Joseph Bryan. "When his gray eyes look at you, they seem to be sighting you along a pistol barrel." Joseph Bryan, the key figure in the history of Sloss Furnaces, is shown here in his prime. Every inch an aristocrat, Bryan exemplified strong connections between the prewar southern slaveholding planter class and Alabama's postwar industrial development. Although Bryan was soon forgotten in Birmingham after his death in 1908, he ranked among the foremost architects of the city's emergence as a major manufacturing center. (Courtesy of Valentine Museum, Richmond, Virginia.)

bankrupt Tanner & Delaney Engine Company and built it into the Richmond Locomotive and Machine Works, the largest firm of its type in the South. He also acquired a struggling newspaper, the Richmond *Times*, and made it a forceful exponent of his political and economic views. When statements in an issue published in 1893 led a reader to challenge Bryan to a duel, Bryan indignantly refused, branding the custom as "absurd and barbarous," and thus helped bring it to an end; his courage was so unquestioned that he was "overwhelmed with congratulations" and made his opponent appear ridiculous.[25] Partly to restrict the power of labor unions, which he believed to be a radical menace, he became the first publisher in the South to adopt the Mergenthaler linotype machine. Ever looking for new opportunities, he speculated in land and bought urban properties in such places as Minnesota and Texas. Hoping to develop foreign markets for southern industrial exports, he became a strong proponent of a Nicaraguan canal.

As Bryan's wealth grew, he became heavily involved, together with the Stewart brothers, in the Danville and Terminal systems. He figured importantly in the "ironclad pool" of 1881, in which he and John Stewart collectively held 6,000 shares. Regarding transportation as the key to southern industrial development, Bryan took a strongly developmental view and spent much of his career opposing opportunistic efforts by northerners and southerners alike to manipulate the assets of southern railroads for quick financial gains. As his son, John Stewart Bryan, aptly said, "his interest was that of a builder . . . not an operator."[26]

After the Georgia Pacific was organized, Bryan became one of its chief promoters. He was named a director in May 1881 and followed the line's progress closely as track was laid across the rugged terrain, mostly wilderness at the time, between Atlanta and Birmingham. As trains moved westward, property values rose and goods began to flow between Atlanta and a chain of new towns along the route, producing the developmental benefits to which Bryan was committed. After the Georgia Pacific had crossed east Alabama's Choccolocco Valley and reached Anniston in February 1883, he participated in ceremonies marking the opening of service to that community, which was soon opened to public investment after previously having been monopolized by the Woodstock Iron Company.[27]

This visit was apparently Bryan's first to Alabama. If so, it was a fateful occasion, because he became deeply involved in the state's development. In addition to investing his own funds, he drew upon those of John and Daniel K. Stewart, over which he gained control by becoming executor of their estates after they died in 1885 and 1889 respectively. He also channeled investments into north Alabama by other Virginians, mostly from Richmond. The social background of these investors was similar to that of Bryan. Contrary to the implications of Jonathan Wiener's thesis involving alleged antipathy between planter values and industrialization, there is no evidence that they had anything but a traditional southern aristocratic outlook. Many had been in the Confederate army; all had strongly supported the war.

Not surprisingly, most of Bryan's associates were heavily involved in the Terminal system and its offshoot, the Georgia Pacific. The group included Thomas and John P. Branch, Richmond bankers and commission merchants; Edward D. (Ned) Christian, a tobacco merchant and president of the Richmond Paper Manufacturing Company; James H. Dooley, a prominent Richmond lawyer and business promoter; James B. Pace, who amassed wealth in the tobacco trade, was reputedly worth $2 million in 1882, and was president of the Planters National Bank of Richmond from that year until 1895; William R. Trigg, a banker who joined Bryan in organizing the Richmond Locomotive and Machine Works and also owned a shipbuilding firm; and Charles E. Whitlock, a banker who made a fortune in lumber.[28]

Dooley was typical of Bryan's associates. He enlisted in the First Virginia Infantry at the outbreak of the Civil War, suffered a shattered right wrist in combat, and spent the rest of the war as an ordnance officer in the reserves. Practicing law after the war, Dooley served for six years in the Virginia legislature before starting a business career. Becoming second vice president of the Danville in 1881, he headed its legal department. Described by Virginius Dabney as "small of stature and ramrod straight, with a keen mind and nonexistent sense of humor," he amassed a fortune and built a large estate, Maymont, on the James River west of Richmond.

Bryan's closest associate was John Campbell Maben. Descended from families that acquired wealth through involvement in tobacco, cotton, and iron, he was born at a Petersburg mansion, Strawberry Hill, in 1837. Receiving his early schooling in Richmond, where he began developing a knowledge of Latin in which he took pride throughout his life, he matriculated at Princeton in 1857 but left in 1861, thus missing graduation, to join the Seventeenth Virginia Regiment, known as the "Richmond Greys." Rising to the rank of captain and serving on the staff of the First Army Corps in northern Virginia, he returned to Richmond after the war. In 1868, he went to New York City, where he joined a brokerage firm, Lancaster, Brown & Company, which was run by Virginians and sold southern railroad securities.[29]

Maben typified an important group of southerners who pursued the region's economic interests on Wall Street after the war.[30] When Lancaster, Brown & Company failed in the Panic of 1873, Maben joined another New York City brokerage, William H. Goadby and Company. Goadby, son of an English merchant who had made a fortune in real estate after coming to Gotham prior to the war, handled the Wall Street accounts of Bryan and the Stewarts. His British origins made him a congenial financial ally; Bryan, an Anglophile, greatly enjoyed his company and entertained him frequently as a house guest.[31] Another Wall Street associate who fit the same profile was Richard Y. Mortimer, a Georgia Pacific bondholder; he was descended from a British merchant who had come to New York from Yorkshire and amassed wealth in wool trading, transportation, and fire insurance.[32]

Among his activities at Goadby's brokerage, Maben handled accounts for such clients as Bryan and the Stewarts, invested time and money in the

James H. Dooley. A wealthy Richmond financier, James H. Dooley was a long-term investor in Sloss Furnaces. "Small of stature and ramrod straight, with a keen mind and nonexistent sense of humor," Dooley, who suffered a shattered wrist fighting for the Confederate cause, typified Bryan's patrician business associates. (Courtesy of Virginia Historical Society, Richmond, Virginia.)

Danville system, became a projector of the Terminal Company, served as a watchdog for Richmond interests on Wall Street, and raised northern capital for southern enterprises, including the Georgia Pacific. Bryan, who was extremely fond of him, corresponded with him ceaselessly, seeking his advice in dealing with northern financial houses.

Another Virginian who worked for Goadby was Thomas Seddon, whose father, James Seddon, had been secretary of war in the Confederate cabinet. Born in 1847 at his family's Sabot Hill plantation in Goochland County,

Thomas suffered a fall in his infancy that resulted in a severe spinal injury, crippling him for life. Prevented by his youth and physical handicap from serving in the Confederate army, he studied law at the University of Virginia, graduated with honors, and practiced his profession for a short time at Baltimore during the postwar years. Finding an attorney's life "too tame for his young and aspiring spirit," he decided upon a business career. After going to Louisiana to manage a sugar plantation, he returned home to become involved in the Richmond wholesale grocery trade and then moved again, this time to New York City. Involving himself in Danville and Terminal affairs, he played a key role in the building of the Georgia Pacific as secretary of the Extension Company. Like Maben, he helped handle accounts for Bryan and the Stewarts. Slight of build and frail in appearance, Seddon compensated for his affliction by having a well-developed sense of humor. Possibly because he had not seen military service, Bryan tended to patronize him. As events would demonstrate, however, Seddon could be a dogged adversary when his personal interests were challenged, and he had a fierce sense of pride.[33]

Bryan visited New York frequently to confer with Goadby, Maben, and Seddon. During the summer of 1882 he was constantly involved in complex dealings as he fought to preserve the power of Richmond interests in the Danville and Terminal systems. A collapse in the price of securities that had occurred in March caused the reputation of both enterprises to suffer badly, and their backers were demoralized. "I am over cropped just now with the affairs of the Danville & Terminal Companies," Joseph wrote at one point to one of his brothers. "The management of the finances of these companies has more devolved upon me, & the burden is very great involving as it does most complicated transactions up in the millions." Complaining of fatigue, he expressed reluctance to assume the first vice presidency of the Terminal Company, tantamount to being put in charge of its day-to-day operations. "It may be that I must accept—but if I can get some parties in N.Y. to take the load I will lighten my shoulders—ease my mind & take some much needed rest—I go back to N.Y. Tuesday Evg—& the matter will soon be determined— whether I am to have more work or less—God grant it may be the latter, as I am tired."[34]

Because of the bickering that had become chronic among investors in Danville and Terminal securities, the rest Bryan craved was elusive. Ten days after writing the letter just quoted, he was back in Gotham, busy as ever. Homesick, he wrote to his children from the Danville's Nassau Street headquarters, telling them that he had been spending his nights on William P. Clyde's private yacht, which was docked behind Coney Island, seeking refuge from an oppressive heat wave. Standing on deck the previous morning, he and Clyde had impulsively taken off their clothes and plunged into the water to cool off.[35] The incident epitomized close friendships with powerful northern backers that lasted throughout Bryan's life. Ruthless in business dealings but nonetheless likeable, Clyde was equally at home in technology and finance. As a designer of seagoing vessels, he was a pioneer in screw propulsion and the use of compound engines and had made millions of dollars in steam-ship lines linking New York and San Francisco via a Panamanian rail line. He was also the biggest investor in the ironclad pool of 1881, owning 5,000 shares.[36]

Another of Bryan's northern allies was Clyde's close friend Walter G. Oak-man. Like Clyde, he came from a rich Philadelphia family and had made a fortune in transportation. He was also well connected politically, being a son-in-law of Roscoe Conkling, Republican senator from New York.[37] Associating with such men as Clyde, Goadby, Mortimer, and Oakman brought Bryan into the inner councils of prominent Wall Street institutions. The most im-portant of these was the Central Trust Company, which was heavily involved in Danville and Terminal affairs.[38]

As the Georgia Pacific pushed westward into Alabama, it was imperative for Bryan, the Stewarts, and other backers of the line to obtain the most accurate knowledge possible about the geology of the region through which it was to be built, for they had to know where best to locate its trackage, acquire mineral lands and town sites, and maximize the value of their investments. One of Virginia's greatest scientists, William H. Ruffner, was appointed to survey mineral lands along the projected route.

Born in 1824 near Lexington, Ruffner was the son of a Swiss immigrant who became president of Washington College and who also acquired an inactive salt works at Kanawha, Virginia. After receiving his bachelor's degree at the institution headed by his father, William reopened the salt mines in 1844. His vocational preference, however, was the Presbyterian ministry. Earning theological degrees from Union Seminary and Princeton, he became pastor of a large Philadelphia church in 1851. Ill health eventually forced him to return to Virginia, where he operated a farm near Harrisonburg and preached on a part-time basis. Despite his opposition to slavery, Ruffner was loyal to the South during the war. After Appomattox, he became Virginia's first superintendent of public instruction. Traveling to every corner of the Old Dominion, he worked unstintingly to establish public schools under a new state constitution that was adopted in 1869. As a result, he became known as the "Horace Mann of Virginia."[39]

Geology had fascinated Ruffner since childhood, and he avidly collected mineral specimens. In 1860, he ran a geological survey across Virginia, from Hampton Roads to the Ohio River. While organizing the state's school system after the war, he lectured on the natural resources and rock formations of the areas he visited. His reputation as a geologist attracted the attention of Thomas M. Logan, a South Carolinian who had become a Confederate general at a remarkably young age and adopted the Old Dominion as his home after marrying the daughter of a Virginia magistrate in 1865. Like Bryan, Logan advocated industrial development and pursued close ties with northern investors. Heavily involved in the Danville and Terminal com-panies, he became an important stockholder in the Georgia Pacific and was keenly interested in Alabama's mineral resources.[40]

William Henry Ruffner. Ruffner, who led the drive for public education in Virginia after the Civil War, was also a geologist whose surveys for the Georgia Pacific Railway made an important contribution to the industrial development of north-central Alabama. Today, one of Alabama's mountains is named in his honor. (Courtesy of Valentine Museum, Richmond, Virginia.)

On Logan's recommendation, Johnston appointed Ruffner to survey the projected route of the Georgia Pacific, stretching approximately 400 miles from Atlanta to the Mississippi River. Resigning his educational respon-sibilities in January 1882, Ruffner made three trips to the region, recording visits to Birmingham and other places in a diary in which he discussed potential mining sites in all three of Alabama's coal fields and commented at length on the state's ore deposits. John L. Campbell, who taught chemistry and geology at Washington College, assisted him and coauthored a book resulting from their findings. The partnership soon ended because the two

men disagreed about classification of coal seams in the Cahaba field, but
Ruffner continued to render useful advice to his fellow Virginians. On his
recommendation, the original line projected for the Georgia Pacific through
the Warrior coal field west of Birmingham was shifted eight miles to the
north, indicating the esteem in which they held him.[41]

IV

Birmingham newspapers contained frequent reports about the progress of
the Georgia Pacific and the benefits it would bring to the city. Excitement
mounted throughout 1883 as construction crews converged on Jones Valley
from opposite directions. Adding to trackage already laid on the earlier
narrow-gauge lines in eastern Mississippi, workers pushed eastward toward
Birmingham. In early April they reached Walker County, Alabama, at the
western edge of the Warrior field. Meanwhile, other construction crews made
their way toward Birmingham through the rugged terrain west of Anniston.

Investors in Richmond and New York kept a watchful eye on developments. Early in 1883, Seddon, who had not ridden a horse for three years,
accompanied Bernard Peyton, an attorney for the Extension Company, on an
inspection tour from Columbus, Mississippi, to Atlanta. The two men went
by rail whenever possible and rode horseback the rest of the way. Continuous
trackage from Atlanta toward Birmingham extended six miles west of Anniston, where workers were anticipating the delivery of an iron bridge that
would cross the Coosa River, scheduled to be reached by the first of May.
Between the Coosa and Birmingham, however, only one-third of the line had
been completed in discontinuous segments. After Seddon and Peyton arrived
in Atlanta, the *Constitution* lauded them for covering the sixty-six miles
between Birmingham and Anniston, "over as rough a country as one would
like to see," in less than a day and a half. The same newspaper predicted that
Seddon would give his superiors in New York an encouraging report on "as
good a new road as was ever built in the south."[42]

Soon after Seddon returned to New York, Johnston completed a significant
transaction that had been rumored around Birmingham for months. After
"slow and tedious" negotiations, the Extension Company bought the Milner
Coal and Railroad Company for a reported $375,000. This enterprise, which
later became a key component of the Sloss Iron and Steel Company, had been
developed by John T. Milner and Henry Caldwell at Coalburg on Five Mile
Creek, nine miles northwest of Birmingham. Ruffner had advised its acquisition by the Extension Company. Among the assets gained by that firm were
extensive coal deposits, an unspecified number of locomotives and railroad
cars, an unfinished spur line that Milner and Caldwell had begun to build
between Coalburg and Birmingham, and access to convict labor that Milner
had used under a contract with the county government.[43]

All parties to the Coalburg transaction were greatly pleased. "Yr acquisition of the Milner property . . . was a triumph," Bryan wrote Johnston. Local boosters were no less jubilant. A newspaper article stated that the deal was "so grand in conception, so vast in wealth, so stupendous in magnitude, that every reader . . . who lives in Birmingham, in Jefferson County, in the State of Alabama, or in this section of the south, may well boast of the future greatness of the Magic City."[44]

Milner and Caldwell's former enterprise was renamed the Coalburg Coal and Coke Company. Incorporated on 9 June 1883, with a capitalization of $1 million, it included Bryan, Clyde, Johnston, Logan, and Trigg among its officers and directors.[45] The venture epitomized the developmental aims of Bryan and his associates; following a practice common among railroad promoters at the time, they were creating their own infrastructure by developing what came to be called captive mines to boost traffic and promote settlement along the Georgia Pacific's right-of-way. Pointing to their ultimate purchase of the Sloss Furnace Company, they also planned to build a blast furnace at Coalburg, which would be the western terminus of the Georgia Pacific until trackage was completed southeastward from Columbus, Mississippi. The outlines of what became a sprawling industrial empire were appearing.

Yet another Virginian, Edward M. Tutwiler, was appointed to manage the Coalburg mines. Born in 1846 at Palmyra in Fluvanna County, he came from a family that already had strong Alabama connections; his uncle had been one of the first four professors at Tuscaloosa when the University of Alabama opened its doors in 1831, teaching classical languages until 1837, and his father had also studied at the same institution. Barely old enough to fight by the time the war began, Edward was a cadet at Virginia Military Institute when its student body was summoned in May 1864 to help block Union general Franz Sigel's advance up the Shenandoah Valley. Taking part in the Confederate victory at New Market, he remained on active duty until the end of the war; afterward, he earned a civil engineering degree at VMI. Traveling widely, he practiced his profession in eight states before becoming assistant chief engineer of the Georgia Pacific in 1881. During the 1880s he built a series of mining camps for the line before organizing his own coal, coke, and iron company in 1889.[46]

On 11 May 1883, the Elyton Land Company gave the Georgia Pacific a right-of-way through Birmingham, along with adjoining property for a depot, shops, and other facilities. By this time, trackage from Anniston had reached the Coosa River, and the Georgia Pacific was within thirty miles of Birmingham. Penetrating the mountainous territory that still remained to be crossed required building McComb's trestle, 800 feet long and 70 feet high, spanning a gorge twelve miles from Birmingham. Construction crews also dug the Cane Creek tunnel, 810 feet long, sixteen miles farther east toward Anniston. Boosters exulted in the benefits that the Magic City would enjoy from the new rail link with outside markets. Work on the right-of-way had already resulted in the first direct telegraphic connections between Atlanta and Birmingham. "Each day brings new evidences that Birmingham will be . . . the greatest railroad center in the south," a local newspaper stated in early September.[47]

In mid-November 1883, completion of the Cane Creek tunnel finally permitted traffic to flow uninterrupted between Atlanta and Birmingham. The first westbound train arrived in Birmingham from Atlanta on 17 November, and eastbound service to the Georgia capital began the following day. Crowds assembled at every depot along the route between the Magic City's Twentieth Street terminal and Atlanta's Whitehall Street station to hail trains passing through during the 117-mile journey, and workers in construction camps shouted as the cars sped by. Aboard the vehicles, the din of "cheering and guffawing" was so loud that ordinary conversation was impossible. When the train reached the Cane Creek tunnel, black laborers who had dug it paid twenty-five cents each to ride through. The train included two cars loaded with Sloss pig iron. Large placards on their sides told where they had started and whither they were bound. After getting behind schedule early in the run, the train gained speed and pulled into the Atlanta terminal "on time to the minute." Soon the Georgia Pacific was helping Alabama furnaces reach northeastern markets. Meanwhile, Birmingham replaced Atlanta as the line's operational headquarters.[48]

Although Bryan did not witness the opening of service between Atlanta and Birmingham, he arrived in the Magic City shortly thereafter to attend the Georgia Pacific's annual meeting. The first of many visits he would make to Jones Valley, it amounted to a personal vindication for his efforts in spearheading the line. Amid the infighting that had become chronic in the Danville and Terminal companies, he had been temporarily removed from the Georgia Pacific's board of directors. Now, he came to Birmingham at the invitation of George S. Scott, a New York banker who, along with Clyde and Oakman, had recently taken control of the Danville through a series of stock acquisitions. At the annual meeting, held on 29 November, Bryan was re-elected to the board, along with such allies as Clyde, Johnston, Oakman, Scott, and Tutwiler. Johnston, who had become president in July after Gordon sold his stake in the line, was confirmed as chief executive. For the Virginians, it was a moment to be savored.[49]

V

Now that the Georgia Pacific had reached Birmingham, the pace of development slowed because of a severe business recession that was gripping the country. By November 1883, the laying of track northeastward from Columbus had been completed to Fayette Court House, and workers were pushing toward Day's Gap, where several coal mines had been opened. It was finally reached in July 1884, but further work was then suspended. It would be a long time before the line reached Coalburg. The Alabama coal trade was adversely affected by the recession, and many mines were either shut down or stayed open only part of the time. Because of the state of the economy, railroad building in the United States plummeted from 9,799 miles in 1882 to 3,652 in 1884.[50]

Under the circumstances, plans to erect a blast furnace at Coalburg were deferred. Construction, however, was begun on a battery of coke ovens. Anxious to have the Georgia Pacific completed through the Warrior field, Bryan chafed at the delay, and Daniel K. Stewart joined him in urging Scott to move ahead. Although the latter signed a contract for 27,000 tons of steel rails, things remained stalled. Bryan also worried about a lack of locomotives and cars but urged Johnston not to despair. "Patience will have its perfect work," he advised.[51]

Scott, Clyde, and Oakman continued to control the Danville system, and Bryan was reelected to the board in November 1884. General Buford, with whom Bryan had close ties, also became a director, signifying a return to power by older Virginians who had dominated the Danville immediately after the war. Stagnation, however, continued into 1885, which opened with increased unemployment, sagging industrial prices, and declining railroad earnings. Besides worrying about the continued economic gloom, Joseph and Belle went through deep personal grief when John Stewart died in March. By becoming executor of his estate, however, Bryan gained added power. Meanwhile, the Danville system continued to undergo retrenchment. It was not alone; trackage laid throughout the nation in 1885 fell again, this time from 3,652 to 2,699 miles.[52]

Under Johnston's management, the Georgia Pacific performed creditably throughout the downturn. Its profits declined only marginally, dropping from $185,481 in 1883–84 to $182,565 in 1884–85. New construction, however, remained in abeyance. At this point, the line had three discontinuous segments, each with different gauges. The first division, from Atlanta to Coalburg, had approximately 186 miles of track with a five-foot gauge, including three spurs in and around Birmingham. The second division, with the four foot, eight-and-one-half-inch gauge that was standard in most other parts of the country, extended for slightly more than 75 miles between Day's Gap and Columbus; it also had two short spurs to the Cane Creek and Patton coal mines. The third division, entirely within Mississippi, had two unconnected segments totaling about 52 miles and used a three-foot gauge. Of the company's total trackage of approximately 318 miles, only 270 miles had steel rails. Scattered throughout the three divisions were 32 locomotives and 950 units of rolling stock, including 498 cars carrying coal and ore. All in all, the Georgia Pacific was badly fragmented and ill equipped, and most of the Warrior coal field remained cut off from Birmingham and the Mississippi River.[53]

In May 1885, the seemingly interminable recession that had lasted three years finally bottomed out.[54] As conditions gradually began to improve throughout the country in the the late spring and early summer, business began to pick up in Birmingham. One sign of impending recovery was the

incorporation in June of the Williamson Iron Company by a group of promoters including Charles P. Williamson, a Union veteran from Ohio. Its
capital, however, was only $150,000, of which $20,000 came from the
Elyton Land Company. Because it owned no coal or iron lands, its modestly
proportioned blast furnace on the corner of First Avenue and Fourteenth
Street, only sixty-five feet high and slightly less than fourteen feet wide at the
bosh, had to depend on outside sources of supply. This requirement proved
beneficial to the Georgia Pacific, which supplied coke from its newly acquired
Coalburg ovens to Willamson under contract. In the future, because it never
achieved vertical integration, the Williamson Furnace would be only a marginal producer, staying in blast when prices were high and shutting down
when they were low. [55]

In August 1885, as economic conditions continued to improve, Bryan and
his associates were at last free to resume implementation of their development
strategy, and the Georgia Pacific finally announced plans to complete trackage
between Coalburg and Day's Gap Station. A $50,000 subscription by the
Birmingham business community, with a like amount promised in the future,
helped make the decision possible. In December, Johnston and officials of the
Mobile & Ohio agreed to share trackage between Columbus and Starkville,
Mississippi. Throughout the early months of 1886, as the Georgia Pacific was
being pushed through the Warrior field and the eastern division anticipated
changing its five-foot gauge to the national standard, Bryan worked hard on a
new bond issue to augment the line's inadequate equipment. In the process,
Daniel K. Stewart and Bryan, now acting both for himself and as executor of
John Stewart's estate, made fresh investments in the Extension Company and
the new equipment trust. The success of the latter enabled the Georgia Pacific
to obtain sixteen locomotives and more than 600 new units of rolling stock
within the next two years. [56]

Throughout 1886, the affairs of the Danville and Terminal companies
became increasingly tumultuous. By this time, the Virginia legislature had
passed a law permitting the Danville to hold stock in noncontiguous railroad
lines, making the Terminal Company legally unnecessary. Partly to econo-
mize and partly to streamline control, Clyde and Scott tried to do away with
it, but this move only infuriated other stockholders. By the end of the year,
their resistance had produced a revolution in which the Terminal Company
not only survived but actually took control of the Danville system, reversing
the relationship that had previously existed between the two entities. As a
result, a new board of directors was elected to govern the Terminal Company.
Clyde, Oakman, and Scott were excluded. [57]

This turn of affairs required adroit action by Bryan, Johnston, and Daniel K.
Stewart. Fortunately, despite their ties to Clyde, Oakman, and Scott, they
were also friends of prominent Virginians in the new controlling syndicate,
including Christian, Logan, and Pace. Another member of the victorious
faction, New York banker John A. Rutherfurd, was also on good terms with
Bryan. As a result, day-to-day control of the Georgia Pacific did not change.

Johnston remained president, Bryan became first vice president, and officials at the Birmingham headquarters remained largely the same. Logan, Maben, and Rutherfurd were also elected to the new board of directors, along with Bryan and Johnston.[58]

Conditions in the Danville and Terminal systems were still favorable, therefore, for Bryan and his associates to pursue developmental aims. So too, were national economic trends as increasingly buoyant conditions in 1886 permitted the Georgia Pacific to return to the vigorous expansionist policies of the early 1880s. The line's net earnings for 1885–86 had mounted to $221,042.64, and things were likely to get even better as new equipment arrived and construction crews connected Coalburg and Day's Gap. Filling the missing link would give coal mines in the Warrior Field access to eastern markets.[59]

VI

As the national economic recovery gained momentum and the price of pig iron rose throughout 1886, Jones Valley began to pulsate with a degree of energy to which even its hyperactive business leaders were unaccustomed. A surge of urban development that began at this time was particularly intense in the interior regions of the country, dramatically increasing the populations of Chattanooga, Lexington, Louisville, and Nashville in the east and Fort Smith, Kansas City, and Wichita in the west. Nowhere, however, was the boom more evident than in Birmingham. A trade report from the Magic City in August boasted that things were "all that could be wished." Demand for real estate had "raised prices away beyond the best figures ever realized before," foundries had "all the work they can do," and furnacemen were so overbooked with orders at good prices that they were "practically out of the market." Stocks of pig iron in furnace yards dwindled throughout September and October; by the end of the year, when the surge had attained the force of a tidal wave, a reporter described "almost feverish activity in real estate speculation, superinduced largely by the recent assurances of the strength of the town's position in the industrial world."[60]

Under these circumstances, there was no further reason to delay a move that had been on the minds of Bryan and his colleagues ever since they had acquired the Coalburg property and service had commenced between Atlanta and Birmingham three years earlier. Like other entrepreneurs who were active in urban planning in the 1880s, the Virginians decided to create an industrial community as part of their overall development plan. Along with his brother Andrew, who was secretary of the Georgia Pacific, Johnston filed on 1 October 1886 for incorporation of the North Birmingham Land Company, aimed at developing a 900-acre site along Village Creek into an industrial boom town. Early in December, three other closely related enterprises were chartered in quick succession: the North Highland Land Company, targeted at acquiring real estate and planning its conversion into manufacturing districts and residential neighborhoods; the North Birmingham Building Association, intended to assist the Land Company in realizing its objectives; and the North Birmingham Street Railway Company, which was to connect the new community to downtown Birmingham. John W. Johnston was president of at least two of these firms; other Virginians, including C. W. Branch, Christian, Dooley, Pace, and Tutwiler, were directors and officers.[61]

Bryan invested heavily in all these enterprises. Late in September 1886, he sent Johnston a check for $9,500 "in payment of first installment due 30th inst on 10 combination blocks of yr land scheme." Soon afterward, Dooley came to Alabama in charge of a delegation that visited the North Birmingham site and reported its findings to Bryan, who stayed in Richmond. After hearing from the group, Bryan told Seddon that "the land we bought will bring $1500 per acre—that is about 600%." Other Bryan associates who invested in the North Birmingham projects included Thomas Branch, Buford, Pace, Daniel K. Stewart, and Thomas Pinckney, a Stewart family relative who lived in Charleston, South Carolina. Convinced that lots in North Birmingham would continue selling at a premium, Bryan was determined to exploit the situation to the fullest by holding out for the highest possible prices. "I am not seeking to sell the land & if any wants it they must bid it up," he told a friend. "I hear the Birmingham boom is still raging."[62]

Meanwhile, the Coalburg Coal and Coke Company received fresh funds and started upgrading its facilities. While Dooley and his group were in Alabama, the corporation held its annual meeting, added $200,000 to its capital stock, and elected a board of directors consisting of Bryan, Buford, Christian, Dooley, Johnston, Oakman, and Pace. Shortly afterward, a local newspaper announced that ambitious improvements were underway at Coalburg, including the installation of an endless wire rope hoist, 2,200 yards long, to speed movement of coal to the surface. The addition of new coke ovens increased the previous 76 units to 190, and a coal-washing machine was acquired. The latter purchase showed that Bryan and his allies, like other local industrialists, were taking steps to cope with lean local resources; besides enhancing the quality of coal for coking, the washer would make it possible to use slack, an inferior grade of coal that had previously gone to waste for lack of a market. Meanwhile, the Coalburg Company still planned to build a blast furnace on a site adjoining Village Creek. It was to have a capacity of 110 tons of pig iron per day.[63] All of these developments epitomized the zeal with which Bryan and his associates were pursuing their developmental approach, now that economic skies had brightened.

VII

The visit of Dooley's group to Birmingham in late October coincided with that of another guest whose arrival drew much more attention from local boosters.

This was Jacob Reese, an inventor from Pittsburgh who had developed a reputation as one of the nation's leading metallurgical experts. As a young man, he had helped introduce Henry Cort's puddling process to the United States. After winning a competition in 1850 to design a nail factory, he became a consulting engineer for the innovative Cambria Iron Company at Johnstown, where John Fritz was revolutionizing the rolling of iron rails. In this phase of his career, Reese designed tanks, stills, and barrel hoops for the infant petroleum industry and made pioneering innovations in rolling, hammering, and extruding iron and steel. [64] It is easy to understand why his visit to Birmingham created so much excitement.

Local boosters were well aware that Reese was the leading American expert in removing phosphorus from pig iron so that it could be converted into basic steel. Here, they believed, was the magic key that could transform Birmingham, confronted by the nagging problem of phosphorus-rich ore, into the steelmaking capital of the world. In 1866, Reese had secured American patents to use his basic steelmaking technique in connection with Bessemer converters. Unfortunately for Birmingham, he had sold these to the Bessemer Steel Company, which did nothing with them. It was not interested in their use because Lake Superior ore, which northern and midwestern steelmakers were exploiting quite profitably, was highly compatible with the acid technology to which they were already committed.

By 1877, Reese had developed a new method for making basic steel in an open-hearth furnace. Not until 1886, however, was he able to prove his claims to priority and secure a viable patent. Meanwhile, he had sold prospective rights to the new process to Andrew Carnegie, thinking that the latter would utilize them when the patent was validated. Carnegie, however, passed them on to the Bessemer Steel Company and the Steel Patents Company, which let them lie dormant. The best Reese could do under the circumstances was to try to recover them and resell permission to use his method in the South, where it was badly needed. He had already visited Birmingham in 1885 with such a project in mind and was in the midst of litigation to regain full control of his patent when he returned in the fall of 1886. Any deals that he might make with local furnace operators would be conditional on his success in court, but Birmingham businessmen were desperate to grasp any possibility that would permit them to make steel from the raw materials at their disposal, including the prospect that Reese would win his legal battles. The stakes were high. One thousand tons of pig iron would yield a profit of $12,000 at prices prevailing late in 1886. Due to high tariffs, the same output, converted into 900 tons of steel, would yield $40,500. [65]

Reese came to Birmingham in 1886, as he had done in the previous year, to assign local rights to his process to entrepreneurs who would pay him royalties to use it. The Pratt Coal and Iron Company quickly obtained a license from him. So, too, did another group of investors. It was led by James W. Sloss, who not only was trying to recoup the large sums of money he had sunk into his own floundering furnace company but also was dreaming of one final accomplishment that would match what he had done for the Magic City by bringing the L&N into Jones Valley.

During Reese's previous visit, Sloss and his associates had already secured the right, conditional upon the successful outcome of the inventor's legal struggles, to apply his basic steelmaking technique in two Alabama counties, Blount and St. Clair. After this agreement was reached, the Alabama legislature had chartered a new firm, the Jefferson Iron Company, in February 1885, "for the production of iron and steel in this State." In addition to Sloss himself, his sons Frederick and Maclin and two other associates, Andrew T. Jones and Arthur Smith, were listed as the incorporators. Activation of the enterprise would not be possible until Reese won control of his patent. As its name indicated, however, Sloss and his partners were aiming to make steel in Jefferson County, not Blount or St. Clair. This was also specified in the charter, which stipulated that the new company would operate in Jefferson, "unless it should be found that it can be carried on more advantageously in one of the adjacent counties."[66]

In 1885, prospects that the earlier agreement with Reese would produce any worthwhile result were cloudy at best. Times were bad, the Sloss Furnace Company was producing pig iron only intermittently, and Reese had not even established the validity of his patent claims. By the fall of 1886, however, Birmingham was in the midst of an economic boom, the Sloss furnaces had been recently remodeled, and Reese had obtained his basic patent even though he had not recovered the right to license its use in Alabama. Soon after Reese arrived in the Magic City, Sloss and his partners moved closer toward implementing their steelmaking plans by negotiating a new agreement under which the Sloss Furnace Company would have the prospective right to use his process in Jefferson County. Shortly after this was signed, the Alabama legislature chartered another corporation, the Sloss Iron and Steel Company, which was authorized to absorb both the Sloss Furnace Company and the dormant Jefferson Iron Company. The new entity received its charter on 29 November 1886.[67]

Because of further developments that had taken place in the previous few weeks, there was no doubt that when Reese came to town the Sloss Iron and Steel Company would soon be activated. This was because Colonel Sloss had enlisted important new backers. One of these was Joseph Forney Johnston, a leading Birmingham banker. He came from Lincoln County, North Carolina, where his family owned a plantation that also included an iron forge, a grist mill, and a sawmill. Johnston migrated to Alabama just before the Civil War. Wounded on four occasions while fighting for the Confederate cause, he returned to the state after Appomattox, studied law at Jacksonville, and established a practice in Selma. Allying himself with George S. Houston, whose accession to the governorship in 1874 ended Republican rule in Alabama, Johnston moved to Birmingham in 1884. In addition to being president of the

Alabama State Bank, which by 1887 was the largest institution of its type in the state, he started a newspaper, the *Herald*, which later merged with another journal and became the Birmingham *Age-Herald*. [68]

Another person who had joined the Sloss syndicate was a prominent Birmingham banker, Edmund Winchester Rucker. A native of Tennessee, he was a former Confederate cavalry officer who had commanded a unit known as Rucker's Legion. Like several of the Civil War veterans who played important roles in the development of Sloss Furnaces, he had been badly wounded in that conflict, losing his left arm at Chickamauga and going down twice in a hail of gunfire at Brice's Cross Roads and Harrisburg in northeast Mississippi. Because he had led a brigade, he was always referred to as "General" Rucker, though his highest rank had been that of colonel. After the war, he had helped build the Selma, Marion and Memphis railroad, assuming charge of operations in 1874 when his former commander, Nathan Bedford Forrest, relinquished control of the line. [69]

Even with the addition of such men as Johnston and Rucker, the Sloss syndicate still needed much more money to implement its steelmaking plans, no matter how well Reese might fare in his legal efforts. It was therefore imperative to find a large source of outside capital. By the time the Sloss Iron and Steel Company received its charter, this had been found, for Sloss and his associates had joined forces with Joseph Bryan and the Georgia Pacific group. It is not clear how the first overtures were made, or from which side they came, but it was a logical marriage of interests. On 19 November 1886, ten days before the legislature granted the Sloss Iron and Steel Company its charter, a preliminary agreement was reached between the two parties. It confirmed the option, mentioned at the beginning of this chapter, that John W. Johnston had recently secured for himself and his friends to purchase the Sloss Furnace Company for $2 million.

"Millions More of Good Hard Cash Brought to Birmingham," proclaimed a headline in one of the city's leading newspapers, announcing the result of what had apparently been a whirlwind courtship between the Sloss and Georgia Pacific groups. "There has been no greater surprise in all the remarkable events of Birmingham's career during the last twelve months," an accompanying article stated, "than that with which people heard the report yesterday that the Sloss Furnace Company had sold their property." All members of the Sloss family who had earlier been involved in running the company, it was indicated, would now retire, bringing a fresh managerial team into power. Noting that negotiations had been in progress for "some weeks," the account stated that "the new company will at once build two new furnaces, and, like their predecessors, have in contemplation the erection, at an early date, of an extensive steel plant." [70]

The transaction constituted a dramatic step in the implementation of the developmental strategy to which Bryan and his fellow Virginians were so strongly committed. In the midst of flush times, nourished by the dream that a new Golconda was about to arise in Jones Valley as the Magic City was transformed into the world's leading producer of basic steel, the Sloss Furnace Company was quickly transformed into the newly chartered Sloss Iron and Steel Company. But the butterfly that emerged from the chrysalis would not be what Birmingham's boosters hoped. From the beginning, the reorganized enterprise would be continually plagued by tension between Alabama stockholders who could not wait to make steel and outside investors who had to reckon with sobering economic realities that conflicted with local visions. Never, in fact, would the new enterprise produce a single ounce of the wonder metal on which Colonel Sloss and his associates had set their hearts. That outcome, however, was still shrouded in the future as the new owners prepared to take control.

5 Takeover, Expansion, and Recession

The economic gyrations of the 1880s, surging upward for a time only to plummet downward with equal rapidity, were epitomized by developments that took place in Jones Valley just after Bryan and his associates took charge of the Sloss Furnace Company. When the new regime began early in 1887, the area was still in a frenzy of growth as the national economy continued to boom. New enterprises were blossoming that would dominate the district for decades, and fresh waves of settlers were swarming into Jones Valley from every point of the compass. Many new residents hailed from Virginia, bearing letters of recommendation from Bryan, and North Birmingham soon became a flourishing community. Within months, however, the national economy cooled off, the Magic City was once more fighting for its life, and the Sloss Iron and Steel Company was engulfed in crisis. As dreams of steelmaking dissolved into sobering realities, a troubled firm struggled merely to survive.

I

In recounting how the Sloss Iron and Steel Company came into existence, Alabama's pioneer industrial historian, Ethel Armes, said that John C. Maben "mustered his forces of credit and influence on Wall Street, and, single-handed, went about raising the capital required. In one day, he raised funds to the amount of three millions of dollars."[1] Though dramatic, this story was not the real one. Far from being the powerful financier that Armes imagined, Maben was at best an agent functioning in behalf of other interests. The amount of money involved in the transaction was $1 million less than Armes reported, and it took much longer than one day to secure it. Most important, much of the cash that made the historic corporate takeover possible was raised not by Maben, but by Joseph Bryan.

Bryan was the key figure in the new enterprise and continued to dominate both it and its successor, the Sloss-Sheffield Steel and Iron Company, until he died in 1908. In the process, he played a role in the industrialization of north Alabama that entitles him to be classed as one of its greatest pioneers. Curiously, Armes made only two passing references to him in her book, *The Story of Coal and Iron in Alabama*, on which historians leaned heavily for more than eight decades after its publication in 1910.[2] Why Bryan received so little attention from Armes is a mystery. Just two years before her book was printed, Bryan's death made the front pages of Birmingham's leading newspapers, and he was mourned in editorials lauding his role in the city's history.[3] For twenty-five years, he had appeared and reappeared on visits to the Magic City, accompanied by investors from Richmond and New York. The private railroad cars that carried him and his partners, including a luxurious vehicle, the Alabama, were frequently seen throughout Jefferson County.[4] Yet Bryan was forgotten, while the names of such subordinates as Seddon and Maben were long remembered.

Why did this neglect occur? The most likely explanation is that Armes was misinformed by persons who wanted to claim more credit for the company's genesis than they really deserved. It is also possible that informants who disliked Bryan downplayed his contributions because they still resented a decision that he and his supporters had made in 1902, dashing the hopes of boosters who believed that Sloss-Sheffield had betrayed the gospel of "salvation through steel" on which the city's faith was perennially grounded. This decision, the most crucial turning point in the history of Sloss Furnaces, will be discussed at length later in this book.[5] Whatever the reason for the ill-deserved neglect that Bryan's many contributions to Birmingham's development have suffered, the city's great indebtedness to him was soon forgotten. Seldom has an industrialist of equal stature received so little recognition for playing a role of comparable magnitude.

II

The origins of the deal that brought Sloss Furnaces into the Georgia Pacific's empire are almost as difficult to trace as it is to explain why Bryan faded so rapidly from local memories. During their visit to Birmingham in late October 1886, Dooley and other directors of the Coalburg Coal and Coke Company could not have escaped the excitement caused by the presence of Jacob Reese in the city at the same time. Letters written by Bryan immediately after Dooley's party returned to Richmond, however, mention neither Reese nor the recent agreement that permitted the Sloss Furnace Company to use his basic steel process in Jefferson County. It appears, therefore, that neither Dooley nor his companions were approached about a takeover of the Sloss Furnace Company by the Georgia Pacific while they were in Birmingham. Evidence is inconclusive, but they probably left before the agreement between the Sloss syndicate and Reese was announced in local newspapers.

The earliest surviving document relating to the takeover is dated 6 November 1886, four days after news of the agreement between Reese and the Sloss group appeared in the Birmingham *Age*. As previously indicated, it is a handwritten proposal to purchase the Sloss Furnace Company "for myself and associates," signed by John W. Johnston. Of the $2 million price that Johnston specified, $1.25 million was to be paid in cash and $750,000 in 6 percent mortgage bonds, which were to mature in thirty-three years. Johnston and his colleagues were to put $500,000 "into the capital of the new company to build two blast furnaces of not less than 100 tons daily capacity each, and such coke ovens and other improvements as may be deemed best." The syndicate was to be organized by 15 December 1886. If it decided to proceed with the transaction at that time, its members were to post a forfeit bond of $10,000, after which they would have until 1 February 1887 to inspect the property and organize the new company.[6]

As he wrote his proposal, Johnston was getting ready to leave for New York City for meetings of the Danville and Terminal systems at which Clyde, Oakman, and other allies of Bryan and the Stewarts would be ousted from their directorships. Johnston had already made arrangements to meet Bryan at the Fifth Avenue Hotel upon arriving in the metropolis. In letters written to Johnston and Seddon on 4 November, Bryan said nothing about the takeover possibility, indicating that he did not yet know about it. The most plausible interpretation that is consistent with surviving evidence is that negotiations between Sloss and his associates on the one hand and Johnston on the other came about suddenly after the signing of the contract between the Sloss syndicate and Reese and that Johnston had prepared his proposal to Sloss, which appears to have been hastily written, just before leaving Birmingham. Apparently Johnston acted on his own authority; at least, there is no evidence that he consulted with business associates in Richmond or New York before making the offer. In any case, it contained enough loopholes to protect him and his backers if the deal fell through.[7]

There is also no evidence that Maben knew about the possibility of the takeover before Bryan did. Because of the close friendship between Johnston and Bryan, the frequency with which they corresponded with each other, and the fact that Bryan was the chief legal representative of the Stewart fortune, it cannot be doubted that Bryan was the person whom Johnston most wanted to contact. Subsequent events make it clear that discussions about the possibility of a takeover took place between Johnston, Bryan, and other Virginians who were in New York for the Danville and Terminal meetings.

Because Bryan and his associates were deeply committed to a developmental strategy, typified by their plans to build a blast furnace at Coalburg and intention to erect two more stacks in North Birmingham, it is not surprising that they jumped at the chance to acquire the Sloss Furnace Company. The alacrity with which they moved is particularly understandable because of the opportunity to build a steel mill if Reese succeeded in his ongoing patent litigation. But the exact way in which events unfolded after Johnston arrived

John W. Johnston. A Virginia-born railroad executive who had been a Confederate
artillery officer during the Civil War, Johnston was one of Bryan's closest friends.
Johnston's leadership in the Richmond and Danville Extension Company and the
Georgia Pacific Railway contributed importantly to Birmingham's growth, and he
played a key role in the purchase of the Sloss Furnace Company by Bryan and his
associates in the winter of 1886–87. A photograph of Johnston taken in the late
nineteenth century suggests the weight of the duties that broke his health and forced
him to retire at the peak of his career in 1890. (Courtesy of Valentine Museum,
Richmond, Virginia.)

in New York is impossible to trace. Because most of the decisions were made face-to-face in New York, the first mention of the takeover possibility in Bryan's letterbooks does not occur until he wrote to Johnston on 15 November after returning to Richmond. By this time, efforts to put Johnston's preliminary offer of 6 November on a firmer basis were underway. In his letter, Bryan indicated that he had made contact with Joseph F. Johnston and Edmund W. Rucker in Birmingham and that he had also consulted with Richmond financiers, including Carter W. Branch, S. A. Ellison, and C. E. Whitlock, who had not been present in New York.

Bryan had already crossed the Rubicon. On 16 November, he contacted a Baltimore firm, Wilson, Colston, and Company, stating that he had "made some engagements touching a transaction in furnace and iron land property in & near Birmingham, Ala. which will require some ready money" and instructing them to sell a block of his Georgia Pacific Car Trust Bonds, even at a sacrifice, to raise cash. He also wrote to John W. Johnston, still in New York, to "put ten blocks subscription to Sloss purchase in my name for self and D. K. Stewart as our interests will not be equal as arranged between us." The closing words of this statement were significant. Sensing an opportunity to establish a substantial fortune independent from that of the Stewarts, Bryan had decided to stake his financial future on the enterprise that he and his syndicate were acquiring.[8]

While Johnston was traveling to New York and parleying with Bryan and other Georgia Pacific investors, Colonel Sloss had been completing arrangements with business associates in Birmingham, Louisville, and Cincinnati without whose consent he could not proceed. On 10 November, he informed Johnston by mail that he accepted the terms set forth in the proposal that the latter had made four days earlier. Agreeing to give Johnston until 15 December to organize his syndicate, Sloss stipulated that, if the deal went through as planned, the new owners must honor the existing contract under which the Sloss Furnace Company obtained its coal supply from the Pratt Coal and Iron Company. In addition, he specified that the two projected blast furnaces in North Birmingham must be completed within fifteen months after 1 April 1887, the date upon which the transfer of property was to become final. The fact that Sloss and his associates would retain a substantial stake in the new company probably accounts for their need to be reassured that the Richmond group would carry out its expansion plans. After Johnston received Sloss's letter in New York, the impending purchase was made public on 19 November 1886.

From mid-November onward, Bryan's letters contain numerous references to arrangements concerning both the Sloss purchase and the ventures that were underway in North Birmingham. Some of the details are so closely intertwined that it is difficult to determine precisely which enterprises were involved in which specific transactions. It is clear, however, that raising the funds necessary for his subscriptions to Sloss Iron and Steel and the four interrelated North Birmingham land and development companies strained

Bryan's resources to the utmost. "I thank you on your congratulations on the 'Sloss turn,' but it hasn't turned yet," he told a friend early in the New Year. "I hope however it will as I need some turns this month—Should 'fickle fortune' make me very poor, I will only be back where I started & where I staid a long time."[9]

Despite Oakman's departure from the Danville and Terminal boards, Bryan's friendship with him remained as strong as ever. He therefore turned to Oakman for help in arranging loans on Wall Street. Notwithstanding his urgent need for cash, Bryan was selective about the terms he was willing to accept. Oakman arranged for Bryan to borrow $80,000 from John H. Inman, a Confederate veteran from Tennessee who made a fortune on Wall Street after the Civil War and became Wall Street's leading investor in southern enterprises, but Bryan did not follow through on this possibility. His later correspondence shows clearly that he distrusted Inman, who had recently become a director of the Terminal Company in the upheaval that had removed Clyde and Oakman.

Subsequent events would show that these doubts about Inman were well founded, demonstrating Bryan's grasp of character and his firm commitment to developmental, as opposed to speculative, goals. Born in Tennessee, Inman had gone to Georgia at an early age and worked in an uncle's bank. Serving in the Confederate army, he came out of the war virtually penniless and went to New York City, starting out by working in a cotton brokerage house. Rising swiftly, he became a partner in three years, and the firm was renamed Austell & Inman. Two years later, Inman and another Tennessean, James Swann, founded Inman, Swann & Company. A financial genius, Inman played a major role in organizing the New York Cotton Exchange and became heavily involved in banks and insurance companies. "In a city crowded with financial titans and self-made men," Maury Klein has written, "Inman took a back seat to no one. He was a millionaire several times over by age thirty-five and had ascended into the best of New York society."

Inman, however, was an opportunist whose unscrupulous tactics contrasted sharply with Bryan's commitment to a patient developmental strategy. There can be no question about Inman's devotion to his native region, into which, by his own estimate, he ultimately funneled more than $100 million of northern money. On the other hand, he was a ruthless operator who could not resist the temptation to play games with the stock of every southern enterprise in which he participated. This eventually led to his downfall. After the Panic of 1893, his affairs were in such disarray that he had a nervous breakdown and died in a Connecticut asylum.[10]

Had Bryan become entangled with Inman at this time, Sloss Iron and Steel might later have been captured by the Tennessee Coal, Iron, and Railroad Company, which had recently invaded the Birmingham District. Inman was its most powerful stockholder and would undoubtedly have coveted Sloss as avidly as other firms that he swallowed in his numerous Wall Street campaigns. In the final analysis, it is impossible to tell exactly how much money

Bryan raised from northern sources that he trusted more than Inman. Most of his support, however, came from below the Mason-Dixon Line, particularly from Richmond and Baltimore. From the beginning, the Sloss Iron and Steel Company was a southern firm, backed mainly by southern capital and credit.[11]

Early in January 1887, Bryan left Richmond, bound for Birmingham with Ellison and Whitlock, to see the Sloss property at first hand and inspect the new equipment that the Georgia Pacific was beginning to acquire as a result of the recent car trusts that Bryan had organized. By this time, the Sloss takeover was a certainty. On 12 January 1887, a mortgage was executed with the Central Trust Company, making it trustee of all property to be acquired by the Virginians and their associates. Returning to Richmond in mid-January, Bryan set out on a long trip to Washington, Chicago, Minnesota, and New York to attend to various business matters, making sure that his other affairs were in order before he plunged into Sloss Iron and Steel. By early February he was back in Alabama, where the new management was in place and preparing to take power.[12]

Far from being arranged single-handedly by Maben in one day, therefore, the Sloss takeover was a much more complex process in which Bryan played a pivotal, though previously unrecognized, role. The original stock registers of the new company, compiled in January and March 1887, show that funding came from three places: Richmond, New York, and Birmingham. The influence of Virginians, however, was paramount. Bryan had the largest single holding, 2,550 shares; John W. Johnston held the second largest block, 1,650. Christian, Pace, and Whitlock each held 1,500 shares, and Maben subscribed for 1,350. Other holdings by Virginians, mostly from Richmond, included those of Thomas Atkinson, 150 shares; A. S. Buford, 900; S. A. Ellison, 600; W. W. Gordon, 300; Sol Haas, 300; A. B. Johnston, 300; Thomas M. Logan, 900; T. B. Lyons, 900; E. T. D. Myers, 150; Thomas Seddon, 100; Daniel K. Stewart, 450; W. R. Trigg, 300; E. M. Tutwiler, 150; and S. W. Venable, 300. James H. Dooley was not an original subscriber but began investing in the company in July 1887, with an initial stake of 268 shares.

Among other stockholders, Colonel Sloss himself continued to hold 1,200 shares, while 1,500 remained in the hands of longtime Louisville associates. Besides Sloss himself, the Birmingham contingent included J. B. Boddie, a leading local banker and real estate promoter; Joseph Forney Johnston; and Edmund W. Rucker, all of whom subscribed 1,500 shares. Andrew T. Jones and Arthur W. Smith, who had joined Sloss in incorporating the Jefferson Iron Company in 1885, each subscribed 300 shares. While these investments were not nearly sufficient to give Birmingham interests control of the enterprise, they were large enough to give them a strong voice in its affairs. Joseph Forney Johnston's appointment as president of the firm indicates that they were given charge of day-to-day operations.

Virtually all the stock was subscribed by southern investors; even most of those listing New York City as their place of residence were really Virginians.

The only major blocks held by northern financiers were those of W. H. Goadby & Co., with 600 shares; Richard Y. Mortimer, 900; Myers, Rutherfurd & Co., 500; and Walter G. Oakman, 300. The fact that Goadby, Mortimer, Oakman, and Rutherfurd were all close friends of Bryan, coupled with the absence of Inman from the list, showed how careful Bryan had been in selecting northern backers. That northern credit played a key role in the transaction is indicated by the fact that the entire subscription was underwritten by the Central Trust Company. In the future it would be Sloss Iron and Steel's main anchor on Wall Street. Northern aid, however, came mostly from loans to southerners, not direct investments.[13]

III

Having completed the takeover, the new owners prepared to assume power amid continuing optimism in the Magic City. As 1887 began, the boom that had started in the previous year was still going full blast, and local furnacemen were extremely choosy about filling orders. "Prices here are to be determined rather from what is refused than from what is taken," stated one trade report. Reflecting the flush times, boom towns were rising on every hand throughout north Alabama, and new furnaces were being built as rapidly as materials could be stockpiled. Of thirty-nine stacks under construction in the entire country, twenty were being erected in Alabama alone, ten of them in Jefferson County. Experienced observers feared that the mania, which was also raging in east Tennessee, was getting out of hand. "The speculative craze that has taken hold of the South is reaching a point where the conservative element in that section should use every means in their power to bring about a return to reason," *Iron Age* cautioned in February, predicting that the bubble would soon burst and warning of "the disaster which must inevitably come."[14]

The number and scope of important corporate reorganizations that took place in Birmingham amid the craze indicates the intensity of the boom during which the birth of the Sloss Iron and Steel Company took place. Within only a few months, no fewer than four major financial transactions occurred that would affect the development of Jones Valley for decades. The Sloss takeover was one of these, but certainly not the most impressive. By the time it happened, three other big new enterprises were already on the scene. One, the Tennessee Coal, Iron, and Railroad Company would become the region's largest industrial enterprise.

TCI burst like a meteor upon north Alabama late in 1886 as the boom surged irresistibly forward. Its early history, typifying the ruthless business tactics prevailing in the age of the Robber Barons, stamped its operations with a bold, often reckless character that affected every local enterprise with which it came in contact. By competing with it, such firms either became victims of its insatiable acquisitive appetite or developed their own individual identities by surviving, just as a blade is shaped by being drawn against another knife.[15]

TCI's corporate personality had been set before the company moved into Alabama. As Birmingham grew from a tiny hamlet to a dynamic boom town in the decade following the 1876 Oxmoor experiment, coal and iron production continued to rise, though less spectacularly, in the mountains of east Tennessee. In the Chattanooga area, the Roane Iron Company displayed characteristic vigor. Its persistent efforts to manufacture steel and its industrial village of Rockwood, an outstanding example of southern paternalism, would hold enduring interest for historians.[16] Much more remarkable, however, was the dynamic venture centered around Tracy City that, as already noted, had begun corporate life in 1852 as the Sewanee Mining Company. Subsequently renamed the Tennessee Coal and Railroad Company, it built an increasingly profitable business in the postwar era by using convict labor and converting its coal into coke.

Amid the financial crises of the mid-1870s, Arthur S. Colyar, the Nashville attorney under whose leadership the company had built the Fiery Gizzard to test the suitability of its coke for iron production, relinquished control to Cherry, O'Connor and Company, which held the contract for its convict labor supply. Under new management, the enterprise not only marketed coke but began to make iron. Reflecting the change, it adopted the name by which it remained known for the rest of its turbulent history.

In 1881, TCI came under Wall Street control for the first time when it was taken over by Inman, who installed a Nashville banker, Nat Baxter, as its president. Under the new regime, the firm quickly manifested the acquisitive appetite and speculative style for which it became famous. Moving into the Chattanooga area with Inman's financial support, TCI absorbed the Southern States Coal, Iron and Land Company in 1882. In the process, it gained the managerial services of James Bowron, Jr., a shrewd, highly articulate young Englishman who had taken charge of the South Pittsburg enterprise in 1877 following the death of his father. Bowron would later become one of Birmingham's most prominent citizens after TCI moved into Alabama.

Inman's hold on TCI was temporarily broken in 1885 after declining iron prices forced him to list its securities on the New York Stock Exchange in order to raise capital. Taking advantage of Inman's plight, Nashville broker William M. Duncan organized a coup and took charge of TCI. In a fateful step with momentous consequences, he ousted its veteran general manager, Alfred M. Shook, who had built the Fiery Gizzard, and put Bowron in his place. Infuriated, Shook went to Birmingham looking for a way to turn the tables on Duncan. Meanwhile, Baxter stayed on as president but felt no loyalty toward the new ownership. Amid the infighting and intrigue that had become chronic by this time, the company was about to enter Alabama.

In Birmingham, Shook found an ally in Thomas T. Hillman, who, along with Henry F. DeBardeleben, had pioneered the development of the Alice Furnace Company. In 1881, Hillman became Alice's operating head when DeBardeleben became convinced that he had contracted tuberculosis and went to Mexico to recover his health by taking up sheep farming. Hillman's power, however, was soon eclipsed by that of Enoch Ensley, a rich Tennessean who came to Birmingham from Memphis in 1881 after Inman had rebuffed his attempts to buy his way into TCI. Ensley, who had inherited a Mississippi plantation from his father and amassed wealth in such enterprises as the Memphis Gas Light Company, was furious about his treatment at Inman's hands. In a transaction that became famous as the first million dollar deal in Birmingham's history—actually, it cost Ensley only $600,000, payable over a six-year period—he bought the Pratt Coal and Coke Company from DeBardeleben. Next, Ensley took control of the Alice Furnace Company, bringing Hillman under his thumb, and acquired the Linn Iron Works. Combining these ventures with the Pratt Coal and Coke Company in 1884, he created a consolidated firm, the Pratt Coal and Iron Company. It was by far the largest business entity that had ever existed in Alabama.

Ensley's huge corporation provided stiff competition for TCI, but Hillman, who had been reduced to the status of an impotent minority stockholder, seethed with resentment. In 1886, he joined Shook in an ambitious scheme to outflank Ensley, take control of the enterprise with Inman's help, and combine it with TCI. Baxter was a willing accomplice, being ready to betray Duncan at the first available opportunity. Meeting in Philadelphia in the spring of 1886, Shook, Inman, and Baxter laid plans for a coup. Playing upon Ensley's ego, they gained his support for the scheme by offering him the presidency of the consolidated enterprise that they had in mind and proposing to build a new boom town that would be named Ensley City in his honor. Meanwhile, Inman mounted a campaign on Wall Street to force down the price of TCI's securities and force Duncan to sell out. Despite some temporary hitches that threatened to derail the plan, a merger of the Pratt Company and TCI was consummated by summer's end. The resulting industrial colossus outclassed anything that had previously existed in the South. It was capitalized at $10 million, an unprecedented sum by regional standards. Ensley was its titular chief executive, Hillman first vice president, Shook second vice president and general manager, Baxter chairman of the executive committee, Bowron treasurer, and Inman the firm's chief financial backer and Wall Street mogul. Duncan, now stripped of real power, remained as a mere director.

Showing a tendency to overextend himself that ultimately caused his downfall, Ensley established a land company to develop the boom town that was to be named after him. Like TCI, it was capitalized at $10 million, a grossly inflated figure even considering the manic growth taking place in Jones Valley. Meanwhile, TCI began building four new blast furnaces just outside Birmingham, marking the first time that this number of stacks had been erected in the United States at one time. By comparison, the $2 million Sloss takeover by Bryan and his friends during the next few months was a modest affair.[17] The contrast between the Sloss Iron and Steel Company and TCI epitomized their underlying commitment to two radically different growth strategies, one developmental, the other speculative.

TCI's invasion of Birmingham was accompanied by an important deal between Inman and Milton H. Smith, who had become president of the L&N Railroad in 1884. A "tall, broad-shouldered man" who had been born in upstate New York, Smith moved to Mississippi before the Civil War and became a telegraph operator, an occupation closely associated with railroading. During the war, showing a pragmatic attitude that would typify his career, he worked alternately on the Southern and Northern sides; by 1865, he was running captured Confederate rail lines for the United States Military Railroads in Alabama, Tennessee, and Georgia. After the war, he stayed in the South as agent for the Eclipse Fast Freight Express Service of the Adams Express Company, headquartered in Louisville. He also became division superintendent for the Alabama and Tennessee Railroad, working out of Selma, where he joined Dallas County businessmen in establishing a cotton press and warehouse. Noticing his talents, Albert Fink brought him back to Louisville as freight agent for the L&N, and he soon became an important force in that line's southward expansion drive. There is reason to believe that he personally drafted the report in which Fink urged the L&N's directors to follow Colonel Sloss's recommendation by moving into Jones Valley in 1871. After Fink left the L&N in 1875 to become commissioner of the Southern Railway and Steamship Association, Smith went north to take part in the freight operations of the B&O and Pennsylvania railroads. In 1882 he rejoined the L&N as third vice president and took command of operations two years later.[18]

Acid-tongued, brutally candid, and contemptuous of people whose talents and judgment were inferior to his own, regardless of their rank or station, Smith became a prime force in southern economic development for nearly four decades. Little interested in passenger service, on the grounds that railroads could not "get a god damn cent out of it," he was what Herman C. Nixon later described as "a hog for freight, especially for long-haul freight." In return for business from north Alabama mines and furnaces, he promised such men as Aldrich and DeBardeleben that he would stay out of coal and iron production, sell land at bargain rates to local industrialists, and keep freight charges as low as possible. Deeply committed to the growth of the Magic City, he spearheaded development by the L&N of what came to be called the Birmingham Mineral Railroad, a complex of feeder lines running eleven miles along both sides of Red Mountain. Completed in 1884, it carried coal, ore, and limestone at cheap rates to furnaces along its right-of-way.

Smith was already committed to expanding the Mineral Road when Inman approached him in 1886 with a deal that he could not refuse. In return for a commitment by TCI to make the L&N its exclusive carrier, Smith agreed to encircle Red Mountain with trackage that would help form a beltway around the entire Birmingham District, from Redding and Graces on the south to Boyle's Station on the north. He also developed a spur line reaching all the way to Woodstock on the west. After the four furnaces at Ensley began production in 1888, Smith made further additions to the system, including the Blue Creek Extension to Blocton Junction in Bibb County. By the time the beltway was finished in 1890, it encompassed 156 miles of rails, laid at a cost exceeding $6 million. The building of this giant "race track," as Nixon later called it, made an enormous contribution to rising production levels throughout the district during the late 1880s. Let local entrepreneurs produce the freight, Smith said, and he would carry it on his back if this became necessary. DeBardeleben expressed the area's debt to the leader of the L&N in his characteristic way: "M. H. Smith is the biggest, broadest man you ever saw in your life!" Ethel Armes later declared that Smith was "the strongest force in the industrial history of Alabama," calling him "the great progenitor."[19]

TCI's entry into the Birmingham District had a more significant impact on the history of Sloss Furnaces than any other development that took place in 1886 and 1887, but other events also had important long-range effects. One of these was the launching of a new enterprise by DeBardeleben, who returned from Mexico in 1882 and, in partnership with Aldrich, developed the Henry Ellen Coal Company. Based in the Cahaba Valley, it was named for DeBardeleben and his wife. DeBardeleben also built Mary Pratt furnace, named for his mother-in-law, on a First Avenue plot adjacent to the two stacks that Sloss had erected. Renewed concerns about his health, however, led DeBardeleben to return to Mexico in 1883, leaving his enterprises in the hands of Aldrich and a Kentuckian, William T. Underwood. In the future, Underwood's brother, Oscar, would become Alabama's best-known politician and perennial favorite son in attempts to occupy the White House.[20]

Stopping in Texas during his second southwestern sojourn, DeBardeleben met a wealthy British businessman, David Roberts, who had become related through marriage to a prominent South Carolina family. With his characteristic flair for words, the voluble Alabamian fired Roberts's imagination with glowing visions of the Birmingham District's future. After DeBardeleben returned home in 1885, bringing Roberts with him, they formed a syndicate and laid plans to develop yet another boom town, to be located about twelve miles southwest of the Magic City. In 1886 they incorporated the DeBardeleben Coal and Iron Company and established a related land and development firm. Both ventures were open only to investors who, in DeBardeleben's words, wanted to "make smoke." Soon a new urban complex was underway, complete with three coal mines, a limestone quarry, two blast furnaces, a battery of coke ovens, and a rail line.

Naming the new community Bessemer after the British inventor, DeBardeleben spared no expense in making it an industrial showcase. The first lots were sold on 12 April 1887; by the end of the first year, more than $2 million had been invested in the town, which already had 3,500 inhabitants. "Magnificent plants of iron furnaces, iron and steel rolling mills, foundry and machine shops, screw works, planing and drying mills and wood working factory, fire-brick works, etc., have been erected," stated a broadside. "Choice business blocks costing from twenty-five thousand to one hundred and twenty-five thousand dollars have been constructed, and nearly four hundred buildings have been completed or are near completion." Officials of the Bessemer Land and Improvement Company had no doubt that "Bessemer is destined to be a large milling and manufacturing center and prosperous city, for its coal and iron fields are inexhaustible, and its transportation facilities are unexcelled in the South; for nearly all the various manufacturing industries and kinds of business enterprises it affords advantages that cannot be rivalled in this country."[21]

With characteristic zeal, DeBardeleben bought an ornate prefabricated structure that had housed Mexico's exhibits at the 1884 New Orleans World Exposition. Carefully taken apart, it was brought to Alabama, reassembled on a hill overlooking the new town, and rechristened the Montezuma Hotel. Other imposing buildings followed, making Bessemer a treasure-trove of Victorian architecture. When the community celebrated its first anniversary in April 1888, George M. Pullman and Robert Todd Lincoln were among the dignitaries on hand. Saluting the venture, the *Manufacturers' Record* stated that "what has been accomplished . . . reads almost like a romance. Where the forest stood fifteen months ago seven railroads now center." Later that year, the first of two original furnaces went into blast, and three more were begun; all were soon consolidated into one enterprise, the DeBardeleben Company. Determined to outstrip every other place in the region in growth and development, citizens petitioned the state government to establish a new county, also to be known as Bessemer. Because of what Bessemer's boosters called "the unholy domination of Birmingham," however, this move was thwarted.[22]

Yet one more large industrial venture and a related boom town got underway in Jones Valley late in 1886. Eighteen years earlier, one of America's leading industrialists, David Thomas, had been introduced to north Alabama by Giles Edwards, a Welsh ironworker who had moved south from Pennsylvania just before the Civil War and built an ironworks at Woodstock in Tuscaloosa County. Thomas, who had come to the United States from Wales in 1839, had taken the lead in transferring hot-blast technology from Great Britain to America and was renowned as the "father of the anthracite iron industry." As a result of his influence, many furnaces had been built in eastern Pennsylvania to utilize anthracite. By the late nineteenth century, the Thomas Iron Company, headquartered at Hokendauqua, Pennsylvania, was the world's largest maker of anthracite iron.[23]

After visiting Alabama, Thomas acquired several ore and coal tracts. In addition to the historic Tannehill property in Roupes Valley, these included Jefferson County deposits. In December 1868, a syndicate headed by Thomas incorporated the Pioneer Mining and Manufacturing Company to develop these holdings. Like the Woodwards, however, the Thomas family did not immediately move into Alabama, even after they bought Williamson Hawkins's 1,774-acre Jones Valley plantation in 1881. Instead, the Pioneer firm remained dormant until 1886, by which time the increasingly successful invasion of the northeast by southern pig iron had shown that the obsolescent anthracite iron industry was doomed. David Thomas was by then dead, and the Thomas Iron Company was headed by his son, Samuel. Resigning as its president in 1887, the latter moved to the Birmingham District. This move delighted local boosters who savored the event as an important victory in the Magic City's relentless war against northern companies.

The Pioneer Company erected a large, up-to-date blast furnace and established a community named Thomas after the family that owned the enterprise. The town, built on a site four miles southwest of Birmingham, was modeled on Hokendauqua; houses with features resembling those of workers' dwellings in eastern Pennsylvania were erected, adding a new element to the increasing architectural diversity of Jones Valley. By the time the first furnace went into blast in May 1888, it was reported that Thomas had decided to build two more, at a cost of $300,000. Only one of these, a duplicate of the first, was actually constructed, beginning production in February 1890. The other did not materialize until 1902, when new owners built a ninety-foot stack, the largest ever erected in the district up to that time.[24]

IV

Due to TCI's appearance in Alabama, the building of DeBardeleben's complex at Bessemer, the Thomas family's move from Pennsylvania, and the takeover of Sloss Iron and Steel by Bryan and his group, the Birmingham District resounded with the din of construction in early 1887. People poured in, seemingly from everywhere, and local newspapers were full of real estate advertisements. Once an object of derision, the Elyton Land Company, which had gone thirteen years without paying a single dividend, was now dispensing steady earnings in cash and securities, ranging from $90,000 to $200,000 per year. This record, however, was dwarfed by a dividend of 340 percent, yielding $680,000, that was declared in 1886. On any given day, the company's offices, located on the corner of Twentieth Street and Morris Avenue, were crowded with speculators who pored over maps, bought unsold properties at inflated prices, and rushed out to sell them at a profit before the title bonds could be executed. One "boomer" realized a 400 percent gain in under three months. Fifty-foot lots that had once sold for as little as $400 now fetched $1,000 per front foot.

Confidence in Birmingham's future was stimulated not only by the nature of the times but also by the opening of beautiful new residential districts. Highland Avenue, 100 feet wide, curved its way through what had previously been "an almost inaccessible wilderness," rising to an elevation of 200 feet above Jones Valley as it ascended from Seventh Avenue South toward an intersection with Twentieth Street. Soon it was lined by imposing residences and served by a street railway whose gaudily painted cars were pulled by mules before the line converted to steam power. Development of both the avenue and the railway was spearheaded by John T. Milner's half-brother, Willis J. Milner, who was the Elyton Land Company's chief engineer.

Meanwhile, the Land Company led a series of projects to enhance the quality of life in the city. Under contract with the Thomson-Houston Company of Philadelphia, electric street lights were installed on a number of thoroughfares in 1885. But the community was still desperately short of water, and municipal officials worried about the possibility of fires that might totally exhaust the small quantities at their disposal. A well that had been bored to a depth of 600 feet just south of Red Mountain soon proved totally inadequate for the city's needs, and a canal was therefore dug to tap springs at the headwaters of Five Mile Creek. When completed in 1887, it delivered 4 million gallons of water per day to residents who had previously endured various restrictions. Because of its burgeoning earnings, the Elyton Land Company was also able, beginning in 1888, to build a belt railroad, twelve miles long, to "move heavy freight within and around the city."

Visitors were impressed both by the youthful vigor of the city's population and by its shifting, transient character. "Birmingham is essentially a young man's town," wrote one observer in September 1886. "During my stay I have not met with more than half a dozen gray and grizzled men, and they were either from the country around, or were prospectors from the North lured hither to satisfy themselves of the truth of the astounding statements concerning this now famous city. . . . Ask any ten persons on the street for a direction in any locality, and eight out of the ten will tell you that they 'have just come.'" Hotels and boarding houses were bursting with new arrivals. "The rush to and fro of men full of business is constant from early morn till late at night," a visitor declared. "Everything is bustle and activity."[25]

In such a frenzied atmosphere, it was hard for even the most conservative managers to avoid expansionist moves that might prove disastrous when conditions inevitably cooled down. In 1887, the Elyton Land Company posted colossal dividends of $4.41 million in cash, bonds, and stock; they amounted to a return of 2,305 percent on the original investment of $100,000. At the annual convention of the enterprise in May 1887, stockholders were told that it was "the most powerful corporation, financially, in the South." Inevitably, confidence bred overconfidence; the Woodward Iron Company stood virtually alone among firms pursuing a safe and steady course, confining itself to completing the furnace it had begun in 1886 after its first stack had proved cost-effective. After the second installation went into blast in January 1887, the Woodwards concentrated on doing what they did best—making money—and went through the rest of the century without adding further capacity. Not until 1903 did they start another furnace. As a result, their operations would not compare with those of such firms as TCI or Sloss in sheer size, but their persistent conservatism, even more pronounced than that displayed by Bryan, yielded compensating advantages at the bottom line.[26] Nevertheless, Bryan, both now and in the future, would pursue a far more deliberate course than Birmingham's boosters wanted him to follow.

V

Boosters predicted great things for the fledgling Sloss Iron and Steel Company. From the beginning, it was expected to proceed rapidly to fulfill the promise of the licensing agreement that Colonel Sloss and his syndicate had made with Jacob Reese. The first move of the new owners, it was predicted, would be to make steel billets and slabs for northern mills. This, however, would be only a beginning; ultimately the firm would make rails, wire, and "many smaller finished forms of steel." Joseph Forney Johnston confirmed this impression, stating that, besides erecting the North Birmingham furnaces, the company would "build a steel converter." It might take fifteen months to complete the North Birmingham stacks, he admitted, but they would probably be finished within a year. By that time, the enterprise would be making 400 tons of pig iron a day.[27]

As preparations were made to build the furnaces, purchasers pored over maps of North Birmingham that were displayed at the Georgia Pacific office. Lots went on sale 3 January 1887. Soon an influx of people was moving into the new community, where the Georgia Pacific's tracks separated a section reserved for whites on the wooded heights to the north from one on the south that was set aside for blacks. Hastily constructed frame structures sprang up in the business district, which fronted what was then known as Seventh Avenue, and the North Birmingham Street Railway Company was soon bringing passengers in and out of town along Twenty-sixth Street. A local newspaper article touted the community as Birmingham's "prettiest suburb." The writer, however, lamented that Indian mounds had been destroyed to provide dirt for a railroad fill and shed crocodile tears about "how this age of utility remorselessly destroys the relics of the past."[28]

Considering the origins of most of the men who now owned Sloss Iron and Steel, it is not surprising that a wave of Virginians poured into the district, carrying with them a rich fund of experience. One of them was Joseph Bryan's brother, St. George Tucker Coalter Bryan. Known to his relatives as Saint, he had entered the Confederate army upon graduating from Episcopal High School in Alexandria, seen duty with the Richmond Howitzers, been badly wounded at the Battle of Sayler's Creek, and served with the haggard, dispirited remnant of Lee's army just before Appomattox. After the war, he studied mining engineering at the University of Virginia. In August 1886, Saint set out for Alabama, where Joseph recommended him to John W. Johnston as a suitable candidate for taking charge of one of the Georgia Pacific's depots. Displaying characteristic rectitude, Bryan told Johnston that he was confident of Saint's "integrity, industry & courage," but added that he lacked faith in his brother's judgment. "What I would desire for him would be a fair chance on his merits . . . but until tested I think some routine place advisable, as he has had no experience in Railroading."[29]

Saint, however, was not satisfied to run a railroad depot. Wanting to use his training as an engineer, he applied to become furnace superintendent. Joseph was uncomfortable about this move, remarking that "we cannot afford to make a mistake in the management of the company" and declining to vote on the matter. Despite Joseph's efforts to avoid nepotism, Saint got the job, at a salary of $2,000 per year. Joseph's reservations about him seem to have been well founded, because the company was soon looking for an experienced furnace manager to work under him.

Like previous superintendents who had tried to deal with local raw materials, Saint quickly ran into trouble for which his academic training had not prepared him. In particular, he was confounded by the enormous quantities of slag resulting from the high percentage of sand and slate in the ore and coal used by the company. It was extremely hard, he complained, to maximize output using ores containing less than 40 percent iron and 28 percent silica and Pratt coal that yielded more than 11 percent ash. "Have read with interest your account of the progress you are making with the slag," Joseph responded. "I wish it was in my power to give you some scientific or practical assistance but I cannot." At the time, the No. 2 furnace was out of blast because of scaffolding. "Your account of the trouble at No. 2 Sloss Furnace was very interesting," Joseph said. "I fear very much the furnace will have to be relined before she can be started again."[30]

Saint was not the only member of the Bryan family who came to Birmingham. Another, who stayed much longer, was John Randolph Bryan, Jr., known to his relatives as Ran. Born in 1841, he had a war record that made him unique among the Confederate veterans thronging into the Magic City. As an aide-de-camp for Col. John B. Magruder in the forces protecting Richmond against McClellan in 1862, he had volunteered for special duty after intercepting a message to Magruder from Gen. Joseph E. Johnston, the commander in charge of defending the Confederate capital. In the message, Johnston requested the services of a soldier capable of distinguishing the character and number of the enemy troops along the front below Yorktown. Such a person would need to be well acquainted with the surrounding countryside. Having been reared on the Bryan family's plantation at Eagle Point in nearby Gloucester County, Ran believed he was well suited for the mission. After questioning Ran closely about his knowledge of local topography and the opposing Union forces, Johnston agreed.

Only after being selected by Johnston did Ran discover that he would be required to ascend in a hot-air balloon to observe Northern troop movements. On 13 April Ran ascended above the protective cover of a forest and found himself temporarily in the danger zone between the treetops and a height at which he would be out of range of federal artillery. After reaching a safe altitude amid enemy fire and carrying out his assignment while air currents slowly spun the craft on a single tether, Ran sketched the Union positions and made a harrowing descent back through the danger zone to report his observations to Johnston. Believing that his duties had been completed, Ran asked permission to return to his unit. He was turned down on the grounds that he was now the only experienced aeronaut in Johnston's army.

Feeling complimented but not elated by the prospects of making additional airborne observations, Ran made two more ascensions, the second coming on a moonlit night just before Johnston withdrew from the Yorktown defenses. Ran quickly found himself in trouble when a soldier who was keeping the balloon tethered got caught in the ropes and a comrade cut him free with an axe. No longer secured to the ground, the balloon careened upward, and the wind carried it back and forth uncontrollably between the opposing lines. After being fired upon by Confederate forces who mistook him for a Yankee spy, Ran lost altitude as the bag cooled and came down in an orchard. Tying the craft to an apple tree, he shinnied to the ground, mounted a horse, and rode to Johnston's headquarters with information that enabled the latter to prepare for a Union attack the following morning.[31]

Ran's postwar experience was disheartening. Returning home after fighting in the battles of Cloyd Mountain, New Market, and Monocacy, he married Margaret R. Minor of Albemarle County, managed a cotton plantation in Louisiana, and then came back to Virginia, where he tried to make a living running a dairy and fruit farm near Charlottesville. He did not prosper, and eked out a precarious income by writing about his wartime experiences. Because of his deepening penury, he suffered the humiliation of having to depend on help from his rich younger brother, Joseph. "I fear that your book enterprise will not last long but you can at least make something," the latter wrote at one point, enclosing a draft for $100. "I will consider what other business you can go into—but would like to have any suggestions you may make."

Shortly thereafter, Ran departed for Birmingham, temporarily leaving Margaret, who suffered intensely from the separation. For many years he was Joseph's business agent in the Magic City, representing him in real-estate transactions and mining deals. Even though he had no formal connection with Sloss Furnaces, he was useful to Joseph as a source of information and Birmingham gossip. Eventually he moved to Tennessee to mine phosphate deposits and seems to have done relatively well until bad eyesight forced him to retire. He never came close, however, to achieving Joseph's wealth.[32]

Far more distinguished among the Virginians who came to Birmingham in the wake of the Sloss takeover was a physician and surgeon, John Randolph Page. Born at Greenway, Gloucester County, in 1830, he received a medical degree from the University of Virginia in 1850, went to Europe, and studied in Paris for several years before returning to America. After the war broke out, he helped organize hospital services for Confederate troops and was a pioneer in antisepsis; among other things, he was credited with originating the use of bichloride of mercury in treating infected gunshot wounds, and he also used "tar water," containing creosote and carbolic acid, for the same purpose. His efforts to combat contagion extended to animals; despite wartime exigencies, he ordered stables to be condemned and horses destroyed to prevent the spread of glanders. After 1865, he taught medicine in Louisiana and Maryland before becoming professor of agriculture, zoology, and botany at the

University of Virginia in 1872. Fifteen years later, he resigned and moved to Birmingham to become chief surgeon for the Georgia Pacific and Sloss Furnaces. Not long after he arrived, Bryan congratulated him for building a hospital for furnace and railroad workers.[33]

Bryan helped a host of Virginians resettle in Birmingham after the takeover. "Reggie Walker . . . about 22 years of age, is anxious to get employment in the South," he informed an official at the furnaces. "His business has been chiefly with brokers in New York, keeping stock books. If you have any place that you could give him to start from and work up, I would be very glad for you to give him a chance." Other recipients of such aid included Cassius F. Lee of Alexandria, a person of "varied business experiences" whom Bryan recommended to John W. Johnston in January 1887; J. A. Burgess, "an experienced housebuilder & reliable man," who received similar assistance at about the same time; and Dr. Edward W. Morris of Hanover County, whose social background and medical abilities were highly esteemed by Bryan.[34]

Bryan also helped Alabamians who wanted jobs with the Georgia Pacific and Sloss Furnaces. A person whom he particularly wanted to aid was a son of Richard Hooker Wilmer, the Episcopal bishop of Alabama. Bishop Wilmer himself was a good example of the Stewart family's impact on the state. Born at Alexandria and educated at Yale, he graduated from Virginia Theological Seminary in 1839 and spent eighteen years ministering throughout the Old Dominion and North Carolina before John Stewart asked him in 1858 to build a church at Brook Hill. Wilmer did so, creating Emmanuel Parish, with which Bryan and his descendants became closely identified. After spending three years at Emmanuel, Wilmer was elected to head the Alabama diocese in 1861, beginning a long episcopate that lasted almost to the end of the century. After John Stewart died, Wilmer wrote an appreciative memoir of his life and piety. Besides wanting to help such an old and distinguished family associate, Bryan had other motives for aiding Wilmer's son. "Ours is an Alabama Corporation & it is a good thing to have friends at home," he told John W. Johnston in asking him to give the matter "as favorable consideration as you can."[35]

VI

Bryan's efforts to help transplanted Virginians were only momentary episodes punctuating an endless series of business decisions in which he became enmeshed. In April 1887, he and his associates took an important step by consolidating the Coalburg Coal and Coke Company with Sloss Iron and Steel. From the outset, they were determined to get rid of their obligation to buy coal from the Pratt Company so that they could benefit, like most of their local competitors, from having independent access to this raw material.[36]

Meanwhile, Bryan's syndicate pushed forward with completion of the North Birmingham furnaces. These were designed by an experienced Pennsylvania engineering firm, Gordon, Stroebel and Laureau, which was currently building eight blast furnaces in the South. The North Birmingham stacks were to be seventy-five feet high and seventeen feet wide at the bosh, considerably larger than the old furnaces erected by Colonel Sloss. New equipment ordered by Bryan and his associates included eight Gordon-Whitwell-Cowper stoves, four blowing engines, and eight boilers. Plans called for the furnaces to be in operation by the spring of 1888.[37]

As construction of the furnaces continued, Bryan and John W. Johnston did their best to complete the Georgia Pacific's right-of-way. In 1887 the territory between Day's Gap and Coalburg was finally crossed, opening the Warrior Coal field from end to end. Construction of a branch line was also undertaken from Woodlawn to Bessemer, and surveys were begun to push westward from Columbus to the Mississippi River. Work on a 141-mile stretch from Columbus to Johnsonville finally got started on 27 July 1887.

In order to finance the North Birmingham furnaces and absorb the Coalburg Coal and Coke Company, capitalization of the Sloss Iron and Steel Company was increased to $4 million in July 1887. Of the 7,000 new shares subscribed, Bryan could take only 440, preserving his status as the single largest investor but showing how badly his resources had been strained by the recent takeover. Daniel K. Stewart took almost as many shares, 420. Pace took by far the largest single block, 1,260 shares, making him the company's second-ranking stockholder; meanwhile, John W. Johnston and his brother, Andrew, subscribed for a total of 936. Whitlock took 532, C. W. Branch and Company 266, Christian 224, and Buford 168. Maben's subscription, only 56 shares, provides further evidence that his role in the emergence of Sloss was less crucial than Armes later indicated. The predominance of Richmond investors, however, was typical of the extent to which the South bore the cost of its own postwar industrialization.[38]

The tight circumstances in which Bryan found himself by mid-1887 were related to hard times again plaguing the country. By that time, the upsurge that had fueled the recent boom in Birmingham was over, and a sharp contraction was underway. Concern about the potential impact of the Interstate Commerce Act, signed into law by Grover Cleveland on 4 February 1887, contributed to declining vigor in the ferrous metals industry; within weeks after the statute was passed, *Iron Age* noted a "halting tendency" in "nearly every department of the iron and steel trade" due to the uncertainty it had created. In late March, the same journal's Cincinnati correspondent called the law "the great bugbear of the business community, and especially of the Iron interest," complaining about its impact on railroad rates and saying, "it is apparently the policy of the railroad fraternity to make the bill as obnoxious as possible, even before it goes into effect."[39]

The new law, however, was only one factor among many that contributed to the recession. What most inhibited business, as Rendig Fels later pointed out, was a shortage of money and credit, stemming from the way in which the feverish pace of development in the preceding boom had outstripped the nation's monetary base. Comments by knowledgeable observers of the iron trade at the time confirm this analysis. "It is the well grounded conviction that the whole country has been going ahead too fast for its own good," stated an editorial in the American Iron and Steel Association's *Bulletin* in November 1887. "Level headed business men believe that this is not a healthy condition of affairs, and banking and other institutions which have control of loanable funds have grown conservative lending them." During the summer of 1887, when the new Sloss stock issue was floated, funds were harder to get than at any time since the depths of the 1884 recession. Building construction, a generally reliable indicator of economic health, was sharply down. Falling agricultural yields and an unfavorable swing in the international balance of trade further darkened the business mood at the end of the year.[40] As always, railroads were implicated in the downturn. Although construction of new track grew because of supplies already on hand, orders for rails shrank in the second and third quarters of 1887, and iron that would have been converted into steel was sold in unaccustomed markets, further depressing prices. Inventories of unsold pig iron rose from 225,629 gross tons at the beginning of the year to 301,913 at its end.[41]

Once again, the South was being hurt by impersonal market forces over which it had no control. In some respects, however, the recession was not as hard on the southern iron industry as it was on furnace operators in other parts of the country. The low cost of southern pig iron kept it in demand. The South's basic problem, as trade reports from Birmingham and Chattanooga indicated, was that the recent boom had resulted in too much construction and manufacturing activity for the region's immature infrastructure to bear.

The most obvious confirmation of this fact was the inability of the Pratt mines to supply enough coal to fuel the Birmingham District's constantly growing number of blast furnaces. Throughout 1887, coke was in desperately short supply, leading local reporters to describe the situation as a "famine." The Sloss Company's two existing furnaces, shut down for thirteen days in December 1886 for lack of fuel, were among five local installations that were hit again in February when the Pratt mines failed to provide coal they needed to stay in operation. The situation led Bryan to complain bitterly to a Baltimore firm that had invested in the company; it was particularly galling because, early in the year, the new management had accepted large orders from western buyers. Now, just as it was trying to get established, its ability to fulfill these contracts was being jeopardized by a lack of raw materials over which it had no control. On April 8, disaster struck again when operations were stopped at Pratt, this time by a fire that also destroyed some of that firm's coke ovens. By late April, the local coke shortage was still acute; as Cincinnati's *Iron Age* correspondent noted, "scarcity of coke in the Birmingham District restricts the furnaces operating there." In some cases, the shortage caused attempts in the district to smelt iron with raw coal.[42]

Compounding these difficulties was a dearth of rolling stock on southern railroads to bring raw materials to factories and furnaces and take their products to markets outside the region. Chattanooga's *Iron Age* correspondent complained about this lack of cars repeatedly, stating in September 1887 that "the inadequacy of the Southern lines of transportation to handle the business that is being offered them, and which is increasing every day, is a source of embarrassment and annoyance. There are instances where manufacturers are very much crippled in their operations for want of coal and coke through delay in transportation. It is true that some of the lines are adding considerable to their rolling stock, but the placing of 200 or 300 new cars on any one line is like placing a drop of water in a pond." Even northern rail lines had trouble finding enough cars, further restricting the export of pig iron to markets on which southern furnaces depended. In November, an analyst in Cincinnati reported that "Southern roads will not issue a through bill of lading beyond the Ohio River for fear their cars will be seized by Northern roads which will forget to return them."[43]

VII

Returning hard times caused labor disturbances that had once been rare in Alabama. The first recorded strike ever to occur in the state took place at Mobile in 1867 among black longshoremen, but its importance was so ephemeral that any trace of its outcome has been lost. But miners began attempting to organize in the Birmingham District from 1876 onward, forming clubs that espoused such nostrums as Greenbackism in response to social and economic ills. Limited initially to whites, the movement expanded as black workers established their own clubs, and interracial cooperation took place as workers became increasingly active in opposition to low wages and the use of convict labor. National ties began to develop as the Knights of Labor, which sent fifteen organizers into the South as soon as recovery was underway from the Panic of 1873, became active in the state late in the decade. Strikes broke out as early as 1878, when one took place at the Eureka Company's Helena mines; another erupted at Coketon in 1880. In both cases, operators fought back by using convict strikebreakers.

As blast furnaces and mines multiplied in the district in the early 1880s, unionism intensified. In 1882, 500 operatives struck at the Pratt mines. That same year, the Amalgamated Association of Iron and Steel Workers organized a lodge and began a series of walkouts, including one in which 837 workers took part without success against the Birmingham Rolling Mills. As strikes in mining camps became more numerous, a Miner's Union and Anti-Convict League was formed in the summer of 1885.[44]

Early in 1887, as the Sloss takeover occurred, shutdowns by organized labor threatened once again. A strike stopped work at the Eureka furnaces but ended early in February. As the economy started to cool down, walkouts began affecting iron and steel producers throughout the country. Business in the Mahoning and Shenango valleys in northeastern Ohio was virtually paralyzed for a time by striking coke workers. Pennsylvania's Connellsville district also experienced a bitter labor dispute lasting eleven weeks.

Ultimately, the spreading discontent reached Birmingham. In late June, that city's *Iron Age* correspondent reported "labor disturbances about the furnaces," asserting that these were "really not serious, being confined to workmen of a cheap class." In July, however, two of the city's largest foundries, TCI's Linn Iron Works and the Williamson facility, were closed down by a "hopeless strike" conducted by molders and machinists belonging to one of the Knights of Labor's local chapters. No prospect of a settlement between operators and workers was in sight. The walkout, which also involved workers at the Birmingham Rolling Mill, was crushed by early September but added to the sense of frustration gripping the city now that boom times had vanished.[45]

Labor leaders particularly resented the leasing of convict workers by the district's leading operators. Slavery had not died, but merely been transformed. As James C. Cobb has indicated, the industrial use of convicts became a distinguishing feature of postwar southern development.[46] It was a natural extension of the low-wage system and the region's massive commitment to servile labor, which could assume many guises.

The industrial utilization of convicts in Alabama had a long history. Prior to the Civil War, because free blacks were rare and slaves who committed unlawful acts were punished privately by their owners, virtually all criminals brought before Alabama courts were white. In 1839, reacting against the vindictive criminal code that had prevailed up to that time, reformers won passage of a law establishing a penitentiary near Wetumpka. In 1842, it received its first prisoners, who worked at such occupations as blacksmithing, coopering, and shoemaking. Within four years, however, the high cost of running the institution, which accumulated a deficit of nearly $40,000, led the state government to lease it to a private contractor who paid a small annual fee for its use and assumed all the risks inherent in operating it for profit. Although state inspectors visited the prison at intervals to see that the inmates were properly treated, critics charged that the system was morally indefensible because any concern about reforming them was lost in an overriding concern for "money, money, money," as historian Jack L. Lerner later stated. Reformers tried to abolish leasing the prison in the early 1850s, but to no avail.

During the war, after a lessee was murdered by a convict in 1862, the state resumed operating the penitentiary and actually earned profits. During Reconstruction, however, its dilapidated condition, overcrowding, a shift in the inmate population from being largely white to heavily black, and lack of state funds led to a new system under which able-bodied inmates were allowed to work for railroads and industrial firms that took responsibility for their custody and paid fees for their use. James W. Sloss and the builders of the South and North Railroad were among the contractors that benefited from this arrangement.[47]

Alabama provided by no means the worst example of mistreatment of convicts in the South during Reconstruction; as penal historian Blake Mc-

Kelvey has pointed out, conditions in such states as Florida were considerably worse.[48] Nevertheless, Alabama's system came under frequent attack by humanitarian reformers. Inspection of working conditions under the new "itinerating system" was virtually nil, and death rates at labor camps throughout the state exceeded 40 percent by 1870. Amid mounting revulsion, some convicts were brought back to the Wetumpka penitentiary, and a prison farm was established at Speigner, on the Tallapoosa River. Partly because of continued demand that the prisons produce revenue, and also because of the burgeoning ranks of black offenders, who outnumbered whites by margins as high as seven to one in the late 1870s, these were only stopgap measures. Following the restoration of home rule, Redeemer governments resorted freely to the contract system. By 1877, as many as 557 state convicts were working for thirteen companies in outside locations under tightened inspection procedures. Workers in top physical condition brought $5 per month, while second-class laborers fetched $2.50; "dead hands" were leased for the cost of their maintenance. Although mortality rates went down as income went up, inspection remained inadequate. In Lerner's words, "the convict suffered; he suffered shamefully; he suffered needlessly; he suffered in the name of punishment for the profits of men whose activities were criminal." Equally bad was the fate of county offenders, who, under laws passed in 1866, were put in the hands of petty contractors. This practice encouraged sheriffs and other local officials to arrest as many people as possible to earn the resulting fees.

Renewed demands for reform were voiced in the 1880s. After becoming warden at Wetumpka in 1881, John H. Bankhead toured fourteen labor camps throughout the state and found them "in most instances totally unfit for the purpose intended. . . . They were as filthy, as a rule, as dirt could make them, and both prisons and prisoners were infested with vermin. The bedding was totally unfit for use. I found that convicts were excessively, and in some instances, cruelly punished; that they were poorly clothed and fed; that the sick were neglected, in so much as no hospitals had been provided, they being confined in the same cells with well convicts. . . . The prisons have no adequate water supply, and I verily believe there were men in them who had not washed their faces in twelve months. . . . The system is a disgrace to the State, a reproach to the civilization and Christian sentiment of the age, and ought to be speedily abandoned."[49]

Bankhead, who was actually working in behalf of powerful vested interests, pointed out to the legislature that no inspection system could be really adequate while inmates were scattered in so many places. As a remedy, he proposed a system that played into the hands of such large corporations as TCI by advocating that able-bodied state convicts be leased to only a few contractors and the rest kept at the Speigner farm. Although not all of his ideas were accepted by political leaders in Montgomery, most state offenders were concentrated at five places by 1886, of which the Pratt Coal and Iron Company's mines were the most important. After taking over the Pratt operation,

TCI won an exclusive ten-year contract for the use of state convicts in 1888 and worked out a system of "dual management" with state officials under which the head of the board of prison inspectors stayed at Pratt City much of the time while TCI provided medical and custodial supervision. TCI also spent approximately $50,000 on new and improved prison facilities at Pratt City. Meanwhile, having lost out in bidding for state inmates, Sloss's new owners continued to use county offenders, as they had already done at Coalburg after acquiring it from Milner and Caldwell. [50]

VIII

Convict labor was one of the most powerful weapons available to Birmingham's furnace operators in their struggle to compete with ironmakers in other parts of the country. Besides the fact that the minimal payments asked by the state for able-bodied workers helped owners to cut costs, the contract system also gave managers an automatic source of strikebreakers and thus helped to hold down wages. Throughout 1887, as the national economy went into a tailspin, these advantages became increasingly vital to survival. "I think consumption will have to improve enormously to consume the product of this & next year," Bryan told Seddon. His forebodings were justified. By October a Birmingham reporter was complaining that "speculation is almost paralyzed and regular business a good deal cramped in some lines for want of money." Banks, he said, had stopped discounting notes. "Three, with aggregate capital of $1,650,000, and deposits in excess of $3,000,000, showed in their statements for October 5 a little less than $350,000 . . . in the vaults."[51]

Within two months, things were even worse. "A continued stringency of money, a painful shortage of rolling stock on most of the railroads and advances in freight rates on some of the others are evident factors of discomfort in the industrial situation of this district," the Birmingham correspondent of *Iron Age* lamented. Pig iron manufacturers had meekly accepted a recent fall in prices "without a protest," forward movement in the market had been "entirely arrested," and there had been "an entire cessation of buying for next year's delivery." Confirming the desperateness of the situation, *Iron Age* printed a harsh estimate of "Ironmaking in the Birmingham District," pointing to the leanness of local raw materials, the unreliability of the labor supply, the difficulty of getting coke, and the dismal prospects of new furnaces now getting ready to add fresh iron to an already overcrowded market. People who had bought lots and acreages on credit from the Elyton Land Company during the boom were now defaulting. Soon, its president, Henry M. Caldwell, had to make liberal extensions to forestall a general market collapse. [52]

Newly arrived enterprises that had been praised as heralds of progress only a year earlier now reeled under the impact of hard times. Continuing to display the internal bickering endemic throughout the company, TCI went through endless episodes of backstabbing that were only partly relieved by the temporary absence of Ensley, who took his wife to France after she became ill. Baxter, enraged by the unscrupulous ways in which Inman had tried to cover up consolidation costs connected with TCI's move into Alabama, bitterly attacked Inman, who in turn offended Bowron by falsely accusing him of misrepresenting earnings. Meanwhile, the value of TCI's securities plummeted. Listed at 118 in December 1886, when the boom was still going strong, they fell to 21 by September 1887, setting off a chain of events that led to Inman's fall from power within two years. [53]

TCI was not the only local firm that suffered as the economic crisis intensified. In January 1887, Joseph Forney Johnston had been in an ebullient mood about the Sloss Iron and Steel Company's prospects. Pointing to the capital that was pouring into the district, he stated that it was practically impossible to stem the flow. "Towns spring up as if by magic," he said. "Streets are thronged with investors. It would take a freight train a day to haul the money into Birmingham sent every twenty-four hours for investment, if it were put in silver dollars." By April, however, Johnston was much subdued and was having nightmares about the "seven lean kine" of biblical days. Anxiety on his part to leave the helm was already forcing Bryan and his syndicate to look for a replacement. As the year ended, the "forlorn visage of bankruptcy" haunted the company as it appeared that interest on the 6 percent bonds it had assumed during the takeover could not be paid. Unable to stand the strain, Johnston resigned to begin a political career that ultimately led to the governor's mansion in Montgomery and a seat in the United States Senate. By the time Johnston left Sloss, Jacob Reese had also suffered a series of reverses in his legal battles to control his patents. Even had he won them, it would have been financially impossible for Bryan and his associates to build a steel mill. With two new furnaces about to be completed in North Birmingham amid bleak prospects that their output could be absorbed by shrinking markets, mere survival was all that could be hoped for. [54]

The recession of 1887–88 cast a long shadow across the future of Sloss Furnaces. Local investors mourned the collapse of their dreams that steel production was about to commence, not simply because of the potential profits that went by the boards but also because they had lost the badge of status that steel production would confer upon them and their city. This, however, was not the only reason for their discontent. After Joseph Forney Johnston resigned, control of day-to-day operations passed from their hands into those of a new chief executive, Thomas Seddon. Unlike his predecessor, he did not come from Birmingham. There was no way for boosters to know that thirty-two years would pass before a local son again presided over the enterprise and that it would never make a single ounce of steel in its entire history. Such was the ultimate legacy of a most distressing year.

6   A Sea of Troubles

Bryan could be glad that courage and determination were among his chief characteristics, because he would need them as an erratic decade came to an end and the nation lurched toward one of the worst depressions in its history. Birmingham was also fortunate that its boosters believed as firmly as ever in the city's future, for this confidence, too, would be severely tested. Despite the unblinking facade that regional prophets presented to the outside world, cracks were appearing in the economic foundations of the New South. Even so, dreams for a better future persisted.

I

When it became apparent in the spring of 1887 that Joseph Forney Johnston did not want to continue as president of Sloss Iron and Steel, the person Bryan wanted most to succeed him was John W. Johnston. The latter, however, did not relish the prospect of administering both the Georgia Pacific and Sloss and was unwilling to give up his post with the railroad. Bryan's second choice was Seddon, who had gained familiarity with north-central Alabama's mineral resources through his 1883 inspection tour of the Georgia Pacific's unfinished line between Atlanta and Columbus, Mississippi. Having represented the interests of Virginians on Wall Street, he was acceptable to key Richmond stockholders including Christian, Pace, and Whitlock. He was also related to the owners of a Baltimore mercantile firm on which Bryan depended for credit.

Seddon, however, was reluctant to assume the challenge of running an enterprise as large as Sloss. As he was frank to admit, he had no experience in either ironmaking or industrial management. He was well aware that the national economy was in a downturn. As an investor in Sloss who was frequently in touch with Bryan and other Richmond capitalists through correspondence and direct personal contact, he was cognizant of the problems facing the recently acquired firm in Birmingham. Personal considerations also influenced his thinking. An acutely sensitive person, he was haunted by the childhood spinal injury that had crippled him for life. Unlike most of the Virginia businessmen with whom he was linked, he had not fought for the Confederate cause and could not share the wartime memories that bound them together. He was thus an outsider.

Perhaps worst of all, Seddon knew that he was not Bryan's first choice, because the latter bluntly told him so in a letter that must have hurt his feelings. Inviting Seddon to go to Birmingham as president of Sloss, Bryan said that he was offering him the job not only because "we must have one of ourselves in the place—a true and tried ally," but also because "there is no one else who as completely fills the requirements as yourself who can possibly be had." Occasionally, Bryan's honesty exceeded the bounds of good judgment. For good measure, he informed Seddon that "it would surely be a good change for you to leave Wall St & engage in *legitimate* work." This may have been meant as a good-natured joke, but it was not the best way to deal with a person as sensitive as Seddon.[1]

Nevertheless, as things went from bad to worse in Birmingham and Joseph Forney Johnston forced the issue by resigning, Seddon was persuaded to reconsider. Agreeing to take command of Sloss on 17 December, 1887, he went to Alabama with many reservations. As he candidly stated early in 1888 in his first report to the directors, he took the job "entirely ignorant of the business, and not fitted for the duties by previous training or experience." Richard Hathaway Edmonds was only exaggerating the truth when he stated that Seddon "scarcely knew a piece of pig iron from a lump of coal" at the time he agreed to go to Jones Valley.[2]

However unfortunate his condescending attitude toward Seddon may have been, Bryan had shown characteristic wisdom in preferring John W. Johnston over the younger Virginian. Both because he had exercised military command and been in charge of a succession of complex undertakings, particularly the Extension Company, Johnston would have brought a wealth of experience to Sloss Iron and Steel. Neither running a Louisiana sugar plantation nor working in a New York City brokerage gave Seddon much of a background for managing a large industrial enterprise that was in deep trouble. Bryan's underlying doubts about Seddon also meant that the latter would not be allowed to run Sloss with the relatively free rein that Johnston, whom Bryan regarded as a peer, might have enjoyed. The fact that Johnston was based in Birmingham, where the Georgia Pacific was now headquartered, also meant that Seddon constantly had an older and much more experienced manager, who furthermore happened to be deeply involved in the same industrial empire, looking over his shoulder. When their judgment differed, Bryan could be expected to side with Johnston. Repeatedly, this turned out to be the case.

Thomas Seddon. President of the Sloss Iron and Steel Company from 1888 until his death in 1896, Thomas Seddon suffered from a spinal injury whose effects are indicated in this photograph taken while he was a student at the University of Virginia. Seddon, the son of a Confederate cabinet officer, was similar to Bryan's other associates in being rooted in Virginia's prewar plantation aristocracy. As the photograph reveals, Seddon's habitual jocularity masked a deep sense of inner loneliness. His badly strained relationship with Bryan, who patronized and sometimes underestimated him, helped make Seddon's long tenure in Birmingham one of the most troubled eras in the history of Sloss Furnaces. (Courtesy of Virginia Historical Society, Richmond, Virginia.)

Seddon's situation was complicated because, by 1887, separation of ownership and control was extremely uncommon in the management of industrial firms and was likely to be found only in the administration of long trunk-line railroads. In his capacity as principal owner, Bryan would not have adopted a hands-off policy toward developments in Birmingham even if Johnston had been in charge of Sloss, but he was certain to interfere constantly with a person whom he held in lower esteem. Bryan's letters in the winter of 1886–87 indicate clearly that he regarded the Sloss Iron and Steel Company as his best hope of winning wealth independently from what he had gained through his marriage to Belle Stewart. It must have grated on his proud nature to know how much of his power came from this circumstance. His father, John Randolph Bryan, Sr., was spending his old age living off his sons because his entire fortune had been wiped out by the war, and two of his brothers, Saint and Ran, were not prospering. Deriving wealth from Sloss became a lifelong quest for Bryan, and it was unthinkable that he would play a passive role in its management. Other directors, most notably Christian, were also extremely strong-minded men. Lacking experience, with Johnston watching him in Birmingham and powerful Richmond investors trying to run things from afar, Seddon was in an unenviable situation.

Managing a big company like Sloss was a tough job. Up to the time of the Civil War, most manufacturing enterprises in America were relatively small, even in the North. As Louis Galambos and Joseph Pratt have stated, they were usually run by "owner-innovators" who oversaw the work force, understood the technology that was used, raised their own capital and credit, purchased raw materials, and marketed their own products. Such enterprises remained numerous after the war but did not have enough capital or administrative know-how to achieve efficient large-scale production and distribution in the merciless, impersonal market economy that the railroads had created. Unfamiliar with any but the most rudimentary accounting procedures, their managers could neither accurately compute unit costs nor control the sources from which they obtained raw materials. Frequently they did not understand new technologies affecting the businesses in which they were engaged. As Galambos and Pratt have observed, such individuals had trouble trying to administer "centralized corporate combines" in which functions that had once been performed by a managerial jack-of-all-trades now had to be divided among persons with specialized expertise; new balances had to be struck in making decisions involving the coordination of raw materials, production, and marketing; and the relationship between an enterprise and its external environment was shifting under changing political, economic, and social conditions.[3]

In choosing Seddon to take charge of Sloss, Bryan apparently felt that the young Virginian's previous experience in running a Louisiana sugar plantation would be relevant to the task of administering an ironworks. In his letters, Bryan repeatedly used plantation analogies in discussing problems that occurred at Sloss. This comparison was not unusual, considering that many American blast furnaces, including those scattered up and down Virginia's Blue Ridge, had still been of the rural "iron plantation" type up to the time of the Civil War and that southern furnaces had been staffed largely by slaves. Far from thinking that large-scale agriculture and manufacturing were dissimilar Bryan saw them as being much alike; to him, Birmingham was simply an urban industrial plantation. Nonetheless, the realities that Galambos and Pratt have analyzed bore heavily upon a person with Seddon's severely limited managerial experience.

Seddon's problems were compounded by his arriving at a bad time. On reaching Birmingham, he found himelf at the head of a demoralized enterprise teetering on the edge of bankruptcy. Nearly $250,000 in unpaid obligations had to be met immediately, and there was no cash on hand with which to settle them. Despite all the time and money that had been spent during the past few years by Colonel Sloss in modifying and repairing the two existing furnaces—Bryan referred to them contemptuously as "old worn-out rattle traps"—they were still not in good working order. Because of its poor condition, the older of the two furnaces had been out of commission for ninety-five days during the previous year; early in January 1888, Seddon had to shut it down again for stripping and relining. With the North Birmingham stacks as yet unfinished, this action left only the second furnace still in blast. It too was in bad shape, because its lining had deteriorated through constant operation since October, 1886. Given the sorry state to which the recession had reduced markets for pig iron, which now sold for about $14 per ton; the undependability of workers; the difficulty of securing an adequate coke supply under the Pratt contract; and the chronic problems that resulted from lean raw materials, it was no wonder that Joseph Forney Johnston had resigned in despair.[4]

Needing cash in a hurry to meet $342,000 in obligations due by the end of January, Seddon sold every last remaining ton of iron on hand for whatever it would bring. This sale, however, was only a temporary expedient, inadequate to meet any but the most pressing immediate needs. As Seddon stated in his first report to the directors, the company was virtually out of money and was desperate for working capital. The original stock subscriptions in the fall and winter of 1886–87 had sufficed merely to assume ownership, and the proceeds of the supplementary issue in July 1887 had been quickly soaked up in meeting day-to-day expenses and building the North Birmingham furnaces. Like Old Mother Hubbard's cupboard, the coffers were bare. Meanwhile, Seddon was so immersed in taking care of emergencies—"everything bad that seemed possible, has happened," he lamented—that he had little opportunity to do anything else. "My time has been so occupied in financiering and trying to get things straight, which should never have been crooked," he complained, "that it has been impossible for me to study and look closely after the current business or physical management."[5] Both for him and his backers in Richmond and New York, disaster loomed.

II

Fortunately, long-range financial planning was in good hands. Mustering his legal acumen and fund-raising expertise, Bryan effected a major transfusion of cash, capitalizing upon the company's only negotiable asset: the further mortgageability of its facilities, equipment, and lands. While T. P. Branch, Buford, Christian, Trigg, and Whitlock provided temporary relief with short-term loans, Bryan got the firm's existing capital stock and other financial obligations consolidated into $2 million in first mortgage bonds and approximately $1.7 million in second mortgage bonds. The entire issue was underwritten by the Central Trust Company, at 7 percent interest per year. In the process, Bryan renegotiated the existing Coalburg mortgage, yielding a windfall of $275,000 for general purposes. "Maben will oppose this on the ground that the company is not to be trusted with so much money," Bryan informed Seddon. "He might as well say, I think, that a carpenter ought not to use sharp tools for fear he will cut himself."[6]

Bryan's plan, which was approved by the directors in March 1888, left the company saddled with heavy fixed expenses in the form of interest payments on its bonded indebtedness but nevertheless enabled it to proceed on a reasonably sound financial footing. Richmonders and their New York allies took up most of the bonds; as always, Goadby and Mortimer provided help from Wall Street, while subscribers in the Virginia capital included such familiar figures as C. W. and Thomas P. Branch, A. H. and E. D. Christian, Dooley, Pace, Trigg, and Whitlock. Bryan held the largest block of consolidated bonds, $171,000. The next biggest holdings were those of Mortimer and Goadby with $100,000 and $82,000 respectively, which in themselves showed an increased reliance on northern capital. Pace, however, was not far behind with $81,000. Birmingham investors also took part; the largest stakes were those of J. B. Boddie ($80,000), Hardy & Company ($38,000), M. G. Hudson ($40,000), E. W. Rucker ($44,000), and James W. Sloss ($72,000). Joseph Forney Johnston, however, virtually dropped out of the picture, subscribing only $2,000.[7]

While Bryan raised fresh financial support, Seddon went through seemingly endless nightmares in Birmingham. The necessity of shutting down the older of his two stacks for a complete overhaul left him with "only one broken-down furnace in blast." Plagued by high costs due to its dilapidated condition, this installation remained operational for only a few months until it, too, had to be shut down for repairs in May 1888, leaving the firm with no iron-producing capacity whatsoever until it returned to blast in July. In early October, however, it broke down again. Fortunately, by this time overhaul of the original furnace had been completed, and production continued uninterrupted. During the fiscal year, however, Seddon got only 439 days of work out of the two stacks. Because of this, pig iron output, only 43,339 tons during the previous year, slipped even further, to 41,820.[8]

Even more disillusioning was the performance of the new North Birmingham furnaces, the first of which went into blast on 18 October 1888. From the beginning, it failed to function properly. According to Seddon, this was at least partly due to the problem of securing an adequate labor supply in a district in which too many new installations were competing for too few skilled workers. "We had to start in a short time, a very large plant, literally in the woods, and it was impossible to get trained men," he stated. "We had to take green men and teach them the business."[9] According to an experienced engineer who later spent much time refitting the North Birmingham plant, the problem actually went deeper. "Although modern in design," he said, the installation "was in many parts built of inferior material and shabby workmanship."[10] It is not clear just why this was the case, but it is possible that Gordon, Stroebel and Laureau, which was building so many blast furnaces in Alabama in the 1880s, had its hands too full of business to do fully satisfactory work for southern entrepreneurs.

After the second of the new furnaces went into operation on 11 February 1889, defects quickly became apparent. The five Babcock and Wilcox water-tube boilers with which they were equipped did not provide adequate steam for full operation, and the water fed to them from a local spring was so hard that their interior surfaces soon became badly scaled with mineral deposits. Attempts to solve the problem by adding two new boilers of a return tubular design were fruitless because they were erected on an ash dump containing unburned coal. When this was ignited by the heat of the boiler furnace, the foundation settled and the boilers became unusable, requiring more modifications.[11] With no previous industrial experience to guide him, Seddon was being badly served by the technical personnel available to him.

Seddon's headaches were multiplied by inadequate control over raw materials. Still bound by the contract with the Pratt mines that had been inherited from the Sloss Furnace Company, he found this arrangement undependable because TCI, which now owned the mines, looked to its own needs first. Partly to deal with this problem, and also to get coal and coke at the lowest possible cost, Seddon tried to develop the Coalburg mines as rapidly as possible. As their production expanded, his desire to use their output conflicted with the Pratt contract. Pending its expiration, he had to sell Coalburg coal and coke to other local furnaces, while continuing to use TCI as a costlier and less reliable source of supply.

A strike at the Pratt mines in the spring of 1888 gave Seddon an opportunity to slip out of TCI's clutches, leading quickly to a situation that might have been comic had it not been so serious. To safeguard its own needs, TCI temporarily interrupted shipping coal to Sloss. Claiming that this break abrogated the contract, Seddon stopped payments to TCI and began using coal and coke from Coalburg. After settling the strike, however, TCI resumed shipments to Sloss and secured an injunction ordering that company to continue using Pratt coal. Seddon countered by taking the coal, coking it in his own ovens, and using it in Sloss's blast furnaces but refusing to pay for it on the grounds that he did not need or want it in the first place and was using it only under duress. Meanwhile, he continued to claim that the existing contract was void because TCI had violated its agreement by interrupting deliveries. The impasse was finally resolved when TCI compromised. Until the contract expired, TCI agreed to furnish coal for only one of Sloss's two furnaces, permitting Seddon to begin phasing in his own internal sources of supply for the other. [12]

Switching to company-owned mines, however, did not assure Sloss a cheap flow of inexpensive coal and coke, because the firm was no more immune from strikes than was TCI. Late in May 1888, while the contract with TCI was still in force, miners at Coalburg demanded higher wages and walked off the job. Never prone to sympathize with malcontents, and stating that the strikers "were acting with their usual folly in pressing an advance at the very time that was dullest and your output more than you could sell," Bryan told Seddon to stand firm. Nevertheless, after a month the company had to accept a new rate of fifty-five cents per ton, ten cents higher than workers received at the Pratt mines. Sloss also agreed to pay a sliding scale for future production, based on fluctuations in the selling price of iron. [13]

Seddon's growing use of coal and coke from Coalburg also resulted in a dispute with John W. Johnston, who wanted what Seddon considered excessively high rates to ship fuel to Birmingham in the Georgia Pacific's cars. Considering the depth of his friendship with Johnston and his own identification with railroading, it is not surprising that Bryan sided with the Georgia Pacific. "I . . . agree with you entirely that our Sloss president is forcing the G. P. to a stand which will only be injurious to both companies," Bryan told Maben. "It is much to be regretted that he fails to understand the sincerity and intelligence of the G. P. management." Trying to keep peace, Bryan arranged a compromise fixing the rates at 20 cents per ton of coal and 22½ cents per ton of coke. Clearly, however, his relations with Seddon were deteriorating. [14]

This situation did not improve. To Bryan's displeasure, Seddon now began copying TCI's own tactics, exposing himself to charges of hypocrisy at the very time that Sloss was trying to break free from legal commitments to the larger firm. As Sloss began shifting to its own coal and coke, Seddon tried to wriggle out of an existing contract under which Coalburg supplied these raw materials to the Williamson furnace, which, as previously noted, did not own its own sources of supply. Displaying a characteristic penchant for moral indignation, Bryan was aghast. "A contract made by so responsible a corporation as ours is, means something more than the mere liability to damages obtained at the end of a law suit," he fulminated, "and whether we make iron or don't make it, if we are able to fulfill our contract with the Williamsons I shall vote to do it." If the behavior Seddon suggested was "a standard of commercial integrity in that region of the country," Bryan admonished, "the sooner it is improved the better." [15]

III

Sloss's contract with TCI soon expired, and Seddon made additional progress toward regularizing its supply of raw materials by developing a growing chain of mining camps in the Jefferson County hinterlands. But the squabbles that had taken place between Bryan and Seddon before the situation was resolved reflected mounting frustration, not only in Birmingham but also in Richmond as the anticipated fruits of the 1887 takeover failed to materialize. Investors were starting to abandon what they plainly regarded as a sinking ship. "Whitlock has nearly sold out his Sloss and I believe is bent on closing out every interest that he has in and around Birmingham," Bryan told Seddon in April 1888. Soon thereafter, he confided to John W. Johnston that "there has been a considerable collapse in the enthusiasm of our stockholders" and complained about fault-finding among investors in Richmond. Things worsened as A. S. Buford sold most of his stock in October 1888; this sale was soon followed by a particularly serious withdrawal when Pace liquidated his substantial holdings, all of which had been sold by February 1889. Many of his shares were bought by Christian and Dooley, who remained faithful to the venture.[16]

Trying to bolster waning enthusiasm among stockholders, Bryan struggled to keep afloat in his own sea of troubles. As always, he was deeply enmeshed in the internal politics of the Danville and Terminal companies, which were becoming increasingly chaotic and bitter. The Terminal system had recently gone through an expansionist phase in which two large railroad lines, the Central of Georgia and the East Tennessee, Virginia & Georgia, had come under its control. In the process, the Terminal had become more than ever a football for the speculative schemes of Wall Street capitalists. At least one of Bryan's allies was among the worst offenders; both now and later, William Clyde had no scruples about manipulating Terminal securities upward or downward when it suited his purposes to do so. Through the continuing power of such Richmond investors as Bryan, Logan, and Pace, the Georgia Pacific was run as a separate operation and was thus to some extent shielded from outside interference, but this situation too was subject to change. The line's autonomy became increasingly tenuous after April 1888, when Inman became president of the Terminal Company after months of deepening involvement in its affairs.[17]

Under these developing storm clouds, Bryan and Johnston spent 1888 trying to complete construction of the Georgia Pacific. Now that the trackage had been laid through the Warrior coal field, they turned their attention to extending the line westward all the way to the Mississippi River. Progress, however, was painfully slow as communities near the projected right-of-way in north-central Mississippi struggled to be included on the route and beggar their neighbors by having nearby towns and villages excluded. Johnston was deluged in 1888 by letters from irate residents of Starkville and Ackerman, protesting a decision by the company that would leave them off the route. Further strife resulted from plans to put the road through Clay and Webster counties, leaving out most of Lowndes and all of Oktibbeha and Choctaw. Later that summer, the building of drawbridges over the Tombigbee, Yazoo, and Sunflower rivers was held up by torrential rains, and a yellow fever epidemic forced contractors to suspend work. Quarantines in August and September interfered with freight and passenger operations over existing trackage. Meanwhile, annual earnings fell to $326,525, down substantially from $396,377 in the previous fiscal year. To make things worse, Bryan and Johnston had only forty miles of new track to show for their efforts.[18]

The strain of constant stress and overwork on both Bryan and Johnston was overwhelming. Taxing his financial and physical resources to the utmost, Bryan was continually engaged in securing loans, arranging mortgages and bond issues, financing equipment trusts, and drawing up legal documents. Inevitably, his health deteriorated. Writing to Maben in April 1888 while finalizing the issuance of second mortgage bonds for the Georgia Pacific and trying to determine where best to locate trackage between Waverly and Bexar, Mississippi, he complained of being "so sick at the Board meeting yesterday" that he was unable to bring up everything needing attention. "I was so ill with one of my raging headaches," he wrote Johnston later the same day, "that I had to leave the Board rather abruptly and go back to the Astor House for such relief as I could get in bed." Still, the press of business could not wait. Before returning to Richmond that same evening, Bryan "received and signed a contract with the Bethlehem Company for 12,000 tons of rail." Meanwhile, Johnston was suffering from recurrent bronchitis.[19]

IV

Fortunately, the national economy began to revive before 1888 was over. During the summer months, markets for pig iron started to pick up in the Midwest, where pipe foundries and rolling mills bought heavily after an anticipated strike failed to materialize. Optimism in Jones Valley was also stimulated by the launching, on 9 June, of an impressive new vessel, the *City of Birmingham*, by the Roach shipyards at Chester, Pennsylvania. Schooner-rigged, with auxiliary triple expansion steam engines that could reach twelve knots per hour at sea, she was approximately 320 feet long by 42 feet abeam and cost $300,000. Designed strictly for freight, with no passenger accommodations whatever, she could carry up to 7,000 bales of cotton or 2,400 tons of pig iron. Soon, she would carry the southern war against northern anthracite furnaces to a higher pitch of intensity.

Henry M. Caldwell, present for the vessel's launching in his capacity as president of the Elyton Land Company, displayed characteristic boosterism in paying tribute to Birmingham's progress and future prospects. When he had moved there twelve years ago, he said, the community had scarcely 2,000 people and did not make a single ton of iron with coke. Now, the district had a population of 50,000, taxable property of $40 million, and twenty-two blast furnaces with a daily capacity of 2,500 tons. "If this much has been accomplished in the last twelve years," Caldwell said, "what imagination can picture the proportions it may assume in the next fifteen years?" Pointing to the *City of Birmingham*, he declared that "even this leviathan of the deep will not be able to do the business for which she was built. At that time it will require one vessel like her leaving Savannah every day in the year to carry the pig iron produced in the Birmingham District. Soon you will have to build her consorts, for the resources of the District upon which she depends for her trade are as boundless as the ocean upon which she floats, and their development as ceaseless as the ebb and flow of its tides."[20]

In northern ports, agents for anthracite furnaces were bracing for a further southern onslaught. Reporting a shipment of 5,000 tons of pig iron from Virginia, a Philadelphia correspondent of *Iron Age* asked, "Can cost be reduced so as to enable the Lehigh and Schuylkill furnaces to retain their supremacy or shall they stand idle while others take their trade?" Not long afterward, the same writer called for threatened Pennsylvania furnacemen to make whatever sacrifices were necessary to meet southern competition. "Sooner or later it will have to be done, and the sooner the better," he stated. "Prompt action is necessary, as the case is serious." Supporting the southern attack, the Central Railroad of Georgia, the Ocean Steamship Company, and other lines tied to the region's interests lowered their rates and expedited shipments. Meanwhile, northern agents of southern firms ridiculed attempts to stop the impending juggernaut. Referring to recent price cuts made by the Thomas Company, whose president was preparing to flee to Alabama, a spokesman representing a New York City mercantile firm predicted that "the desperate effort to 'shut out the South' will not avail, for the great ironmasters of that region, with their modern furnaces of immense capacity, can lay their iron on dock in New York at less than its cost of production in the Lehigh or Schuylkill valleys and still have a margin of profit."[21]

Late frosts, good harvests, and heavy sales of agricultural implements buoyed the market later in the year, and southern iron began finding increased markets in Ohio, Indiana, and Illinois. As winter approached, demand rose along the East Coast as imports of pig iron from abroad declined; British producers were enjoying a domestic price rise and could get more for their output at home than they could earn by shipping it to America. A sharp contraction set in after the start of the New Year, triggered in part by a continuing decline in steel rail production, which had shrunk by approximately 675,000 tons in 1888 despite overall gains in other areas of business. But demand from this sector recovered later in 1889, and for the second straight year buoyant home markets in Great Britain, especially in shipbuilding, helped keep foreign iron out of American ports. By year's end, despite the fact that more and more southern furnaces were coming on line, pig iron was selling for about $2 per ton higher than it had twelve months earlier.[22]

To Bryan and his beleaguered colleagues, the improved business conditions were providential. Although the North Birmingham furnaces continued to operate far below rated capacity, their combined production of 46,771 tons in fiscal 1889–90 was substantially more than the company's entire output for the previous twelve months. Meanwhile, the older pair of stacks that Colonel Sloss had erected—now formally identified as the City Furnaces—yielded approximately 62,561 tons. All told, production for the fiscal year climbed to 109,332 tons, a gain of 67,511 over 1888–89.[23]

Efforts to control costs also seemed to be succeeding. Although the cost per ton of iron made at North Birmingham, $10.60, was still markedly higher than that achieved by the two City installations, $9.12, both pairs of furnaces were operating more efficiently than they had done during the last fiscal year. This improvement was partly because the company had now broken completely free from its obligation to purchase coal from the Pratt mines, which was finally ended by mutual agreement in November 1888.

Buoyed by brightening circumstances, Bryan and his partners continued to pursue the developmental strategy to which they were committed. To maximize the usefulness of Sloss's resources and advise the owners about their future exploitation, Ruffner, who had bought a small stake in the company, returned for further prospecting in Alabama. While president, Joseph Forney Johnston had acquired 15,000 additional acres of mineral lands, and the directors wanted to determine their true value. "I would be glad," Bryan wrote Seddon as Ruffner's work proceeded, "if you could give me from time to time some brief statement of the results of his examination. Some, I am sure, must be disappointing, while others no doubt confirm our expectations, and for one I shall be entirely satisfied if just one-half of all we have heard of the mineral value of the Sloss properties turns out to be true." After Ruffner had finished his work, Seddon recommended the purchase of another 800 acres of coal lands to consolidate the company's properties and augment its existing resources.

During 1888 and 1889, Seddon undertook various projects aimed at improving and expanding operations at the coal and ore mines. These facilities were of two types: drift mines, which were driven laterally into a seam, and slope mines, which penetrated a seam at an angle and required complex hoisting apparatus. Under Tutwiler's direction, two new drift mines were opened on Drinking Branch at Coalburg, expanding the number in operation there to five, with a total output approaching 2,000 tons per day. A pair of additional drift mines were opened at Brookside, where the Warrior coal seam was thicker than at Coalburg. Seddon hoped to raise output at Brookside from an 1888 level of approximately 772 tons per day to more than 4,000 tons per day, or 1,254,300 tons per year. Approximately 100 new beehive coke ovens were constructed, sixty-three of which were a new design permitting coke to be automatically withdrawn by machine and loaded into railroad cars without human aid. They were built under direction of their inventor, Richard

Thomas, who believed they would decrease costs by half. All told, coke-making capacity at the various facilities along Five Mile Creek rose to 236,100 tons per year.[24]

During the late 1880s and early 1890s, Bryan and Seddon would devote much attention to the Brookside mine, showing that southern owners and managers were willing and able to engage in a considerable amount of mechanization if this approach was necessary and the expense could be justified. The coal deposits at Brookside were embedded in fire clay that was too hard for easy hand picking and in pyrites that impeded the saw-like operation of conventional chain-driven cutters. Overcoming such obstacles by using advanced mechanical apparatus made sense because the deposits at Brookside were sufficiently extensive to be mined for two or three decades. By contrast with other company-owned workings, where traditional pick-and-shovel methods were used to dig coal that was hand loaded onto mule-drawn tramcars, the Brookside mine was equipped with Ingersoll-Sergeant machines that used compressed air to drive reciprocal pick-like cutting heads into a seam repeatedly and with great force. Coal was brought out of the mine by an air-powered "mole engine," built in Pittsburgh, that was capable of pulling twenty cars. The mine was ventilated by a Crawford and McCrimmen fan, twelve feet in diameter and modern by all prevailing standards. Outside the entrance was an up-to-date engine house, a large tipple, and, ultimately, an advanced mechanical coal washer of an English design that had been adapted to local conditions by TCI's leading engineer, Erskine Ramsay, who had come to Alabama from Pennsylvania's famed Connellsville coal field. As Jack R. Bergstresser has stated, "Brookside would have compared favorably with any mechanized drift mine in the United States."[25] Although an operation like this was untypical of the Birmingham District as a whole, it made sense because of local geological conditions and the degree to which Sloss Iron and Steel depended on selling coal for its survival. In such cases, selective mechanization was used within a larger context of labor-intensive technology.

Meanwhile, a new slope was opened at the Sloss ore mines near Bessemer, which produced 136,351 tons of red ore in fiscal 1889. Along with output from the Irondale mines, this brought the company's annual ore production to 225,168 tons. During the same twelve months, ending 1 February 1890, coal production rose from 240,960 to 451,723 tons, and coke tonnage climbed from 57,402 to 138,281. Such developments typified what was taking place in Alabama's coal industry generally as production rose from 1,568,000 to 5,529,000 tons between 1883 and 1893. Seddon spent $239,749 on new land purchases and physical improvements to existing ore and coal mining operations in fiscal 1889 alone. His efforts seemed to be paying off. One sign of improvement in the company's finances was an increase in gross earnings from coal, which rose from $11,198 in fiscal 1888 to $102,581 for the next twelve months.[26]

Things were also going better for the Georgia Pacific. By May 1889, operations in the Mississippi Division had been extended from West Point to

Malbin, 54 miles west of Columbus. Grading and track laying proceeded rapidly across the flat Mississippi delta. On 17 June 1889, the last spike was driven as the line reached its western terminus at Greenville, permitting 1,110 miles of uninterrupted travel via the Danville system from Washington, D.C., to the Mississippi River. Negotiations with the Louisville, New Orleans, and Texas Railway led to the use of that line's trackage from Greenville to Huntington, on the Arkansas side, where connections were made with the Missouri Pacific.[27]

V

The familiar pattern of boom and bust continued to hold true, and the good times could not last indefinitely. For Bryan, personal losses were interwoven with his business problems. On 11 August 1889, Daniel K. Stewart died after a lingering illness that had kept family members alternately hoping and despairing for weeks on end. His passing increased Bryan's economic power, because the latter now became executor of Stewart's estate. Still, the old financier would be greatly missed.[28]

Joseph and Belle had hardly had time to recover from their grief before Johnston, worn out by pushing the Georgia Pacific to completion, resigned as chief executive of the railway in November and stepped all the way down to a fourth vice presidency. After doing everything he could think of to persuade Johnston to stay in office, Bryan had no choice but to acquiesce and to assume the presidency of the line to protect both his own stake and the investments of his friends. This position added greatly to his responsibilities, but whether it increased his power was another matter. Although the Georgia Pacific continued to have its own officers and directors, it was now being operated by the Danville under a twenty-year lease. At the time this instrument had been executed, George S. Scott, a friend of Bryan and Johnston, was still president of the Danville. In December 1889, however, Scott yielded his post to Inman, who wanted to extend trackage west of the Mississippi despite the opposition that this expansion would provoke from such powerful magnates as Collis Huntington. Bryan, who did not share Gordon's dream of a transcontinental line, wanted no part of this idea, wanting instead to develop the economic potential of the areas already served by the Georgia Pacific. Once more, conflict was brewing between speculative and developmental goals.[29]

This threat became a reality in 1890. As the year began, Bryan was locked in combat with Inman over the latter's expansionist policies and was feeling increasingly deserted by friends upon whom he had previously depended. Having accepted the presidency of the Georgia Pacific only under duress, he wanted Johnston to remain as actively involved as possible in its management. Johnston, however, was full of foreboding about how things would go under Inman's control and wanted to get out entirely. After some behind-the-scenes

Richmond & West Point Railroad System in 1890, Showing Constituent Rail Lines.
Note particularly the route of the Georgia Pacific Railway, connecting Atlanta and
the Mississippi River via Birmingham. The size and complexity of the sprawling
network typified the often-reckless expansionism that was prevalent throughout much
of the late nineteenth century, resulting in shaky enterprises that were highly
vulnerable to sudden shifts in the business cycle. After going bankrupt in 1892, the
system shown above was reorganized as the Southern Railway under the aegis of J. P.
Morgan. (Originally published in *Investor's Supplement, Commercial and Financial
Chronicle*, March 1890.)

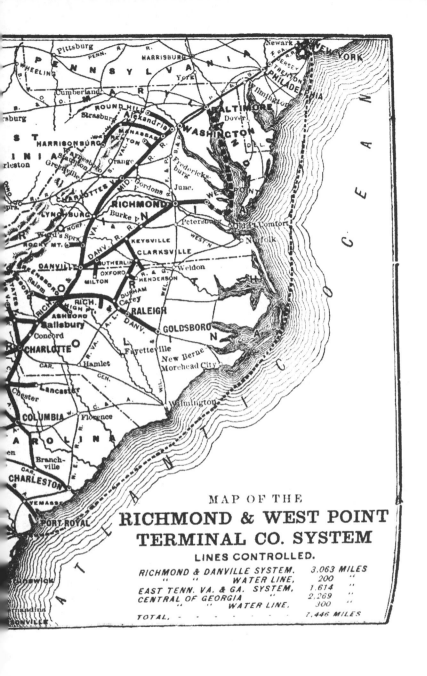

MAP OF THE

RICHMOND & WEST POINT
TERMINAL CO. SYSTEM

LINES CONTROLLED.

RICHMOND & DANVILLE SYSTEM.	3,063	MILES
" " WATER LINE,	200	"
EAST TENN. VA. & GA. SYSTEM,	1,614	"
CENTRAL OF GEORGIA "	2,269	"
" " WATER LINE.	300	"
TOTAL, - - - - - - -	7,446	MILES

maneuvering by Bryan, he halfheartedly agreed to stay on as third vice president and resume limited activity if his health improved during a three-month leave of absence. After a short rest at a New York City hospital, he sailed for France in February 1890. Bryan, who saw him off, told his brother Ran that "the old man's heart was very heavy and he took it out in attacking the French violently for their pronunciation, to which he said he would never give up, but would pronounce words only as they were spelt." The vacation did not change Johnston's mind about wanting to be free from his railroad responsibilities. To Bryan's regret, he retired shortly after returning home from Europe. [30]

Such developments typified what was happening to an enterprise that had once been dominated by development-minded Virginians and their New York allies but was now facing increasing interference from Inman. Only a few weeks after Johnston bowed out, I. Y. Sage, another Bryan associate who had long been general superintendent of the Georgia Pacific, announced that he too would retire after having threatened for months to leave the company. Late in July, he left Birmingham after a tearful sendoff in which a throng of friends gave him a gold watch. [31] Increasingly, allies like Johnston and Scott were leaving the board of directors while the line became dominated by Inman's friends, including Ohio speculator Calvin S. Brice and George I. Seney, a New York banker who was known for his stock-watering propensities and his tendency to play games with other people's money while carefully protecting his own. "The fundamental difficulty about the Georgia Pacific Company," Bryan told Johnston, "is that its destinies are controlled by men who have no substantial interest in the property itself." [32] Nor could Bryan ignore Inman's deep involvement in one of Sloss's chief competitors, TCI. More and more, he was swimming with sharks, not friends.

Under these circumstances, Bryan became increasingly frustrated in trying to pursue developmental goals. In the interest of stimulating traffic on the Georgia Pacific, he worked tirelessly to secure backing from two Baltimore firms for an issue of $1.5 million in equipment bonds that would permit purchase of badly needed additional locomotives and rolling stock. He succeeded, only to have Inman veto the deal. Furious, Bryan told Inman that he was "ashamed of being the President of the Georgia Pacific Railway." But for his proprietary interests, he told a friend in Baltimore, he would "immediately resign from any participation in the management." Because of the support he received from such allies as Oakman, who was still active in Danville and Terminal affairs, Bryan finally managed to secure an equipment bond issue of $1 million. He was glad to see that the Georgia Pacific was building new repair shops in Birmingham, also furthering developmental aims. Profits, however, were down, and fresh setbacks took place as a rash of accidents plagued the rail line, at least one of which seemed due to malicious mischief. As if this were not enough, spring floods along the Mississippi burst through levees protecting the right-of-way, inundating trackage and delaying planting of the annual cotton crop. Mother Nature herself seemed to have turned against Bryan, leaving him only headaches and frustrations in return for all his efforts.[33]

VI

Meanwhile, the Sloss Iron and Steel Company became embroiled in crisis. At Brookside, the new Thomas coke ovens were a disappointment, yielding no advantages over conventional beehive ovens.[34] Worse yet, the North Birmingham furnaces, already a persistent source of difficulty, became virtually inoperable. Together, they produced less than 24,000 tons during the entire 1890–91 fiscal year, and the cost per ton of what little pig iron they yielded was more than $13. In view of the fact that the average price received by the company had dipped to approximately $11.48 per ton, this outlay created an impossible situation. If something could not be done to bring expenses down, the company would have to keep the stacks out of blast. Because of the heavy investments that had already been made in their construction and repair, however, such an outcome was unthinkable.[35]

By March 1890, drastic action was required. Desperate, Bryan, Christian, and Maben turned for help to Henry Morton, a distinguished engineering educator who had recently become an investor and director in Sloss. Morton's credentials offered hope that he might find some way out of the company's dire straits. Born in Philadelphia in 1836, he had been secretary of the prestigious Franklin Institute, had made the first successful photographs of a solar eclipse, was a member of the National Academy of Sciences, and had been president of Stevens Institute of Technology in Hoboken, New Jersey, since its founding in 1870. A long-time friendship with Andrew Carnegie intensified his interest in ferrous metals.[36]

On Morton's recommendation, Bryan and his associates commissioned William W. Dashiell, a Stevens alumnus who had headed the Red Star Steamship Company's repair shops in New Jersey and served as an engineer for Standard Oil, to investigate the debacle that had taken place in Alabama and report his findings. Unfortunately, the document that Dashiell produced has been lost. The only surviving evidence about its contents is a statement in unpublished reminiscences by Edward A. Uehling, a Stevens graduate and Morton protégé who also became associated with Sloss at this time. According to Uehling, the report "was quite favorable as to the possibilities of the company's properties, if properly managed." The final three words, however, show clearly that the document was critical about the way Seddon was running things. This criticism was also apparent from a letter in which Morton told Uehling that Sloss's management "has proved very far from satisfactory and a change will very probably be made within a few months." Confident that the company could be a "bonanza" if its present "very bad management" were changed, Morton planned to put such protégés as Dashiell and Uehling in "entire charge of the property in its various branches," converting it into "a model iron-producing concern."[37]

The assertive spirit of Morton's letter, lacking in the passive role played by such northern investors as Goadby and Mortimer, was in itself a new development with which Bryan and his Virginia associates would have to reckon in the future. Southern control of Sloss might be lost if Morton were permitted to go too far. For the time being, however, Bryan had decided that the company was not being "properly managed" and that he had no choice but to act, regardless of the consequences. From July 1888 onward, his letters showed his increasing frustration with developments in Birmingham. He was apparently instrumental in removing his own brother, Saint, as furnace superintendent, and his dissatisfaction with the way things were going at the North Birmingham furnaces was probably a factor in the sudden departure of their manager, Kenneth Robertson. "To say that I am not satisfied with the management would be putting it mildly," he told Rutherfurd in July. "I am satisfied that some important changes will have to be made, but what they are, and exactly how they will be brought about, I am not determined."[38]

Late in 1889, as national economic conditions brightened again, Bryan became temporarily more hopeful about Sloss. Being preoccupied at the time with the problems of the Danville and Terminal systems and with new responsibilities that he had assumed in becoming publisher of the Richmond *Times*, he was indisposed to force an immediate crisis in Alabama. It did not take much prodding, however, to make him cross the line. Frustrated beyond endurance by his struggles with Inman, the loss of Johnston's help in running the Georgia Pacific, and the fiasco in North Birmingham, he reached the limits of his patience early in 1890 and turned to Morton as the only person who could help him find a way out.

Even more quickly than Morton had predicted in his letters to Uehling, therefore, things came to a head. After reading Dashiell's report, the directors scheduled a meeting at Birmingham in March 1890 to decide what must be done to save the enterprise. Morton took part, traveling with Bryan to Alabama after keeping in close contact with him before the trip. Little can be gleaned from Bryan's outgoing letters about what took place after the party had reached its destination. Repeatedly, at critical junctures he did not put much in writing about key decisions, perhaps merely because he was in frequent face-to-face or telephone contact with the people involved, but possibly also because he wanted to preserve his options. Nevertheless, every clue that exists—the fact that the deliberations lasted three days, a total lack of coverage about them in Birmingham newspapers, and the tenor of Uehling's later reminiscences about the circumstances surrounding the meeting—points to heated discussions that were kept private because of their sensitive nature.

The outcome was bitter for Seddon. Although the directors permitted him to remain as president, they appointed Dashiell to a newly created post, that of general manager, and gave him a mandate to carry out such changes and reforms as he deemed necessary. In a jubilant message to Uehling, who had been invited to come to Birmingham but was unable to make the trip, Morton stated that "my programme . . . was carried out to the letter" and indicated that Dashiell would be in Birmingham by the beginning of April.[39]

Uehling, one of the two greatest engineers to work for Sloss in its entire history, was named furnace superintendent at the meeting and was given a handsome salary of $5,000 per year. He came highly recommended by Morton, who stated that he bore "the highest character for ability and diligence." Because Uehling wrote an unpublished autobiography, his background is easier to trace than that of Dashiell. Born to German immigrant parents in Dodge County, Wisconsin, in 1849, he grew up on a farm and displayed mechanical precocity at an early age, inventing an improved land roller patented in November 1869. Given the best public and private education that his father could provide from limited earnings, he developed strong interests in music, literature, and various political, social, and philosophical issues that concerned him for the rest of his life. He thought deeply about such topics as unemployment insurance, municipal government, constitutional reform, the place of women in society, and religious intolerance. Clearly, he was a person of depth.[40]

Fluent in both English and German, Uehling studied for two years at Northwestern College in Watertown, Wisconsin. He then matriculated at Stevens Institute of Technology in 1873 as a result of seeing an advertisement about its programs in *Scientific American*. His academic performance won Morton's admiration. After graduating from Stevens in 1877 with a degree in mechanical engineering, Uehling assisted one of the nation's greatest engineering educators, Henry Thurston, in a study of the properties of cold rolled iron and structural steel. In 1879, he became master mechanic for the Douglas Furnace Company of Sharpsville, Pennsylvania, gaining practical experience in blast furnace operations. Studying chemistry, he became adept in the scientific analysis of ferrous metals and later managed five blast furnaces for the Bethlehem Iron Company. Disappointed when that firm's acceptance of a large government order for armor plate forced it to build an open-hearth steel furnace and cancel plans under which he was to build a battery of hot-blast stoves, Uehling left Bethlehem in 1890 to accept the new post that Morton had secured for him at Sloss Iron and Steel Company. A sizeable increase in pay helped focus his mind; continuing to follow a practice already well established in the antebellum period, Bryan and his associates were combining high-priced technical assistance from outside the South with cheap, severely exploited black labor.[41]

Dashiell's background is harder to trace than that of Uehling. It is perhaps significant that he was a Marylander, which may have made him more palatable to Bryan as a general manager than a person from a more distinctly northern state would have been. In any case, Morton's rescue operation capped an emerging pattern, present in Sloss's development since the inception of the enterprise, that bears out Gavin Wright's later emphasis on "the absence or small size of the indigenous regional technological community" as a limiting factor in southern industrialization.[42] Repeatedly, in contracting with such northern firms as Gordon, Stroebel and Laureau or dealing with such individual engineers or inventors as Hargreaves, Gutilius, Reese, Dashiell, and Uehling, Sloss's management had been forced to look beyond the South for technological expertise. St. George Bryan's failure as superintendent of the City Furnaces provides further support for Wright's argument.

It is true, as Jack R. Bergstresser has suggested, that northern or European technical help was relatively easy for southern firms to obtain. On the other hand, outside assistance had not always been as effective as Sloss's owners might have wished. Furnaces and stoves designed to be used with raw materials from other places had not worked well under southern conditions, as developments in Birmingham and North Birmingham had shown. Furthermore, even though technicians could be imported from outside the region, there were never enough skilled workers to build the facilities that the district needed in a timely manner; the chronic coke shortages that plagued local ironmasters provided good evidence of this lack. For such reasons, affairs at Sloss were once more in desperate straits.

VII

As Dashiell and Uehling prepared to go to Birmingham, the city stood at a crossroads. Boosterism remained rampant as its leaders clung to grandiose dreams, based on their continuing faith that iron could be produced in Jones Valley more cheaply than anywhere else. Utilizing vertical integration to the fullest possible extent, the Woodwards had proved that the cost advantages claimed by such promoters actually existed. Low costs, however, were not in themselves enough to turn the district's dreams into reality. In particular, Birmingham had already discovered that its overwhelming commitment to two basic industrial commodities, coal and pig iron, rendered it critically vulnerable to every gyration of cyclical trends in distant markets over which it had no control.

There was still much about which local go-getters could take satisfaction. By 1890 Birmingham had solidified its position as one of America's fastest-growing cities. Preliminary figures released by federal census takers in 1890 indicated that its population was approximately 27,000, compared with only 3,086 a decade before. Boosters insisted that this was too low and were chagrined when the correct figure turned out to be 26,178, but Jefferson County as a whole had 88,501 persons living within its boundaries, and the district was clearly burgeoning.[43]

Such growth reflected the district's phenomenal progress in making pig iron. By 1889, Alabama ranked fourth among all American states in the fabrication of this commodity—Pennsylvania, Ohio, and Illinois were the top three—with a total output of 791,425 tons. Most of this production came from up-to-date furnaces that used coke. Ten years earlier, Alabama had made only 49,841 tons of iron, nearly 65 percent of which came from outmoded charcoal furnaces. Capital invested in iron production with mineral fuel had risen from $955,800 to $12,394,757 during the decade, and the value of the state's total output of ferrous metals had ballooned from $554,162 to $8,374,816. Almost all of this achievement was attributable to developments in the Birmingham District.[44]

Further signs of Birmingham's emergence from a raw outpost on the nation's urban frontier to a young city attracting much notice at home and abroad were the paving of its downtown streets, beginning in 1887; completion of the ornate five-story Morris Office Building in 1888, followed within a year by the elegant Caldwell Hotel; the growth of the city's public school system under the leadership of a dedicated educator, John Herbert Phillips; and the building by the Elyton Land Company of a system that brought water to the city from a pumping station located in the Cahaba Valley, nine miles from the business district.

The new water system was an impressive achievement; Caldwell was justified in calling it a "stupendous work." Its completion required digging a tunnel under Red Mountain, took two years, and cost half a million dollars. The expenditure of time and money was imperative, because the city's need for water had outrun the capacity of the reservoirs previously built at the head of Five Mile Creek. Alfred F. Brainerd, an analytical chemist, stated in 1888 that the system would assure "a good and adequate supply of water . . . for five hundred thousand people." A decision by the American Institute of Mining Engineers to hold its annual meeting at Birmingham in the same year was also hailed by local residents as an indication of the status that the Magic City had attained as a focal point of industrial growth.

Evidences of mushrooming development were everywhere. Lots in the fashionable area of South Highlands were selling for $5,000 in January 1890. The four-story Steiner Bank, made of Vermont marble and pressed bricks from St. Louis, was going up at the corner of First Avenue and Twenty-first Street on a lot that had been sold for $25,000 in cash. Typical of statistics avidly collected by boosters was the fact that the Birmingham Mineral Railroad was handling 4,000 cars, loaded with coal, ore, limestone, and pig iron, per week. On any given day, 700 cars rumbled over its tracks.[45]

To a large degree, however, such quantitative measurements merely reflected the tiny statistical base from which the city and district had started. However impressively pig iron production had grown in ten years, Alabama's output was dwarfed by the 4,181,242 tons made by the nation's leader, Pennsylvania, more than three times as much as that of all the southern states put together. Furthermore, the cooling off of the national economy since the collapse of the 1886–87 boom had dashed the prospects of many Alabama boom towns. Ensley, created to flatter the ego of the entrepreneur after whom it was named, was a case in point. By 1890, its development was going so poorly that the Tennessee capitalist had taken off for Alabama's northwest corner to pursue another of his perennial development schemes.[46]

Even TCI, despite its great size, was not living up to prospects that had seemed bright when it moved to Alabama in 1886. As in the case of the Georgia Pacific, its troubles could be traced partly to Inman, who pocketed speculative profits during the late 1880s by refunding its bonded indebtedness at high cost to its financial health. To the disgust of such persons as Bowron and Baxter, Inman also damaged TCI's reputation among investors by organizing alternate buying campaigns and bear raids, causing violent fluctuations in the value of its stock. Bitter infighting continued to plague the firm, leading to such episodes as the dumping of Baxter from the presidency in 1889. Clearly, Bryan had been lucky to have had the foresight not to borrow from Inman in 1887, even though he had not been able to escape the results of Inman's influence so far as the Georgia Pacific was concerned.

Inman's mishandling of TCI ultimately backfired. In 1889, Duncan, who had nursed a grudge against him ever since TCI came to Alabama, helped organize a coup that removed Inman from power. In the process, however, control of TCI passed to Thomas C. Platt, a New York politician who had assumed leadership of the powerful Republican faction formerly dominated by Roscoe Conkling. By manipulating TCI's securities as ruthlessly as Inman had done, Platt touched off a financial panic that seriously threatened the interests of his fellow investors, giving Inman a chance to regain control. Allying himself with some of the men who had previously opposed him, including Baxter, Inman launched successful Wall Street raids that soon brought more and more TCI stock back into his hands. In December 1890 he achieved equal board representation for himself and his supporters against Platt and his allies. TCI's profits, which were partly spurious because of faulty bookkeeping methods that were used in their calculation, suffered from these gyrations, plummeting from $1,251,300 in 1888 to $327,573 in 1889 and sinking further to $209,902 in 1890. Everything that had happened since 1887 revealed the extent to which the company had become a football to be kicked about by Wall Street speculators.[47] To some extent, the same thing was becoming true of the Georgia Pacific, as Bryan had learned in trying to preserve what was left of his dwindling power in that enterprise.

VIII

While boosters continued to dream about endless vistas of growth and danger signs became increasingly obvious, the person who had done more than anyone else to launch Birmingham on its quest for greatness passed away. On the evening of 4 May 1890, James W. Sloss died of heart failure. Two days later, an "immense concourse" of local residents gathered at the First Methodist Church to attend his funeral, at which the Reverend Dr. C. B. Riddick preached on the text, "David, after serving his generation, by the will of God, fell asleep." Stating that "the history of Colonel Sloss for the past twenty-five years was the history of Alabama," Riddick characterized him as a fitting role model for "every citizen, from the laborer to the banker." Laid to rest at Oak Hill Cemetery, Sloss was mourned by the Birmingham *Daily News*, which declared in an editorial that "his memory will be revered by all who love in a man simplicity and goodness, sterling integrity and manly ability, the gift of getting and the grace of giving."[48]

Praising Sloss, an obituary in the national trade journal, *Iron Age*, stressed "his farseeing discernment, indomitable energy and modern ideas." Stating that Sloss had "done more probably to develop the South since the war than any man in it," the article underscored the fallen industrialist's hope that Birmingham might become a great steelmaking center. "Since his retirement from active business life he had his heart very much set upon the question of producing steel from Alabama irons," it stated. "He took a lively interest in everything looking to the culmination of successful steel making in his section, believing that when this was accomplished the last remaining obstacle would be overcome that stood between the iron producing District of the South, and the most unbounded industrial success." Accompanying the obituary was a striking picture of Sloss, showing him looking like Moses, with a long, flowing beard. Just as Moses had died before he could lead his followers into Palestine, Sloss had passed away before he could guide Birmingham into the promised land of steel.[49]

STEEL: Here, both to *Iron Age* and to local go-getters, was the key to Birmingham's future. As Sloss lay dying, a group of local businessmen, including his son Maclin, staged the city's latest effort to produce this wonder alloy and open the road to Golconda. Apart from the fact that profits from steelmaking were potentially much greater than those available to pig iron manufacturers, civic leaders were painfully aware that the district had lost lucrative industrial opportunities because of its inability to produce steel. In 1881, soon after the Birmingham Rolling Mill had been established, its superintendent, Arthur Moxham, had been visited by a friend from Louisville, Tom Johnson, who had just designed a promising girder rail and wanted Moxham to devise ways of rolling it at a reasonable cost. Known as the Jaybird because of a distinctive appearance resulting from two offset flanges, Johnson's brainchild would later prove phenomenally successful for streetcar trackage. Despite his best efforts, Moxham had not been able to secure good results in producing the Jaybird with wrought iron made in Birmingham. Giving up, he resigned his Alabama position and went back to Louisville, where he found that only steel was adequate to withstand the lateral stresses that the rail had to sustain. Ultimately the Johnson Steel Street Railway Company, which was chartered in Kentucky in 1883 to produce this distinctive product, established a plant in Johnstown, Pennsylvania, because of the presence there of the Cambria Iron Company, which made Bessemer steel.

In 1886 and 1887, electric street railways began to spring up in various parts of the country. Southern cities, including Montgomery and Richmond, took the lead in installing them. By 1889, more than 180 similar systems had been built or were being constructed. Birmingham had one in operation by 1892;

other places that were building them included Atlanta, Charlotte, and St. Louis. Cashing in on the superior features of the Jaybird, the Johnson Company was well on its way to becoming the country's largest and most successful manufacturer of the rails, frogs, turnouts, and other components required for streetcar systems, and Moxham had devised all sorts of imaginative techniques for the fabrication of these items. But for the peculiarities of southern raw materials, the Johnson plant could very well have been located in Jones Valley.[50] Such things could not help but gall Birmingham's boosters.

Because of the ever-present problems stemming from the presence of phosphorus, silica, and sulfur in red hematite and Alabama coke, fulfillment of Birmingham's dreams continued to rest on the possibility that one or another form of basic steel manufacture could be established in Jones Valley as a rival to the acid technology prevailing in the North. In February 1888, as hopes that had been placed in Reese's inventions passed into oblivion, a small run of steel, the first specimen ever made in the district, was produced in an open-hearth furnace that had been brought all the way from Boston to North Birmingham. Pig iron from the Mary Pratt Furnace was used in the charge.

The technology underlying this pilot project had been developed by James Henderson, a European-trained inventor who had visited Birmingham in 1887 to take part in a meeting of metallurgists and mineralogists and attracted the attention of local entrepreneurs because he had spent a quarter-century in steelmaking and had once worked for the great German industrialist Friedrich Krupp. Henderson called this method the "flame process," converting iron into steel by heat alone, and stated that it would yield savings of $10.75 per ton over the Bessemer technique. Utilizing an open-hearth furnace operating at extremely high temperatures, it depended upon high-quality refractory bricks that were made according to a special formula that Henderson had patented and also utilized ferromanganese or spiegeleisen after phosphorus, silica, sulfur, and excess carbon had been removed from the charge. "Come around in sixty days and I will show you as fine steel . . . as can be produced in this country," Henderson boasted.

Typifying the district's dearth of funds for ventures that did not promise immediate profits, the total capital invested in the Henderson Steel & Manufacturing Company was only $20,000 of the $1 million authorized by its charter. Promoters hoped, however, to "erect one or more steel furnaces of 100 tons each daily capacity, and make ingots for sale to the four rolling mills here and in the vicinity, and erect a works to make steel castings and tool steel under the Henderson patents." The project was spearheaded by W. H. Hassinger, who had studied engineering at Rensselaer Polytechnic Institute and been employed as a chemist by steel and iron companies in Pittsburgh and Youngstown. Waiting in the wings to provide additional seed money if initial developments proved encouraging were Joseph Forney Johnston and Edmund W. Rucker, who had earlier joined forces with Colonel Sloss in trying to import Jacob Reese's basic steelmaking methods into the district. The modestly successful outcome of the Henderson experiment created much momentary excitement, raising hopes that Birmingham's emergence as a major steelmaking center was finally at hand. The first trial, in March 1888, utilized a mixture of pig iron, ore, and fluorspar on a dolomite hearth. According to the Birmingham *Herald*, it "worked like a charm" and produced a small quantity of "the very finest of soft steel." A souvenir penknife was made for Henderson from the metal, and Rucker was given a set of razors.[51]

Like so many other recent dreams, however, hopes that the Henderson process would yield a bonanza for the district were quickly dashed. In January 1890, after issuing $50,000 in first mortgage bonds, promoters attempted further development at an expanded facility in North Birmingham, equipped with three converters and having a capacity of twenty tons per day. Once again, the small amount of funding raised—only $40,000 worth of bonds were actually subscribed—typified the difficulty of obtaining local capital for truly pioneering efforts based upon relatively untried technologies. The renewed attempt took place "within a pistol shot" of the new blast furnaces that had been erected by Bryan and his associates, who watched its progress closely. The fires of the expanded facility were ignited on 29 January 1890, but not until 24 April did the first test, "a successful and glorious run of ten tons of best grade of soft steel," take place before a crowd of onlookers that had braved a rainstorm to witness the event. Some of the spectators came from as far away as New York and Chicago.

Plans were quickly announced for a new plant, to be located on a 140-acre site served by three railroads. The installation was to be equipped with a blooming train and a rolling mill. Such visions, however, were premature. Using pig iron from the furnaces that DeBardeleben had recently built in Bessemer, the promoters made runs ranging up to thirty-five tons of steel and at one point attracted a trainload of European experts to visit the works. Conscious that a similar experiment was being conducted at the same time by the Southern Iron Company in Chattanooga, using a basic process invented by Sidney Thomas Gilchrist in Great Britain, local boosters fervently hoped that the Birmingham venture would lead to greater things. Its excessive operating costs, however, gave it little chance for commercial success, and experienced investors refrained from risking the large amounts of capital that would have been required for further development. The Henderson process itself was difficult to control, producing such a violent reaction that it sometimes shook the furnaces in which it was utilized and burned out their linings. It also required steel scrap, a commodity that was scarce in the South because it was a predominantly agricultural region. In short, Henderson's methods were not the answer to Birmingham's needs. By the end of the year, the mirage of steelmaking in the Magic City had faded once again.[52]

IX

Into such a setting, still full of hope for the future but increasingly marred by frustration and dependence upon outside resources that local boosters could not control, Dashiell and Uehling came in April 1890. Upon their shoulders rested Morton's hope of rescuing the Sloss Iron and Steel Company from its persisting problems and transforming it into a model enterprise. Embarrassed by the events of the previous month but still clinging to his presidential office, Seddon did not wait to extend the newcomers a welcome that would have been insincere. Leaving for a vacation the day before Dashiell arrived, he did not return to Birmingham for six weeks.

In Richmond, Bryan was too busy coping with the latest problems of the Georgia Pacific to do more than watch events at Sloss from the sidelines. As floodwaters poured through broken levees along the Mississippi and damaged its recently completed tracks, he could only hope that he and his fellow directors had found at least a temporary solution to the problems that sprang up so endlessly in Jones Valley. Even in letters to Maben, with whom he shared news of the inundation as it was relayed from the scene by Captain Sage, who was still a few months away from retirement, Bryan said nothing about recent developments in Alabama. If he was discouraged, his correspondence did not show it. Now, as always, courage and the will to persist remained deeply ingrained in him. Surely, however, he must have found the past few years sobering. It was fortunate that he could not foresee how short Dashiell's tenure would be, predict what would happen to the national economy in the period of financial panic and depression that lay just ahead, or know how much strife would take place in Birmingham during the crises that were yet to come. Had he possessed such foresight, his mood might have been even more somber.[53]

Turmoil and Tenacity 7

The sudden economic upswings and downturns of the 1880s were only a prelude to a devastating depression that began in 1893. Before it hit, Bryan and his associates did their best to put Sloss on a firmer footing. Seddon, however, clung obstinately to power despite the effort that had been made to unseat him, and the company was plagued by internal divisions. Nationally, a wave of corporate consolidations got underway as the American economy began to be restructured in the wake of the turbulence that the 1880s had produced. Reflecting the times, Sloss became involved in merger talks with TCI that proved fruitless after momentarily raising the hopes of local boosters. A reorganization of the Danville and Terminal systems by J. P. Morgan did take place, foreshadowing growing northern domination of the southern economy. Still, Bryan and his fellow investors in the Georgia Pacific came out of the situation as well as they had any reason to hope. As the depression wore on, Sloss responded creatively to its challenges. Mounting strife with organized labor, however, required state intervention to sustain the low-cost strategies on which southern competition with northern producers depended. It was a time of turmoil, but the Virginia syndicate had the tenacity to meet it. In 1896, as the company looked back on a crisis involving its use of convict workers, Seddon died, giving Bryan a welcome opportunity to choose a new president.

I

During the spring and summer months of 1890, while Seddon nursed his wounded feelings and considered his options following his loss of power, Dashiell and Uehling did their best to make Sloss more efficient and profitable. Heartened by their progress, Bryan compared Uehling to a physician who was "removing the causes of clots in our machinery." After reading one of Uehling's reports, transmitted by Morton, Bryan exuded confidence in the future of the enterprise and was full of praise for the young engineer. "He writes like a straightforward, true man, and I am very favorably impressed with his letter," Bryan wrote Morton. "Mr. Christian read it in my office this morning, and turned around and said, 'We must give . . . Dashiell and Uehling . . . our full support,' and I say 'Amen.' From Uehling's account, the mechanical management at the Sloss must have been worthy of an old, run-down plantation."[1]

Eager to implement Morton's plans for revitalizing the company, Dashiell and Uehling attempted to improve the quality of the furnace fuel at their disposal by washing coal before it was coked. Bryan's support of the effort demonstrated his continued willingness, already displayed in connection with the modern equipment recently installed at the Brookside mine, to blend selective mechanization with labor-intensive production methods. Eliminating impurities like sulfur and slate from Warrior coal would result in stronger coke with less ash content, enhancing the smelting process and yielding better iron.

Mechanical coal washing had been used in the United States for only a few decades. Only sporadic attempts to implement the practice had been made in Alabama, including the previously mentioned construction of a washery at Coalburg by Bryan and his associates in 1886. Various cleaning methods were employed, all of which agitated coal in a stream of running water to remove impurities. Some innovative firms were experimenting with jig washers, equipped with plungers or other pulsating devices that caused waste particles to pass through filters or screens and settle at the bottom of a tank while coal rose to the surface in boxes or troughs.

Judging from the Sloss Iron and Steel Company's need to conduct a fresh attempt at coal washing after Dashiell and Uehling arrived in Birmingham, the apparatus used at Coalburg in 1886 had not been effective. In 1890, while Bryan and Morton kept track of developments from Richmond and Hoboken, Sloss became the first firm in the United States to install the Luhrig jig washer, a model that had recently been invented in Germany. A British emigrant, Alexander Cunningham, erected the washery on a site adjacent to the coke ovens at the City Furnaces. Visitors came from as far away as Colorado to inspect the new system, which was later adopted successfully in other American locations. Unfortunately, the Luhrig filtration process required elaborate presorting of coal into specific sizes and chemical categories. For this reason, it could not be used economically with fuel from the Warrior field. Sloss soon abandoned the project, but the failed experiment indicated the vigor with which Dashiell and Uehling were trying to solve the company's problems.[2]

By summer's end, attention was shifting to other concerns. After vacationing at White Sulphur Springs in West Virginia, Seddon returned to the Magic City in a fighting mood. Together with secretary-treasurer W. L. Sims and the company's most powerful local stockholder, Rucker, he launched a counterattack upon Dashiell and Uehling. Now, as later, Seddon made the most of cross-purposes existing between investors in Birmingham and Richmond. It is easy to see how he and Rucker, one crippled by a childhood accident and the other maimed at Chickamauga, formed an instinctive bond. Rucker's interference in personnel matters soon led to the resignation of John Dowling, a foreman at the North Birmingham furnaces whom Uehling held in high esteem. This development increased managerial tensions and indicated that a crisis was brewing.[3]

As autumn began, Seddon was sufficiently sure of himself to give the directors an ultimatum: either he or Dashiell would have to go. Uehling regarded this as a bluff, but Bryan did not. Besides being deeply rooted in the Virginia aristocracy and having many friends in the Danville and Terminal systems, Seddon had other powerful business connections, particularly in Baltimore. In addition, he was strongly backed by Birmingham investors who were already disappointed by the failure of the company to move into steel manufacture and whose support Bryan needed. Characteristically, because Bryan rarely discussed sensitive matters in writing and could easily deliberate with his associates in Richmond and New York on a face-to-face basis, his letterbooks reveal nothing about whatever considerations weighed most heavily on his mind in deciding what course to take. The outcome of the crisis, however, is clear. Confronted by Seddon's intransigence, a majority of the directors backed down. Bowing to Seddon's demands at a special meeting held at Birmingham late in September 1890, the board voted to dismiss Dashiell, who soon left and took a position as a consulting engineer in the petroleum industry. In 1893, he founded a successful enterprise, the New York Lubricating Oil Company, with which he remained the rest of his career.[4]

Learning from Morton about Dashiell's dismissal, Uehling decided that he, too, would resign. Morton, however, persuaded him to stay on as furnace superintendent. Though infuriated that the directors had "not had the backbone" to stand by Dashiell, Morton was not ready to abandon hope in the company's future and told Uehling that he had won Seddon's promise to cooperate "in every way to make the blast furnace department a success." It is not clear precisely what agreements had been made at the meetings. Apparently, however, a compromise had been reached putting Seddon in overall charge but leaving the day-to-day supervision of smelting operations in more capable hands.

Putting the crisis behind him, Bryan worked hard in the next few months to arrange financing for repairs and modifications that Uehling desired. Now that Seddon was again in charge, Bryan took a conciliatory attitude toward him. "I fully appreciate the troubles that you have had, and am not disposed to bear down unreasonably upon you," he told Seddon at the end of the year. On the other hand, he stated, the firm's future still hung in the balance, and "it is a fact that we must make every edge cut."[5]

Despite such gestures, unresolved tensions continued to plague operations in Birmingham. Even before the board had tried to limit Seddon's power,

Bryan had complained about receiving too little information from him. This now became a chronic problem. Seddon also displayed underlying resentment in other ways, as by requiring Bryan to justify in a meticulous way expenses that he incurred in periodic trips to Alabama. Seddon liked to make sardonic jokes about his relationships with Bryan and other board members. After one protracted board meeting in Birmingham, a black janitor commented that "Dem correctors o' your'n sho' do set long, Marse Tom!" Relishing the unintended appropriateness of the remark, Seddon habitually used the term "correctors" in referring to the board from this time on. He also boasted that he was immune from dismissal because it had cost his superiors too much to train him. "They never could afford to turn me off, you know, no matter how much they might want to," he was fond of saying. "It would have been too big a loss for 'em to stand. . . . No, sir, no danger of my losing my job." Later, in her history of the coal and iron industry in Alabama, Ethel Armes recounted such jibes as an indication of Seddon's good humor, apparently not understanding the context within which they were made.[6]

II

Fortunately, persisting tensions did not prevent the company from making a comeback in 1891 after the turmoil of the previous year. Despite foot-dragging by Seddon, who objected to the costs involved, Uehling repaired the North Birmingham furnaces and restored them to operation. After being put back in blast, they performed much better than before, producing 40,083 tons of pig iron in fiscal 1891 at an average cost of approximately $9.83 per ton. During the same twelve months, the City Furnaces yielded 68,796 tons at even lower unit expense, resulting in a combined output of 108,879 tons at an average cost of $9.30. Demand throughout much of the year was sluggish, and the average profit per ton for the first six months was only ninety-two cents. Nevertheless, improved yields and lower costs helped produce a net profit of $190,340, easily the company's best performance since the 1887 takeover. Uehling was obviously doing a good job, and Bryan continued to praise him in letters to Morton.[7]

Notwithstanding the enhanced functioning of the company's blast furnaces, profits from coal, coke, and ore sales still accounted for most of its income. During the first half of 1891, they produced nearly 56 percent of earnings from sales. Sloss's dependence on these sources of revenue underscored the importance of the growing network of mining camps that were spawned by the company and other local firms in outlying areas of Jefferson County. Besides yielding income in the form of rents and commissary receipts, such settlements also provided a valuable form of social control to management. Their very isolation, combined with the fact that they were completely owned by the enterprises that maintained them, helped tighten the grip of such firms over labor.[8]

Typifying the rough-and-tumble aspects of the industrial order that emerged in Alabama in the late nineteenth century, Jefferson County's mining camps and company towns consisted mainly of rude frame dwellings with two to four unplastered rooms, built and owned by various local firms and rented to employees; the Sloss Iron and Steel Company charged $2 per room. None of the houses had running water, and the use of outdoor privies was universal. Working conditions were equally primitive. Walking or crawling through dark passageways deep under the hills with the aid of kerosene lamps attached to the backs of their caps, the miners emerged at day's end covered with mud and coal dust. Working hours were from sunrise to sunset. Even free workers—most of the miners were convicts—did not start earning their pay until they reached the face of the subterranean coal seams and generally had nothing but survival to show for their efforts after years of backbreaking toil. Numerous deductions were made from their pay for powder, fuses, tool repairs, and other charges. Whatever remained of their earnings after they had settled monthly balances at the company store was likely to be claimed by bill collectors who always showed up on payday. No fringe benefits eased the prospect of accidents, lung diseases, or old age. The miners themselves thus bore much of the unrelenting effort made by the owners to hold down costs in order to compete with northern producers. Despite such conditions the camps afforded better living conditions than those existing in Alabama's rural districts, where abject poverty was so prevalent that sharecroppers and agricultural laborers gladly left for industrial jobs in search of a better life.[9]

As had been true in the South's extractive industries during the slave era, most of the miners were black. Perpetuating the white myth that they enjoyed their lot, Alfred F. Brainerd, who presented a paper about their use at the Chattanooga meeting of the American Institute of Mining Engineers in 1885, took a patronizing view by comparing them to children. "The colored laborers are fervently religious, intensely superstitious, improvident, and usually good natured and happy," he said. "That strong religious emotions do not involve necessarily a moral sense highly developed in the direction of ethics is shown in the propensity of the colored population for chicken-stealing, and in the brutal fights and other outbreaks of which they are not infrequently guilty, upon any slight provocation." While some observers might think they were lazy, he continued, they were actually more efficient than they appeared to be on the surface, for "what they may lack in speed they make up in muscular efforts and in longer hours of labor." Frequently, he added, they "enliven the monotony of their task by singing some melody, keeping time with their hammers, picks, and shovels to the music. One never hears as yet of strikes among them. Yet they hang together with exceptional loyalty. In case of mishap due to either accident or carelessness, the management fails to get at the cause by examining any number of them. They always manage to get together and agree as to the story they will tell if called upon. Among white miners, on the contrary, there is usually some one to turn 'States' evidence,' either for the sake of getting out of difficulty himself, or of getting some one he dislikes into trouble, or from the higher motives of truth and duty."[10]

By far the most important of Sloss's mining communities was Coalburg, where Frank Hill, an engineer with Belgian training, took charge of operations in the early 1890s. Boasting a hotel, a schoolhouse, a telegraph office, a railroad station, a company store, and a church where services were conducted twice a month by an Episcopal clergyman, Coalburg was characterized by a visitor in 1888 as a "first class railroad town." By this time, it already had a population exceeding 1,000, including 150 convicts who occupied a two-story prison built along Five Mile Creek. Free workers lived in houses scattered throughout the forest covering the 13,000-acre tract. During the next few years, the population grew rapidly; by summer 1890, convicts alone numbered more than 1,000.

Coal and coke production at Coalburg escalated along with the number of inhabitants. The mines exploited the "seemingly inexhaustless" Pratt seam, as one analyst described it. Running two miles under Light's Mountain, it had veins ranging from three to four feet thick. The galleries, like those at most other company mines, consisted of horizontal tunnels served by tramcars. These vehicles were of one-ton capacity and were impelled by ropes and pulleys at ten to fifteen miles per hour. Steam engines provided motive power, but an electrical signaling system was used. During the first half of 1891, Coalburg produced 230,532 tons of coal and 42,904 tons of coke. At this time, it was Sloss's single most important asset.[11]

Brookside, the company's other main mining camp, was located seven miles west of Coalburg on the Georgia Pacific. By mid-1890 it had a population of about 2,000, was producing 650 tons of coal per day, and had a battery of beehive coke ovens. Before the decade ended, its coal production outstripped that of Coalburg, aggregating 386,116 tons in 1896 as opposed to the older installation's 296,974. As was true of the firm's other mining settlements, Brookside's population included both blacks and whites, living in segregated quarters. Racial tensions sometimes got ugly. In June 1890, after what a local newspaper described as a "race war," a posse of twenty-four deputies was sent to the community to restore order and "scour the woods for the ring leaders of the negroes who had fired upon the whites." The fugitives were not found; had they been apprehended before the deputies arrived, the article stated, they would have been lynched. Later in the decade, Brookside's ethnic diversity increased when it attracted eastern European immigrants who built what was for a time the only Russian Orthodox church in the Southeast. Dedicated to St. Nicholas, it featured an onion-shaped copper dome in Byzantine style. Brookside was only one of the mining camps in the vicinity that was becoming inhabited by immigrants at this time; by 1890, about 17 percent of the district's miners and 12 percent of its ironworkers came from foreign countries. As demand for workers increased, arrivals from such places as Greece, Hungary, Italy, Poland, and Russia made the Birmingham District "unique in the South," as a recent study has indicated.[12]

Blossburg, a few miles west of Brookside on a branch of the Georgia Pacific near that line's Fishtrap Tunnel, was another of Sloss's mining camps. Founded in 1889, it grew from nothing to a booming community of 1,000 persons within six months. Starting with the first shipments in December 1889, coal output increased rapidly to 1,000 tons per day by mid-1891. "The Blossburg mine has been fitted up with the very best sort of machinery and boasts of being the finest equipped one in the South," said a reporter who visited the camp. By 1897, output had risen to 130,517 tons of coal and 43,636 tons of coke per year.

Tutwiler, previously in charge at Coalburg, supervised Blossburg's early development. A stockade was built for convict workers, adding to an unstable population of blacks, Scots, Welsh, and Germans who were prone to outbreaks of sporadic violence and vigilante justice. Early in 1890, amid drunken revelry on payday, one black worker shot and killed another in a dispute over an unpaid debt. After the murderer defied arrest, a group of white hunters came on the scene. Sizing up the situation, they took the law into their hands and made short work of the culprit. "His body was riddled with bullets," a newspaper account stated laconically. "No further trouble is feared."[13]

Cardiff, another settlement built for Sloss under Tutwiler's direction, was a few miles north of Brookside and Blossburg. By the end of the century, it had more than 500 inhabitants, mostly British and Irish immigrants, and output had spurted to 4,300 tons of coal per day. Brazil, another hamlet in the same area, completed the roster of early mining communities developed for Sloss by Tutwiler, who left in 1893 to start his own enterprise, the Tutwiler Coal, Coke, and Iron Company.[14]

The Sloss Iron and Steel Company's industrial empire also included two ore-mining communities. One of these, Irondale, was a few miles east of Birmingham near the Shades Valley site where McElwain had built Oxmoor furnace in 1863. Some nearby mines had been acquired by Colonel Sloss before he sold out to the Bryan syndicate; others had resulted from Ruffner's prospecting activities and were later named for him. By 1897, ore deposits in this area were yielding more than 130,000 tons per year. Nearby, quarries at Gate City supplied the firm with limestone.[15]

The company's other ore-mining community was Sloss, named for the founding father. It adjoined Bessemer on a site where rich seams of red hematite, fifteen feet thick in some places, outcropped on Red Mountain at one end and extended deep under Shades Valley at the other. Manned by black workers, the mine produced more than half the company's ore supply by 1897, with an output of 223,740 tons. Much of the ore held so much carbonate of lime that it was self-fluxing, lowering furnace costs.[16] Production at Irondale expanded less rapidly than at Sloss, reaching 132,555 tons in fiscal 1896 but falling to 116,939 the following year. This output was partly due to high costs associated with topographical features that required cutting through solid rock and constructing a long tram line to deliver ore to the tipple. The fact that deposits in the area were not as extensive as those near

Sloss furnished another reason for "husbanding this ore as much as possible," as an adviser counseled Seddon in 1896.[17]

Profits derived by the company from its sprawling network of settlements were not based solely on their output of coal, ore, and limestone. Rents from employee housing and the earnings of company stores were crucial in keeping the firm afloat throughout the 1890s and beyond. Commissaries, a characteristic feature of antebellum industrial slavery, were equally typical of the social and economic order prevailing in Jefferson County after the war. Employees were not forced to buy food and other necessities at the commissaries, but usually did so, particularly because it was convenient for them to obtain goods on credit and have the amounts that they owed deducted from their wages. The use of scrip was common among the miners, who lived on $3.50 to $7 per month. "At the end of the month, rent, doctor's bill, and the amount of scrip drawn, or money advanced, are deducted from the amount due for wages, and the balance paid in cash," Brainerd reported. "Most companies employ their own physicians, and the employees are taxed to pay the doctor's salaries and the cost of medicines used. A few of the colored miners lay up a certain amount every month from their earnings. Most of them keep in debt to the storekeeper, or simply draw enough to support themselves as they go along, and on pay-day receive the remainder and spend it within a short time in some foolish manner."[18]

During the early 1890s, commissaries at the City and North Birmingham furnaces were run by independent merchants who paid the company a commission on sales, but those at the mining camps were owned and operated by the firm itself. During the six months ending 31 July 1891, profits from the stores at Coalburg, Brookside, Irondale, and Sloss yielded $17,780.47; commissions from merchants running the commissaries at the City and North Birmingham furnaces added $5,012.49. Rents from company-owned housing in the same period totaled $5,600.45. Considering that profits from all sources for the six months were $113,957.62, the $28,393.41 that accrued from rents and store receipts was an important part of the firm's operating income.[19]

Company administrators depended heavily upon the ability and advice of the superintendents who ran the mining camps. One of these, John David Hanby, had supervised the Sloss mines before the 1887 takeover and eventually came back to work for the new owners. Sporting a walrus mustache, about which friends liked to tease him, he strongly resembled Grover Cleveland. He was a grandson of David Hanby, who had sent coal by flatboat to Mobile in the 1840s. Earlier forebears had come from the Old Dominion, making Hanby yet another example of the company's strong Virginia connections.

Another key official, Thomas C. Culverhouse, supervised the coal mines throughout much of the 1890s. He was a strong advocate of the company's policy of relying chiefly upon black labor. In 1896, in a report to Seddon, he expressed regret that the work force at Brookside, Brazil, and Cardiff, which had more than doubled from 539 to 1,101 men in twelve months, was slightly more than half white. In his opinion, the best ratio was 70 percent black to 30 percent white; this was partly because "our negroes trade 65% to 70% of their earnings in the Company store," whereas whites were less dependable in this regard. One of the main problems that Culverhouse faced in recruiting black miners was his inability to provide adequate housing for them when operations were expanding rapidly. He told Seddon that, unlike blacks, who were highly family oriented, "the whites look out for themselves, locating among the natives out in the country within two or three miles of the mines." He was sorry that "the increase of whites has kept pace with that of the negroes in the face of my efforts to locate negroes and my indifference to employing whites."[20]

In securing workers, Culverhouse and Hanby depended heavily on black subcontractors who recruited operatives and paid them after taking a percentage of their earnings as a commission. The system was similar to the use of trusted bondsmen as slave managers during the antebellum period and helped employers exploit divisions within the black community.[21] Samuel Dooley and Paul Thompson were typical subcontractors for Sloss. Born at Madisonville, Tennessee, in 1867, Dooley worked at ore mines and blast furnaces in his native state before coming to Alabama in 1886. Four years later, he signed on with Sloss. After leaving the district in an unsuccessful attempt to improve his lot in Oklahoma and Colorado, he returned to become a contractor at the Sloss ore mines, furnishing the company up to forty operatives per day and earning commissions of $200 or more per month. Thompson, from Tuscumbia, went to work for the Sloss Furnace Company in 1884 and lost his right eye the same year when a spark from a miner's lamp ignited a gunpowder magazine. Well liked by company officials because of his steady work habits, he was hired as a contractor and earned "hundreds of dollars monthly."

Labor contracting was one of only a few avenues of upward mobility open to blacks in Alabama in the late nineteenth century. Photographs of Dooley and Thompson appeared in a Bessemer newspaper in 1901. Both were elegantly dressed, befitting the successful entrepreneurs they had become. The accompanying account of their careers emphasized their "industrious habits and . . . abstention from the vices too often affected by the young men of the colored race." It also noted approvingly that both were active in the Red Mountain Baptist Church. Declaring that "worth and merit command success, irrespective of color or previous condition," the article argued that the achievements of the two men showed how the South provided a "fair competitive field" for all persons, "unbiased and unfettered by caste, color or previous condition."[22] Such rhetoric had to be taken with more than a few grains of salt. A careful study of upward mobility among Birmingham workers in the late nineteenth and early twentieth centuries has shown that, while one in every two or three whites improved their economic condition after living in the city for twenty years, only one black worker in six did so. Indeed, most blacks went from better to poorer jobs. "For the city's black workers," the study concluded, "the road to success too often ran in the wrong direction." Dooley and Thompson were clearly exceptions to the rule. [23]

Besides establishing mining camps in the Jefferson County hinterlands, the Sloss Iron and Steel Company maintained quarters near the City and North Birmingham furnaces where housing was provided for its overwhelmingly black work forces. Such areas were analogous to the slave quarters on ante-bellum plantations, emphasizing the degree to which the New South was rooted in the Old. A statistical analysis of the workers who lived in the quarters has shown that more than 94 percent either rented their dwellings or boarded with employees who did. Because blacks were steadily squeezed into areas near the furnaces while upwardly mobile whites moved to the suburbs, Sloss did not force its workers to live in company-owned housing. Indeed, more than half the persons who rented houses near the furnaces were not employed by the firm. Workers were still extremely transient, just as they had been when Colonel Sloss had earlier testified to senate investigators. This made it as hard as ever for the company to secure a dependable labor supply, forcing it to hire far more operatives than would otherwise have been needed. Black workers, however, were more firmly rooted in the district than were whites, because the latter had more of a chance to move upward and onward. This stability gave the firm an added reason, besides the fact that blacks would take lower wages, to prefer hiring them. [24]

III

Although the company's blast furnaces earned less income than the coal mines during most of the 1890s, they were crucial to its survival. Their output continued to be marketed mainly through commission merchants in large cities, particularly Cincinnati and Louisville. Matthew Addy & Company, in Cincinnati, was the most important dealer, handling more than half of Sloss's pig iron business during a six-month period in 1891. During the next year it received nearly $460,000 worth of iron, much more than any other agent. The company's most important dealer in Louisville was George H. Hull & Company, to which it consigned $108,367.72 worth of iron in fiscal 1891. A creative force in the industry, Hull organized schemes, in which Sloss took part, to bolster prices by restricting output during slack times. He also enlisted Sloss and other firms in a system enabling them to withhold iron from sale by putting it in storage and having negotiable warrants issued against it. Rogers, Brown & Company of Cincinnati, which paid $711,445.40 to Sloss over a three-year period in the early 1890s, and Hugh W. Adams & Company of New York City, which handled $437,339.97 worth of Sloss pig iron in fiscal 1891 and 1892, were other key agents. [25]

In the early 1890s, Sloss sold only small amounts of pig iron directly to individual customers. It earned less than $20,000 this way during the first six months of the 1891 fiscal year. Coal, coke, and ore, however, were frequently sold directly to specific individuals or firms. Many of these were railroads, particularly lines belonging to the Danville and Terminal systems. Other customers included iron works, cotton mills, ice companies, and a brewery.[26]

Despite the improved financial health that Sloss was enjoying by 1891, earnings reported by Seddon and Sims were suspect due to inaccurate bookkeeping. Like most industrial managers at this time, they had no experience in cost accounting. In December 1891, as chairman of the finance committee, Maben sent an auditor, William Kent, to Birmingham to inspect the books. Kent concluded that these had been "honestly and faithfully kept" but noted many irregularities, particularly because inadequate allowance had been made for depreciation and the cost of "extraordinary renewals and repairs." Because of such errors, even modest earnings reported for four fiscal years prior to 1891, averaging $33,000, were misleading; in fact, Kent stated, the firm had lost almost $224,000 in this period. Even though fiscal 1891 was admittedly a good year, he continued, Seddon and Sims had gone overboard by calculating that Sloss had earned roughly $190,000 after paying interest on mortgage bonds. Kent computed the true figure at $67,241.08.[27]

IV

Notwithstanding faulty bookkeeping, Sloss's improved financial performance encouraged its owners to persevere when they might otherwise have given up and sold out. During the winter of 1890–91, Bryan gave serious thought to liquidating when he was approached by a group of businessmen led by J. J. Newman of North Carolina.[28] Nothing came of this idea, but a much more serious proposition arose early in 1892 when TCI's general manager, Shook, proposed a merger with Sloss. Adding Sloss to TCI's existing properties, Shook pointed out, would give the surviving company virtual control of iron manufacturing in the Birmingham area. It would own twenty-one blast furnaces, coal mines capable of producing 18,000 tons a day, coke ovens with a daily capacity of 5,000 tons, and more than 80 percent of the coal and ore lands in the district. Shook estimated that savings of nearly $2 million could be effected by the consolidation and calculated that the surviving firm would yield profits exceeding $2.5 million per year.[29]

Shook's proposal was characteristic of the time. During the late 1880s and early 1890s, the United States was undergoing what Leslie Hannah has called "a minor merger boomlet," fueled by the desire of industrialists to gain monopoly control over markets, achieve economies of scale, and enhance efficiency by integrating operations.[30] Shook's suggestion led to intense discussions between representatives of Sloss and TCI at the Fifth Avenue Hotel in New York City, beginning on 1 March 1892. DeBardeleben, who arrived the following day, joined the parley, fearing what might happen if he were left out, and reporters speculated that Pioneer and Woodward would also be drawn into the proposed corporate marriage.[31] The prospect that most of the furnace companies in the Birmingham District might unite gave boosters fresh visions of the marvelous destiny awaiting the area. The *Age-Herald* proclaimed that the projected merger would create "the strongest company in America."[32]

The merger talks coincided with another crisis in the Danville and Terminal systems that had brought Bryan, Christian, and other investors from Richmond to New York. Because of damaging revelations about the financial condition of these rail networks that the New York *Herald* had published in August 1891, Inman's control had been seriously weakened. In December, he was temporarily reelected president of the Terminal Company while a committee headed by New York financier Frederic P. Olcott prepared to sort out the system's tangled affairs and draft a reorganization plan. This action, however, only bought time. Inman's removal was imminent by early March 1892.

As TCI, Sloss, and DeBardeleben negotiated, rumors about the fate of the Terminal Company spread like wildfire throughout Wall Street. Speculation became even more intense when it was learned that minority stockholders of Central of Georgia, an important railroad leased by the Terminal, had gotten that line placed in receivership and were challenging the constitutionality of the lease under Georgia law. Gossip about the situation dominated financial news in such papers as the New York *Times*, overshadowing the merger talks between TCI, DeBardeleben, Sloss, and other firms at the Fifth Avenue Hotel. Everything pointed to a speedy end of Inman's tenure at the helm of the Terminal and his replacement by Walter G. Oakman, who was far more congenial to Bryan.[33]

Shuttling back and forth between New York, Richmond, and Baltimore while attending to personal and financial matters, Bryan functioned like a chief of staff while associates represented him in the Wall Street discussions.[34] Because of Maben's deep involvement in the Terminal crisis, Christian, John W. Johnston, and Seddon represented Bryan in the talks with TCI and DeBardeleben. As rumors flew that the Terminal was about to go into receivership, Maben got attention from reporters by joining Samuel Thomas in issuing public denials.[35] Meanwhile, daily front-page reports in Birmingham and Nashville newspapers predicted an imminent consummation of the impending merger between Alabama's largest ironmaking enterprises. "SIGNED: THE CONSOLIDATION SCHEME NOW A FIXED FACT," blared a headline to a story asserting that TCI's current chief executive, Thomas C. Platt, would lead the new firm and that Seddon would be a vice president. Sounding a familiar theme, the Birmingham *Age-Herald* assured its readers that the deal would result in construction of a "mammoth steel plant" and exulted, "The world is ours!"[36]

Behind the scenes, however, things were not proceeding as smoothly as readers of the stories were led to believe. Not realizing that TCI was in bad financial condition, DeBardeleben readily accepted an exchange of stock proposed by that firm's chief negotiator, Bowron, but Bryan and his fellow Virginians were not fooled. Writing in his diary on 4 March, Bowron complained that Seddon was "kicking and demanding everything in sight." "Sloss people very stubborn," he reiterated three days later after wearisome wrangling had produced an "apparent break off of everything." On 9 March, however, Bowron seemed to be on the verge of success. After three sessions of hard bargaining that had started the previous afternoon and lasted into the wee hours of the morning, he wrote triumphantly that he had reached a preliminary agreement with Johnston and Seddon. Bryan expressed satisfaction with its terms in a letter to Maben the next day.

To Bowron's intense disappointment, the pact fell through. Newspaper reports of its consummation led to a sudden rise in the price of TCI's stock. Protesting that shareholders in Sloss had gained nothing from this upturn, Christian, possibly following instructions from Bryan, objected to ratifying the agreement and demanded that an entirely new corporation be formed to buy out all three member firms instead of allowing TCI to emerge as the surviving corporation. This demand killed the deal.[37]

Despite some last-minute hitches, TCI and DeBardeleben finally did join forces, with TCI emerging as the surviving corporation. A merger was made public on 14 March and was signed by representatives of both companies four days later. By the end of the month, Aldrich's Cahaba coal mining company also joined, completing the creation of an industrial giant unprecedented by any previous southern standards. With a capital stock of $21 million, land holdings of 400,000 acres, coal mines with a capacity of 13,000 tons per day, and seventeen blast furnaces, TCI became the third-largest pig iron manufacturer in the United States, outranked only by Illinois Steel and Carnegie. As usual, DeBardeleben was not at a loss for words. "I had been the eagle eating the crawfish," he quipped. "Now, a bigger eagle than I had ever been came along and swallowed me!"[38]

While the merger between TCI and DeBardeleben was still pending, Bryan's forces persisted with bear raids to drive down the price of TCI's stock. After this effort failed, they made a futile last-ditch attempt to reopen negotiations. Upon the announcement of the merger, however, Sloss went its own way. This was a good thing for the firm. DeBardeleben and Aldrich had not known that, prior to absorbing them, TCI was teetering on the verge of bankruptcy. Within a few months, despite its enormous size, the giant combine was virtually out of money. Loaded with fixed expenses, a huge bonded debt, and heavily watered stock, it floundered badly in the ensuing national depression. While Sloss continued to follow Bryan's patient developmental path, TCI pursued its speculative way.

As usual, TCI was plagued by internal warfare as factions headed by DeBardeleben and Inman struggled for control. In February 1893, DeBardeleben launched a drive on Wall Street to corner TCI stock but was soon outwitted by Inman. By May, Inman had scooped up DeBardeleben's holdings in TCI. In the process, DeBardeleben lost most of his fortune, once estimated to have been $2.5 million. Although he was no longer one of TCI's owners, he clung to the office of first vice president for more than a year through the influence of friends in Birmingham and even managed to organize some new ore mining and ironmaking enterprises. Increasingly, however, he became estranged from even some of his closest former associates, including Roberts. In October 1894, he withdrew from TCI after winning an agreement that it would purchase ore from one of his new companies. A few days after his departure, the directors double-crossed him by reneging on the deal.[39]

V

Characteristically tight-lipped when things did not turn out as he had hoped, Bryan said nothing in his letters to indicate how he felt about the collapse of the attempted merger that would have brought Sloss within TCI's ambit. But he was highly pleased with the immediate outcome of the Terminal crisis. On 16 March 1893, after the details of the Olcott plan were made public, Inman resigned as president of the Terminal Company and was replaced by Oakman. Two other friends of Bryan, Clyde and Scott, were also elected directors of key lines in the Terminal network. The publication of Olcott's plan, which proposed consolidating the Danville and Terminal systems into one corporation, set off a furious battle between rival groups that hoped to profit from the reorganization that now seemed imminent. Whatever might happen, Bryan and his friends could be sure that their interests would be well represented.[40]

Caught in a crossfire of conflicting interests, the Olcott plan was soon scuttled. After another plan proposed by Samuel Thomas met the same fate, it looked as if the entire Terminal network would fall apart. Facing chaos, frightened investors turned to Drexel, Morgan and Company, hoping that this prestigious investment house might succeed in finding an acceptable solution. J. P. Morgan, however, refused to get involved unless financiers who had long made the Terminal a speculative plaything, including William Clyde, would promise him their cooperation and give prior consent to anything he might care to decide. This medicine was too hard to swallow, and the Terminal system continued lurching toward disaster. During the summer of 1892, things got worse as various lines connected with the troubled network, including the Georgia Pacific, went bankrupt and were placed in receivership pending resolution of the crisis.[41]

As the Georgia Pacific entered bankruptcy, Bryan stayed at the helm and continued to fight for developmental goals. He was particularly determined to spearhead the creation of a barge line between Greenville and New Orleans that would break the stranglehold enjoyed by mine operators in the upper Ohio Valley who supplied coal to the Crescent City. Securing detailed information about existing barge traffic on the Ohio and Mississippi, and collecting reams of statistics on the riverine coal trade, he bombarded Clyde, Oakman, and other financiers with pleas to support the project. In August 1892, he informed the Georgia Pacific's receivers, Frederick W. Huidekoper and Reuben Foster, that he had "$30,000 in sight" from three Alabama coal companies to implement the plan. Along with Sloss itself, two of the largest enterprises in the Warrior field, the Corona and Patton mines, were involved. But the time was anything but ripe for the idea, and Bryan got nowhere.[42]

Meanwhile, Bryan, Maben, and other Virginians fought doggedly to defend their financial interests as the Terminal system moved toward reorganization. Like other investors trying to flee from the collapsing house of cards, they cared less about the network as a whole than they did about the welfare of lines in which they were specifically involved. Seeking the best terms that they could get, they joined Clyde and Oakman in procrastinating as long as possible to avoid handing Morgan a blank check. In February 1893, having exhausted all possible delaying tactics and demonstrated their power to create mischief, Clyde, Maben, and other persons extended a fresh invitation to Morgan to step into the deadlocked situation. This time, they managed to negotiate terms agreeable to him. Maben played a key role in the dickering. During the following year, under Morgan's watchful eye, a committee headed by one of his associates, Charles H. Coster, worked out a reorganization plan favoring such senior bondholders as Bryan and carefully limiting fixed charges to safeguard future funding for orderly expansion and maintenance.[43]

Bryan gladly endorsed the plan, which replaced the bankrupt Danville and Terminal systems with the Southern Railway Company. From Bryan's developmental perspective, Morgan, who was deeply engaged in efforts to impose discipline upon large enterprises whose unruly behavior had produced endless gyrations in the business cycle, was infinitely preferable to Inman and his ilk. Bryan's cooperation with Coster and Morgan resulted in his appointment as a director in the Southern Railway upon its organization in June 1894 and a continuing role for the Bryan family in its corporate boardroom long after his death in 1908.[44] In a letter to a beneficiary of the Stewart fortune, he wrote that "the reorganization of the Southern Ry will be of very great benefit to the property interests of the 2 estates and in time I look for a large increase in your income. We have seen the worst I hope for many years to come."[45] He also lauded Maben for his part in the protracted horse-trading with Coster.[46]

Led by Samuel Spencer, a Confederate veteran from Georgia who had won Morgan's admiration for his outstanding work as an executive in the Baltimore and Ohio, the Southern became an extremely successful enterprise. Along with the L&N, it provided most of Sloss's future freight linkage with the outside world. Its senior officials included Alexander B. Andrews, a North Carolinian who had been one of Bryan's warmest friends for many years and soon became a director of Sloss.[47] Still, the new dispensation reduced Bryan's involvement in transportation. No matter how much he might prefer the leadership of Morgan and Spencer to that of Inman, he was now merely a director and no longer the president of a rail line. Whatever impact he might now have on the Southern would come from the sidelines. In the future, his role in Alabama's industrial development would stem chiefly from being the principal owner of Sloss.

Because of his troubled relationship with Seddon, who contacted him only when necessary, even this role would be hard to play. In a letter to Maben, Bryan complained that he never heard from Seddon "except indirectly."[48] "First and foremost a Virginian," as he liked to call himself, and increasingly occupied by his roles as publisher of the Richmond *Times* and head of the Richmond Locomotive Works, Bryan was much too deeply attached to the Old Dominion to entertain the slightest thought of moving to Birmingham or functioning as anything but an absentee owner of installations in Alabama.[49]

"Generally considered Richmond's leading citizen," as Virginius Dabney described him, Bryan kept track of Sloss as best he could from Laburnum, a Victorian mansion that he built on a property that he had purchased in 1883. Named for a tree that produced brilliant yellow blossoms, the ornate structure was widely regarded as the city's showcase. Because of poor drainage on his land, Bryan invested heavily in underground tiles and became expert in subsoil water management; as always, his industrial activities were deeply rooted in his agricultural background. Keeping constantly in touch with Goadby, Maben, and Mortimer in New York, he regularly visited Birmingham with them and other investors on inspection tours of Sloss's various properties. He also kept his ear to the ground about Jones Valley happenings through his brother, Ran. Meanwhile, he relentlessly prodded Seddon for information, fretting and fuming about unanswered letters.[50]

Regardless of discouragements, of which there were many in the 1890s, Bryan continued to believe in Birmingham's future. Tenaciously, he played the role of a long-term investor, spurning quick, speculative gains and keeping his eye on long-term developmental goals. Late in 1894, while successfully resisting a move by Seddon and his Birmingham allies to list Sloss's securities on the New York Stock Exchange, Bryan remarked to a friend that he had never sold a single share of his holdings in the company. Loathing everything Inman had stood for, he adamantly opposed making Sloss a target for bulls and bears. Whatever advice Seddon might get from such people as Rucker and Sims to depart from the model of a closely held corporation, Bryan was determined to protect the firm from opportunism of the type that had destroyed the Danville and Terminal systems.[51]

VI

Economic developments in the mid-1890s tested Bryan's resolve to the utmost. The reckless business tactics that had made the preceding period of boom and bust such a nightmare for a person with his developmental instincts had reached a tragic denouement. Early in May 1893, Wall Street was swept by a panic that led swiftly to one of the worst depressions in American history. By year's end, 573 banking and loan companies were out of business, and more than 15,000 firms of all sorts had gone bankrupt. As the crisis intensified, output shrank, prices plummeted, and wages were cut.[52] Waves of discontent rolled over the country, resulting in the march of Coxey's Army, the Pullman strike, and capture of the Democratic Party by free silver forces in 1896.[53]

Like every other segment of the economy, the iron and steel industry underwent sharp contraction throughout the last half of 1893 as blast furnaces, foundries, rolling mills, and steel works shut down. By the end of the year, only 137 of the nation's blast furnaces remained in operation. The immediate impact of the panic in the South, however, was less severe than in the North and Midwest. As in the past, high-cost northeastern operators were the first to suspend operations. By 1897, the moribund anthracite iron industry had been almost wiped out. Trusting their ability to compete under virtually any price conditions, large-scale, low-cost southern installations not only remained in blast but actually expanded their operations. By 1895, the average daily output of furnaces in the Birmingham District was 175 tons, up sharply from 90 tons in 1891–92.[54]

Nonetheless, the depression had a profound impact upon the iron industry in Alabama. Even before it struck in 1893, falling prices had forced eleven furnaces to put out their fires since the beginning of the decade. After the onset of the panic, ten more installations went to the wall as prices plunged to abysmal levels and inferior grades of foundry iron sold for as little as $5.75 per ton. Most of the casualties were economically marginal or technologically obsolescent operations. Some, like two of Anniston's four Woodstock furnaces, still used charcoal in making pig iron for such specialized products as railroad car wheels. Others, such as a stack built at Fort Payne by New England capitalists, used low-grade raw materials or, like Birmingham's Mary Pratt and Williamson furnaces, lacked access to their own coal and ore. A few plants, including the Hattie Ensley and Lady Ensley furnaces in Sheffield, were potentially profitable but underfunded. Some secondary processors, including the Bessemer Rolling Mills, also stopped production. Despite the region's low labor costs and other advantages on which boosters had counted heavily, the extent of the shakeout reflected the overconfidence that had prevailed in the previous decade.[55]

In such an environment, only large firms that owned their own raw materials and were capable of achieving maximum economies of scale could remain in production through drastic cost-cutting and cutthroat competition. Even the biggest of these were hard-pressed to survive. Already in shaky condition before 1893, TCI weathered the storm only because Inman and other directors extended personal loans and creditors refrained from demanding payment when cash was short. Despite such help, the firm could not meet all of its payrolls. Although the crisis bore hardest upon workers and families who took severe pay cuts or lost their livelihood, executives also felt the strain. Bowron, who moved to Alabama in 1895 when TCI shifted its operating headquarters from Nashville to Birmingham, was typical. "My health began to suffer from the constant financial strain and intense anxiety," he wrote later. "I was bilious, suffering from indigestion and insomnia, and . . . was notified by my physician that I was threatened with diabetes."[56]

Although the depression had a severe impact on Sloss, the company got through it better than most local enterprises. In fiscal 1893, after paying interest on its mortgage bonds, it incurred a net loss of $9,470.45; during the next twelve months, it went $83,868.14 deeper in the red. In fiscal 1895, however, it posted a net profit of $20,540.96 and bettered this figure with earnings of $22,120.88 the following year. Sales of coal, coke, and ore to outside customers, commissary receipts, and rents from company-owned housing helped it to survive. Income from such sources aggregated $256,080.42 in fiscal 1893, $204,832.98 in 1894, $229,045.93 in 1895, and $241,024.07 in 1896. [57]

By contrast, Sloss lost a total of $37,881.06 on pig iron production in fiscal 1893 and 1894. After producing these losses, however, the furnaces began to yield small profits, earning $26,625.97 in fiscal 1895 and $45,871.57 in 1896. [58] High fixed costs imposed by heavy bonded indebtedness left the company no choice but to keep its furnaces making as much iron as possible, depression or no depression, however savagely this course of action might force it to compete with other operators fortunate enough to stay in business. Large-scale producers throughout the country, North and South alike, tended to do the same; as Andrew Carnegie said, "the condition of cheap manufacture is running full." [59] Following the advice of J. C. Brooks, a Philadelphia foundry owner who had urged him early in the decade to "sell iron at any price it would bring," Bryan pressed Seddon to keep the furnaces going steadily while cutting costs to the bone. "If we can find somebody to buy our products," he said grimly, "we will soon learn to make them cheap enough." [60]

The company's use of a low-paid, largely black labor force also constrained it to avoid shutdowns. As Culverhouse stated in a report to Seddon in 1896, blacks tended to "scatter to the farms" when they were not steadily employed, and it was hard to get good operatives back once they had been lost. As a result, one had "to figure to give them enough work at all times to make a living, or you will lose them." Culverhouse was referring to miners, but his logic applied equally well to furnace crews. [61]

Efforts to sustain high levels of iron production during the depression led to a major shift in marketing strategy at Sloss. Though continuing to ship most of its output to commission merchants, it sold pig iron directly to sixty-four individual customers in fiscal 1894. Some of these, including the Oregon State Stove Foundry and the San Francisco Stove Company, were far from Alabama. Sales to such firms aggregated $55,012.08, as against only $2,401.62 in the previous twelve months. Having long practiced what economists would later call "backward integration" by controlling access to raw materials, Sloss was now starting to use "forward integration" by taking distribution and sales into its own hands. [62]

Sloss followed a national trend in making this change. During the late nineteenth century, as Glenn Porter and Harold Livesay have indicated in a detailed study of marketing practices, many large-scale producers of iron and steel found that it cost less to establish sales offices in large cities than to pay commission merchants for their services.[63] Continuing this policy in the future, Sloss developed a far-flung system of sales representatives in large urban centers, both in the United States and abroad. Surviving records do not shed light on the extent to which the company created such a network in the mid-1890s, but the increasing numbers of customers who were receiving direct shipments show that the change was starting to take place at this time.

Sloss also played a pioneering role by becoming the first iron producer in the Birmingham District to exploit foreign markets. Bryan, whose Richmond Locomotive Works was shipping engines to such places as Chile, Finland, and Sweden, and who strongly advocated the creation of a Nicaraguan canal, was probably responsible for this innovation. In March 1894, Sloss began shipping pig iron to Great Britain, sending 100 tons to New Orleans via the L&N for consignment to Liverpool. By 1896 its exports of iron had jumped to 26,250 tons, followed by further increases to 50,866 tons and 66,927 tons in 1897 and 1898 respectively. At one point the firm was shipping approximately one-third of its pig iron to foreign countries, not only to such relatively nearby places as Canada and Latin America but also to such distant countries as Australia, India, and Japan. Not until domestic markets revived late in the decade did exports fall off, shrinking to 30,300 tons in 1899. Copying Sloss, TCI also began developing a large foreign trade.[64]

Sloss also found a new market for iron in the United States Navy, which was expanding rapidly in the 1880s and 1890s as the nation began breaking away from isolationism and scanning the rewards of imperialism. Among other vessels, the battleships *Oregon* and *Texas* were made partly of iron from Sloss Furnaces. Its use in the first of these two ships brought Sloss momentarily into the public eye after a critical battle in the Spanish-American War. As the prospect of armed conflict with Spain became imminent in March 1898, the *Oregon*, the newest and most advanced ship of its class in the American fleet, was dispatched to sail as quickly as possible from San Francisco to the Caribbean. Racing around South America, it reached Florida late in May after an epic voyage of sixty-eight days, enabling it to share in the American victory at Santiago Bay. Asked to explain the ship's phenomenal endurance, the navy's chief engineer, Adm. George W. Melville, said, "After all, you know, she was built of Sloss iron."[65]

VII

Uehling deserved much of the credit for the ability of the company's blast furnaces to stay competitive despite plummeting prices. Early in the decade, with Bryan's support, he conducted furnace modifications that paid off in lower operating costs. Shortly after Uehling came to Birmingham in 1890, the older of the two City furnaces burst its jacket. Over the opposition of Seddon, who objected to the cost, the directors approved a proposal by

Uehling not simply to patch up the chronically ailing facility but also to alter its original design by lowering the bosh and installing a new two-pass stove of his own design. Because Uehling had trouble getting parts from local machine shops—another sign of the district's inadequate infrastructure—the project went slowly, and the furnace remained out of blast longer than he anticipated. Its subsequent performance, however, was praised by the owners.[66]

The directors next permitted Uehling to overhaul the North Birmingham furnaces by installing new return tubular boilers, feedwater purifiers and heaters, and a central draft stack. In addition to exceeding the original cost estimates, these repairs were less successful than those performed on the City furnace. According to Uehling, this failure was partly because the existing stoves, which were retained, "would not stand up under the blast pressure necessary to produce the expected output of iron." Problems persisted, however, even under a reduced rate of driving. Uehling traced them partly to defects in the original jacket design that prevented him from modifying its proportions along lines he had followed in reconstructing the older City furnace. In this way, mistakes made by the original designers, stemming from unfamiliarity with local raw materials, continued to plague the firm. Uehling also complained that, whereas the City Furnaces had their own coke ovens, those at North Birmingham were supplied from Coalburg, where coking operations were not as well managed. His efforts to get Seddon to rectify this problem went unheeded.[67]

Seddon's opposition to the cost of Uehling's modification program intensified after the depression struck. Bryan, however, continued to support Uehling, helping him secure authorization to finish the program by remodeling the second City furnace. After that project was completed, Bryan praised the results and reminded Seddon pointedly that "we would not be making money of any kind today if we had not heretofore spent money to keep up & improve our plant." At the same time, he repeated earlier complaints about being continually kept in the dark by Seddon and Sims. "I have to depend upon information accidentally received about matters of the greatest interest to me," he declared.[68]

Bryan's annoyance reflected renewed conflict within the board of directors about Seddon's capacity to provide effective leadership as the depression wore on. Because of the firestorm that would have arisen among Seddon's supporters in Birmingham and his friends and relatives in Richmond and Baltimore had his dismissal been proposed, the old idea of appointing a general manager was revived. In Birmingham, Seddon's allies lined up behind veteran ironmaster J. H. McCune, who had rendered effective service to the Woodwards, as a suitable candidate. Although he liked Uehling, Bryan also admired McCune.[69]

Morton had other ideas. Still a member of the board of directors, he predictably favored promoting his protégé, Uehling. He also thought well of William Kent, who was not likely to appeal to Seddon and Sims because he

Edward A. Uehling. A graduate of Stevens Institute of Technology, Edward Uehling was one of the two greatest engineers in the history of Sloss Furnaces. While living in Birmingham, Uehling invented an automatic pig-casting machine that was ill-suited for the labor-intensive southern approach to ironmaking but well fitted for the highly mechanized style of blast furnace practice emerging in western Pennsylvania. A photograph of Uehling taken in his prime suggests the creative genius that enabled him to conceive and design a highly complex apparatus that existed only in his fertile imagination before its successful adoption by Andrew Carnegie. (Courtesy of S. C. Williams Library, Stevens Institute of Technology, Hoboken, New Jersey.)

had recently criticized their accounting methods. Writing to Uehling in early February 1894 about rumors that were flying around Birmingham as possible changes were being discussed, Morton cautioned his former student to "be on your guard" and avoid hasty action. "As you know," he stated, "I have never had any confidence in Mr. Seddon and believe that if he thought he could do so he would put in McCune as Gen. Manager and get rid of every one who was not directly dependant [*sic*] on himself." Uehling's best policy toward Seddon, he advised, was to "keep on the best of terms with him but never to *trust* him and to be careful not to say or do anything which he can use against you." Morton assured Uehling that "the members of the board here have the highest opinion of your ability . . . and I believe that they would be glad to get rid of S. if they could do so easily, but they have little back bone." Asking for Uehling's opinion of Kent, he added, "Do you not *now* feel that you would make a good General Manager? If not why not?"[70]

Morton, however, was outmaneuvered by Seddon, who had never accepted Uehling despite the promises he had made to cooperate with the young engineer at the time of Dashiell's dismissal. Brazenly disregarding the esteem Uehling enjoyed among the directors—indeed, partly because of it—he asked Uehling to resign. Tired of the obstacles that Seddon and his friends had continually put in his way, Uehling complied. Confronted with a fait accompli, the board acquiesced, and McCune was hired as furnace superintendent. "I am as disgusted with what has happened as you can be," Morton wrote Uehling, "but I want to vindicate you when the chance comes and my own advice and efforts in getting you down there." Once more, Seddon had shown how hard it was for outside directors to control events in faraway Alabama. Writing to Maben, Bryan acknowledged that Seddon had made a number of mistakes but approved his "general administration of the property" and concluded that his advice should be followed "as long as he is the responsible officer in charge."[71]

VIII

After resigning, Uehling stayed in Birmingham temporarily to work on several projects that he had already begun at Sloss. The most important of these was a new way of casting pigs mechanically. Uehling had devoted much thought to the problems inherent in time-honored methods of molding pig iron by hand in sand beds. "The product is rough and irregular and the operation is very wasteful, producing large quantities of scrap, which must be remelted, and carries much sand with the iron, which reduces the value of the iron," he wrote. "Moreover, this method of running the iron from the furnace into the sand is under very poor control. It sometimes runs so slowly that the iron chills in the long runner before it reaches the lower pig-beds. At other times the running of the iron is with such a rush that the cores, which form the partitions between the pig-molds, are washed away. . . . In either case the iron must be taken up from the sand beds and remelted to make it marketable, besides the expense of having to break it up and remove it from the beds." Sometimes, two or more grades of iron resulted from the same melt. This was because some parts of the melt had too little silica and graphitic carbon and others too much.

Bad working conditions also affected the quality of the product, particularly in the South, where discipline among unmotivated black workers was poor.

Not only because of malingering and inefficiency, but also because of increasing levels of output resulting from larger blast furnaces that now ranged up to seventy-five feet in height, it was becoming harder to clear casting beds of pig iron resulting from one melt before the next was ready to be emitted from the crucible. Sometimes, if the flow of molten iron got out of control during the searing heat of an Alabama summer, a near-catastrophe might take place in the casting process when the beds become "sheeted." Jumpers—multiple pigs that were formed when the ridges of sand that would ordinarily have separated them got washed away—would become attached to one another by layers of iron accumulating to a depth of six inches or more. As the air rang with curses, all available hands would be summoned to break the coagulated mass apart while the blast was turned off to avert a possible breakout of the next melt. Half a day's production might be lost before normal operations were resumed.[72]

Uehling's desire to mechanize pig-casting was not motivated solely by a desire to improve output and cope with technical problems. Running through an unpublished autobiography and an article that he later wrote on his automatic pig-casting system is an unmistakable current of moral earnestness, showing that he was emotionally repelled by the sheer drudgery of the work required in the casting sheds. "The task of breaking and carrying out the iron from the casting beds of even a moderate sized furnace is not a fit one for human beings," he stated. "If it were possible to employ horses, mules, or oxen to perform this work, the Society for the Prevention of Cruelty to Dumb Beasts would have interfered long ago, and rightfully so."[73]

For both technological and humanitarian reasons, therefore, Uehling devised an automatic system, worked out on paper but never implemented in practice during his years in Birmingham, to replace hand-molding techniques. His invention was elegantly conceived and was all the more remarkable because he had no means of making a prototype or model of his apparatus. The large size of the equipment required for his casting system, which had to work as a totality or not at all, was essential to its proper functioning. Building a prototype would have necessitated space, materials, and levels of funding that the Sloss Company could not have justified, even had its owners and managers been sympathetic to Uehling's ideas and purposes. That the machine existed only in its inventor's mind until after he left Birmingham testified to his creative genius.

In Uehling's system, molten iron from a blast furnace was tapped through a runner into a large mixer of forty or fifty tons capacity. Because the surface area of the iron in the mixer was small by comparison with what it would have been had the same substance been distributed throughout a maze of channels in a casting-bed, the iron remained hot for several hours and could "be treated in any manner that may seem expedient to improve its quality or change its character before tapping it into molds," as Uehling wrote in a subsequent patent application. After the iron had been mechanically stirred and blended, the mixer was tilted forward to discharge its molten contents through a spout

E. A. UEHLING.

APPARATUS FOR AND METHOD OF CASTING AND CONVEYING METALS.

No. 548,146.   **Patented Oct. 15, 1895.**

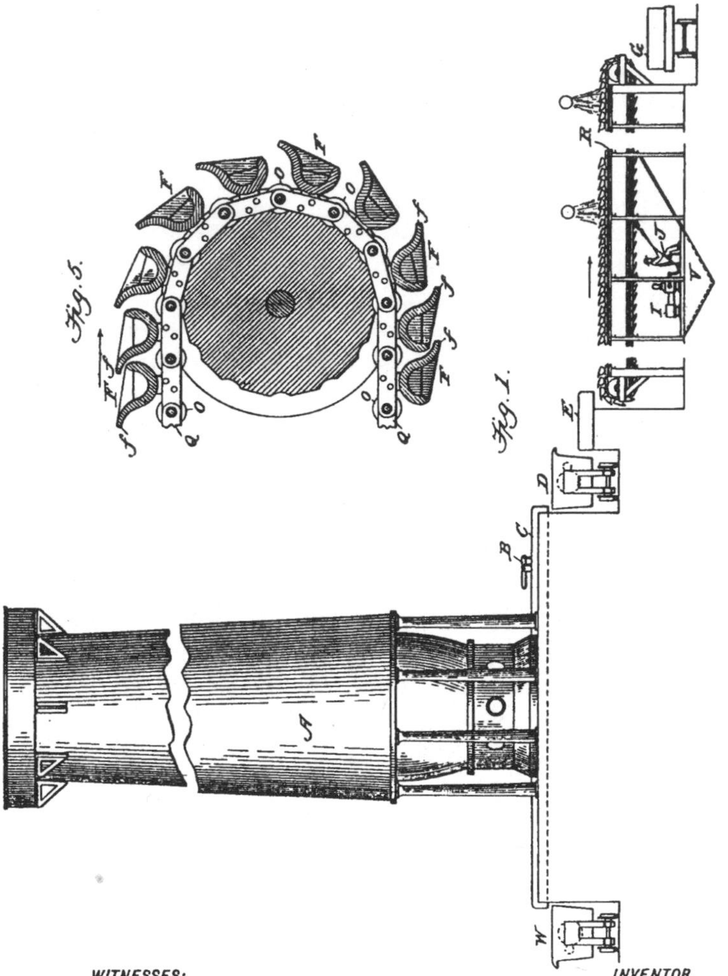

Fig. 5.

Fig. 1.

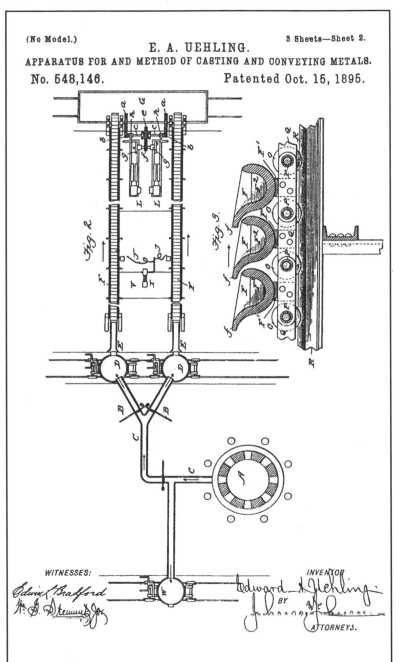

Uehling Pig-Casting Machine, 1895. These patent drawings show Edward A. Uehling's pig-casting machine. Every detail of this remarkable invention epitomizes the careful thought Uehling bestowed upon it while living in Birmingham, where he had no chance of building a prototype to test his ideas. Note, for example, how the overlapping containers in which pig iron was molded are designed to prevent spillage, and the way in which the spray rig was conceived to prevent wastage of graphite. The partitions in the moving containers were designed to make it possible to mold two pigs at once, but this feature was not used in most applications of the system. (U.S. Patent Office, Washington, D.C.; courtesy of Ralph Brown Draughon Library, Microforms and Documents Department, Auburn University, Alabama.)

into an endless chain of bucket-shaped molds. Because these containers were also made of iron, Uehling saw that their interior surfaces would have to be coated with a material that would prevent fusion with the metal they were carrying. He selected graphite as a suitable material for this purpose, but it was later found that a mixture of coal dust and powdered limestone would serve just as well and be much less expensive. In Uehling's original conception of the machine, each mold was partly divided by a transverse rib that enabled two pigs to be cast at once in the same container, connected by a thin coupling so that they could be readily knocked apart. This, however, was an unnecessary refinement.

The molds traveled forward on an endless chain made of interconnected links held together by axles whose projecting ends terminated in flanged wheels that ran on rails situated on either side of the links. The rails were in turn mounted in a supporting structure so that the flanged wheels would be held in place regardless of whether the molds were facing downward or upward. At each end of the structure were steam-powered drum shafts. The system was designed so that two or more strands of mold-conveyors could be operated simultaneously. Every detail of the drive mechanism and the molds themselves exemplified the care that Uehling took to make sure the equipment functioned smoothly. For example, the rear side of each individual mold was curved in such a way that the lip overlapped the front of the mold immediately behind it. This feature prevented spillage of molten iron that would otherwise have dripped down onto the endless drive chain carrying the molds forward, making the system inoperable.

Because the traveling molds were open to the ambient air on all sides and were completely exposed at the top, heat in the molten iron that they were carrying dissipated much more rapidly than was possible in beds of sand on a casting floor. Depending on the rate of speed at which the machinery was operated, Uehling calculated that the contents of the molds might solidify within a distance of fifty feet, because it would become safe to spray the iron with water as it began to harden. Moving ahead and cooling as it traveled, the metal in the molds hardened into billets of pig iron. As the endless conveyor reached a revolving drum at the opposite end of the line, the molds tilted downward and the billets fell into a waiting rail car. As the molds emptied, they became inverted for the return journey to the mixer.

Now facing downward, the molds passed above a reservoir filled with a graphite-water mixture. The water served as a medium to distribute the graphite on the inner surfaces of the molds. Using a pressurized hose with a nozzle specifically designed to insure even distribution, a worker coated the molds with the water-graphite solution as they passed above a spray rig. Because of the cost of the graphite, it was essential to conserve as much of that material as possible. Uehling therefore designed a dashboard that channeled splashing liquid back into the reservoir for reuse. Uehling also saw that heat radiating downward from unsolidified iron in the forward-moving molds passing overhead would evaporate water in the inverted molds as they gradually approached the mixer, leaving only a film of graphite remaining by the time they turned upward to be refilled and began another cycle of operation.

With great ingenuity, Uehling had contrived a series of components that would cast pig iron automatically in identical units with uniform chemical characteristics. The system was patented on 15 October 1895. Collaborating with Alfred Steinbart, a German engineer who came to Birmingham in 1893, Uehling also invented a pyrometer that accurately measured the heat of air supplied to a blast furnace by stoves, thus enhancing efficiency. Patented on 11 February 1896 and tested at Stevens Institute at temperatures up to 2,500 degrees Fahrenheit, this instrument became the standard device of its type.[74]

Despite his difficulties with Seddon, Uehling continued to work for Sloss as a consultant. One of his assignments was to evaluate Fortimolite, a supposedly improved form of pig iron involving the use of graphite-treated sand molds. Rights to the process had been offered to both TCI and Sloss by two promoters who were typical of the visitors attracted to Birmingham by the prospect of making money from new ideas, most of which were of little value. Nothing came of the Fortimolite project, because costs outweighed whatever benefits might be gained, but Uehling was still bargaining with Sloss for possible royalties resulting from its use by prospective licensees at the time he left Alabama to start producing his pyrometer in partnership with Steinbart at Newark, New Jersey. Morton tried to work out satisfactory contractual arrangements for Uehling in case Fortimolite caught on within the industry, but he failed to make headway with other board members, whom he called "economical cranks."[75]

It is not clear precisely when Uehling left Birmingham, but he soon returned to the North. By letting him go, Sloss lost one of the two most creative engineers ever to work for the company; not until James P. Dovel joined the firm in the twentieth century would it be served by a person of comparable technical ability. Uehling's experience at Sloss invites speculation about what might have happened had things turned out differently. This is particularly true with regard to his mechanical pig-casting method, which was not adopted by Sloss when it was invented but would ultimately be adopted in a modified form by its successor, Sloss-Sheffield. Had it been implemented by Sloss in the 1890s, it could have revolutionized the company's operations and enabled it to rival TCI, which was then about to introduce the South's first effective steelmaking technology, as an innovative force in the southern ferrous metals industry.

Even if relations between Uehling and Sloss had been ideal, however, Sloss would not have adopted the young engineer's pig-casting technique. Despite the fact that Bryan greatly admired Uehling, there is no evidence that he gave the invention so much as a passing thought. It would have been strange if he had, because it was antithetical to the entire economic and social milieu within which southern blast furnace practice existed. As had been true from the inception of ironmaking in the Birmingham District, the strategy governing production there was predicated upon taking advantage of low-cost black labor, close proximity to all the raw materials required for smelting, and short interior supply lines to compete with firms outside the region. This strategy compensated for one of the basic problems facing southern industrialists: their lack of the large amounts of investment capital that would have been needed for the degree of mechanization that northern iron and steel producers had been implementing since the 1870s. In the early 1890s, northern capitalists were still limiting their southern investments mainly to railroads. Raising the money required for untested machinery of the type that Uehling had designed would have been impossible to begin with and would have imperiled control of the firm by Bryan and his fellow Virginians even if adequate funding could have been found. In addition, the peculiarities of southern raw materials still prevented the region from exploiting growing markets for pig iron that would in turn be converted into steel. Because outlets for southern pig iron were still limited to foundries and rolling mills that made products from cast and wrought iron, ironmasters in places like Birmingham could not aim at production levels compatible with a high degree of mechanization. Instead, they mechanized only selectively, as Bryan had done by authorizing the use of advanced equipment at the Brookside mine when this step was demanded by the nature of the coal seam at that specific location.

The very nature of southern society made it impossible for Sloss to give serious consideration to Uehling's pig-casting machine. Despite the presence in Alabama of such northern-born engineers as Uehling, the South also lacked enough native technicians to consider extensive mechanization as an alternative to labor-intensive technologies utilizing cheap raw materials. Even more important, southern iron producers remained committed to using unskilled and semiskilled black workers as a matter of social policy. Employment practices remained predicated on the idea that black workers were particularly well suited for hot, dangerous, and dirty operations requiring brute strength and that using them in such a way left whites free to engage in other, more desirable types of work. This policy avoided mixing blacks with whites as equals in the workplace, a practice to which southern entrepreneurs were unalterably opposed.

Economic and social conditions were different in the North. Since the 1870s, furnace operators in Pittsburgh and other cities had been committed to making lavish outlays to bring raw materials long distances in huge quantities to enormous plants where vast investments in mechanization would reduce unit costs of production to the lowest possible extent. Unlike southern raw materials, the natural resources available to northern producers from such places as the Lake Superior region and the Connellsville District were highly suitable for steel manufacture. Blast furnaces capable of making 300 or more tons of pig iron per day for sprawling installations like Carnegie's Edgar Thomson Steel Works fitted naturally into the northern strategy for high-volume ferrous metals production. Uehling was correct to state in his autobiography that it was becoming impossible by the early 1890s to clear casting sheds of the volume of pig iron that the largest northern furnaces could emit before the next melt took place, but he neglected to point out that this was only one feature of a much larger system that had become diametrically different from the southern approach, based on totally different economic assumptions.

Furthermore, the racial overtones present in southern ironmaking were not present in the North. Furnace operators in such places as western Pennsylvania were shifting to immigrant work crews composed of men who often spoke a myriad of languages, but they had no shortage of technical personnel to service complex machines, and had no overriding need to preserve a racially segregated society. By the early 1890s, northern and southern blast furnace practices had become two different worlds. At its Duquesne furnaces in western Pennsylvania, where production was much larger than at Sloss, the Carnegie Company was already using electrically powered traveling cranes and heavy mechanical pig-breakers to cope with some of the problems with which Uehling had been wrestling in Birmingham.

Even before Uehling left Alabama, Carnegie had caught wind of what he was doing and "sent a representative to Birmingham to negotiate for the purchase of my patent," as Uehling later wrote. Uehling asked $50,000 for the right to use his machine any place in the United States, $20,000 for its use in Pennsylvania's Allegheny County, and $10,000 at a single furnace. Ultimately, he settled for a $2,000 payment from Carnegie for a ninety-day option to test the device at the latter's Lucy furnaces. Amply funded by Carnegie, engineers and technicians used Uehling's patent sketches to improvise the apparatus needed to carry out the experiment.

Lucy's superintendent, James Scott, had already been thinking along lines analogous to what Uehling had been doing in Birmingham. Scott, however, had been pursuing a method that Uehling later described as "even more radical than mine," by pouring a thin stream of pig iron into water, causing it to granulate, and converting the resulting granules directly into steel. As Uehling observed, the technical problems involved in the process were formidable, because "iron in granular form is exceedingly difficult to handle, and . . . cannot be stored for any length of time, because it rusts up very rapidly due to the excessive surface exposure to atmospheric moisture."

As soon as he saw Uehling's patent sketches, Scott realized that their underlying conception was superior to his own. "Uehling has got what we want," he said. Scott's judgment was confirmed by the Lucy trials. "Due principally to the practical ability and honest effort of Mr. Scott," Uehling later wrote, "the casting machine was, from the start, an unqualified success. The Lucy furnaces became a real Mecca for doubting iron men, as well as interested blast furnace managers. They came and were convinced that the output of the largest blast furnace could be successfully handled over a machine from the furnace to the railroad car without being touched by hand."

Concentrating for the time being on producing pyrometers at his shop in Newark, Uehling assigned a representative, J. W. Miller, to negotiate the details of a new agreement with Carnegie; the deal involved Carnegie's paying royalties to Uehling instead of the flat fee that the inventor had originally suggested. Soon, Uehling's device was adopted by other pig iron producers in western Pennsylvania and Ohio's Mahoning Valley under license from the American Casting Machine Company, which was established to control Uehling's patent. Later, the system was installed by European furnace operators, led by owners in Germany and Austria-Hungary. Four decades later, Uehling stated that his apparatus had "stood the test of time, and is now in use at practically all blast furnaces."[76] By that time, the roster of corporations using automatic pig-casting would include the Sloss-Sheffield Steel and Iron Company, the successor of Uehling's former employer.

IX

Added to all the other circumstances that prevented Sloss from considering the adoption of Uehling's pig-casting machine in the early 1890s was the fact that it was still primarily a coal-producing, not an ironmaking, enterprise. In his autobiography, Uehling stated that Seddon regarded the company's coal mines as his "pet department." There was every reason for Seddon to feel this way; sales of coal and coke, which were already the main sources of Sloss's profits in 1891, remained so throughout most of the decade. In fiscal 1893, Sloss lost $19,392.29 on pig iron production but sold 695,466 tons of coal for a net gain of $90,157.11 and 113,938 tons of coke at a profit of $66,594.25. Combined with receipts from company stores, rents from housing, and other miscellaneous income, these sums enabled the company to come out of the year with net earnings of $51,693.30 despite the onset of the depression. Over the next two years, earnings from coal and coke far exceeded those from pig iron.[77] Because of its high output and the profits it generated, Coalburg was the chief jewel in the company's crown throughout most of the 1890s. Its production was much greater than that of the mining camps at Blossburg, Brookside, and Cardiff and was not exceeded by that of the Brookside facility until late in the decade. But Coalburg was also an extremely controversial asset, because it was run largely with convict workers. Partly because of this practice and also because of the company's need to keep costs as low as possible, Sloss became deeply embroiled in labor disputes as the depression that began in 1893 became increasingly severe.

Sloss's failure to secure state convicts in 1888, when it had lost out to TCI in bidding for these inmate workers, forced the company to rely on county and local offenders, for which it paid an average of $9 per month. The social and legal system that permitted this practice to exist was rife with corruption and inequity. Sheriffs made their living from fees exacted from persons found guilty of petty offenses. These delinquents were often too poor to pay fines or court costs and were therefore put to hard labor with large companies as a way of collecting the amounts involved. Pocketing part of the payments, law enforcement officers had every incentive to arrest as many people as possible. Largely consisting of blacks and indigent whites, the victims of this self-perpetuating system were sent to places like Coalburg. Most of the inmates who were sent there worked in the mines, but a few lived on a farm that supplied food for the prisoners. A few female offenders, for whom the company paid about $4 per month, lived in a separate area, doing washing, sewing, and light agricultural work.[78]

The use of convict labor by Sloss and TCI aroused indignation among Alabama's free coal miners, who protested that the system deprived them of jobs. Complaints mounted in 1893 as the depression produced a wave of layoffs and wage cuts. Strikes against both TCI and Sloss took place in June and July of that year; the following April, an even larger walkout involved about 6,000 miners. TCI was the primary target, but Sloss and other companies were also affected. In addition to free miners at Coalburg, the strikers included workers from the Blossburg, Brookside, and Cardiff mines.[79]

Confronting more determined opposition from organized labor than ever before, administrators at TCI and Sloss looked for help to elected officials who controlled the state government in Montgomery. Known as Bourbons because of their strongly traditional outlook, the latter were already worried about the rise of populism in the state's poverty-stricken agricultural districts. Concerned that discontent among farmers and laborers might overturn the political order that had been dominant since the restoration of home rule in the 1870s, Bourbon leaders were also fearful of rebellion by rural sharecroppers and haunted by the possibility that blacks and whites might join forces, destroying the prevailing political order. This coalition might in turn lead to a new situation much like the Jacksonian order that had prevailed during much of the antebellum era but would actually be far worse in that it would include blacks as equal participants. That powerful downstate planters would join hands with Birmingham industrialists against such an eventuality was only natural. It is misleading to argue, as Jonathan Wiener has done, that the Big Mule–Black Belt alliance that proved to be an effective force against social discontent, both in the 1890s and later, was a sudden development. The fierce struggles that erupted near the end of the century were deeply rooted in the same cleavages between Broad River planters and small farmers that had figured prominently in Alabama politics before the Civil War.[80]

Closing ranks with upstate industrialists against the radical menace, Bourbon officials prepared for what loomed as a potentially bloody confrontation. While governor Thomas Goode Jones readied troops to quell violence, TCI and Sloss fired free workers who dared to take part in the strike, evicted them from company-owned housing, and imported substitutes whom the strikers called Blacklegs. At Brookside, Sloss hired a hundred such persons and protected them with fifty deputies; critics charged that the coal they mined cost more than $2 per ton, an exorbitant figure at the time. Building temporary huts in the woods, workers and their starving families could not help but resent the strikebreakers, who came from as far away as Kansas and were mostly black, intensifying racial tensions already prevalent in southern society. Leaders of the United Mine Workers of Alabama, under whose aegis the strike had been called, officially deplored violence, but individual strikers who were threatened with the permanent loss of homes and jobs could not be restrained. Dynamiting a mine at Horse Creek in May 1894, they also attacked bridges and trestles on the trackage of the L&N and the Georgia Pacific, trying to prevent them from hauling coal.[81]

After another outbreak of violence at the Pratt mines on 9 July, Jones intervened with armed force. Many miners were arrested, and the attitude of such companies as TCI and Sloss hardened. The last hopes of the miners died early in August when William C. Oates, a Silk Hat Bourbon Democrat who was even more conservative than Jones, won election to the governorship over Reuben F. Kolb, the reform candidate supported by Wool Hat farmers and strikers. Later that month, most miners striking against TCI accepted a compromise settlement and returned to work despite having previously pledged not to make a separate settlement that did not include Sloss employees. The latter capitulated to Seddon and Rucker, who, with Bryan's support, had been unyielding in negotiations with union representatives.[82]

Besides failing to gain their wage demands, the strikers lost their battle to abolish convict labor. In 1895, however, a protest movement, led by Birmingham physician Thomas D. Parke, was launched against the system on humanitarian grounds. Conditions at Coalburg furnished all the evidence that the reformers needed. With the exception of females, the convicts who worked there were confined in a prison made out of rough planks, located on the bank of Five Mile Creek near a battery of coke ovens. Working from nine to eleven hours a day, six days a week, they almost never saw daylight. Efforts by the company to obtain the highest possible production of coal at the lowest attainable cost resulted in terrible suffering among the inmates. Only one fan was used to ventilate the mines; ditching and pumping were inadequate; brattices—partitions used to control air circulation—were allowed to deteriorate without proper repair; long tramways led through low-topped passageways along which the convicts pushed carts of coal on wooden rails to the surface. A rude hospital, made from the same materials as the prison and described by Parke as inadequate, was the only structure provided for the many offenders who became ill. Comparable to buildings that had been maintained for a similar purpose by industrial slaveholders in antebellum times, it housed convict miners who suffered from such maladies as tuberculosis and erysipelas. The latter disease, known as sore leg, resulted from cuts or abrasions that, if not properly treated with carbolic acid to kill streptococcal bacteria, could result in extremely high fevers and incapacitating infections. The death rate for the two years before 1895 was 90 per 1,000—much greater than comparable rates at prisons in such states as Illinois, Michigan, Ohio, and Pennsylvania. Parke did not merely confine himself to advocating better medical care at Coalburg, but argued that the convict lease system itself was responsible for the scandalous conditions he described.[83]

Embarrassed by Parke's claims, Sloss disputed his findings in a pamphlet written by one of Seddon's representatives, J. W. Castleman, and by prison physician F. P. Lewis. According to Lewis, Parke had reached his conclusions through "intuition, not . . . facts." Admitting that mining was an inherently unhealthful occupation because it was necessarily conducted underground in the absence of sunshine, the company protested that it could not be expected to install skylights. Nevertheless, as state prison inspectors began to relocate some convicts to Lee County's Chewacla Lime Works and the state farm at Speigner, public indignation forced Sloss to take limited action to quiet the mounting clamor. In March 1896, Castleman reported to Seddon that changes had been made "to improve the mine and prison conditions, on the theory that the *ultimate* business interest of the company was to surround the convicts with such conditions as would keep them healthy and in shape to do the largest amount of work possible without breaking them." The ventilation system was remodeled, giving each mine a separate air intake; bratticing was repaired; air courses were cleared; drainage was improved; convicts were now taken to the surface by rail instead of being forced to trudge long distances through muddy walkways in a stooped position.[84] As the controversy abated, Sloss began to increase production at its other coal-mining communities so as not to be so dependent on Coalburg. Meanwhile, its use of convicts continued, for the firm never accepted Parke's claim that the lease system itself was inherently bad.

X

As Seddon cleaned up the worst aspects of the Coalburg situation early in 1896, there seemed little reason to think that his long and often stormy administration, now entering its tenth year, was about to end. On 9 May 1896, however, he became ill at the home of a friend, was put to bed, and died the following day. "Only a few of his acquaintances knew that he was at all indisposed," stated his obituary in the Richmond *Times*, "and they thought him only temporarily affected by a bilious attack." In fact, Seddon had been suffering from Bright's disease and had contracted pneumonia. Because he was so widely admired in the Magic City, whose interests he had supported in his battles with absentee owners, the local press heaped praise on him. The *State Herald* called his death a "public calamity," declaring that "his was the magic hand which made everything in and around the property of the company flourish and improve."

A grieving Edmund Rucker helped escort the body of his fallen friend to Richmond for burial. After it had been laid to rest, Bryan composed a glowing tribute that was unanimously adopted by the directors. In it, he referred to the dark days through which the industry was still passing. "Through this period of trial, which called for the exercise of the highest qualities of judgment, courage, and executive ability, Mr. Seddon successfully conducted the affairs of this great corporation," he stated, attributing the doubling of iron production and tripling of coal output that had taken place at Sloss in the past decade to the fallen executive's guiding hand. Bryan also lauded Seddon's personal attributes. "Ever easy of access," he stated, "whether sought by those who enjoyed power and influence or by the day laborer, by his unvarying kindness and absolute impartiality he gained our confidence and won our love."[85]

In fact, as every member of the board knew, Bryan could have felt little but relief as he extolled Seddon's achievements, for the latter had been anything but a model of successful leadership. Whatever gains Sloss had made in iron production were largely attributable to Uehling, and Bryan himself was more responsible than any other person for the strategies that had guided the firm through the depression. Nonetheless, the satisfaction that Bryan took in the way the company had weathered a severe economic crisis that had killed off many other southern enterprises was genuine. Within a year of Seddon's death, economic conditions would brighten, and the enterprise would enter a fresh phase of expansion and growth. It would carry with it, however, a heavy burden of ill feeling stemming from the way in which local forces in Birmingham and absentee owners in Richmond and New York had disagreed about what goals it should pursue and how it might best attain them. Under another president who shared Seddon's willingness to lock horns with the owners, this would ultimately produce renewed conflict and a major decision involving market orientation. Such things lay in the future as the firm emerged from a turbulent decade and became involved with TCI in a scramble to buy up a number of Alabama firms that had failed to survive the storm.

Brown Ore, Basic Steel, and the Emergence of Sloss-Sheffield 8

In his classic study, *Origins of the New South*, C. Vann Woodward pointed out that northern investment below the Mason-Dixon Line accelerated after the Panic of 1893, producing a "colonial economy." Morgan's reorganization of the former Danville and Terminal systems into the Southern Railway was followed by expansionist moves into other parts of the South as Yankee capitalists including Henry M. Flagler and Henry B. Plant created vast rail networks to previously undeveloped places such as Miami and Key West. After the Spindletop Gusher of 1901 was brought in, northern interests began exploiting southern oil deposits; soon, enormous reserves of bauxite, manganese, phosphate rock, and sulfur also fell into Yankee hands. On another front, New Englanders bought up textile mills that had once belonged to southern investors.

Before this time, the South, though hospitable to northern capital, had funded much of its own industrial development. Now, devastated by the depression, it was losing whatever command of its economic destiny it had once possessed. Sympathetic northerners warned southerners what to expect. "Our capitalists are going into your country because they see a chance to make money there, but you must not think that they will give your people the benefit of the money they make," a correspondent from Massachusetts wrote to the *Manufacturers' Record* in 1895. "Your people should not be dazzled by the glamour nor caught with the jingle of Northern gold, but they should exact terms from these men that will be of benefit to the communities in which they may select to establish themselves." The cost of ignoring such advice was severe. Partly because it went unheeded, Dixie became, in Woodward's words, "a raw-material economy, with the attendant penalties of low wages, lack of opportunity, and poverty."[1]

No part of the South would pay more heavily for its dependence on northern decision-makers than Jones Valley. Nonetheless, boosters of the Magic

City clung doggedly to dreams of future greatness in ferrous metals manufacture because of Birmingham's access to cheap raw materials and ability to marshal low-cost black labor. As the national economy began to improve at the turn of the century, TCI took the lead in pursuing the will-o'-the-wisp that had up to now eluded the district's grasp: steelmaking. After failing to convince absentee owners to furnish the necessary funds, local executives displayed remarkable initiative by mustering the support necessary to build a steel mill. Boosters hailed the first batches of the wonder alloy as a triumph that would swiftly unlock Golconda's doors.

Bryan pursued a radically different course. Frustrated by the difficulties caused by dependence on Alabama's lean red hematite deposits, he and his associates took advantage of discoveries in the Cahaba Valley that revealed sizeable reserves of richer brown ore with which red ore could be blended, thereby reducing the amount of coke necessary for smelting. The fact that Cahaba ore could be strip-mined at much lower cost than red ore, which required underground mining, made the case for such a move all the more compelling. Coupled with a decision to emphasize pig iron over coal for the first time as the company's chief mainstay, the new strategy of mixing brown ore with red had important consequences for the future of the enterprise.

As the century neared its close, Sloss and TCI became engaged in a fierce struggle to acquire furnaces, ore lands, coal mines, and other northwest Alabama assets that had once belonged to firms that had not survived the recent depression. Part of the incentive to obtain these properties was a desire to secure market control through horizontal integration, but a major goal was to capture rich reserves of brown ore in the northwest corner of the state. Having already seen the benefits that came from mixing red and brown ore, Bryan was particularly pleased to acquire, near Russellville in Franklin County, large deposits of brown ore that could be hydraulically strip-mined.

As a result of the gains it scored in its battle with TCI, Sloss became a much larger enterprise known as Sloss-Sheffield. Local boosters were sure that it would follow TCI into the magic world of steel. They were wrong, but this error was not immediately obvious; even Bryan and his associates did not realize what a fundamental shift was coming. As one of the most tumultuous decades in American history came to an end, bringing down the curtain on the nineteenth century, Sloss groped its way toward the most important turning point in its entire history.

I

Because of Bryan's long involvement in the Danville and Terminal systems, it was logical for him to choose a person with the same background, Solomon Haas, to become president of Sloss after Seddon's death. Born in Germany in the early 1840s, Haas came to America at a young age and settled in Georgia. After fighting for the Confederacy in the Civil War, he began a railroad career in 1868 as a soliciting agent for the Atlantic Coast Line. He then became a freight agent for the Danville, rising to the rank of traffic manager by 1887. In 1894, he assumed the same position in the Southern Railway after it absorbed the Terminal system.[2] Known to his friends as Sol, he corresponded frequently with Bryan before the latter put him in charge of Sloss in 1896.[3]

Haas took command at Birmingham less than two months after Seddon's death. "I am glad you have become so actively engaged & installed & hope you will find the occupation agreeable to you," Bryan wrote shortly after his arrival.[4] Showing unaffected cordiality and no sign of the tensions that had surfaced so frequently in Bryan's correspondence with Seddon, the letters that followed indicate that Bryan had chosen Haas mainly because of the marketing expertise that he had gained in mass distribution of goods throughout his many years as a traffic officer. From this time onward, under Haas and other executives who later followed in his footsteps, effective salesmanship became an enduring hallmark of Sloss's corporate style. "What we want is a *market* above everything," Bryan told Haas early in 1897. "I hope and trust that you will spread your market as rapidly as you have your reputation as a first class advertiser," he stated not long afterward while congratulating Haas for shipping iron to Australia and receiving an inquiry from Russia. "You have certainly gotten the Sloss Company before the public in a way that would make a patent medicine man green with envy."[5] Bryan also told Maben what good work Haas was doing. His only concerns about the company involved problems with two agents who had saddled it with a $20,000 loss. Blaming this on Seddon's bad judgment, Bryan remarked that it "only showed how deep poor Tom had gotten us in the mire."[6]

Haas responded quickly to requests for information from Bryan, who no longer complained about not knowing what was happening in Birmingham. An announcement in June 1897 that William L. Sims had resigned as secretary-treasurer of the company to "engage in other business" was another sign of change, indicating that Haas was dismantling the structure of alliances that Seddon had built to control day-to-day operations.[7] Shortly before Sims stepped down, another important change took place when Henry Morton, whose relationship with the company had become badly strained, resigned from the board of directors. He was replaced by William E. Strong, a New York banker and former Danville stockholder who had worked with Bryan, Maben, and other friends of Clyde in reorganizing the Terminal system. Haas now occupied Seddon's former place on the board, ex officio. The other directors were Bryan, Christian, Goadby, Maben, Mortimer, Rucker, and Rutherfurd. With the sole exception of Rucker, who was now the lone representative of Birmingham's interests, the board presented a solid front of Virginians and longtime northern allies who could be expected to work together in close harmony. Bryan, who had previously been caught in a crossfire between Rucker and Seddon on one hand and Morton on the other, had reason to be pleased.[8]

Business conditions were reviving in Alabama when Haas arrived in Birmingham. Late in 1895, the state's governor, William C. Oates, boasted that "mines are putting out every ton of coal possible; the factories, mills and foundries give forth the hum of engines, wheels and hammers; the glare of acres of coke ovens and the furnaces light up the country for miles around, both day and night, while their tall chimneys with their splendid plumes of black smoke ascending heavenward proclaim to the world that there are thousands of busy men there and no enforced idleness." Despite such rhetoric, which bespoke the continued enthusiasm of Bourbon politicians for industrial development, the pace of recovery was slow, and Sloss was lucky to post a profit of $22,120.88 for the 1896 fiscal year after paying interest on bonded indebtedness. Bryan's concerns about markets were underscored by the fact that more than 21,000 tons of iron remained unsold by 31 January 1897. At that time, only two of the company's furnaces were in blast.[9]

Still, Bryan was optimistic about the future. Loathing his namesake, William Jennings Bryan, whose "Cross of Gold" speech he derided in the Richmond *Times* as a "studied piece of sophomorical rodomontade," he had presided over a Richmond convention of Gold Bugs immediately after the 1896 Democratic convention. In the ensuing campaign, he had supported the Gold Democrat, John M. Palmer, despite steep losses in circulation and a threat of mob action against the newspaper's main office. After the November balloting, heartened by defeat of the Nebraska congressman, Bryan was relieved that McKinley had won and sure that better times lay ahead. Upon receiving encouraging news from Haas just before Christmas about October's operations, he stated that "if that report had been in Nov. or Dec. we would have thought it McKinley prosperity." As 1897 began, he saw "many indications . . . that the present collapse will not last indefinitely."[10]

Bryan was correct. By June 1897, an economic upswing, which lasted about two years, was in progress. After losing impetus in the final two quarters of 1899, it resumed after the turn of the new century, peaking in September 1902. Unexpected events in faraway places, having to do particularly with the gold supply, lay behind the upswing. Gold rushes occurred in Alaska, Australia, and South Africa, and a new cyanamide process made it possible to extract gold from lean ores. Poor harvests in Europe and India coincided with bumper crops in America, producing massive exports of wheat and an influx of funds from abroad to pay for them. Businessmen, already pleased by McKinley's election, received further encouragement when the victorious Republicans won passage in 1897 of the Dingley Tariff; it established protective rates averaging 52 percent on dutiable products, the second-highest level in the nation's history. By the end of the year, the depression was almost over.[11]

Despite reviving markets for iron, Sloss's financial health did not improve. Indeed, it got worse; the annual report for the fiscal year ending 31 January 1898 showed a net loss of $50,683 after bond interest had been paid. As in the past, sales of coal and coke continued to generate more income than the marketing of iron did. Even the earnings of the company stores were more than twice those of the furnaces.

Settling bad debts left over from Seddon's administration at a loss or writing them off completely accounted in part for the company's dismal financial showing, but problems connected with its perennially troublesome equipment also helped prevent it from taking advantage of improved conditions. The basic cause of these troubles was the constant damage done to the firm's apparatus by undesirable elements in the raw materials that were fed into the furnaces, but some of the technical difficulties stemmed from Uehling's having not performed anything but routine repairs on the original stoves that supplied hot air to the stacks. Kenneth Robertson, who returned as furnace superintendent after having served in the same capacity in the late 1880s, reported to Haas in 1899 that the interior brickwork of the stoves at City Furnace No. 1 had "never been renewed and the shells are in bad condition, are leaking, and have been patched till the possibility of patching has passed." The boilers at the City and North Birmingham furnaces were too small, and the power of the blowing engines at North Birmingham did not match the size of the furnaces, as Uehling had already noted. For all sorts of reasons, pig iron output was extremely irregular. In fiscal 1897, the City Furnaces produced only 45,967 tons during the entire year because they had to be shut down for frequent repairs. [12]

Compounding the firm's mechanical problems were unduly high costs, stemming chiefly from the large amount of coke needed to smelt Red Mountain ore. Excessive coke consumption and cutthroat competition with other producers, not only Alabama furnaces but also midwestern ones that catered to markets in cities like Cincinnati, Louisville, and St. Louis, resulted in extremely slim profit margins on pig iron sales. During the 1897 fiscal year, Sloss sold approximately 130,195 tons of iron, but the anticipated net profit was only $20,624.56, less than sixteen cents per ton. One way to deal with the situation, at least locally, was to form pools with other southern producers, as had already been done during the depression. Such arrangements, however, were unenforceable in the courts, violated the Sherman Antitrust Act of 1890, and had never worked well to begin with. As a temporary expedient, Bryan approved when Haas entered into a pool with TCI and the Alabama Iron and Railway Company in February 1898. He was just as happy, however, when Haas withdrew from a more ambitious one with the Southern Iron Association. [13]

II

As recovery from the depression got into full swing, Bryan demonstrated his continuing commitment to a developmental strategy by launching what amounted to a sea change at Sloss. Up to this time, the firm had been only secondarily a pig iron manufacturer, profiting more from coal sales than from anything else. From now on, this emphasis was going to change: for the first time, Sloss would become chiefly an ironmaking enterprise. As Bryan knew, the shift would cost a lot of money, but he was determined to go ahead.

As a first step, the company had to update its existing equipment. Production at the older of the two City Furnaces was suspended throughout fiscal 1898 while the stoves were thoroughly overhauled. For the future, however, Bryan's hopes rested primarily on completing a project that occupied more and more of his time and energy in 1897 and 1898: acquiring and exploiting large brown ore deposits that had recently been discovered in the Cahaba Valley near Leeds by Seddon's former assistant, J. W. Castleman.

Mixing red hematite with brown ore was the best available way to remedy the high costs connected with the excessive amounts of fuel that the hematite required proportionate to the quantity of pig iron that it yielded. This problem was becoming aggravated by the early 1890s because most of the red ore deposits that had lain close to the surface had now been mined—"skimming the cream," as one analyst later put it—and expensive underground tunneling was required to reach what was left. Red ore was also harder at deeper levels than it was at surface levels. Although the greater lime content gave it valuable self-fluxing properties, it also had adverse side effects because it wore out furnace linings unless the ore was first concentrated and purified through expensive crushing and washing.

Seeking answers to the problems resulting from exclusive reliance on red ore, owners and managers of furnaces in the Birmingham District turned to brown ore, which was available in large quantities throughout north Alabama and had once been the mainstay of the charcoal iron industry. Brown ore, when washed to eliminate impurities, chiefly clay, yielded up to 50 percent metallic iron and could be scooped from open pits at far less cost than was required for mining red hematite. For these reasons, Birmingham ironmakers began in the late 1880s to mix brown ore with red in their furnaces. The trend accelerated until by 1912 Alabama ranked first in the nation in use of brown ore, with a whopping 46 percent of the national total. Whereas only 379,334 tons were mined within the state in 1889, the figure soared to 1,114,836 tons by 1909. Because of this increase, some areas in Alabama began to resemble parts of Minnesota's Mesabi Range.[14]

After geologist William Battle Phillips confirmed the value of Castleman's discoveries in the Cahaba field, Bryan moved quickly to acquire the tract in which the ore was located. By using brown ore, Sloss could make a ton of pig iron with half a ton less coke than was needed with red, saving seventy-five cents per ton. Furthermore, because of the higher iron content of the brown ore, production would rise. Bryan ordered Haas to get as many options as possible on tracts near Leeds. "Under no circumstances permit any of your options on the Cahawba Brown ore to fail," he said. "It means everything to us."[15]

Under a grant of "informal authority" from the directors, Haas bought more than 1,200 acres of Cahaba ore lands. In a meeting at Birmingham's Morris Hotel in November 1897, the directors ratified his purchases and authorized issuing $1 million worth of new capital stock to finance and exploit them. Underwritten by the Central Trust Company, the plan permitted current stockholders to buy for $20 a share new stock equaling one-fourth of their existing holdings. By mid-December, almost all the stock had been taken, except for a small amount retained in the treasury. Bryan took the largest block, subscribing for 1,222 shares in his own name and 424 as trustee of Daniel K. Stewart's estate. At a board meeting in New York City in December, Haas was ordered to proceed with development of the new property and pursue further prospecting in the Cahaba Valley.[16]

Constantly goaded by Bryan to pursue the Cahaba project with maximum speed, Haas negotiated with the Southern Railway to build a spur line to the new mines. When matters hit a snag in March 1898, Bryan, protesting that "we are losing every day by the delay," ordered Maben to intercede with the Southern's president, Spencer, and secured prompt remedial action. Meanwhile, the new mines were developed as fast as circumstances would permit, and equipment was purchased to exploit them. The new apparatus included two pumps, one of which had a capacity of 1,000 gallons per minute, for stripping away the overburden and a large air compressor, built by the Hardie-Tynes Company of Birmingham, for drilling. By early June, Haas was finally able to report that limited quantities of Cahaba ore would be shipped to City Furnaces by mid-July. "The importance of this ore development is so immense," Bryan responded, "that I am thankful whenever you give me a word on the subject." Underscoring the need for haste, he told Maben that coke consumption at Birmingham was "simply terrible."[17]

Late in July, Bryan attended a Confederate reunion in Atlanta. Traveling in a private railroad car provided by his longtime friend Alexander B. Andrews, who was now an officer in the Southern Railway, Bryan proceeded from the Georgia capital to Birmingham and inspected Sloss's older properties. Then he visited the new Cahaba mine, about which he wrote a long letter to Maben. Because of heavy rains, he said, ore shipments to Birmingham had not yet begun. The new spur line from Leeds to the mine, however, was in "excellent condition," and a "large body of ore" had been uncovered by hydraulic stripping. A flume had been built and was partly lined with iron plates; soon it would carry ore from the workings to waiting rail cars. "Castleman says that the development of ore far exceeds anything he had expected," Bryan told Maben. Meanwhile, because of new handling facilities at Greenville, his old dream of shipping coal from Alabama to New Orleans via the Southern's Georgia Pacific branch was about to be realized. "A deputation of Louisiana planters has been gotten up on the Georgia Pacific," he wrote, "and they now believe that there is some other coal besides that of Pittsburgh."[18]

Bryan's letter, which reported a "generally better feeling in Birmingham," reflected intensified optimism in the business community as recovery from the depression continued and the nation relished its recent victory in the Spanish-American War. Although the problem of finding enough workers continued to delay full-scale exploitation of the Cahaba mine, it had attained a modest production of about 100 tons per day by October, and shipments finally began to reach the furnaces. Reporting to the Birmingham Industrial and Scientific Society, Castleman predicted that the Cahaba project would trigger a revolution in southern ironmaking. Previously, he said, the North had utilized expensive high-grade ores, and the South had relied upon cheap low-grade deposits. Now, with cheap high-grade ore at its disposal, the South could drive northern competitors to the wall, and "no other resource will remain to them but to follow their cotton mills south."[19]

Though complaining to Maben that brown ore output cut "a small figure" and that the furnaces were "either in bad order—or are badly managed," Bryan believed that things were finally turning around at Sloss. Estimated earnings for fiscal 1898 showed a tiny but nonetheless welcome net profit of $15,536.40 after payment of bond interest. Although one of the City furnaces remained out of blast pending completion of repairs, the North Birmingham stacks had produced more than 100,000 tons of pig iron during the year, and the overall cost per ton, $6.45, was highly satisfactory. Earnings from pig iron, $58,682.53, were also somewhat closer to those from coal and coke, $109,760.19.[20]

Looking ahead, Bryan was intrigued by the prospect that Sloss might someday become part of the growing American steel industry by making basic iron from southern raw materials and shipping it to northern open-hearth furnaces to be converted into steel. Noting the recent formation of Federal Steel Company, a $200 million enterprise organized by J. P. Morgan and Elbert H. Gary, he told Goadby that he foresaw a time when both Sloss and TCI might be absorbed by Federal. He thought that the chances for selling out on favorable terms would be better if Sloss could make further progress in exploiting brown ore. This possibility was one reason he was pushing development of the Cahaba mines so vigorously.[21]

III

Bryan's willingness to contemplate the assimilation of Sloss into a larger enterprise was characteristic of what was taking place throughout the country as businessmen welcomed the first glimmerings of returning prosperity. Anxious to recoup losses caused by the depression, many owners were eager to liquidate. The formation of Federal Steel was a sign of the times. Until the late 1890s, northern investors had been mainly interested in railroads and utilities, not manufacturing concerns. This situation was one reason why Sloss, like many industrial enterprises, had not listed its securities on the New York Stock Exchange. Until recently, even such large enterprises as the Carnegie Company had been unincorporated. Now things were changing. Even before the depression, a merger boomlet had taken place in the early 1890s, and industrial trust certificates had begun to be traded on Wall Street as unlisted securities despite their highly speculative reputation. The passage by New Jersey in 1891 of the nation's first modern incorporation law had contributed to this trend by starting to clear up questions involving the legality of creating large holding companies as a way to suppress competition. As the depression started to lift in 1897 and investment capital became increasingly available, a wave of industrial mergers got underway. In 1898, there were sixteen consolidations; the next year, sixty-three more. Eleven major reorganizations or consolidations occurred in the ferrous metals industry alone, producing such giant enterprises as Bethlehem Steel, Federal, and Republic. By the time the movement ended in 1904, it had changed the basic structure of American manufacturing. [22]

Besides being lured into consolidation by the prospect of eliminating the cutthroat competition that had prevailed during the recent depression, firms participating in the merger drive did so hoping that larger size would give them financial muscle with which they could gain greater control over economic variables affecting profits. Unlike such classical economists as Adam Smith, who had taught that public welfare was best served by trusting in an "invisible hand" to regulate the ill effects of competition between many small business entities, more and more entrepreneurs were now trying to maximize control over economic development with a "visible hand." One tactic for achieving such control, horizontal combination, involved acquiring enterprises that made the same type of goods, thus enhancing control over output. Another, vertical integration, involved two complementary strategies. By practicing "backward integration," firms tried to gain more favorable access to raw materials by acquiring ownership of the sources from which they came and controlling transport systems that shipped them to manufacturing facilities. By using "forward integration," they tried to control distribution and marketing by taking these functions into their own hands. [23]

None of these methods were new. Because of the close proximity of all essential raw materials needed for iron production, a high degree of backward integration had long been common in Alabama; firms like Sloss and TCI had customarily owned their own mines and quarries and been closely affiliated with the rail lines that delivered coal, coke, iron ore, and limestone to their furnaces. TCI had also practiced forward integration by acquiring such installations as the Linn Iron Works, which used pig iron in making boilers, blowing engines, and other products. Similarly, TCI's absorption of DeBardeleben's ironmaking enterprises and Aldrich's Cahaba coal company in 1892 represented horizontal integration in the former instance and backward integration in the latter.

During the depression, competition had intensified in Alabama, and the pace of consolidation had slowed. Nevertheless, tendencies toward integration of operations continued. Sloss's switch from using commission merchants to selling iron directly to individual customers through its own sales department was one example, representing forward integration. As the merger movement gained momentum later in the decade, the urge to consolidate reappeared in the Birmingham District, just as it did everywhere else. A series of transactions involving the Thomas family's Pioneer Furnaces exemplified this trend. In 1898, these installations were acquired by William Pinckard, a former associate of Aldrich and DeBardeleben; the following year, Pinckard sold out to a larger northern-based combine, Republic Iron and Steel. Because Sloss had already become greatly enlarged, this sale left Woodward the only independent iron producer in the district still unaffected by the trend. Two local rolling mills, along with large coal and ore reserves, were also absorbed by Republic.[24]

Another national tendency that accelerated because of the greater availability of capital at the end of the century was an intensified use of technological innovations that reduced costs, increased dependability of operations, and maximized output, thus raising profits. This strategy, which had already been used before the depression by firms handling liquid or semiliquid products, such as petroleum or grain, had also been adopted in the long-distance transport of iron ore from the Lake Superior area to the furnaces of eastern Ohio and western Pennsylvania. Highly rationalized techniques of iron manufacture, such as those adopted by Carnegie, offered an alternative or accompaniment to consolidation. Uehling's continuous pig-casting technique, conceived at Sloss but used by Carnegie, was a case in point; other related developments included streamlined furnace design, intensified hard driving, mechanical charging, and new rolling techniques.[25]

Southern producers lagged behind northern counterparts in adopting innovations, partly because the South had little in the way of a machine tool industry, fewer trained engineers, and, above all, less money than the North.[26] Southern owners and managers had to play a different game with less capital, lean raw materials, and intensive use of cheap labor while trying to make the most of their superior proximity to natural resources. Within the limitations imposed by these realities, they too were responding creatively to new opportunities; Sloss's acquisition of Cahaba ore deposits to benefit from strip mining and ore mixing was a case in point. Seen in this perspective, southern practice was a mirror image of its northern counterpart. But some southern managers clung to the belief that the region could play the northern game. At the turn of the century, the South's largest firm, TCI, took a long step toward meeting northern producers on their own ground by escaping from its previous limitation to the manufacture of pig iron.

IV

TCI was the district's showcase example of modernization in the late 1890s, finally succeeding in making basic iron and steel where other southern firms had failed. Birmingham was desperate for such a development. "WHAT WE NEED MOST," began a section in a promotional booklet issued by the Commercial Club in 1896. "We need a company with ample capital and experienced men to erect a Steel Plant in or near our city, which would give an unprecedented impetus to the growth of the city by the development of all kinds of diversified industries. There is no more promising field for investment than money spent in the erection of a steel plant large enough to supply ingots not only for our rolling mills, but other industries that would locate here at once were we able to supply them with the steel they would need for their products."[27]

As always, the chief problems that had to be overcome in order to make steel in Jones Valley were the harmful results of the phosphorus, silica, and sulfur present in southern raw materials. Because the Bessemer technology that remained dominant in America utilized converters with "acid" linings of siliceous materials mixed with clay, any phosphorus in the molten iron that was decarbonized remained chemically unaffected by the blow, to the detriment of the resulting steel. To deal with this problem, two British inventors, Sidney G. Thomas and Percy C. Gilchrist, had mixed lime with the molten pig iron and devised a "basic" lining containing limestone, dolomite, or magnesite; this would absorb phosphorus oxides that formed as slag during the blow. In the United States, Jacob Reese, whose visit to Birmingham in 1886 had encouraged Sloss and other industrial leaders to believe that the advent of steel manufacturing in Alabama was at hand, had developed similar techniques. But legal difficulties prevented southern operators from adopting such advances. Partly to keep the Thomas-Gilchrist patents from benefiting low-cost southeastern firms, Carnegie had helped the Bessemer Steel Association acquire the rights to them. Similarly, after protracted litigation, the Bessemer Association got control of Reese's patents and was highly restrictive in permitting their use. Other new European techniques, ironically, demanded even more phosphorus than was present in southern ores.[28]

The best way of solving southern problems, which was developed in Great Britain and transferred to America in the late 1880s, was to use open-hearth furnaces with basic linings. Once again, however, northern manufacturers, including Carnegie and Samuel Wellman, captured the American rights to the patents involved. In 1886, the Otis Iron and Steel Company of Cleveland, Ohio, of which Wellman was an owner, made the first experimental run of basic open-hearth steel in the United States. It produced only a small amount before reverting to the acid process. Carnegie, however, saw that, because of its capacity to make more scientifically determinable grades of steel, the open-hearth process would sooner or later replace Bessemer's technology, regardless of whether acid or basic technology was used. Adopting the new method, in March 1888 he inaugurated the first significant American production of basic open-hearth steel at his Homestead works. Within the next few years, Otis and other northern steelmakers began using the process, but the Southeast was left out.[29]

Even basic open-hearth technology, however, did not completely answer the South's problems. Besides needing to solve the difficulties inherent in phosphorus-rich ores, the region's ironmakers also had to cope with complications caused by excess silica. As the main heat-producing element in southern pig iron that might be converted into steel, it was a troublesome element in either Bessemer or open-hearth manufacture, causing the shells of conversion chambers to overheat and requiring cooling techniques that adversely affected the driving rate. The silica content of southern pig iron could be reduced by adding extra limestone to the furnace charge, but this addition lowered operating temperatures and increased scaffolding, which was already a major problem in the region. Another difficulty was that scrap metal, a major ingredient of open-hearth burdens, was scarce in the South. [30]

Despite such obstacles, TCI's local management, led by Baxter, Bowron, and Shook, persistently urged their company's absentee owners in New York City to support experiments aimed at making basic steel. In the spring of 1893, after conducting an intensive study, a select committee confirmed that use of the acid process in the South, whether in connection with Bessemer or open-hearth technology, was hopeless because of the high phosphorus content of regional ores. On the other hand, the report urged, basic open-hearth technology was a potentially successful alternative. This would require utilizing brown ores, minimizing scrap, and eliminating undesirable elements like silica and sulfur from pig iron used in the charge. As a way of dealing with the last of these problems, TCI's local administrators were already experimenting with the Barton-McCormack method of passing sintered ore through a magnetic separator. For further progress along the same line, they urged the adoption of a process developed by British inventor Benjamin Talbot, in which passing molten iron through a bath of liquefied basic slag removed silica from the iron. This had already been tried by the Southern Steel Company in experiments at Chattanooga, but the effort had been unsuccessful because of the small size of the furnace used. [31]

Leasing the abandoned Henderson plant in North Birmingham, TCI's local leaders experimented with the Talbot process. They achieved enough success for Shook to recommend securing rights to it and building a steel mill at an estimated cost of $1 million. The onset of the 1893 depression, however, prevented either of these steps from being authorized by Inman and other absentee owners, who were more interested in making money through stock speculation than by investing in new technologies. Undeterred, local officials were buoyed by finding that dolomite, large deposits of which existed in the district, enhanced the liquidity of slag and permitted a burden heavy enough to minimize silica content in the resulting iron without increasing scaffolding. They also found that running molten iron through a bath of calcium chloride could eliminate sulfur resulting from contact with Warrior coal. [32]

The best way to deal with sulfur, however, was to eliminate it at the source. Attaining this objective required washing coal as it came from the mine, a process that no furnace company in the district had yet been able to master. Developing effective mechanical washing methods would not only eliminate the problems posed by the presence of sulfur in Warrior coal but also remove slate, silica, and other impurities that caused an excessively high ash content in coke made from this troublesome mineral resource. Only by significantly enhancing the quality of its furnace fuel could TCI hope to produce basic pig iron that would meet the exacting standards of northern steel mills, let alone make basic steel in Birmingham itself. Because of the high stakes involved, TCI's local executives made an all-out effort to solve the technical problems that had aborted earlier coal-washing experiments conducted in the district by Sloss and other firms.

Erskine Ramsay, TCI's superintendent of mines and chief engineer, found the solution for which the company was looking by redesigning a transplanted English machine, the Robinson washer, that had already received experimental use in eastern Tennessee. Unlike the Luhrig model that had proved impractical when tried by Sloss, the Robinson apparatus did not require presorting and classification of coal. This particular feature made the British device seem like a promising answer to TCI's needs. But when it was brought to Birmingham and used with Warrior coal, it became clogged with fine particles of slate and sulfur removed in the washing process. This gritty sludge caused frequent breakdowns and severely damaged the pumping system.

Ramsay's adaptation of the Robinson machine was a classic example of what historians of technology would later call development: redesigning an invention that performed only problematically so that it would work effectively and dependably in day-to-day service. Differing accounts of the modifications that Ramsay made to the British apparatus create the impression that he made a series of piecemeal changes before arriving at a remarkable machine that he patented late in November 1894. The model that he had developed by that time featured a large washing vessel built in the shape of two converging cones. The slanting walls, which narrowed as they approached the upper and lower ends of the container, formed a shell within which rotating vertical stirring arms agitated a mixture of coal and water. Coal was fed downward into the interior of the vessel through a screw conveyor the drive shaft of which actuated the stirring mechanism. A perforated baffle plate prevented the uncleaned fuel from descending to the bottom of the chamber. As coal came out of the conveyor and entered the washer at its widest point, water was pumped upward through the emerging mass from pipes located below the baffle plate. The force of the flow was carefully adjusted so that the coal, which was lighter than the impurities it contained, would rise to the top of the vessel. Meanwhile, heavier waste particles were whirled around the interior of the vessel by the vertical stirring arms, impelled toward the bottom of the lower cone by gravity and centrifugal force, carried over the rim of the circular baffle plate, and discharged into a drain pipe. Accumulating grit was released through valves, helping eliminate the clogging that had made the Robinson prototype ineffective.

Borne upward by the pressure of the water, coal was agitated by the stirring arms, cleaned, carried to the top of the washing vessel, and dumped onto a downward-sloping screen conveyor that carried the fuel away from the container. Water dripping from the coal, still wet from the drenching it had received, fell into a settling tank equipped with an ingenious sludge-removal system that eliminated particulate residues before the water itself was recycled back to the washing vessel. Descriptions of Ramsay's work that appeared in

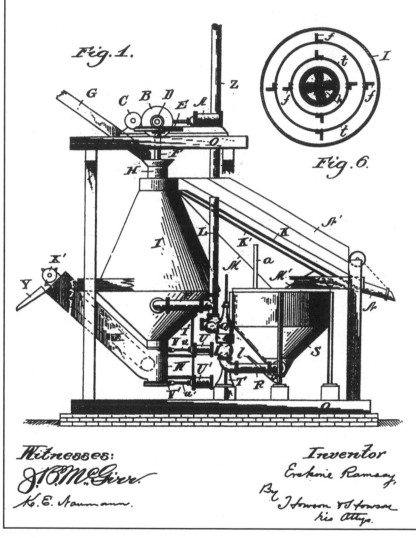

E. RAMSAY.
COAL AND MINERAL WASHER.

No. 528,803.   Patented Nov. 6, 1894.

Fig.1.

Fig.6.

Witnesses:
J. P. McGivern.
K. E. Naumann.

Inventor
Erskine Ramsay,
By Howson & Howson
his Attys.

Robinson-Ramsay Washer. These drawings show coal-washing machinery patented by Erskine Ramsay in 1894, based on an original device invented by Robert Robinson in Great Britain. By eliminating sulfur and other impurities from Warrior coal, Ramsay's apparatus played a major role in TCI's successful drive to manufacture basic steel. It also helped make Alabama a focal point of coal-washing technology in the United States. (U.S. Patent Office, Washington, D.C.; courtesy of Ralph Brown Draughon Library, Microforms and Documents Department, Auburn University, Alabama.)

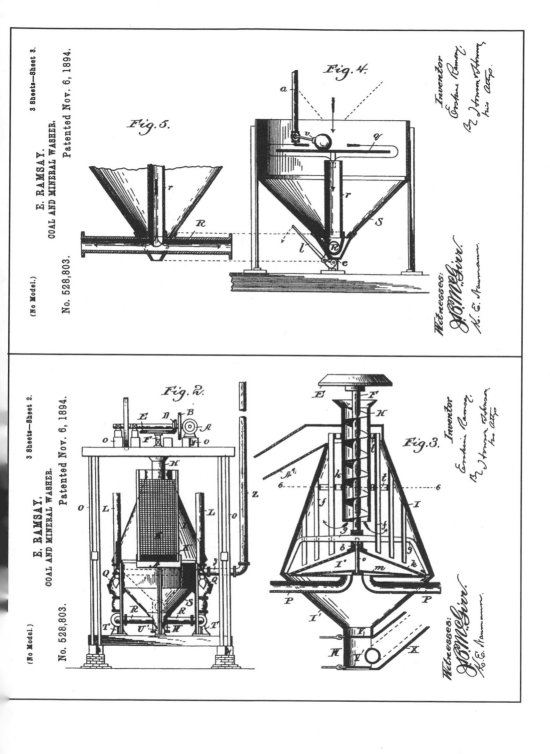

(No Model.)

3 Sheets—Sheet 3.

E. RAMSAY.
COAL AND MINERAL WASHER.

No. 528,803. Patented Nov. 6, 1894.

Fig. 4.

Fig. 5.

(No Model.)

3 Sheets—Sheet 2.

E. RAMSAY.
COAL AND MINERAL WASHER.

No. 528,803. Patented Nov. 6, 1894.

Fig. 2.

Fig. 3.

trade publications paid special attention to the tank and the apparatus it contained as being among his most original contributions in adapting the Robinson washer for American use. Upon reaching the end of the conveyor, the coal was dumped into rail cars and transported to a furnace for smelting.

Ramsay's contrivance, which became known as the Robinson-Ramsay washer, was a major engineering benchmark in Alabama's industrial history. It was now possible to eliminate impurities from Warrior coal that had plagued mine and furnace operators in the Birmingham District for many years. From this time onward, Alabama became a leading center of advanced practice in coal-washing technology. The most immediate result of Ramsay's work, however, was to remove a major roadblock impeding TCI's local officials in their ongoing efforts to surmount problems connected with making basic iron and steel. By 1895, TCI was operating three Robinson-Ramsay washers at its Pratt mines and planning to build additional installations in the near future.[33]

Still unable to gain support from the absentee owners to build a steel mill, Baxter, Bowron, and Shook succeeded in making small batches of basic pig iron at one of TCI's Alice furnaces. Late in August 1895, they sold a sample shipment to the Carnegie Steel Company, which signed an option to buy an additional 24,000 tons if the first lot proved satisfactory. Early the following month, TCI also shipped 20,000 tons of basic iron to Illinois Steel. Another customer, Otis Steel, reported good results with a trial sample; in October, the Carnegie Company, after having achieved success with its initial order, took up its option on the remainder. Meanwhile, yet another northern firm, Jones and Laughlin, began using basic iron supplied by TCI. Both Alice furnaces now started making the new product, and it was widely marketed throughout the United States.[34]

Elated by the success of their experiments, TCI's local officials intensified pressure on New York financiers to fund construction of a steel mill in the Birmingham area, but to no avail. Pointing to the failure of the Southern Company's previous experiments at Chattanooga, northern ironmaster Abram S. Hewitt cautioned TCI's directors not to repeat the same mistake. Estimates of the cost of the projected facility, submitted by outside engineers, ran as high as $850,000, further dampening prospects. Obstacles continued to multiply. Late in 1895, when it seemed that a powerful director, W. S. Gurnee, would commit $500,000 to the project, Grover Cleveland's declaration that the United States would support Venezuela in opposing British border claims set off a war scare, and Gurnee lost interest.

In the spring of 1896, after fears of war had subsided, the directors finally agreed to build a steel mill, but only if capital could be secured elsewhere. Bowron went abroad to obtain it, but failed. After he came home, concern that William Jennings Bryan might win his bid for the White House dried up potential domestic sources of investment. Although local supporters pledged to back a bond issue with which to finance a steel mill, the fact that the company was having a hard time meeting its payrolls made prospects for a potentially expensive experiment in steelmaking seem hopeless.[35]

Early in 1897, hope revived when John W. Gates, who was president of the Illinois Steel Company and a leading maverick in the industry, showed an interest in supporting the project. Known as "Bet a Million Gates" because of his penchant for taking risks, he was intrigued by the possibility of combining an open-hearth steel plant with facilities for producing steel bars, plates, and wire in the Birmingham area and coupling this with a shipyard at Mobile or Pensacola to build oceangoing vessels. The total investment required for such a complex undertaking would be at least $3 million. Little of this funding was likely to come from Alabama, but local promoters secured a subscription of $75,000 so that the Birmingham Rolling Mill Company could build a thirty-ton experimental basic open-hearth furnace. Meanwhile, Baxter sent one of Shook's closest associates, George B. McCormack, to Montgomery to secure a charter for a new enterprise, the Alabama Steel and Shipbuilding Company. Aided by downstate legislators who were enthusiastic about the possible rewards for Mobile, which was now enjoying an economic renaissance after spending decades in the doldrums, McCormack succeeded.

As construction of the experimental facility got underway in Birmingham, contracts were negotiated through Matthew Addy & Company of Cincinnati to deliver trial shipments of basic steel to Pittsburgh. As the *Age-Herald* noted, "it will be a great source of satisfaction to Birmingham to know that the first ton of steel manufactured will be sent right into the heart of her rival manufacturing district in Pennsylvania. Those who have doubted the possibility of Birmingham making steel cheaper than Pittsburgh or any other city in the United States can no longer substantiate their arguments." On 22 July 1897, the first heat of basic steel, about 1,400 pounds of ingots, was poured at the Birmingham Rolling Mill. Boosters lost no time in hailing the achievement. In the words of a Birmingham newspaper, "The chrysalis has opened and from its hidden interior has come forth the butterfly of prosperity, its wings goldenhued and its life sufficiently strong to last for unheard of ages."[36]

Local backers realized, however, that small experimental runs of steel from a pilot plant were no more conclusive for realizing Birmingham's hopes than were the earlier batches that had been produced by the Henderson works. Enthusiasm was quickly subdued by the outbreak of a yellow fever epidemic that spread northward from the Gulf Coast, causing Governor Oates to move his office temporarily from Montgomery to Birmingham. A quarantine blocked the movement of freight cars in and out of the city and prevented ships loaded with Alabama pig iron from leaving Savannah and Charleston. Once again, things were at an impasse.[37]

Late in 1897, one of Birmingham's most devoted champions broke the stalemate.[38] Recognizing, as always, how deeply the fortunes of the L&N were tied to those of north-central Alabama, Milton H. Smith came to the Magic City in December to have what Bowron called "a long heart to heart conversation" with Baxter. Earlier in the decade, Smith had poured yet more money into the state by throwing an enormous rail loop, the Alabama Mineral Railroad, around most of the area east of Birmingham. Consolidating two smaller lines, the Anniston & Atlantic and the Anniston & Cincinnati, Smith connected Calera and Sylacauga. Then, absorbing other properties, he started filling the final gap between Attalla and Champion. The work would not be completed until 1905, but it showed how deeply Smith remained committed to Birmingham's future.[39]

Smith was characteristically blunt in his talk with Baxter. "He said that the Birmingham District was practically 'broke,' and that it could not go on any longer selling half its pig iron at the base cost of manufacture with no allowance for depreciation, wearing out of plants, and exhaustion of minerals," Bowron later recalled. In Bowron's words, Smith said "that the L&N Railroad had fifteen million dollars invested in the District and that if the District collapsed with all its heavy traffic, it would be ruinous to the L&N. He said it was the duty of the T.C.I. Co. as the largest local enterprise to lead the way into the manufacture of steel, and that he had come to say to us that if we would not do so the L&N Railroad, which was already a stockholder in the Birmingham Rolling Mill, would put enough additional money into that concern to enable it to go ahead."[40]

TCI's local officers had no intention of seeing the Rolling Mill Company march off with a prize that they coveted intensely. After Smith went back to Louisville, McCormack drafted a plan under which he, Baxter, Bowron, Ramsay, and Shook each offered to commit $10,000 to a fund for building a basic steel works if other Birmingham residents would put up a matching sum, $50,000. After this sum had been raised, Smith and Spencer would each be invited to subscribe $100,000 on behalf of the L&N and Southern railroads. With the resulting $300,000, Bowron wrote, "we would put a small plant up, small as to the number of its units, but modern as to their size and capacity, where we could deliver molten pig iron very cheaply and with the profits expand the plant unit by unit."

Before playing this final card, Baxter wrote a long message to TCI's executive committee in New York, conveying the substance of Smith's comments and asking once more for support in building the proposed steel works. Again, the results were negative. After delivering the letter personally to New York, Shook returned just before Christmas with a reply from the owners "that the company would not go into the manufacture of steel; that it had no money for the purpose, but that the officers were authorized to subscribe to the capital stock of the Birmingham Rolling Mill Co. $100,000 payable over one year in iron and coal at market prices to assist that company in enlarging its steel equipment." The directors also urged that TCI promise, for a specified period of years, not to compete with the Birmingham Rolling Mill Company by making steel on its own.[41]

Sickened by the owners' willingness to abdicate to the Rolling Mill Company and determined not to see TCI withdraw from the field and lose everything for which they had been fighting, the local executives submitted a counterproposal to the owners, embodying McCormack's plan. Written by Bowron, it pointed out that many of the TCI's furnaces were currently shut down for lack of business and stated that "if we had to sell pig iron and nothing else the more furnaces we had the worse off we were and we would drag along at the tail end of the procession instead of being at the head as our size and property ownership would suggest." Instead of meekly accepting such an outcome, Bowron and the other signers asked that the owners permit them to raise, on their own, the capital needed to build a basic steel plant. While awaiting the reply, Bowron attended a convention in Tuscaloosa and gathered support by painting a glowing verbal picture of "the Warrior river traffic, Birmingham at one end furnishing steel plates and shapes, Mobile at the other building them into ocean steamships loading out with Birmingham coal." Responding favorably in a long editorial, the Mobile *Register* closed ranks with the upstate group. In early January 1898, after TCI's executive committee finally acceded to their proposal, the local officers swung into action with a decision "to proceed on the basis of a plant to handle 400 tons of furnace metal per day at a cost of $500,000 for construction and $200,000 for working capital."[42]

Going to Louisville, Baxter won Smith's consent for a $200,000 subscription by the L&N. Early in March, after securing estimates that a basic steel plant would cost $419,000 ($609,000 if a rail mill was added), the Birmingham group wrote to August Belmont, Jr., a key Morgan associate and American representative of the Rothschild interests in London. Telling him about Smith's pledge, they asked for an equal commitment from the Southern Railway. Soon afterward, an "extended interview" with Spencer in the latter's private railroad car "practically clinched the steel works business," as Bowron later wrote. As a result of the meeting, the Southern joined the L&N in subscribing $200,000 to the bonds of the Alabama Steel and Shipbuilding Company. Additional commitments by individual investors followed through TCI's New York brokerage house, Moore and Schley.

Alarmed by the speed at which the project was moving, TCI's executive committee demanded an independent audit of the estimates that Bowron and his colleagues had received. After this had been made, the committee agreed to have the projected mill built at Ensley, with a capitalization of $1.1 million. It demanded, however, that subscriptions totaling $150,000 be raised in Birmingham itself as a condition of going ahead. The money was raised in two days; Bowron himself committed $25,000. Despite further snags, final agreement with the executive committee was reached by the end of June, and ground was broken for the new plant at Ensley on July 14.[43]

As TCI's executive committee had feared, the cost of building the new steel plant greatly exceeded the original estimates. It reached approximately $1.8 million by the time the first billets were made. Because of the dramatic improvement in the national economy throughout 1898 and 1899, however, the increase never became an issue. Indeed, as things brightened, TCI provided the additional funds itself and leased the new facility from the Alabama Steel and Shipbuilding Company, running it as a subsidiary within its own corporate structure.

The new plant, designed and built by the engineering firm of Wellman, Seaver and Company, would have ten basic open-hearth furnaces mounted on foundations that permitted them to be rocked back and forth under hydraulic power. Carloads of iron from the nearby Ensley stacks would be charged into arched conversion chambers along with dolomite, scrap, and other materials and raised to temperatures in excess of 2,000 degrees Centigrade with a mixture of superheated air and gas. Each chamber would have a fifty-ton capacity and be equipped with basic linings to facilitate removal of phosphorus and other undesirable chemical elements. As flames swept across the materials that were being converted into steel, an intense state of combustion would be maintained by continuously recycling the heat from exhaust gases back and forth between regenerative ovens located on either side of the conversion chambers. Each oven would contain masses of firebricks stacked in a honeycomb arrangement to trap heat from spent gases before residual fumes passed up tall chimney stacks. Circulating through the staggered layers of firebricks, a fresh mixture of incandescent air and gas would be blown back into the furnace from one of the ovens, come out the other side of the conversion chamber, and pass into the adjoining oven to maintain a continuous process of heat exchange. As valves were opened and closed, the draft would be repeatedly reversed so that the flames in the conversion chamber could be kept as hot as possible.

One of the great advantages of the open-hearth method over the older Bessemer technology was that the conversion process could be closely monitored. In TCI's new steel plant, samples of molten metal would be tapped from the furnace at intervals and tested until the desired metallurgical quality was reached. At that point, the furnace would be rocked backward to discharge slag and then slowly tilted forward to pour a continuous stream of liquid steel into giant ladles, each of which would be capable of holding twenty-two tons of the white-hot brew. After the steel had been tempered with a dash of ferromanganese, the ladles would be lifted by an electrically powered crane and inverted to pour their contents into vertical ingot molds mounted on rail cars. After cooling, the ingots, each weighing three tons, would be stripped from the molds, reheated to insure uniform internal consistency, and taken to a blooming mill for repeated passes through rolls. The resulting billets, measuring about four inches square, would constitute the end product of the facility and would be hydraulically sheared into various lengths for shipment to market.[44]

Completion of the plant was delayed by problems connected with procuring materials at a time when demand for structural iron and steel was rising throughout the country. As a result, only three of TCI's new open-hearth furnaces were ready for operation when the long-awaited initial runs of steel took place. Shortly after noon on Thanksgiving Day, 30 November 1899, the new installation produced its first batches of the much-prized alloy. Hedging against the possibility that something might go wrong, TCI kept the event as simple as possible; as a newspaper account stated, "no invited guests, champagne, or speeches graced the day." Only a small group of executives, including Shook, who had been injured in a recent accident and came on crutches, watched as a lever was thrown and one of the furnaces was tilted forward to discharge its burden of molten steel. For the time being, unnecessary complications were avoided by dispensing with ladles and running the white-hot metal into forehearths with nozzles leading directly to the ingot molds.

Much to everyone's relief—Shook, it was said, had "almost lived on the site" during the final preparations—the test came off triumphantly. "Every part of the plant worked like a charm and gave no indication that it was the first trial," stated an account that appeared in the New York *Times*, showing that outside observers were keeping a watchful eye on developments in Jones Valley. Predictably, the local press was ecstatic. "It was a great day for Birmingham and Alabama," the *Age-Herald* declared, "and presages the coming of that greater day when this district will control the steel as well as the iron trade of the world." Hundreds of additional runs took place in December, with good results, as the new plant was readied for full-scale production. After decades of frustration, the Magic City had finally entered the Steel Age.[45]

V

TCI's successful production of basic steel on a commercially profitable scale overshadowed every other development that took place at Birmingham in 1899. Nevertheless, it was also a landmark year for Sloss. Instead of being absorbed by a larger combine, as Bryan had anticipated, the company spearheaded the most important consolidation to occur in the district since TCI, DeBardeleben, and Aldrich's Cahaba mining firm had merged seven years before. By acquiring a group of properties that had gone out of business during the depression, Bryan and his associates created a much larger enterprise, epitomizing expansionist tendencies transforming the American economy at the turn of the century.

Most of the properties acquired by Sloss were located in Alabama's northwest corner. Like so many enterprises that failed to survive the Panic of 1893, they had been founded during the flush times of the 1880s. Lured by the area's extensive brown ore deposits, the proximity of the Warrior field, and the potential commercial advantages of cheap water transportation to St. Louis and New Orleans via the Tennessee River, a group of speculators bought a town site in Colbert County and extensive mineral properties in Franklin, Walker, and Winston counties in 1883. Among the entrepreneurs were E. C. Gordon of Clarksville, Tennessee, and Walter S. Gordon of Atlanta, who had long been active as railroad promoters in northwest Alabama and northeast Mississippi. Along with their brother, John B. Gordon, they had also figured prominently in the early development of the Georgia Pacific. Alfred H. Moses, a lawyer and real-estate promoter from Montgomery, was also one of the leaders, together with three other investors from the capital city. Because of the advantages of northwest Alabama for combining rail and water transport, these promoters believed they could undersell Birmingham pig iron in St. Louis.[46]

Like Birmingham's founders, the men who launched this new growth frontier adopted the name of a famous British industrial center to express their visions. Organizing the Sheffield Land, Iron, and Coal Company, they created the city of Sheffield on a limestone bluff overlooking the Tennessee River. A crowd estimated at between 5,000 and 10,000 people gathered for the first day of auctions that were held to sell lots in May 1884. "The first lot offered brought $1,000," the *Manufacturers' Record* reported. "The bidding on others was rapid and even excited, and at the end of the first day it was found that sales aggregated over $100,000. By the following day the demand was so great that 50 per cent advance was in some cases offered and refused for lots purchased the day before. On the second day $8,900 was paid for one lot 150 feet square, on which to build an opera house."

Besides calling for two blast furnaces, plans for the site included building a large hotel, a bank capitalized at $50,000, a grain elevator, two sawmills, a planing mill, a sash and door factory, and a brick works. "That Southern men have planned this new town, and that Southern men have been the heaviest investors in it, while it is Southern money that is to build most of these many manufacturing establishments," said the *Manufacturers' Record*, "is sufficient proof that the people of the South are awake, and that the development of the South is not, as some seem to think, due entirely to outside men and money."[47]

Calling the city site "beautiful beyond comparison," the promoters issued broadsides stressing the prospective boat landings it afforded, the fertility of the surrounding land, the salubrious climate ("no chills, fever or miasma exist there"), the abundance of local building materials, and the low cost of transporting pig iron by water to St. Louis, which was estimated at $1 per ton as opposed to $3.55 from Birmingham. Experts from outside the area praised its advantages. "It beats anything along Lake Superior," said one of them. "You walk along and stumble against whole hills of coal and iron."[48]

No sooner had the venture been launched, however, than the sharp economic contraction that hit the country in 1884 set in. "Even those who had been most eager to make investments became thoroughly frightened," stated an account written only a few years later. "Immediately the entire property began to depreciate. . . . Companies which had entered into agreements to build smelters and furnaces, refused to fulfill them until affairs brightened. The stock of the Sheffield Company could find no purchasers. On all sides the enterprise was held to be a failure; by many it was looked upon as a swindle."

Within two years, boom times returned. In 1886, E. W. Cole, a Nashville capitalist who had figured prominently in railroad-building in east Tennessee, invested more than $200,000 in Sheffield. "I have absolute faith in Sheffield's future," he said. "In two years from now you will see 300 carloads of coke being delivered daily at my furnaces here. You will see 100 carloads daily of pig-iron being exported from the same furnaces. You will see the Tennessee River alive with Sheffield shipping, and there will not be a river in the great Mississippi valleys which will not be coursed by Sheffield's pilots." Under Cole's leadership, a furnace company was formed. During the summer of 1886 it began constructing a stack, seventy-five feet high and seventeen feet wide at the bosh, on a site immediately west of the planned community. Another new venture, the East Sheffield Land Company, was also launched. It had a capital stock of $400,000, of which $200,000 was soon paid in.

Put into blast on New Year's Eve 1887, Cole's new furnace had a capacity of 125 tons. It was built under the direction of Edward Doud, a civil engineer from Vermont. Because no coke ovens had yet been built and the 100-mile Sheffield and Birmingham Railroad was little more than half finished, the Cole installation used fuel from Virginia's Pocohontas coal beds. Two more furnaces were built on a nearby tract by the Alabama and Tennessee Coal and Iron Company, which was soon consolidated with the uncompleted railroad. Epitomizing the same optimism that was also rampant in Birmingham at the time, the new firm was capitalized at $8.25 million. As president of the expanded enterprise, Cole said that "in the way of transportation and consequently cheap raw material, there is no company in the United States with such advantages." The new stacks, which were named the Cole Furnaces, went into blast in September 1888 and October 1889 and were operated by workers who lived in a nearby village known as Furnace Hill. The owners also built 1,000 coke ovens to use Warrior coal as soon as the railroad was finished, aiming to add 2,000 more in the future. Rounding out their scheme was a barge line to link their properties with outside markets via the Tennessee and other rivers.[49]

In 1888, another stack was begun on the Sheffield Company's tract by Enoch Ensley, whose hopes for the Jefferson County boom town that bore his name had collapsed in the recession that had ended the flush times of 1886–87. Resigning the presidency of TCI in April 1887, he sold his properties in and around Birmingham and concentrated on Sheffield. In addition to building his new installation, which went into blast in April 1889, he also acquired the original furnace that the Sheffield Company had completed in 1887, naming the two stacks the Hattie Ensley and Lady Ensley after his two daughters. In 1891, in partnership with a young coal operator, Walter Moore, he organized the Lady Ensley Coal, Iron, & Railroad Company, which also owned brown ore lands in Franklin County, and acquired the Horse Creek Coal Mine in Walker County. A battery of 200 coke ovens at the Horse Creek facility produced fuel for Ensley's Sheffield furnaces.[50]

Meanwhile, across the river, two furnaces were built in Florence, the chief town in Lauderdale County and chief entrepot of the entire area from its founding in 1818 up to the time Sheffield was established. Despite being on the Memphis and Charleston Railroad, it had long been stagnant. "Five years ago," a reporter wrote late in the decade, "Florence was as dead as a town can be killed, but now it is brim-full of energy and enterprise." Aside from the desire to compete with Sheffield, the main force behind the change was the Florence Land, Mining & Manufacturing Company. It began during the 1886 boom and was capitalized at $3 million, of which $400,000 had been subscribed by the end of that year.

After hundreds of workers had prepared the site on which new dwellings would be erected, the company began selling property in April 1887. "People who have visited this historic old town recently have been surprised to find what an immense amount of work was being done in cutting new streets, grading and improving them, and putting everything into such a shape that the work of building houses can go on very rapidly," the *Manufacturers' Record* reported. Calling the city "Fair Florence," a broadside stated that "There is no Place on the Continent more Healthful and Salubrious," praised its existing population as "moral, social and hospitable," enumerated its colleges, schools, churches, and lodges, and hailed the promise of "vast beds of iron ore" and "immense coal fields" located nearby.

Headed by William B. Wood, a steamboat and railroad promoter, the company quickly signed a contract to build "the most complete furnace plant in Alabama." It was to have a daily capacity estimated at 100 to 150 tons, but the recession that had such a severe impact on Birmingham in 1887 had a similar effect in Florence. Not until 1891 was the stack completed by Alabama and Virginia capitalists. Using brown ore from Tennessee and coke from the Old Dominion, it shipped pig iron on barges to foundries outside the region. A second furnace, slightly smaller than its counterpart and begun before the recession hit in 1887, had a similarly tangled history. It was eventually finished with the help of northern and midwestern capital by an enterprise known as the North Alabama Furnace, Foundry and Land Company after an earlier firm, the Florence Coal, Coke & Iron Company, had failed to complete it. Like the group led by Wood, the owners of the furnace supplied iron to northern foundries by the river route.[51]

The recession of 1887 halted the growth of Florence and Sheffield only temporarily. Horace Ware, who had organized the Shelby Iron Works four decades earlier, joined investors who swarmed into Colbert and Lauderdale counties as various enterprises, including a pipe works and a stove company, sprang up in Florence and Sheffield despite the gyrations of the business cycle. Ware's visions for the area epitomized the boosterism rampant at the time. "Sheffield, as a grain and provision distributing point, is bound to become the Southern St. Louis," he stated in an interview published by the Atlanta *Constitution*. Coming "from every point of the compass," rail and water routes would "pour into the lap of Sheffield the iron, coal, grain, and other products of a continent." Unable to compete, Birmingham would "be compelled to ship the greater part of her products from the docks of Sheffield, thus giving us her tonnage."

Alfred Moses and other promoters of the area were equally hopeful. Estimating that shops to be built by the Memphis and Charleston Railroad would employ between 400 and 500 skilled mechanics, they heralded the district as the "Iron Manufacturing Center of the South." A furniture factory, a bottling works, a bakery, five brick plants, a water works, three banks, a printing establishment, and a "first-class semi-weekly newspaper, printed by steam," were listed among enterprises "already in successful operation." Planned for the future were "Electric Light Works, Paint Works and Agricultural Implement and Machine Works, Large Stove Works, Machine Shops, additional Iron Blast Furnaces, an extensive Charcoal and Chemical Plant, a large Rolling Mill, and other manufacturing establishments." Sites for such ventures would be donated by the Sheffield Land, Iron and Coal Company, along with lots for free public schools and churches. "No 'Old Fogy' element here," boosters assured prospective investors.[52]

Completion of additional railroad lines in and out of the area, supplementing east-west connections with Memphis and Chattanooga via the Memphis and Charleston Railroad, provided further encouragement. By 1888 the Sheffield and Birmingham Railroad had been put through to Jasper, eighty-five miles to the south, where a junction was effected with the Kansas City, Memphis, and Birmingham. Speedy progress was expected to permit linkage with the Georgia Pacific and the Danville and Terminal systems with which it was associated. The L&N anticipated completing a subsidiary, the Nashville, Florence & Sheffield, by July of the same year, providing direct northern connections with Louisville and Cincinnati. Other lines were in the planning stage. Boosters reasoned that the outstanding accessibility of the region by water would keep freight rates low on these rail systems. Pointing out that "in 1883 the site of Sheffield was a corn and cotton field," a correspondent of the Cleveland *Iron Trade Review* stated that "it is impossible for nature to have done more in the way of a fitting location for a large inland city."[53]

Despite such inflated rhetoric, the area was ill prepared to withstand the economic storms of the 1890s. Ensley died in 1891, and his properties went into receivership. Hattie Ensley Furnace was blown out in 1892 and was foreclosed the following year by its builder, James P. Witherow, from Pennsylvania. It operated sporadically throughout the decade, but the Lady Ensley remained idle for nine years after it, too, was blown out in 1892. Things went no better in Florence. The furnace that had been built there by Wood and his partners went out of blast soon after the onset of the depression in 1893 and was foreclosed by a Philadelphia bank to which it was indebted. The smaller North Alabama Company furnace had already shut down in 1890. After undergoing major repairs in 1893, it returned to blast the following year under the ownership of a Tennessee firm, the Spathite Iron Company. This enterprise, however, failed in 1895, leaving only the Cole Furnaces at Sheffield still operating. They, too, went through legal difficulties and at least three changes of ownership in the first half of the decade. In 1895, their owners bucked prevailing trends by completing a third stack and putting it into blast. Little more was left of the industrial complex for which such high hopes had been entertained. [54]

VI

Despite the heavy blows experienced by northwest Alabama in the 1890s, conditions were ripe for its bankrupt facilities to be reanimated as buoyant conditions returned late in the decade. Only buyers willing to acquire them and invest the funds necessary to return them to working order were required to start another surge of development. Sloss and TCI took advantage of this opportunity; by early 1899 they were competing vigorously with one another to buy as many properties as possible in and around Sheffield and Florence. Having only recently aspired to outstrip the Birmingham District as the South's leading industrial center, Alabama's northwest corner soon became a satellite of the Magic City.

Because Bryan's letterbooks for most of 1899 have been lost, it is impossible to cover in detail the way in which Sloss acquired three furnaces and many other valuable properties in northwest Alabama or fully to assess the motives that led the company to engage in this frenzy of expansion. But Bryan's developmental ardor was obviously as strong as ever, overriding his flirtation with the idea of selling Sloss to a larger enterprise like Federal Steel. The success of his recent foray into the Cahaba ore field had opened his eyes to what might be accomplished by intensifying his commitment to brown ore. High-grade deposits near Russellville, not far from Sheffield and Florence, offered the same substance in copious quantities, exploitable at low cost by strip mining and now readily accessible to the Magic City via the completed Sheffield and Birmingham Railroad. [55] Combining the rich Russellville beds with newly built furnaces in Florence and Sheffield and adding Warrior coal deposits north of Jasper offered an abundance of opportunities for horizontal and vertical integration. It is not hard to see why even such a prudent entrepreneur as Bryan became excited by the potential rewards of a major expansionist effort.

As business improved nationally throughout 1899, Sloss's increasingly buoyant condition also helped convince Bryan and his associates that a major expansion of their Alabama operations would pay off. All four of the company's existing furnaces were in blast. By the end of the fiscal year, pig iron production had mounted to more than 172,000 tons. Because of the Cahaba Valley deposits, yields from ore were increasing substantially. Despite heavy outlays for repairs and new equipment that boosted production costs to about $8.10 per ton, prices advanced faster, reaching about $17 per ton. Benefiting from Haas's sales experience, the company now supplied most of its iron directly to foundries, rolling mills, pipe works, stove companies, and machine producers without going through middlemen. Never had the enterprise been in such good condition. [56]

Buoyed by the return of prosperity, Bryan and Maben were negotiating as early as March 1899 for the purchase of properties sufficiently extensive to require the organization of a "proposed New Company." Among the assets involved were the abundant brown ore deposits near Russellville; in one of the greatest bargains in the history of the firm, these beds were acquired from J. D. Lacey of Chicago for about $178,000. Located on what had once been a cotton plantation, the deposits would quickly become a bonanza, providing vast quantities of limonite to be mixed with Jones Valley's red hematite in different proportions for the needs of specific customers. The old slave quarters were still standing, providing housing for workers who would soon operate steam-powered engines to wash impurities out of ore scooped from giant pits where antebellum landowners had once maintained a racetrack.

Because the Virginia syndicate did not have sufficient funds to effect the anticipated consolidation, a committee consisting of Bryan, Maben, and Rutherfurd was appointed to secure subscriptions to an underwriters' fund from such Wall Street allies as Oakman and Olcott. Meanwhile, TCI scored a victory by paying $850,000 to acquire all of the Sheffield Iron Company's properties. "This gave us 19 blast furnaces," Bowron later wrote. He added that "in the light of calm, sober after thought it was a mistake, because there was not sufficient trade or demand for foundry iron in the entire country for us to run nineteen furnaces steadily year in and year out."[57]

Such hindsight manifested itself only after the passions of the moment had died away. With merger mania running at a fever pitch as high in Alabama as in the rest of the country, Sloss and TCI continued scrambling for new acquisitions. Having lost the battle for the Sheffield Iron Company, Sloss won another prize on 15 June, when Maben paid $107,750 to owners in Philadelphia to acquire the foreclosed furnace in Florence that had originally been started in 1887 by the W. B. Wood Company. The pace quickened at a meeting of Sloss's board of directors at Goadby's New York City brokerage on 12 July, when Maben, Rutherfurd, and Rucker were appointed to negotiate the purchase of the Lady Ensley Company, which had been in receivership since 1891. The assets involved were under litigation between Ensley's widow, who claimed that $1 million worth of bonds had been illegally canceled, and a group of defendants including some Memphis banks, Walter Moore, and the Horse Creek Coal and Coke Company.

TCI had already acquired $220,000 worth of bonds from litigants in the Ensley case and thought it would soon amass the remaining securities. This anticipated gain, however, did not occur. As Bowron later wrote, "the Sloss Co. was better advised than we, for it bought the stock of the bankrupt company from Mrs. Ensley and paid the company out of bankruptcy, paid off the bonds and took possession of the property and left us wondering 'who hit Billy Patterson.'" By outmaneuvering TCI, Sloss acquired $2 million in mortgage bonds, $2 million in stock, and "large amounts of judgments" held by Mrs. Ensley at what Maben later described as "greatly depreciated prices." Complications growing out of the struggle kept Sloss and TCI engaged in legal battles until Sloss finally gained title to the entire enterprise in 1903 for an additional expenditure of about $620,000.[58]

By mid-July, the results of the battle that TCI and Sloss had been waging behind the scenes surfaced in local newspapers as stories appeared that Sloss was in the midst of a major reorganization. The news was confirmed at the end of the month with an announcement from New York that a new entity, the Sloss-Sheffield Steel and Iron Company, would soon be incorporated in New Jersey. This development became official on 16 August 1899, when formal chartering took place. The consolidation added an impressive list of properties to Sloss's previous holdings. The new corporation controlled six blast furnaces and two-thirds of another, 63,603 acres of coal lands, 48,005 acres of ore deposits, 1,100 beehive coke ovens, 1,400 tenement houses, and a host of other assets. Twelve enterprises were acquired in whole or in part. These holdings included the American Coal and Coke Company; the Colbert Iron Company; the Franklin Mining Company; the Hamilton Creek Ore Company, which owned a limestone quarry and additional ore tracts in Franklin County; the Lady Ensley Coal, Iron, & Railroad Company, which included 32,400 acres of coal and ore lands in six Alabama counties and the Hattie Ensley furnace; the Lady Ensley Furnace Company; the Lost Creek Coal Company; the Miss Emma Ore Mining Company; the North American Furnace Company; the Philadelphia Furnace Company, which had purchased the former Wood installation in Florence; the Russellville Ore Company, with its rich reserves of surface brown ore deposits; and the Walker County Coal Company. The total authorized capital stock of the new corporation was $20 million, consisting of equal amounts of preferred and common. Only $6.7 million worth of preferred and $7.5 million of common stock, however, were actually issued, the remainder being kept in the treasury.

Understandably, the Virginia syndicate lacked sufficient funds to finance such a large undertaking, even with help from such traditional allies as Goadby and Mortimer. Investors from New York now became for the first time numerically preponderant in the new company, holding more than 40,000 shares of its common stock and approximately 33,000 of preferred. Bryan, however, remained the largest single stockholder, with 6,122 shares of common and 7,858 of preferred. Other Virginians included C. W. and Thomas Branch, T. B. Lyons, Scott, and Trigg. As in the past, some New York investors were actually from the Old Dominion; these included Maben, who held 1,401 shares of common. In addition, large blocks of stock were held by such Bryan allies as Goadby, with 5,536 shares of common, and Mortimer, with 2,350. Stockholders residing in Birmingham included Haas, John W. Johnston, and Rucker, all with relatively small investments. The list also included other southerners who had been friends of Bryan for many years, such as Andrews, with 610 shares, and Henry O. Seixas, a Louisiana capitalist who held 1,500. The roster of owners, therefore, continued to include many familiar names.[59]

Nevertheless, for the first time in the history of the enterprise, major holdings were now owned by prominent Wall Street investors who had not been previously active in the company. Prince & Whitely, for example, now held 2,899 shares of common; Frederic P. Olcott, president of the Central Trust Company, 2,415; George Parsons, a New York financier who may have been related to Henry C. Parsons, a prominent promoter of southern industries, 2,293; Rogers, Brown & Company, important commission merchants in the iron trade, 1,975; Kean, Van Cortlandt & Company, 1,323; Ladenburg, Thalman & Company, 1,517; and J. W. Seligman & Company, 1,216. Another new investor was Edward O. Hopkins, an Indianian who was involved in the tangled Lady Ensley business and held 1,231 shares. In the future, he would become a pivotal figure in the history of Sloss Furnaces.[60]

Reflecting the fresh capital that Bryan and his allies had been obliged to raise, the composition of the new board of directors differed significantly from that of the Sloss Iron and Steel Company. There were, of course, many holdovers, including Bryan, Goadby, Haas, Maben, Mortimer, Rucker, Rutherfurd, and Strong. The inclusion of Scott as a representative of minor Richmond stockholders further bolstered the Virginia contingent. Although not previously a director, Oakman, a key figure in the Central Trust Company, was a familiar and reassuring presence in his new capacity as chairman of the executive committee. The same firm's chief attorney, Adrian H. Larkin, was also on the board, as was George Parsons. Another new member, Archer Brown of Rogers, Brown & Company, was well known to Sloss because of his firm's previous role in handling pig iron shipments to northeastern destinations. Some of the new faces, however, were not so familiar. Besides Hopkins, they included veteran Wall Street operator Moses Taylor, representing Prince & Whitely, and R. B. Van Cortlandt, representing Kean, Van Cortlandt & Company.[61]

How the firm's new directors would fit in with the old remained to be seen. Among other possibilities, Rucker, in a lonely minority since Seddon's death, might now find fresh opportunities to forge alliances between the Birmingham investors he represented and board members who did not see eye to eye with Bryan and his closest associates. The events of the next few years would demonstrate just how real this possibility was. In any case, the emergence of Sloss-Sheffield in 1899 marked the end of an era and the opening of a major new phase in the history of the enterprise that Colonel Sloss had founded nearly two decades before. Looming just ahead was a crisis that would have a dramatic impact upon the future of the firm, launching it on a trajectory radically different from the one to which TCI had committed itself. Out of it, however, would emerge a company that remained deeply rooted in the past and still dominated by Bryan.

9

The Turning Point

As the twentieth century opened, Birmingham's boosters continued to look upon steel as a magic key that would unlock endless vistas of growth for the city. To such people, TCI's recent success in producing basic steel overshadowed every other development that had taken place in the history of Jones Valley, culminating a quest to which the district's pioneers had been dedicated for almost three decades. Only in this perspective is it possible to understand the dismay many local residents felt in 1902 when Sloss-Sheffield abandoned the dream of making steel and, in a historic turning point, shifted its focus in another direction. From this time on, Alabama's two largest enterprises diverged sharply, becoming mirror-image twins.[1]

I

One by one, as an old century closed and a new one opened, Birmingham's pioneers continued to pass from the scene. Nothing epitomized this transition better than the death of John T. Milner on 18 August 1898. During the last decade of his life, he had served eight years in the state senate while continuing to run the Milner Coal and Railway Company, described by a local newspaper as "one of the most successful mining companies in the state." During the winter preceding his death, he had suffered a stroke. He had another apoplectic seizure about a week before he died but had seemingly recovered and was "riding around in his buggy with usual good health" just prior to his third and fatal attack. A large throng, including about fifty black servants, crowded First Presbyterian Church to hear an old friend, Dr. J. H. Nall, eulogize the fallen engineer; afterward, his body was interred at Oak Hill cemetery. "With this ceremony was buried . . . one who had done as much or more than any of his contemporaries in the upbuilding of the District," declared the *Age-Herald*.[2]

Despite the passing of a generation, the city that Milner had helped to create remained a raw urban frontier. Swarming into the city from poverty-stricken rural areas that had been hard hit by the recent depression, an army of badly dressed, poorly paid, ill-disciplined industrial workers tried to adjust to an unaccustomed way of life. Lured into brothels and saloons in squalid enclaves known by such names as Beer Mash, Buzzard's Roost, Dry Branch, Hole-in-the-Wall, and Scratch Ankle, they tried to forget, if only for a while, long hours of hard labor in dirty, physically exhausting, and dangerous occupations. Violence was endemic, causing the community to be known as "Bad Birmingham, Murder Capital of the World." In 1902, it had by far the most arrests, 9,626, of any city of its approximate size. Spokane, Washington, which placed second, had 5,117.[3]

Birmingham's business district and fashionable residential sections, however, reflected its status as the foremost industrial center of the Deep South. The arrival of its first electric streetcars, beginning in 1892, was only one indication of the way in which power surging through overhead wires was transforming the city; the "immense iron stack" of the Consolidated Electric Light Company, reaching 150 feet into the sky, became one of its most prominent landmarks. Twenty-seven dynamos, mostly of Thomson-Houston design, supplied current for commercial use, and six Edison generators, using a three-wire system, were utilized for home lighting and other residential purposes. By the end of the decade, a few horseless carriages were appearing in streets that were still dominated by equine traffic.

Even in the depths of the depression following 1893, construction had not stopped. After the six-story Caldwell Hotel burned to the ground in July 1894, the former Morris office building was turned into a suitable place to accommodate important visitors from out of town. Designed by Edouard Sidel, a French-born architect, it was an ornate five-story structure in Second Empire style, with a glass-roofed arcade, two mansard towers, and an elegant dining room paneled in mahogany and oak. Mercantile firms were also creating impressive establishments in the downtown district. In 1895, Louis Saks built a fashionable department store directly across from the Morris House on the northeast corner of First Avenue and Nineteenth Street, competing for patronage with nearby counterparts like Blach's, Caheen Brothers, Loveman's, Pizitz, and Yielding's. As business picked up late in the decade with the return of prosperity, the Bromberg family moved in from Mobile and contested a growing jewelry trade with such earlier arrivals as Ash Brothers and Jobe-Rose. Founded in 1836 and reputed to be the oldest business enterprise in Alabama, Bromberg's catered to families living in opulent areas such as the South Highlands. Stores like these typified a revolution in retailing that had taken place after the Civil War as department store buyers and independent operators, profiting from the market system that the railroads had made possible, undermined old wholesale distribution networks and dealt with an army of jobbers and drummers who sold brand-name goods directly from producers to urban outlets. Specialization of function was fundamental to the change.[4]

Numerous cultural, charitable, and philanthropic institutions reflected the city's dynamism. It now boasted a conservatory of music. James Bowron, who moved to Birmingham from Nashville in 1895 to become TCI's highest-ranking local officer, was an avid musician who did much to advance the city's artistic horizons by supporting various organizations including the Mendelssohn Society. In 1899, he played a prominent role in bringing the Boston Festival Orchestra to Jones Valley for two days of concerts. But Bowron was unusual among businessmen in promoting the arts. Most of his fellow executives were too busy making money, abdicating cultural leadership to women who founded literary societies, among them the North Highlands Chautauqua and the Clioian Club.

Institutions of higher education were springing up. Howard College, a Baptist institution that moved to the city in 1887, completed a new campus in the East Lake District by 1892, and the North Alabama Conference College for Men, a Methodist school later to become known as Birmingham-Southern, began admitting students in 1897. A business college—"there is not a single graduate of the Bookkeeping, Shorthand and Typewriting departments but what has a remunerative position," one description stated—occupied quarters in the Potter Building on First Avenue. In addition to the public schools, there were also such private institutions as the Pollock-Stevens Institute for young ladies and the Taylor School for boys.

St. Vincent's Hospital, built by the Catholic Sisters of Charity at the turn of the century, was an important addition to the city's medical facilities. Another sign of increased social concern was an industrial school for boys, established by Julia S. Tutwiler and other reformers to provide separate treatment for youthful delinquents instead of confining them with hardened offenders. Born in 1841, Tutwiler was the daughter of Henry Tutwiler, one of the first professors at the University of Alabama, and a cousin of Edward M. Tutwiler, providing one more illustration of the strong Virginia connection in the development of the Mineral District. One of Alabama's greatest humanitarians, she was well known for her severe criticism of convict labor, arguing that it embodied all the evils of slavery without any of its redeeming virtues.[5]

Another of Birmingham's prominent citizens was a prolific novelist, Mary Johnston. As the daughter of John W. Johnston, she exemplified the cultural contribution that the Georgia Pacific had made to the district. Born in the Old Dominion at Buchanan, Botetourt County, in 1870, her health was too frail to endure regular schooling, so she obtained her early education from relatives and governesses. After the Johnstons came to Birmingham she was sent to school in Atlanta, but illness forced her to return home after only three months of instruction. This was the only formal education she ever received. When her mother died shortly afterward, she and her father took care of the other five children. Accompanying him to France in 1889 when his health broke under the strain of overwork, she ultimately returned to the Magic City with him and began an outstanding career as a writer of historical novels, mostly set in Virginia during the Revolutionary and Civil wars. In 1898, she achieved her first success when *Prisoners of Hope* was published by Houghton Mifflin after attracting attention from Walter Hines Page; two years later, *To Have and to Hold* was even more widely acclaimed, setting her on a path that led her to write twenty-five novels, some best-sellers. Most of her later work was done in Virginia, to which she returned after her father died in 1905. She lived there until she died in 1936.[6]

Besides yielding cultural custodianship to women, captains of industry played little or no role in Birmingham's governance unless there was a threat to their vested interests, which seldom happened. "No mayor came from the major upper-ranking manufacturing, mining, utility, or banking firms," stated political scientist Carl V. Harris after surveying political leadership in the city between 1871 and 1953, "and no member of these firms ever ran for mayor." Instead, according to Harris, holders of this office were "typically young, aggressive downtown business or professional men who had demonstrated administrative skill by nurturing their own small firms to prominent success." Such individuals personified the booster spirit that pervaded the community. Not a single mayor represented working-class interests; the least affluent was a grocer whose store was well above average in the size of its operations. Aldermen also tended to be people of middling income, usually businessmen; as Harris stated, "lower-ranking groups, which embraced the lower 80 percent of the population, were always underrepresented on the board." Political campaigns seldom if ever involved fundamental economic issues. Usually they revolved around such concerns as saloon regulation, control of vice, and Sunday observance.[7]

One thing on which white residents agreed was that blacks, who made up about 40 percent of the city's population and an overwhelming majority of its unskilled work force, had to be kept firmly in subjection. As a newspaper editor observed, "The negro is a good laborer when his labor can be controlled and directed, but he is a very undesirable citizen." Long before Jim Crow laws were enacted at the turn of the century, blacks had been segregated by custom and feared for supposedly criminal tendencies. Even though they seldom harmed whites, acts of violence perpetrated by them were luridly reported in newspapers, reflecting widespread anxiety that such behavior would affect fashionable areas of the city. "When criminal negroes carve up, stab and kill other criminal negroes, they rapidly develop into human tigers," the *News* declared. "The taste of blood makes them reckless. They will readily, surely give vent to their hatred of some white men, and sooner or later kill white men." The word of a white person, no matter how low his or her status might be, was always superior to that of a black, and most of the social ordinances that were enacted by the board of aldermen were aimed at keeping blacks under control.[8]

The 1900 census showed that 38,415 persons lived in the Magic City, representing a gain of 12,237 above the number that had resided there ten years before. Jefferson County had 140,420 inhabitants, an increase of 51,919 since 1890. Boosters protested that the city's importance was obscured by the fact that so many persons lived outside its boundaries and conducted a vigorous campaign to create a "Greater Birmingham" by annexing outlying areas. Among other benefits, they argued, this would produce an integrated sanitation system and also yield a fairer apportionment of taxes, particularly because the plants belonging to such companies as Sloss-Sheffield and TCI were not located inside the city limits. In an atypical instance of attacks being made against vested industrial interests, a bill to include suburbs and manufacturing installations inside Birmingham's boundaries nearly passed the state legislature in 1899. Managers and their attorneys thwarted it.[9]

Despite the protracted depression of the mid-1890s, the city bustled with industry. A promotional booklet issued by the Commercial Club in 1896 indicated that there were about 180 manufacturing plants, including "Furnaces, Basket Factories, Bicycle Factories, Blank Book Manufactories, Boiler Works, Bolt and Nut Works, Bottling Works, Brass Foundries, Breweries, Brick Works, Iron Bridge Works, Broom Factories, Wagon and Buggy Works, Candy Factories, Cabinet and Pattern Works, Cigar Factories, Cider Works, Cotton Compresses, Foundries and Machine Shops, Furniture Factories, Grist Mills, Harness Factories, Lithographers, Mattress Factories, three Daily Newspapers and Ten Weeklies, Packing Houses, Pipe Works, Planing Mills, Rolling Mills, Saw Works, Stove Factories, Trunk Factories, Wheelbarrow and Truck Factories, Wood Turning and Scroll Works, Yeast Factory, Car Wheel Works, Cultivator Works, Pump Works, Cotton Seed Huller Works, and others." Five railroads served Jones Valley; Jefferson County had almost 600 miles of track.

As always, mineral industries constituted the heart of the local economy. In 1896, the district had approximately 3,400 furnacemen, 3,500 ore miners, 5,000 coal miners, 1,500 coke oven employees, and 8,000 workers in rolling mills, pipe foundries, and other metal-fabricating installations. Coal production in 1900 was more than twice what it had been a decade earlier, and pig iron production had climbed by one-third. Textile manufacturing also flourished, most notably at the Avondale Mills of Braxton Bragg Comer, a Barbour County native who moved to Jones Valley from Anniston in 1897 to escape high freight rates and built a plant employing about 400 poor whites, many of them children. By the turn of the century, as the American economy surged forward, the entire area was experiencing a boom outstripping even the flush times of 1886; in the twelve months preceding August 1900, more than $11 million was pumped into new industrial and mining enterprises throughout the district. According to the *Age-Herald*, this was "twice or three times the amount ever invested here within any other similar period." Noting that approximately seventy-five new companies were operating, boosters rejoiced not only because of the number of these enterprises but also because many of them made finished goods, encouraging local leaders to believe that the district was at last breaking free from its almost total reliance on the extraction of raw materials and manufacture of pig iron. "Today there is no gloom, no uncertainty, no doubt, no fear," the *Age-Herald* boasted, "but instead confidence, push and aggressiveness."[10]

Overshadowing everything else was TCI's recent success in producing the first steel ever made in the area in significant quantities at commercially competitive prices. "This great corporation has probably done more toward the industrial development of the South than any other agency," the *Age-Herald* declared. Although not all of its facilities were in the Birmingham District, the great majority were located there. The company's three Tracy City coal pits in eastern Tennessee, producing 300,000 tons per year, were dwarfed by its fourteen Jefferson County mines, yielding 2,700,000 tons. Similarly, TCI's three blast furnaces at South Pittsburg in the Chattanooga District, with an annual capacity of 210,000 tons, cut a small figure alongside its fourteen furnaces in the Birmingham area, with a potential yearly output of 830,500 tons.

Everything about TCI was enormous by southern standards. Its twenty blast furnaces, including three stacks that it had recently acquired in northwest Alabama, produced 3,550 tons of pig iron per day. Its thirty coal mines, mostly in the immediate vicinity of Birmingham but also including its Blue Creek, Blocton, Henry Ellen, Sheffield, Tracy City, and Whitwell collieries, had an average daily output of 19,000 tons. It owned 3,722 coke ovens, with a daily capacity of 6,000 tons; more than 400,000 acres of mineral lands; and four large quarries that yielded 100 railroad cars of limestone and dolomite per day. Its payrolls numbered 14,000 employees, and its company stores handled $2 million worth of merchandise per year. Its total output of pig iron, coal, coke, ore, and flux in 1899 aggregated 7,809,927 tons. "To haul this gigantic output of raw material and finished product would require over three hundred thousand railroad cars, making a train over 1,500 miles long," the *Age-Herald* reported. During the same year, TCI sold 767,220 tons of iron, which was shipped to thirty-five American states and eight foreign countries.[11]

II

Despite its greatly expanded size, Sloss-Sheffield cut a modest figure beside its mammoth rival. Nevertheless, as the *Age-Herald* stated, it, too, was "a most powerful developing agency" in the district's progress, "successfully controlled by men of public spirit, energy and brains." As it had demonstrated in November 1899 by temporarily acquiring 4,480 acres of mineral lands from The University of Alabama at a small fraction of their true value, the firm had less public spirit than this ringing endorsement indicated. Joseph Forney Johnston, a past president of the Sloss Iron and Steel Company who had become governor of Alabama in 1896, was a party to the attempted grab, as was his perennial business associate, Edmund W. Rucker. The deal, made in secrecy with the university's board of trustees, led to a furor that the *Age-Herald* itself did much to create, but the tumult died down after the board rescinded the transaction under pressure from irate alumni. Later, historian John Craig Stewart stated that "no evidence of corruption was ever produced," but Johnston's critics pointed out that he owned stock in one of the coal and coke companies that Sloss-Sheffield had absorbed in the recent consolidation. In any case, the unseemly episode was soon forgotten, even by the *Age-Herald*, amid enthusiasm about Birmingham's growth and future prospects.

Boosters had no doubt that Sloss-Sheffield would soon follow TCI into the production of basic iron and steel. At the time of the 1899 merger, it was reported that the newly consolidated firm would soon build a 1,000-ton steel mill and that the $5.8 million worth of unsold common and preferred stock that had been retained in its treasury would be used partly for that purpose. The steel plant, it was said, would be built in North Birmingham, along with other new installations including a pipe works.[12] Sloss-Sheffield's financial performance did nothing to discourage such speculations. Fiscal 1899 was a halcyon year in which net earnings after allowance for depreciation and payment of bond interest swelled to more than $500,000. Two historic break-throughs highlighted the gains registered by the enterprise. For the first time, earnings from the sale of pig iron, aggregating $530,539, far outstripped those from the marketing of coal, coke, and ore, which totaled only $179,539.13. The second development was equally gratifying; in March 1900, a dividend of 1¾ percent was declared on capital stock, marking the first time in the firm's history that this had ever happened. As a newspaper account stated, the company had made "a splendid showing," particularly because much of its pig iron had been sold under contracts signed early in 1899, when prices were still relatively low.[13]

Behind the scenes things were not going as smoothly as Sloss-Sheffield's profits seemed to indicate. In the last quarter of 1899 and the first few months of 1900, Bryan went through a period of intense anxiety. Beginning in June 1899, the expansion of business activity that had been proceeding so strongly throughout the country since 1897 momentarily faltered. The downturn was brief, hitting bottom before the end of the year, but was severe while it lasted.[14] As winter approached, money became extremely tight. Caught in the crunch, Bryan was short of cash just when he needed it most to make good on his stock subscriptions in Sloss-Sheffield, the organization of which was still underway. He also had to raise a great deal of money so that the Richmond Locomotive Works, which lost more than $200,000 in 1899 because of intense competition with northern rivals like Baldwin, could fill about $2.5 million worth of recently accepted orders.

The resulting crisis, one of the worst in Bryan's entire business career, was particularly vexing because he could not understand why things were suddenly so bad on Wall Street. "There is nothing in this 'rattle' in New York that affects the substance of our properties, and there is no reason that I can see to justify the commotion," he wrote at one point to a banker in Birmingham. "The business of the country has never been in better shape." Blaming the contraction on speculators and recent financial developments in England, he complained to Maben that "there is nothing to justify any such condition of things."[15] Increasingly short of funds, Bryan drew heavily upon Goadby and Oakman late in 1899 to keep his locomotive works afloat and fulfill his subscriptions to Sloss-Sheffield stock. He was also forced to borrow money in Birmingham, where, perhaps because of the euphoria surrounding TCI's recent success in making steel, credit was easier to get than in New York or Richmond. Galling as it was to become dependent upon sources that might come back to haunt him in the future, he managed with Haas's help to borrow $20,000 from the Alabama National Bank, of which Rucker was president. He also got a loan of $30,000 from the Birmingham Trust and Savings Company. Meanwhile, his normally excellent relationship with Maben showed signs of strain as he pressed his longtime friend to settle a long-standing obligation of $12,000, apparently connected with the settlement of Daniel K. Stewart's estate.[16]

By mid-December, having exhausted every other source of collateral at his disposal, Bryan was forced to take a step that would previously have been unthinkable: sell some of his holdings in Sloss-Sheffield. On December 19, he reluctantly authorized Goadby to sell 200 shares of preferred stock, "at the best price you can get for it." Consoling himself that he would "have a plenty left when this is gone," he told Goadby that "my indisposition to sell Sloss has almost become a disease, which I had just as well begin to get over." Three days later, to meet a $24,500 payroll at the Locomotive Works, he drew on Goadby for $10,000 and authorized him to sell yet more of his Sloss-Sheffield stock, 700 shares of preferred and 400 common. The outlook for 1900, he stated, was "simply superb . . . but that is not ready cash." Two days after Christmas, he wrote that he had disposed of $20,000 worth of Sloss mortgage bonds, while expressing regret at prices Goadby had secured in the previous sale. "Of course it was very unpleasant to me to give up the stock at these figures," he said, "but I had other matters to provide for." Writing that he would need another $15,000 from Goadby to keep the locomotive works going, he soon sold another 100 shares of Sloss-Sheffield. "I have need for all the cash in sight," he told Goadby.[17]

Early in 1900, Bryan wrote a reproachful letter to Maben, who had not responded to his recent request to settle his debt. "For lack of spot cash I had most reluctantly to sell some Sloss-Sheffield last week," he stated. "It hurt me to do so. Now I would like if you can do so to convert the $12,000 you owe me into Sloss-Sheffield so as to recoup in some degree my loss." By this time, Bryan was under pressure from the Central Trust Company to fulfill yet more of his stock subscription in Sloss-Sheffield and had to ask that institution for an extension of an existing $30,000 loan. Acknowledging a letter from

Goadby reporting the sale of ten Sloss mortgage bonds, he had him sell thirty more. His aversion to taking such steps was intensified by his high hopes for the firm. In January, just before going to Birmingham on an inspection tour, he answered a request for information about the company by stating, "we believe that the Sloss-Sheffield Co. has the most valuable coal and iron property in the South and that it will make the cheapest iron of any concern."[18]

Although he reported to Goadby that he was "not at all well" after getting back to Richmond, Bryan's letters, as usual, revealed little about his trip to Alabama. During February the financial pressures he was facing seem to have abated, but when his $50,000 debt to Birmingham banks came due in early March, he could not liquidate it and renewed the notes for another three months. By late March he was again deep in crisis. The Central Trust Company demanded a $70,000 installment on a subscription that he had made to Sloss-Sheffield for Daniel K. Stewart's estate, and he was hard pressed to comply as it became increasingly evident that the Richmond Locomotive Works was a millstone around his neck. When the Central Trust Company, to which he already owed $100,000, turned down his request to finance his stock subscription in Sloss-Sheffield by borrowing yet more money, he wrote a stinging letter to Maben, who had still not paid his overdue debt. "I write again to tell you that I will absolutely need the $12,000 this week," he said. "I have a payment to make which *makes* me call on you. Please let me hear from you."[19]

At this point, Sloss-Sheffield came to Bryan's aid, showing the degree to which its financial health was superior to that of his locomotive works. Because of the 1¾ percent dividend declared by Sloss-Sheffield at the end of March, Bryan received a check for $27,786.50. With this infusion of cash and the help of Ladenburg, Thalman and Company in New York, he managed to come up with the rest of the funds required by the Central Trust Company. In the process, he also got credit for a disputed 1,728 shares of Sloss-Sheffield stock that he claimed as a bonus for helping organize the new company. Hurt by the strain in his relationship with a firm that had been so crucial to him for so many years, he wrote a conciliatory letter to a Central Trust Company officer stating that he had been doing business with that enterprise for two decades, "and this is the solitary instance of anything approaching a disagreement between us." In April, Maben finally settled his $12,000 obligation, with interest. "I return the note herewith marked paid in full," Bryan responded. Considering how long he had waited, effusive gratitude would have been inappropriate.[20]

Soon thereafter, having searched for markets in such faraway places as Australia and South Africa, Bryan bowed to the inevitable and sold the Richmond Locomotive Works at a loss to a group of northern businessmen who formed the American Locomotive Company by combining it with seven other enterprises. Bryan continued to be a director in the new entity and served for a time as chief operating officer of its Richmond plant. From this time on, however, his entrepreneurial energies were devoted principally to his publishing activities and Sloss-Sheffield. In 1903, he acquired a competing Richmond newspaper, the *Dispatch*, and combined it with the *Times*. After surviving a few more years, the locomotive shops closed down soon after he died in 1908. According to James M. Lindgren, they had apparently been acquired by their new owners "only to eliminate his competition."[21] The colonial status about which perceptive observers had warned southern entrepreneurs in the 1890s was becoming a reality.

By unloading his locomotive business, Bryan stanched the drain it had made upon his financial resources, enabling him to retain control of Sloss-Sheffield. Meanwhile, prodding investors who held stock in the Sloss Iron and Steel Company to exchange their securities for certificates in Sloss-Sheffield occupied much of his time. Maben bore the brunt of this activity, but Bryan worried constantly about it and was relieved when Maben reported in February that "substantially all" of the transfers had been made. Completion of the project made Sloss-Sheffield the official operating entity that emerged from the recent consolidation, reducing Sloss Iron and Steel to a dormant subsidiary. Soon thereafter, Sloss-Sheffield began listing its shares publicly on the New York Stock Exchange. Although this practice had become customary among large industrial corporations, it was the first time that Bryan had been forced to take such a step. Coupled with the financial distress through which he had been going, it could not have helped but raise fears in his mind of seeing his power in the increasingly profitable enterprise slip from his grasp as it attracted attention from ever-watchful investors.[22]

Throughout the winter of 1899–1900, as Maben gathered in the outstanding shares of the Sloss Iron and Steel Company, legal tangles connected with the Lady Ensley properties further muddled efforts to complete the organization of Sloss-Sheffield. At one point, these complications became so difficult as to require a trip to Alabama by the Central Trust Company's leading attorney, Adrian Larkin. In the scramble for Enoch Ensley's former assets, a welter of conflicting obligations and shareholdings had fallen into the hands of local and outside interests. A particularly large stake had been acquired by Edward O. Hopkins, a railroad promoter from Evansville, Indiana, who was also involved with a syndicate that had taken over a moribund furnace company in Florence. One of four sons born to John S. Hopkins, a native of upstate New York who went to southern Indiana as a boy and made a fortune in banking, steamboating, and railroads, Edward inherited wealth from his father and increased it through involvement in such rail lines as the Peoria and Pekin Union and the Peoria, Decatur, and Evansville. In trying to deal with him, Sloss-Sheffield had the choice of buying him out or taking him into the firm as a part owner and director. When satisfactory terms for a buyout could not be reached, Bryan yielded reluctantly to the second option rather than go to court. "I do not like this kind of partnership business, as it is very apt to breed trouble," he remarked prophetically in a letter to Haas, "but it is probably the best we can do and will help along the settlement of the Lady Ainsley matter, I think."[23]

III

Among other problems facing Bryan in this difficult period were managerial and technological issues connected with restarting the welter of ill-assorted mines, coke ovens, blast furnaces, and other facilities that had recently been acquired in northwest Alabama. Getting the mines back into production was especially urgent; despite its recent Cahaba acquisitions, Sloss-Sheffield needed as much brown ore as it could get and was eager to benefit from the rich Russellville deposits. Only an experienced person with strong technical credentials could undertake the task, which was beyond Haas's capabilities.

Following Rucker's advice, Bryan and his advisors dealt with the problem by recruiting one of the most prestigious figures in the Birmingham District: Truman H. Aldrich, who had resigned from TCI several years earlier because of disagreements with DeBardeleben. As a trained mining engineer with a degree from RPI, he was an ideal person to help Sloss-Sheffield exploit its new assets. "I am sorry that you have such a burden of labor upon you and I wish I could help you more," Bryan wrote Haas early in December 1899. "The management of the Sloss Co. is a big thing, and I hope you will not fail to get the benefit of Mr. Aldrich's assistance if you possibly can. Don't stand upon the salary question but get the man." Much to Bryan's satisfaction, Aldrich was hired within a few days as general manager and second vice president at a yearly salary of $10,000. "I am sure Aldrich will help our cause greatly," Bryan wrote Maben shortly after the beginning of the new year. "He seems practical & energetic & judicious."[24]

Although Bryan did not realize it when he wrote these words, Haas, who had worn himself out in his efforts to run the greatly expanded enterprise, had contracted tuberculosis. Learning about this in early April, the executive committee gave Haas a ninety-day leave of absence, and he went to the mountains of western North Carolina in search of a cure. Bryan, still struggling to extricate himself from financial difficulty, encouraged Haas by telling him that a relative had recently gone to the same area to make a successful recovery from illness. "I really envy you the rest which you are going to have, especially as you are going to have Mrs. Haas with you," he wrote. "It will be like another honeymoon."[25]

Pending Haas's return, Aldrich took charge at Birmingham as acting president. Apparently hurt that he had not been chosen for this role, Rucker turned against Aldrich and questioned his fitness for it. Even before Haas departed from Birmingham, Rucker began bombarding Bryan with letters critical of Aldrich. "I was not prepared for your opinion of Mr. Aldrich," Bryan responded. Reminding Rucker that Aldrich had been hired upon his recommendation, Bryan expressed hope "that your first judgment of him will be confirmed." Writing to Haas as the latter prepared to leave Birmingham, he confided that "Gen. Rucker is restless and anxious about the situation in your absence. I don't know how far he thinks his own services should be availed of. I am sure it would be a great delight to him to get his hand on the helm." After Rucker wrote again to convey "a very decided opinion that Mr. Aldrich is not equal to the emergency," Bryan told Oakman, with evident sarcasm, that the general's aspersions had resulted partly from a "desire to have his own services more used." For the present, he recommended that Aldrich move slowly in implementing any changes that he might have in mind, but supported him fully. "I think Aldrich should receive all the help which our joint experience can give him," he told Maben. [26]

Within a few days, an unexpected event brought the already volatile situation to a boil. At about 9:40 p.m. on Saturday, 6 May, while workers were preparing for a run of pig iron at City Furnace No. 1, an explosion within the stack blew out the rear wall, releasing a torrent of molten metal that set fire to everything with which it came in contact. Spreading rapidly, the flames incinerated anything that was combustible, including coal bins, an ore lift, the stock house, and fifteen freight cars loaded with raw materials. Fire engines from the Northside and Southside departments fought the blaze for two hours while a crowd that had been attracted by the blast gawked at the spectacle. Not until midnight was the fire under control. Even then, the flames were still not totally extinguished. According to the *Age-Herald*, the scene at daybreak the next morning "was one of appalling ruin." [27]

Forced to shut down both of the City furnaces, Aldrich moved quickly to restore production. With the aid of the L&N and Southern railroads, which rushed supplies, damage was quickly repaired. City Furnace No. 2 was back in blast within four days, and even No. 1 was returned to operation by 16 May. "Prompt Action of Acting President Aldrich Saved the Company from Heavy Losses," a headline in the *Age-Herald* proclaimed. The accompanying article praised Aldrich and his associates for their "vigorous and capable management" and the "almost incredible rapidity" with which the repairs had been effected. "The Sloss company may well feel proud of such an achievement," the newspaper stated, "and has reason to congratulate itself." [28]

Bryan supported Aldrich strongly throughout the crisis. "I am very sorry that you have had your administration inaugurated with such an accident . . . but I know that you will do what is best and quickest to remedy the trouble," he wrote Aldrich the day after the conflagration. Nothing but full cooperation with Aldrich, he told Maben, would be "just or sensible." Aldrich, he bluntly informed Rucker, must be given "a fair show and full support before we conclude that he is not equal to his task." A few days later, he again told Rucker that he was "heartily in favor of giving Mr. Aldrich all the authority necessary to make his administration effective." His confidence in the future of the enterprise remained high as repairs continued. "I am more than ever persuaded that we have a magnificent property," he wrote Oakman—"much more valuable than we have any idea of." [29]

During the next few weeks, Haas reported gradually improving health. After earlier admonishing him to stay inactive, Bryan now began using his assistance in connection with some private business dealings in Birmingham in which the two men were jointly engaged. Nevertheless, Bryan worried incessantly about Haas. "For God's sake get well," he urged him in late May, scolding him for not having revealed his illness before it had become so bad that he had lost the use of one lung. "I have been distressed ever since you have been sick, and if I had had any idea of how you fooled me about your sickness I would have had you jerked up and carried to a doctor months ago. But I tell you I will never be fooled by you again. If I get you straight this time I will see that you keep straight."[30]

Haas not only improved, but survived for nine more years. His tenure at the helm of Sloss-Sheffield, however, was over. Early in August he returned to Birmingham looking much better, but the duties of the presidency were too "onerous and confining" for him to bear without "absolute rest for a time." On 3 August 1900, the *Age-Herald* reported his resignation and speculated that Aldrich would succeed him. Noting that the latter had been "actively connected with the coal and iron industry of the District for over twenty years," it asserted that "there is no better informed man in Alabama on the resources and conditions in this District, and his elevation to the presidency . . . would be a good stroke for the company."[31]

But things did not happen as expected. Within a week, at a meeting in New York City, the directors chose the company's newest director, Edward O. Hopkins, to succeed Haas. At the time, Hopkins was so little known in Birmingham that the *Age-Herald* referred to him as "E. O. Haskins" in a headline announcing his appointment. Haas did not sever connections with the company; he was put on the executive committee, which continued to be chaired by Oakman and also included Bryan, Maben, and Van Cortlandt. Aldrich temporarily stayed on as second vice president.[32]

Because Bryan's letterbooks between June 1900 and August 1902 have been lost, there is no clue why the directors had acted as they did. It is logical to assume that Rucker's opposition to Aldrich hurt the latter's chances of succeeding Haas. The presence of several new directors on the board as a result of the recent expansion also created a degree of volatility previously lacking. Given Bryan's enormous influence, however, it is inconceivable that Hopkins could have been chosen had the Richmond industrialist actively opposed him. It is possible that Hopkins was a compromise candidate, preferable in Bryan's view to the first vice president, Rucker. Haas's inclusion on the executive committee, coupled with his improving health, suggests that some of the directors believed he might be able to return as president in the future. A newspaper article reported that Haas and his wife greatly enjoyed living in Birmingham and intended to continue making it their home after going to California in the hope that Haas's health might improve there.[33] Hopkins's elevation may therefore have been regarded as a temporary move.

Hopkins remained in Evansville for a few weeks to wind up the affairs of the

Peoria, Decatur, and Evansville Railroad, a bankrupt line that was being absorbed by the Illinois Central. After this task was done, he moved to Birmingham in early September. Haas came down from North Carolina to help him settle in. Aldrich resigned within a few weeks to devote himself to other business interests. Soon thereafter, Hopkins assembled his administrative team. Some members, including M. J. Lewis, assistant to the president, and A. G. Palmer, general traffic and sales agent, were new to the company. Most, however, were not, including J. H. Means, a previous furnace superintendent who now returned to his former job; Henry Hiden, retained as purchasing agent; James W. McQueen, a longtime employee who stayed on as secretary-treasurer and marketing head; Priestley Toulmin, who had been with the engineering and mining department for several years and became general manager; and C. H. Schoolar, who continued as auditor. Everything indicated that a smooth transition was underway.[34]

A few months after taking office, Hopkins presented an annual report that was mostly a recapitulation of what Aldrich had already accomplished. During the past year, Sloss-Sheffield had launched a major program of physical improvements to its newly acquired northwest Alabama properties. In Florence, Philadelphia Furnace had received $60,000 worth of renovations, including new boilers and a new stove. The Hattie Ensley and Lady Ensley furnaces in Sheffield had been even more thoroughly overhauled, at a cost of $134,000 and $146,000 respectively. Among key additions were an Allis blowing engine and a new pumping station on the Tennessee River that was capable of providing boiler and cooling water to both installations. Another $285,000 had been invested in improvements at Russellville to facilitate strip-mining. These included hydraulic washers, steam shovels, and mine locomotives, aimed at making this installation "the largest and best equipped plant of its kind in the South." Newly acquired coal lands in Walker County had received less costly improvements, new openings had been made to increase coal production at Blossburg, and various changes had been made at the Sloss mines near Bessemer to insure enough red ore for future operations at all four Birmingham furnaces. Despite making these improvements and expenditures, the firm had earned a net profit of $546,099.53.[35]

During fiscal 1901, Hopkins continued an aggressive program of renovations and improvements. Turning his attention to the old City furnaces, he decided that Sloss No. 2 was in such bad condition that it needed to be torn down and "built new from the ground up." In the process, its height was increased to eighty-five feet, thirteen feet higher than the previous stack. Though Sloss No. 1 was not dismantled, it was relined and fitted with new base plates, crucible and tuyére jackets, bosh plates, bustle pipes, increased stove capacity, and other improvements. Further renovations were made to the Hattie Ensley and Lady Ensley furnaces, and large sums were again invested at Russellville, where brown ore output rose from 129,502 to 234,661 tons in 1902. All told, $512,172.85 was spent in improvements at company sites, and profits increased to $666,443.25.[36]

TCI was also enjoying banner profits, outstripping those of Sloss-Sheffield. In 1899 it earned an unprecedented net gain of $1,287,873, followed by $1,634,745 in 1900. Despite its prosperity, however, the company was in turmoil as usual. Quarrels between outside owners and local administrators over management of the new steel plant led Bowron to resign as treasurer in October 1900. The success of the steelmaking venture also produced an increase of interest among northern investors. This situation brought into power a new group of businessmen, largely connected with the Hanover Bank of New York. Early in 1901, Don H. Bacon, a Minnesota mining magnate, became chairman of the board. Heads rolled in Birmingham as Bacon replaced TCI's chief local executives with men of his own choosing, Baxter and Shook being among those removed. Understandably, the displaced leaders were deeply resentful. Despite enthusiasm among local boosters, the new steel works of which Birmingham was so proud had led to intensification of outside control.[37]

IV

Sloss-Sheffield, too, was soon deep in crisis despite its increasing prosperity. Pressure to emulate TCI by following its giant rival into steelmaking was the primary cause, placing Hopkins, who favored such a move, at odds with members of the board who thought differently. Among physical modifications reported to the directors by Hopkins at the end of the 1901 fiscal year were changes to one of the North Birmingham furnaces so that it could "produce either basic or foundry iron as desired by the management." Behind these words lurked an issue creating intense debate on the board as the year came to an end: the intent of the modifications was to enable Sloss-Sheffield to begin making basic iron for sale to companies that made open-hearth steel. In November 1901, Hopkins submitted to the executive committee a proposal from E. T. Schuler, president of the Alabama Steel and Wire Company, "on the subject of constructing a steel plant in the Sloss District." According to the minutes, this "was deemed worthy of consideration and referred to the Chairman, who was requested to communicate with Mr. Schuler and obtain further details." Two months later, on 9 January 1902, William Garrett and Horace W. Lash of the Garrett-Cromwell Engineering Company "were present by request, and estimated the cost of constructing a plant of open hearth furnaces for the production of about three hundred tons of steel billets per diem, with a view to the possible construction of such a plant by this Company."[38]

Schuler had previously been involved with TCI in converting billets from its Ensley steel works into rods and wire.[39] So far as local boosters were concerned, the adoption of his proposal by Sloss-Sheffield would represent the logical culmination of rumors that had flown around the district, at the time of the 1899 consolidation, about the impending construction by Sloss-Sheffield of a steel mill in North Birmingham. The very fact that the enterprise called itself the Sloss-Sheffield Steel and Iron Company, putting the magic word "steel" before "iron," was further evidence that, at least originally, it had regarded steelmaking as its ultimate goal.

During the past three years, however, controversy had arisen within the firm about whether this intention should be carried out. On 20 March 1902, matters came to a head at the annual stockholders' meeting in New York, and the most significant turning point in the history of the company was reached. Although the minutes reveal nothing about the arguments that took place, the debate must have been highly charged. Henceforth, instead of moving into the manufacture of basic iron or steel, Sloss-Sheffield would concentrate on foundry pig iron.

As the meeting began, there was no suspicion in Birmingham that anything was wrong. "No special changes are contemplated," the *Ledger* reported. "President E. O. Hopkins will be reappointed." Soon after the meeting, however, the *News* indicated that Hopkins would resign. Interviewed by a reporter, Hopkins neither affirmed or denied the story. The next day, the *Age-Herald* confirmed it and said that Maben would replace Hopkins. Friction within the board, the report stated, had begun "as far back as last summer," resulting from "disagreement over policies touching certain large questions." Maben, it continued, was Hopkins's chief adversary on the board; indeed, it referred to members who opposed Hopkins as the "Maben faction." Although Hopkins had won election for another year, the article stated, the situation had become so bad that he had decided to step down.[40]

However vague the story was concerning the matters at issue among the directors, it was true that Hopkins had relinquished his post and that Maben would replace him as president. The gravity of the crisis was indicated by changes that quickly took place. Within twenty-four hours, a wave of resignations occurred within Sloss-Sheffield's Birmingham hierarchy; among those leaving were Gentry Hillman, manager of the City Furnaces, and C. H. Schoolar, who had been auditor for eleven years. Articles in the *Age-Herald* and the *News*, praising Schoolar's "fine record" and emphasizing the local popularity of Hopkins himself, left no doubt that powerful members of the community were angered by what was taking place. Within a short time, local newspapers announced that Hopkins and his family would leave in June for a trip to Europe and that Maben would arrive by mid-April to take charge, but other stories speculated that Hopkins had received "a flattering business offer from a strong southern corporation" and that his impending European trip might not occur. "His friends here," stated the *Age-Herald*, "hope that he will see his way clear to accept any position that will keep him in this part of the country." The *Ledger* underscored prevailing sympathy for Hopkins by saying that "probably no man who has been so short a time connected with the development of the Birmingham District will leave . . . with more genuine and general regret attached to his departure." Calling Hopkins a "pusher," the article said that "his personal following in Birmingham is strong and wide-spread. It will be difficult to replace him."[41]

Maben was by far the most powerful figure yet to become president of the company. The fact that he had been sent from Wall Street to Birmingham to take the reins indicated the gravity of the directors' decision. In light of what happened after Maben took power, it is clear that Schuler's proposition had created the crisis by bringing the issue of basic steelmaking to a head. Various changes in the slates of directors and officers elected at the company's annual meeting reflected the outcome of the debate, which had triggered Hopkins' decision to bow out. One of these was the resignation from the board of Archer Brown, whose support for steelmaking was indicated by the fact that he had wanted Bowron, who was well known as an enthusiast for basic iron and steel, to leave TCI in 1899 to become president of Sloss-Sheffield. H. O. Seixas, a Louisiana capitalist and longtime friend of Bryan, was selected to fill the seat that Brown had vacated. In a related move, indicative of waning strength on the part of Birmingham interests, Bryan assumed formal office in the company, becoming first vice president while Rucker was demoted to second vice president. Things had obviously gone to Bryan's liking, while Birmingham's deep desire for the firm to emulate TCI had been rejected.[42]

Events moved swiftly as the incoming Maben administration prepared to take control. Accompanied by Goadby and Seixas, Maben arrived in Birmingham on 27 April 1902 and checked in at the Hillman Hotel, a six-story structure that had recently replaced the Morris House as the city's foremost place for accommodating visiting dignitaries. Bryan and other directors soon followed. When reporters asked Maben what his policies would be, he replied evasively that he was "not prepared to be interviewed." The *Age-Herald* noted that he was "thoroughly familiar with all the details of the company's business methods and operations," but the paper displayed none of the warmth that had been so obvious in its treatment of Hopkins. The latter, it was announced, would sail to Europe, "to be abroad a year or more."[43]

Two days later, Hopkins turned his office over to Maben and left the city. Before departing, he was given a going-away party that showed how sorry many local leaders were to see him go. As the *Age-Herald* reported, "many prominent citizens called on Mr. Hopkins to say good bye and express their regret at his quitting Birmingham." At five o'clock in the afternoon, heads of the various departments in the firm presented him with "a token of their esteem in the shape of a large diamond." Hopkins responded with a speech in which he "said many nice things about Birmingham" and predicted a bright future for the city. Mine superintendent John Hanby gave Mrs. Hopkins "a box of beautiful flowers." Schoolar was also given "a costly diamond of the same size and setting as that which Mr. Hopkins received," and "made a brief but feeling speech of acceptance." That night, Hopkins and his family boarded the L&N for Evansville, where they planned to stay until they left for Europe.[44]

V

Despite the fact that newspaper accounts did not mention precisely what was at issue, it is not hard to understand why they treated Maben's arrival with coolness and emphasized the high esteem in which Hopkins was held. Everything that had happened was a keen disappointment to boosters who equated the words "steel" and "progress." In aspiring to achieve the same result at Sloss-Sheffield that Baxter, Bowron, and Shook had accomplished at TCI by moving that company into the production of basic iron and steel, Hopkins and his local backers had met the same type of resistance from absentee owners. Unlike their counterparts at TCI, however, they had failed to overcome the opposition.

As Maben took the helm, Sloss-Sheffield did not formally announce its future plans. Clearly, however, it was not prepared to move into the production of basic iron and steel. Early in 1903, Bryan showed where he stood in a letter to the company's newest director, Seixas. "As a general line of policy," he said, "I am distinctly and affirmatively opposed to experiments in steel. There is not a man in our crowd that understands anything about it." The art of making basic steel from Alabama's raw materials, he continued, had not advanced to a point at which the company could be sure of producing it profitably. Success in such an undertaking required "the eye of the master," and the firm had access to no such technological expertise. Therefore, Bryan concluded, "I am disposed to believe that our wisest plan would be to accumulate our money and see what the course of business will be during this year before we begin spending any serious amount in new construction." Bryan did not completely dismiss the possibility of moving into steel production later. It might be advisable, he conceded, to have a committee study "the best form and cost of two new furnaces to produce not less than 250 tons a day, preferably 300, and to consider the best location should the company determine to develop in that direction." The study should be taken, however, only for informational purposes, "and not with the expectation that we would be satisfied that we could now conduct the operation satisfactorily in Alabama."[45]

Left unspoken in Bryan's letter were other considerations that dampened his enthusiasm for following TCI's example. Sloss-Sheffield was already earning excellent profits without embracing a risky proposition. In organizing the new company, Bryan had nearly failed to complete his stock subscriptions and thereby risked losing the power that he had held for so long. Only by selling the Richmond Locomotive Works had he managed to salvage his leadership role. Memories of that crisis were still fresh in his mind. Moving into basic iron and steel production, even on an experimental scale, would require far more capital than he and his most trusted associates possessed. It was far better to pursue a safe and steady course than lose control of an enterprise that had every prospect of success.

This was particularly true because traditional markets for southern pig iron were no longer as crowded as had once been the case. Although TCI continued to make foundry iron at Bessemer, it was now putting most of its efforts into basic iron and steel. Under Bacon's leadership, the new Ensley mills were being expanded and modernized.[46] Why should Sloss-Sheffield return to cutthroat competition with its giant rival when this was no longer necessary because the two firms were becoming complementary? Now that Pioneer had been absorbed by Republic, Sloss-Sheffield and Woodward were the only firms in the district that remained oriented exclusively toward pig iron production. Sensing the opportunities inherent in the changed situation, Woodward was conducting its first significant expansion since 1887 by building a third blast furnace, scheduled to go into operation in 1905.[47] Under the circumstances, it was logical for Sloss-Sheffield to follow Woodward into exploiting the widening market niche that both firms were now moving to fill: the foundry trade.

Recent trends in the foundry industry made it an inviting target for new marketing strategies by southern entrepreneurs like Bryan. Basic open-hearth steel producers and firms that made finished products from steel billets were mainly concentrated in the North. By contrast, foundries were already migrating southward, placing them within easy reach of plants in Birmingham and northwest Alabama. Prior to the late 1880s, many American enterprises had been scornful of southern iron. A St. Louis stove manufacturer told Bowron in 1879 "that he could not dream of putting such an inferior material into his cupola." But attitudes began to change, however, as it became known that silica helped soften castings. Because this chemical constituent was so prevalent in southern ores, high-silica metal from the region, known as "bright" iron, won increasing markets among foundries that had previously preferred iron from such places as the Hanging Rock region of Kentucky. Because the high phosphorus content of southern pig iron enhanced fluidity, it also gained increasing favor among producers of agricultural castings, hollowware, piping, radiators, and stove plates as an alternative to imported Glengarnock or Scotch iron, which had long been esteemed among American foundrymen for the ease with which it could be molded.

In 1888, Bowron had analyzed shipments of foundry iron in America and discovered "that the product was shipped over 30 different States, and that the entire consumption in the South amounted to only 13½ per cent." Conducting a similar study seven years later, he learned that "while the product was then distributed over 39 different States and 5 foreign countries, the Southern consumption had risen to nearly 22 per cent." Within this short period, Alabama had risen from seventh to second rank among American states as a market for foundry pig iron. Even in England, where Middlesboro iron had been renowned for ease of pouring and molding, Alabama iron had won markets during the export drive of the 1890s. Meanwhile, demand for what was known as forge iron, previously used for plates, bars, and nails, was decreasing under what Bowron called "the relentless march of steel." As a result, at least one Alabama producer of forge iron, at Helena, had gone out of business. During the late nineteenth and early twentieth centuries, the proportion of southern pig iron that was sold to rolling mills declined steadily, plunging steeply after 1904. Clearly, the foundry trade was the biggest remaining refuge for firms that did not want to produce steel. Given the extent of the market that beckoned, it made little sense for Sloss-Sheffield to throw away the natural advantages of southern raw materials for the foundry trade and spend money that its owners could not afford on costly steelmaking experiments. All of Bryan's deeply conservative instincts impelled him to avoid such a move.[48]

Concentrating increasingly on foundry pig iron also appealed to Bryan and Maben for another reason. During the depression of the 1890s, while Sloss-Sheffield and other southern producers had relied on cutthroat competition as their main strategy for survival, many American manufacturers had chosen a different approach, that of product differentiation. Emphasizing quality rather than quantity and concentrating on winning as good a reputation as possible in specialized markets, such enterprises had escaped the worst effects of the economic downturn. Following this strategy, makers of diverse products the likes of oatmeal, ledger paper, lubricating oils, household starch, terne plate (a specialized variety of tinplate used in roofing), and fine whiskeys had learned that they could maintain relatively high price levels. A growing use of brand names was part of this approach, distinguishing the products of quality-oriented companies from those of firms that sold their output in bulk. "In general," as Naomi R. Lamoreaux later remarked, "the more successful a firm was in differentiating its product from potential substitutes and staking out its claim to the upper end of the market, the more freedom from general price trends it was able to secure."

Several northeastern pig iron producers had adopted this strategy to good effect in the 1890s while southern producers had clung to more traditional types of competition or, in the case of TCI, moved away from mere pig iron manufacturing toward steelmaking. By emphasizing the production of high-quality foundry iron, some furnaces were able to postpone or avoid price changes that adversely affected firms remaining committed to a more indiscriminate approach. Information about such trends was readily available in the weekly bulletins of the American Iron and Steel Association, furnishing Bryan and Maben with potential ammunition for arguments in the executive committee meetings that took place in the winter of 1901–02 preceding Hopkins' resignation.[49]

The contrast between the diverging courses that Sloss-Sheffield and TCI were following was great. By 1902, TCI had spent millions of dollars in what would later be called a "high-tech" solution to the problems of the Birmingham District, based on *compensating* for the peculiar characteristics of its natural resources. That same year, Sloss-Sheffield chose a far less costly and much more traditional approach in an attempt to *take advantage of* the same peculiarities. Despite the anguish of boosters who remained mesmerized by the lure of steel, it was a recipe for success. Quickly, Sloss-Sheffield would reap its rewards.

The situation was also ironic. In their unsuccessful effort to emulate TCI's move into steel production, Sloss-Sheffield's local investors had been led by a northerner, Hopkins, who had become a hero in Birmingham by fighting a losing battle to attain an objective that symbolized the community's fondest aspirations. Conversely, the people who had sent Hopkins back to Evansville in defeat were not Yankee investors like the New York financiers who had resisted the dreams of Baxter, Bowron, and Shook, but Confederate veterans who represented the old plantation order in which the Magic City had been conceived. Epitomizing stubborn resistance to change, the decision that was made in 1902 to concentrate on foundry pig iron aptly manifested the enduring southernness of Sloss Furnaces. Yet no move that was taken in the entire history of the enterprise was more locally unpopular. The outcome must have been bitter indeed to Rucker and other investors who had hoped that the Virginia syndicate to which Colonel Sloss had sold his furnace company in 1886 would fulfill the old industrialist's dreams for the district by manufacturing the wonder alloy that would open the gates to Golconda. One cannot help but think that from this time onward Bryan was regarded as a Judas Iscariot by many of Birmingham's boosters. Perhaps this is why his memory faded so quickly after his death, which took place six years later and which many people in the city had no particular cause to mourn.

Progress and Paradox   10

As the national economy remained vibrant in the early years of the twentieth century, Birmingham continued to be confident about the future. The creation of a mammoth statue of Vulcan epitomized its aspirations, but the ultimate fate of the colossus betrayed the immaturity of a young community that lacked a sure sense of its heritage. Birmingham's hopes were naive, and the visions of its leaders were unrealistic. Growth strategies that seemed self-evident were misleading, and opportunities that were soundly based upon regional assets, not fully appreciated. In no case was this naivete more evident than in the contrast between TCI, which committed itself to the glamorous goal of becoming a major steel producer, and Sloss-Sheffield, which concentrated on the less exciting role of making foundry pig iron. By 1907, it was not clear where the contrasting paths of the district's two largest firms would take them, but it was obvious that they were pursuing different destinies. Late in that year, the acquisition of TCI by the world's largest corporation amid a financial panic caused boosters to believe even more strongly that Birmingham was destined to become the world's foremost steelmaking center. But future events would show that Sloss-Sheffield's focus on a product for which regional resources were ideally suited was more congruent with the South's potential for long range growth.

I

Maben's move to Birmingham in 1902 to take command of Sloss-Sheffield launched that company on the most profitable period it had yet experienced. Having argued for concentrating on the production of foundry iron, he soon justified the wisdom of that course. For an enterprise that had come through so many crises, it was a welcome change.

Whatever reservations local leaders may have had about Maben's role in the events that removed Hopkins from power and dashed hopes that Sloss-Sheffield might follow TCI into steelmaking, Birmingham's citizens regarded him as a personification of "big business." At the time he took office, he was credited with being "many times a millionaire" by a reporter who knew nothing about the difficulty with which he had recently retired a relatively small debt to Bryan. Unaware that Bryan had been the key figure in the buyout of the Sloss Furnace Company in 1887, the *Age-Herald* credited Maben with arranging it, lending credence to a myth that Ethel Armes would later perpetuate. Maben did nothing to disabuse the locals about the formative role in which he had been cast. His connections with Virginia's colonial aristocracy, going back to such families as the Spotswoods; his background as a Confederate officer; and his years on Wall Street all added to a commanding public image. Dignified, impeccably dressed, and patrician in bearing, he was the very model of an industrial titan, even though he was merely Bryan's chief lieutenant. [1]

The fact that such a person as Maben would consent to move from Wall Street to Birmingham only confirmed the Magic City's sense of self-importance. Fittingly, his arrival coincided with the beginnings of modern high-rise construction in Jones Valley. In 1902, the first steel-frame structure ever built in Alabama, the Woodward Building, rose at the corner of First Avenue and Twentieth Street. Ten stories high, it contrasted with everything around it, because of not only its size but also the forthright way in which it embodied the plain and unadorned commercial style pioneered in Chicago during the 1880s by architects Daniel Burnham and John W. Root. Its "bare vigor," to use the apt words of Marjorie L. White, epitomized both the spirit of Birmingham and the capitalist values undergirding its rise. [2]

The new building was named for the person who had financed it, iron magnate William H. Woodward. Because of its size, skeptics doubted that it would ever be fully occupied. Soon, however, it attracted a swarm of tenants. Sloss-Sheffield was one of these, moving its executive headquarters to the prestigious new address in 1903. Here, in an oak-paneled office overlooking the heart of the city, Maben presided over the South's second-largest enterprise.

The Woodward Building was only temporarily Birmingham's most elite business structure. Soon, it was eclipsed by a much larger edifice that went up diagonally across the street on the site where Charles Linn's bank had previously stood: the Brown-Marx Building, named in honor of real estate executive Robert A. Brown and one of the community's most distinguished Jewish families, that of bankers Otto and Victor H. Marx. Its construction took place in two stages, starting with a tall but relatively narrow building erected in 1906 and ending in 1908 as a monster three times its original size. The largest edifice ever put up in any part of the South, it had 800 rooms and was sixteen stories high, dwarfing anything that had previously existed in the Magic City. TCI soon moved its executive offices there, and Sloss-Sheffield was again overshadowed by its larger rival. [3]

John C. Maben. John Campbell Maben represented the interests of Bryan and other Virginians on Wall Street for many years before becoming president of Sloss-Sheffield from 1902 to 1917. His advocacy of specializing in foundry pig iron during the greatest crisis in the company's history was perhaps the most important of his many contributions to its success. A photograph of Maben published in 1914 by the Birmingham *Ledger* epitomizes the patrician bearing and deeply ingrained conservatism of a man who was second only to Bryan in his impact on the development of Sloss Furnaces during the long period in which Virginians dominated the enterprise. (Courtesy of Birmingham Public Library, Department of Archives and Manuscripts, Birmingham, Alabama.)

Despite an increasing number of automobiles that were gradually replacing horse-drawn vehicles on its busy streets, Birmingham remained dependent for its existence on the railroads linking it to the outside world. By the turn of the century it was apparent that Union Depot, which had been built in 1886 on Morris Avenue between Nineteenth and Twentieth streets at a cost of $134,000, was no longer adequate for the city's needs. Agitation for a new facility capable of handling Jones Valley's growing volume of rail traffic resulted in the construction of a new terminal, begun in March 1906 on a site covering twelve and one-half acres, bounded by Second and Seventh avenues and Twenty-sixth and Twenty-seventh streets, North. The main building, 720 feet long, had two 130-foot towers and a central dome rising to a height of 96 feet. Made of "pressed brick, stone, and reinforced concrete, with trimmings in terra-cotta and marble," it was an impressive structure when it was opened in April 1909 with appropriate ceremonies and lauded by the press as a "dream come true." The central section, with an area of 20,400 square feet, was "devoted to the use, comforts, and conveniences of white passengers exclusively." Blacks occupied much less space—5,200 square feet—in the south wing, most of which was set aside for baggage and mail. The north wing contained dining facilities, offices, and express rooms. Ten tracks, with eighty-pound rails rolled at TCI's new Ensley steel mill, ran through the train shed; they were used daily by forty-four passenger trains operated by seven systems serving the community. Including its midway and train sheds, the new facility, which cost about $1.75 million, was more than three times the size of the older depot, befitting what one writer called "the wonderful growth of Birmingham and the expansion and development of its contiguous interests." The L&N continued to use the old station, but the rest of the lines serving the city gladly adopted the new one, making the most of its superior advantages.[4]

II

Although it did not take long for new buildings to get besmirched by the smoke and soot that continually hung over Jones Valley, Birmingham's changing skyline epitomized the city's pride in being the South's greatest industrial center. An opportunity to express its civic spirit materialized in 1903 when Alabama was invited to sponsor an exhibit at the Louisiana Purchase Exposition, to be held at St. Louis the following year. When the state government had to decline for lack of funds, the Commercial Club of Birmingham—later to become the Chamber of Commerce—stepped into the breach. It had already helped bring steelmaking to the Magic City by donating $40,000 to the pilot venture at the Birmingham Rolling Mill in 1897 and had aided textile manufacturer Braxton Bragg Comer that same year by investing $150,000 in his Avondale plant, the first in the city to make cotton cloth. Now, two of its leading members, Frederick M. Jackson and James A. Mac-

Knight, conceived the idea of creating a colossal iron statue symbolizing Birmingham's industrial might.[5]

Jackson, president of the Commercial Club, was general manager of the Alabama Consolidated Coal and Iron Company, one of the greatly enlarged firms that had resulted from the recent merger wave. A series of social events that had taken place in the past few years probably suggested to him and MacKnight the identity of the figure that was chosen to represent the city at the St. Louis fair. During the late 1890s, at Birmingham's annual Mardi Gras celebrations, a prominent local citizen had been disguised as "Rex Vulcan" before being unmasked at a lavish ball. In October 1903, the Commercial Club chose the Roman fire god as a suitable subject for a "giant advertisement" for Alabama at St. Louis and began soliciting subscriptions to defray the estimated $15,000 cost. A Vulcan Committee was appointed to coordinate the project. Its thirteen members included two representatives of Sloss-Sheffield: secretary-treasurer and sales director James W. McQueen and mine superintendent Jones G. Moore.

In November, MacKnight went east in search of a sculptor. He was advised in Boston that casting a statue of the type wanted by the Commercial Club might take two to three years, but he had better luck in New York, where he met an expatriated Italian artist, Guiseppe Moretti. Born in Siena, Moretti had received early instruction in sculpture there before completing his training in Florence and Carrara. Working primarily in marble, he lived in Vienna and Budapest and created pieces including a representation of Austro-Hungarian emperor Franz Josef II. By the time the likeness of the Hapsburg emperor was shown at the Paris Exposition of 1889, however, Moretti had departed, having come to America during the previous year after military officials had frustrated his plans to develop quarries in Transylvania on the grounds that the rail connections required for their exploitation would have opened a possible invasion route from Russia. After arriving in the United States, he designed statuary for William K. Vanderbilt's mansion at Newport, Rhode Island; executed several impressive bronze, granite, and marble monuments in Pittsburgh's affluent Schenley District; and became a partner in a New York City bronze foundry.[6]

Because of business reverses in the last of these ventures and the failure of his main Pittsburgh patron, Edward M. Bigelow, to win reelection as director of that city's parks and public works, Moretti was ripe for a new challenge when MacKnight approached him on behalf of Birmingham's Commercial Club in 1903. He agreed to produce the preliminary plaster cast required for the Vulcan statue for $6,000—other sculptors had demanded twice that much—and complete the entire project in forty days. After building a wooden framework and making an eight-foot clay prototype of the statue at his studio on West Thirty-eighth Street in Manhattan, Moretti and sixteen assistants produced a full-sized clay model at an unfinished church in Passaic, New Jersey. Because it would have collapsed of its own weight if built as a whole, it was made in sections around wooden armatures. The legs alone were so enormous that a horse and buggy could be driven between them. Visitors to the church piously crossed themselves as the work proceeded, thinking that Moretti was creating a gigantic figure of a saint.

The height of the colossus, originally set at fifty feet, was soon increased by six feet to make it larger than a bronze Buddha in Tokyo that had been found to be the world's biggest comparable statue. Because of the limited load-bearing capacity of cast iron, the expanded dimensions represented a formidable challenge, emphasizing Moretti's audacity. Because he could not view the entire structure as a whole before work was completed, he did not know until too late that the foreshortened torso was out of proportion to the rest of the body. Perhaps it was just as well, because he had also found it hard to persuade his Birmingham patrons that the lame fire god should not be over-idealized but rather depicted as "a strong but none-too-handsome colossus."

Time was short in which to meet the deadline, for the St. Louis fair would begin in April 1904. Beginning in late January, as soon as they were completed, sections of the plaster cast were wrapped in thick padding to prevent any possible damage and shipped by rail to Birmingham. The first car, containing the waist and thighs, arrived via the Southern Railway on 11 February 1904. Moretti himself came to the Magic City within a week to supervise casting the statue in metal. When nobody showed up to meet him, he and two assistants went to the Linn Building and introduced themselves to red-faced officials of the Commercial Club before registering at the Hillman Hotel.

The casting was to be done by the Alabama Steel and Iron Company's new foundry at 1421 First Avenue, North. Its president, James R. McWane, had come to Birmingham in 1904 at the invitation of the Commercial Club, leaving Lynchburg, Virginia, where his family had made plows for three generations going back to an association between his grandfather and Cyrus McCormick. Because of his eagerness to have the honor of casting the statue, McWane had made as low a bid as possible. After it was accepted, he created an enormous pit in the floor of his foundry for the gigantic molds required. Moretti set up a studio directly across the street to supervise the work. Pig iron from Sloss-Sheffield was to be used in the project, which was described by the *Age-Herald* as "the largest task ever undertaken in Birmingham."[7]

On 26 February, two days after the statue's feet and legs had arrived from Passaic, a gala event was held at the Jefferson Theatre to raise money for the project. Musical performances were rendered by such local artists as Mrs. Solon Jacobs, who sang "Ben Bolt," and speeches were delivered by dignitaries including Bowron, who spoke about "The Resources of Alabama." Moretti, himself an accomplished musician, sang an operatic aria and was called out for "several encores."

The last of the plaster parts arrived on 8 March; three days later, the first cast, that of the waist, which alone weighed more than 12,000 pounds, took place under the watchful eyes of a team of experts headed by McWane. Much to everyone's relief, the operation was successful. The molding process, however, proved difficult. At one point, it took Moretti three days to repair a plaster pattern for the head, which had been dropped and badly damaged. To replicate the clay model in iron, workers fashioned 150 individual pieces, called drawbacks, of brick and loam, making them capable of withstanding molten metal by packing them against larger sections brought from Passaic and baking them until they had hardened. Without these smaller units, features like the fire god's nose and ears or the folds in his tunic could not have been cast in iron.

After all the patterns for the components had been finished, they were assembled in fifteen larger pieces, each of which was bound in a two-box frame known as a cope and drag. Working sixty hours a week for thirty-five cents an hour, founders cast each piece in the pit, pouring molten iron into the boxes, allowing it to set, opening the molds, and taking out the cores so that the metal would not break while cooling and shrinking. After the iron had cooled, the molds were taken away, rough edges were smoothed off, and the iron was painted. After the components were bolted together, the completed statue weighed 250,000 pounds. A photograph taken during the project showed Moretti, wearing a fedora, standing with seven workers in front of one of the massive legs, which was hanging from wooden scaffolding over the casting pit. One of the founders claimed that he had not left the plant even once during the entire process. Proud of their work, he and his companions declared that "Birmingham is the only place in the world this could have happened."

As molding proceeded, fund-raising events took place. A visit to Birmingham by the New York Giants baseball team, on its way north from spring training, provided an opportunity on which local leaders were quick to capitalize. Like Maben's arrival the year before, it signified growing connections between the Magic City and the nation's largest metropolis. Led by manager John J. McGraw, the Giants were scheduled to play a three-game exhibition series with Birmingham's minor-league team, the Barons, at the local ball park, which had been built along First Avenue near the slag pile of the Alice Furnaces.[8] Later that year, McGraw's team would win the National League pennant and refuse to meet the American League champions, the Boston Red Sox, in a World Series.[9]

Scoring almost at will, the Giants won the first game by a score of thirteen to nothing and also took the second contest by a margin of twelve to seven. Boosters, however, were particularly primed for the third game, which was to be played on what was officially designated "Vulcan Day" to raise money for completing the statue. In honor of the occasion, the Barons were temporarily renamed the Iron Men. Ace pitcher Joe McGinnity of the Giants, who was nicknamed "Iron Man" not only because he had once been a foundry worker but also because he had performed such feats as winning three double-headers within one week, agreed to take the mound for the visitors. As a publicity stunt, he also took part in the casting operations. Reporters made much of a celebration held by the Giants on McGinnity's thirty-third birthday, 19

March, four days before the big game took place. "All the New Yorkers will drink a libation of copperas solution," stated the Birmingham *News* in a preview of the party. "A pie of iron filings well sauced with emery dust will top the festivities. Ferrous sulphate, three centimeters to the man, will be an important item on the bill of fare. Steel ingots buttered with silicon will be served on toast. . . . Vive l'homme de fer!"[10]

Even though the local team lost the Vulcan Day game to the big-league visitors, nothing could dampen the satisfaction that boosters took in the way they played, showing that the Magic City could offer worthy competition to a top team representing America's largest metropolis. To the delight of "nearly two thousand baseball cranks in fine fettle," the Iron Men, batting first despite their hometown status, scored a run in the opening inning. The Giants, however, quickly countered with two runs of their own when future Hall of Fame outfielder Roger Bresnahan, the "Duke of Tralee," singled, Dan McGann scored him with a triple, and McGann came home on Sam Mertes's grounder to second. Mertes added another run by stealing two bases and scoring on a Birmingham error in the fourth inning, and outfielder George Brown crossed the plate in the fifth when McGann singled him home. The Iron Men scored two runs in the top of the eighth, but the Giants clinched the game in their half of the same inning when Bresnahan doubled, went to third on a fielder's choice, and scored on a sacrifice fly by Mertes.

The New Yorkers thus won, five to three. Nevertheless, local supporters took consolation in the fact that the score was close, that the Iron Men had outhit the Giants, and that McGinnity had struck out only five batters before yielding the mound to lefthander George ("Hooks") Wiltse in the late innings. Moretti himself was in the stands, shouting in Italian, and the Vulcan fund netted approximately $750. "It was a fine exhibition," the *News* declared. The *Age-Herald* agreed, saying that the home team had "fought their big opponents to a standstill and whacked the ball into the daisies at all stages of the game."[11]

By early April, only a few parts of the statue remained to be cast, and the others were being assembled. On 18 April, the feet and legs were shipped to St. Louis. The rest of the components followed, requiring eight freight cars. Not until mid-May did all the parts arrive at the exposition. Meanwhile, the colossus was shown in incomplete form and reportedly attracted "more attention at the Fair than any other exhibit." On 7 June 1904, standing on a pedestal of coal and coke, it was dedicated to the accompaniment of music by the Philippines Band, christened with Cahaba River water, and acclaimed in speeches by Jackson and other dignitaries.

Throughout the summer, Vulcan was admired by thousands of fairgoers. In September, the colossus won a grand prize from a jury chosen by the exhibition's mineral department. MacKnight, McWane, and Moretti received silver medals for their parts in the project. Alabama Day was observed at the fair on 10 October; a parade was held, and representatives of Alabama cities laid flags at the statue's feet. Back home, Republican politicians secured the right to use

Vulcan as the party's state emblem on the November ballot in return for giving a substantial amount to help pay remaining costs, which had been much higher than anticipated. In February, after the fair was over, the giant figure was disassembled and brought back to Birmingham, free of charge, by the L&N Railroad.

While the freight cars that brought the colossus home rested on a siding, community leaders debated what to do with it. Now that it had served the advertising function for which it had been created, local enthusiasm for it began to wane. Llewellyn Johns, a Welsh mining engineer who had come to America and risen to prominence with Pratt, Republic, and TCI, offered to pay all remaining debts on condition that the statue be prominently displayed between the tracks in the railroad reservation and dedicated to Henry F. DeBardeleben. The latter, however, declined the honor. Now that the symbol had served its purpose, he said, it should be sold for scrap.

As apathy continued to set in, the L&N demanded the return of its freight cars, and the statue's components were scattered in a field on Red Mountain behind a barbed-wire fence. After months of wrangling over what should be done with them, individual donors and firms that had contributed to the project voted to put the colossus in Capitol Park, but unexpected opposition arose when women's groups protested that the fire god was so scantily clad as to offend public morals. A member of the Commercial Club finally came to the rescue by suggesting that the statue be exhibited at the 1906 Alabama State Fair. This compromise was agreed upon, but there was no rail line to the grounds, which were located far outside the city limits, near Bessemer. A solution was found by carrying the segments on streetcars belonging to the Birmingham and Bessemer Dummy Line.

A succession of unseemly events emphasized the disrepute into which Vulcan had fallen. Hastening to reassemble the colossus in time for the opening of the fair, workmen carelessly positioned its right hand backwards, preventing it from clutching the arrow it had been designed to hold aloft. After the statue was erected, the Weldon-Jenkins Ice Cream Company used a hole in its left hand, designed to accommodate the handle of its massive hammer, as a convenient receptacle for what a local historian later described as "a huge ice cream cone made out of plaster and chicken wire." Disassembled at the end of the fair, the fire god lay in pieces on the grounds like an unwanted Humpty Dumpty and was not put back together again for almost three decades. Its left hand, however, was still used periodically for advertising purposes, holding pickle jars and Coca-Cola bottles.

III

Vulcan's devolution from a proud symbol to an assortment of discarded junk epitomized the commercialism that obscured Birmingham's sense of its civic heritage. Things might have been different, however, had the colossus been made of steel, upon which the imagination of the city's boosters was so strongly fixed. By contrast, cast iron was unexciting, like the unwanted relic into which it had been molded.

Steel was the wonder metal of America's industrial age. The ability to produce it in large quantities at low cost, made possible by Bessemer and open-hearth technologies that had been developed since the mid-nineteenth century, had become a hallmark of modernity. Steel's remarkable range of qualities, combining great strength, resistance to wear, toughness, ductility, and other valuable properties, diverted attention from the continuing significance of other ferrous metals that had been crucial to human material progress but were now regarded by a misinformed public as old-fashioned and, hence, inferior. This impression was as prevalent in Alabama as anywhere else. In the eyes of boosters, Birmingham could never attain its just status in the industrial world until it had joined the ranks of Pittsburgh and other steel-producing centers.

Under Maben's leadership, Sloss-Sheffield concentrated after 1902 on catering to the needs of foundries that made various products molded from a much less esteemed product. Using either a cylindrical furnace known as a cupola or a reverberatory air furnace, founders converted pig iron into an alloy known as cast iron, containing varying percentages of such chemical constituents as carbon, manganese, phosphorus, silica, and sulfur. Despite its Cinderella image, cast iron was a highly important commodity. Its continuing significance provided an excellent example of a phenomenon that might aptly be called technological persistence: the tendency for a traditional and supposedly outmoded material to retain its usefulness long after its public image has been undermined by dramatic innovations. [12]

During the late nineteenth and early twentieth centuries, the manufacture of cast iron became not only increasingly scientific but also more and more crucial to industrial and economic growth. In 1915, Richard Moldenke, the dean of American foundry metallurgists, summarized its applications and uses as "agricultural machinery, air, ammonia, steam and hydraulic cylinders, automobile cylinders, pistons and rings, annealing boxes, bed plates, brake shoes, car wheels, crusher jaws, electrical castings, fly wheel and general machinery castings, gears, grate bars, cannon, hardware, ingot molds, locomotive and car castings, ornamental work, fittings, plow points, pulleys, pumps, chilled and sand rolls, water and soil pipe, stove and furnace castings, valves, etc." Due to such versatility, Moldenke stated, "the melt in the United States alone runs above 6,000,000 tons in every normal year." [13]

Cast iron, however, had little impact on public consciousness. Scholarly literature continues to mirror this fact. A number of careful studies document the development of Bessemer, open-hearth, and other steelmaking technologies, but few historians have paid much attention to the evolution of the foundry trade or to that of merchant pig iron manufacturers, such as Sloss-Sheffield and Woodward, that catered to it. [14]

By 1914, there were 6,507 foundries in North America, ranging from what

Moldenke called "tumble down barracks" to highly rationalized installations that he described as "perfect wonders of efficiency." Most foundries used molding machines placed so that workers could transfer completed molds to traveling conveyors that received molten cast iron, moving slowly enough to allow it to set and then dumping the finished castings over a grate so that excess sand could fall through. Remaining hot sand was then re-tempered, fresh sand added, and the resulting mixture returned to the molding machines for another run.

The sand was of three basic varieties: greensand, drysand, and loam, the last having a high clay content and being particularly difficult to mold. Cores—mold parts that shaped the interiors of hollow castings—were made in ovens in special rooms where great care was paid to the chemical composition of binders so that the cores would have an even internal consistency instead of being dry on the outside and damp in the interior.[15] The making of malleable castings, dating back to the time of Reaumur and introduced into the United States in the early nineteenth century by New Jersey ironmaster Seth Boyden, was still an arcane procedure as the twentieth century opened, requiring complex annealing, baking, melting, pouring, sorting, grinding, and chipping techniques to obtain proper consistency, freedom from excess shrinkage, and tensile strength. It was becoming scientific, however, under the prodding of such persons as Moldenke.[16]

Cast iron had just as remarkable a range of properties as steel. As Moldenke stated, "iron castings may be soft enough to be machined with the greatest ease, and *per contra* as hard as glass. They may be so weak that they are unfit for anything better than sash weights, and on the other hand almost as strong as ordinary steel castings."[17] Although the graphite flakes that interlaced its matrix structure rendered cast iron brittle, making it inferior to steel in ability to withstand abrasion, its relatively low melting and pouring temperature, combined with high fluidity in a molten state, made it ideal for assuming intricate shapes. In its characteristic form of "gray iron," the staple product of most foundries to which such firms as Sloss-Sheffield catered, it was easily machinable, provided its carbon level was not excessive. Its surface qualities made it highly suitable for conveying liquids, gases, and other fluid or semi-fluid substances from one place to another. It was an excellent radiator of heat and also had a tendency to absorb, rather than transmit, sound; it was therefore a good dampener of noise and vibration, making it the material of choice for engine components that did not have to withstand abrasion. Its compressive strength made it ideal for bearing quiescent loads, and its massive quality made it an obvious material for flywheels. Depending on elements in the alloy, it was highly resistant to corrosion. Being cheaper than steel, it could be used in large quantities at less expense. For these and other reasons, cast iron not only survived the onslaught of steel but actually proved superior to that alloy for many industrial and commercial applications.[18]

Southern pig iron was particularly suitable for conversion into cast iron because of its high phosphorus content. This chemical element had a bad name among nineteenth-century metallurgists, who sometimes referred to it as "rank poison."[19] In reality, phosphorus enhanced the fluidity of iron, increasing its ability to conform to the inner surfaces of molds and making for sharper, cleaner castings. It also retarded a phenomenon known as growth, the tendency of iron to increase in volume under repeated heatings and coolings, and improved surface finish by countering tendencies toward pitting associated with excess graphitic carbon, or kish. Because of its phosphorus content, southern foundry pig iron chilled less rapidly than northern varieties, retaining superior softness. Such properties, confirmed through painstaking research by a number of European and American metallurgists, explain why the South in general, and Alabama in particular, became the heart of the American foundry trade.[20]

During the late nineteenth and early twentieth centuries, as different types of alloy steels began to replace wrought and charcoal irons for a variety of uses, specialized steel castings appeared with mechanical properties that had previously been unachievable. The slowness with which some foundries learned how to make good malleable iron accelerated this trend.[21] Because of its high phosphorus content, southern pig iron was not well suited for malleable iron, but enough large markets remained for a wide range of cast-iron products that favorably located merchant pig iron producers could benefit from the resulting demand. Sloss-Sheffield and Woodward were in exactly the right place to occupy this niche, just as the Southeast was well situated to attract fabricators of cast-iron products. By 1900, Alabama was already producing slightly more than 25 percent of the nation's foundry iron; it had overtaken Pennsylvania as first-ranking state in that category by a narrow margin, 883,208 as opposed to 856,472 tons. This lead widened steadily in the early twentieth century. By 1915, Alabama produced 1,221,476 tons of foundry iron, against Pennsylvania's 871,718. It was sometimes said that two categories existed insofar as production of foundry iron was concerned: Alabama on the one hand and the rest of the United States on the other.[22]

Because of the suitability of southern iron for molding, plants that fabricated various products from foundry pig iron, which had originally been centered in the states of New York, New Jersey, Pennsylvania, and Ohio, moved toward the southern blast furnaces that increasingly became their main source of supply. By the 1920s, the foundry trade was concentrated either in the South or around the periphery of the region in cities like Cincinnati, Louisville, and St. Louis. Birmingham became a magnet for firms that made various types of foundry products, as did such nearby Alabama cities as Anniston and Gadsden.

As Birmingham's boosters were well aware, cast iron commanded significantly smaller markets than steel. By the eve of World War I, American foundries melted about 6 million tons of iron per year. The country's annual production of steel ingots at the same time was more than 30 million tons. Nevertheless, the versatility and indispensability of cast iron guaranteed a thriving business to the relatively few companies that continued to cater to the foundry trade. Sloss-Sheffield and Woodward were located in the right place at the right time.[23]

By far the most important foundry product that stimulated the growth of Sloss-Sheffield, and the emergence of the southern foundry trade in general, was cast-iron pipe, used for conveying water, sewage, chemicals, fuels, gases, and other substances. Two general classes of pipe made up this key segment of the foundry industry: soil pipe, through which water or other substances flowed under the force of gravity, and pressure pipe, through which, as its name implies, substances were pumped under force.

The use of cast-iron pipe for drains, sewers, water delivery systems, and public fountains was well established in Europe by the seventeenth century. Large cities, such as London, began replacing water pipes made from lead or bored-out logs with cast-iron piping. Ultimately, this practice spread to the United States; in 1817, Philadelphia became the first major American city to adopt cast-iron pipe for drinking and sanitary purposes. New York, Pennsylvania, and Ohio dominated the pipe industry throughout the period before the Civil War. Starting in 1867, however, it began moving to the South, where two cities, Anniston and Birmingham, gradually won preeminence in its manufacture. A pioneer pipe mill, the Birmingham Iron Works, was established in 1881. It soon became the largest producer of water, gas, and steam pipe in the region.

By 1890, in which year alone five pipe foundries were established in Birmingham, Avondale, Bessemer, Pell City, and Gadsden, the industry had become a mainstay of Alabama's economy. The rapid pace of urbanization throughout the nation, creating a steady demand for water conduits and sewage systems, guaranteed markets for a growing number of installations. The natural advantages of the South for dominating the trade became particularly compelling after a court decision in the Addyston Pipe Case disallowed pools that northern installations had formerly used to maintain market share, leading to increasing concentration within the industry. In 1898, a number of shops in Anniston, Bessemer, and east Tennessee became consolidated into the American Pipe and Foundry Company. The following year, this in turn became involved in an even larger entity, the United States Cast Iron Pipe and Foundry Company. From this point, virtually all growth in pipe manufacturing occurred in the South. Of twelve plants built throughout the nation between 1900 and 1914, seven were located in Alabama alone.[24]

The entrepreneurs who built the Alabama pipe industry came from a variety of backgrounds. Hamilton T. Beggs, an English immigrant who learned the foundry trade as a teenager and came to Virginia in 1849, moved to Alabama during the Civil War to help make armaments. After the war, he played a key role in the development of southern pipe manufacture, establishing foundries in Birmingham, North Birmingham, and Bessemer. Robert Campbell, born at Inverness, Scotland, in 1852, helped establish the pipe business during the same period in Gadsden. At Anniston, Samuel Noble, whose family had come to the United States from Cornwall, as previously noted, built a six-pit foundry in 1888. At that time, it was the largest installation of its type in the world. In the 1890s, pipe works proliferated in Anniston as the city diversified in response to the depression that struck during that decade. After the turn of the century, a further expansion of the industry took place in and around the community under the leadership of Henry Bascom Rudisill, a native of Georgia who came to Alabama in the early 1880s as an employee of the Georgia Pacific Railway. In 1905, he and two other entrepreneurs established the Anniston Foundry and Machine Company. As the business grew, the firm moved to a fifteen-acre site in the western part of the city in 1910. Within the next decade, it and other pipe works made Anniston the largest soil pipe manufacturing center in the world, with an annual production of 140,000 tons. By 1928, one Anniston-based enterprise, the Alabama Pipe Company, controlled thirteen pipe works in four cities. Administered by Charles Anglin Hamilton, a self-made man from a humble rural background, it was the largest firm in the industry. [25]

Meanwhile, Birmingham won preeminence in production of pressure pipe. One of the pioneers who built this industry, William B. Stockham, came to the Magic City from Chicago in 1903 and established the Stockham Pipe and Fittings Company. By 1928, it had 1,500 employees and made 5,000 different types of pipes, valves, and fittings. Another important producer was the American Cast Iron Pipe Company (ACIPCO). Its founder, John J. Eagan, was born in 1870 in Griffin, Georgia, the son of a store owner who died of tuberculosis that he had contracted during the Civil War in a northern prison camp. Never completing high school, Eagan inherited $6,000 from a grandmother and through shrewd investments turned it into $73,000 before he was thirty years old. Soon thereafter, he inherited most of the $750,000 estate left by his stepfather, William A. Russell, an Atlanta businessman. Deciding to invest his fortune in the cast-iron pipe industry, Eagan founded ACIPCO, which, as Rebecca L. Thomas has stated, was "financed entirely by southern capital." Although it was incorporated in Georgia, Eagan built his plant in Birmingham "to take advantage of the low-cost pig iron in the District." Knowing nothing about foundry operations, he hired James R. McWane to run the installation. McWane had lost so heavily in the Vulcan project that he had been forced to sell the Birmingham Steel and Iron Company.

Under McWane's management, ACIPCO became a thriving enterprise. Beginning with a payroll of $151 per week, it bought a tract of land in 1906 at North Birmingham, where it erected a small frame office building, foundry, commissary, and chemical laboratory. Ultimately, it became one of Alabama's largest and most progressive industrial establishments, known particularly for a profit-sharing system rooted in Eagan's strong religious beliefs. Partly because of hearing a sermon by Josiah Strong, a key figure in the Social Gospel movement, Eagan "believed that Christ gave certain men special business abilities" and that his own mission was to "see God's Providence divided fairly among the people of the world." Paying the highest wages in the Birmingham District, ACIPCO implemented a "Golden Rule plan" under which it was administered by a "Board of Operatives" and a "Board of Management," acting as joint trustees of the firm's capital. Prevented by Alabama's Jim Crow laws from meeting in the same room with white employees, black workers, who comprised about 70 percent of the firm's 1,000 operatives, were constituted as an "Auxiliary Board." Eagan voluntarily limited what he could earn on his own shareholdings, and most of the profits went to the employees. When Eagan died in 1924, his will bequeathed all his common stock to ACIPCO, which continued as a cooperative venture making high-quality castings. Eventually, it became "the largest pipe shop under one roof in the United States."[26]

McWane left ACIPCO in 1921 to establish his own enterprise, the McWane Cast Iron Pipe Company. Its best-known products were precaulked watertight joints that eliminated the need for making joints after pipe had been laid in underground trenches, thereby saving labor costs. McWane also departed from standard foundry practice by casting pipe in horizontal molds instead of vertical pits. Later, he abandoned hand-ramming techniques for an assembly-line technique that featured sand conveyors and overhead cranes for moving molds and cores. In 1926, he founded the Pacific States Cast Iron Pipe Company in Provo, Utah, which made fire hydrants and various types of pressure pipe and valves. For the rest of his life, he managed both its facilities and those of his Birmingham firm. By his death in 1933, he had won seventy-five patents for his innovations.[27]

Yet another Birmingham pipe-making enterprise was the Dimmick Pipe Company, founded by two brothers from Massachusetts who had migrated to Anniston before moving to Birmingham at the beginning of the twentieth century. Charlotte Blair, who had been a stenographer for a Virginia pipe manufacturer, took part in the enterprise and was reputedly the only woman in Alabama to serve on the board of a corporation at that time. After leaving the Dimmick Company because of a dispute in 1904, she assisted Eagan in founding ACIPCO. Bowron also invested heavily in the Dimmick enterprise, which was badly managed, and lost all of his stake after its sale in 1911. Later, however, one of the two brothers, Dan B. Dimmick, became a successful manufacturer of spiral corrugated cast-iron culvert pipe for drainage of roads and railroad rights-of-way.

By the 1920s, 60 percent of all cast-iron pipe in the United States was made in the South, mostly in Alabama. About 75 percent of the 2 million tons produced annually was pressure pipe, the rest being soil pipe. Sloss-Sheffield and Woodward both catered heavily to this trade.[28]

Steam engines also used large amounts of cast iron, which normally accounted for about three-fourths of their total weight. The only components of such engines that were made from wrought iron or steel were those that had to bear unusual stress, such as piston rods and connecting rods. Taking advantage of cast iron's relative cheapness and its ability to assume all sorts of intricate shapes, all other parts, including cylinders, pistons, flywheels, cranks, and steam chests, were molded in foundries.[29]

A Birmingham firm that built large numbers of steam engines was the Hardie-Tynes Manufacturing Company, incorporated in 1903. It traced its beginnings to the Birmingham Iron Works, founded in 1882 by John T. Hardie, an Alabama native who had been a cotton factor in New Orleans before the Civil War. Hardie's son, William D. Hardie, studied mechanical engineering in Scotland, worked for a time on the technical staff of the Sloss Iron and Steel Company, and eventually became general manager of his father's enterprise. Upon John T. Hardie's death in 1895, William established the Hardie-Tynes Foundry and Machine Company in cooperation with a partner, William Tynes, who ran the financial and marketing side of the business. For a time the Hardie-Tynes plant was located directly across from Sloss Furnaces on First Avenue. After their plant suffered heavy damage from a fire in 1901, its owners built greatly expanded facilities eight blocks away, on the eastern side of Birmingham. They reorganized their firm under its enduring corporate name two years later. By 1912, its sprawling facilities covered an entire city block.

During the early twentieth century, Hardie-Tynes was particularly well known for building Corliss steam engines, of which it made sixty-three in 1902 alone. Sold primarily in the South, these power plants were bought by producers of cotton oil, furniture, bricks, chemicals, and cement. Another major customer was the Louisiana Rice Milling Company, one of the largest enterprises of its type in the United States. Hardie-Tynes also made slide-valve engines, which were marketed as far away as Michigan, and hoisting engines for coal mines. During World War I, the firm built marine engines, cast-iron ship propellers, and artillery shells.

Hardie-Tynes responded resourcefully to shifts in demand resulting from technological change and altered business conditions affecting markets for cast-iron products. When steam engine sales fell drastically in the 1920s due to the growing use of electricity and diesel engines, the company emphasized the production of air and gas compressors, ammonia stills, hoists, pumps, vacuum pans, and valves. It also made heavy iron gates and other equipment for Boulder (later Hoover) and Grand Coulee dams and for hydroelectric facilities that were built by the Tennessee Valley Authority. Still later, Hardie-Tynes became an important supplier of air compressors for the United States Navy, particularly those to be used in launching torpedoes from submarines. For a time, it was said that "there wasn't any navy vessel built that didn't have a Hardie-Tynes compressor." Eventually, the firm closed its foundry in 1962 after becoming briefly involved in making components for atomic reactors and housings for missile canisters. After 1962, Hardie-Tynes survived as a producer of fabricated steel shapes for heavy industrial equipment, dams, and naval ordnance. Throughout its history, up to the time its foundry closed, it had been a major purchaser of pig iron from both Sloss-Sheffield and Woodward.[30]

Another important foundry product, with a history even longer than that of cast-iron pipe, was stove iron, which capitalized on the ability of cast iron to radiate heat. One of the most important buyers of Birmingham pig iron in the city's early years was a stove works in Nashville, Tennessee, the Phillips and Buttorff Manufacturing Company. Starting as a small foundry in 1881, the year in which the Sloss Furnace Company was born, it expanded steadily as its markets grew; by 1887, it had three cupolas and was making 4,500 stoves per month with a force of 190 molders. A ten-year contract between it and DeBardeleben in 1884 was a key development in Birmingham's early conquest of outside markets. Two years later, Phillips and Buttorff bought its first order of pig iron from Sloss. Partly through the ability of Benjamin E. McCarthy, who joined its bookkeeping staff as a boy in 1892 and became president in 1915 after the death of H. W. Buttorff, the firm continued to prosper. By 1927, still one of Sloss-Sheffield's most important customers, it was making more than 100,000 stoves, ranges, space heaters, and hot-air furnaces per year.[31]

Nashville was only one important southern stovemaking center; another was Rome, Georgia, which had been the home of a well-known Civil War munitions foundry. After the war, beginning with establishment of the Rome Stove and Range Company in 1868, the city became famous for stove manufacture as other similar enterprises appeared, including the Southern Cooperative Foundry Company in 1897, the Hanks Stove and Range Company in 1901, the Eagle Stove Works in 1919, and the Standard Stove and Range Company in 1926. By the late 1920s, Rome was producing about 125,000 stoves per year and was billing itself as "The Stove Center of the South." Even more important as a nerve center for the stove trade was St. Louis, which by 1929 claimed to make "three times as many Stoves and Ranges as any other city in the world." Its enterprises included the American Stove Company, the Charter Oak Stove Company, the Orbon Stove Company, and the Willard Range Company. All were important customers of Sloss-Sheffield.[32]

Alabama itself became an important producer of stoves. One firm, the Birmingham Stove and Range Company, started as a small producer of hollowware in 1905 but became best known for making gas ranges under its Majik Baker trade name. In the state's northwest corner, the King Stove and Range Company at Sheffield and the Martin Stove and Range Company at Florence employed several hundred workers by the 1920s. Headed by William H. Martin, an English immigrant, they advertised a wide range of products including "Ranges, Stoves, Box Heaters, Laundry Heaters, Charcoal Furnaces, Holloware, Sugar Kettles, Dog Irons and Sad Irons." Another important Alabama stove producer was the Agricola Furnace Company, founded by Otto Agricola, a key figure in the growth of Gadsden and Attalla.[33]

Another type of heating apparatus that provided large markets for Sloss-Sheffield was the cast-iron steam radiator. It was developed in the 1890s by an inventor from Connecticut, Robert Calef, whose patents were acquired

by James B. Clow and Sons, iron founders of Coshocton and Newcomers-
town, Ohio. Clow ultimately established a subsidiary, the National Cast Iron
Pipe Company, at Birmingham, as part of a nationwide network of plants
producing Gasteam radiators. The American Radiator Company, another
Birmingham firm, was also prominent in the industry. [34]

Kitchen and bathroom fixtures, including sinks, bathtubs, and water clos-
ets, also utilized large quantities of cast iron. The modern bathtub appeared
in the 1880s. Made of molded iron castings coated with glass, enamel, or
vitreous china, it replaced older fixtures made of wood, tin, or lead. Among
southern installations prominent in this field was the Chattanooga plant of
the Crane Enamelware Company. By 1930, it was making 1,200 bathtubs per
day, plus large quantities of sinks and lavatories. Another producer, the Standard Sanitary Manufacturing Company of Louisville, Kentucky, started as a
small enterprise in the 1860s. Originally a pipe foundry, it began making
sanitary and plumbing fixtures in the 1880s. By 1930 it had grown into one of
the industry's leading installations, with a fifty-acre plant employing several
thousand workers. [35]

The coming of the automobile in the 1890s led to another class of products
using large amounts of cast iron. Engine blocks, piston rings, and exhaust
manifolds were all made from this highly versatile material. The Packard
Motor Car Company, founded at Warren, Ohio, in 1900, was among auto-
motive firms that bought large quantities of pig iron from Sloss-Sheffield.
Like most American car manufacturers, it moved to Detroit as the industry
grew. Many makers of automobile components, however, moved into or near
the Southeast in order to be as close as possible to southern pig iron producers. One, the Chance Company of Centralia, Missouri, made piston rings
and became an important customer of Sloss-Sheffield. [36]

IV

By concentrating most of its efforts on making basic steel after 1899, TCI gave
Sloss-Sheffield and Woodward an increasingly clear field to exploit growing
markets for foundry pig iron. By the time Maben came to Birmingham in
1902, conditions were optimal to grasp this opportunity. Because of its acquisitions in northwest Alabama, Sloss-Sheffield now had inexhaustible supplies of brown ore to mix with Jefferson County's red hematite, completing
what Bryan had begun with his move into the Cahaba Valley. In the future,
the Russellville strip mines would become one of the company's most important assets.

Sloss-Sheffield's foundry orientation required careful attention to the
quality of the raw materials fed into its furnaces. Coal washing was now a
normal part of operations. A survey published in 1904 showed that the company was using Robinson-Ramsay washers at six mines. By this time, however, it had begun to adopt a new jig washer invented by Elwood Stewart, a

midwesterner who had moved to Birmingham because of the pioneering role that the district was now playing in coal-cleaning technology. Stewart's machine, introduced at one of the Alabama Consolidated Coal and Iron Company's mines in 1900, was superior to earlier jig washers that used plungers to send intermittent bursts of water through stationary boxes with perforated bottoms, causing coal to rise to the top while waste material sank to the bottom. The new Stewart device had movable jig boxes that went up and down in water as a drive mechanism imparted an eccentric motion that eliminated waste particles from coal. The Stewart apparatus took up less space, cleaned more coal in less time, and needed less water than the Robinson-Ramsay washer, which became obsolete. Sloss-Sheffield was one of the earliest firms to implement the improved design, installing four Stewart washers at a mine that went into operation at Flat Top, northwest of Birmingham, in 1902. This facility, staffed mostly by convict workers using manual methods of extraction, soon had the highest output of any mine in the district. Once again, the company had proved adept at selective mechanization, just as it had done earlier by adopting modern equipment at Brookside while retaining older techniques at other mines.[37]

Because of the renovations begun by Aldrich and Hopkins, Sloss-Sheffield's blast furnaces were in the best shape ever. At the City Furnaces, a new brick building had been built to house three Allis-Chalmers blowing engines of cross-head design. Ninety years later, the same structure and two of the engines would still be standing, mute reminders of a period when the firm was about to enter one of the most prosperous periods in its history. New stoves of Whitwell or Gordon-Whitwell-Cowper design were also obtained. In 1904, City Furnace No. 2 was rebuilt and provided with a new casting shed. Six years later, before the pace of improvement slackened, a battery of Rust vertical water-tube boilers, rated at 400 horsepower, was added. Designed by E. G. Rust, chief engineer of the Colorado Fuel and Iron Company, apparatus of this type was described by blast furnace expert J. E. Johnson in 1917 as "the latest developed of any boiler in extensive modern use," having five rows of straight vertical tubes for minimal dust collection and easy cleaning. Because flames did not come into contact with the tubing, such boilers were ideal for continuous operation with little maintenance.[38]

Things were also in good shape financially. Despite periodic downturns and crises on Wall Street, the economic recovery that had begun in 1897 gained momentum throughout most of the following decade. During this ten-year period, the volume of America's foreign trade doubled; deposits in national banks nearly tripled; the volume of money in circulation increased by approximately $1.2 billion; and the value of assets held by the country's financial institutions rose to more than $21 billion. Prosperity was worldwide, shared not only by the United States and major western European nations but also by such countries as Argentina and Japan.[39]

Taking advantage of these conditions, Bryan and Maben did not hesitate to increase output and spend money. During fiscal 1902, production was increased at the Sloss and Russellville ore mines. More than $500,000 was spent on capital improvements, "all of which," Maben stated proudly, had been "provided for out of earnings, no additional securities having been issued." During the year, profits on sales of pig iron alone soared to $932,492.00. After adding returns from coal sales and other sources, total net profits aggregated $1,128,391.35, enabling the company to issue a 7 percent dividend on its preferred stock.[40]

In 1903, following protracted negotiations, an adjustment of land boundaries, and an outlay of more than $620,000, years of wrangling came to an end when a settlement was reached in legal disputes with TCI over the Lady Ensley coal and ore properties. Maben placed a conservative estimate of $2 million on the value of Sloss-Sheffield's acquisitions. During the year, the firm completed the development of a major new coal-mining facility by building 200 coke ovens at Flat Top, on the line between Jefferson and Walker counties. Despite an interruption in the prevailing boom, which caused prices to fall after September 1902, profits on pig iron sales mounted to $1,139,146.00, and total net profits hit an all-time record, $1,585,640.68. Despite lamenting falling prices in a letter to a large holder of common stock, who naturally wanted to share in the company's increasing good fortune, Bryan admitted that "the earnings of the company for the past year have been very large." Though inclined to be cautious, he said that the firm could "easily make dividends on its preferred stock" and start paying a "very considerable" return on common once prices had picked up.[41]

Maben described the 1904 fiscal year as "in some respects one of the most disappointing and trying in the history of the company." The downturn that began late in 1902 did not bottom out until August 1904. Markets for iron were weak, and more than 40 percent of the blast furnaces in the country shut down. Sloss-Sheffield did not blow out any of its own furnaces except for routine repairs, but profits on sales of pig iron fell drastically to $303,092.88, and total net profits plunged to $483,637.14. Despite the situation, the directors felt comfortable enough about the surpluses that had been registered over the past few years to declare another 7 percent dividend on preferred stock. Bryan's hope of paying a dividend on common, however, was temporarily deferred.[42]

Late in 1904, the economy revived, beginning another upturn that continued for almost three years. Sloss-Sheffield's earnings for fiscal 1905 reflected the surge, rising to $1,205,079.36. Encouraged by the excellent showing, the directors not only declared another 7 percent dividend on preferred stock but took the unprecedented step of adding a 5 percent dividend on common. They also declared a 33⅓ percent stock dividend but were still able to add $361,079.36 to the company's surplus account, finishing the year with a total accumulated surplus of $2,691,478.81.[43]

In his annual report for 1905, Maben referred to several new mining projects that were underway. The most important of these was the development of a mine at Bessie, near Flat Top, named for a daughter of John T.

Milner. "Here everything is being done in the most substantial way," Maben stated. "The machinery installed is the strongest and the heaviest of its kind." A slope, twenty-one feet wide, was driven into the mountain and a double track, with forty-pound steel rails, was laid. All in all, Maben declared, the facility would be "one of, if not the best in the State." By this time the Ruffner mines at Irondale had been closed for some years because the Sloss mines were capable of providing all the red ore needed by the firm's seven furnaces, but rising demand for iron production had led to a decision to reopen them. The fact that a post office in the northwest part of the county, just east of Littleton, had recently been named Maben in the president's honor buoyed his mood as he looked back on a banner year.[44]

Having settled with TCI its legal difficulties involving the Lady Ensley coal and ore lands, Sloss-Sheffield took further steps in 1906 to gain clear title to its own share of properties that had been acquired in complex dealings in 1899 and 1900 by purchasing the remaining assets of the North Alabama Furnace Company. In the process, a block of securities formerly owned by E. O. Hopkins fell into its hands. Meanwhile, the firm's already impressive physical plant was further improved. Construction of facilities at the Bessie mine was virtually completed. The Ruffner mines at Irondale were reopened as planned, and additional housing was built for employees who worked in them. Profits were down slightly from the previous year but still topped $1 million, leading the directors once more to declare dividends on both pre-ferred and common stock.[45] In 1907, the company had another prosperous year. Although it spent more than $660,000 on repairs and improvements to its facilities, it posted an annual profit of $1,479,662.76 and added more than $500,000 to surplus despite paying nearly $1 million in dividends on pre-ferred and common. For the first time in its history, accumulated surplus topped $3 million.[46]

Notwithstanding these gains, Maben saw room for improvement. In addi-tion to lamenting problems connected with labor disturbances that had plagued the company in recent years, he complained repeatedly about the inadequacy of freight cars to move its products to market. Because it cost nearly as much to move an empty car as to move one that was full, rail lines followed a policy of keeping cars where they had been unloaded until they had taken on enough freight for a return trip. This often resulted in an accumulation of rolling stock along the heavily populated East Coast and a dearth in places like Birmingham, which shipped such bulky commodities as pig iron and steel billets outward and received lighter manufactured goods in return. "The profits from business for the last year would have been consider-ably greater had the railroads been able to furnish cars for the transportation of our iron," Maben declared in a characteristic protest. Fluctuations in the price of iron that made it hard to gauge when to accept orders and when to hold back created further difficulties.

Nevertheless, the company's overall performance in this period was the best in its entire history. For the first time, it was earning repeated seven-digit profits after paying interest on bonded indebtedness, yielding healthy dividends to both classes of stockholders. Whatever its problems, it was enjoying halcyon times. Given the near-certainty of losing control of the enterprise that they would have risked had they tried to raise the capital required to move into basic iron and steel manufacture, the Virginia syndicate and its northern allies had made a wise decision by opposing Hopkins and his supporters.[47]

V

As Sloss-Sheffield prospered, TCI went through another stormy period in its perennially troubled development. Upon taking command in 1901, Minnesota mining magnate Don Bacon was dismayed by the company's condition. "I found an empty treasury, and a property that needed millions for upbuilding," he later recalled. The open-hearth operation of which local boosters were so proud was plagued by problems stemming partly from the inexperience of the engineers who had built it with making steel under southern conditions but also from the desire of the previous owners to erect the steel mill as cheaply as possible. Visiting Birmingham in 1901, a British expert, J. S. Jeans, learned that, because of a gas shortage, only seven of the ten new steel furnaces could be run at one time. The facility was cramped and had no mixer or ladle for adjusting furnace burdens. By northern standards, TCI's coal and ore mines were woefully inefficient, as were its Ensley blast furnaces, all of which would be torn down after U.S. Steel acquired the company in 1907. "There was scarcely any of the property that was right, if it was possible for it to be wrong," Bacon said. Compounding these physical and technological problems was the undependability of its still mainly black workers, who were often in poor health. Having no incentive to do otherwise, they were careless about showing up on time and so inefficient that, as always, much larger numbers had to be hired than were theoretically necessary to run operations.[48]

Making wholesale changes in TCI's local management, Bacon brought in northern engineers who developed an ingenious "duplex" process for producing basic steel, using an acid Bessemer converter to remove excess carbon and silica from molten iron before transferring it to an open-hearth furnace to eliminate phosphorus, which could then be used in fertilizer. Bacon also installed blooming and rolling mills to convert steel ingots into various shapes and began building a plant to make steel rails. Partly because of the very avidity with which he embraced technological change, however, Bacon's tenure at the helm of TCI contrasted markedly with Maben's smooth and successful assumption of power at Sloss-Sheffield. Although Maben had his share of problems with organized labor, Bacon's were much greater. In his five years in office, no fewer than six strikes took place; one of these, which broke out in 1904, temporarily shut down the new steel plant at Ensley. Meanwhile, Bowron and other former TCI officials did their best to discredit Bacon's leadership by publicly accusing him of waste and mismanagement. Despite TCI's much greater size, its profits as a percentage of capital were much smaller than those of Sloss-Sheffield. It paid no stock dividends between 1900 and 1905. Despite his modernization program, Bacon estimated that $25 million more was needed to bring TCI's operations up to northern standards; twelve Mesta vertical blowing engines purchased in 1905 could not be efficiently used because of weak blast mains and stove shells. Despite the propaganda of local steel enthusiasts, the firm lacked Sloss-Sheffield's financial health.[49]

As Bacon's grasp became increasingly tenuous, John W. Gates, the colorful and somewhat reckless financier who had earlier played a key role in nudging TCI toward steel production by urging creation of the Alabama Steel and Shipbuilding Company, made repeated attempts to take control. Displaying a driving ambition that marked his entire career, he tried in 1904 to bring about a merger of TCI, Sloss-Sheffield, and the Republic Iron and Steel Company, which had acquired the Pioneer Mining and Manufacturing Company in 1899 and added a third blast furnace to the two earlier ones at Thomas. Though intrigued by the possibilities, neither Bryan and his associates on the one hand nor Republic's owners on the other were willing to exchange securities on the basis of capitalization and property holdings alone, insisting that any buyout must take note that both firms were more profitable than TCI. Gates's scheme collapsed almost as quickly as it had begun.

In 1905, Gates again managed to interest Sloss-Sheffield in an even larger merger with the Alabama Consolidated Coal and Iron Company, the Alabama Steel and Wire Company, Republic, and TCI. This time, the bait was more enticing to Bryan, who was ready to sell out for $125 per share. When the bidding passed $120, Goadby counseled Bryan to accept. "You know how markets go," he said. "Why wait for an extra point or two? Sell now while you can, and get out." Bryan, however, displayed characteristic stubbornness. "That stock is worth $125, and I won't take less," he declared. "Making money hasn't made a fool of me, and losing will not disturb my equilibrium." Gates had made TCI's inclusion a prerequisite for closing the deal, and that firm's New York bankers also vetoed it. After the negotiations collapsed, quotations for Sloss-Sheffield's stock turned downward, but Bryan took his usual long-range developmental attitude. "If I had had one-tenth the sense in selling I had in buying," he mused at one point, "I would have had all the money in the world; but . . . when I get into a thing I like to stay and help work it out."[50]

Gates finally gained control of TCI in 1906 by arranging a buyout through a prominent New York brokerage firm, Moore and Schley. Among other investors, James B. Duke of the American Tobacco Company and Oliver H. Payne of Standard Oil took part in the transaction. Despite the results of Bacon's modernization efforts, the new owners were appalled by the state of the operation. "Its physical condition was very poor, and the property was run down at the heel," one manager observed. After taking command,

Gates retained Bacon briefly as operational head but soon replaced him with Frank H. Crockard, who was also an executive at Republic. Although TCI and Republic remained separate, Crockard administered the two enterprises jointly.

With funding arranged by Gates, a major expansion of TCI's steelmaking capacity took place. Ingot production surged to 400,000 tons in 1906, and plans were made to expand this to 600,000 in 1907. Intensification of Bacon's earlier move into rail production won TCI increasing markets for basic open-hearth rails, which it now made in larger quantities than any other firm. Early in 1907, it secured an order of 157,000 tons of steel rails from railroad magnate E. H. Harriman for the latter's Union Pacific and Southern Pacific lines.[51]

VI

TCI's progress alarmed an extremely dangerous adversary that had dominated the steel industry since 1901. In that year, fearing that the Carnegie Company would start making finished products and throw the entire steel industry into chaos, J. P. Morgan bought it out for $480 million. Adding fourteen other firms, including Federal Steel, he organized the United States Steel Corporation and soon placed Federal's chief executive, Elbert H. Gary, at its helm. Capitalized at the staggering sum of $1.4 billion, U.S. Steel was by far the largest industrial corporation in the world, dwarfing such firms as TCI. At the time it was formed, it controlled 161 different manufacturing facilities, 1,000 miles of railroads, 112 steamships, and landholdings containing enormous deposits of coal, limestone, and iron ore. It made half the coke and pig iron produced in the entire country. With a yearly capacity of almost 8 million tons of finished goods, it had a virtual monopoly of American barbed wire, steel tube, tinplate, and wire nail output in America and also fabricated about 60 percent of the nation's structural steel.[52]

Despite the vast disparity in size between TCI and United States Steel, TCI did have an edge in one respect: its basic open-hearth rails, as Harriman's 1907 order indicated, were superior to the Bessemer rails produced by its gigantic competitor. U.S. Steel was building new open-hearth rail mills in Pennsylvania and Indiana, but these were still two years from completion, increasing Morgan's desire to acquire the Alabama-based firm. Morgan also disliked Gates, who had made large profits at his expense in a railroad deal in 1902. Viewing Gates as a reckless speculator and "an unfit custodian of other people's property," Morgan had been waiting to pounce on him as soon as he had a chance.[53]

This opportunity came late in 1907. Early that year, a succession of overseas financial crises created a jittery mood on Wall Street. Economists worried that the heady economic growth taking place in the United States was outstripping gold supplies. The trust-busting activities of Theodore Roosevelt also caused concern, particularly when Standard Oil was fined nearly $30 million after being indicted on more than 500 charges of taking illegal rebates. As the year wore on, financial institutions became increasingly insecure, and corporations found it harder to raise funds for expansion. Events in the iron and steel industry reflected prevailing anxiety as blast furnaces were shut down and steel production was curtailed. TCI and Republic, however, seemed impervious to the trend. Demand for their open-hearth rails remained high, and they actually increased output. [54]

Under the circumstances, anything could trigger a panic on Wall Street. In October, the breaking point came with the collapse of the United Copper Company, which had been engaged in a reckless scheme of stock manipulation aimed at aggrandizing the fortunes of its rapacious leader, Frederick A. Heinze. Summoned to New York from Richmond, where he was attending a convention of the Episcopal Church, Morgan organized a campaign to shore up the market. Costing him and his financial allies many millions of dollars, it succeeded, and the panic began to dissipate in the early months of 1908. In the process, however, even Morgan's enormous financial resources had been strained. [55]

Morgan exacted a price for his rescue operation. During the Wall Street crisis, TCI's New York brokerage firm, Moore and Schley, became seriously overextended. A large amount of its assets consisted of TCI stock that it was using as collateral on loans it could not pay off. Faced with bankruptcy, Grant B. Schley, a major stockholder in TCI, turned for advice to another TCI shareholder, Oliver Payne, who recommended a buyout by Morgan. This created the opportunity for which the latter had been waiting. Knowing that the already strained atmosphere on Wall Street would only worsen if Moore and Schley liquidated its holdings in TCI at a sacrifice, Morgan arranged to have U.S. Steel acquire the troubled brokerage firm's controlling stake in the Alabama-based enterprise.

The idea of including TCI in U.S. Steel had been broached in 1901, but Gary had vetoed it because he knew that the Alabama company was unprofitable and took a poor view of its management. Along with another key leader in U.S. Steel, Pittsburgh millionaire Henry Clay Frick, Gary still opposed acquiring TCI in 1907 on the grounds that its costs were too high and that the deal might prompt federal antitrust action. Morgan, however, argued forcefully that U.S. Steel could acquire TCI for a fraction of its real value. His persistence led Frick and Gary to acquiesce on condition that the federal government formally agree not to intervene. Taking a special night train, consisting of a locomotive and a single private Pullman car, to Washington, they held a hastily arranged conference with Theodore Roosevelt at the White House on the morning of 4 November 1907. So quickly was the meeting convened that Roosevelt did not wait for his attorney general, Charles J. Bonaparte, who had spent the night in Baltimore, to attend. Recognizing, however, that it would be advisable to have a lawyer present, he had Secretary of State Elihu Root sit in as a substitute.

Roosevelt was under strong pressure, both from Old Guard Republicans and from members of the national business community, to tone down his image as a trust-buster and exponent of governmental intervention in economic affairs. Throughout 1906, his hostility toward John D. Rockefeller and vigorous backing of such measures as the Hepburn, Pure Food and Drug, and Meat Inspection acts had alarmed powerful conservative legislators including Nelson Aldrich and "Uncle Joe" Cannon. In 1907, as economic conditions deteriorated, prominent bankers warned the president that his attitude toward Big Business was partly responsible. As James MacGregor Burns later indicated, Roosevelt "needed such a banker friend" as Morgan in 1907. The White House visit by Frick and Gary gave him a welcome opportunity to play up to the financial magnate. As a result, Birmingham became a pawn in a larger game.[56]

Without revealing the identity of Moore and Schley, Frick and Gary told Roosevelt that an important Wall Street firm with large holdings in TCI was in jeopardy and that the acquisition of a majority of TCI's stock by U.S. Steel was needed to prevent a market collapse. Such a move, they assured Roosevelt, would increase U.S. Steel's share of the industry only marginally and therefore would not create a monopoly. It did not take long for Roosevelt's visitors to get what they wanted. Despite posing as an enemy of trusts, the chief executive believed that the increasing degree of concentration that had taken place in American industry in recent years was both necessary and irreversible. He was confident in his ability to distinguish between "good" and "bad" trusts, and looked upon U.S. Steel as being in the former category, particularly because of his esteem for Gary, from whom he had recently received an admiring letter. Fearing the effects of a Wall Street debacle, he told Gary and Frick that he "felt it no public duty of mine to interpose any objections" to their proposal and dictated a letter to Bonaparte to that effect. The entire matter, so crucial in its bearing upon the future of Birmingham, was decided in less than an hour. Shortly before ten o'clock, Gary telephoned word of the president's commitment not to interfere in Morgan's plans to one of the latter's representatives in New York. The news was quickly transmitted to the New York Stock Exchange and "received with the proper bullish sentiment."[57]

Morgan had already done everything necessary to effect the takeover. Feeling that he deserved a reward for the way he had dealt with the 1907 panic, he had developed a plan under which a consortium of New York bankers would underwrite the deal, sparing him the trouble. Locking these men in his tapestried library, he forced them to sign the requisite documents. Less than $700,000 of the money that financed the buyout was actual cash; the rest was composed of $35,470,000 in 5 percent gold bonds issued by U.S. Steel. This was a rare bargain. Julian Kennedy, a respected metallurgical engineer, valued TCI's assets in 1907 at $90 million to $100 million; Gary himelf later testified to a congressional committee that the property was worth at least $200 million.[58]

Gary and Frick had not been candid with Roosevelt. While telling him that the acquisition of TCI would give U.S. Steel only about 60 percent of the industry, they neglected to point out that in many significant categories, including ore reserves and open-hearth rail production, the share controlled by their corporation would be much higher because of the deal. Furthermore, as William Glenn Moore has indicated, acquisition of TCI by U.S. Steel was not really necessary to save Moore and Schley; it could easily have been rescued by loans from the same interests that had seen a chance to buy out an increasingly serious competitor at distress sale prices and leaped to make a killing.

Gates, aboard an ocean liner returning from Europe at the time the transaction took place, was furious about the way his interests had been betrayed. After the Wall Street panic had dissipated, public opinion was also outraged. Two later probes by congressional committees found the sale monopolistic and contrary to the antitrust laws. From the time the takeover was announced, however, it was greeted with naive enthusiasm in Birmingham by boosters who believed that it would fulfill their aspirations for the district by making much more capital available for developing its resources. W. P. G. Harding, president of the First National Bank, predicted that it would open a "new era in Alabama's industrial development." The Birmingham *News* said that the deal was "full of splendid possibilities" and declared that it would make the Magic City "the largest steel-making center in the universe."[59]

VII

Events would soon demonstrate that U.S. Steel's entry into Alabama was a blessing for many local residents who, for the first time, began to receive decent medical care, a rudimentary education, and other social and recreational benefits as a result of welfare programs imposed by the new regime. Otherwise, however, the takeover had unfortunate long-range consequences, particularly by removing any possibility that TCI could ever compete effectively with northern steel mills.[60] Hugh W. Sanford, a Tennessean who wrote a perceptive letter, "The South and the Steel Corporation," published by the *Manufacturers' Record* in 1908, saw what was coming. Asserting that "the Steel Trust holds the iron manufacturing interests of the South in the hollow of its hand," he rejected predictions that Birmingham was about to become a second Pittsburgh.

Sanford recited the familiar list of sectional advantages that Birmingham's boosters were fond of presenting. "The South is making the best of steel rail by the basic process," he stated. "The high-phosphorus ores in the South are adapted by the basic process. Iron ore, coal and limestone flux being in abundance and close to one another, it seems as though the future of the South was just as bright in the steel as in the cotton business. . . . Pig-iron can be made cheaper south of the Ohio River than anywhere else in the

United States. The raw material can be obtained at less expense. The railroad hauls of ore, fuel and flux are much shorter in the South than in the North, because we have all these things here close together. Labor is cheaper here, and this is guaranteed for all time to come by the mild climate, cheaper houses and buildings to stand this climate, cheaper food and living expense in every way." There seemed to be "no reason in the world," therefore, "why bar iron, bar steel, plate steel, bolts and all such . . . articles cannot be made in the South at the same cost advantage over the North as is found in the pig-iron."

Why, then, was there cause for gloom about the potential results of the recent takeover? Because, as Sanford said, "United States Steel Corporation has an arbitrary rule for the price of steel bars and plates." By selling these uniformly at the Pittsburgh price, currently $1.40 per hundred pounds, plus freight rates from Pittsburgh regardless of where the shipments originated, competition with any other part of the country was forestalled. Sanford calculated that steel bars and plates could be made at least $3 per ton cheaper in Birmingham than in Pittsburgh, but this cost advantage meant nothing under the prevailing system. U.S. Steel's policy, he stated, would have "a boa-constrictor effect on Southern concerns manufacturing iron and steel goods," making their products competitive with those of Pittsburgh only within severely localized markets. As a result, "the Lilliputian has no chance to grow. Pittsburgh is the Lord and Master."[61]

Sanford's words were prophetic. Although U.S. Steel would initially pump large sums of money into the Birmingham District, seemingly confirming the rosy visions of the area's boosters, the long-run impact of "Pittsburgh Plus" would be to strangle TCI's chances of competing on equal terms with the rest of the steel industry. The pricing formula would have a particularly devastating impact upon the South's ability to attract secondary steel processing industries whose interests were better served by staying in areas where transportation costs made steel bars and plates from such places as western Pennsylvania more competitive than those from the Birmingham District. U.S. Steel's takeover of TCI would drive home a lesson that boosters should have learned from the defeats that the Barons had suffered at the hands of McGraw's Giants in 1904: it was a heady experience to play with major leaguers, but the Magic City was likely to lose in the process.

But the takeover was a fait accompli, and the courts later upheld it in two bitterly contested antitrust proceedings that dragged on until 1920.[62] In the future, the southern ferrous metals industry would grow principally by attracting pipe mills, stove works, and other foundry-based industries unaffected by Pittsburgh Plus and therefore able to capitalize upon the special qualities of the region's natural resources. In the process, Sloss-Sheffield and Woodward became the world's largest producers of foundry pig iron, while TCI was subjected to freight rate restrictions that impeded its growth.

More than a century before, one of the world's greatest political economists had provided a suitable prologue for the situation now unfolding in Alabama.

In *The Wealth of Nations,* Adam Smith stated that "the natural advantages which one country has over another in producing particular commodities are so great that it is acknowledged by all the world to be in vain to struggle with them."[63] This statement was equally applicable to regions within a country. Given a clear competitive field, the South was so well fitted by its natural endowments to produce foundry pig iron that it attracted a constellation of industries based on the cost advantages that they conferred. Whether it could have derived similar advantages from TCI's ability to manufacture basic steel, once it had solved the technological problems involved, is moot. Gavin Wright, Kenneth Warren, and other analysts have emphasized other enduring problems, including the region's deficient skilled labor supply, its shortage of scientists and technicians, and its lack of capital, that would have compromised its hopes of competing successfully in the national arena.[64] U.S. Steel's basing point pricing policies, however, gave it no chance from the outset.

TCI's decision to produce steel benefited both Sloss-Sheffield and Woodward by leaving the market for foundry iron largely in their hands. Had TCI remained in this arena instead of gradually withdrawing from it, their future would have been different. By not following TCI's example, the two companies also unwittingly insured their ultimate survival, because otherwise they too might have been tempting targets for Morgan's acquisitive instincts and wound up being absorbed by U.S. Steel. By limiting themselves to the production of foundry iron, they presented no threat to northern steel producers. Despite their healthy earnings, it was not worth the effort that would have been required to take them over, especially given the implications that this would have had for federal antitrust policy. They therefore survived as independent entities, reaping the benefits of technological persistence in an Age of Steel that remained critically dependent upon cast iron.

Better than they could have known at the time, Bryan and his supporters had made a wise decision in 1902. Following the course they charted, Sloss-Sheffield continued to pursue a separate destiny under the leadership of the Virginia syndicate and its northern allies. Their deeply conservative policies would contrast greatly with the program of welfare capitalism that TCI's new owners were about to launch. After 1908, however, Sloss-Sheffield would no longer be under the guiding hand of the Richmond industrialist who had been a major force in Alabama's development for many years. For Bryan, time was about to run out.

11 Divergent Paths

As Sloss-Sheffield reached the end of the period in which Bryan was its principal stockholder, increasing profits did not bring about any basic change in long-standing labor policies. The firm continued to use traditional casting methods and clung to an extremely conservative attitude so far as employee welfare was concerned. In this respect, Sloss-Sheffield diverged from TCI just as markedly as it had done in becoming oriented toward foundry pig iron instead of steel. While Sloss-Sheffield clung to long-accustomed southern ways, TCI launched impressive new employee welfare programs. After the 1907 takeover by U.S. Steel, the contrast between the two firms became steadily more pronounced.

Although TCI's new policies were rooted in hard-headed business considerations, they were congruent with a wave of reformism that was sweeping the country at the time. They also resulted in much improved living and working conditions for thousands of Alabama's most disadvantaged citizens. Basking in a honeymoon atmosphere amid great expectations following the 1907 takeover, TCI largely escaped the restrictive impact of municipal regulations that were spearheaded by local reformers who were much less inclined than before to regard industrial growth as an unmixed blessing.

Sloss-Sheffield was not so fortunate. Out of step with forces of change that were sweeping away the Gilded Age, it became subject to increased taxation and restrictions on smoke emissions that urban reformers pushed through the city council. As international skies began to darken with the approach of World War I, the company entered a period in which its problems would begin to mount. Ultimately, ill winds blowing in from the future would bring the Virginia dynasty to an end.

I

During the early twentieth century the United States reached a historic watershed. With returning prosperity in 1897 and a swift, decisive victory over Spain in 1898, the Gilded Age ended, and the Progressive Era began. Despite dramatic technological achievements, economic growth, and America's rising status as a world leader, many citizens felt ambivalent about the direction of events. Political corruption, urban blight, the power of huge corporations, immense disparities between lifestyles of the rich and the poor, and other intractable problems set off a wave of reformism at every level of national politics—municipal, state, and federal.[1]

Neither the South in general nor Alabama in particular was immune from this ferment. Southern Progressives, mostly from urban and middle-class backgrounds, joined forces with rural Populists who had returned to the Democratic fold by the early twentieth century. Together, they made common cause against vested interests, particularly railroad companies. After being elected governor of Alabama in 1906, Birmingham industrialist Braxton Bragg Comer vigorously attacked freight rate discrimination. Congressmen from the region, including Alabama's favorite son, Oscar W. Underwood, spearheaded drives for tariff reform and a graduated income tax. Southerners fought for antitrust laws against oil and insurance companies. The commission and city-manager plans of municipal administration originated in the South in response to urban malaise. Rural and urban prohibitionists fought side by side; six southern states, including Alabama, outlawed the sale of intoxicants between 1907 and 1909. Religious leaders, including a beloved Birmingham clergyman, James A. ("Brother") Bryan, campaigned for social justice. Feminists, including Alabama's Patti Ruffner Jacobs and Julia S. Tutwiler, crusaded for woman suffrage and humane treatment of criminals.[2]

Southern progressivism, however, was selective, as was true in the nation as a whole. Throughout the region, racism actually intensified. During much of the late nineteenth century, as indicated by the way in which James R. Powell had mobilized blacks to help Birmingham supplant Elyton as the Jefferson County seat, white leaders had vied for black support, giving black people a degree of political leverage. Under the depressed conditions of the 1890s, however, mounting discontent among the lower classes created a threat that poor people might unite across racial lines, convincing white politicians that things must change if a social revolution was to be averted. As a result, new state constitutions almost totally disenfranchised blacks through grandfather clauses and other legal subterfuges. Poll taxes were also imposed, and the "white primary" became the only election that really counted. Pseudo-scholars vied in attributing bestial characteristics to persons of color, and segregation laws multiplied. In the words of C. Vann Woodward, "racism was conceived of by some as the very foundation of Southern progressivism."[3]

Alabama was no exception. Despite opposition from populist strongholds, even such moderates as Reuben Kolb, who led the plebeian Wool Hat faction of the state's Democratic party, approved a new 1901 state constitution creating what Sheldon Hackney has aptly called "a most elaborate maze through which one had to grope to claim the privilege of becoming an elector." Besides being disfranchised, blacks were lynched in increasing numbers. Jim Crow laws proliferated as blacks were compelled to use separate public lavatories and toilets, drink from their own fountains, ride at the rear of streetcars, and submit to other humiliations. "The city of Birmingham denied blacks access to many public institutions," states one account of the period. "They were kept out of the seventeen public playgrounds. . . . The zoo and the library were also off limits. Most hotels and restaurants were closed to blacks, unless they worked there as a maid, waiter, or cook. Blacks could shop in downtown stores but were not allowed to try on a coat or a pair of shoes before purchasing. . . . Stores refused to take back merchandise from blacks."[4]

Although united in upholding white superiority, ex-Populists and Progressives disagreed on many issues affecting industrial growth. Child labor was one of these topics. Edgar Gardner Murphy and other Alabama reformers fought against the practice and won passage of a state law restricting its use, but the statute was full of loopholes and came under attack by a reformist governor, Braxton Bragg Comer, who used large numbers of children in one of Birmingham's biggest industrial plants, the Avondale textile mills. Fear that northern capital would stop flowing into the South if child labor was restricted had much to do with its retention.[5]

Comer was typical of Alabama industrialists in supporting some reforms and resisting others. Because of self-interest, Sloss-Sheffield opposed the pernicious practice of paying arrest fees to law-enforcement officials, which encouraged policemen to detain black workers for a host of petty offenses. The company also tried to persuade the Interstate Commerce Commission to end freight rate discrimination against southern firms.[6] On most social issues, however, it was deeply conservative, remaining true to its Old South heritage.

Sloss-Sheffield clung to ingrained ways partly because there was no compelling necessity for it to change. Unlike TCI, it was profitable. By the early twentieth century it had solved its technological problems by acquiring extensive brown ore lands and modernizing its furnaces. It was also oriented toward customers who preferred traditional techniques of making pig iron. Because of the wide variety of characteristics needed in cast iron, depending upon the nature of the products into which it was molded, foundrymen preferred sand-cast pig iron that was made in relatively small batches. Because it was difficult to combine a high degree of mechanization with such market expectations, both Sloss-Sheffield and Woodward clung to techniques that emphasized manual labor.

Labor-intensive technology was in turn compatible with traditional southern social and racial attitudes. Bryan and Maben had shown great astuteness in 1902 by opting for a style of operation that was highly consistent with their patrician values. Although Sloss-Sheffield's black workers lacked incentive and were poorly disciplined, they were well accustomed to the tasks they performed. "Blast furnaces require little skilled labor," economist John A. Fitch stated in 1911 in a study of the ferrous metal trades. "There is no delicate machinery—it is simply a question of feeding the furnace with the raw material, and tapping out the molten product."[7] Operations in the casting shed required more finesse, but through decades of practice Sloss-Sheffield's workers had learned to perform them properly. Unlike TCI, which had moved into a rapidly changing area fraught with technological complexities, Sloss-Sheffield and Woodward had chosen to remain in an industrial niche in which they could earn steady profits without having to alter basic production methods or rely on a different labor force. Under such circumstances, expensive programs of welfare capitalism such as those upon which TCI would soon embark did not make sense to entrepreneurs who, like Bryan, did not have the capital to keep up with U.S. Steel and were too imbued with a traditional outlook to see any need for social innovation.

The aging process also contributed to Sloss-Sheffield's increasing resistance to change. From the time Hopkins resigned in 1902, the company's board of directors became more and more ingrown as investors who had wanted to move into steelmaking withdrew and older members tightened their grip on power. Bryan's death in 1908 did nothing to change matters; for almost a decade afterward, his longtime associates remained in control. Meanwhile, as northern influences became increasingly dominant in its management, TCI moved steadily in a different direction. Even before the 1907 takeover, it instituted new labor policies under executives drawn from outside the region. Events after 1907 climaxed this trend as U.S. Steel put a dynamic young president in charge of TCI and adopted policies that contrasted greatly from those of Sloss-Sheffield's ruling gerontocracy.

II

Sloss-Sheffield's decision to concentrate on foundry iron led to an exodus of directors who had wanted to pursue a more adventurous path. One by one, such men as Archer Brown, Moses Taylor, and R. B. Van Cortlandt left; so, too, did a group of Canadian investors, led by A. E. Ames of Toronto, who became briefly involved in the firm after 1902 but soon departed following frustration with the company's unwillingness to pay dividends on common stock, even though it was increasingly profitable. Within a few years, the faces in the corporate boardroom were essentially the same as they had been prior to 1899. The vacancies that occurred were filled by persons drawn from the same tightly knit club. When the Canadians withdrew, J. N. Plummer, who had represented them, was replaced by one of Bryan's oldest friends, James Dooley. The same thing happened when William E. Strong died shortly after being elected to the executive committee in March 1905; he was replaced by James N. Wallace, a protégé of Frederic Olcott who had succeeded his mentor as president of the Central Trust Company. Olcott, who himself did not have long to live, became chairman of the executive committee when Oakman, who was heavily involved in other business ventures, left shortly thereafter. Increasingly, Sloss-Sheffield was playing a game of musical chairs in an atmosphere that was hostile to new ideas.

As the company became more and more ingrown, conflict disappeared. In 1908, Rucker withdrew from the board, thus ending any chance of disagreement between absentee owners and investors from Birmingham, of whom he was the last remaining representative. When other members died or departed, their places usually remained unfilled. From his finely paneled office at the Woodward Building, Maben faithfully carried out Bryan's wishes. [8]

As always, Bryan visited Alabama regularly to inspect the company's properties. He now did so, however, in truly regal style, befitting his status as the chief investor in one of the South's largest corporations. In 1902, the board of directors appointed Bryan, Goadby, and Maben to buy a company-owned vehicle for such visits, something that it had not previously possessed. Leading the search, Bryan found that a seventy-foot passenger car, the Cleopatra, was available from the Pullman Company. [9]

Like boys who had discovered a new toy, Bryan and Goadby had the Cleopatra brought to Jersey City in January 1903 for inspection by one of the Southern Railway's technical experts. Their excitement in refurbishing it reflected company priorities at a time when no fundamental changes were likely to occur. "If you are in doubt about accepting the car or not," Bryan wrote Goadby from Richmond, "wire me and I will come on. My fear is that it is too large for our purposes, as it will have to go on spur tracks and some very troublesome places." Maben, however, warned not to buy "too short and light a car, as on these crooked and sometimes rough roads it makes very hard riding, and it costs no more to maintain a good sized car than a small one." [10]

Bryan's letters during the next few months reflected the delight with which he and his associates savored every detail connected with their new status symbol. In mid-January, Goadby received from the Southern Railway a favorable report about the Cleopatra estimating the car to be worth about $6,500, plus an additional $700 for china, silverware, and bedding to be included in the deal. The report also recommended improvements bringing the total cost to about $8,000. After offering this sum to the Pullman Company if it would agree to perform the modifications, Bryan and his friends settled for $7,350 and had the Southern do the repairs at its Manchester, Virginia, shops. [11]

In February, amid a flurry of letters and telegrams, Bryan had the Cleopatra taken from a siding in Weehawken, New Jersey, and brought to Manchester. "I shall have new carpet put down, new linoleum, the whole car cleaned and varnished inside, painted and lettered outside . . . a new platform and vestibule extension put on the kitchen end, with arrangements for extra boxes of provisions," he wrote Maben soon after the car had arrived from New Jersey. Ordering sheets, pillowcases, and towels, he pondered installing acetylene gas lighting but assured Maben that he would "leave enough of the old lamps to light us in case of trouble." He estimated that finishing the job would take at least thirty days. "We are very fortunate in having the car located in Manchester where I can give attention to it and where they are best fitted to do such work than at any other point on the Southern Railway," he declared. By this time, the committee had decided to give the car a new name: the Alabama.[12]

By April, the car was ready for its maiden trip to Birmingham, but because of a sudden stock market crisis the New York directors postponed a planned visit. "We have gotten a car for the convenience of the directors," Bryan lamented to Haas, "and now the directors have no heart to travel in it." In late April, however, Bryan, Goadby, and Seixas finally set out for the Magic City. "The car was a perfect success on our trip and was greatly admired," Bryan reported after he had returned to Richmond. During the trip, he had decided that the car needed an added dressing room and "a place for the servants to sleep," and it was sent back to Manchester for modifications. In October, before next coming south, Bryan informed Maben that a new partition had also been added to provide a separate washroom and toilet, so that company leaders might be spared the indignity of sharing the ones used by the crew.[13]

No matter how well things were going, it was not in Bryan's nature to adopt a passive attitude toward the company and its affairs. In 1903, he demanded semiweekly sales reports from secretary-treasurer J. W. McQueen, tried to clear up accusations of mismanagement against a furnace superintendent, strongly advised Maben to get out of pooling arrangements with other iron producers and warmly approved when this was done, worried about the performance of the Hattie Ensley furnace and delays in completing new coke ovens, and played an advisory role in purchasing boilers from a Chicago manufacturer. No detail was too small to escape his attention. "If there is anything I would like to know it," he told McQueen in one of his letters.[14]

Many of Bryan's other activities were focused on the past. In 1903, he spent much of his time raising money for John S. Mosby, who was now living in poverty. By seeking a job for his old commander in the federal justice department and trying to bolster his waning spirits, Bryan became "the warmest friend Mosby ever had," according to one of the latter's biographers, even though he disagreed sharply with many of the political opinions held by the former guerilla leader, who was now a Republican. Bryan also attended Confederate reunions, raised money for equestrian monuments that were proliferating throughout Richmond, and spearheaded historic preservation projects. He took special pleasure in recalling his own role in the Lost Cause, never wavering in his convictions about its justice. "I fully sympathize with your sorrow that you did not do more to repel the invaders of Virginia in 1861–1865," he told his brother, Saint. "I wish I had been 100,000 men. I would have put every one of them on Grant's flank in '64, after the battle of Cold Harbor, and wiped them off the face of the earth."[15]

One by one, cherished links with the past began to snap. William R. Trigg was stricken with paralysis early in 1902 and died of blood poisoning from a kidney ailment in February 1903. An even more painful blow to Bryan was the death of John W. Johnston, who had become increasingly frail, in May 1905. An editorial in the Richmond *Times-Dispatch*, probably written by Bryan himself, recounted Johnston's role as president of the Georgia Pacific Railway and recalled that it was he who had "secured the first properties from Colonel Sloss and the Coalburg Coal and Coke Company, and arranged the financial scheme which created the Sloss Iron and Steel Company."[16]

Despite keeping busy, Bryan found time for a few leisure activities. Having long envied such friends as Johnston for being able to visit foreign countries, he began to travel abroad. In July 1903, he and Belle went to Europe for a three-month vacation. Letters that he wrote after their return did not indicate where they went or what they saw, but more information is available about a later trip to Europe. Because surviving correspondence is scanty, Bryan's activities are hard to follow after 1903. Travel notes kept by Belle, however, show that he continued visiting Alabama to inspect Sloss-Sheffield's properties. By this time she apparently accompanied him wherever he went. Their last trip to the Magic City took place in 1908. Arriving aboard the Alabama on 24 April, they spent a few days inspecting facilities at North Birmingham, Flat Top, Bessie, Sheffield, Florence, and Russellville. "Altogether a most satisfactory trip," Belle noted after their return home.[17]

On 30 May 1908, Joseph and Belle sailed for Liverpool aboard the S. S. *Carmania* to begin a long vacation. In her diary, Belle traced their wanderings through London, Paris, Geneva, Rome, Naples, Florence, Venice, Cologne, Amsterdam, Antwerp, and Brussels. Returning from Belgium to England, they went to Scotland, where they visited the Stewart family's ancestral home at Rothesay and saw Glasgow and Edinburgh before turning southward to York. After worshiping in York Minster on 16 September, they entrained for Liverpool and sailed for America the next day. "Laus Deo," Belle wrote in her diary on 30 September. "After weeks of travel returned well to find all safe and well at home."[18]

"Home" was no longer the ornate Victorian mansion, Laburnum, into which the Bryans had moved in 1885. In the wee hours of 6 January 1906, Belle had awakened suddenly to find that the house was on fire. The flames were so intense that Laburnum was reduced to ashes within two hours. Everything was destroyed except for a few pieces of furniture and some books, papers, and art objects, but members of the family felt fortunate to escape with their lives. Faulty electrical wiring was the apparent cause of the blaze. After the brick walls had cooled and the ashes were raked, Bryan found what was left of a steel sword that Robert E. Lee had carried at Appomattox. The damaged blade continued to be a treasured family heirloom.

Undaunted, Joseph immediately laid plans for an even more impressive structure to replace his former residence. Given the same name as its predecessor, it had thirty-five rooms and was built in Georgian style, with a classical portico adorned with six limestone Corinthian columns. Because it was believed that the fire that had destroyed the old Laburnum had broken

Belle Bryan. Isobel Lamont Stewart (Belle) Bryan, wife of Joseph Bryan, came from an upper-class background and met her future husband at a plantation owned by one of the clients whom he served as a rising young attorney. She was one of seven daughters born to John Stewart, who along with his brother, Daniel K. Stewart, amassed a fortune in the tobacco trade that helped nurture southern industrial development after the Civil War. The sustaining role that Belle played in her husband's career was particularly evident in the closing years of his life, when she seems to have accompanied him wherever he went. An undated photograph suggests the same lively intelligence revealed by the journal she kept of their travels to Europe and other places. (Courtesy of Virginia Historical Society, Richmond, Virginia.)

out in a closet, the new mansion had only three such compartments. To minimize the danger of future conflagrations, it had a steel frame, and no wood was used in the structural members. The most impressive room was a walnut-paneled library copied from an Italian design. When completed in January 1908, the house was "considered to be the most handsome residence in Richmond." Some of the Bryan children and their families stayed there later that year while waiting for Joseph and Belle to return from Europe.[19]

The diary that Belle kept during the European trip shows that Joseph's health was declining. On 16 July, he complained about "a sharp pain in his chest" but insisted on attending a performance of *H. M. S. Pinafore* despite Belle's protests. He had chills and a fever afterward. After rallying for a short time, he had a "sharp attack of gout" just before they crossed the Channel to France; was "sick and tired" when they reached Rome; had another "slight gout attack" in the Italian capital; and complained of chest pains during the homeward voyage in September. Bryan had angina. Not long after returning to Richmond, his condition worsened, and he was told by his attending physician that death was imminent. On 20 November 1908, after spending much of his remaining time in severe pain, he died.[20]

Bryan's death prompted an outpouring of tributes to his memory. "I had known Mr. Bryan for forty years, and do not believe that Virginia has ever produced a more devoted son or that any community has ever had a more useful citizen," stated Richmond's mayor, David C. Richardson. Such a multitude was anticipated at the funeral, held on 22 November, that the Richmond and Chesapeake Bay Railroad ran special trains to a siding near Emmanuel Church, where the service took place. Only a small fraction of the thousands of people who came could be squeezed into the building after the cortege had wound its way from Laburnum to the house of worship by way of the Brook Hill Turnpike. Surviving veterans of the Richmond Howitzers and employees of the Richmond Locomotive Works lined the drive as the casket was carried into the sanctuary, where the Episcopal burial service began; because of the size of the throng, however, as much of the ceremony as possible was conducted at the gravesite immediately outside the building, close by the plots where John and Daniel K. Stewart rested. Former comrades-in-arms wept silently as Bryan was interred in his Confederate uniform, in a coffin draped with the colors under which he had fought.

Soon thereafter, officers and employees of the Richmond *Times-Dispatch* commissioned a stained glass window for Emmanuel Church to honor Bryan's memory. Depicting him in the armor of a crusader, it was dedicated at a service on 27 June 1909 at which the Right Reverend Alfred M. Randolph, bishop of the Episcopal Diocese of Southern Virginia, praised his "noble Christian manhood." Later that year, Belle donated to the city a 262-acre tract on the north side of Richmond, to be known as "Joseph Bryan Park," for the "use and benefit of all its citizens." On 10 June 1911, a large bronze statue of Bryan was unveiled in Monroe Park before an "immense crowd" that included his old commander, Mosby; the inscription at the base of the monument ended with the statement, "The Character of the Citizen is the Strength of the State." A choir of 1,000 school children sang at the ceremony, and the band of the Richmond Light Infantry Blues played "Dixie" as the throng dispersed. In a newspaper editorial published prior to the event, Bryan was lauded as "the greatest citizen of the Commonwealth in his day and generation."[21]

Richmond was not alone in mourning Bryan's death. In Birmingham, newspapers gave it front-page attention. "Bryan was one of Birmingham's most ardent prophets," the *Age-Herald* stated. "Few men had so unfaltering a faith in the Birmingham District as did Joseph Bryan," echoed the *News*. "Back in the dark days of the nineties when many feared the bottom would drop out of Birmingham, Mr. Bryan continually prophesied that this city would rally and come to the front as the South's great industrial center. He backed his faith by his substance, and invested in Birmingham when things looked blue to many others. He inspired others in Birmingham, and never doubted its future."[22]

III

Soon after Bryan died, one of Birmingham's greatest citizens passed away. Early in December 1910, Henry F. DeBardeleben had a massive heart attack while inspecting a mine near Acton, about fifteen miles west of the city, that he had recently discovered in one of his perennial prospecting ventures. Most of his fortune was now gone; his estate, when probated, was found to be worth only $84,000.

News of the event sent waves of disbelief through the Magic City as residents found it hard to conceive that a man of such prodigious energy could somehow be dead. DeBardeleben's body was brought to his residence, built in the mid-1880s on the west end of town near Alice furnaces and the Birmingham Rolling Mill. As years went by, friends had urged him to move to the fashionable South Highlands, but the aging industrialist had protested that he wanted to stay close to the "enterprises in which he was interested." The corpse lay in state in a front room while wagonloads of flowers were brought to the home, accumulating in a mountainous pile around his bier. Three heavily laden vehicles took them to the gravesite at Oak Hill Cemetery for his burial on Thursday morning after a long line of "railroad leaders, merchants, capitalists, industrial leaders, corporation heads, judges, engineers, doctors, and miners" had filed past his casket to view his remains.[23]

Ex-mayor Walter M. Drennen was among those who paid tribute to the fallen industrialist. "If we had 25 men in this state of the energy, capacity, and enthusiasm of Mr. DeBardeleben, Alabama's population would soon reach 2,500,000," he said. "Just imagine what 25 citizens of his type could do in building up the state, adding rapidly to her wealth and attracting population."[24]

Drennen, scion of a family that owned one of Birmingham's largest department stores, was a machine politician whose administration had been marred by the lust for boodle that was strongly prevalent in leading American cities at the turn of the century.[25] His tribute to DeBardeleben conveyed no hint that the quality of life in Birmingham might be at least as important as quantitative measures of its ceaseless development. From such a perspective, the district that Bryan and DeBardeleben had done so much to build left much to be desired. As a growing cadre of reformers recognized, social conditions there were becoming more and more desperate. Unfortunately for the district, even Bryan's careful developmental strategy was unleavened by any sense of concern for the long-range consequences of values deeply embedded in the Old South heritage.

Things were particularly bad in north Alabama's squalid mining camps. Many had no public water supply, forcing residents to use springs or wells that often ran dry in the summer. Even where better systems existed, public taps often served five or six houses. People drank water from outdoor pools into which workers spat as they rinsed their hands. Reeking privies, overflowing with human excrement, were seldom cleaned, and pigs scavenged for refuse in the absence of garbage collectors. Outbreaks of typhoid, described by one commentator as "a greater economic burden than amputated fingers," swept over the camps as swarms of flies carried germs from outhouses and refuse heaps to overcrowded hovels where miners and their families shared living space with boarders, dogs, and chickens. Hookworm, lung disease, malaria, and pellagra were rampant.[26]

Conditions were no better in many of Birmingham's own neighborhoods. Garbage collection was inadequate, and roving gangs of convicts dumped piles of decaying refuse in open lots. Until the turn of the century, the sewer system poured human waste directly into Village Creek; even after a new disposal plan was implemented in 1905, many parts of the city remained unprotected from contagion. As late as 1910, 10 percent of Birmingham's population continued to use outdoor privies, many of which served multiple dwellings. More deaths occurred from typhoid—59.70 per 100,000 people— than in any other American city. Birmingham also had the highest death rate from tuberculosis of any place except Denver, which was atypical because victims of pulmonary disease went there only to die. Smallpox epidemics swept through the Magic City eight times from 1899 to 1910. Pneumonia was also prevalent, particularly during raw, damp winters. Outbreaks of measles, diphtheria, diarrhea, and dysentery were equally common. Partly because blacks who could not afford physicians used midwives, one of every ten pregnancies ended in stillbirth.[27]

Dust and soot from blast furnaces, foundries, and rolling mills blanketed the city, raising the incidence of respiratory disease. When Sol Haas returned to Birmingham in 1903, Bryan warned him to live as far as possible from "the smoke and dirt of First Avenue and Twentieth Street." Even the pristine facades of such newer structures as the Woodward and Brown-Marx buildings were quickly discolored by layers of grime. The *News* said that smoke caused $3 million worth of damage to goods and property every year.[28]

Trying to escape the fumes, wealthy citizens moved toward the crest of Red Mountain, where they built large residences. Typical was an imposing house on Fourteenth Avenue, built in 1905 by lumber magnate Frank H. Lathrop. Adorned with Ionic columns, it had ten-inch-thick walls, leaded glass windows, a grand entrance hall with a large Victorian chandelier, and a spacious porch surmounted by a balustrade with urn-shaped finials. In 1906, developers began creating a luxurious enclave, Forest Park, two miles from the heart of town. A noted landscape architect laid out terraces ornamented with flowering trees, vines, and imported European shrubs; baronial estates lined Crescent and Glenview roads. Even the rich, however, could not escape the epidemics that plagued the city, partly because of germs brought into their mansions by servants who arrived in the morning from some of the city's most noisome areas and returned to them at night. [29]

Unable to move elsewhere, the poor lived near the center of town. South of the main business district, a sprawling section of ramshackle shotgun houses, so called because it was said that their layout enabled adulterous visitors fleeing out the back door to be shot by irate husbands from the front porch, housed most of the black population. Here, occupying hovels that harked back to Birmingham's Old South plantation origins, powerless victims of a racist society were regularly winnowed by disease. In 1912, an article in *Survey* described their dismal shanties, crowded fifty or sixty into a city block, as "nests of infection." In many cases, ten families used the same outhouse. Blacks were five times more likely than whites to die of tuberculosis, the chief cause of mortality in the slums. [30]

Even for white children, public education had a low priority in Birmingham; things had degenerated since the 1880s, when good schools had been a matter of civic pride. Expenditures per pupil now ranked near the bottom among American cities. In the 1903–1904 term, only an emergency bank loan kept schools from closing. Provisions made for educating blacks were particularly bad and got even worse after the mid-1890s when a school in East Birmingham that had been attended by 1,000 students burned down and was not replaced. During a sixteen-year period ending in 1907, not one school was built for black pupils, who were limited to three dilapidated frame buildings. White students, by contrast, had eight elementary schools and one high school, all made of brick. In 1905, the teacher-student ratio in white schools was forty to one; in black schools, seventy-three to one. Two years before, the *News* had commented that "The east end Negro school is in miserable quarters. It has about five hundred pupils who are packed in five rooms without light, without heating facilities, with a distressing ventilation, and without desks." [31]

Sloss-Sheffield was prominent among the firms that condoned and perpetuated such conditions. Most of the predominantly black work force at the City Furnaces lived in what was familiarly known as Sloss Quarters, clustered near the stacks. The fact that such a name was given to them indicated that they, and other similar enclaves throughout the city, were the early twentieth-century analogues of similar plots of ground that had been inhabited by slaves on antebellum estates. John Fitch found Sloss Quarters particularly appalling when he visited the Magic City in 1911 to write an article for *Survey*. With "a slag dump for a rear view, blast furnaces and beehive coke ovens for a front view, railroad tracks in the street, and indecently built toilets in the back yards," the area was in his words "an abomination of desolation." Residents peered outside through gaping holes where boards were missing in unpainted walls. An accompanying photograph showed squalid shanties and privies perched near the edge of the slag pile. Questioned about such conditions, Maben answered that he did not believe in "coddling workmen."[32]

After Fitch and other writers published their findings in the 6 January 1912 issue of *Survey*, Maben and other industrialists denounced them for having committed "the most inexcusable injustice that had ever been done a District." Responding to criticism of garbage-seeking pigs in the mining camps, Maben declared that swine were "the natural and logical scavengers" of such places. "Pooh," he said, "I'd rather have twelve hogs than fifty men cleaning up my camps." Another local business leader, Walter Moore, was equally exasperated. "Why must the Birmingham District be entered in the night, as it were, and lambasted by men whose minds never light upon anything concerned with the good, the beautiful, and the true?" he demanded. "There is a handful of men in Birmingham who have withstood the fever's blistering breath; they have conquered the untrampled depths of this District; they have worked until the blood almost spurted from them; they have conquered the elements and have brought wealth and happiness to many score of thousands. Yet they are crucified before the world by a staff of alleged philanthropic men whose whole life is spent in turning up the muck."[33]

A few local leaders responded differently. George B. Ward, for example, said that most of the charges leveled by *Survey* were true. Ward was one of the greatest figures in Birmingham's history. His life revealed much about how a sensitive and compassionate person reacted to the harsh realities of the industrial age. Reputedly the first male ever born in Birmingham, he was really a native of Georgia. His parents had come to the Magic City when it was still a jumble of shanties and boxcars to run the Relay House, the two-story frame structure that served as its first hotel. Ward had vivid memories of the community's formative years. In 1888, he had nearly lost his life when a sheriff fired a bullet through his hat as he stood surrounded by a mob gathered in front of the municipal jail to lynch a murderer. During his early life as a bank runner and teller, both before and after completing his education at a preparatory school in Lebanon, Tennessee, he had seen the best and worst of the city. Elected as an alderman at the turn of the century, he ran unsuccessfully for mayor against Drennen in 1903 before swamping a member of the latter's machine for the same office two years later.

An idealistic person with large, soulful eyes, high forehead, and Adonis-like face, Ward broke the mold in which the rest of Birmingham's mayors were cast. Epitomizing the high tide of progressivism in the Magic City, he spearheaded codification of laws and regulations, expanded the public parks, enlarged the police and fire departments, fought for cleaner streets and better public health, imposed higher taxes and stricter regulations on utilities, and tried to combat crime and violence by making it harder for dives and saloons to secure licenses.

Ward, however, ran afoul not only of political hacks who hated everything he stood for but also of moralistic prohibitionists who were offended by his realistic approach to saloon regulation. When he left for a European vacation in 1907, his opponents, taking advantage of a new municipal code passed during his absence by the state government to separate executive and legislative powers, seized control of vital committee chairmanships on the municipal council. After returning, he was prevented by local foes, state legislators, and the judiciary from recouping the ground he had lost. Through skillful maneuvering in Montgomery, his enemies also blocked his efforts to establish effective supervision of the police department, which was colluding with gambling house operators for rakeoffs. Recognizing the pitiful state of education for black children, Ward started a high school in rented quarters for them before his second term expired, but his reform efforts were largely hamstrung and he retired from politics after losing a race for sheriff in 1910.[34]

Local industrialists, who normally stayed aloof from municipal politics as long as their interests were not threatened, helped block Ward's reform efforts. Fighting vice through regulation instead of wasting time in unrealistic efforts to outlaw it, he tried to confine brothels and saloons—particularly those frequented by blacks—to a special red-light district. After this measure was defeated by special interests that profited from the unrestricted sale of alcohol, he reluctantly joined other reformers to secure outright prohibition of intoxicants but was strongly opposed by such firms as Sloss-Sheffield and TCI. "Skilled, as well as ordinary labor is scarce and in demand all over the country," their managers declared, "and we know that if a sweeping prohibition law should be enforced in this District, large numbers of our best workmen would leave." Ultimately, suburban and rural voters imposed prohibition upon Jefferson County as a whole, and the state legislature passed a general law banning the sale of alcohol. Neither move, however, had much effect in Birmingham, where semisecret "blind tiger" clubs and saloons served thirsty blacks and whites in the absence of effective law enforcement. This situation vindicated Ward's assessment of political moves that did more to symbolize the value systems of the respectable than solve the problems of the needy.[35]

While opposing prohibition, Maben and other industrial leaders supported laws aimed at reducing vagrancy because it served their interests to do so. At the turn of the century, Sloss-Sheffield and TCI still had to keep 50 to 75 percent more men on their payrolls than they could expect to report for work on a given day. Anti-vagrancy ordinances, dating as far back as 1891, helped keep this situation from getting even worse. Ostensibly, such laws applied to both blacks and whites but were aimed chiefly at the former. Under pressure from industrialists, state legislators passed severe vagrancy codes in 1903 and 1907. Apart from giving such companies as Sloss-Sheffield a better supply of floating labor, these measures increased the ranks of convict work gangs, mainly black, who shuffled chained and manacled along Birmingham's streets under the eye of local police.[36]

IV

Labor unions fought against vagrancy laws. They were particularly concerned that such legislation would strengthen the convict lease system, which was strongly supported by Maben and other industrialists. It is not hard to see why large companies resisted abolishing what one historian later called a "Monument to Shame." In addition to being cheaper, convicts were also more dependable than free workers, black or white, who might take off at any time to attend social events or merely to loaf. Events in the depression-ridden 1890s had shown that convict labor was an effective weapon against strikes, permitting Sloss-Sheffield, TCI, and other large firms to continue operating during walkouts. Industrialists, therefore, fiercely opposed efforts by the United Mine Workers and other labor organizations to limit the use of convicts to repairing roads.[37]

Maben showed his strong opposition to unions and firm support of convict labor soon after coming to Birmingham. At the time he took charge of Sloss-Sheffield in 1902, the company was embroiled in renewed controversy over conditions among inmate workers at Coalburg. During the first three months of the year, ten state offenders and twenty-two county convicts had died there from pneumonia, tuberculosis, and other causes. Because this rate was much higher than the incidence of mortality at any other comparable Alabama facility, Gov. William D. Jelks had ordered a special investigation shortly before Hopkins left office.

Even before Maben moved to Birmingham, plans had already been made to abandon the Coalburg mines. Because these workings had been depleted by constant exploitation, Sloss-Sheffield was developing a new mine on Flat Top Mountain, west of Coalburg near the Walker County line, and planning to move all of its inmate labor force there as soon as it was ready for operation. About thirty or forty convicts were working at Flat Top by the time conditions at Coalburg became a public issue, but neither the stockade nor a spur to the new mine from Littleton, a nearby village on the Southern Railway, had been completed.[38]

As public indignation mounted, Maben completed the Flat Top facility. Shipments of coal from the new mine began in September 1902, and the last remaining convicts from Coalburg were transferred there by the end of October. By March 1903, a slope had been driven into the mountain to a depth of about 2,000 feet, and 200 coke ovens had been built. Predicting that Flat Top would be "the most profitable of the Company's coal mines," Maben estimated that it could produce 1,500 tons of coal and 1,400 tons of coke per day, enough to supply all seven of the company's furnaces.[39]

Unwilling to lose revenue, however wretched conditions at Alabama's industrial labor camps might be, officials at Montgomery negotiated a new three-year contract with Sloss-Sheffield for the use of state convicts. Under a new law resulting from the earlier furor about conditions at Coalburg and other places where such persons worked, state prisoners would now be segregated from county inmates and supervised by state officials. The coal they extracted was to be sold by the ton to the companies owning the mines. From this point on, the two types of offenders were housed in separate wings at Flat Top. State convicts dug coal on the left side of the slope under overseers appointed in Montgomery. County inmates worked on the right side under company supervision. A mining journal admitted that the resulting output was inferior to that of free workers but said that the system was "very efficient because of the steadiness of work and the immunity from strikes, lock-outs, etc."[40]

At Flat Top, a labyrinth of rooms, called chambers or breasts, fanned out from entries or headings that branched at right angles off the central slope. Here, as at other Alabama mines, workers labored in what two historians, describing similar conditions in northeastern Pennsylvania, have aptly called "Black Hell."[41] Coal was extracted by means of the room and pillar technique, leaving columns intact to support the sandstone roof.[42] The work was done entirely by hand and illuminated by carbide lamps that had replaced the smoky oil or kerosene lamps of the late nineteenth century. Water or saliva flowing from a chamber at the top of the lamp created a gas when it came into contact with carbide in a receptacle at the bottom; this vapor was in turn ignited by sparks from a friction device. Although the resulting flame was steadier than the flickering glare from oil or kerosene lamps, it was just as deadly if it set off an explosion of gases, such as methane, that were always present in a coal mine.

Regardless of whether free or convict labor was used, mining was a dangerous and unhealthy occupation. Kneeling or lying on their sides and using picks to open the base of a seam, workers undercut the coal, which, at Flat Top, was about ten feet thick. After this task was done, the coal was detached by wooden wedges driven upward into it, or, if necessary, with blasting powder. While working under the seam, miners used wooden blocks, called sprags, to support the mass until it was ready to be brought down. To prepare for blasting, they made cartridges by wrapping newspapers around a stick. After withdrawing the latter, they filled the cartridge with powder and fastened it to the end of a rod known as a miner's needle. Drilling a hole into the coal while lying on his back, a miner pressed against the mass above him with a curved breastplate to gain leverage. After the hole had been drilled, he inserted the rod and cartridge and ignited the latter with a fuse, or squib, made of waxed paper sprinkled with powder. Because the flame went quickly up the hole, a miner left a room in a hurry after lighting the fuse. Following the explosion, he returned and dug out the fallen coal by hand.[43]

The danger of setting off an uncontrolled explosion was not the only peril confronting miners. Rock falls sometimes resulted in death or serious injury by crushing them or burying them alive. Constant exposure to coal dust also led to pneumoconiosis, the dreaded "black lung disease." Returning to the surface after a day's work, the men emerged covered with mud and slime, hardly looking like human beings. Because of this, they were often called "lizards." Such conditions were shared by free and inmate workers alike, but adding to them the privations associated with incarceration made them only worse. Maben and other advocates of the system, however, praised it for giving the offenders an "honest trade" that they could follow after serving their time.[44]

Union claims that convict labor gave management an effective weapon against strikes were validated anew in 1904, when the district had its first major walkout in ten years. Sloss-Sheffield, TCI, and other large corporations had helped make conditions more stable in 1901 by forming the Alabama Coal Operators Association (ACOA), which bargained collectively with the United Mine Workers (UMW).[45] After the passage of a 1903 state law prohibiting boycotts, however, operators refused to continue negotiating with unions. Using a momentary downturn in economic conditions as an excuse, they demanded substantial wage cuts, triggering a strike that lasted more than two years before the miners were finally defeated in August 1906.

Maben's tactics in dealing with the stoppage were typical of those used by local industrialists. Although Sloss-Sheffield's coal and coke output was initially curtailed, it weathered the crisis by using convicts and strikebreakers. "While this strike has been expensive and in many ways trying," Maben commented, "the ability of the Company to control its own business . . . must ensure greatly to its benefit in the future." By winning the struggle, Sloss-Sheffield was able to establish an unyielding open shop policy.

In 1908, the United Mine Workers called another strike in Alabama's coalfields. Sloss-Sheffield's mining community at Blossburg was among the camps affected. Feelings ran high in the wake of the Panic of 1907, and speakers at a public meeting in Birmingham demanded that labor leaders be lynched. When the walkout continued, Governor Comer called out the National Guard, but many of the troops, who were drawn from rural areas and became sickened by the conditions they witnessed, openly sympathized with the workers. Despite such fellow feeling, the strike failed. Comer drove his car to a place where miners had gathered and, speaking from the running board, ordered them back to work. Coupling his words with force, he ordered troops to tear down a tent city that the strikers had erected. His success in ending the walkout virtually ended the power of the UMW in Alabama.[46]

As usual, racism was evident in the tactics used by operators to quell the strike. Unlike most unions, the UMW had no restrictions on black membership; its constitution specified that its aim was to "unite in one organization, regardless of creed, color, or nationality, all workmen employed in and around coal mines." Throughout the walkout, spokesmen for management took the position that it had been fomented by outsiders who wanted to force racial equality upon Alabama and its people. Denying that they opposed white supremacy, leaders of the UMW asserted that they had organized blacks only "for purposes of self-defense, since the union knew that if it did not, the operators would use them as strikebreakers and to keep down the wages of white miners." Such disavowals were in vain; as a searching analyst of the strike later commented, "The skillful use of racial prejudice by the operators completely alienated the union and its members from Birmingham society, and made any kind of victory impossible."[47]

By taking a hard-boiled approach toward labor and using racism as a tool to divide the working class, Birmingham's industrial leaders perpetuated attitudes that had prevailed since the birth of the city, but reformers were becoming increasingly vocal about the need for change. In "The Spirit of the Founders," an article published in the same issue of *Survey* that offended Maben and other managers in 1912, Ethel Armes, who had come to Alabama to join the staff of the *Age-Herald* after previous service as a reporter with the Chicago *Chronicle* and Washington *Post*, indicted the way in which the city's pioneers had played the "game of self-interest," heedless of social consequences. Armes epitomized a rising wave of ameliorative journalism that was bent upon refashioning American society during the Progressive Era. While admiring the "go-ahead, grit, gumption, and great business capacity" of Birmingham's early promoters and admitting that the city could not have been built without their enterprise, she found much to criticize about the social order they had created. Although she did not demand that it depart from white supremacy, she believed that it had to manifest something beyond merely pecuniary aspirations.

Armes flayed the lack of public spirit shown by local industrialists, charging that they had "let the town go hang." Interested only in profit, they had neglected the arts and refused to back philanthropic activities. As a result, she said, the city had "no central public library building, no first-class theater nor concert hall. It has not yet been possible to provide and maintain any very large parks or playgrounds worth mentioning. There are few places for people to go to in the evenings. Such public buildings as are here . . . were designed and built some years ago, without the slightest reference to such minor considerations as air, light, heat, ventilation, or any hygienic or sanitary arrangements deserving of the name."

Having just published a history of coal mining and iron manufacturing in Alabama made Armes particularly conscious of the district's shortcomings. In her zeal to gather facts at first hand, she had contracted typhoid; ignoring a taboo that bad luck would result from a woman's presence in a mine, she had gone deep into the earth to see for herself the conditions under which lizards toiled. Nevertheless, despite her concern she concluded her essay with an optimistic vision of Birmingham's destiny. After four decades of laissez-faire, she said, a group of new leaders had fastened their eyes on a different future. Through their influence, the spirit of Aphrodite, goddess of love and beauty, might one day be joined with that of Hephaestus, god of smoke and fire.[48]

V

Armes's faith in Birmingham's futue was based partly on recent developments at TCI, which had fared much better than Sloss-Sheffield in the *Survey* articles. Although continuing to fight unionization, employ convict labor, and take harsh action against strikers, TCI was diverging from Sloss-Sheffield in significant ways. Late in 1907, a new president, George Gordon Crawford, was put in charge of TCI by U.S. Steel's high command. Though southern-born, he was committed to a program of social action that differed radically from Maben's hardbitten approach. Armes's regard for Crawford as "a man of great constructive power" and "marked personal force" had much to do with her belief that Aphrodite and Hephaestus might eventually join forces in Jones Valley.[49]

Like Sloss-Sheffield's indifference to employee welfare, TCI's new attitude reflected hard-nosed business motives. Steel production was a high-volume operation requiring steady, uninterrupted output that was hard to achieve with the ill-disciplined work forces characteristic of Jones Valley. By adopting steel-making, TCI had also entered an arena in which skills were far more neces-sary than was true in the blast furnace operations to which Sloss-Sheffield remained committed. Speaking of the operatives who manned the Bessemer and open-hearth installations of the Pittsburgh District, John Fitch called them "craftsmen in heat," stressing the extent to which their work involved "judgment rather than physical labor," as for example in deciding when ingots should be drawn from soaking pits to be rolled into blooms. Paradoxically, the skill required for many tasks in a heavily mechanized industry with a huge output of standardized goods was greater than that demanded of furnace workers in a much less heavily mechanized sector of ferrous metals produc-tion that made pig iron in relatively small batches for foundries with spe-cialized needs.

Even in the most basic operations, Fitch pointed out, production standards in the steel industry were difficult to attain, particularly in plate rolling, where workers had to use micrometers to determine proper degrees of thickness. From then on, every step up the line became more exacting. Structural and rail mills were in a class by themselves; "their product is more nearly finished than are rods, bars or plates, because it does not require any further manufac-turing process," Fitch reported. "Moreover, the shapes to be rolled are more intricate and the process is more difficult. It is a simple thing as far as the construction of a mill is concerned, to roll a round bar or rod, and compara-tively so to roll an angle, but it is evident that a more difficult technique is involved in rolling a bar for a T-rail or an I-beam so that it shall be thinner in the middle than at the edges."

Writing about conditions prevailing in the steel industry in the early twen-tieth century, David Brody later underscored the same realities. "Even where mechanization went furthest," he said, "gaps remained." Various types of operatives, including "oval and diamond roll hands . . . straighteners in rail mills, pit cranemen in blooming mills, chargers in open-hearth furnaces, and top fillers in blast furnaces," still possessed skills that were in short supply. Of these, Sloss-Sheffield had to worry about only the last, but TCI was dependent on all. For this reason, U.S. Steel and other steelmaking firms tried to foster loyalty among such men by keeping them in continuous service and providing jobs for them even in slack times. [50]

In embracing the gospel of steel, local executives including Baxter, Bowron, and Shook had gotten more than they bargained for, not realizing that making this glamour alloy might require something beyond conquering technical problems involving silica, sulfur, and phosphorus. In particular, these men had not anticipated potential conflict between the requirements of steelmaking and the consequences of the regional low-wage labor system. This lack of foresight helps explain why they encountered unexpected problems soon after 1899, resulting in their dismissal and the beginning of a new order under Bacon. Well before the 1907 takeover, TCI had been obliged to take a new approach to labor relations, while Sloss-Sheffield and Woodward clung to more traditional southern ways.

VI

TCI's new approach had deep historical roots. Beginning with Samuel Slater in the late eighteenth century, some American industrialists had concluded that they could get more work from their employees by providing decent housing, schools, hospitals, and other social welfare services than by using repressive policies. After the strike-plagued 1870s, a few corporations including National Cash Register, Procter and Gamble, and the Pullman Palace Car Company began practicing what came to be known as "welfare capitalism." Such programs typically featured profit sharing, model tenements and planned communities, libraries for workers, and other innovations aimed at replacing conflict with cooperation. Benevolent paternalism was expected to reduce labor turnover. It was also aimed at deterring unionization, which welfare capitalists abhorred. [51]

TCI took its first steps toward welfare capitalism under Bacon. Grasping the larger implications of the company's shift into basic steelmaking, he recognized that the scarcity of skilled labor in Alabama, coupled with excessive employee turnover and absenteeism, reduced TCI's chances of competing with northern and midwestern mills. One way of coping with this was to import immigrants in the hope that they might prove more competent and efficient than unmotivated and badly educated blacks. In the early years of the century, TCI recruited 5,000 steelworkers and coal miners in Italy alone.

Besides breaking the color line, TCI also started hiring American-born whites, who formed more than 37 percent of its work force by 1913. By that time, blacks comprised only 55 percent. Undertaking these moves, which clashed starkly with traditional southern customs, indicated the depth of the labor problems caused by TCI's switch to steelmaking. Despite the change, however, performance levels remained poor. Immigrants reported for work about twenty days per month, blacks not quite seventeen. For operatives as a whole the average was just over eighteen, indicating that TCI's native whites were not well disciplined either. All told, TCI had to employ 18,567 operatives in order to get 13,622 days' worth of work. A company official characterized its employees as "shiftless, thriftless, sloppy and dirty." Things had not changed much in Birmingham since Sloss had given similar testimony to senate investigators in the 1880s.[52]

In attempting to deal with such intractable difficulties, Bacon became the first manager in the district to try to cope with them by embracing welfare capitalism. Less inclined than his predecessors to see workers as mere "mine and furnace fodder," he set out "to increase the loyalty and efficiency of the employees; to add to their comfort and decrease the dangers to which they were subject; and to adopt the methods that had stood the test at those mills, furnaces, and mines that were recognized as being the best."[53]

Bacon was particularly eager to improve housing conditions in TCI's mining camps. He imported technical experts from Minnesota to supervise construction of new dwellings. Many of these persons remained in the district after he was gone and played important roles in TCI's later employee welfare programs. Conceiving his program partly as a weapon against unions, Bacon used improved housing to accommodate strikebreakers during the 1904 coal walkout.

Believing that improving the health of TCI's workers might also help combat absenteeism, Bacon brought physicians from Minnesota to staff emergency treatment facilities at the company's Fossil and Ishkooda mines and erected a small hospital at Muscoda. He also contracted to hire local doctors to provide supplementary medical services, based on the number of workers at a given furnace or mine. Because such physicians bore the cost of their expenses, they had an obvious interest in skimping to maximize income. Even so, the new system was far better than the previous one of neglect.[54]

Soon after the Gates syndicate acquired TCI, Bacon left Alabama, but his program was continued under John A. Topping, a New York financier who became chairman of TCI's executive committee. Frank H. Crockard, an engineering graduate of Lehigh and Michigan College of Mines who became vice president and general manager, also helped spearhead the campaign. For a time, these men ran TCI and Republic as joint ventures. As had been true under Bacon, no-nonsense business motives continued to prevail. "The main thing we're driving for right now in the South is to get good men, skilled labor," Topping stated. "To do that . . . we have to present such advantages as will attract the higher order of workingmen. That means we have got to regard the aesthetic and the human side a trifle more. It's business to be human."

Accelerating Bacon's housing program, Topping and Crockard built schools and better quarters for employees who lived near the Ensley steel mills. Until recently a sparsely populated pine barren, Ensley grew rapidly from a village of less than 600 people in 1899 to a town of approximately 20,000 in 1910 as workers flooded into the area. By 1908, dusty, unpaved streets that had been lined with primitive storefronts and rude hovels at the turn of the century now featured brick schoolhouses, a library funded by Andrew Carnegie, a post office, a city hall and fire station, and neat employee residences, all served by modern water and sewer systems. In an effort to retain immigrant strike-breakers who were brought from Ellis Island in 1904, TCI also built improved housing in other company towns such as Wylam, a coal-mining community on Birmingham's western fringe. Many of the newcomers came from eastern Europe. To meet their religious and social needs, the company built a church, dedicated to St. Stanislaus, in 1906. Because its priest, Father Paul Kilch, knew eight languages, TCI hired him as a translator. [55]

The program started by Bacon, Crockard, and Topping reached its climax under managers appointed by U.S. Steel after the 1907 takeover. Under Elbert Gary's leadership, that giant firm had become the nation's foremost exponent of welfare capitalism. To counter unionization, Gary believed, "we must make it certain that the men in our employ are treated as well as, if not a little better than, those who are working for people who deal and contract with unions." Assisted by George Perkins, who had instituted welfare programs at International Harvester, Gary started a profit-sharing plan permitting U.S. Steel's workers to buy stipulated amounts of stock in proportion to their earnings. Wage incentives, safety programs, improved medical care, and retirement benefits were also used to raise morale. [56]

After U.S. Steel absorbed TCI in 1907, it used the same approach in Alabama. Welfare policies mandated by Gary were imposed by George Gordon Crawford, who became president of TCI in November 1907. Born in 1869, Crawford had been reared on a Georgia plantation, a background suggesting once again the close connections between southern agricultural and industrial development. Later, as one of the region's most prominent industrialists, he did his best to promote scientific development of its crops and rationalize its use of land. For his career, however, he chose engineering; he matriculated at the newly established Georgia School of Technology when it opened in 1888 and was in its first graduating class. He then studied metallurgy for two years in Germany, where social welfare programs had been developed to combat the spread of Marxism.

Ironically, in view of what happened later, Crawford served briefly as a draftsman at Sloss after returning to America in 1892. Finding Seddon's management crude and inefficient by comparison with what he had seen in Germany, he soon left to take a position with the Carnegie Steel Company in Pittsburgh. Rising rapidly as he mastered various facets of steelmaking, he became superintendent of the Edgar Thomson Works, then the most modern installation in the industry. Later, he established an outstanding managerial reputation with the National Tube Company, which became one of U.S.

Steel's three largest components in 1901. Gary's high regard for him and his work resulted in his appointment to a succession of powerful committees and, in time, to the presidency of TCI.[57]

With Gary's backing, Crawford established a multifaceted program of welfare capitalism at TCI that attracted widespread attention both within and outside the South. At the time he took command, the labor turnover rate was still 400 percent despite the programs instituted by Bacon, Topping, and Crockard. As Marlene H. Rikard has indicated, this reflected both the prevailing state of public health in Alabama and "the tendency of many of the men to work only long enough to get cash for their immediate needs."[58]

Retaining Crockard and other managers whom he had inherited from previous administrations, Crawford hired professional medical and social workers to administer a growing network of clinics, schools, and recreational facilities. In 1908, he appointed a prominent black banker and educational reformer, W. R. Pettiford, to develop an instructional program in housekeeping, sewing, and other domestic skills at TCI's mining camps. The following year, Crawford built segregated bathhouses, known as drys, for black and white miners, providing hot showers and lockers for employees who had previously stayed home on rainy days rather than work in wet clothing. After starting relatively modest educational programs, he hired a social worker, Marian Whidden, to create an entire school system, complete with kindergartens, home visitation, supervised playgrounds, and company-sponsored troops of Boy Scouts and Campfire Girls. A lavish pageant, *Wenonah*, that Whidden organized in one of the mining camps in 1916, leading to further such festivals each year thereafter, was typical of her contributions to enriching the cultural life of TCI's employees. Whidden was in turn a precursor of other female professionals who came to Birmingham to implement Crawford's programs from offices in the Brown-Marx building. She was followed by Winifred Collins, a social worker from Chicago who stayed with TCI for twenty-four years; Sue Berta Coleman, a black woman who graduated from Fisk University and worked tirelessly to improve the quality of life in the mining camp of Muscoda, particularly among its children; and Mary Dolliver, a South Dakotan with a degree from Morningside University in Iowa who, after being prevented by illness from going to Paris to study voice, came to TCI to teach music, organize study groups, provide counseling, and improve morale among employees in many other ways.[59]

Much of Crawford's long-term effort at TCI went into building model industrial communities, starting in 1909. Harking back to Robert Owen's New Lanark, towns of this sort had been created in Europe and America by Alfred Krupp, George Pullman, Sir Titus Salt and other industrialists. In Alabama, Anniston had been formed in the same mold, but Pullman, Illinois, was the best known example of the type in the United States. Established in 1880 by the railroad-car magnate after whom it was named, it was located on the south side of Chicago and included a library, a theater, an interdenominational church, a public park, and more than 1,400 dwellings built to house workers at affordable rents. Its chief drawback, searchingly identified by such social critics as Richard Ely even before the Pullman strike of 1894 discredited it throughout the country, was the heavy-handed paternalism with which it was run. This was typical of welfare capitalism and would be a prominent feature of the practices that Crawford instituted at TCI.[60]

As Pullman's visions collapsed at the turn of the century, no corporation did more than U.S. Steel to keep the idea of model towns alive. Gary's interest in their potential was stimulated when he acquired a rolling mill that had already set up a community of this type at Vandergrift, Pennsylvania, in the 1890s. When he built Gary, Indiana, begun in 1906 amid sand ridges and swamps at the southern end of Lake Michigan, it was immediately hailed by progressive reformers as a showcase of welfare capitalism.[61]

Such communities as Vandergrift and Gary gave Crawford models to follow. The first model town that he built in Alabama was located on Birmingham's western outskirts. Originally named Corey, in honor of William E. Corey, who was president of U.S. Steel when the community was founded in 1906, it was later renamed Fairfield, after the Connecticut home place of Corey's successor, James A. Farrell. Fairfield offered a variety of attractive, up-to-date houses for different classes of employees. These dwellings were situated on wide, tree-lined streets that avoided the monotonous gridiron pattern characteristic of earlier prototypes. Single-family cottages with indoor plumbing, hot and cold running water, and central heating were provided to employees for reasonable rents or for purchase on easy terms. Other amenities included a municipal park, a YMCA, a public library, and churches of several denominations. A bank and other business structures were ornamented with elaborate brickwork and imaginative terra-cotta motifs, and a complex of civic structures in classical style was built around a tree-lined mall, looking much like a college campus. Other attractions included playgrounds, an herb garden, and a plaza where ex-president Theodore Roosevelt praised TCI's initiative in 1911 in a speech that he made while visiting the Magic City as a delegate to a National Child Labor Convention. Progress in building the community was set back by a recession that hit the steel industry in 1910, but Crawford loaned one of its developers $30,000 from his own pocket to see the project through the emergency.[62]

TCI ultimately created three other model communities to demonstrate that mining camps could be attractive. The first was a village for workers at the company's No. 12 mine, about seven miles northwest of Pratt City. Here TCI built Docena, named for the Spanish equivalent of the number twelve. Conforming to the consistent use of Iberian terminology, it was laid out around a public square called the Prado. Neat four-room bungalows, locally known as "square tops" because of a distinctive configuration featuring a hipped roof and central chimney, were provided for workers and their families. Supervisory personnel were quartered in larger residences. A brick commissary, known as the Mercado, faced the central square; other public buildings included Baptist and Methodist churches and a Masonic lodge. Black miners lived in a separate section on the north side of town, and segregated schools were provided for black and white children.

Despite having to use outdoor privies, Docena's residents enjoyed such amenities as kitchen sinks with running water, effective garbage disposal, and sanitary conditions far superior to those prevailing in the district's other mining camps. A public dispensary provided basic medical care, and the educational program was also much better than that available in most of the state's public schools. Miners and their families had varied social and recreational opportunities, such as dances, pageants, study groups, membership in fraternal organizations, and free movies on Wednesday nights.

Granting workers the privilege of living in Docena gave TCI an important means of social control. Mine superintendents could force residents whose behavior did not measure up to company standards to leave after a single day's notice. Company-appointed welfare workers and medical practitioners swarmed throughout the community, conducting home visitations, making sure that children brushed their teeth, seeing that they changed their underwear, and enforcing the use of communal bathhouses. Such paternalism was the price that workers had to pay for the improved conditions they were allowed to enjoy.[63]

TCI also built two other model villages at Edgewater, a few miles west of Ensley, after a new mine was opened there in 1911, and at Bayview, a short distance west of Docena, where the company created a man-made lake by damming Village Creek. Before Docena was built, Edgewater was regarded as "the best and the most up-to-date mining camp in the District," praised by *Survey* as "revolutionary" because of the roominess of its houses and the sanitary nature of its waterproof privies, which were built with "hinged rear doors to facilitate cleaning and prevent exposure to flies." Bayview was the last of TCI's town-building ventures. Partly because of the increased experience that TCI had gained in planning new communities by the time it was created, and also because of the natural advantages of its site, it was especially attractive. The street layout was tastefully varied, the road to the mines was lined with shrubs and flower beds, and the lakeside location provided many scenic vistas. According to one admiring observer, "not many resorts of the wealthy have a more delightful setting."[64]

The most impressive result of welfare capitalism at TCI was the dramatically improved medical care that it brought to a region long known for appalling deficiencies in public health. After providing subsidized nursing care and taking steps to improve sanitation in facilities and communities maintained by the company, Crawford set up a centralized health program in 1913 and appointed Lloyd Noland, a young physician who had gained distinction in nine years of medical work in the Panama Canal Zone under Gen. William C. Gorgas, to direct it. With a generous budget starting at $750,000 per year, Noland launched a sweeping inspection program that greatly reduced the incidence of malaria, hookworm, and other diseases by reducing water pollution, enforcing effective sewage disposal, killing insects and other pests, controlling the quality of food sold in company stores, and systematically disposing of garbage and industrial waste.

Abandoning the contract system that Bacon had established, Crawford hired a large staff of physicians and set up modern company-owned dispensaries, emergency treatment facilities, and hospitals. Instituting regular physical examinations among employees, he started a payroll deduction plan under which they and their families received medical care at little expense. In 1915 he added a dental department, beginning with two traveling practitioners and then establishing clinics throughout the district. He also began an industrial safety program to reduce work-related accidents and provided ambulances and first-aid services for emergencies.

In 1919, after a protracted campaign for funding from U.S. Steel, Noland fulfilled a cherished dream by opening a seventy-eight-bed hospital on Flint Ridge, overlooking the Fairfield steel plant. Modern in every respect, the four-story building, with six wings "so arranged as to furnish ample air and light to every room," offered surgical facilities, a bacteriological laboratory, an obstetrical ward, and a radiology department. It also maintained a nursing school. The hospital became Noland's "command post" throughout the remainder of his career, which lasted for thirty more years until he died in 1949. Admitting nearly 4,000 patients in its first year of operations, it became the first institution in the state to use radiology in treating cancer victims, to provide electrical therapy, and to hire a full-time specialist in anesthesiology. Gary paid it the ultimate accolade at one point by checking in as a patient.[65]

VII

In Alabama, as elsewhere, welfare capitalism proved to be an effective defensive strategy for corporations in the Progressive Era. Large American corporations were by no means hostile toward all the political, economic, and social reforms that took place at this time and were in fact instrumental in initiating and implementing many of them. Some industrialists regarded economic regulation as a welcome way to combat cutthroat competition, prevent unfair business practices, and impose discipline upon a disorderly social and political system. On the other hand, corporate attitudes toward regulation and their reform goals were selective, sympathetic when such measures advanced specific business objectives but hostile when vested interests were threatened. Although the activities of industrialists who founded model communities or improved public health by establishing clinics and hospitals were consistent with the Progressive spirit, Stuart D. Brandes has made an apt assessment in stating that "welfare capitalism was affected by the progressive milieu . . . while managing to remain apart from it." TCI's social welfare program fits this characterization.[66]

By manifesting industrial statesmanship, welfare capitalism in Alabama produced results that contrasted greatly with the neglect shown by firms that followed more traditional policies. Sloss-Sheffield was one of these companies. While developments at TCI became increasingly congruent with progressivism, Sloss-Sheffield's aging, ultraconservative leadership marched out of step with it. Maben and other corporate officials not only refrained from embracing welfare capitalism but actively opposed it. James W. McQueen, one of Maben's lieutenants, spoke scathingly about the "new-fangled ideas" that TCI had implemented.[67] Such sentiments were probably much more typical of southern business attitudes than those of Crawford and his associates, who drew heavily on the support of northern owners to whom they reported. On the other hand, Sloss-Sheffield was demonstrably less successful than TCI in warding off reforms that were contrary to its interests. Even in Alabama, welfare capitalism paid off.

The fact that Crawford's approach could yield tangible benefits beyond achieving a healthier and more efficient work force became apparent in its resourceful response to municipal changes that were pushed through the state legislature with Governor Comer's support. This steel-eyed, square-jawed industrialist adroitly capitalized on the reformism of the period by challenging the vested interests of railroads and posing as a champion of change even as he continued to exploit child labor in his Avondale mills, baited blacks with racist rhetoric, and used troops to crush the 1908 coal strike. His split personality as a politician was in no case better revealed than by his alliance with local forces that wished to impose taxation on the district's manufacturing concerns, of which he was one of the owners. Yielding to exponents of the Greater Birmingham movement, he agreed to most features of an annexation bill that brought seven outlying communities within the city limits, giving it an area of more than forty-eight square miles and boosting its population to 132,685.[68]

TCI fared far better from this action than Sloss-Sheffield, whose City and North Birmingham furnaces and their appurtenant structures came under municipal jurisdiction for the first time. As a result, company property valued at more than $664,000 became subject to city taxes. Although some of TCI's properties also fell within Birmingham's new boundaries, its iron and steel plants, valued at $3,778,239, remained outside the lines. In lobbying for this exclusion, Crawford successfully threatened municipal councilmen that U.S. Steel's absentee owners would refrain from further investment in the district if they believed that their Alabama subsidiary was being unfairly treated. Sloss-Sheffield's less powerful owners could not muster such convincing arguments. In addition to assuming new tax burdens, Maben and his firm also became liable to municipal regulation, while most of TCI's installations were spared.[69]

While Sloss-Sheffield waited to see what further results the passage of the annexation bill would bring, another series of events dramatized the growing difference between the contrasting responses that it and TCI were making to changing events in an increasingly reform-charged atmosphere. On 8 April

1911, a catastrophe suddenly crystallized public indignation against the conditions endured by all the state's miners, both convict and free. In an explosion at Pratt Consolidated Coal Company's Banner Mine in northwest Jefferson County, 128 men, mostly black convicts, died. Capitalizing on the outcry, union leaders and social reformers won passage by the Alabama legislature of a Mine Safety Law that increased the number of state inspectors and set forth a list of provisions aimed at reducing the chance of accidents. Even more threatening to industrial interests was a determined effort by organized labor and Progressive reformers to outlaw the use of convicts in company-owned mines. Strongly opposed by Sloss-Sheffield and other firms, it was beaten back on the final day of the 1911 legislative session. Soon afterward, the state began operating the Banner Mine, assuming direct control of its convict workers.[70]

Seeking to improve TCI's public image by cooperating with labor, despite continuing to resist unionization, Crawford announced in December 1911 that his company would abandon convict leasing and operate its mines exclusively with free workers. A dramatic demonstration of the new policy took place on New Year's Day 1912, when 300 inmates who had previously worked for TCI, handcuffed and chained together in pairs, were transferred to the state-operated Banner Mine in railroad cars while heavily armed guards and sixteen bloodhounds stood by to track down potential escapees. Thus, while TCI cleansed itself from association with the system, the inmates themselves continued to suffer. Nevertheless—and even if it is true that TCI had taken this step "grudgingly and without any outcry of moral indignation against the system," as Robert D. Ward and William W. Rogers later stated—the move was a striking departure from earlier company policy and a fresh affirmation of TCI's commitment to welfare capitalism. Soon Docena was constructed on the site where a prison stockade had recently stood, further emphasizing the contrast between the old order and the new dispensation.[71]

Sloss-Sheffield remained as committed as ever to its traditional southern ways. Continuing to use state convicts, it negotiated a new three-year contract for their employment at Flat Top. In an article in the *Age-Herald* that appeared soon after the agreement went into effect, Maben strongly defended Sloss-Sheffield's policy, using the familiar argument that mining taught convicts a trade that they could use after their sentences expired. This was far superior, he said, to putting them in a penitentiary or making public spectacles of them in repairing roads. Citing the favorable impressions of a visiting northern clergyman who had recently toured Flat Top, Maben denied that conditions there were objectionable. Indeed, he declared, the inmates often finished their tasks by noon and had the rest of the day to themselves. "They have spacious halls to frolic in," he said in apparent seriousness, "and . . . are well satisfied as a general thing."[72]

Soon after Maben's views became public, Gov. Emmet O'Neal came to Birmingham with the head of the state convict board, James G. Oakley, to discuss the new contract, which was being publicly criticized because it specified a lower rate of compensation to the state than the one stipulated by the previous agreement. It was also suspect because it permitted Sloss-Sheffield to pay a sliding figure to the state for inmate-mined coal, depending on its slate content. Newspaper coverage of the visit and the negotiations that took place between Sloss-Sheffield and state officials indicated clearly that the matters at issue were merely financial, having nothing to do with convict working conditions. A compromise was quickly reached, the modified agreement went into effect, and the dispute was laid to rest. As in the past, officials in Montgomery were much more concerned about the monetary rewards of the leasing system than about inmate welfare, and Sloss-Sheffield's vital interests were left undisturbed. [73]

Not so easily terminated, however, because it involved the health and well-being of taxpaying voters, was a bout of legal wrangling between Sloss-Sheffield and municipal officials over smoke emissions from the company's furnaces. Reformers in the city council were also determined to eliminate other ways in which the company's operations fouled the local environment, such as its failure to remove slag piles that prevented streets from being extended through areas near the furnaces. While TCI was immune from remedial action because its facilities lay mostly outside the redrawn city boundaries, Sloss-Sheffield and other less fortunate firms faced burdensome new regulations.

Late in 1912, a citizen's group brought suit against Sloss-Sheffield to remove its slag pile at the City Furnaces and other obstructions that prevented the extension or use of nearby streets. Protesting that "the Sloss company has proved to be one of the best advertisements that Birmingham has ever had," Maben strongly opposed this action and predicted that his firm would prevail in court. Meanwhile, the company came under increasing public pressure to abate its smoke emissions. According to the *News*, the City Furnaces, located only eight blocks from the heart of the business district, were "a dirty splotch upon the fair face of Birmingham," and should be moved "to a more isolated site." Maben professed willingness to deal with the problem if he could be shown how to do so without hurting the firm's legitimate interests, but Sloss-Sheffield took a hard line. Furnace superintendent John Shannon pointedly reminded the city how much it stood to lose if demands for reform forced a plant with a $2 million payroll to close.

In December 1912, city commissioner James Weatherly won enactment of an ordinance requiring local industries to install equipment to cut smoke emissions and limiting the number of hours in which smoke-producing operations could be carried out. Two inspectors were hired to enforce the regulations, which specified that responsible officials would be subject to arrest and fines for noncompliance. Within a short time, one of Sloss-Sheffield's vice presidents was arrested under the ordinance. [74]

Weatherly's crusade for cleaner air aroused a chorus of opposition from local businessmen, who won the support of the *Age-Herald* and the *Labor Advocate* in efforts to rescind the ordinance. Fifty firms, including Sloss-

Sheffield, issued a petition protesting that it would wreck the local economy and discourage new enterprises from coming to the city. The fact that the signatories had payrolls aggregating $2,705,000 was pointedly mentioned. Leading industrialists warned that they were not issuing idle threats. W. N. Crellin, owner and operator of the Birmingham Boiler Works, declared that he would move his plant to Chattanooga. "The people who want pure air, free from smoke and other ingredients which go to make up a city, should move to the country and take up farming," protested another industrialist, W. H. Kettig, who recalled that residents of the district had longed to see smoke during the depression of the 1890s. Anybody seeking to ban it from the air at that time, he said, would have been deemed fit for the state insane asylum at Tuscaloosa. "Smoke is not harmful nor injurious to health," stated a prominent Alabama physician, Cunningham Wilson. "No one was ever harmed by breathing pure carbon such as is contained in coal dust and coal smoke. The miners in the coal mines of the Birmingham District are some of the longest lived people on earth. It's because of the coal dust and smoke they breathe."[75]

The counterattack was successful. Within three months, the new ordinance was a dead letter, being watered down with so many amendments, drafted by manufacturers and adopted by a majority of the commissioners, that smoke inspectors resigned rather than try to enforce it. In 1915, at the behest of Sloss-Sheffield and other manufacturers, the state government removed the power of cities in Birmingham's class to regulate smoke emissions, on the grounds that such action would drive away corporations.[76]

Meanwhile, grudgingly, Sloss-Sheffield took steps to deal with demands for removal of its slag pile and other obstructions to local traffic. Settling the issue required months of legal wrangling, during which the company brought its chief corporate attorney, Adrian Larkin, from New York to Birmingham to assist a prominent local law firm, Tillman and Morrow, in defending its interests. Eventually, the parties reached a compromise. Though arguing that it would be forced to stop operations at the City Furnaces if all the demands facing it were upheld, Sloss-Sheffield consented to remove the slag pile and other traffic obstructions on two streets if municipal officials agreed to the continued blockage of others. Implementation of the plan required the company to move some existing railroad tracks and relocate several buildings. As part of the settlement, a concrete viaduct was built to carry First Avenue, North, across the area between Twenty-fifth and Twenty-second streets. To quiet public indignation about smoke emissions, Sloss-Sheffield also agreed to shut down the 288 beehive coke ovens at its City Furnaces. The outcome satisfied no one but saved the city an economic asset while providing some relief.[77]

Like so much else that had happened in the past few years, the smoke emission episode highlighted the ever-growing contrast between Sloss-Sheffield and TCI. By threatening that U.S. Steel would withhold future investment in the district, Crawford had adroitly kept most of his company's property outside the new city limits and consequently free from taxation and municipal regulation. In addition, he had defused hostility among free workers by abandoning convict labor and created a local image far superior to that of Sloss-Sheffield by pushing through the social welfare programs conducted by such charismatic leaders as Noland. As the Progressive Era reached its climax in the last few years before the outbreak of the First World War, the pragmatic policies of Crawford and Gary, combining hardheaded business sense with humanitarianism, were more in tune with public sentiment than were those of Maben and his fellow directors, which were still anchored in the economic, political, and social attitudes of a bygone age.

Yet another important difference now existed between the local managements of the two firms. By 1913, Maben was not even living in Birmingham. He had returned to the more congenial atmosphere of New York City and placed the conduct of day-to-day operations in the hands of subordinates. It was the first time since the birth of the Sloss Furnace Company that the president of the firm had not lived in the Magic City, emphasizing how dramatically things had changed since the days of such locally popular managers as Seddon and Hopkins and epitomizing the degree to which absentee ownership had prevailed over Birmingham's interests in recent years. Increasingly, decisions were made on Wall Street by superannuated businessmen whose periodic visits aboard the Alabama could not keep them abreast of shifting conditions in the district. Meanwhile, amid the dirt and smoke of Jones Valley, a divided group of subordinates ran the show. This state of affairs would soon have unfortunate consequences.

The End of an Era **12**

The dynasty died with the man. The collapse of the control that Virginians and their northern allies had exercised over Sloss Furnaces since 1887 came only by degrees after Bryan passed away, but the result was no less certain as the grip of his lieutenants weakened under the stress of changing times. The aging process was partly responsible for the growing malaise; so, too, were the natural desires of a septuagenarian father, Maben, to spend his remaining years in comfort and throw his presidential mantle over the shoulders of a trusted son and heir. More fundamental problems, however, resulted from the outbreak in 1914 of the worst war that the world had seen in almost a century. After Archduke Francis Ferdinand was shot to death in Sarajevo, things would never be the same.

Changing technology helped spin the web in which Sloss-Sheffield became enmeshed. As the United States was drawn deeper and deeper into the international maelstrom because of its expanding role in supplying financial credit and war materiel to the Allied powers, the company came under increasing pressure to abandon its antiquated beehive coke ovens and replace them with more costly equipment capable of reclaiming chemical by-products that were needed for the manufacture of explosives and dyes. Facing loss of control over the enterprise because of the heavy outlays of capital that modernization would require, its leaders temporized until the pressure became too great to resist. Early in 1917, as the United States prepared for war, power finally slipped from their grasp.

I

Sloss-Sheffield's reluctance to invest fresh capital in modernization and expansion helped set it apart from TCI after 1907. Despite its success in developing markets for foundry iron and earning steady profits in the early years of the new century, its resources did not begin to compare with those now available to its much larger rival. During the first year of Crawford's administration alone, U.S. Steel spent nearly $3.5 million updating and expanding TCI's facilities. Work was begun to replace three of the original open-hearth furnaces at Ensley, build a new chemical laboratory, install a 2,000-kilowatt electrical generator, and provide a new coal and coke laboratory for the Pratt mines. In addition, the Docena Coal Mine and a new limestone quarry were opened. Hailing these steps, the *Manufacturers' Record* declared that the takeover had advanced Birmingham's fortunes by fifteen to twenty years. By 1913, when new mines had been opened at Edgewater and Bayview, two more open-hearth furnaces had been put into operation at Ensley, and five other blast furnaces had been rebuilt, the benefits of the merger seemed all the more evident.[1]

One of the most significant steps taken by TCI was a determined effort to deal with the water shortage that had plagued industrial operations in Jones Valley ever since Birmingham's founding. After commissioning a study by Morris Knowles, a civil engineer who had been a hydrological consultant for Boston, New York, and Philadelphia, Crawford built a dam across Village Creek at a point where it had cut a deep gorge in the hills near the mouth of one of its tributaries. Construction of the massive structure, five hundred feet wide and ninety feet high, was described locally as "the greatest engineering project in the history of the state." Built of concrete-reinforced boulders, a technique known as "cyclopean masonry construction," it formed a man-made lake covering more than 500 acres, with a capacity of approximately 2.5 billion gallons. Pumped from a station four miles above the dam at the rate of 84 million gallons per day, water was carried by a steel pipeline and concrete-lined tunnel through one and one-half miles of solid rock to a reservoir, holding approximately 15 million gallons, on a hill above the Ensley plant. From there it flowed by gravity to the steel mills below. Water that was not lost through evaporation or otherwise wasted was recycled to Village Creek.[2]

While working steadily to improve the quality of life in TCI's mining camps, Crawford modernized the technology used in the mines themselves. Electrical equipment for undercutting and hauling coal replaced picks, shovels, and mule-drawn tramcars. Ore mining was also updated by standardizing machinery and adopting interchangeable parts. Tramcars that had previously dumped their cargo individually into tipples now deposited ore into twelve-ton skip hoists. After carrying ore to the top of the tipples, the hoists automatically dropped it into gyratory crushers, which deposited it in railroad cars for transport to the blast furnaces. Besides being safer than older methods, the new system was more efficient. Before being fed into the furnaces, ores from different mines were chemically analyzed and blended to assure uniform quality. Slag from the furnaces, previously wasted, was now processed for conversion into Portland cement and other building materials. Phosphorus-rich by-products were used to make fertilizer at an Ensley plant opened in 1914.[3]

Crawford wanted to develop an infrastructure of consumer-oriented industries in his native region so that it might escape the limitations inherent in merely supplying raw materials to other parts of the country. Acquiring 1,700 acres of land for TCI between Wylam and Dolomite in the Opossum Valley, he persuaded one of U.S. Steel's subsidiaries, the American Steel and Wire Company, to erect a plant to make wire and rods on the tract. At the new mill, which opened in 1909, steel from TCI's Ensley works was converted into barbed wire, nails, and wire for telegraph and telephone lines. Plans to house the 3,000 workers employed at the new facility provided the impetus for developing Corey, the model community that later became known as Fairfield. Crawford was disappointed that U.S. Steel did not put the plant under his direct supervision, but it did ultimately become a division of TCI, known as the Fairfield Wire Works.[4]

II

Not until the 1920s, when fresh capital was pumped into the company, would Sloss-Sheffield undertake a modernization program in any way comparable to the one that took place at TCI after 1907. While Crawford carried out his sweeping changes, Maben pursued an increasingly conservative course by conducting routine repairs and installing only a modest amount of new equipment including additional boilers and stoves, an $87,000 system of rope haulage at the Ivy coal mines, and a new ore washer completed at a cost of $124,000 at Russellville. After flooding at the Sloss ore mines caused heavy losses and shut down production for several months in 1910–11, two centrifugal electric pumps were installed to forestall future disasters. Nothing, however, was done to modernize extraction.

Little was done, either, to update the company's furnaces, which were showing signs of age. One of them, the Hattie Ensley stack at Sheffield, compiled a phenomenal record by staying in blast for almost eight years on the same lining before being shut down for refitting in 1911. But the company suffered a major disaster in January 1912 when the lining of City Furnace No. 1 collapsed, causing the stack to career sideways. As preparations were made to rebuild it, Maben assured his fellow directors that the project would entail "only a modest expenditure," estimated by the *Age-Herald* at "between $50,000 and $75,000." Typically, nothing new or different was planned in the reconstruction.[5]

Sloss-Sheffield's conservatism was underscored by the contrast between what it and its arch-rival, the Woodward Iron Company, were doing. Early in 1914, Woodward blew in a new blast furnace, replacing the first stack that had been lighted in 1883. Ninety feet high, the installation had a capacity of between 400 and 450 tons per day, was designed for rapid loading, and had many up-to-date features that were lacking in Sloss-Sheffield's existing stacks, including a high-pressure turbo-blower built by General Electric. Maben's firm was doing nothing comparable at the time.[6]

Maben's reluctance to invest fresh capital in major improvements after Bryan's death is understandable. Since its creation in 1899, Sloss-Sheffield had conducted an extensive program of modernization designed to bring its methods and equipment up to date with standards prevailing in the largely traditional foundry trade toward which it was oriented. The impulse to digest the fruits of the successful transition that Sloss-Sheffield had made was stronger than the urge to innovate as the firm adjusted to the passing of its chief stockholder, Bryan.

Hard times affecting the American economy intensified Maben's desire to retrench. After panic swept Wall Street in 1907, the price of pig iron plummeted by nearly $4 per ton, and Sloss-Sheffield was forced to shut down several of its furnaces for the first time in many years. In Maben's words, 1908 opened with a "gloomy outlook, following closely upon the banking crisis of the preceding October, and serious fears were entertained of a demoralization in the iron and steel trade that might prove disastrous to many." Nationally, business remained in a slump until June. The return to normal levels of production at Sloss-Sheffield was correspondingly slow, and net profit after payment of dividends fell precipitously from $1,689,662.76 in fiscal 1907 to $69,960.00 in 1908. Attempting to console the stockholders, Maben pointed out that the firm was "entirely free from floating debt," but such reassurances could not hide his low spirits.[7]

Despite a nationwide recovery that began in early 1909, demand for pig iron lagged behind other economic indicators. Buoyed by prices that rose $3 per ton by November, Sloss-Sheffield restored idle furnaces to production and was doing a brisk business as the year ended. At this point, however, flooding shut down two slopes at the Sloss ore mines, forcing the company to turn to outside sources of supply and incur higher costs. The opening months of 1910 brought another national economic downturn. Despite deteriorating markets, Maben and his advisors made a serious mistake by maintaining high levels of production for several months. When demand failed to pick up, 74,000 tons of unsold pig iron remained on the yards by the end of November, and the company assumed its first floating debts in six years. Continuing to believe that prosperity would soon return, the directors erred again by declaring dividends totaling $844,000 on both preferred and common stock, creating a shrinkage of $172,477.45 in the surplus fund. Without Bryan's hand at the tiller, the enterprise was floundering.[8]

Markets remained depressed throughout 1911, forcing Sloss-Sheffield to sell, at low prices, much of the high-cost iron that it had accumulated during the previous year. As earnings continued to sag, the directors stopped paying dividends on common stock but continued to issue the customary 7 percent dividends on preferred. The resulting outlay of $469,000 caused a net deficit of $64,351.25, further shrinking surplus. Meanwhile, the firm continued to pile up floating debts, which totaled at least $500,000 by the end of the year, and its inventory of unsold pig iron climbed to 88,000 tons.[9]

III

Compounding the company's deepening problems, Maben precipitated a crisis at this point by deciding to return to New York, though still remaining president. Because of a related decision that smacked all too obviously of nepotism, he left behind him a situation guaranteed to create friction between executives who continued to reside in Birmingham. This, added to the other predicaments that the company was encountering, hastened the ultimate collapse of the company's ruling regime.

Maben's decision to return to New York was related in part to his basic managerial outlook. From the beginning of his administration, he had delegated as much control over day-to-day operations as possible to key subordinates. "I'm not a technical man, and I don't pretend to be one," he told Ethel Armes. "All I want to take care of are these cost sheets right here and look after results, and see that results show up. That's what I'm here for. Why should I be a mining engineer or a coal and ore expert? . . . If anything is wrong it shows right here in these reports, and I call up the men in charge and see what's the matter and have it corrected. That, in my opinion, is the only way to run a company."[10]

Maben had moved to Birmingham only reluctantly, to checkmate Hopkins's steelmaking plans. Basically a financier, he had never managed an industrial enterprise. His hands-off attitude was consistent with trends toward separation of ownership and control that had become increasingly prevalent in large American industrial corporations. Gradually, Sloss-Sheffield was becoming a firm in which managers managed while directors directed, but embracing the new style of operation would be difficult for a firm that was deeply accustomed to older ways.

Under Maben's passive managerial approach, power gravitated into the hands of such persons as James W. McQueen, who handled central administration and marketing; Jones G. Moore, who supervised the mines; and John Shannon, who ran the blast furnaces. Of these, McQueen was by far the most important and took charge whenever Maben was away from Birmingham. "Owing to his long term of service, nearly twenty years all told," Armes stated in 1910, "Mr. McQueen has a more intimate knowledge of the company's affairs than possibly any other of its officers." In her words, he was "the company's main influence and strongest force." McQueen's prowess as a marketing expert was legendary throughout the ferrous metals industry. It was said that he had "sold more Southern foundry pig iron than any other man living or dead, and is deservedly credited with being the best pig iron salesman in the United States." If Sloss-Sheffield had one indispensable executive, it was McQueen.[11]

Because he delegated so much responsibility, it was logical for Maben to believe that he could function just as effectively as president by living elsewhere. Had he followed company traditions, he would have yielded that title to someone exercising managerial authority, but he did not. Forthcoming events were to indicate why; because of McQueen's influence, the time was not ripe for Maben to place his mantle on the man he had in mind. Instead, he tried to make that person—his eldest son—an heir apparent. This effort had unfortunate results.

Personal considerations precipitated the timing of Maben's impending move, which was announced in December 1911. He had suffered a recent fall, and his wife, Virginia, was in poor health. Having lived in New York City for many years, Maben understandably preferred its many attractions to the dirt and smoke of Jones Valley. Just before Christmas, he stated publicly that he would return to Manhattan because of his wife's condition.[12]

Prior to returning north, Maben secured the appointment of his scion and namesake, John C. Maben, Jr., as vice president, putting him on a par with McQueen. Born in New York City in 1871, Maben, Jr., had graduated from Columbia University with a B.S. degree in 1897 and studied for two years at Colorado School of Mines in preparation for what a misleading account of his life later characterized as a "brilliant industrial career." Moving to Alabama, he ultimately became president of the Davis Creek Coal and Coke Company. Based in Tuscaloosa County and founded in 1902 to exploit what was known locally as the Jagger seam, its board of directors included many of the same men who dominated Sloss-Sheffield. Through hard work, the young engineer developed it into a profitable concern; the *Age-Herald* called the venture "a mine that he opened against the advice of many experts and . . . made into one of the best propositions in this District." In August 1911, he was unanimously elected to the board of directors of Sloss-Sheffield to fill a vacancy created when Henry O. Seixas committed suicide earlier that year. It is clear from later events that the subsequent elevation of Maben, Jr., to a vice presidency was already in prospect. At the time of his promotion to that post, it was stated that he had been giving "practically all of his time" to Sloss-Sheffield in recent months. It was also evident at the time of his appointment that he was destined for elevation to the company throne. Symbolically, he was given his father's office at the Woodward Building.[13]

These steps affronted McQueen, who, despite having been a vice president since 1902, had never been made a director. Highly popular in the Birmingham area, he had long aspired to assume the top spot in the company whenever Maben stepped down. Now, suddenly, his hopes were jeopardized, sending shock waves throughout the company, alarming employees whose career aspirations were linked to his, and angering his many friends in the local business community. Latent conflicts and resentments traceable to earlier struggles between Birmingham interests and absentee owners during the Seddon and Hopkins administrations were quickly rekindled.

This was not the first time that McQueen's rising star had been threatened. In December 1905, Bryan had resigned the first vice presidency that he had assumed during Rucker's fall from grace in 1902. It was a largely meaningless position, because his power in the company did not derive from his title but from his role as its dominant stockholder. At Bryan's behest, the directors elected M. M. Richey to succeed him at an annual salary of $10,000. Richey had previously been assistant general superintendent of the Southern Railway system, with headquarters in Birmingham. McQueen continued as second vice president, at a salary of $7,500.[14]

Bryan had been trying to solve two problems at once by helping Maben go back to New York as soon as possible and grooming a likely successor as his replacement. Ever since Maben's arrival in 1902, observers in Birmingham had assumed that he would remain in the Magic City only temporarily until a younger man took over. Richey's appointment indicated that Bryan was planning to replace Maben with a person whose background in railroading was similar to that of Haas. At the time of Richey's appointment, the *Age-Herald* reported that he would relieve Maben of "all details of the executive office, and . . . have full charge of operations." His outstanding record indicated that he was worthy of the challenge; as assistant general superintendent of the Southern, he had administered its Atlanta, Birmingham, and Knoxville divisions, the three largest parts of the system. His duties had required him to travel thousands of miles each year in a private railroad car to make on-the-spot decisions of vital importance to the line.[15]

Despite his impressive credentials, Richey failed to derail McQueen's ambitions. In September 1906, upon granting Maben a leave of absence to take a European vacation, the executive committee directed that the company be jointly administered by Richey, McQueen, and Moore while he was away. "Should any differences of opinion arise as to the management of the Company," the committee stipulated, a majority decision of these men would "govern in all cases." Obviously, Richey had not succeeded in taking "full charge of operations." After Maben left for Europe, matters deteriorated. In mid-October, while Maben was still abroad and Frederic Olcott was serving as president pro tempore, Richey notified the directors "that he was prepared to offer his resignation as an official of the Company and to annul the contract which he has with it, provided that he be paid the sum of $10,000." His offer was accepted, and he left at the beginning of November. At the company's next annual meeting, in March 1907, McQueen was the only person elected to a vice presidency, and the distinction between first and second vice president disappeared. Nothing was said in local newspapers about what had happened. A plausible inference is that, during Maben's absence, McQueen had joined Moore in making Richey's position untenable and thereby regained momentum in his drive to succeed Maben as chief executive.[16]

Five years later, Maben's impending return to New York and the appointment of his son as vice president confronted McQueen with another challenge. In an obvious indication that a crisis was brewing, the executive committee sent Goadby to Birmingham in mid-November 1911, "for consultation with the President, as to the organization and conduct of the Company's business." Soon thereafter, McQueen suddenly left for the northeast and was gone for two weeks. Local newspaper reports did not mention his specific destination, but it is not hard to guess that he was bound for Wall Street, conveying remonstrances from business associates in Birmingham.

McQueen succeeded in repairing at least some of the damage that had been done to his position. On December 9, a few days prior to being formally appointed as vice president, Maben, Jr., resigned from the board of directors, putting him more nearly on a par with McQueen, who remained a vice president as before. When the executive committee confirmed the appointment of Maben, Jr., on December 12, it took care to specify the differences between his duties and those of McQueen. Under job descriptions drawn up by Larkin, McQueen was put in charge of marketing and finance, while Maben, Jr., was given "supervision and control of all the physical operations including furnaces, coke ovens, ore and coal mines."[17]

Nevertheless, the relative status of the two men in the corporate pecking order remained ambiguous. When the board's actions were ratified by the stockholders at their annual meeting in March 1912, the resulting slate of officers listed McQueen ahead of Maben, Jr., but refrained from referring to one as first vice president and the other as second. Apparently, they were regarded as equals, but this arrangement would prove hard to maintain as they struggled for power and preference in years to come. Meanwhile, insisting that his annual salary as president be reduced from $25,000 to $15,000, Maben vacated his residence on South Twenty-first Street in Birmingham and moved back to Manhattan to occupy a suite of rooms with his wife at the Waldorf-Astoria. From time to time he returned to the Magic City aboard the Alabama, often in Goadby's company, to inspect company properties and grant interviews to reporters on such controversial subjects as its use of convict labor. His role in local affairs, however, was conducted mostly at arms' length. Having reached his mid-seventies, it was hard for him to leave New York's creature comforts.[18]

IV

Despite the directors' disinclination to invest fresh capital in technological improvements, Maben, Jr., plunged vigorously into his new responsibilities. Minutes of executive committee meetings held in New York City show that he frequently requested substantial outlays for physical improvements. At a meeting on 12 April 1912, Maben presented a letter from his son asking for "about $50,000" to install "compressors for the Sloss and Bessie Mines, hoisting engine for #2 mine, and cylinders at Flat Top Mines etc." This was approved. Over the next several months, Maben, Jr., secured funding for such projects as new ore washers at Russellville, a small electric power plant and pumping station for the Hattie Ensley furnace in Sheffield, fifty new houses for employees, and "a fire proof building to be located on the City Furnace property . . . to be used for storing the records of the Company." All told, such projects cost the firm approximately $300,000. Although this sum was small compared to the funds that U.S. Steel was pouring into TCI, it was still impressive in view of the directors' wish to conserve money. Maben, Jr., also undertook numerous repairs. Late in December, for example, he received an appropriation of $27,633.50 for maintenance on "tipples, track work, ties, frogs, switches, etc." at the Brazil, Brookside, and Blossburg mines. [19]

For the time being, there was no sign of friction, at least partly because financial conditions improved temporarily after Maben's return to New York. As 1912 began, a national economic upswing got underway, ending the downturn that had begun two years earlier. Steel, however, benefited from the trend more quickly than foundry iron, the price of which did not rise until the second half of the year. Even then, it went up only ten cents per ton above 1911 levels. Still, Sloss-Sheffield managed to reduce its inventory of unsold pig iron to 25,000 tons and liquidated its floating debt. Net profits after payment of dividends on preferred stock were only $84,254.75, but at least the company was once again operating in the black. Encouraged that things were getting better, Maben reported early in 1913 that further improvements to the company's physical plant would be undertaken. Characteristically, however, he indicated that these would be paid for strictly out of earnings, which "has been the fixed policy of the company—no securities of any kind being issued to cover expenditures." [20]

But the upturn did not last long. Whereas 1912 began with low prices and ended with higher ones, 1913 showed the opposite tendency. By late December, pig iron was selling for $3 less per ton than it had at the beginning of January. Because of relatively high prices prevailing at the beginning of the year, Sloss-Sheffield managed to post a net surplus of $209,466.05. This figure was deceptive, however, because the company had been forced again to assume about $500,000 in floating debts. The difficulties that it was experiencing with local reformers in the debate about smoke emissions only compounded its financial worries. Because of the need to replace the City Furnace coke ovens under the agreement it made with the municipal council, other improvements that had been planned were put on hold. [21]

A darkening international climate added to the gloom engulfing Sloss-Sheffield. The outbreak of war in the Balkans, beginning in October 1912, created uneasiness, both in foreign financial capitals and on Wall Street. Anxiety intensified when J. P. Morgan died in Italy from a heart attack in March 1913. While flags flew at half-mast above the New York Stock Exchange, American bankers, now without an acknowledged leader, worried about the impending passage of the Federal Reserve Act. As 1914 began, the business slump that had begun early in the previous year showed no signs of ending. Things got worse by summer as the outbreak of a general European war created a temporary cessation of foreign demand for American exports and increased uncertainty about the future. Fear that Great Britain, France, and other Allied powers would drain gold from America to finance military operations led the New York Stock Exchange to suspend operations at the end of July 1914. Not until April 1915 was unrestricted trading resumed. [22]

Increasingly, Sloss-Sheffield felt the impact of the prevailing malaise. Pig iron prices, falling more rapidly than the price of steel, dropped below $10 per ton for the first time in a decade. Concerned about the long-range impact of the war, southern ironmakers also complained about a reduction of protective rates under the recently enacted Underwood Tariff. Such circumstances made 1914 a bad year for Sloss-Sheffield. Despite efforts to cut back on outlays for new equipment, the need to replace the coke ovens at the City Furnaces resulted in heavy expenditures, and the floating debt ballooned to $1.3 million. Company officials also continued to make mistakes in their efforts to balance supply and demand. Guessing that prices would rise late in the year, they maintained high levels of production but withheld iron from the market, resulting in a staggering unsold inventory of 149,000 tons. Even though the new coke ovens produced lower costs, net profit for the fiscal year sagged to $490,189.41, leaving only $21,139.41 to add to surplus after payment of the usual $469,000.00 in preferred stock dividends. Even worse, dividends that came due in January 1915 had to be paid in scrip.[23]

As matters deteriorated, new and ominous themes began surfacing in deliberations of the executive committee, reflecting deep divisions within the company. They involved two types of irregularities: improvements made by Maben, Jr., without prior appropriations, and cost overruns exceeding estimates approved by the committee upon his request. Indications that a campaign against Maben, Jr., was underway became evident at an executive committee meeting on 29 December 1913, when Goadby submitted a letter from an unnamed vice president complaining about cost overruns, resulting in a supplementary grant of funds to cover a list of "expenditures without appropriations heretofore made." This was a sensitive matter, because the difficulty Maben, Jr., had in staying within his budget added to the floating debts that Sloss-Sheffield was continually incurring. Early in 1914, the company was forced to borrow an additional $500,000 from the Central Trust Company, using warehouse receipts for unsold pig iron as collateral.[24]

Although the executive committee's minutes failed to specify which of the two vice presidents had written the letter presented by Goadby, it was obviously McQueen, acting in his capacity as chief financial officer; Maben, Jr., would hardly have taken a step guaranteed to bring discredit upon himself, and McQueen had every motive to embarrass his rival. As matters turned out, the episode would be merely the first in a series of similar occurrences that continued to take place in the next few years.

There were also subtle indications that the conflicting interests of McQueen and Maben, Jr., had rekindled long-standing power struggles between investors in Birmingham and absentee owners in New York. At the annual meeting in March 1914, after a seven-year absence, Rucker, the stormy petrel of previous administrations, suddenly returned to the board of directors; for the first time since 1907, a spokesman for the Magic City again had a voice in determining policy. The following year, further unrest was signaled when the board was reduced from twelve to ten members and Dooley, who had represented the Bryan family's interests on the executive committee since 1908, temporarily dropped out, only to be just as suddenly reinstated two days later. Amid these unexplained developments, when the annual election of officers took place in March 1915, Maben, Jr., was listed as second vice president, whereas no distinction had previously been made between his rank and that of McQueen. A year later, however, the practice was resumed of listing the two men as equals. The unfolding instability indicated that the balance of power within the company was subtly shifting and that McQueen's fortunes were brightening. [25]

Unfortunately for Maben, Jr., his accuracy in estimating the cost of improvements did not get any better. He was caught in a cruel dilemma, faced on one hand by the need to keep the company's increasingly obsolescent equipment functioning properly and on the other hand by the onset of a war-related inflationary spiral that made it harder and harder for him to calculate costs. At an executive committee meeting early in 1915, auditor Hugh Franklin reported that recent expenditures for improvements had exceeded estimates by $123,115.77. Once more, Franklin indicated, a number of modifications had been made without preliminary appropriations. At this time, when the company was considering a general reduction in salaries and wages because of continuing financial difficulties and was paying its dividends in scrip, such revelations were embarrassing. Occasionally, Maben himself began voting against his son on such things as providing a better water supply at the Russellville mines and improving the performance of the Hattie Ensley furnace. [26]

Meanwhile, the composition of the board of directors continued to change. A. E. Ames, the Canadian investor whose desire for common stock dividends had vexed Bryan a decade earlier, started to purchase company securities once more and returned to the boardroom in 1912. An important link with the past snapped in April 1915 with the death of Alexander B. Andrews, who had been one of Bryan's closest friends; his place was taken by a New York investment banker, Theodore C. Camp, who had been previously connected with Bryan through their mutual involvement in the Lanston Monotype Company, a supplier of printing equipment to the Richmond *Times*. Camp's appointment represented continuity with the past, but any fresh face on the board added further unpredictability to an increasingly volatile situation. [27]

Early in 1915, amid indications of dissatisfaction concerning the substitution of scrip for cash in payment of dividends, two auditors were dispatched to Birmingham to investigate the way costs were being computed. By June, the normally antiseptic quality of minutes taken at meetings of the board of directors assumed a sharper tone by referring to "interrogations" of Maben in an effort to elicit information from him. Matters deteriorated further as cost overruns connected with projects undertaken by Maben, Jr., continued to occur, leading to acerbic statements at an executive committee meeting in February 1916 "that the appropriations should be so accurately prepared that no excess should result" and demanding that methods of "covering up excess appropriations or work not authorized by special appropriation be completely discontinued in the future."[28]

V

Fortunately for the company, a brightening economic picture began to cushion the friction. As the European war dragged on, it became increasingly apparent that earlier fears among business leaders about its impact on the American economy had been unjustified. While Allied and German forces became bogged down in endless trench warfare, Great Britain and France turned to the United States for enormous quantities of munitions. Instead of being drained from America, gold began flowing into it, and New York started to replace London as the world's financial capital. As Robert Sobel later said, "the economic forecasters were wrong: the war would not destroy American business, but rather would prove to be its greatest opportunity for expansion since the turn of the century." Following the trend, iron and steel manufacture became increasingly profitable, and production smashed all previous records. During 1915 alone, pig iron output in the United States rose from 19 million to 38 million tons. Throughout the next year, shipments of iron and steel from Birmingham to foreign ports escalated sharply. More than 40,000 tons of metal, mostly pig iron, were shipped abroad from the district in October 1916 alone.[29]

After enduring so many hardships, Sloss-Sheffield now began to prosper. Throughout 1915, as the war became an increasing bonanza for American industry, the company's condition improved correspondingly; its inventory of unsold iron decreased to 59,000 tons, and the floating debt declined sharply to $800,000. Net profit for the fiscal year rose to $522,387.95, and $351,750.00 was paid in cash dividends for the final three quarters. In his annual report in March 1916, Maben took pleasure in announcing that the scrip that had been issued earlier in place of dividends could now be redeemed. Things might have been even better, however, had the president and his aging advisors been less conservative. Having guessed wrong the previous year by augmenting production, they now limited the number of furnaces in blast, shut down many coke ovens, and stopped work at some coal and ore mines. As a result, even greater profits went by the board.[30]

As massive orders for munitions poured in from Allied powers in 1916, American pig iron production soared. Belatedly adjusting to the new situation, Sloss-Sheffield increased its output by 57 percent. Buoyed by prices that increased by $7 per ton, it earned a net profit of $1,990,647.21 for fiscal 1916 and added $1,521,674.21 to surplus even after paying cash dividends of $469,000 on preferred stock. Despite paying nearly $800,000 for capital improvements to enhance production, it came close to liquidating its floating debt. Meanwhile, cumulative surplus reached an unprecedented $4,465,070.13.[31]

VI

Sniping about cost overruns by Maben, Jr., disappeared as economic conditions improved. The directors, however, were now wrestling with an even more important issue that would have a major impact upon the company's future. This involved whether or not it should begin replacing its antiquated beehive coke ovens with more modern ones capable of recovering valuable by-products that, up to now, had been irretrievably lost. Debate among the directors about this issue mirrored intense national concerns that had become more and more pressing since the outbreak of the war, as shrinking imports from Germany produced severe shortages of coal-tar derivatives.

These key products, based on scientific and technological advances in the late nineteenth century, were critical for the manufacture of synthetic dyes, which became extremely scarce in the United States after 1914. Fumbling attempts were made to return to natural color sources that had long been unused—indigo, madder, and saffron, among others. What was needed, however, was mastery of the same technologies that had enabled Germany to turn coal into a multitude of hues. By-products of coal distillation were also urgently needed to make explosives that were currently in high demand in Great Britain and France and were also crucial to national preparedness in case the United States formally declared war. As diplomatic crises, like those involving the loss of American lives to German submarine warfare in the sinking of the *Lusitania* in May 1915 and the *Sussex* a year later, multiplied, progress became more and more imperative. [32]

Up to the late nineteenth century, there had been little American demand for coal tar derivatives, which were produced mostly by city gasworks for use in roofing, wood preservation, and, to a limited extent, road building. Under such circumstances, there was no incentive for manufacturers of coal or iron to derive such little-appreciated by-products as anthracene, benzol, naphthalene, or phenol from coke. Even a much more fundamental chemical, ammonia, could be produced in adequate quantities by other means. By-product ovens, first developed in Europe, used various techniques to carry gases through mazes of pipes and flues to coolers, condensers, saturators, separators, scrubbers, and stills to recover and refine potentially useful chemicals. Such large, complex installations required much capital and skilled labor and needed large markets to be economically justifiable. As a result, beehive coke ovens, which were cheap and easy to contruct, were built in enormous numbers in the United States despite the fact that the gases they produced went to waste. [33]

During the 1880s and 1890s, things began to change. New developments, particularly the growth of the electrochemical industry, resulted in mounting demand for such alkalies as soda ash and caustic soda, which could be most economically produced by means of an ammonia-soda conversion process invented in 1863 by two Belgians, Albert and Ernest Solvay. Textile manufacturers also recognized the potential value of making soda in the United States instead of importing it from Europe. William B. Cogswell, an American engineer, and Rowland Hazard, a textile producer, secured rights to the Solvay process and chose Syracuse, New York, as the site for the process's implementation because of the city's proximity to raw materials the likes of coal, limestone, and salt. Here, in 1893, the first by-product plant in America was built to supply coke for limestone burning and synthetic ammonia for soda conversion. It used an oven design developed by a European inventor, Louis Semet.

After Hazard's death in 1898, his son, Frederick, organized the Semet-Solvay Corporation to make, sell, and install by-product coke ovens. A rival design, invented by Heinrich Koppers of Germany, was brought to America during the same period. The resulting firm, the Koppers Company, was initially based in Chicago but subsequently moved to Pittsburgh, where it received infusions of capital from the Mellon family, which was heavily involved in the manufacture of a major electrochemical product, aluminum. After the Cambria Steel Company installed sixty by-product coke ovens in 1895, the new technology began to be used in iron and steel manufacture. One of the first places where this occurred was Alabama. In July 1898 a significant pilot project began when TCI and Semet-Solvay started bringing on line a battery of 120 by-product ovens that had been under construction for thirteen months on a site near TCI's Ensley furnace complex. The new installation was in full operation by the end of October. Semet-Solvay owned the ovens, using coal that TCI supplied from its Pratt mines. Semet-Solvay in turn delivered coke to TCI's stock houses, after which TCI's workers hand-loaded it into the Ensley furnaces. TCI paid Semet-Solvay twenty-five cents per ton for the coke, plus "one half of the value of the reduction in coal used, as compared with its bee-hive practice." In addition, TCI received all surplus gas from Semet-Solvay's ovens at no charge, burning the fuel at its boiler plant and thus reducing the amount of coal previously needed for operating its steam engines. Gas from the new plant was also used in the manufacture of creosote, fertilizers, roofing pitch, tar paper, and other products that were needed in Alabama's predominantly agricultural economy. Along with smaller establishments located in Pennsylvania and West Virginia, the facility was among the first three batteries of by-product ovens to be built at blast furnace sites in the United States by Semet-Solvay, which gained valuable experience from being able to experiment with several northern and southern varieties of coal. The success of the Ensley experiment was indicated by the fact that, in 1902, TCI installed 120 more by-product ovens at Ensley as part of Bacon's modernization program.[34]

By 1913, by-product coke output in the United States had reached 12,714,700 tons—about 27 percent of all the coke manufactured. Even so, it was estimated that $71 million per year was still being lost because of the continued use of beehive equipment. Among the firms that had become increasingly committed to by-product technology was TCI, which, under Crawford's leadership, installed a fresh battery of by-product coke ovens in

1912. Its 280 Koppers-type units had a daily capacity of 5,500 tons of coke, yielding such by-products as ammonia, benzene, and naphthalene. Some recovered gases were also used to heat the ovens themselves or supply fuel for the Ensley steelworks; further residues went into making motor fuels, paint, and road-building materials. The fact that Woodward was also installing Koppers-type ovens at the time created a situation from which Sloss-Sheffield would ultimately benefit; the commitment of two of the Birmingham District's three largest firms to Koppers left Semet-Solvay in need of finding a suitable local manufacturer to install its latest equipment. When the time came to act, there was little doubt where Sloss-Sheffield would find it advisable to turn, but this still lay in the future as war broke out in Europe in 1914.[35]

As Jack R. Bergstresser has pointed out, TCI had good reason to be in the vanguard of American ferrous metal producers in adopting by-product coke manufacture. By-product ovens yielded significantly more coke—by a margin of 5 to 10 percent—than beehive types. Because Alabama led the nation in coke consumed per ton of iron produced, the cost savings that firms like TCI could gain by converting to by-product ovens were substantial. The coking chambers of by-product ovens also permitted less expansion of the coal with which they were charged than was true in the case of beehive ovens. As a result, by-product coke was stronger and harder—or, in metallurgical terminology, less friable—than beehive coke. The strength and hardness of by-product coke was particularly desirable in the Birmingham area because Red Mountain ore was becoming harder as relatively soft surface deposits gave out and operators had to dig deeper into the mines honeycombing the hills along Jones Valley. Because by-product coke was stronger than the beehive variety, it did not crumble as readily when used in combination with hard red ore. This resistance to compression helped avoid an accumulation of small coke particles, known as "fines," that could cause clogging in a blast furnace. The additional revenue that could be earned from selling chemical by-products of the conversion process made the case for adopting the new technology even more compelling. For all these reasons, as Bergstresser has stated, the Birmingham District ultimately became "the first major iron and steel-producing region in the country to convert fully to by-product coke." The fact that Sloss-Sheffield was one of the last firms in the area to make the switch showed plainly how regressive its management had become in the period immediately following Bryan's death.[36]

By 1915, when Sloss-Sheffield's board of directors first began to consider switching to by-product technology, nearly 34 percent of all coke manufactured in the United States was being produced in this manner, and the number of by-product ovens in operation had swollen to more than 7,000. As war-induced pressures mounted for America to intensify its production of such munitions-related chemicals as benzene for conversion into picric acid and toluene for nitration into TNT, more and more by-product ovens went on line with each passing week. While the price of benzene rose to $1.25 per gallon, that of toluene skyrocketed to $4.00, making for enormous potential financial rewards. In the words of a leading historian of the chemical industry, Williams Haynes, such prices "touched off the wildest explosion of chemical activity this country has ever seen."[37]

Despite the lure of such inducements, Sloss-Sheffield's aging leaders moved cautiously as they contemplated moving into by-product coke production. Their slowness to act was all the more glaring because of an ambitious program of expansion and modernization that had been carried out by their arch rival, the Woodward Iron Company. In 1912, Woodward had acquired two stacks, the Vanderbilt Furnaces, that had been built in North Birmingham. The transaction, by means of which Woodward added valuable coal and ore mines to its holdings, created a stir in the district because it featured the writing of the largest single check, for $1 million, that had been made out in Alabama up to that time. The extremely modern new blast furnace that Woodward put in blast in 1914 was another milestone in the firm's continued progress. Among its other features, it utilized by-product coke from 170 Koppers regenerative ovens.[38]

The minutes of the meetings at which the issue of switching to by-product coke was discussed by Sloss-Sheffield's board from 1915 onward provide few details about the nature of the arguments that were advanced on either side. It is not difficult, however, to see why the superannuated directors were reluctant to endorse a fundamental departure from the company's traditional methods. In addition to requiring skilled labor, the new technology was extremely expensive. Raising the capital required to adopt it inevitably would have swept away what was left of the power wielded by longtime investors whose financial resources were already strained.

Still, talking about possibilities cost nothing. Discussions about the advisability of installing by-product coke ovens began among Sloss-Sheffield's directors in September 1915, resulting in requests for information from the Koppers and Semet-Solvay companies. A special meeting was held early in February 1916 to consider preliminary information received by Maben, Jr., acting in his capacity as the company official responsible for physical operations. One director, Harry Bronner, who was concerned about the possible repercussions that the matter might have on future ownership of the company, noted that the thirty-year bonds that had financed the takeover of the Sloss Furnace Company in 1887 would soon mature. In view of the outlays that would be required if by-product ovens were built, he said, the two matters should be considered at the same time. A resolution was therefore passed directing the executive committee to appoint subcommittees to "report separately on the two propositions." On 28 February, four directors, including Goadby and Wallace, agreed to study the issue of building by-product ovens and to probe the impact that such a change would have on renewal of the bonds.[39]

In mid-May, after the four directors had visited Birmingham, they submitted their findings to the executive committee. The nature of the report was not specified by the minutes, and no action was taken. The matter was again discussed later in the month but was tabled pending a report from Maben, due in two weeks, about "certain recommendations" that were once again unspecified. Judging from a lack of references to the matter in the minutes of meetings that took place throughout the summer and into the fall, nothing further was done. Instead, the members busied themselves with building a store at Brookside, providing a filtration system for the water works at the Bessie mine, installing a slag washer at Russellville, and suing the Pratt Consolidated Company for mining 290,000 tons of coal on Sloss-Sheffield's properties. In no hurry to modernize, the directors marked time.[40]

VII

Early in November, a terrible incident at Sloss-Sheffield's Bessie coal mine—an installation employing only free though nonunionized labor—temporarily diverted attention from everything else. At approximately 2:45 in the morning of 4 November, while heading a crew loading coal from one of the seams, miner John McGowan "heard a rumble, and felt a sort of a quiver—that came with a sort of blowing and a sucking," as he later described it. Suddenly, McGowan continued, "fire flashed before my eyes. The whole mine seemed to be aflame. It seemed to last four or five minutes. I don't remember what happened then. The air was black with dust, bits of coal and, Lord, but it was black."

Almost overcome by the heat—"it seemed as if I would die when I took my first quick breath of that thick, smoky dust," McGowan later recalled—he and four companions stumbled blindly through a series of passageways to an underground pumping station. Soaking their clothes in a pool of water and putting them over their heads, they made their way past timbers "blazing with fire" until they reached an entry where they could smell fresh air. "It was then," McGowan said, "that we knew that we were saved." Coming out of the mine, the men saw a starlit sky at the mouth of the slope. "I never saw anything that looked so good to me," McGowan stated. "We were so happy and glad that I don't recollect that we even said anything to each other."

John Shell, a fire boss reputed by the *Age-Herald* to have "the most level head of any miner in the deep passages of the earth," led two other workers to safety by plugging a small area near an air pipe so that fumes from firedamp could not penetrate it and staying there with his companions until rescuers from Sayreton found them fifteen hours after the blast. Eight men, three white and five black, thus survived the tragedy, but many others were not so fortunate. All told, thirty miners died in the disaster. One by one, their mangled bodies were brought to the surface. Exhausted teams of searchers carried them to a "rude washhouse that was converted into a morgue." Aided by "generous tanks of oxygen," the searchers had battled afterdamp by plumbing the depths of the mine in a "vigil that extended nearly 20 hours." So fierce was the subterranean blast that two miners had been beheaded, either "by the force of the explosion or by the fury of flying wreckage."

Declaring that the Bessie mine had been "inspected quite recently and pronounced safe," the *Age-Herald* was unable to explain what had happened. "It seems that no matter how elaborate the precautions taken," an editorial stated, "there is no absolute safeguard against explosions." Experts led by C. H. Nesbitt, state inspector of mines, later conducted an official investigation. As usual, little came of it. "The extent of damage to the mine is expected to be rather slight," the *Age-Herald* said. "The two car tracks in the slope were not battered by the explosion, but the brattices were torn into infinitesimal bits and roof timbers and electrical apparatus must be replaced." Meanwhile, amid "deep gloom," residents of the mining camp at Bessie prepared "to commit the bodies of the victims with last rites to the earth." Two weeks after the disaster, a less tragic incident occurred when a fire of unknown origin destroyed a bathhouse at Sloss ore mines. There were no casualties, but the mine was shut down temporarily to replace a pole carrying electricity to the pumps.[41]

VIII

However dramatic their immediate impact, these episodes only briefly diverted attention from the pressures that were forcing Sloss-Sheffield to move toward modernizing coke production. Just before the Bessie tragedy occurred, the *Age-Herald* reported that outside interests were planning to invest approximately $15 million in "gigantic new plans" for the Birmingham District, including "a steel plant, several blast furnaces of immense proportion and of the most modern type, a number of batteries of by-product coke ovens, benzol plants and other auxiliary industries." The Semet-Solvay Corporation had recently acquired 7,000 new acres of coal lands and several ore properties in the area. Its existing agreement with TCI, which would soon expire, was likely to be discontinued, for Crawford's company, because of its ties with Koppers, was competing with Semet-Solvay in marketing coke. Meanwhile, as Anglo-French casualties in four months of intense fighting in Flanders mounted to a reported 600,000 and renewed German submarine warfare resulted in five vessels being sunk off the New England coast, the United States was being pulled ever closer to war, causing demands for explosives and other coal-tar derivatives to escalate.[42]

As pressure to act became increasingly insistent, Sloss-Sheffield finally announced plans in late November 1916 to build a battery of by-product ovens in North Birmingham, at an outlay of $2.5 million. The project was characterized as tentative, and the time at which it would commence was left purposely indefinite. According to the press release, four directors, including Maben and Goadby, would soon come to Alabama to begin a two-week tour of inspection. Maben, Jr., told the *Age-Herald* that work on the new facility could be "begun immediately were it not for the abnormal prices that machinery, equipment and labor demand now." Echoing this cautionary note, the paper predicted that the new plant would not be built until such conditions were "materially altered."[43]

The very manner in which this tantalizing but inconclusive announcement was made indicated that the board, though closer to a final decision, remained uncertain about how to proceed. Interviewed by *Age-Herald* reporters in New York before leaving for Birmingham, Maben "flatly denied" that changes were impending in the ownership of the firm. That the question was posed, however, indicated that the outlay required to build a by-product plant had led astute observers to think that the current owners would soon lose control. Maben dismissed this prospect as a joke, boasted about Sloss-Sheffield's latest earnings, and further obscured the situation by saying that his upcoming trip would deal only with routine issues.

Maben and Goadby reached Birmingham on 4 December and proceeded to tour the district along with the other members of their party. As the *Age-Herald* said, the main question they faced was "whether or not the battery of by-product coke ovens should be installed soon or whether it is better to abide time and wait until materials that will be used in the construction of the ovens and by-product recovery plants . . . can be purchased more cheaply and delivered with greater dispatch." After the directors had spent more than a week in the district, local observers were still no wiser than before about when, if ever, the new ovens would be built. Speaking to reporters on 13 December, just before the group left for New York, Maben declared that the firm's properties were in "splendid condition" and asserted that its production of foundry iron, which he claimed was now the largest in the world, would reach nearly 600,000 tons by the end of the year. Responding to questions about "big improvements," however, he indicated only that they would not take place in 1917. "We will be governed by conditions as they arise," he said. "I can't say what we shall decide upon later, but at present we have no definite plans. The price of materials is too high to make extensive developments."[44]

Minutes of meetings that took place after the delegation returned to New York provide no clues about what its members said to the other directors, but no action resulted. Meanwhile, the war neared a turning point. On 1 February 1917, amid mounting desperation among powers on both sides, the Germans ended indecisive peace negotiations by declaring unrestricted submarine warfare, gambling that they could win before the United States, now certain to enter the conflict, could send reinforcements across the Atlantic in time to avert Allied defeat. After German U-boats began sinking American ships and the British intercepted a telegram from German foreign minister Arthur Zimmerman to Venustiano Carranza, president of Mexico, attempting to embroil that country in armed conflict with the United States, Woodrow Wilson saw no alternative but to act. On 2 April 1917, he asked Congress for a declaration of war, and America entered the holocaust.[45]

Amidst these momentous developments, the old order at Sloss-Sheffield began to collapse. At the board of directors' annual meeting on 15 March 1917,
just three weeks before America entered the war, Maben relinquished the
presidency that he had held since 1902 and became chairman of the board.
Pending designation of a permanent successor, James N. Wallace, who was
president of the Central Trust Company and had been a director since 1906,
assumed temporary leadership of the firm. McQueen and Maben, Jr., remained vice presidents as before. Eight days later, a new administration took
shape as two additional directors were named. One of these, John S. Sewell,
of Gantt's Quarry, Alabama, had made a recent survey of the company's
properties. He joined Rucker as one of two Alabamians on the board, indicating continued resurgence of local influence in the enterprise.

The other new director, Waddill Catchings, was much more important
than Sewell, becoming president as Wallace resigned. Like such earlier presidents as Seddon and Maben, Catchings was a "southerner on Wall Street."
Born in Sewanee, Tennessee, he had grown up in Mississippi, where he
obtained his early schooling. Matriculating at Harvard, he graduated in 1901
and received a law degree in 1904. Working for the New York City law firm of
Sullivan and Cromwell, he became adept at advising companies that went
into receivership during and after the Panic of 1907; these included Milliken
Brothers, a steelmaking enterprise, and the Central Foundry Company,
which owned a number of plants in Alabama. Becoming president of the
latter firm, he restored it to solvency and then became chief executive of the
Central Iron and Coal Company, which operated blast furnaces and coal
mines at Holt and Kellerman, near Tuscaloosa. After managing these facilities
for nearly eight years, he moved back to New York City and became active in
the purchasing and export departments of J. P. Morgan and Company. Still
under forty years old by the time he became president of Sloss-Sheffield, he
was already a rising star on Wall Street.[46]

It is not clear whether Maben had stepped aside voluntarily or given up his
office under duress. Certainly, however, his leadership position had become
increasingly untenable due to his indecision about modernization, and his
resignation was a foregone conclusion. His advanced age alone may have led
him to decide that he lacked the vigor to steer the company through the war
that lay ahead and that the time had come to turn his office over to a younger man.

It is also uncertain how or why Catchings had been chosen as Maben's
successor. The geriatric character of the board of directors, however, had
made it practically inevitable that the company would have to look outside for
a new president, particularly in view of the impasse still prevailing between
Maben, Jr., and McQueen, the most likely inside candidates. The fact that
Catchings had southern roots and close connections with the foundry trade
made him a logical man for the post. At the time of his appointment, a

headline in the *Age-Herald* pointed out that he was well known in Birmingham and throughout Alabama; a cousin, Ben Catchings, had previously lived in the Magic City. Through his growing prominence on Wall Street, the new president must have been equally familiar to the absentee owners in New York. Because of the simultaneous appointment of Sewell as a director and the way in which later events would reveal close concord between Catchings, McQueen, and the Birmingham business community, it is possible that forces allied with McQueen had played a role in Catchings's selection. The *Age-Herald* indicated that Maben, Jr., was among local company officials who had been caught by surprise when Catchings's election was announced. The newspaper, however, did not go into further specifics, indicating merely that "new interests" had "determined that the company shall be re-energized."[47]

Noting Catchings's relative youthfulness, the *Age-Herald* predicted that major changes were in store for Sloss-Sheffield. Inevitably, speculation was rife that Sloss-Sheffield would finally embark upon the long-deferred dream of making steel. At the very least, the erection of by-product coke ovens was regarded as a certainty. A flurry of recent surveys by accountants, auditors, and engineers who had visited the district on special assignment for Sloss-Sheffield, it was rumored, had been made for the purpose of determining whether current assets were undervalued, thus providing a basis for a new stock issue to fund major capital improvements. On the other hand, Catchings, a bespectacled, soft-spoken man with dark hair, brown eyes, and "firm mouth," sounded a cautionary note after making a four-day visit to the city with his wife immediately after his appointment. "I would advise you not to use your imagination much about any developments that the company has projected," he told reporters. "It is likely that they would be amiss."[48]

These were prophetic words. People who had hoped that Sloss-Sheffield was about to launch a sweeping program of modernization were quickly disappointed. So, too, were residents of the Magic City who had wanted Catchings to reside in Birmingham. By the time he returned to New York in late March, it was increasingly apparent that Catchings had no plan to institute dramatic changes and that neither he nor his wife was in a hurry to move to Alabama. While predicting confidently that modernization would ultimately occur, a "reliable source" who was interviewed by the *Age-Herald* sounded the by-then-familiar note about high construction costs and said that the directors would want to study the company's future prospects with great care before committing themselves to heavy outlays. "There is no doubt . . . that the Sloss-Sheffield company will order improvements and construction that will mean the expenditure of money," the *Age-Herald* stated shortly before Catchings returned to Birmingham for a second visit early in April. On the other hand, it added, "those in position to know say that it will not be a very large amount."

Catchings repeated the same cautionary theme upon arriving in the city on 8 April, two days after the United States entered the war. Denying that recent property surveys portended a new stock or bond issue, he also indicated that the chances of his moving to Birmingham were slim. For the time being, he said, "the company's efforts will be directed toward maintaining a full and uninterrupted output from its existing plants. . . . there will be no material difference from the operations in the past."[49]

Under the goad of war, environmental concerns were soon forgotten. After negotiations between Catchings and the municipal government, Sloss-Sheffield received permission to repair and relight at the City Furnaces the 288 beehive ovens that had been inoperative since the smoke emission crisis of 1914. This move undid the last vestige of the pollution abatement that such reformers as Weatherly had fought to secure. The agreement was to run for three years. Sloss-Sheffield reaffirmed its earlier intention to build a new by-product coke plant at North Birmingham, but nothing in the agreement indicated when. Indeed, the accord specified that if the new facility was under construction but not yet operative when the contract expired, the old beehive ovens would be allowed to remain in operation pending their completion. Things might drag on for years.[50]

It is not hard to understand why Sloss-Sheffield was still unready to move ahead with modernization. Catchings's appointment had not been accompanied by an infusion of new capital, and the way in which the company's existing thirty-year bonds were to be retired remained undecided. Inflation had been mounting rapidly prior to America's entry into the war, and reports in local newspapers frequently indicated that Sloss-Sheffield was not ready to assume rising construction costs. As Catchings visited Birmingham in March and April, it was also unclear how the federal government would finance the staggering war costs it was facing, or how potential war bond issues that were contemplated by treasury secretary William G. McAdoo would affect the availability of capital for industrial purposes. Plans for coordinating the war effort were also in a chaotic state. Since August 1916, these matters had been vested in the hands of the purely advisory Council of National Defense, which lacked statutory authority to do its work. It was uncertain at best whether another advisory body, the General Munitions Board, could perform the task any better after it was created in April 1917.[51]

Pending further developments, Catchings did his best to conserve Sloss-Sheffield's cash reserves. Early in 1917, because of the company's increasing profits, Maben had been under pressure to reinstitute dividend payments on common stock. Catchings now persuaded the directors to renounce the idea, which, if implemented, would have shrunk the firm's accumulated surplus by as much as $600,000. Disgusted by this latest setback, Ames once more sold his stake in the company, and Canadian representation on the board came to an end.[52]

Sloss-Sheffield's continuing failure to modernize contrasted with what was happening at TCI. While Crawford was away on one of his periodic visits to U.S. Steel's New York City headquarters in July 1917, it was announced that the parent corporation would invest $11 million in building a massive new facility at Fairfield, designed to convert ingots produced at Ensley into plates and other structural components for building merchant vessels at Mobile. Three new subsidiaries were created to implement the arrangement. C. J. Barr, who had previously been general superintendent of TCI's Ensley division before moving to another position in Canada, was brought back to Birmingham to direct the project. "Mr. Barr will launch right into his new task with the idea of making every moment count and before many days it is expected that dirt will be flying and building materials will be rapidly transformed into structures that will house the immense additions," the *Age-Herald* stated after Barr returned to the city in late July.[53]

X

Although Sloss-Sheffield continued to delay modernization, it moved rapidly toward getting its house in order by resolving tensions that had festered since 1911 due to the conflicting ambitions of McQueen and Maben, Jr. For a short time after Catchings became president, the two executives continued to divide supervision of day-to-day affairs in Birmingham. But there was no question about which one ranked higher on the corporate ladder. At a special meeting of the board of directors, held in New York just prior to his visit to Birmingham in April 1917, Catchings urged that one of the existing vice presidencies be formally classed as senior to the other in order to eliminate "divided authority." McQueen, he said, should be given precedence. Accordingly, the veteran marketing head was named first vice president and Maben, Jr., was designated second. This action signaled final victory for McQueen and indicated that days with the company were numbered for Maben, Jr. He hung on until December 1917, when he was given a paid leave of absence. Shortly thereafter, the board forced him to resign, awarding him $5,000 in severance pay.[54]

Despite temporarily retaining his official status as chairman of the board, the elder Maben was now often absent from corporate meetings and played no further role of consequence in the company. At the annual meeting held in March 1918, he was relegated to the rank of director and stripped of his previous title, which had evidently been intended to be ceremonial from the outset. Meanwhile, two veteran directors and former Bryan associates, Dooley and Mortimer, had also resigned from the board. Dooley's place was taken by M. C. Branch of Richmond, who was named to the executive committee. Nevertheless, the power that Virginians had once held in the firm was waning.[55]

While McQueen ran day-to-day operations in Birmingham, Catchings became more and more active in the war effort. The federal labor department appointed him to its advisory council, and the U.S. Chamber of Commerce named him head of its war committee. The second of these posts was particularly prestigious, because the chamber was assuming increasing importance in the industrial mobilization program being pursued by Bernard

Baruch as head of a newly constituted War Industries Board that had been created in July 1917. In December of that year, the naming of Catchings as a partner in the powerful Wall Street firm of Goldman, Sachs and Company indicated how swiftly his stature had grown. There was no longer any idea that he would move to Birmingham, and it was obvious that he could not continue to serve as president of Sloss-Sheffield under the pressure of his other responsibilities. Both now and in the future, he would be far more valuable as a fund-raiser and lobbyist in New York and Washington, where his ties with such persons as Baruch and McAdoo were of great potential benefit to the company. On 15 February 1918 he resigned his presidency at a special meeting of the board of directors but agreed to stay on as a director, at a yearly salary of $25,000. At the annual meeting in March, he succeeded Maben as chairman of the board, indicating that he would continue to play a powerful role in the firm. [56]

There was no question who would succeed Catchings as president. Having spent twenty-six years in the company's service, McQueen now received the promotion for which he had waited and struggled so long. His election as chief executive took place immediately after Catchings resigned and was enthusiastically hailed in Birmingham. Echoing Horatio Alger, the *Age-Herald* declared that McQueen was "typical of the men of the south who, starting at the bottom of the ladder, have by pluck and capacity risen to the top." His election, the article said, would "cause his host of friends to rejoice and shower him with congratulations." [57]

In heaping praise upon McQueen, Birmingham was also acclaiming itself. The *Age-Herald* displayed its boosterism by saluting the new president as a "home product." Abroad, the war raged on while casualty lists that appeared daily in Alabama newspapers testified to its tragic cost. At home, nagging questions persisted about the role that Sloss-Sheffield would play in munitions production as the by-product coke ovens that had been deferred for so long remained in limbo. Despite everything, Birmingham could take pride that, for the first time since Joseph Forney Johnston had resigned the presidency of Sloss Iron and Steel in 1887, a local son was in charge. As he reveled in congratulations, McQueen was no doubt repeatedly reminded of this fact by his legion of supporters and admirers.

McQueen in Command 13

McQueen's administration of operations at Sloss-Sheffield showed how far the company had come in permitting managers to run day-to-day affairs without interference from absentee owners. Long gone were the days when Bryan and other directors had interfered with Seddon, Haas, and Hopkins. Watching events from his Wall Street office, Catchings was secure knowing that the firm was in good hands. Separation of ownership and control had become an established fact at what, for the first time, was a truly modern corporation.

I

In 1917, as the United States formally entered the war that had been raging for almost three years, Birmingham's civic leaders were in a confident mood. For the first time, according to the estimates of *Polk's City Directory*, the district's population had passed 200,000. Not all the changes taking place were quite so heartening to local boosters; many worried about a northward migration of black workers that threatened coal production late in 1916 and was causing labor disturbances all over Dixie. Some analysts argued that the exodus was a good thing because it might "stamp out indolence" and "entice foreign immigrants to the south," but others feared that it would upset the region's long-standing advantage in labor costs over competition from other parts of the country. Nevertheless, boosters could not help but be encouraged by the stimulating effects of war-related orders that poured into north-central Alabama as the nation girded for its role as a declared belligerent in the worldwide conflict. "Merchants are jubilant over the fact that a vast amount of the material needed by the government . . . will be drawn from this District," the *Age-Herald* stated in May 1917 as local payrolls began to exceed $1 million per week. "There is more money in Birmingham now than ever before in her history."[1]

One development that contributed greatly to local optimism was the prospect that the city would soon enjoy the benefit of water transportation directly to Tuscaloosa, Mobile, and New Orleans by way of the Warrior and Tombigbee rivers. Since 1871, when steps were first taken to remove obstructions to navigation on these waterways, work had proceeded intermittently on channels, dams, and locks to facilitate achievement of this long-range goal. By 1915, the spreading network of improvements had reached Attwood's Ferry, not far from the city. Plans for further canalization were underway, looking toward the eventual opening of Birmingport, a complex of docks, railheads, and other facilities at the confluence of Short Creek and the Warrior River's Locust branch. Although this would not be finished until 1920, barges bound for Mobile were already being loaded with coal, iron, and steel at Attwood's Ferry as the inability of the district's overworked rail system to meet wartime needs became more and more vexing to Alabama industrialists. [2]

Boosters were also proud of the city's newest hotel, the finest structure of its type yet to be built in Birmingham. Construction began in 1913 after Crawford complained that the district lacked lodgings worthy of accommodating visitors from U.S. Steel's New York headquarters and might therefore lose potential investment capital. Led by local developer Robert Jemison, Jr., a consortium of the city's elite purchased a site on the corner of Fifth Avenue and Twentieth Street and erected the Tutwiler Hotel. It was named after its largest stockholder, Edward M. Tutwiler, who survived from the early days of the district as one of its most venerated industrialists.

A twelve-story steel-frame brick and terra-cotta edifice designed in ornate Renaissance Revival style by architect William L. Walton and built by Wells Brothers and Company of New York City, the Tutwiler possessed what the *Age-Herald* proudly called "every modern feature of high class hotel construction." It had such amenities as an orangery with a leaded glass ceiling and fountain of Caen stone, a sumptuous dining room with an Italian balcony garden, reception rooms in Louis XIV style, bedroom corridors in "French verdure design," and a large open-air roof garden floored with "red promenade tile" and adorned with latticework trellises and flower boxes. Sparing no expense, the promoters arranged to have all tanks, vents, and other features that might mar its appearance tastefully housed in such a manner that it would be "highly ornamental" when viewed from any side. [3]

A growing cluster of skyscrapers crowded Birmingham's urban core. Seven of these had been built in the decade between 1903 and 1913. Two recent additions, the sixteen-story City National Bank building and an even higher twenty-one-story structure erected by the American Trust and Savings Bank, flanked the Woodward and Brown-Marx buildings at the intersection of First Avenue and Twentieth Street. Citizens called the crossing "the heaviest corner on earth."

Completion of the American Trust and Savings Bank building in 1912 gave Sloss-Sheffield a chance to assert itself vis-à-vis TCI, which had upstaged it earlier by setting up headquarters in the Brown-Marx. Vacating its old head offices in the Woodward Building, Sloss-Sheffield moved into the new edifice, a handsome neoclassical structure faced with white terra-cotta panels and adorned by stately pilasters crowned with Corinthian capitals. The following year, the twenty-five-story Jefferson County Savings Bank building on Second Avenue became Birmingham's tallest skyscraper. Despite having been so quickly upscaled, and notwithstanding that it was already becoming discolored by the pall of smoke and dust continually overhanging the city, the new corporate home was an impressive setting for McQueen's assumption of power when he took charge early in 1918.[4]

Sloss-Sheffield shared Birmingham's sense of importance. Although it was not a corporate giant compared with many other firms in the ferrous metals industry, McQueen had climbed a mountain on his way to becoming chief executive. At the time America entered the war, the company's total assets of $27 million ranked 28th among primary metal manufacturers in the United States and 206th among all industrial firms. Its corporate structure and way of doing business were typical of a modern, vertically integrated enterprise. Having long ceased relying on general commission agents, it marketed its products through sales agents in such major American cities as Boston, Chicago, Cincinnati, Los Angeles, New York, Philadelphia, St. Louis, and San Francisco. It also had a foreign sales office in Glasgow, Scotland. Its seven blast furnaces were rated at a total annual capacity of approximately 600,000 net tons. Except under unusual circumstances, it relied on its own raw materials from coal and ore lands comprising 62,500 and 46,400 acres respectively. Gross current assets had grown from $1,867,866 in 1911 to $3,613,078 in 1917, and net working capital from $935,736 to $2,398,445. By regional standards, at least, Sloss-Sheffield was in every respect a "big business," and its new president well fitted the model of a "captain of industry."[5]

II

The sudden transfer of power from Maben to Catchings and McQueen epitomized the final stages of a generational shift in the management of southern corporations as younger men who had not taken part in the Civil War replaced a large number of former Confederate officers to whom the region had turned for industrial leadership after Appomattox. Except for his being southern, McQueen's background typified that of many executives who had risen to positions of power in the American iron and steel industry in six northern cities during the late nineteenth and early twentieth centuries. John Ingham has shown that the great majority of these individuals did not climb from "rags to riches" but came from solid middle- or upper-class backgrounds.[6] A comparable study by Justin Fuller indicates that the same generalization can be made about Alabama's business leaders.[7] McQueen's origins, like those of Crawford, support such findings. Although his life and career resembled the plot of a Horatio Alger novel, especially in that he had been reared by a widow who had lost most of her possessions in the Civil War, Sloss-Sheffield's new president had a formidable ancestry. His father, Gen. John McQueen, was descended from one of the greatest of all Scottish kings, Robert Bruce. His mother, Sarah Pickens McQueen, was a granddaughter of one of South Carolina's most illustrious Revolutionary War heroes, Gen. Andrew Pickens. After serving in Washington as a representative from South Carolina prior to the Civil War, John McQueen became a Confederate congressman and a close friend of James Seddon, secretary of war in Jefferson Davis's cabinet. This relationship would stand McQueen's youngest son in good stead when Seddon's son, Thomas, gave him a job at Sloss Iron and Steel in 1891.[8]

Born in 1866, James William McQueen came to Alabama with his mother and two brothers the following year, after his father died. Their family home in South Carolina had been destroyed by Sherman's forces, and the children were reared in genteel poverty. Forced to fend for himself early in life, McQueen took a job at age fifteen as a telegraph operator for the Alabama Great Southern Railroad at Eutaw, southwest of Tuscaloosa on the way to Meridian, Mississippi. In 1886, he became a joint agent for that line and the Cahaba Coal Company, working in such places as Woodstock, a village between Tuscaloosa and Birmingham. There his lifelong association with the iron industry began when he met his future wife, Lydia. Her father, Giles Edwards, was a blast furnace superintendent who had migrated to Pennsylvania from Wales in the 1840s, moved to Chattanooga in 1859, and come to Alabama three years later to rebuild the Shelby Iron Works at Columbiana. McQueen, a ruddy-faced, robust young man known to his friends as Will, married Lydia in 1889. A year later, he and his bride moved to Birmingham when the Alabama Great Southern sent him there as a train dispatcher. Soon, Seddon attracted him to Sloss by offering him higher pay.

Beginning as a clerk, McQueen quickly moved upward in the company hierarchy, taking over the transportation department and helping organize Sloss's move into the export trade after depression struck in the mid-1890s. Winning promotion to auditor, he became secretary-treasurer in 1896 and was put in charge of sales. Along with mine superintendent Thomas Culverhouse, he represented the company in the district's Coal Operators Association and took part in wage negotiations with the United Mine Workers. The national trade journal *Iron Age* later commented on his "great capacity for work and a painstaking attention to the details of duty," stating that these qualities, coupled with his "wise handling and marketing of the company's output," were major factors underlying the firm's progress. His ability attracted the attention of Bryan, who corresponded with him more and more frequently as time went by, receiving information about local developments and using McQueen's help in pursuing business deals in and around Birmingham.[9]

Bryan's confidence in McQueen resulted in his elevation to second vice president in 1902 during the crisis surrounding Hopkins's forced resignation. Consolidating his growing power during Richey's unsuccessful stint as first vice president in 1905–1906, he played an increasingly dominant role in day-to-day operations under Maben's loose managerial reins. Shrewdly, he paid to have a full-page picture of himself included in Ethel Armes's book, *The Story of Coal and Iron in Alabama*; it contained an admiring account of his life and career when it appeared in 1910.[10] Over the years, McQueen acquired large real-estate holdings in the district and became a director in the American Trust and Savings Bank. Proud to be descended from "old Scotch fighting stock," he proved himself more than a match for John C. Maben, Jr., after the latter joined Sloss-Sheffield in 1911 as heir apparent. Seven years later, having won the ensuing power struggle, McQueen stood at the pinnacle of his career. By this time, he had also become chief of the MacQueen Clan in the United States.

Besides being a home-grown product, McQueen was unique among Sloss-Sheffield's presidents to date in having ascended through the ranks. Like four of his predecessors—Seddon, Haas, Hopkins, and Maben—he had a background in railroading. He differed from these men, however, in having had hands-on involvement in industrial management, a distinction that he shared only with Catchings. In being able to claim familiarity with day-to-day operations, McQueen was similar to his counterpart at TCI, Crawford, though he lacked the latter's record of formal academic training, European travel, and northern conditioning. Nor did he have the Wall Street background of some of Sloss-Sheffield's earlier chief executives or the patina of having studied at such outstanding institutions as Princeton and Harvard that Maben and Catchings could claim. Whatever he lacked, he made up in ability and drive.

Continuing the company's tendency to put railroad men in leadership posts, McQueen selected Landers Sevier as his right-hand man in a move that ironically recalled Bryan's earlier choice of Richey. Born in Canton, Mississippi, in 1866, Sevier went to work with the Alabama Great Southern at an early age. In 1907 he became vice president of the Seaboard Air Line Railway, following which the Southern tapped him in 1909 to become executive general agent, with headquarters in Birmingham. Sloss-Sheffield made him vice president soon after McQueen took office. Appropriately, Sevier began his new duties by taking charge of transportation; after performing this job, the *Age-Herald* stated, he would take control of sales and other functions, giving McQueen "more time to devote to the financial and purely executive branches of the company's activities." Sevier was also elected to the board of directors, becoming one of four members from the Birmingham area. Never before had the Magic City enjoyed such representation in the firm's inner councils.[11]

III

Despite a dramatic surge of demand for foundry pig iron that took place under wartime conditions, and a consequent rise in the price of iron, McQueen had a tough first year as president. Military service claimed approximately 700 of

James W. McQueen. Long the central figure in the day-to-day operations of Sloss Furnaces, James William McQueen was a highly effective marketing executive whose strong leadership as president of the Sloss-Sheffield Steel and Iron Company from 1918 until his death in 1925 marked him as the best chief executive in the history of the enterprise up to that time. This photograph of him at a relatively young age suggests the winsome characteristics that helped make him a superb salesman. Nevertheless, as he showed by repeatedly thwarting the ambitions of rivals who vied with him for power, he possessed formidable skills as a corporate infighter. (Courtesy of Sloss Furnaces National Historic Landmark, Birmingham, Alabama.)

Sloss-Sheffield's employees; in trying to replace them, McQueen was severely handicapped by a labor shortage throughout the district, partly because of black migration northward and also because wartime construction projects tended to soak up what workers were left. As a result, pay scales escalated rapidly, hiking costs. "Never has there been in the history of your Company a more unsatisfactory condition of labor as to its scarcity, its wage and its efficiency," McQueen lamented in his annual report for 1918, stating that the

National War Labor Board and other governmental agencies had mandated pay raises that the firm was ill prepared to grant. As demand for production intensified, the company was also forced to turn to outside sources for some raw materials costing more than ones from its own properties. Meanwhile, freight rates climbed steeply as local firms competed for scarce rolling stock. Under these circumstances, the cost of producing pig iron rose to an unprecedented $32 per ton.

Because of labor shortages and the difficulty of getting raw materials, Sloss-Sheffield could not keep up with demand. Indeed, output actually shrank. Part of the reason was the same labor inefficiency with which local managers had been struggling for decades. Try as the company might, McQueen told reporters after returning from a trip to New York in May 1918, it could not get anything approaching optimum performance from its work force. "Production is only about 75 per cent of what it should be with the number of men employed," he said. Within two months, personal grief added to his troubles when his wife died and was buried in Oak Hill cemetery. Rucker, Sevier, and Tutwiler were among the distinguished citizens who served as honorary pallbearers. [12]

During 1918, Sloss-Sheffield made only 387,497 tons of pig iron, as against an average of 475,000 for the preceding two years. Coal production fell to 1,568,019 tons from a previous two-year mean of 1,900,000, and coke production to 585,413 tons from an average of 600,000 in 1916–17. Despite its problems and the need to reserve funds to meet a wartime excess profits tax liability estimated at about $1 million, the firm posted a net profit of $1,972,071 for 1918 before paying $1,519,000 in stock dividends. Nevertheless, the $453,000 remaining for surplus was far below the figure for 1917. [13]

IV

Despite these difficulties, 1918 produced at least one major accomplishment. Public anxiety about what was happening in Europe helped goad Sloss-Sheffield, at long last, to begin building a battery of by-product coke ovens. The collapse of Russian resistance in the winter of 1917–18 enabled the Germans to transfer large numbers of troops to the western front, where Ludendorff soon launched a furious offensive against Allied positions. The onslaught produced heavy casualties and serious losses of territory before it was finally checked early in April. Later in the spring, as American forces under Pershing began to brace crumbling British and French units, another German offensive reached the Marne not far east of Paris before losing impetus. Renewed attacks followed in mid-July, centered around Rheims. [14]

The German drive intensified pressure upon American firms to produce munitions vital for the war effort. In Alabama, the federal government speeded efforts to build two large nitrate plants at Muscle Shoals following its selection late in 1917 by President Wilson and Secretary of War Newton D.

Baker to be the site of these facilities under legislation passed the previous year. Development of this area, rich in hydroelectric potential, had long been urged by promoters who now foresaw that cheap nitrates might fuel a postwar industrial boom in the Tennessee Valley. Demand also soared for such coal-tar derivatives as ammonia, phenol, and toluol. In April 1918, TCI announced that it would add two batteries of Koppers by-product ovens to the four it already operated at Fairfield. Reporting that the cost might run as high as $4 million, the *Age-Herald* hailed this project as yet another example of U.S. Steel's "patriotic purpose" and "large way of doing things."[15]

TCI's announcement stimulated rumors that Sloss-Sheffield would construct comparable facilities, reportedly near Sheffield so as to be close as possible to the nitrate plants under construction at Muscle Shoals. In mid-July, McQueen ended the suspense by announcing a contract recently negotiated with officials in Washington. Under its terms, Sloss-Sheffield would install 120 Semet-Solvay by-product ovens at its own expense. Following their completion, the government would buy their entire output of coal-tar derivatives, estimated at approximately 8,250 short tons of ammonium sulphate and 250,000 gallons of toluol per year, for two years. It was anticipated that the plant would be ready for production about 1 September 1919. The total cost, McQueen told reporters, would exceed $5 million. Much to the delight of local interests, North Birmingham was selected as the site in preference to northwest Alabama, where skilled labor was in short supply and housing costs had gotten out of hand as a result of the frenzied pace with which that area was being transformed by the nitrate project. Construction was soon underway in North Birmingham, and the *Age-Herald* reported steady progress as work proceeded.[16]

The project, however, had hardly started before the war took a sudden turn for the better. An allied offensive that began on 18 July rapidly overran German positions in France. American troops, now present in force, added weight to the initiative at such places as the St. Mihiel salient and the Argonne forest. By mid-August, Germany's situation was growing desperate; in September and October, it became even worse with the collapse of its allies, Austria-Hungary, Bulgaria, and the Ottoman Empire. Efforts to secure peace at almost any price intensified as morale crumbled on the home front, the navy mutinied, and the Kaiser abdicated. On 11 November 1918, an armistice signed in a railroad car near Compiegne in eastern France formalized Germany's surrender, ending hostilities.[17]

Suddenly, the emergency conditions under which Sloss-Sheffield had agreed to build its new by-product coke ovens vanished. Instead of pressing American firms to undertake large war-related projects, federal officials began a wave of contract cancellations. In mid-December, McQueen was told by the government to suspend construction of the by-product plant, pending negotiation of a supplemental contract under which the company would be compensated for losses due to cancellation of the July agreement. Bowing to this demand, Sloss-Sheffield began meetings with a federal adjustment board in Cincinnati to reach a fair settlement based on building costs already incurred and loss of projected sales.

The discussions at Cincinnati took account of the fact that the project had begun at a time when construction costs were abnormally high due to the large number of facilities that were being built throughout the country under emergency conditions. Represented by Adrian Larkin from the New York headquarters and by Birmingham attorney Hugh Morrow, Sloss-Sheffield proposed three alternative plans for compensation. Under the first, aimed at producing a settlement estimated at $3 million, construction would end, and the firm would receive all the costs it had incurred, plus overhead. Under the second, involving a potential award of approximately $2.5 million, construction would continue, and Sloss-Sheffield would secure damages based on a differential between prewar costs and price levels prevailing in 1918. Under the third, which would yield an estimated $2 million, the company would receive an award based on a complex formula that took into account the difference between the past and present prices of commodities specified in the contract, the projected capacity of the North Birmingham installation, and profits recorded by "other and similar constructed plants."[18]

The final agreement, reported to the directors by McQueen in June 1919, netted Sloss-Sheffield only $1,525,207.30. Although this was much less than the firm had hoped for, it permitted construction of the North Birmingham plant to resume. Because of repeated delays in the unsettled postwar period, however, ten months elapsed before operations finally got underway. The startup took place in two stages, beginning in mid-April 1920, when sixty of the new ovens went on line; the remaining units were functioning by the end of May. Anticipating that substantial profits would accrue from lower raw material costs and sale of by-products, McQueen took satisfaction in telling the directors that the project had stayed within its $6 million budget.[19]

Far from playing a pioneering role in putting the new plant in operation, Sloss-Sheffield was bringing up the rear; production of by-product coke in Alabama had exceeded the beehive variety every year since 1913. During the last year of the war, 2,233,210 tons of by-product coke had been made, as opposed to 1,164,528 of the beehive type. Within less than a decade, no beehive ovens would remain in operation in the entire state.[20]

Still, the new plant was an impressive facility, hailed by the *Age-Herald* as typifying "the indomitable spirit of the industrial captains, both of today and those that are gone." The sprawling complex covered about twenty-five acres. Besides the ovens itself, it included administrative offices, a boiler plant, an engine house, a laboratory, a machine shop, a six-story ammonia concentration building, two pump houses, a large reservoir, and two spray ponds. Because a lack of water was one of the Birmingham District's most enduring problems, great care was taken to assure an ample supply of this critical liquid from three separate sources and to provide for systematic recycling. All power came from electricity. Three railroads, including the L&N and Southern, served the facility, bringing washed coal from company mines. Stopping above a funnel-shaped receptacle, cars dropped their loads onto an inclined belt conveyor that carried the coal to bins located above and between the twin batteries of ovens. Receiving coal from the bins, sixteen-ton hopper cars ran back and forth to deposit it into the coking chambers.

Unlike Koppers installations, which had a vertical design, Semet-Solvay ovens featured horizontal flues that increased in size as they descended from the top of a heating chamber. Combustion began at the top, and air and fresh gases were added to maintain uniform temperature as coking proceeded. Ovens were separated from one another by division walls to permit individual units to be repaired while others continued operating. As was customary in oven design from the time of Sir William Siemens onward, the regenerative principle was employed, and the flow of heated air was periodically reversed. This was done in such a way that an entire battery of sixty or more ovens was swept at one time, preventing "hot spots" and ensuring consistent temperatures.

After the coking process was completed, the incandescent fuel was conveyed by rail cars to a quenching station that used recycled hot water, again demonstrating the care that was taken to reduce liquid consumption. As an article in *Iron Age* pointed out, Semet-Solvay's engineers had found that "a smaller quantity of hot water is required for quenching a car of coke than when cold water is used." After cooling had taken place, requiring about fifty seconds per carload, the coke was dumped on a sloping wharf, delivered by gravity to a rotary feeder, loaded onto a belt conveyor, and taken to a screening station. Fine particles of coke, known as "breeze," passed through the screens and were returned to mechanical stokers at the boiler plant, providing further evidence of the way in which the entire facility had been designed to conserve materials and control costs. The remaining coke fell into rail cars and was taken to the blast furnaces.

Meanwhile, gases produced by coking passed through a maze of washers, pipe coils, decanters, and other apparatus for further processing. The gases first yielded tar, then ammonia, then light oils that could be fractionated into a variety of chemical compounds. Residual by-products were used to provide fuel for Sloss-Sheffield's own boilers and dynamos and supply energy to outside customers. The plant had two 2,000-kilowatt electric turbines the output of which was supplemented by that of a 2,000-horsepower generating facility at the nearby North Birmingham furnaces. Ultimately, the company planned to use the resulting capacity to supply all the electricity needed by its blast furnaces, mines, and quarries after these properties had been modernized. "It is difficult for the human mind to conceive a more complete use of a raw material," stated an admiring newspaper article after recounting the ways in which a single substance, coal, was transformed to meet an almost endless series of industrial and commercial requirements. "The only thing comparable is the meat packing plants, where it is claimed that everything is utilized except the pig's squeal."

The new plant had a maximum capacity of about 62,000 tons of coke per month, requiring about 85,000 tons of coal. Its output would supply five blast furnaces, as many as the company normally operated at any given time. This yield enabled the firm not only to retire once and for all the beehive ovens at the City Furnaces that had so long been a public nuisance but also to abandon other obsolete batteries at its coal mines. Meanwhile, it could profit from the sale of by-products derived from gases that had previously gone to waste. The recovered gases had a potential daily yield of 20,000 gallons of crude tar, 60,000 pounds of ammonium sulphate, and 7,000 gallons of light oils or other fractional compounds, including benzol, naphtha, and toluol. These chemicals were marketed for Sloss-Sheffield by the Barrett Company, which handled up to 90 percent of the entire American output of by-product coal tar not directly consumed by producers themselves. Even this list of benefits did not exhaust the rewards that would accrue from the new plant, for there would remain enough fuel gas to furnish 7 million cubic feet of marketable surplus each day after the company's own electrical needs had been met.[21]

Bursting with pride, the *Age-Herald* declared that Birmingham was "especially pleased to have the Sloss-Sheffield company take this forward step because it is Alabama's oldest and largest strictly foundry iron producer and has attained its prestige in the industrial world under the management of strictly southern men." Despite such self-congratulation, funding for the new installation had come not from "strictly southern men" but from Wall Street and the federal government. Shortly after McQueen took office, the company completed arrangements to liquidate $4 million worth of mortgage bonds, mostly dating back to the takeover by Bryan and his friends of the Sloss Furnace Company in 1887. Half of these obligations were paid on 1 April 1918 from surplus cash; the rest were discharged on 1 February 1920 after the company had floated $6 million worth of 6 percent gold notes underwritten by three leading New York investment banks. One of these was the Central Union Trust Company, as the Central Trust Company was now known after a merger with a former competitor. The second was Goldman, Sachs and Company, reflecting the role Catchings now played in that Wall Street investment house. The third was Lehman Brothers, stemming from Sloss-Sheffield's recently established connection with Semet-Solvay. Besides providing the means to retire old mortgage bonds, much of the new bond issue went into paying for the by-product coke ovens. The $1.5 million settlement from the federal government took care of the remaining cost of building the North Birmingham facility.[22]

V

As the postwar period got underway, continuity was mingled with change. Reaffirming established tradition, Sloss-Sheffield retained the Central Union Trust Company as its chief financial agent. A new director with strong ties to that institution, James N. Jarvie, was elected to the board in 1919 following the death of Richmond banker M. C. Branch, an event that severed yet another link with Bryan's former group of Virginia associates. Two other directors passed away during the same year. One was James N. Wallace, who had served briefly as president in 1917 before handing the reins to Catchings. "My God, I am dying," he cried as he collapsed from a sudden heart attack at his home near Nyack, New York. His place on the board was taken by George W. Davison, who had earlier succeeded him as president of Central Union Trust. Meanwhile, Hugh Morrow, who had ably represented Sloss-Sheffield as legal counsel at Cincinnati, became a vice president in 1919 and was named a director at the same time. His presence on the board affirmed the trend toward giving Birmingham strong representation. [23]

The succession of deaths that was steadily removing persons who had figured importantly in development of the firm continued during the early postwar years. Maclin Sloss, who had come from Athens, Alabama, to Birmingham with his father and brother Frederick during the formative years of the district to prospect for ores in the surrounding hills, and had served as one of the first officers of the Sloss Furnace Company, died in January 1919. In March 1922, Oakman, who had been a constant source of financial support to Bryan, died in New York City. Eight months later, Dooley, who had suffered two strokes since his resignation as a director, passed away at his Maymont mansion near Richmond. [24]

VI

Having finally resolved the by-product coke oven debate and successfully rearranged its finances, Sloss-Sheffield plunged into a difficult period of postwar readjustment. It was widely feared that a recession would take place as soon as the war ended. This anticipated development did not immediately occur, partly because the government continued to spend large amounts of money for European emergency relief and supported the building of a large merchant fleet and also because treasury officials failed to restrain credit for fear that this would hamper industrial production and limit jobs for returning veterans. Sloss-Sheffield benefited from these policies, which were in effect while it was liquidating its old mortgage bonds and floating the new $6 million note issue. [25]

But the respite was only temporary. Despite continuation of the wartime boom, a number of important sectors of the economy, including the coal, textile, iron, and steel industries, soon ran into trouble. Crippling strikes at Pittsburgh and in other steelmaking districts in 1919 had an unsettling impact on the ferrous metal industry. Because of the unavailability of coal and the effect of concurrent strikes by their employees, rail lines on which iron and steel producers depended also shut down. The lifting of wartime controls further destabilized iron and steel prices, and the Justice Department compounded prevailing uneasiness by deciding that efforts by the secretary of commerce to set new guidelines violated antitrust laws. Amid these conditions, the national output of ferrous metals fell precipitously, reaching the lowest point since 1908. Conditions continued to fluctuate unpredictably during the next few years as the industry searched for an elusive stability.[26]

As the South's leading industrial center, Birmingham could not escape the turbulence of the times. TCI, hard hit in 1918 by an unsuccessful nine-week walkout called by the local Metal Trades Council, was threatened once more in 1919 when U.S. Steel refused to grant a series of demands made by the American Federation of Labor (AFL). In New York, corporate officials trusted that continued implementation of welfare capitalism would allay discontent, but this attitude proved delusory. Beginning in late September, a wave of strikes broke out in Pittsburgh, Gary, South Chicago, Youngstown, and other steelmaking cities; by the end of the second week, 365,000 men were idle. The disturbance soon spread to Alabama as about 500 men walked out at TCI's Ensley plant. Night after night, Crawford anxiously scanned the sky from his Red Mountain home for a fiery glow on the horizon, reassuring him that the mills were continuing to operate. Much to his relief, the strike ended quickly.

Trouble in the traditionally militant Alabama coal fields proved to be more of a problem. Here, during the war, federal officials had tried unsuccessfully to persuade TCI to agree to a series of labor demands without conceding recognition to the United Mine Workers (UMW). Amid persisting tension, a strike broke out in September 1920; it quickly produced violence, threatening to stop work at TCI's Bayview and Edgewater mines. Despite attempts by Gov. Thomas E. Kilby to put down the strike with armed force, it dragged on for five months. As in previous battles between management and labor, the walkout ended in humiliation for the unions. As violence continued and public opinion turned against the workers, the latter were maneuvered into accepting arbitration by a board whose members were appointed by Kilby himself. After it rejected their demands, the UMW withdrew support for the strike, which quickly collapsed. Crawford, who had sympathized with the workers to some degree but kept quiet due to U.S. Steel's stalwart anti-unionism, was conveniently away from Birmingham during much of the dispute and remained discreetly aloof. Predictably, Sloss-Sheffield's stance toward the UMW was hostile; in a speech to the Birmingham Kiwanis Club, one of its officers, Russell Hunt, called for vigilantes to take the law into their own hands against Van Bittner, a labor organizer who had been sent to Alabama by UMW president John L. Lewis and was depicted by local operators as a champion of racial equality.[27]

McQueen did not even mention the coal strike in his annual report for 1920; apparently he had weightier problems on his mind as he looked back on a troubled time of wartime adjustment that had given way to an equally precarious period of peacetime uncertainty. In 1919, responding to the sharp contraction of markets throughout the ferrous metals industry as a whole, pig iron production at the company's furnaces had shrunk from 387,000 to 250,000 tons. Output of ore, coal, and coke had also plunged as the price of pig iron plummeted from $34 to $26.75 per ton during the first three months of the year alone. Ordinarily, the company kept five furnaces in blast; for almost six months beginning in April, however, it operated only two. Demand revived late in the year, causing pig iron to rise to $38 per ton, but gross profits fell to approximately $1.8 million, less than one-third of what they had been in 1918. Although the firm posted net earnings of $2,514,826 and added $1,025,826 to surplus after paying dividends on both common and preferred stock, this was possible only because of the $1.5 million settlement negotiated with federal officials in Cincinnati.[28]

Demand remained strong for a time during the first half of 1920. An article in the *Age-Herald* at the end of May stated that the local iron trade was "throbbing with record business" as orders poured in from such places as England and Scotland. "All the pipe plants are as busy as can be," the paper said, and "new ones are rushing construction so as to be in the game as soon as possible." But markets contracted again early in the fall as the country was hit by a severe recession caused by a steep decline in government spending since the end of the war, an increase in business and personal taxes, and restrictions on credit as key banks belonging to the Federal Reserve System raised rediscount rates. By 1921 these conditions produced what George Soule later called "one of the most violent crashes of prices that the nation has ever experienced." Like other iron and steel producers, including TCI, Sloss-Sheffield curtailed output drastically, shutting down all but one of its blast furnaces and canceling previously scheduled physical improvements. Idle workers roamed the streets of Birmingham, and empty railroad cars became common throughout the district.[29]

Luckily for Sloss-Sheffield, the start of operations at the by-product plant cushioned the impact of the crisis. After the new ovens went on line, the company agreed to supply the city of Birmingham with 6 million cubic feet of gas per day. Abandoning beehive ovens also yielded cost reductions, not only at the City Furnaces but also at the Blossburg, Brookside, Dora, and Flat Top mines. As a result, Sloss-Sheffield earned a net profit of $2,293,779.07 in 1920 despite the contraction.

As hard times intensified in 1921, however, the downturn became a rout, and Sloss-Sheffield posted the smallest output of pig iron in its entire history. Production shriveled to 15 percent of the previous year's level. Net profits sank to $514,356.85, and dividends were paid on common stock only during the first quarter. Had it not been for operations at the by-product plant, which went on uninterrupted, the situation would have been worse. This continuation of coke production, however, caused fresh problems. By making large amounts of coal-tar derivatives, the company accumulated a large surplus of coke that it could not use in its idle furnaces. This commodity was later sold to foundries in the Middle West, but carrying excessive inventory was temporarily worrisome.

Prosperity began to return in 1922 as the postwar roller coaster continued. Stimulated by pent-up postwar demand for housing, residential construction helped fuel the recovery, as did rising consumer spending. Sloss-Sheffield, however, did not feel the effects of the upturn until summer, and net profits for the year rose only slightly to $578,893.68. As markets revived, five furnaces were kept in blast from June throughout the remainder of the year, but income from sale of iron and other products merely made up for previous losses, and payment of dividends stopped. Instead of sharing the firm's meager earnings with the stockholders, McQueen took advantage of the recovery to retire floating debts that had accumulated during the recent downturn.

Sales of pig iron became so brisk in March and April 1923 that experts feared a runaway market, but demand subsided in May as a minor recession began and remained dull until December, when another upswing commenced. Despite such ups and downs, net profits for the year soared to $2,491,019.55, and preferred stockholders received dividends in April, July, and October. Still, much of the improvement was due not to iron sales but to surging demand among local residents for the company's by-product gas. In 1923 alone, city authorities purchased 1,396,000,000 cubic feet from Sloss-Sheffield. Clearly, the company was benefiting from having built the North Birmingham installation.

Another mild recession occurred in 1924. Because automobile production went into a temporary tailspin, it hit the iron and steel industry harder than most. As a result of the downswing, net earnings fell to $1,516,276.67, but steady dividends were paid on both common and preferred stock as the by-product plant continued to pick up the slack. As the decade neared its midpoint, McQueen could take satisfaction in having pulled the firm through a series of difficult postwar adjustments. [30]

VII

McQueen showed courage and prudence in curtailing dividends during the worst of the hard times that the company suffered in the early 1920s. At the same time, he also showed faith in the future by continuing to press for modernization. Conditions were ripe in the postwar era for such effort; not only in the South but throughout the nation as a whole, manufacturing firms were achieving significant productivity gains. Investment capital was available at low interest rates, and the cost of equipment was declining relative to that of labor, inducing managers and owners to adopt new technologies to increase efficiency.

Under McQueen's leadership, Sloss-Sheffield began to modernize blast furnace operations for the first time in more than a decade. In 1919, as labor conditions remained unsettled and pig iron prices were beginning what could turn into a steady decline, McQueen began a series of modifications at the North Birmingham stacks. Among other things, he authorized furnace manager James P. Dovel to design and install an overhead crane and a large air-driven hammer in the casting shed to hoist sows into place and knock away individual pigs, replacing decades-old methods of breaking and carrying by hand. By April 1921, the new system was in place at both of the North Birmingham furnaces. Although only an incomplete form of the automatic pig-casting method that Uehling had invented in the 1890s, it lowered costs significantly. McQueen also installed modern hoisting equipment to replace earlier charging methods, another step toward increased efficiency and control of operations. The renovations foreshadowed a much more ambitious program of modifications that would take place at Sloss-Sheffield's City Furnaces later in the decade. [31]

During the postwar period, McQueen also began modernizing mining operations, which had previously used electricity only for pumping. He did so despite costs initially estimated at $1.5 million for transmission lines, motive power, compressors, drills, and other equipment, again showing that the company was willing to invest heavily in rationalization. Because of the need to assure a better coal supply for the by-product plant, this fuel was the first raw material affected by the change; various improvements in coal mining technology took place at Bessie, Blossburg, Brookside, and Flat Top, ending in April 1921. The rest of the project was concluded in August 1922 with the installation of transmission lines to the Sloss ore mines. By this time all major mines belonging to the company were being supplied with electrical energy generated by the central power station at the by-product plant. The wisdom of making the financial outlay needed for this effort, McQueen told directors, was "constantly reflected in the cost sheets." [32]

VIII

Sloss-Sheffield followed yet another prevailing postwar trend by absorbing two firms through merger. During the war, there was a resumption of the movement toward larger industrial units that had been so strong at the turn of the century. This trend continued after 1921 under Republican presidents who had little interest in enforcing antitrust laws or fostering competition. As secretary of commerce, Herbert Hoover promoted policies aimed at cooperation and consolidation. Throughout the decade, firms seeking increased control of raw materials, production facilities, and markets pushed steadily toward higher levels of concentration, and investment bankers who relished the resulting underwriting commissions were glad to oblige. Between 1919 and 1929, 1,268 mergers took place in the United States among mining and manufacturing firms alone, resulting in the absorption and disappearance of 7,256 formerly separate enterprises. By the end of the decade, almost half of America's corporate wealth was held by only 200 companies. [33]

Sloss-Sheffield consummated a merger in April 1923 by acquiring, through a foreclosure sale, all properties belonging to the Sheffield Iron Corporation. During the war, a syndicate headed by James Gayley, a former partner of Andrew Carnegie, had formed that enterprise by purchasing the three Cole

Furnaces in northwest Alabama. Because of the cost of bringing coking coal to the area, the stacks had been only intermittently in blast for many years. Efforts by the new owners to return them to operation were hampered by litigation and a lack of blowing equipment. As a result, only one furnace went into blast until unsettled business conditions in 1919 forced the Gayley syndicate to shut it down. Not long after the war, the Bankers Trust Company of New York foreclosed a $1.1 million mortgage on the property, which remained idle until Sloss-Sheffield bought it for $265,000 in 1923. By this time, only one of the old stacks was still standing. Increasing its current levels of iron production was not Sloss-Sheffield's motive for acquiring the dormant enterprise; it did so chiefly to gain 7,000 acres of additional coal and ore lands in northwest Alabama and west Tennessee, significantly augmenting its mineral reserves. McQueen concentrated on this aspect of the deal in his next annual report, not even mentioning the furnace.[34]

More important than taking over the Sheffield Iron Corporation, particularly because of the extent of the properties involved, was the acquisition in October 1924 of another ironmaking enterprise, the Alabama Company. This purchase, which added $3,675,000 to Sloss-Sheffield's bonded indebtedness, was made for a variety of motives, including getting more mineral reserves, securing sales benefits from two popular brands of foundry iron, and gaining experience in operating the only automatic pig-casting equipment owned by any merchant iron producer in the state. The transaction also prevented the Alabama Company from falling into other hands, particularly those of Woodward, which had embarked upon its own expansion program during the war.[35]

The Alabama Company had been organized in 1913 by Baltimore capitalists. It included a large number of properties that had previously belonged to the Alabama Consolidated Coal and Iron Company, which had been formed in 1899 during the wave of mergers that had produced Sloss-Sheffield itself. Some of the holdings that were acquired by the Baltimore syndicate from Alabama Consolidated in 1913 were located in Talladega County, including valuable brown ore lands and two stacks that had been known at various times as the Clifton or Ironaton furnaces. These installations had originally been built to use charcoal but were later modified for smelting with coke, with a rated capacity in 1924 of 200,000 tons per year. Their "Clifton" brand of pig iron, high in manganese and low in phosphorus, commanded premium prices. Other assets acquired by the Alabama Company were located in Etowah County, including red hematite deposits supplying a furnace that had been built in the late 1880s. Like many other marginal installations, this stack had been blown out in 1893 when hard times struck the industry. By the time the Alabama Company bought the property twenty years later, it had been reconstructed and another furnace now stood on the same site. Further properties taken over by the Alabama Company in 1913 that ultimately gravitated to Sloss-Sheffield included coal mines in Tuscaloosa County and a tract of land at Gate City containing red ore, limestone, dolomite, refractory clay, and sandstone.

After World War I broke out, the Baltimore syndicate launched a major expansion program and profited temporarily from the high price of iron. With the return of peace, the owners ran into trouble as markets gyrated erratically, explaining why they were willing to sell out by 1924. Although the takeover by Sloss-Sheffield increased its rated annual capacity for pig iron production from 600,000 to 900,000 tons, some of the furnaces that it acquired in the deal were of only marginal value. As Gary Kulik has pointed out, they were obsolete, lacked ready access to good coking coal, and were located in small cities, thus being vulnerable to labor shortages as more and more black workers left the region. [36]

Nevertheless, the Alabama Company was worth acquiring, if only to keep it out of Woodward's grasp. Despite uneven conditions prevailing in 1923, the Alabama Company's output for that year included 121,171 tons of pig iron, 938,256 tons of coal, and 293,153 tons of coke. Its 62,500 acres of coal and ore properties comprised four coal mines, two brown ore mines, five red ore mines, and two limestone quarries. But its most notable assets were the two furnaces at Gadsden, which constituted the most technologically advanced merchant ironmaking facility in the state. Among its brands of pig iron was a product, marketed under the trade name Etowah, that was highly regarded by foundrymen. One of the stacks, which had been extensively rebuilt in 1921, was equipped with a single-strand Heyl & Patterson pig-casting machine that had reduced the number of men required to operate the plant from 305 to 160. In addition, the apparatus produced smaller, cleaner, and more uniform pigs that were increasingly wanted by some foundrymen who were beginning to appreciate these characteristics. Similar in many respects to the equipment invented by Uehling in the 1890s, the mechanism was the first automatic casting device to be used by an Alabama merchant furnace. Its acquisition by Sloss-Sheffield was another step toward modernization, permitting the firm to familiarize itself with techniques that later became standard at its City and North Birmingham furnaces. [37]

IX

The takeover of the Alabama Company was the chief topic of the annual report that McQueen planned to give at Sloss-Sheffield's yearly meeting at Jersey City in April 1925. The same document contained tributes to no fewer than four directors who had recently died, including T. C. Camp and Henry Evans, who had served on the board since 1916 and 1917 respectively. Both men were also members of the executive committee. [38]

The other two deaths were more important, because they involved individuals who had played extraordinary roles in the company's development. One, General Rucker, had been in bad health for four years. Still, nothing suggested that his death was imminent as he strolled on the terrace of his Birmingham home on Sunday afternoon, 13 April 1924. According to the next day's *Age-Herald*, "he seemed to experience a heating sensation" later that evening, and "asked for fresh air." Suddenly, he was dead. Had he lived until 22 July, he would have been eighty-eight years old.[39]

At their next meeting, the directors adopted a resolution recounting Rucker's long record of service to the firm and describing his life as one that "began in hardship and ended in splendor." After mentioning Rucker's valor in the Confederate cause, which had cost him an arm at Chickamauga in 1864, the statement touched upon his later business career in such a way as to obscure the stormy nature of his relationship with other persons who had been involved in Sloss-Sheffield's sometimes turbulent history. "Severely maimed and wounded on the battlefield, he returned at the close of the war to a devastated home, to begin as it were, life anew," the tribute declared. "Nature had added to the strong, rugged character which distinguished him as a soldier, a splendid judgment of business and of men. These talents, which were given by him freely to this company, redounded greatly to its success."[40] Eighteen years later, in 1942, the U.S. Army made a lasting contribution to his memory by naming an important military base Fort Rucker in his honor. Located near Enterprise, Alabama, it became known as the "Home of Army Aviation" after August 1954, when the U.S. Army Aviation School was transferred there from Fort Sill, Oklahoma. From that point on, the rolling hills surrounding the base reverberated to "the steady whop-whop-whop of helicopter rotor blades," as one writer observed, forming "a backdrop to the normal cadence of life."[41]

Less than five months after Rucker died, Maben was next to go, dying in Atlantic City, New Jersey, on 31 August 1924 after a long illness. Like Rucker, he was eighty-seven years old. Lamenting his death, the *Age-Herald* stated that he had been "one of the pioneer iron and industrial men of this District, believing always in the paramount position of Birmingham among industrial communities." Along with such persons as DeBardeleben and Ensley, the paper said, he had never doubted "that the District was destined ultimately to become the leading industrial section of the world." By heading the story "Pioneer Steel Man Succumbs At Atlantic City," the newspaper misstated the role that Maben had played in Birmingham's history, especially in view of his successful opposition to Hopkins in the 1902 crisis that oriented Sloss-Sheffield toward the foundry trade. Like Joe McGinnity of McGraw's New York Giants, who had pitched in the Vulcan Day game of 1904, Maben was emphatically an "Iron Man."[42]

The deaths that winnowed the company's boardroom at this time wiped out most of the remains of the old order that had been instituted by Bryan and his associates in the mid-1880s. As the aging process continued exacting its toll, a cadre of new directors, later to be identified and discussed, became dominant. Their financial ties reflected changes that were taking place throughout the American economy. Whereas previously the company had been directed by railroad men, it was now controlled by Allied Chemical and Dye, a holding corporation formed in 1920 in a merger of the Barrett, General Chemical, National Analine, Semet-Solvay, and Solvay Process companies. The new regime reflected Sloss-Sheffield's heavy involvement in the by-product chemical field and its close connections with both Barrett and Semet-Solvay.[43]

McQueen was ill as he boarded the train that took him north to the 1925 annual meeting, where he was also to preside at a conclave of the MacQueen Clan. Upon arriving in New York and registering at the Waldorf-Astoria, he had such a high fever that he was put under the care of a physician, who diagnosed his condition as a mild case of influenza. Friends and relatives who hastened to his bedside departed after a few days when it appeared that he was out of danger, but this impression compounded their shock when he had a relapse and died suddenly on Monday, 20 April, six days after arriving in the city. While his body was taken back to Birmingham for burial from the Episcopal Church of the Advent, the *Age-Herald* paid tribute to him as "one of the District's outstanding leaders who had been a conspicuous factor in Birmingham's industrial development during a period of more than thirty-five years." In that span of time, the article continued, the company that he served had grown "from a struggling pioneer concern to one of the nation's outstanding coal, iron and by-products producers." The national trade journal, *Iron Age*, echoed these remarks; it stated that McQueen "had had a part in every development of the Southern pig iron market, from the days when most of the Alabama product was sold in the North, often in a profitless competition . . . to a time when the industrial development in the South was sufficient to absorb nearly all the output of Alabama furnaces." Shortly after McQueen's passing, Sloss-Sheffield's directors commissioned the creation of a leather-bound booklet, adorned with ornate calligraphy, paying tribute to his memory.[44]

X

McQueen left an impressive legacy. In every respect, he had been an effective president, the best in the firm's history to date. Ending years of indecision, he had resolved the nagging by-product coke oven issue. In addition to lowering costs by eliminating antiquated beehive ovens that had wasted natural resources and constituted a public nuisance, this resolution had diversified the company's markets, thus moderating the impact of periodic downturns in demand for its traditional products. McQueen had also ended another protracted debate by electrifying the company's mines, further enhancing efficiency. At North Birmingham, he had made limited but significant progress toward the mechanical handling of pig iron, beginning a shift from long-standing policies that had favored labor-intensive strategies over adoption of newer and more efficient methods. By acquiring the Sheffield and Alabama companies, he had increased Sloss-Sheffield's mineral reserves, strengthened its market position by acquiring new brands of foundry iron, gained an up-to-date pig-casting operation, and forestalled acquisition of these enterprises by the firm's chief competitor, Woodward. In addition, he had steered the firm through a difficult period of postwar adjustment, during which the country had experienced a severe recession and a succession of sudden economic shifts. Navigating such troubled waters had required remarkable managerial skills.

Immediately ahead, under an able new chief executive, Sloss-Sheffield would continue implementing the program that McQueen had begun. Meanwhile, looking back on his administration, the local business community could take satisfaction in knowing that a person who had lived so long in Alabama as to be considered a native son had performed with great ability at the company helm. Like the Tutwiler Hotel, it was something in which the Magic City could take pride as it anticipated a period in which Sloss-Sheffield would be headed by yet another home-grown product.

14 Morrow and Modernization

During the 1920s, Birmingham's boosters did not realize that they were heading into an economic maelstrom and remained as confident as ever. Within two decades, the Chamber of Commerce predicted, the Magic City would rival Chicago in size. John R. Hornady's *Book of Birmingham*, published in 1921, envisioned vistas of future growth that would produce "one of the world's greatest centers of population, and thus fulfill to the utmost the vision of that little group of men who, fifty years ago, entered a barren waste and there laid the foundations of a great city." Four years later, a writer in *World's Work*, saying that the people of Birmingham were "keenly conscious of their opportunities, and believe firmly in their destiny," called the community "the next capital of the steel age."[1]

Despite such optimism, evidences of malaise abounded. The environmental costs of industrialization were multiplying. There were obvious indications that the dream of salvation through steelmaking had been illusory. New political leaders, drawn from the middle class but now representing white industrial workers, were failing to deal with the problems of a fragmented social order. Members of the old elite, frustrated and discouraged, were retreating from the activist role that such wealthy reformers as George B. Ward had played in the Progressive Era. For the most part, such signs of decay went unrecognized or were ignored.

Both in Birmingham and throughout the country as a whole, few people questioned that science and technology were magic keys that would unlock future progress. Sloss-Sheffield marched in step with this faith. Under the leadership of Hugh Morrow, a native son who assumed the presidency of the firm after McQueen's death, modernization was vigorously pursued, and strong support was given to the inventive efforts of James Pickering Dovel, who along with Uehling was one of the two greatest engineers in the history of the enterprise. Preaching a scientific gospel, the company began to bring the entire foundry trade abreast of modern standards through a new publication aimed at replacing rule-of-thumb practices with more rigorous standards and methods. As economic disaster struck the nation at the end of the decade, the company's technological progress had put it in a good position to survive the ensuing crisis.

I

Until the Great Depression hit Birmingham in 1929, boosters shrugged off the city's problems and exulted in its future prospects. Statistics seemed to validate their visions. In 1920, Birmingham's population reached nearly 179,000, and that of the surrounding district surpassed 310,000; both figures indicated that Jones Valley was growing faster than most other American metropolitan areas. On 24 October 1921, a mammoth celebration at Avondale Park observed the Magic City's fiftieth birthday, featuring a "Pageant of Birmingham" in six scenes. It traced the evolution of the district from prehistoric times to the present and had a cast of 2,000 people. John C. Henley, scion of one of the city's oldest families and a descendant of its first mayor, impersonated James R. Powell and delivered an oration echoing the promotional rhetoric at which the latter had once excelled. Birmingham, Henley claimed, was "the Magic City of the World, the marvel of the South, the miracle of the Continent, the dream of the Hemisphere, the vision of all Mankind!"[2]

Highlighting the anniversary festivities was a visit by Pres. Warren G. Harding. Arriving by special train, he was taken by motorcade to Woodrow Wilson Park, as the former Capitol Park was now known, to address "the largest crowd ever assembled in Birmingham." Harding found much about which to praise the city, but the event turned into a debacle when he urged that black people be given equal voting rights with whites. As blacks in a segregated section cheered, whites fell silent; angry, Harding turned to the latter and said, "Whether you like it or not, unless our democracy is a lie, you must stand for that equality." After he returned to Washington, a local newspaper called his remarks "tactless," an "untimely and ill-considered intrusion into a question of which he evidently knows little." As Carl V. Harris later said, "Harding had appalled Birmingham by highlighting at its moment of most visible triumph its most tense and distressing social division."[3]

"Birmingham is the *nouveau riche* of Alabama cities," wrote Carl Carmer, an upstate New Yorker who had come south to teach English literature at the University of Alabama after studying at Harvard and serving as a gunnery officer in the war. "With an arrogant gesture she builds her most luxurious homes on a mountain of ore yet unmined. . . . She has no traditions. She is the New South. On one side of her rises a mountain of iron. On another a mountain of coal. She lies in the valley between, breathing flame." Carmer scorned the local aristocracy for collecting second-rate art, decried the lack of money spent on books for the public library, and coupled the popularity of service clubs with the hypocrisy of a community that had more Sunday school students per capita than any other place in America, yet allowed Catholics, Jews, and black residents to be beaten by white-hooded night riders. But what particularly aroused his imagination was the sheer spectacle of the city's industrial might. "The valley of the furnaces is an inferno," he wrote. "Molten steel, pouring from seething vats, lights the night skies with a spreading red flare. Negroes, sweating, bared to the waist, are moving silhouettes. On the top of a big mold they tamp the sand in rhythmic unison—a shambling frieze. Steel cranes, cars of the juggernaut, screech above the simmering red pools of spitting, rippling metal."[4]

Carmer was wrong about a number of things. Taking the concept of a New South at face value, for example, he failed to realize that Birmingham did have traditions that were deeply rooted in the old plantation order. Basically, his essay typified a revulsion against commercial values that was common among American intellectuals during the 1920s; the book in which it appeared, *Stars Fell on Alabama*, came from the same mold as Sinclair Lewis's *Babbitt*, F. Scott Fitzgerald's *The Great Gatsby*, and H. L. Mencken's "Sahara of the Bozart." Still, despite their trendiness, Carmer's words, like Harding's critical remarks, rang true. Betraying Ethel Armes's hopes for a marriage of Aphrodite and Hephaestus, and notwithstanding Crawford's achievements in welfare capitalism at TCI, Birmingham suffered from a failure on the part of its ruling classes to blend technological progress with human fulfillment. As Glenn Eskew has argued in a probing analysis of the city's politics after World War I, its elite bore much of the blame for this situation by having abandoned responsibilities that should have accompanied the exercise of economic power.[5]

Some of Birmingham's wealthiest citizens continued to live on or near Highland Avenue, winding its way up Red Mountain through some of Alabama's most affluent neighborhoods. Many, however, moved over the crest and beyond the city limits into Shades Valley, where developer Robert Jemison, Jr., created one of the most beautiful residential areas in the United States: Mountain Brook. For years, it had the distinction of being one of the nation's ten highest communities in per capita income. Like other affluent Americans who were fleeing to suburbia in the 1920s, its residents abdicated involvement in the problems of the urban core by taking refuge in a world of their own.

Jemison was the greatest developer in Birmingham's history. Like many of his clients who bought homes in Mountain Brook, he was descended from a tradition of wealth and political power. His father had been a prominent antebellum slaveholding industrialist, an associate of James R. Powell in the stagecoach business, a Confederate senator, and a real-estate executive and streetcar pioneer in the postwar era. Besides heading the district's most prestigious development company and publishing his own magazine, Jemison, Jr., was president of such firms as the Iron and Oak Insurance Company, the East Lake Land Company, and the Birmingham Railway, Light and Power

Company. His sprawling rural estate, Spring Lake Farms, became a showcase of the district; people drove cars from near and far to gape at its floral vistas, created by Boston landscape architect Warren Manning and teeming with such plants as "Azalea, Mountain Laurel, Gray Beard, Wahoo, Flowering Dogwood, Callicarpa, Huckleberry, Cardinal Flower, Blue Lobelia and Cornflower."

In Mountain Brook, Jemison wanted to provide affluent home buyers with an escape from the ugly realities of industrial society that were obvious just across the brow of Red Mountain. In his oasis, limited by restrictive covenants to "white persons of the Caucasian Race," residents could plant gardens with sundials, fountains, and pools "where lilies bloom in the shade of drooping ferns," and children could romp in bucolic safety. "Jemison was selective in the types of businesses that could be conducted in Mountain Brook," said a local historian. "No business that was obnoxious by the emission of fumes, gas, odor, dust or noise was acceptable. No hospital, asylum or sanitarium of any kind was allowed, nor undertaking parlors, livery stables or veterinary hospitals. No garages for repairing automobiles could be within the corporate limits." One of many local executives who moved into the sanctuary was Russell Hunt, Sloss-Sheffield's marketing head.[6]

Like Jemison, a few of the district's wealthiest citizens were able to afford an even more magnificent type of seclusion. Among the palatial homes built during the decade, none were more grand than the one erected by Birmingham's erstwhile chief executive, George B. Ward. Epitomizing the withdrawal of the local elite beyond the city's boundaries, it was built to provide solace from wounds that Ward had suffered in his disappointments as an urban reformer.

Ward had done his best for Birmingham, but his spirit was broken. Despite announcing his retirement from politics in 1911, he had been unable to stay aloof from the Magic City's problems. By 1913, it was approaching bankruptcy because enthusiasts for the "Greater Birmingham" movement had been unable to foresee the staggering financial burdens that would result from providing municipal services to all the areas that had been annexed in its recent expansion. Trying to eliminate the corruption that had been chronic under the old mayor-aldermanic frame of government, Progressives had managed to have it replaced by a board of three commissioners. Running successfully for the presidency of that body, Ward took office in 1913.

Unfortunately for Ward, the problems that he encountered in the second phase of his political career were similar to those that had frustrated his earlier efforts as mayor. When he failed to raise taxes to provide better social services, the already abysmal education budget had to be slashed, the police and fire departments reduced, and the zoo closed. His only success was bringing the "City Beautiful Movement" to the district. Calling reporters to his office, he said, "Boys, I want you to write stories about this movement, day after day. If you don't, I am not going to give anybody any more news." Creating an exhibit of flowers and vegetables at City Hall, he distributed free seeds; on his initiative, residents planted gardens and spruced up their neighborhoods. After Ethel Armes left Birmingham in 1917 to write for the Boston *Herald*, she sent him a heartfelt Christmas greeting: "No one ever tried to make a whole city lovely—as hard as you tried."[7]

Armes sent the card in an effort to console Ward for a bitter setback that he had just suffered. Running for a second term, he had been defeated by a nativist who called him "a tool of a Roman Catholic conspiracy to take over the government." The election marked a turning point in Birmingham's political life, signaling the end of attempts by elite reformers to provide civic leadership and launching a long period in which a "new middle class," dominated by white industrial workers, ran City Hall. Becoming a "very private citizen," Ward devoted himself from then on to his brokerage firm and his personal tastes. For a haven in which to spend the rest of his life, he built the most spectacular home in Alabama.

Ward's creation was begun in the spring of 1924 and was finished in the following year. Named Vestavia, from "Vesta," Roman goddess of the hearth, and "via," Latin for roadway, it was patterned after shrines of Ceres and Sibyl that he had seen in Italy. Circular, like ancient sanctuaries devoted to domestic deities, it was surrounded by twenty columns supporting a frieze ornamented with classic relief work. Measuring 182 feet in circumference, 60 feet in diameter, and 58 feet high, the house was made of dark pink sandstone, "shading from old rose to the Pompeiian Red of oriental granite, or mellowed in rich madder brown and sienna tints." Standing majestically at the crest of Shades Mountain, far from Ensley's steel mills and Sloss-Sheffield's blast furnaces, it was the first thing that Carmer mentioned in his essay on the district.

To Carmer, it was grotesque to find such a structure in the heart of an area pervaded by smoke and grime; it made him wonder "about mirages and the tricks the sun's rays play on a man's vision." Ward acknowledged that he had built it as an escape from adversity, telling one visitor that "a majestic building and the seclusion of a picturesque location has the effect of making events and human affairs seem unimportant and petty and not worth worrying over." His comments provided both a suitable epitaph for the passing of progressivism and an apt commentary on the fate of reformism in the Roaring Twenties. Although Ward lived a scant few miles from Birmingham, he was almost as absent from its everyday concerns as Bryan had been at Laburnum.

However incongruous it may have appeared alongside the industrial squalor that lay so close at hand, Vestavia was a magnificent place for a defeated urban statesman to nurse his wounds. The main floor was an enormous living room, "luxuriant with rich, soft-toned rugs, couches, and furnishings." In homage to the goddess for whom the house was named, the area was dominated by a huge stone fireplace. Greek and Roman statuary filled niches in delicately tinted gray walls; large bookcases with rows of varicolored volumes, tastefully keyed to hues featured in the rest of the decor, hung from the ceiling, suspended by chains. Between the bookcases were long windows, "in

the summer curtained in filmy gold, in winter draped with rich red velvet."
Above, reached by a semicircular wrought-iron stair, was the master bedroom;
below, a dining room in which as many as forty people could be served. Other
facilities included a spare bedroom, bathroom, pantry, and kitchen.

The grounds were no less impressive. In the estimation of one admirer,
they were worthy to be compared with "the famous gardens of Adonis, and of
Hesperides, or the Persian gardens of Cyrus." A three-basined fountain contained tiers of plants, vines, and flowers, among which floated "tiny boats and
other little conceits." It was illuminated at night by jade and blue lights
submerged in the water. "The glow from these," it was reported, "gives an
eerie, beautiful light, similar to that of the Blue Grotto at Capri." A Roman
ornamental garden had busts of Vergil and Sappho, and a statue of Vesta
herself lurked "under a bower of cyprus, laurel, and chinaberry trees." Figures
of Nidia, a blind maiden of Pompeii, and her swain, Glaucus, stood near "a
fern-edged rockery and a pool on which floats the Egyptian pink lotus of the
Nile, surrounded by valley lilies, purple iris, and yellow orchids." Flocks of
peacocks strutted across emerald grass, preening in the sun and showing
off enormously elongated tails, glittering bronze-green and blue with large
eyelets to attract nearby harems of peahens. Three kennels—named Villa
Cleopatra, Villa Nero, and Villa Plato after Ward's dogs—were built in the
style of the Parthenon, with diminutive pillars and porticoes. A radio was
hidden in a bird house. Across the road, a circular Temple of Sibyl jutted
from the mountainside. "Here," it was said, "one may sit in calm repose and
meditation, far from the hurrying world of industry, and watch the ever-
changing scenes of interest and charm." Underneath was a crypt in which
Ward wished to be buried.

Brightly illuminated at night, Vestavia "glittered like a huge constellation
above the city." After moving into the house in 1925, Ward delighted in
throwing parties. Black servants in white jackets embroidered with yellow
Roman insignia, featuring the letters SPQR, received guests; other attendants
wore helmets and carried spears. Wanting to share his excitement about
classical civilization, Ward allowed visitors to tour the grounds on stated days.
Arriving in Chevrolets and Fords, they wandered among the shrubs and
hedges, "marveling at the splendors, but tired and oppressed by the over-
powering heat." With characteristic liberality, Ward allowed black people to
tour Vestavia, but deferred to segregationism by making them "welcome on
certain days set aside for them."[8]

As a necessary concession to modernity, Vestavia had a basement garage;
the automobile helped make it possible for the rich to escape Birmingham
and its problems. Although Jemison would not allow cars to be repaired in
Mountain Brook, he had to make provision for gas stations. Automobile
ownership made it easy for those who could afford cars to escape from the
smoky downtown area at the end of a working day. Despite playing less of a
role in municipal affairs, many of the same people put pressure on City Hall
to pave streets, draw up stricter traffic codes, hire extra policemen, and

provide adequate downtown parking. Safety considerations and a wish to protect streetcar companies from allegedly unfair competition led them to urge restrictions on a growing swarm of jitneys used by persons too poor to drive cars. Although organized labor tended to support the same move to protect streetcar workers, such journals as the *Labor Advocate* and *Southern Labor Review* defended jitneys as "the poor man's automobile" and complained that restricting their use would "deprive the poor of the little pleasures of an automobile ride and a quick trip to and from work . . . while the more fortunate ride in privately owned cars." By the end of the decade, the jitneys had been forced out of existence.[9]

Automobile traffic worsened the already polluted atmosphere of Jones Valley. "A thick, oily blanket of smoke hung over Birmingham Wednesday—the heaviest by far that the city has experienced in years," the *Post* stated just before Christmas in 1928. "Up on the top and sides of Red Mountain the sun shone brightly, but was unable to penetrate the smoke in the valley below. Automobiles and streetcars were using their brightest lights, while pedestrians flitted in and out of the shadows of the smoke blanket like wraiths. . . . The moral of our black mornings is that a city may become so engrossed in enjoying its prosperity and good fortune that it will permit an ugly and insidious evil to rob it of the fresh, clean and wholesome air, which is the elementary right of every man."[10]

While Ward fled to Vestavia and other wealthy residents retreated to Mountain Brook, demagoguery increasingly set the tone of Birmingham's political life. Throughout the 1920s, the Ku Klux Klan was stronger in the Magic City than in any other southern municipality of comparable size. Its Robert E. Lee Klan No. 1, founded in 1916, had the distinction of being "the first klavern in Alabama and for many years, the most powerful." Among its members was a newly elected United States senator, Hugo L. Black. Huge rallies were held at East Lake Park in 1923 and 1924 to initiate thousands of new members; at one point, the Klan had at least 10,000 members in Birmingham itself and up to 10,000 more in Jefferson County as a whole. Believed to control about half of the eligible voters, its rolls included two judges, the county sheriff, most Birmingham policemen, and a score of other public officials. The True American Society, an anti-Catholic organization dating from the war years, was also active. James M. ("Jimmy") Jones, who headed a local trucking firm and had held a succession of elected offices since 1908, won election as commissioner in 1925 with Klan support and dominated City Hall until his death in 1940. Despite refusing from time to time to knuckle under to the Invisible Empire's demands, as in failing at one point to issue masked marchers a parade permit, he usually followed its wishes.[11]

Birmingham became proverbial among American cities for the virulence of its racism and nativism. A reporter for the *Nation* called it "the American hotbed of anti-Catholic fanaticism." A leading historian later asserted that there was "certainly no question that Birmingham's Protestant churches supported the Ku Klux Klan," and the head of the Robert E. Lee Klavern estimated that more than half of the local white Protestant ministers belonged to the organization. Cross burnings and lynchings became routine, and night riders burned a convent. After a Catholic priest was murdered by a Methodist clergyman for marrying the latter's daughter to a Puerto Rican paperhanger, a jury found his assailant, who was defended by Hugo Black, "not guilty by reason of temporary insanity." After 1927, when hooded vigilantes were convicted for abducting and beating a teenage boy for his drinking habits, the Klan's power rapidly waned; by 1930, it had fewer than 1,500 members statewide. But it left an ugly legacy of hatred and violence.[12]

Caught between the Klan and business leaders who ignored local politics unless their interests were directly challenged, elected officials became expert in what Edward S. LaMonte later called "nondecision making." As a result, the city ranked at or near the bottom of national rankings having to do with virtually every aspect of social welfare. Its per capita tax rate in 1928 was the lowest in the country among cities of its class; high license fees were imposed to avoid equitable assessment of property. Alone among all American municipalities, Birmingham spent less on charitable, correctional, and medical institutions in 1928 than at the beginning of the decade. Its public school system, deprived of its greatest leader when J. H. Phillips died in 1921, ranked last in expenditures per pupil among thirty-five major cities surveyed by the United States Bureau of Education in 1927–28. The city maintained no public parks for black residents; had far fewer hospital beds for charity patients than Atlanta, which had roughly the same population; paid municipal employees much less than they would have earned elsewhere; and had nearly 50 percent fewer policemen per thousand than the national average.[13]

Despite the city's failings, local authors sang its praises, confident that industrialism and technological ingenuity were synonymous with progress. Hornady filled his *Book of Birmingham* with chapters entitled "Iron and Steel to the Rescue," "Turning Ore into Ships," "Blazing New Trails," and "Aladdin's Lamp Surpassed." Lauding local inventors, he called them "hard-headed and dynamic individuals, determined to make their particular mine, furnace or mill more effective than the mine or furnace or mill of the other fellow." Comparing operations in the American Steel & Wire Company's roughing mill to "such serpentine performances as no snake charmer ever dreamed of," he evoked a vivid spectacle of "rushing, squirming rods of red hot steel" that were converted into a myriad of products, such as nails, fabricated by "great batteries of machines . . . amid a perfect inferno of noise." Hornady admitted that "in a community where trained minds are in demand, and where technical knowledge is so essential, it is surprising that there exist no great institutions specializing along these lines." Apart from a business college, the only institutions of higher education in the city were small private ones maintained by the Baptists and Methodists.[14]

Enthusiasm for aviation epitomized the prevailing worship of technology. A flying club was organized in 1919 by a World War I ace, James A. Meissner. The following year, coal gas from Sloss-Sheffield's new by-product plant made it possible to hold an international balloon race, which attracted nearly 50,000 people; it was won by a Belgian craft that flew all the way to Vermont after ascending from North Birmingham. Glenn Messer, a stunt pilot, helped build the Dixie Flying Field in 1920. Six years later, he started a plant to manufacture the "Air Bass," a two-seater plane with a short fuselage based on his own theory of balance. In 1922, the city built Roberts Field, named for a local flyer who had been killed in France during the war. Pilots landing after dark spotted the runway by looking for the fires of the Republic steel plant and the glow of its slag pile. In 1927, Birmingham became a stop on the federal airmail route between New Orleans and Atlanta. In October of the same year, Charles A. Lindbergh touched down at Roberts Field aboard the *Spirit of St. Louis* and was feted at a banquet in the municipal auditorium. Responding to his pleas for better airmail facilities, officials quickly agreed to build a new municipal airport. [15]

II

Despite Henley's oratory and Hornady's predictions, technological development and progress were not marching hand in hand. In sober fact, gaping cracks were beginning to appear in the foundations upon which the Magic City had built its faith. Its economy was sluggish, betraying the dreams of boosters who had predicted that U.S. Steel's 1907 takeover of TCI would make Birmingham the world's leading steelmaking center. Instead, the impact of Morgan's deal with Roosevelt had been just the opposite, freezing the district in a permanent second-class status.

Unfortunately for Birmingham, it was in the perceived best interest of U.S. Steel to restrict competition rather than to capitalize on any advantages that the district might have vis-à-vis other parts of the country. During the first four decades of the twentieth century, American steel ingot production exceeded 90 percent of capacity only twice, in 1916 and 1917. Believing that high fixed costs and market inelasticity made it necessary to stabilize the industry, Elbert Gary and his advisers had tried to control demand accordingly. The Magic City's visions fell victim to this policy. In 1905 the South made 14.2 percent of American pig iron; by 1929 its share had declined to 11.6 percent. The fact that the region had experienced an extraordinary growth in its output of foundry iron indicated clearly enough what accounted for the overall decline: steel production was shrinking relative to other parts of the country. By 1931, the South as a whole was producing only 3 percent of the national output of this alloy. Golconda had failed to materialize. [16]

The reasons for Birmingham's plight had much to do with the desire of powerful financial interests to end the bitter competition that had prevailed in the 1880s and 1890s, producing sharp upswings and downturns in the business cycle. Paradoxically, it was because of Birmingham's cost advantages in such economic warfare that its leaders had counted on gaining supremacy over northern and midwestern centers of ferrous metals manufacture. What local boosters had not realized in 1907, when they had hailed Morgan's coup in bringing TCI under the control of U.S. Steel, was that the rules of the game had changed, to Birmingham's disadvantage.

Unlike Andrew Carnegie's chief lieutenant, Charles Schwab, who became U.S. Steel's first president in 1901, Morgan was not interested in maximizing profits through vigorous competition. This difference, as well as Schwab's flamboyant lifestyle and penchant for gambling at such places as Monte Carlo, had led to increasing friction between Schwab and Morgan. In 1903, Schwab departed to take command of Bethlehem Steel, which he soon built into a successful firm in his own aggressive image. His replacement at U.S. Steel, Elbert Gary, believed in being politically careful; dreaded the thought of antitrust prosecution; thought it prudent to allow competitors to make a fair profit; knew little about technology; shunned innovation; and was so loath to arouse public criticism that he even made an admirer of muckraker Ida Tarbell, who wrote a flattering biography of him.[17] No matter how much local boosters tried to curry favor with Gary at ceremonial banquets whenever he came to Birmingham, even going so far as to build the Tutwiler Hotel to impress him and other visiting executives, U.S. Steel had no intention of sacrificing the interests of northern and midwestern steel producers to fulfill the Magic City's aspirations. Gary was not hostile to local interests; he was simply opposed to playing the brutally competitive game to which the city had become accustomed in its formative age.

As a means of suppressing competition in the steel industry without running afoul of existing federal antitrust laws, U.S. Steel used a practice known as basing point pricing. As Sanford had pointed out in the *Manufacturers' Record* in 1908, this tactic involved designating Pittsburgh as the reference point for prices of steel ingots throughout the entire country and adding uniform freight charges to shipments from other areas, regardless of where they might originate. The imposition of this policy, known as Pittsburgh Plus, made the price of steel at Pittsburgh lower than at anywhere else in the country, regardless of whatever potentially competitive advantages such places might enjoy. U.S. Steel employed the practice even though it handicapped some of its own subsidiaries, including TCI.

In an effort to protect TCI's interests, Crawford persuaded his northern superiors to substitute a "Birmingham Differential" of $3 per ton for the normal Pittsburgh Plus formula whenever this alternative enabled TCI to charge lower prices for its products than would have been otherwise possible. Even this modest concession, however, did not apply to U.S. Steel's other local subsidiary, American Steel and Wire, which had to charge customers in the Birmingham District up to $15 per ton more than they paid for steel from Pittsburgh. Compounding the problem, U.S. Steel ultimately raised TCI's Birmingham Differential to $5 per ton. Besides its impact on TCI, basing point pricing made it hard for secondary steel processors to gain a foothold in the district, further hurting local interests.[18]

The results of freight differentials were apparent in Jones Valley by the early

1920s as steel production failed to keep pace with other, more favored, parts of the country. Basing point pricing put TCI and other southern steel producers at a disadvantage, even in competing for nearby markets. A typical protest was lodged by a Tennessee plowshare manufacturer whose business shrank after the war because of the way in which Pittsburgh Plus gave an unfair advantage to northern steel mills and firms that used their products. "We have lost all our carload business in the Carolinas," he said. "With a Birmingham base equal to the Pittsburgh base we could have retained that business as against Pittsburgh and Ohio competitors and we could have enlarged it." But for the Birmingham Differential that Crawford had secured, his plowshare firm would probably have gone bankrupt.

Because he was under Gary's thumb, Crawford could not publicly protest the situation, but other local interests were free to do so. As the restrictive impact of basing point pricing became increasingly obvious, a chorus of southerners denounced its effects. After holding extended hearings, the Federal Trade Commission found in 1924 that U.S. Steel was in violation of the Clayton Antitrust Act and ordered it to stop imposing Pittsburgh Plus. Refusing to admit the justice of the decision, the giant corporation brazenly evaded it by instituting a new "multiple basing point pricing" scheme. Although this included Birmingham as a basing point, it was still harmful to the city's interests, because the price of Alabama steel remained $3 per ton higher than steel from Pittsburgh.

Southern criticism of U.S. Steel's policies was later validated in a detailed study by a Vanderbilt University economist, George W. Stocking. After careful analysis, Stocking concluded that "basing point pricing in steel has contributed to the South's poverty by curbing the expansion and utilization of its steelmaking facilities and by retarding the growth of steel-consuming industries." Stocking's cost figures proved that steel could be produced more cheaply in the Birmingham area than anywhere else. Nevertheless, as he showed, discriminatory pricing policies not only robbed the South of the share it would otherwise have enjoyed in steel ingot production, but also effectively limited or completely suppressed the manufacture of such commodities as hot-rolled sheets, steel plate, tin plate, and steel tubing. The substitution of multiple basing point pricing for the earlier Birmingham Differential did not help matters, for the district still supplied only about half of the limited range of steel products that U.S. Steel permitted TCI to produce in what Stocking characterized as its "natural market area . . . that is, the area within which Birmingham could lay down steel at a lower average cost than its rivals."

Stocking highlighted the contrast between TCI, hobbled by discriminatory freight differentials, and the district's merchant pig iron producers and pipe foundries, which were unaffected by basing point pricing. By the 1930s, Alabama alone made roughly 50 percent of all the cast-iron pipe manufactured in the entire country. Because of the natural advantages inherent in phosphorus-rich southern ores, Sloss-Sheffield and Woodward enjoyed market dominance in pig iron production without any need to employ artificial methods of price manipulation. Because they did not possess their own foundries, they could not use the arbitrary tactics U.S. Steel imposed on plants that made various products from steel. But for U.S. Steel's policies, Stocking charged, Birmingham's locational advantages might have enabled it to become the nation's leading center of steel pipe manufacture, just as it dominated cast-iron pipe production. [19]

III

U.S. Steel and its basing point pricing policies were not solely responsible for Birmingham's failure to become the world-dominating steel center about which boosters had dreamed. It is easy to forget that U.S. Steel pumped millions of dollars into the district after 1907, updating facilities and launching an expensive program of welfare capitalism in an effort to rationalize and expand TCI's operations. During World War I, the parent corporation built a new plant, the Fairfield Steel Works, in the Birmingham area to supply steel for shipbuilding at Mobile. After the war, this facility was kept in operation to make and repair railroad cars. In 1924, TCI built a steel bar mill at Fairfield. The following year, it constructed a new railroad across Jones and Opossum valleys to convey ore to Ensley and Fairfield, running up to twenty feet above ground level to avoid grade crossings and requiring a fifty-three-foot cut through Flint Ridge. In 1926, TCI completed building the first steel sheet mill in the South. In 1928, two new blast furnaces and sixty-three new by-product coke ovens went into operation at Fairfield. In 1929, a mill for making steel strips to tie around cotton bales went on line. Clearly, even though Gary and his fellow executives were not willing to aggrandize Birmingham at the expense of other steelmaking centers, they were not opposed to the district's growth. [20]

Furthermore, U.S. Steel's basing point pricing policies were not the only factors that adversely affected the Birmingham District's ability to compete with other parts of the country after the turn of the century. Compounding its problems were new regulations that the Interstate Commerce Commission imposed on shipping ferrous metals by rail and water. Earlier, the Southern Railway and Steamship Association, which had a vested interest in promoting regional industries, had cooperated in establishing rates that helped Birmingham iron sell at low prices in northern markets. Assistance of this type had aided the southern onslaught against northeastern anthracite iron producers in the 1880s. In 1907, however, the Interstate Commerce Commission, using authority gained under the Hepburn Act, instituted new rate structures that not only prevented such practices but in fact discriminated against southern ferrous metals in long-haul interregional markets. Pressure on the ICC from firms in Alabama and Tennessee, including Sloss-Sheffield, brought some measure of relief by 1914, but things became worse during World War I, when railroads were nationalized. Regional interests deteriorated further when private ownership returned under the Transportation Act of 1920, which produced a wave of railroad consolidations and higher rates that impeded access by southern ferrous metals to outside destinations. Because so much of the foundry trade had moved below the Mason-Dixon Line by this time, Sloss-Sheffield and Woodward were spared some of the damage that they might otherwise have sustained, but southern manufacturers as a whole were hurt. Indeed, as Robert J. Norrell has indicated, higher freight rates may have had a more restrictive influence on southern steel production after this time than the multiple basing point pricing scheme that U.S. Steel adopted in 1924 when the Federal Trade Commission outlawed Pittsburgh Plus.[21]

Birmingham and the South were also partly responsible for their own problems. To some degree, as Gavin Wright has argued, the Magic City's failure to attain greatness as a steelmaking center was due to the southern low-wage strategy and other disincentives that were imposed upon the district's workers. Despite the social welfare programs that Crawford had instituted in an effort to upgrade the local labor force, and the importation of immigrant workers to supplement black operatives, absenteeism at TCI was still 57.4 percent in 1923. Although this was better than 400 percent in 1912 and 145.3 percent in 1919, it was appalling by northern standards, providing living testimony of the obstacles to industrial discipline in a racist society that socially humiliated the southern steel industry's still heavily black labor force, reacted to immigration with nativism, and maintained wage scales lower than those prevailing in other parts of the country. Because of the inefficiencies that resulted from trying to combine an unmotivated work force with the technology required for steel manufacture, real southern labor costs were actually higher than they appeared to be, as Kenneth Warren later pointed out. Because of the southern low-wage strategy, the region also never developed the consumer purchasing power that might otherwise have nurtured the growth of secondary processing industries and tertiary service enterprises based upon steel. A detailed study published by the University of Alabama in 1953 argued that the low incomes characteristic of the South, attributable in part to the "predominance of agriculture" but equally traceable to the long-standing reliance on cheap labor, constituted "an unfavorable factor of the greatest importance" in limiting markets for both iron and steel throughout the region.[22]

In combination with segregationism and racism, the low-wage strategy on which boosters of the Birmingham District had counted for competitive advantage since the founding of the city had gone awry. The exodus of black and immigrant workers that was already beginning in World War I indicated what was happening, draining the region of skilled labor that was needed more in the steel industry than in the highly traditional merchant blast furnace and foundry trades. As Wright has stated, "educated, experienced miners and steelworkers of the Birmingham District were among the first ones to leave for the better-paying jobs in the North." Nothing happened after the war to reverse the trend; needing no Underground Railroad to escape a region where they were subjected to every form of discrimination that the white power structure could impose upon them, black workers whose talents might have benefited the region in a different educational and social atmosphere voted with their feet for the dream of a better life in the North. Simultaneously, any hope that regional labor problems could be alleviated by immigration ended as Congress moved toward imposing a quota system that was finally enacted in 1924.

Because skills were more crucial in steelmaking than in pig iron manufacturing and effective industrial discipline in Alabama was so hard to attain, the potential rewards to U.S. Steel for pouring money into the Birmingham District beyond what was necessary to keep TCI's facilities marginally competitive were not worth the effort, as Wright has suggested. Boosters who had believed that cheap, docile labor formed part of an unbeatable combination for southern steelmaking supremacy had miscalculated just as badly as they had done in thinking that U.S. Steel would capitalize upon Birmingham's proximity to raw materials and other locational advantages to compete vigorously with other parts of the country.[23]

IV

While Birmingham's share of national steel production began to decline, Sloss-Sheffield continued to build on the gains it had made under McQueen's leadership. Not only because of the diversification it had achieved by beginning to manufacture coke by-products, but also because of the limited degree of modernization that had occurred in mining and blast furnace operations earlier in the decade, the company was in an improved position to exploit the business opportunities remaining in the last few years of prosperity before the Great Depression hit. That it not only took advantage of these possibilities but actually enhanced its competitive edge by completing a thorough overhaul of its blast furnaces was due to the leadership supplied by McQueen's successor, Hugh Morrow.

From the moment McQueen died, it was obvious who would succeed him. Within a few days, Morrow was named acting president; shortly thereafter, the directors made his elevation permanent. These developments confirmed the trend, already evident in McQueen's administration, toward local control of day-to-day operations. As the *Age-Herald* commented, Morrow's appointment was highly welcome in the Magic City because it placed "a purely Birmingham product, a boy born and reared in the city of Birmingham, in charge of one of the District's great industrial concerns." Morrow's roots were embedded in the earliest days of Birmingham's history, and it was fitting for him to head Sloss-Sheffield as it approached its fiftieth anniversary. His father, John Calhoun Morrow, had presided over the incorporation of the Sloss Furnace Company in 1881 during one of his six terms as Jefferson County's probate judge.[24]

Hugh Morrow was born in Birmingham in 1873, only two years after the first lots had been sold by the Elyton Land Company. He was educated in the local schools and earned pocket money as a delivery boy for the *Age-Herald*. Matriculating at the University of Alabama, he won renown for his athletic ability; besides quarterbacking the 1893 football team, he was an ace hurler in baseball, credited with introducing curveball pitching in the South. His academic performance was equally outstanding; winning election to Phi Beta Kappa, he was valedictorian of his class on graduating in 1893. Indicating the high regard in which the university held him, he was appointed private secretary in the president's office while staying in Tuscaloosa to earn a law degree.[25]

In 1894, after graduating from law school, Morrow was admitted to the Alabama bar. Returning home at a time when Birmingham was still reeling from the Panic of 1893, he established a law office and capitalized on his forensic and oratorical skills in building a thriving practice. Ultimately, he became a partner in the city's most prestigious law firm, Tillman, Bradley, and Morrow. A strikingly handsome young man with a high forehead, large piercing eyes, and firm jaw, he exuded confidence and determination. In 1897, he married Margaret Julia Smith, who bore him five daughters and a son. His father-in-law was a prominent local physician and surgeon, Joseph Riley Smith, who had made a fortune by turning 600 acres of real estate into one of Birmingham's choicest suburbs, Smithfield. Morrow also became a pillar of the Presbyterian church. As the depression lifted and the twentieth century began on a wave of prosperity, his growing wealth and ability made him one of the brightest stars in Alabama society.[26]

Entering politics as a Democrat, Morrow served two terms as assistant solicitor of Jefferson County and was elected to the state senate in 1901. The speed with which his impact was felt was remarkable considering that he was still under thirty years old on taking office; by the time his four-year term ended, he was chairman of the judiciary committee. Returning to the upper chamber seven years later for a second term, he became president pro tempore. In 1908, the University of Alabama selected him as a trustee, in which capacity he served for eleven years. He was an avid sportsman, and his hunting and fishing activities contributed to the steady expansion of his contacts among the rich and powerful. His popularity was further enhanced by a marked sense of humor—he liked to refer to himself as "Humoro"—and a talent for spinning yarns. After his death in 1960, the Birmingham *News* called him "the man with a million stories and a million friends."[27]

Morrow was unswervingly conservative in his political and economic outlook, and swiftly found a place among the Big Mules who dominated the district. His reputation as spokesman for such enterprises as the Birmingham Railway, Light, and Power Company and Sloss-Sheffield became firmly established early in his career. In 1903 he helped kill legislation to create a "Greater Birmingham" by expanding the city's boundaries, arguing that it would drive manufacturers out of the district. In 1911, both before and during

Hugh Morrow. Morrow, who succeeded McQueen as president of Sloss-Sheffield in 1925, completed modernizing its facilities and guided the enterprise with a firm hand through the Great Depression and World War II. After resigning as president in 1948 and becoming chairman of the board, he spent much of his time fighting desegregation until he died in 1960. This photograph, taken in 1911 when Morrow was a rising star in Alabama politics and a highly successful lawyer in Birmingham, shows the poise, determination, and self-assurance of a young man who had already attained considerable wealth and power. (Courtesy of Alabama Department of Archives and History, Montgomery, Alabama.)

debates aroused by the Banner Mine tragedy, he played a key role in making sure that efforts by reform-minded legislators to establish tighter control over convict labor would not hurt the interests of Sloss-Sheffield and other clients.[28]

Morrow ably represented Sloss-Sheffield in disputes with local officials. He helped blunt the impact of the 1912 antismoke ordinance and settle quarrels arising from the way in which the firm's operations had blocked street expansion near the City Furnaces.[29] His role in securing an acceptable settlement from federal officials at Cincinnati in 1919 after the war's sudden end had interrupted construction of the by-product coke plant in North Birmingham clinched his appointment as a vice president later that same year.

Landers Sevier, Sloss-Sheffield's other vice president, left the firm in 1920 to become president of the Associated Industries of Alabama, an arch-conservative lobbying organization devoted to protecting the interests of Big Mules in Montgomery.[30] Morrow then became McQueen's chief policy adviser and heir apparent. His status was evident shortly after McQueen's death, when Birmingham staged one of its periodic banquets to honor Gary and other visitors from U.S. Steel's Broadway headquarters. Chairing the event, Crawford introduced Morrow as "soon to be made permanent president of the Sloss-Sheffield Steel and Iron Company." Never at a loss for words, Morrow expressed regret that Crawford did not belong to his board of directors. As everybody knew, the post would go to Morrow, and the appointment became official in a few days.[31]

<h2 style="text-align:center">V</h2>

Like every other previous chief executive of the company, Morrow had no background in industrial technology. His administration, however, broke fresh ground by giving the title of vice president, rather than the less prestigious one of superintendent, to two persons who possessed such knowledge. Up to this time, Maben, Jr., whose appointment had been due mainly to family connections, had been the only engineer ever elevated to vice presidential rank. By conferring high status upon technical men, Sloss-Sheffield showed that it was marching in step with an increasingly rationalized age.

One of the new vice presidents, James P. Dovel, was an engineer who had already begun modernizing furnace operations under McQueen. The other, Frank Miller, a Semet-Solvay employee, headed the by-product coke plant in North Birmingham. By the time Morrow became president, Russell Hunt, a veteran employee with a strong record in marketing, had been elevated to secretary-treasurer in addition to retaining supervision of sales. For the first time, therefore, the executive roster revealed a clear and consistent emphasis upon allocating authority according to functional responsibilities. The days when executive titles had been given to financiers like Bryan and Rucker were long past.[32]

Birmingham remained well represented in the corporate boardroom. Morrow headed a group of four local businessmen who served as directors. The identities and affiliations of the rest showed the guiding hand of Catchings, whose wartime career had solidified his role in the growing movement toward cooperation by industrialists in national trade associations. Whereas Bryan and other nineteenth-century industrialists could only rely on such undependable extralegal arrangements as pools to deal with cutthroat competition, Catchings and his fellows preferred to work out "trade practice codes" within the cozy surroundings of chambers of commerce and other semifraternal business associations. As Robert F. Himmelberg has pointed out in a study of the roots of the National Recovery Administration, the Republican-dominated Federal Trade Commission was happy to endorse these codes in the 1920s as "clarifying the law and intensifying its enforcement," despite the way in which they eroded antitrust legislation. Heading the Department of Commerce, Herbert Hoover aided and abetted the process.[33]

Because of their wartime experience on boards linking the growing Chamber of Commerce movement with the federal government, Sloss-Sheffield's local directors were well accustomed to such a milieu. John L. Kaul, a Pennsylvania-born entrepreneur who had been in the Alabama lumber business since 1890, was typical. Through participation in the Chamber of Commerce movement, he had become involved with the War Industries Board, and his close identification with cooperative business groups was shown by his activity in the National Federation of Construction Industries and the Southern Pine Association. Another board member from Alabama, William H. Kettig, was also involved in Chamber of Commerce activities. He had lived in Birmingham since moving to the district from Kentucky in the 1880s to help organize a foundry that was later absorbed by the Crane Company of Chicago. He had been one of the most vocal opponents of the antismoke ordinance enacted in 1912. Among other posts that he held, he was board chairman of the Birmingham Federal Reserve Bank, vice president of the Redd Chemical and Nitrate Company, and a director of the Protective Life Insurance Company.

William Webb Crawford, who had been appointed to the board in 1918 to succeed J. S. Sewell, rounded out Birmingham's group of directors. He had been active in the local banking fraternity for almost forty years and was president of the American Trust and Savings Bank. During his youth, he had become a close friend of McQueen while working for the Alabama Great Southern Railroad. He was also a Presbyterian, as Sloss-Sheffield's executives and directors now tended to be. This affiliation was another indication that the Virginia dynasty, which had been dominated by Episcopalians, was over.[34]

By the time Morrow took office, three directors who were active in the New York financial community had filled vacancies caused by the series of deaths that had occurred in 1924. These newcomers exemplified Sloss-Sheffield's ties with the Allied Chemical and Dye Corporation and the contacts with key government officials that Catchings had made during the war. The most important new director was Bernard M. Baruch, a Wall Street financier who had been chairman of the War Industries Board and the guiding force in Woodrow Wilson's Supreme Economic Council. A strong advocate of cooperative business organizations and champion of cartelization, his activities as industrial statesman, public servant, and advisor to presidents brought him growing national and international fame.[35]

Another new Wall Street representative was lawyer Francis H. McAdoo. His father, William G. McAdoo, was a perennial power in national politics, had served as Woodrow Wilson's secretary of the treasury and railroad czar during the war, and was a top contender for the Democratic presidential nomination in 1924. Also new as a director in the corporate boardroom was Matthew Scott Sloan, an important figure in the electric utilities industry who had been president of such firms as the New York Edison Company and the United Electric Light and Power Company. He was also closely associated with the Irving Trust Company and the National City Bank of New York.

Despite their strong connections with northern finance, all of the new board members came from below the Mason-Dixon Line. Like Catchings, they exemplified Sloss-Sheffield's enduring tradition of maintaining close ties with "southerners on Wall Street." Baruch was a native of South Carolina; he still owned a large estate there, the Hobcaw Barony, where he threw lavish parties for his friends. McAdoo was a Tennessean with an Ivy League education; he was descended from prominent families in the Volunteer State on his father's side and Georgia on his mother's. Sloan, a native of Mobile, had earned engineering degrees from Alabama Polytechnic Institute at Auburn and was married to a daughter of a Confederate general, James Henry Lane. Steeped in southern culture and social institutions, such men assured Morrow and other company executives of support and sympathetic understanding as they ran day-to-day operations in accordance with deeply ingrained local traditions of racial segregation, opposition to labor unions, heavy-handed paternalism, and other regional ways of doing business.[36]

Still, the board was different from any previous group of directors in the company's history. The roots of its members were largely in banking, chemical production, electrical distribution, and governmental service, not railroading. These men were highly receptive to technological innovation. Although none belonged to the Republican Party that now held power in Washington, most had been directly involved in profitable contacts with federal bureaucrats and were committed to a hands-on approach to governmental relations through politically active trade associations. In essence, they were Hoover Democrats.

Though highly supportive of technological modernization and steeped in northern ways of doing business, the new generation of southerners who ran Sloss-Sheffield had no intention of emulating the employee welfare program that had been implemented at TCI. Unlike its larger competitor, Sloss-Sheffield was southern to the depths of its corporate soul, upholding values and attitudes going back to the antebellum planter aristocracy and the Broad River group. In this respect, as in their contrasting size and market orientation, Sloss-Sheffield and TCI presented strongly contrasting images.

While long-standing traditions endured, personal connections with earlier times continued to snap. On 22 April 1925, Edward M. Tutwiler died at sea two days after visiting Honolulu. News of his death, reaching Birmingham by radio, produced a chorus of tributes to the role he had played in the district's history. "He trudged the hills and climbed the mountains," the *Age-Herald* declared. "He studied coal and iron deposits until the field was like an open book to him." His body, brought back to Jones Valley, was buried after services at the Episcopal Church of the Advent, where he had long been junior warden.[37] Less than three months after Tutwiler's death, Sloss-Sheffield's oldest director, William H. Goadby, died at his Fifth Avenue home in New York City. A venerable figure on Wall Street, he had lived to become the third oldest member of the New York Stock Exchange, on which he had taken his seat in 1870. His death left Larkin as the only remaining director who had been closely affiliated with Joseph Bryan.[38]

VI

Despite having diversified by building the by-product coke plant in North Birmingham, Sloss-Sheffield remained deeply committed to the production of foundry pig iron. The wisdom shown by the directors in 1902 in deciding to concentrate on this commodity instead of producing basic steel became more and more apparent during the 1920s, as large steel producers became increasingly self-sufficient and many independent merchant blast furnaces that continued to make pig iron for a general market went to the wall. Twenty-two furnaces owned by such operators—one-fifth of the stacks remaining in the entire nation—were torn down in 1927 alone.[39] The only large market that remained for firms specializing in pig iron was the foundry trade, still a key component of the ferrous metals industry because of the continued usefulness of cast iron for many technological applications. The fact that foundries were becoming concentrated in the South worked strongly in favor of Sloss-Sheffield and Woodward.

Morrow's administration began on an encouraging note. As rising demand swept the yards virtually clean of accumulated inventory, pig iron production for 1925 broke previous records. Prices remained unsteady as suppliers fought to maintain market share, but Sloss-Sheffield's high volume of operations, combined with the limited degree of modernization that McQueen and Dovel had achieved at the North Birmingham furnaces, drove down unit costs and kept the firm competitive with Woodward. Together with the sale of coal-tar derivatives from the by-product coke plant, this advantage resulted in a net profit of $1,978,941, up from $1,516,276 earned the previous year. By the end of December, Sloss-Sheffield's inventory of unsold iron was the lowest in its entire history.[40] Although pig iron production declined in 1926, earnings climbed to $2,106,759. Increasingly, however, diversification helped keep the company in the black, while declining prices for pig iron cut into returns from sales to the foundry trade. At the end of the year, prices for Birmingham pigs were holding firm at about $20 per ton. This was far below levels prevailing in 1920, when the price of southern foundry iron had reached a postwar peak at more than $40 per ton, or even in 1923, when it averaged nearly $24.[41]

The crunch got worse in 1927. By the end of December, the price of Birmingham pig iron had sagged to $16 per ton, and the company's net profit of $1,150,934 was down by nearly half from the previous year. A recession that began late in 1926 did not end until November 1927, causing industrial production to drop nationally by a percentage point. Even after business picked up, demand for ferrous metals remained sluggish. In 1928, the price of Birmingham pigs sank to $15.50, the lowest level since 1915, and net earnings for the year fell to $1,079,857. Without remedial action, the company's situation was bound to get even worse.[42]

Well before 1928, Morrow had taken stock of the situation and resolved to push ahead with modernization. One important result of this decision was a steady departure from the company's traditional reliance on labor-intensive methods of production, leading ultimately to the adoption of mechanical pig casting. Writing at a time when Sloss-Sheffield's corporate records were unavailable, Gary Kulik argued that northward migration of black workers was responsible for the firm's belated adoption of this technique and other features of modern blast furnace practice that had been common since the 1890s in places like Pittsburgh. It is clear that unsettled labor conditions after World War I contributed to the adoption of Dovel's pig-breaking machine and new hoisting equipment at the North Birmingham furnaces, but evidence to support Kulik's hypothesis is otherwise inconclusive. Birmingham's black population actually grew dramatically in the 1920s, from 70,230 at the beginning of the decade to 99,077 at the end, as African-Americans continued to flock to the city from depressed rural areas.[43] Although large numbers of skilled black steelworkers may have migrated to northern cities, adversely affecting TCI, there remained a large pool of unskilled workers upon which Sloss-Sheffield could draw.

A different perspective emerges on the company's belated adoption of pig-casting machines and other devices that had long been used elsewhere if one examines this development as part of a larger rationalization process that took place under McQueen and Morrow and considers the conflicting priorities that these executives had to weigh as they decided what to do next. Insofar as concern about labor influenced their thinking, it was mostly because they saw that they could no longer rely on manual work done by undependable personnel at a time when the national economy was constantly in flux and the price of pig iron was declining. The main thing on their minds, to which labor and every other aspect of the company's operations was related, was the need to cut costs by maximizing efficiency. McQueen's annual reports in the early 1920s indicate that he began to modify operations at the North Birmingham furnaces at that time primarily to reduce expenses and lessen Sloss-Sheffield's dependence on inefficient workers. But other priorities were even more pressing, such as completing the by-product plant, shutting down the company's antiquated beehive ovens, and finishing the electrification of its mines to assure an adequate supply of coal to the expensive new coking facility. McQueen therefore deferred a major overhaul of the North Birmingham stacks until these needs had been met and did nothing at all to modernize the outmoded City Furnaces. [44]

But this neglect could not continue indefinitely. By the time Morrow became president, the steady erosion of pig iron prices had made it imperative to attend to the poor condition of the firm's increasingly obsolescent furnaces. Frequent references to cost-cutting in Morrow's annual reports throughout the late 1920s underscored the urgency of the situation. It was within this context that a series of steps began to be taken that ultimately led to the adoption of pig-casting machines. [45]

Fortunately, conditions were ripe in 1925 for a prompt and vigorous response by the directors. Led by Catchings, members of the board had access to capital and favored keeping up with technological change, to which they were well accustomed through their involvement in science-based industries. In Birmingham itself, technically oriented executives held greater power than ever before. Quickly, therefore, the push for modernization gained momentum.

Beginning in 1925, the North Birmingham furnaces were rebuilt, new boilers were installed, and the blowing apparatus was improved. In the process, the height of the stacks was increased to eighty feet, ten inches, and their individual capacities were boosted to 118,050 tons per year. In October 1925, the directors authorized Morrow to spend up to $625,000 to begin a massive rebuilding program at the City Furnaces in Birmingham. Starting early the following year, Sloss No. 2 was completely dismantled and reconstructed. The new stack, completed on 25 July 1927, went into blast eight days later. Supported by cast-iron columns that were anchored in a concrete foundation lying deep under the surface of the ground, the installation was eighty-two feet high and twenty-one feet wide at the bosh, about seven feet higher and five feet wider than before. Starting in March 1928, after some debate about whether to spend the $450,000 required to complete overhauling the entire plant, Sloss No. 1 was torn down, rebuilt along comparable lines, and returned to blast in January 1929. Each of the new furnaces had an average daily capacity of 400 to 450 tons, compared to a previous maximum output of about 250 tons. Increased size, however, was not the main difference between the old and new stacks, which were still much smaller than many comparable northern installations. More important was their basic design, incorporating modifications aimed at increased efficiency and economy in the handling of Southern raw materials. [46]

The person responsible for implementing modernization of the furnaces was Dovel, whose recent elevation to vice president signified Sloss-Sheffield's commitment to technological change. His career climaxed the long evolutionary development of a distinctly southern style of merchant blast furnace, adapted to the specific requirements of regional raw materials and economic conditions. Interior furnace design, not mechanization, was the key element in this style; without recognizing this fact, it is impossible to appreciate what Dovel accomplished at Sloss-Sheffield in the late 1920s and early 1930s.

VII

James P. Dovel was born in 1868 at Pickerington, Ohio, not far from Columbus. Like Uehling, he grew up on a farm. Because his father was an invalid, Dovel had to take responsibility for his family at an early age. Like Uehling, he was mechanically precocious. When only seventeen years old, he devised a corn harvester that was later copied by several manufacturers of agricultural equipment.[47]

Hard times in the 1890s forced Dovel, who now had a wife and children, to leave the farm and move to Ohio's capital city, where he took a job for $9 a week with the Columbus Steam Boiler and Manufacturing Company. On his first night at work he conceived an idea for improving boiler efficiency; within a week, he had been elevated to shop foreman and given a pay raise. Unlike Uehling, however, he was never financially able to secure formal education at a technological institute; despite having had early aspirations to become a lawyer, he never finished high school. Instead, Dovel pursued another avenue that still provided access to an engineering career by gaining on-the-job experience in what Monte A. Calvert later called the "shop culture."[48]

Dovel moved restlessly from one employer to another. Leaving the Columbus boilermaking firm that had first hired him, he took a job with the Jeffrey Manufacturing Company, which made various products from steel sheets. Here, he increased efficiency by devising methods of interchangeable manufacture that made it unnecessary to take individual measurements when riveting plates or sheets together. About 1903, he shifted to yet another Ohio firm, the Portsmouth Machine and Foundry Company, where he began a preoccupation with blast-furnace technology that lasted through the rest of his career. This move also brought Dovel to Alabama, because his new employer had contracted to build a furnace for the Woodward Iron Company. Dovel came to Birmingham to supervise its construction.

Though northern-born, Dovel identified strongly with the South because of ancestral ties with Virginia's Shenandoah Valley. Unlike Uehling, he did not come to Alabama feeling like an outsider but rather like someone who had come home where he belonged. He adjusted easily to the region's way of life.[49] Remaining in Alabama after the Woodward furnace was completed, he became superintendent of the Birmingham Engineering Company in 1905. Four years later, he began working for Sloss-Sheffield. Sources differ about whether he started as a draftsman or superintendent of construction; in any case, he worked for the firm from 1909 until arthritis forced him to retire in 1930. Becoming general manager of furnaces and winning promotion to vice president in 1925, he produced a stream of inventions that gained him international recognition and career achievement awards from such organizations as the National Association of Manufacturers.

Most of Dovel's creative activities were focused on improving the efficiency and reliability of blast furnaces and eliminating the need for frequent repairs that had plagued Sloss-Sheffield since its inception in 1883. Despite persistent

James P. Dovel. James Pickering Dovel, whose contributions to the handling of southern raw materials underlay modernization of Sloss-Sheffield's City Furnaces in the 1920s, was one of the two greatest engineers, along with Uehling, in the history of the enterprise. This photograph, taken in the late 1920s, suggests the inner strength of an immensely gifted man who remained active as an industrial consultant even after becoming bedridden with severe arthritis. (Courtesy of Dr. Dudley Dovel Shearburn, Winston-Salem, North Carolina.)

efforts to improve the quality of raw materials with which furnaces were charged, Sloss-Sheffield, like other companies in the district, continued to struggle with the buildup of deposits on the inwalls of its blast furnaces and resulting scaffolding. Because of the importance to the district of reducing the need for constant relining, Dovel was already working on inventions connected with this problem before he started work at Sloss-Sheffield. His early efforts, aimed at cleaning the heated air blast blown into a furnace, resulted in two types of gas-purifying apparatus. He also tried to improve the quality of the ore with which the furnaces were charged; in 1914, he and another inventor patented an elegantly conceived ore concentrator in which the contraction of mercury within annular grooves governed the rotation of a table revolving transversely to the line of flow of the ore passing over it.[50]

The abnormally high temperatures and blast pressures required for smelting lean Alabama ore were prime causes of furnace deterioration in the Birmingham District. Much of Dovel's work was aimed at alleviating this distinctively regional problem. Typical of the innovations that resulted from his efforts was a new method of cooling blast furnaces, jointly patented by Dovel and Sloss-Sheffield's furnace superintendent, John J. Shannon, in 1914. It featured a carefully aligned network of vertical passages, metal plates, valves, and pipes that conducted heated air upward and discharged it from the top of the stack.[51]

Improved cooling techniques continued to preoccupy Dovel as he tried to extend the useful life of furnace components. In 1919, he patented a new method of reinforcing a furnace with a superstructure through which water circulated downward through a series of channels before being discharged at the base of the stack. A year later, he patented a new way to surround a tuyére with water circulating through a cooling box that was equipped with a threaded inlet port and corresponding outlet. Unlike some of his earlier projects, which were jointly developed with other persons, both of these inventions were his own independent conception.[52] They were benchmarks in the evolution of an inventive style that would be climaxed in the late 1920s by furnace modifications that drew visitors to Birmingham from as far away as Japan and the Soviet Union.

During World War I and the immediate postwar period, Dovel conceived other new ways to improve the quality of raw materials and rationalize ironmaking operations. In 1918, he patented several new stationary and portable ore-washing devices. In May 1921, seeking further to enhance the quality of raw materials with which blast furnaces were charged, he filed five claims involving a water purification system with two elongated horizontal tanks and a transferrable rack with narrow baffles that shifted from the top of one tank to the other. This apparatus, based on the settling of impurities as water flowed in one end and was discharged out the other, won patent protection in 1924.

Almost all of Dovel's patents involved furnace design and the processing of raw materials. A few, however, were different. Ever since Bryan had spearheaded the acquisition of the Cahaba and Russellville brown ore deposits because of their relatively high metallic iron content and strip-mining potential, achieving economies in the extraction of raw materials had been of great importance to Sloss-Sheffield; this background also explains why McQueen had placed such strong emphasis on the electrification of coal and ore mines early in the 1920s. As a contribution to lowering extractive costs, Dovel invented a power shovel with a pivoted horizontal boom that swung upward and downward as a hydraulic cylinder imparted reciprocal motion.

Yet another of Dovel's patents involved mechanizing casting methods to a limited degree. In 1921, at North Birmingham, Sloss-Sheffield installed a pig-breaking machine for which Dovel filed a patent application involving twenty-six separate claims. A modified version of similar devices developed elsewhere since 1890, it was deployed from a traveling crane running on a track high above the floor of the casting shed. Moving back and forth inside a cylinder was a vertical hammer consisting of a piston, piston rod, and chisel-shaped breaker head. Guided by metal plates bolted both to the cylinder and to each other, it came down repeatedly, with great force, to break pigs off a sow. As Dovel indicated in an article that he published in *Iron Age*, reducing labor costs was one of the motives underlying this innovation but was by no means the sole reason for it. Sales officials of such companies as Sloss-Sheffield, he stated, received more complaints from foundrymen about iron being improperly broken than was true with regard to any other problem. Such customers wanted to feed their cupolas with smaller pieces of pig iron "to insure a quicker and more uniform melt." This objective, he said, was "impossible by the old method of breaking by hand" but could "be met nicely by the pneumatic hammer system of breaking."[53]

After the directors decided to proceed with an extensive program of furnace modernization in 1926, all of Dovel's long-accumulated experience went into rebuilding the City Furnaces, which today remain substantially as he left them. His constant concern for cooling was evident in their design. At the base of each furnace, water circulating through pipes cooled the hearth, which had a capacity of 100 tons of molten iron. A well filled with water that filtered into an array of firebricks located directly beneath the crucible protected that critical area from being damaged by excessive heat. The water-cooled tuyéres were also protected from overheating with bronze plates designed for easy replacement. A jacket of cooling plates and pipes protected the bosh, which was angled in a way that Dovel had found effective in smelting southern ores. Water for cooling purposes came from a massive storage tank that towered high in the air on metal supports. After passing around the extremely hot areas located near the base of a furnace, the water was conveyed through pipes to a pond where pumps sprayed it into the air in cascading jets. This process, analogous to the action of a radiator in an automobile, caused the heat that the water had accumulated to dissipate. Upon settling back into the pond, the water was returned to the overhead storage tank for further use.

Recycling was also evident in the way the furnaces were designed by Dovel to clean and conserve the vast amount of gas—up to 80 million cubic feet per day—that was produced by the smelting process. This gas resulted from complex chemical reactions that took place as hot air was pumped into bustle pipes that encircled the bosh, entered the blazing crucible through the tuyéres, interacted with incandescent raw materials in the roaring interior of the stack, and ascended to the top of the furnace at velocities exceeding 300 miles per hour. The resulting fumes contained about 15 percent carbon dioxide, 24 percent carbon monoxide, 4 percent hydrogen, and 57 percent

nitrogen. Each upward journey through the stack also produced an accumulation of about 12 tons of flue dust that had to be removed before the gas
could be sent back to the stoves and boilers for another round of operations. A
brick-lined pipe called the downcomer took the uncleaned effluent from the
top of the furnace to the dust catcher, a domed receptacle that was designed to
reduce the speed with which the gas was moving. Large bits of soot were
trapped in a vertical chamber before the flue gas entered the dust catcher
itself, where remaining particles settled into a funnel-shaped area at the
bottom. This accumulating residue was dumped into a rail car and carried
away.

The remaining gas, which still contained small bits of dust, proceeded to a
washer that ranked among Dovel's most remarkable inventions. Many decades
later, the mechanism largely unchanged, still lurked among more imposing
units, a relatively small but nonetheless intriguing artifact. Its most conspicuous component was an elongated tank in which the gas was cleaned. Viewed
from either end, this vessel resembled a tooth with a large crown and two
truncated roots. The tank sloped gently downward to facilitate water flow. In
the interior were three vertical baffle plates. Two troughs, filled with moving
water, formed V-shaped protrusions at the bottom of the chamber. Entering
the vessel through an inlet, uncleaned gas was agitated as it moved under and
over the baffles and mingled with water in the troughs. In the washing
process, dust particles were thrown against the surface of the water by centrifugal force and carried off through an outlet. Purified gas was discharged
through another passage and returned to the boilers and stoves.

A key feature of the apparatus was the way in which it removed sludge
formed by dust particles that accumulated at the base of the tank. Reciprocating scrapers imparted scouring action by moving back and forth along the
walls of the V-shaped troughs. Endless cables, attached to the scrapers by
clamps, received motive power from reversible electric motors and rotating
drums located directly under the washer. Grooved wheels guided the cables,
which entered the cleaning chamber above the water line at opposite ends of
the troughs. (See illustrations on pp. 400–401.)

Dovel's gas-washer was significant not merely because it prevented clogging
more effectively than other devices of its type but because it was the latest in a
series of similar contrivances developed in the Birmingham District to cope
with the dirty condition of local raw materials. Dovel was following in the
footsteps of Ramsay and Stewart, who, in developing pioneering apparatus for
washing coal, had dealt with problems analogous to those involved in cleaning gas. These men had helped create a distinctively southern tradition of
mineral and metallurgical processing in which the district could take great
pride.

Equally a part of the same tradition was the ingenious way in which Dovel
designed the exterior and interior walls of Sloss-Sheffield's new blast furnaces.
The twin stacks featured a main shell and an upper shell, based on a two-
tiered concept for which Dovel sought patent protection in November 1925.

These shells were lined with bricks up to the point where abrasion from the stock commenced and with metal plates from there to the top, forming a combination that Dovel described as being "practically indestructible." Application of this principle required about half the number of bricks—140,000 as against 280,000—needed in conventional furnaces of the same size. The twin-shell concept also made it possible to modify the pitch of the inwalls, which were protected by rows of bronze cooling plates. Catwalks encircling the stacks enabled workers to climb around their outer surfaces to inspect and replace the plates during periodic overhauls. Dovel's careful study of optimal proportions and his ever-present concern for cooling permitted the furnaces, each of which had a gross burden of 25,199 tons, to hold more raw materials than had previously been possible in Alabama stacks that were necessarily shorter than their northern counterparts because of the limited mechanical strength of Warrior coal. The enhanced capacity of the new City Furnaces not only increased production but also, combined with a recuperator of Dovel's own design, promoted heat retention, reducing coke consumption to 2,690 pounds per ton of iron. This was 15 percent less than that required in previous models, providing a good illustration of the types of costs that Morrow was trying to save.

One of the most ingenious aspects of the new furnaces, embodying ideas for which Dovel filed a patent claim in 1927, was the way in which the uniform upward passage of gases reached a velocity along the inner walls as high as that prevailing in any other part of the stack. Taking advantage of the extremely high durability of the linings he had created, Dovel also redesigned the charging bell at the top so that it threw coarse materials along the inwall perimeter instead of fine particles that had previously been necessary to protect the same surfaces from abrasion in older furnaces. Because the air blast could readily penetrate the granular materials, tendencies toward clogging that had previously resulted from the coagulation of fine particles were eliminated. Dovel was particularly proud of another feature of the new stacks. Instead of being lined in a conventional manner, their inner walls were covered with agglomerated particles composed of the same stock with which the furnace was charged. Because of the continuous and uniform travel of the gases, the interior surfaces were continually renewed by the action of the furnace itself. This concept was patented in February 1929.[54]

Despite all of Dovel's efforts, the four furnaces that Sloss-Sheffield operated in Birmingham and North Birmingham in 1930 continued to require half a ton more coke than the national average per ton of pig iron produced. As Jack R. Bergstresser has stated, "if each furnace had an average output of about 300 tons of pig iron per day, the resulting 1,200 tons would have required 600 tons more coke than was needed for the same output in other parts of the country." Thus, even with Russellville brown ore mixed in, a combination of what Bergstresser has aptly called "low-grade, relatively inaccessible ore and coal only marginally suited for metallurgical purposes" continued to have an adverse effect on the development of Sloss-Sheffield.[55]

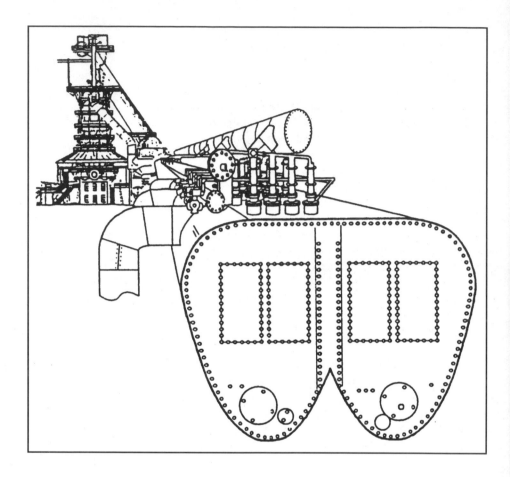

Dovel Gas Washer. These two sets of patent illustrations and a line drawing of the gas washer that Dovel designed for Sloss-Sheffield's remodeled City Furnaces show the operation and exterior appearance of a mechanism designed to cope with the large quantities of flue dust generated by smelting iron with raw materials from the Birmingham District. The patent illustrations show the distinctive configuration of the vessel within which washing took place, the baffle plates, the scrapers that effected sludge removal, and the underlying drive mechanism that imparted reciprocating motion to the scrapers. The line drawing shows the position of the washer relative to other parts of the furnace to which it was attached. (Patent illustrations: U.S. Patent Office, Washington, D.C.; courtesy of Ralph Brown Draughon Library, Microforms and Documents Department, Auburn University, Alabama. Line drawing: Courtesy of Sloss Furnaces National Historic Landmark, Birmingham, Alabama.)

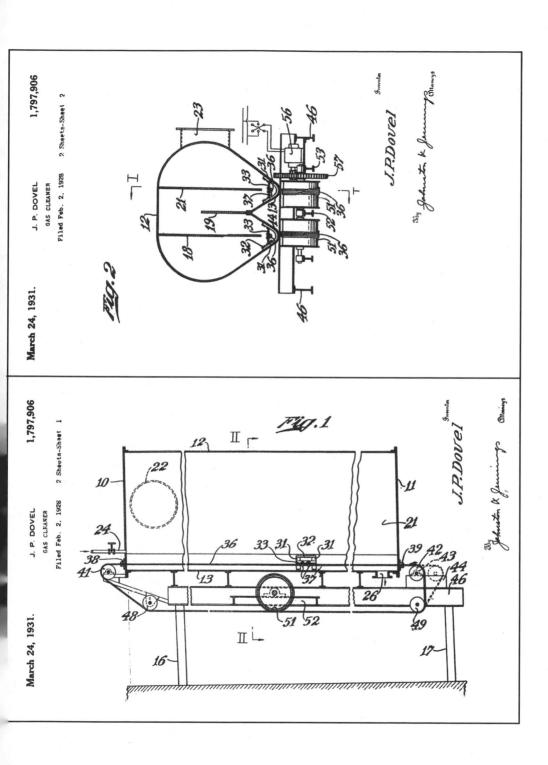

March 24, 1931.

J. P. DOVEL

GAS CLEANER

Filed Feb. 2. 1928   2 Sheets-Sheet 2

1,797,906

Fig.2

J.P.Dovel

By *Johnston & Jennings* *Attorneys*

March 24, 1931.

J. P. DOVEL

GAS CLEANER

Filed Feb. 2. 1928   2 Sheets-Sheet 1

1,797,906

Fig.1

J.P.Dovel

By *Johnston & Jennings* *Attorneys*

Regardless of this fact, Dovel's new City Furnaces, which were substantially different from any stacks previously built in the South, represented the high-water mark of merchant blast furnace design in the region up to this time.

VIII

Less original than the interior and exterior design of the new City Furnaces was the way in which various aspects of their operation became mechanized. For the first time, Sloss-Sheffield began to adopt equipment that had been standard for decades in other parts of the country. The process, however, took place haltingly, in an incremental way. This was not because the firm was backward but because it, like Woodward, was part of a distinctive southern syndrome with its own specific characteristics and demands.

The most visible sign of change at the rebuilt City Furnaces was the presence of skip hoists, inclined tracks on which cars loaded with ore, coke, and flux materials ran diagonally up the sides of the stacks. A steam engine housed in a building located between the hoist and the dust catcher provided power for this method of charging, which had long been common in leading steel-making centers and had already been adopted by Sloss-Sheffield at North Birmingham in the early 1920s as part of the selective remodeling effort during which Dovel had developed his pig-breaking machine. Skip hoists were much more efficient than older vertical hoists that were hand-loaded from wheelbarrows. Railroad cars filled with raw materials dumped coke, ore, and fluxing stone from a stock trestle into large storage bins with doors opening downward into a tunnel. Running along a narrow-gauge track in the tunnel, weighing cars received carefully measured quantities of these raw materials from the overhead bins and dropped them into the skip cars, which commenced their upward journey to the apex of the stack from a point deep underground. Upon reaching the peak, the skip cars discharged their cargoes into McKee tops, ingenious revolving devices that combined a receiving hopper with two charging bells, one small and the other considerably larger. Filling and discharging the bells in alternating sequence permitted the top of the stack to remain closed at all times, so that constant pressure could be maintained within the interior. Because the new system did away with handloading, it reduced labor costs. By controlling pressure and facilitating more accurate handling of the ingredients that were fed into the stack, however, it also promoted Dovel's more basic aim of utilizing raw materials more scientifically. In adopting long-established northern techniques, Sloss-Sheffield also discarded its previous use of antiquated ladles, known as hot pots, that had carried molten waste to slag piles. Now, slag went directly to pits, from which it flowed by gravity into rail cars.

Sloss-Sheffield also installed equipment that eliminated much of the grueling labor formerly required in casting and molding pig iron. This was a piecemeal process, beginning with the installation of mud guns at the re-

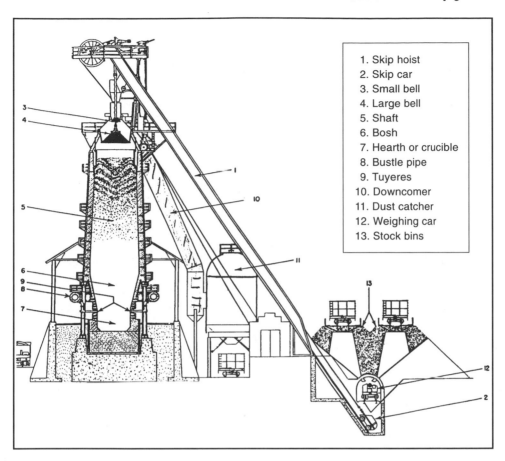

1. Skip hoist
2. Skip car
3. Small bell
4. Large bell
5. Shaft
6. Bosh
7. Hearth or crucible
8. Bustle pipe
9. Tuyeres
10. Downcomer
11. Dust catcher
12. Weighing car
13. Stock bins

Furnace Schematic. This line drawing of one of Sloss-Sheffield's two City Furnaces after the modernization drive of the late 1920s shows the stock trestle, the skip hoist, the McKee revolving top with double charging bells at the top of the furnace, the reconfigured stack, the raw materials at various stages of reduction, the bustle pipe and tuyéres, the subsoil foundations and cooling system, the downcomer, the dust catcher, and other features of the installation. (Courtesy of Sloss Furnaces National Historic Landmark, Birmingham, Alabama.)

modeled furnaces in 1927 and 1928. Long common in other parts of the country, these devices fired clay plugs into the tapping hole to close it after a melt had been released into the sand molds on the floor of the casting shed. Although this process saved some manual labor, a more important reason for its adoption by Dovel was that it assured that the furnace would be returned to operating condition as soon as possible after a melt, with less disturbance to the pressures and temperatures prevailing inside the stack.[56]

Significantly, the company did not immediately proceed to abandon hand molding. The fact that the full adoption of mechanical casting was delayed for about two years after 1929 is a significant clue that mechanization was less important in the company's scale of priorities than earlier modifications having to do primarily with achieving control over raw materials. This was not only because labor costs were less crucial than the cost of obtaining and handling raw materials but also because of what Louis C. Hunter called "the influence of market upon technique." The company's situation was like that of ironmasters in western Pennsylvania before the Civil War, when railroads had not yet opened up vast eastern outlets, the age of cheap steel had not yet begun, and furnaces in the Allegheny Mountains were still catering to farmers and other customers who needed the wide range of characteristics that charcoal iron could provide.[57] Sloss-Sheffield confronted a similar state of affairs in the late 1920s; instead of serving steel mills that used vast quantities of standardized pig iron, it sold its output to foundries that had a traditional orientation. These installations required relatively small quantities of pig iron with a wide range of chemical properties suitable for many potential uses in products varying from stove iron to piston rings; a veteran furnace superintendent likened the situation to that of a pharmacist asked to compound unique prescriptions for extremely specific illnesses.[58] Still dominated by a craft-oriented outlook, most foundry operators were also suspicious of anything but sand-cast pig iron; furnaces that changed to mechanized methods risked rejection at the hands of such customers. It was for this reason that not only Sloss-Sheffield, but also its chief competitor, Woodward, clung to hand-casting methods. Commenting that Woodward adopted mechanized casting only in 1930, *Fortune* mistakenly attributed this to "the sloth and indifference of the South." From a northern point of view, it was hard to recognize that southern merchant blast furnaces and the markets to which they catered were worlds unto themselves.[59]

Until late in the 1920s, the only machine-cast pig iron produced by Sloss-Sheffield was its Etowah brand, made at the Gadsden furnaces that had once belonged to the Alabama Company. Early in 1929, however, a transitional step in the direction of mechanical casting took place at the City Furnaces when a uni-pig machine was installed that made pre-formed pigs of identical size and weight. In deference to tradition-minded customers, the pigs were still allowed to solidify in sand molds. Because of this method, individual pigs were not absolutely unvarying in chemical composition, however uniform they might otherwise be.[60]

Not until early 1931 did Sloss-Sheffield finally complete mechanization of pig-casting. In February of that year, it installed two 125-ton mixing ladles— known as bull ladles because of their great size—at the City Furnaces. Running through curved troughs from the furnaces into the ladles, which were located immediately outside the sheds that had previously been used for casting operations, melts of iron were homogenized before being poured into an endless train of individual pig molds, impelled by a chain drive. Cooling and solidifying as they traveled up an incline, the finished pigs, weighing from forty-five to fifty pounds each, were shaken loose as they reached the top and dumped into waiting rail cars. Ironically, this technique, which eliminated sand casting entirely, utilized Heyl-Patterson equipment similar to that which Uehling had conceived in Birmingham in the 1890s. Mixing itself was also an old idea, having been used for decades by more mechanized northern furnaces after the concept was introduced in the late nineteenth century by one of Andrew Carnegie's chief lieutenants, William Jones.[61]

Sloss-Sheffield's selective mechanization of its furnaces in the 1920s was consistent with its previous record of making similar changes whenever it became necessary to do so. Earlier examples of such adaptation were the use of modern equipment at the Brookside mine and the installation of Ramsay and Stewart coal washers at several locations including Flat Top. The recent effort to modernize had taken much time; McQueen had started the process by introducing skip hoists and an experimental pig-breaking system at North Birmingham in the early 1920s, and Morrow did not fully implement mechanical casting at the City Furnaces until roughly ten years later. The slowness with which action took place was partly traceable to competing priorities, but in the case of mechanical pig casting there was a particularly long delay because many foundries continued to demand sand-cast pig iron. Over a ten-year period, however, mechanization gradually won out, even in casting, because it paid off at the bottom line. By abandoning hand-loaded vertical hoists, manual ramming of plugs into tapping holes, and sand-casting techniques in favor of skip hoists, mud guns, mixing ladles, and automatic casting apparatus, Sloss-Sheffield gained ever-increasing control over the handling of raw materials. Such command became more and more crucial at a time when the price of foundry pig iron kept steadily falling.

It is not clear precisely what considerations impelled the company to undertake mechanical casting, the final step in its long rationalization process. The most likely explanation is that the augmented capacity of the new furnaces made it extremely difficult to clear pig iron from sand beds, even with the help of Dovel's pig-breaking apparatus, before the next melt took place. Uehling had commented on this problem in explaining why it was necessary for furnacemen in places like Pittsburgh to adopt his casting machine in the 1890s; the same situation may ultimately have made older methods impracticable at Sloss-Sheffield. A new skimming technique for removing excess graphitic carbon (kish) from machine-cast pig iron had also been developed by the time Sloss-Sheffield abandoned sand casting, satisfying all but the most stubborn foundry operators that the time had come to accept the inevitability of change.[62] Whatever the reason, the protracted modernization drive that McQueen had launched soon after the end of World War I finally reached its climax as the firm embraced a technique that one of its own employees had conceived under the hostile eyes of Thomas Seddon.

In adapting to changing circumstances, Sloss-Sheffield had reached a historic juncture in a corporate evolution that now encompassed half a century. By developing a remarkable series of inventions aimed at securing better blast furnace performance, Dovel enabled the firm to make a concerted attack on one of the Birmingham District's two greatest long-range problems, that of coping with the peculiar requirements of southern raw materials. Similarly, by embracing mechanical pig-casting, the company climaxed a long tradition of selective mechanization by dealing with a second persistent difficulty, that of securing dependable performance from an undependable, because severely disadvantaged, labor force. Up to now it had been impossible to exercise effective control over operations because of inefficiency resulting from the use of unmotivated black workers. In the 1920s, inexorable market forces finally forced the company to assume this type of command.

But some things remained the same for a company that had always been highly resistant to change. Adopting mechanical devices that had long been common elsewhere did not require Sloss-Sheffield to stop using ill-paid black operatives. Unlike some of the equipment used in steel production, the new apparatus that the company had installed was compatible with traditional southern labor policies. As a result, the firm could still recruit inexperienced African-Americans, hire fewer employees more selectively, put them on the job with minimal training, maintain low wage scales, and preserve rigid practices designed to keep black workers and white bosses in separate occupational roles. Sloss-Sheffield could thus embrace modernity and at the same time affirm its enduring commitment to southern values. Under such circumstances, mechanization was an irresistibly compelling technological fix.

IX

Despite its southern ambience, the dramatically changed installation that now stood along the First Avenue Viaduct reflected larger trends taking place in the United States during the decade in which it was built. Control of operations had become a watchword in the 1920s among corporate officials who embraced a gospel of efficiency preached by Frederick Winslow Taylor and other champions of scientific management. Taylor's disciples worked out ingenious schemes for studying industrial processes and eliminating unnecessary motions in the belief that applying rigorous empirical methods would enhance productivity and mitigate labor conflict—exactly what Morrow and Dovel were trying to accomplish. Southern executives were as active in the efficiency crusade as their counterparts in any other region of the country. One of Taylorism's foremost exponents was Henry L. Gantt, who was born below the Mason-Dixon Line, grew up in Maryland, and idolized "Stonewall" Jackson.[63] It was only natural that simple technological solutions to complex industrial problems would appeal strongly to such southern-born financiers as Baruch, Catchings, and Sloan, who were heavily involved in big modern enterprises like Allied Chemical and Dye. Just as TCI's experiments in welfare capitalism reflected the reformism of the Progressive Era, Sloss-Sheffield's impressive new City Furnaces were characteristic products of a postwar period in which Americans had grown weary of social experiments but remained deeply enamored of science and engineering.

The huge casting sheds flanking the furnaces constituted a distinctive feature of the remodeled plant and added greatly to its imposing bulk. But much of the floor space in the sheds became unnecessary after the company adopted mechanical casting and no longer needed the sprawling sand beds in which pig iron had formerly solidified. Relatively short, curving channels now led from the crucibles to areas directly behind the sheds where the mixing ladles and pig-casting machine were located. As a result of the new layout, spectators could get a better view of the company's operations from atop Red Mountain than was possible from the viaduct running in front of the installation. Looking down from the crest of the ridge, one could see a vivid panorama of industrial activity as railroad cars rumbled back and forth across the stock trestle, skip hoists carried raw materials to the tops of the stacks, and bull ladles poured their fiery brew of white-hot iron into the endlessly revolving pig molds. The scene was particularly breathtaking at night as flames lit up the sky.

Less dramatic, but just as typical of Sloss-Sheffield's commitment to scientific and technological progress, were testing procedures conducted in a laboratory that the company built in 1930. Here, metallurgists painstakingly examined sample billets of pig iron to make sure that they were absolutely uniform in their chemical composition. Previously, the exact amount of key nonferrous elements in pigs from a single cast had depended on when the iron had come out of the furnace; the amount of silica, for example, might vary by as much as 50 percentage points between the first and last iron produced in a 100-ton cast. Now, customers could be guaranteed absolute accuracy. To dramatize this unvarying consistency, the company adopted a new advertising logo to indicate that its pig iron was "as uniform as peas in a pod." Significantly, this emblem was part of an educational campaign designed to overcome prejudices against machine-cast pig iron that were still widespread throughout the foundry industry.[64]

In November 1926, Sloss-Sheffield launched a bold new marketing venture to help it compete with such firms as Woodward. Known as *Pig Iron Rough Notes*, a name based on the expression "as rough as pig iron," it began as a small monthly publication edited by Russell Hunt and sent to a list of 600 subscribers. Conforming to the argot of the industry, issues were designated by "casts" and "heats" instead of "volumes" and "numbers." Within less than a decade, the little journal was being mailed to almost 5,000 readers. Despite its modest size, it became one of Sloss-Sheffield's greatest contributions to the advancement of the foundry trade, playing an educational and technical role far transcending that of a mere advertising tool.[65]

One of the most significant things about *Pig Iron Rough Notes*, easy to miss unless one understands the rationale underlying Sloss-Sheffield's modernization drive, was the timing of the magazine's introduction, which speaks volumes about the need the company felt to launch an educational campaign as it began updating its furnace practice. Morrow and Hunt, well aware that foundrymen were suspicious of change, created *Pig Iron Rough Notes* not

Sloss-Sheffield No. 1 City Furnace in 1906. This photograph shows how the installation looked before the modernization drive that began two decades later. Particularly noteworthy are the manually filled vertical hoists and charging apparatus, ultimately replaced by mechanical skip hoists and McKee tops; the relatively small dimensions of the stack, which were considerably enlarged in the late 1920s; and the location of the casting shed, which jutted outward toward First Avenue. The enhanced capacity of the furnaces after modernization in the 1920s, creating a need for more casting space than was available in front of the stacks, explains why it became necessary to relocate the sheds so that they flanked the two furnaces on each side, giving the plant a more horizontal layout. (Courtesy of Henry Ford Museum and Greenfield Village, Dearborn, Michigan.)

simply as a promotional tool but as a medium for showing foundries that they, too, needed to modernize and would benefit from the increasingly rigorous standards to which the firm's products were beginning to adhere. Implicit in the strategy was the prestige that science and technology commanded in the 1920s.

From the beginning, advertising and education went hand in hand. As modernization went forward, Hunt constantly emphasized the high quality of the iron that resulted and the scientific specifications that Dovel's innovations had made possible. At the time the magazine appeared, the company made four brands of pig iron. These were known respectively as Sloss, produced at Birmingham and using red ore from the Sloss mines, blended with selected brown ores; Noala, smelted at Sheffield and particularly esteemed because of its suitability for stove plates, stemming from its high percentage of Russellville brown ore; Etowah, a machine-cast brand from the recently acquired Gadsden furnaces; and Clifton, a blend of red and brown ore produced at Birmingham, though based upon a popular brand formerly identified with the Alabama Company. Each carload was shipped with a detailed analysis of its precise silica, sulfur, phosphorus, and manganese content.[66]

Knowing his audience, and displaying the same flair for salesmanship that Haas and McQueen had shown in their time, Hunt blended the firm's increasingly modern, scientific image with an emphasis on time-honored, old-fashioned virtues. Sloss-Sheffield, he maintained, was "a corporation with a soul," deeply valuing its customers and strongly committed to prompt, personalized service. To prove that this dedication was genuine, the company appointed a graduate chemist, W. O. McMahon, as a full-time representative to visit foundries and assist them in solving whatever technical problems they might experience. Given the title of "Sales Engineer, Cupola-Foundry Ser-

Sloss City Furnaces after Modernization. This photograph of the City Furnaces of the Sloss-Sheffield Steel and Iron Company after their modernization, taken at an undetermined date, shows the side of the plant that faced Red Mountain. It gives a good idea of the hustle and bustle of industrial operations as raw materials were brought to the site by rail, deposited into bins beneath the stock trestle, taken to the top of the furnaces by skip hoists, and discharged into the stacks. What the picture does not reveal are less obvious ways in which the furnaces had been redesigned by Dovel to optimize use of local raw materials. Though not apparent to the eye, these hidden modifications were crucial features of the modernized facility. (Courtesy of Sloss Furnaces National Historic Landmark, Birmingham, Alabama.)

vice," he was described by Hunt as having had "as much actual experience in melting iron as any man in Alabama. . . . He will be glad to come to your plant for a day or a week, as the circumstances may demand." To emphasize that McMahon and other sales representatives whom the company hired were "not afraid of soiling their hands and clothes," Hunt coined the term "Shirt-Sleeves Service" to describe the help they rendered.[67]

Knowing that prominent foundrymen would appreciate seeing their own names or information about their companies in print, Hunt devoted one or two pages of each issue to a leading entrepreneur, enterprise, or community that was prominently associated with the foundry trade. An early issue featured the life story of Joseph Lodge, once a poor orphan of immigrant parentage but by 1927 a leading soil pipe and hollowware manufacturer in the Chattanooga area. Another article contained a brief history of the Maytag Company, which used cast iron in its washing machines, and recounted the saga of its founder, Frederick L. Maytag, whose contributions to the American household were compared by Hunt to those of Henry Ford to the automobile industry. Yet another issue was devoted to Rome, Georgia, the "Stove Center of the South," with information about its leading firms and pictures of their products. *Pig Iron Rough Notes* thus became a valuable source of free advertising to Sloss-Sheffield's customers, something that its subscribers could appreciate.[68]

Reflecting the spirit of industrial cooperation to which the company's directors were devoted, Hunt declared that progressive firms no longer tried to keep business secrets, but exchanged information freely for the good of all. In this vein, *Pig Iron Rough Notes* promoted the activities of various associations connected with the foundry trade in particular and American industry in general. These included the Alabama-based Soil Pipe Association, founded in Birmingham in 1925 to "increase the knowledge, utility and use of Cast Iron Soil Pipe, through scientific research and public education," and the Gray Iron Institute, headquartered in Cleveland, Ohio, which fostered the adoption of uniform standards and business practices throughout the foundry industry. The magazine also encouraged readers to attend the annual meetings of organizations like the National Association of Manufacturers, which held its 1927 conclave in Chattanooga.[69]

Hunt never lost sight of the mission of *Pig Iron Rough Notes* to show foundrymen that it was in their self-interest to buy Sloss-Sheffield's increasingly standardized products. The magazine constantly exhorted its readers to aspire to higher standards of excellence and in the process taught them not to be suspicious of Sloss-Sheffield's steady departure from time-honored methods. Each issue contained a special section aimed at promoting good foundry practice and keeping subscribers abreast of recent scientific discoveries affecting the trade. Early issues included short articles by Y. A. Dyer, a consulting metallurgist, dealing with such subjects as cupola furnaces, the proper location of tuyéres, the chemical composition of cast iron, and the nature of fuels and fluxes.[70]

Soon the section became even more authoritative with the appearance of regular contributions by Richard Moldenke, a distinguished scientist and engineer. Moldenke's academic training, painstaking research, prolific publications, and contributions as a consultant to numerous industrial firms and professional organizations had established him as the leading foundry expert in America and the key figure in transforming what had previously been a traditional craft into an increasingly systematic industrial discipline. Known as the dean of American metallurgists, his role in foundry engineering was similar to the one played earlier by Alexander Lyman Holley in bringing Bessemer steel technology to America.

Born in Wisconsin of East Prussian immigrant stock, Moldenke had received a degree in mining engineering from Columbia School of Mines at age twenty-one and earned a Ph.D. from Columbia two years later. During the academic phase of his career, he rose to become head of the mechanical engineering department at Michigan School of Mines. His passion, however, was foundry technology. This abiding preoccupation resulted in repeated trips to Europe to gather information about the latest developments in advanced foundry practice and bring such data back to the United States, producing significant productivity gains. In 1925, the American Foundrymen's Association conferred its highest accolade on him by making him the first recipient of its Joseph S. Seaman Gold Medal. "He is known all over the world wherever iron is melted," Hunt declared in announcing his appointment to write articles for *Pig Iron Rough Notes*.[71]

Beginning with an initial essay in the January 1928 issue, in which he meticulously analyzed common defects in castings and advocated systematic record keeping as a way to overcome them, Moldenke contributed frequently to the magazine until his death in 1930. His articles, dealing with virtually every aspect of foundry practice, combined practical insight with rigorous scientific methods, providing readers with a wealth of insight and information. Meanwhile, other experts published articles on problems common to the industry, producing a growing body of industrial and scientific literature that had an incalculably significant impact on the improvement of American foundry practice.[72]

X

In with the new, out with the old. As modernization proceeded at Sloss-Sheffield, obsolete equipment was discarded. In August 1928, the directors approved the abandonment of a large number of antiquated facilities and much old, worn-out machinery. Steam-powered apparatus at the Sloss, Bessie, Flat Top, Searles, and Lewisburg mines that had been retained on a standby basis after electrification was now written off, together with beehive ovens that were no longer needed because of the North Birmingham by-product plant.

Not just equipment, but entire facilities, were discarded. Coal and ore mines were abandoned because they were either worked out or uneconomical to utilize. So, too, were mining camps, such as Brookside and Blossburg, that had once been home to thousands of employees.[73] Over the years following their abandonment, their squalid shacks would gradually be overgrown with weeds and brush or disappear entirely, leaving little trace of what had once been busy settlements. By the late 1980s, scraggly trees and bushes covered slag heaps, making them look like natural hummocks; decaying tipples and gaping mine portals peered out forlornly amid tall grasses, barely visible from nearby roads.[74]

Changing times also forced the company to give up its use of convict labor. Throughout the 1920s, reformers pushed vigorously for an end to this blot on the state's reputation, only to be frustrated by delays smacking of deliberate collusion between the industrialists and the Alabama legislature. A law passed in 1919, making it illegal after 1 January 1923 "for any person to lease or let for hire any state or county convict," was so badly written as to be unenforceable; hairsplitting over the difference between a lease and a contract rendered it effectively null and void. An amendment in 1923 extended the use of the system for four more years unless the governor found it financially possible to discontinue it earlier. But the incumbent, William W. ("Plain Bill") Brandon, who had campaigned on an "economy without new taxes" platform, saw no reason to take such a step. Such foot-dragging subjected the state to increasing scorn in other parts of the country; in July 1923, the *Nation* flayed Alabama's continued use of convict labor in a stinging editorial.[75]

Sloss-Sheffield inadvertently helped hasten the system's demise by providing irate citizens with a glaring example of the consequences to which it could lead. In August 1924, James Knox, a forger convicted in Mobile County for falsifying a thirty-dollar check, died at Flat Top. Company officials explained that he had taken his life by swallowing bichloride of mercury, but this allegation was shown false after rumors that Knox had actually been lashed with a steel wire and immersed in a vat of hot water prompted an investigation by the state attorney general's office. Its report, issued in March 1926, created a sensation. According to a physician who examined the corpse, Knox had indeed died in the vat. His death, the report stated was "a result of heart failure, which probably was caused by a combination of unusual exertion and fear acting upon an abnormally small heart which in turn was weakened by an extra large load of fat." Afterward, the physician charged, a "discoloring poison" had been injected into the victim's stomach to simulate suicide.

Publication of the report set off a firestorm of criticism throughout the state. Leading newspapers excoriated the convict lease system; even the Montgomery *Advertiser*, which had previously defended it, now changed its tune, branding it as "humiliating" and "disgraceful." A Jefferson County attorney, declaring that the situation at Flat Top recalled the "savage and inhuman treatment of ancient days when thumbscrew and rack were used," pledged that he would not rest until all parties guilty of wrongdoing had been punished. As shock spread throughout the country, the Washington *Post* called the episode a "hideous affair," *Literary Digest* compared the Alabama penal system to the Spanish Inquisition, and the New York *World* sent an investigative reporter, Donald Ewing, to Montgomery. Interviewed by Ewing, Brandon declared that he found "nothing wrong with the prisons of Alabama" but later promised to dismiss any state employee found guilty of mistreating inmate workers.

After being empaneled in Jefferson County, a grand jury heard testimony from a parade of convicts who reported being confined in tiny "dog houses," whipped, beaten with clubs, and knocked senseless by being struck on the head with heavy rubber tubes. Outraged reformers resurrected the idea of using inmates to work on the roads instead of forcing them to labor in mines. Called into special session late in 1926 after the election of a new governor, Bibb Graves, the state legislature responded by calling for a constitutional amendment permitting a $25 million bond issue for highway maintenance, to be repaid through the imposition of gasoline taxes.

During his campaign, Graves, a progressive politician who had withdrawn from earlier membership in the Ku Klux Klan, had pledged to eliminate convict labor once and for all. After taking office, he began removing state inmates from industrial prisons, including the one at Flat Top. But he had no authority to remove county convicts from mines, and Alabama's Big Mules fought tenaciously to preserve what was left of the system as the legislature debated how best to kill it. Responding to the delaying tactics, Graves extended a projected termination date of 1 December 1927 "so as not to cripple any industry by taking too many men from any one operation" and also to provide more time to build stockades for inmates assigned to work on the roads. After a nasty floor fight, the legislature finally agreed to end the system by 1 July 1928. On that date, Sloss-Sheffield's last convict workers were transferred to such places as the state farm or to hastily constructed road camps.[76]

The death of the convict lease system accentuated changes that were sweeping away familiar aspects of the industrial landscape and forcing managers to adjust to new situations. One by one, furnaces that had once blazed with activity were decommissioned and leveled, leaving hardly a trace of their previous existence. Despite the outstanding reputation of their Clifton brand of pig iron, which was henceforth produced in Birmingham, the Ironaton stacks that had formerly belonged to the Alabama Company were abandoned by 1929. Plagued by a deficient ore supply and the high cost of supplying it with fuel, the Etowah installation at Gadsden also went into oblivion; its pig-casting equipment was cannibalized for use at the City Furnaces in Birmingham, and the rest of the plant was torn down. Another casualty of modernization was Philadelphia Furnace at Florence, which had been out of blast since 1923 and was dismantled. A similar fate awaited the Cole Furnaces at Sheffield, only one of which had been in operation since 1920; it went out of blast for the final time in August 1927, after which the plant was demolished. Similar things were happening all over the country as moldering piles of crumbling masonry and rusting metal that had once teemed with industrial life were reduced to scrap. By the onset of the Great Depression, economic imperatives had forced Sloss-Sheffield to consolidate all its smelting operations at the four Jefferson County furnaces whose continued viability had been assured by the protracted overhaul they had received in the preceding decade. [77]

XI

As Sloss-Sheffield's new furnaces, mechanical charging and casting equipment, and scientific testing facilities came on line from 1927 through 1931, Hunt kept readers of *Pig Iron Rough Notes* well informed about the firm's latest technological achievements and how they contributed to producing "the best Foundry Pig Iron on the market." By the time modernization had been completed with the total mechanization of pig-casting in 1931, this message was more and more crucial, not just to maintain market share but also to help Sloss-Sheffield survive the worst economic crisis in the nation's history.

As early as 1928, key indicators of economic activity had begun to dip. Even though most business experts remained optimistic about the future, a few forecast trouble ahead. One disturbing indicator was a declining level of funding for capital improvements, such as the new equipment that Sloss-Sheffield had brought on line throughout the decade. Instead of being used to finance such projects, more and more money was being thrown into gambling on the stock market. As in Bryan's day, the same speculative tactics at which Inman had been so adept prevailed over developmental growth strategies, with the same result that had occurred in the depression-ridden 1890s.

Sloss-Sheffield itself was implicated in the speculative trend through its connections on Wall Street. As stock prices climbed steadily higher from 1927 through most of 1929, investors were willing to pay as much as 20 percent interest on call loans to buy securities on credit, making it more profitable for banks to funnel holdings into highly leveraged investment trusts rather than industrial improvements that might produce only half as much financial yield. Goldman, Sachs and Company, to which Sloss-Sheffield was closely tied through Catchings, helped fuel the mania after December 1928 by organizing a number of get-rich-quick schemes under the aegis of the Goldman Sachs Trading Corporation, the Shenandoah Corporation, and the Blue Ridge Corporation. One of Sloss-Sheffield's directors, Bernard Baruch, was making large profits by taking advantage of inside information to buy stocks at reduced prices before they went on sale. [78]

Throughout 1929, the stock market index kept rocketing upward despite fears expressed by Herbert Hoover, who had just moved into the White House, that the bubble would soon burst and bring about an economic disaster. Concern that the situation was getting out of hand finally drove Baruch out of the market during the summer, but he continued to make reassuring public statements about the country's financial health. His actions proved wiser than his words. Late in October, Wall Street spun out of control as a stock market crash of unprecedented magnitude took place.

Economic historians have debated ever since about the nature of the connections—or, indeed, whether or not connections existed—between the crash and the Great Depression that followed. For a time, the market seemed to recover, leading business and governmental leaders to assume that the worst was over. Despite such optimism, the crash proved to be only the beginning of a deepening crisis pervading the entire world as investment dwindled, output of goods and services became more and more feeble, unemployment became increasingly widespread, and consumption plummeted. By July 1932, business and industrial securities that had seemed unassailable at their peak in 1929 had sunk to unimaginably low levels. General Motors had fallen from 73 to 8, Montgomery Ward from 138 to 4, and U.S. Steel from 262 to 22. Based on a 1929 index of 100, American industrial production stood at 54 by 1932. The value of securities listed on the New York Stock Exchange plunged from $87 billion in 1929 to $19 billion in 1933.[79]

It did not take long for Sloss-Sheffield's financial statements to reflect the harshness of the times. Nevertheless, because of the diversification and modernization that had been effected under McQueen and Morrow the picture was not as dark as might otherwise have been the case. In November 1929, as Wall Street reeled from the debacle of the previous month, Morrow negotiated a ten-year contract under which the Birmingham Gas Company agreed to purchase 2.5 billion cubic feet of coke oven gas per year from Sloss-Sheffield at fifteen cents per thousand units. Due to competition from other suppliers, the agreement was slightly less lucrative than the one it replaced. Still, it guaranteed dependable income for the company, regardless of what it earned from other operations.[80]

This cushion was badly needed. By the end of 1930, pig iron production in the United States was down by 25 percent from the previous year, and market quotations had sagged to $14 per ton, the lowest figure since 1915. At times, Sloss-Sheffield sold iron for as little as $10 per ton for northern delivery. Still, Morrow could report to the directors in March 1931 that diminished price levels had been "more than offset by lower production costs," demonstrating the wisdom of modernization. Indicating its faith in the future, the firm had pushed ahead with the installation of automatic pig-casting machinery at both the City and North Birmingham furnaces. Due to the depression, however, it had already been forced to stop paying dividends on its common stock and had paid only three dividends instead of the usual four on preferred. In large part because of the sale of by-product coke gas, it continued to operate in the black. Though materially lower than its average earnings throughout the 1920s, its net profits for the year were $576,741.30. Many corporations were doing far worse.[81] There was no way, however, for Sloss-Sheffield to escape the intensifying impact of the depression. Despite stopping all dividends and declaring a moratorium on purchases of new equipment, it posted a net profit of only $79,342.34 in 1931.[82] Directly ahead loomed a succession of annual deficits and other hardships that would force the company to dip deeply into accumulated surplus in order to survive.

Sloss-Sheffield celebrated its fiftieth anniversary in 1931 amid deepening gloom. Half a century after the founding of the Sloss Furnace Company, the future seemed bleak, but the situation was not as desperate as it could have been in the absence of effective leadership in the previous decade. Because of its creative response to changing times under McQueen and Morrow, the firm was prepared to ride out the crisis. Over many years, it had weathered a succession of panics and depressions. Now, under Morrow's tightfisted leadership, it would do so again.

From Hugh Morrow to Jim Walter

Completion in 1931 of the modernization program that McQueen had begun thirteen years earlier brought the formative period in the history of Sloss Furnaces to an end. During the next four decades, the enterprise that Colonel Sloss had founded in 1881 would survive the Great Depression, help the United States win World War II, be sold to new owners, earn the largest profits it had ever recorded, and climax its technological evolution by building the most advanced blast furnace in the history of the American foundry pig iron industry. Then, swiftly, nine decades of development came to an abrupt end as drastic economic and technological changes undermined the market for foundry pig iron. Meanwhile, Birmingham itself experienced a shattering crisis, traceable at least in part to the lack of vision that generations of industrialists had shown in their heedlessness for its civic and social welfare. In 1970, while the Magic City was being reborn under new leadership, the fires of the last furnaces to make iron on the site where the first stacks had been built in the early 1880s went out forever.

I

The Great Depression was a watershed in Birmingham's history, dashing hopes that it would become the country's leading industrial center. Devastated by the worst economic disaster ever to hit the United States, it gained a reputation as "the hardest-hit city in the nation," confirming a local adage that "hard times come here first and stay longest." Edward S. LaMonte later characterized it as "a community overwhelmed by an economic catastrophe totally beyond its control."[1]

Conditions in Birmingham were deteriorating long before the stock market crashed in October 1929. Unemployment mounted throughout 1927, causing a wave of applications for public aid that soon overwhelmed the city's anemic welfare system. Banks began failing by June 1929. Despite such warning signs, civic and business leaders thought that the crisis, when it finally struck, would not last long. They were mistaken. Testifying to a senate relief committee in 1932, congressman George Huddleston estimated that 100,000 of the city's 108,000 workers were unemployed or working only part-time. "My people are desperate," he said. "They're in an agony of distress and starvation. They are likely to do anything. They have almost lost the power of reasoning. The situation is full of dynamite."

Even in such affluent suburbs as Mountain Brook, some residents survived the depression only by "exercising the severest restraints in their life style." Many boarded up their homes and left. As the economy went into shock, George B. Ward's brokerage business suffered badly. Before he died in 1940, he had to sell his peafowl and cancel his membership in the Red Mountain Garden Club to make ends meet.[2]

Exploiting the catastrophe, the American Communist Party moved its regional headquarters to Birmingham in 1930. Shaking off submissive attitudes, black slum dwellers organized block committees, and even women's sewing clubs seethed with revolt. In May 1933, a melee broke out when police attacked thousands of black demonstrators who had gathered in front of the Sixteenth Street Baptist Church to protest unemployment, poll taxes, disenfranchisement, and lynching. Even old men and women resisted; one elderly female clubbed a policeman across the face with her parasol. "This was the first time that Negro women came to know that they could stand up and fight back and defend themselves," a black leader later recalled.[3]

Confronting the worst crisis in its history, Birmingham had little in the way of local leadership. The "nativist progressives" who had taken charge of the city in the 1917 municipal election retained their grip on local politics throughout the 1930s and long thereafter. Adept at penny-pinching, dodging responsibility, and perpetuating ethnic and racial cleavages, they were far more disposed to temporize than to meet the challenge of change. Disclaiming any responsibility to offer public assistance, municipal officials turned as always to private agencies, which soon ran out of funds. Fragmentation within the city contributed to paralysis; reeling under mounting burdens, the Community Chest and other private agencies refused to give blacks or labor unions a voice in determining welfare policies. County and state agencies also failed to provide adequate help, wrangling about how or whether to dispense pathetically meager relief payments while people died from starvation. Only the Red Cross Family Service and federal intervention through New Deal agencies including the Civil Works Administration (CWA), Public Works Administration (PWA), Reconstruction Finance Corporation (RFC), and Works Progress Administration (WPA) prevented the situation from becoming hopeless. During the mid-1930s alone, Jefferson County received $361 million in federal aid.[4]

Soon after the depression hit, the district lost one of the few local industrialists who might have made some difference in coping with its problems when George Gordon Crawford resigned as president of TCI in January 1930 to become chief executive of the Jones and Laughlin Steel Corporation in Pittsburgh. Throughout his twenty-three years in Birmingham, he had done his best to promote the best interests of his native region despite his inability to exert more than a moderating influence on U.S. Steel's discriminatory pricing policies. During the 1920s, he had spearheaded the development of modern port facilities at Mobile, completed in 1927; persuaded Goodyear Tire and Rubber to establish a plant at Gadsden; and worked tirelessly to promote agricultural diversification. Leaving TCI because Jones and Laughlin had offered him better pay, he picked a bad time to lead a projected expansion program at Pittsburgh and returned to Birmingham in 1935 broken in spirit by the reverses he had encountered. Too exhausted to resume his former leadership role, he died in March 1936 after surgery at TCI's hospital from which he never recovered.[5]

Under Crawford's successor, Herbert C. Ryding, TCI made deep cuts in production. Rather than dismiss employees, it tried to spread what little work was available among as many people as possible. After the worst years of the depression had ended, the sheet mill added capacity, and a tin plate mill, the first in the region, was finished in 1938. If TCI could have done more for the district, it could also have done less.[6] On the other hand, the company was unquestionably a weak reed as Birmingham faced the greatest economic crisis in its history. As ever, it reflected the restrictive influence of attitudes prevalent in the front office of its parent corporation. There, executives who were more interested in stability than competitiveness had long pursued what two writers have aptly characterized as a "drift toward technical mediocrity." In 1936, a study published by a team of industrial experts listed a series of innovations that U.S. Steel had either failed to embrace or adopted only haltingly, including continuous strip mills, stainless steel, and cold-rolled products. At the time the report was released, a leading business magazine stated that "the chief energies of the men who guided the corporation were directed to preventing deterioration in the investment value of the enormous properties confided to their care. To achieve this, they consistently tried to freeze the steel industry at present, or better yet, past levels."[7]

In 1931, an article in *Fortune* highlighted the fate that had overtaken Birmingham, a city that had once expected much from the 1907 takeover, now only a distant and increasingly bitter memory. Pointing to the district's natural advantages, the unidentified author stated that "the God of Steel could not have laid out his domain more beautifully. It is the world's masterpiece of natural assembly. Down from the Red Mountain rolls the ore, down from the west rolls the coal, meeting at the valley's furnaces. Chicago, a Lake port, is a cheaper assembly point than Pittsburgh, but neither will match costs with Birmingham." A photo of the Magic City's skyline was captioned, "Nowhere in the world can iron be produced more cheaply. Yet Birmingham's development as a steel center has been slow." A map of American ore deposits, coal deposits, and blast furnaces reinforced the point by showing that the district made only 3 percent of the steel produced in the United States.

Seeking the cause for such a paradoxical situation, the writer asked why Alabama had produced only a "paltry 1,700,000 tons" of steel in 1929. "With the Steel God's incomparable largesse, with pig iron dropping into the freight cars at $9 or $9.50 a ton, why does the valley not glow with open hearths? Why were 1,000 miles of steel pipe sold last year within the Birmingham territory, and not one foot of it from a Birmingham mill?" After considering a variety of explanations, including "the sloth and indifference of the South itself," a dearth of markets within the region, and the effects of the multiple basing point system that U.S. Steel had adopted after being forced in 1924 to abandon Pittsburgh Plus, the article concluded that eastern financial interests had too much invested in the Mesabi Range, in ore fleets plying the Great Lakes, and in installations at such places as Pittsburgh, the Mahoning Valley, and Chicago to want to increase output in Jones Valley beyond relatively modest levels. For consolation, it predicted that U.S. Steel would ultimately have to turn to Alabama when Mesabi gave out. "Birmingham will have its day," the author said, "but there is still much to be worked out in the North before this spacious, easy-going southern city becomes our second Pittsburgh."[8] This, too, was a false prospect. By the time the Lake Superior region was yielding only low-grade ore, foreign sources had made Alabama's remaining deposits commercially valueless.

By donating a parcel of land overlooking the city, TCI did assist in a project to rescue Moretti's statue of Vulcan from the state fairgrounds and erect it atop Red Mountain as a symbol of Birmingham's industrial prowess. For the most part, however, even this effort was financed by the New Deal; the WPA paid $38,874 of the $44,062 required to complete the undertaking. At some point, the scattered components of the colossus had been reassembled, colored with flesh tones, given black eyebrows and rouged cheeks, and adorned with a pair of painted overalls. In the summer of 1936, workers took it apart, removed the paint and rouge, and transferred the cast iron segments from the fairgrounds to the mountaintop site provided by TCI. Three years later, the project was completed; in May 1939, the statue was dedicated at ceremonies highlighted by a pageant with episodes recounting Vulcan's—actually Hephaestus's—role in making a suit of armor for Achilles, Hernando de Soto's 1540 visit to what later became Alabama, and the birth of Birmingham in the 1870s. The *News* declared that the giant symbol was the city's "greatest single attraction," and civic leaders looked forward to the tourist dollars it would bring to the district.[9]

Whether tourists would want to come was another matter considering the bad publicity that Birmingham was now receiving. During the 1930s, it was repeatedly scorned by critics who pointed out that it continued to rank first in the nation in illiteracy and venereal disease and last in per capita income among cities of its class. In 1937, New York journalist George R. Leighton published a trenchant analysis of its problems in *Harper's Magazine*. Deriding Birmingham as a "city of perpetual promise," he decried the contrast between the living conditions of the rich and the poor, the indifference of local leaders to social problems, the hypocrisy of clergymen who denounced Sunday movies while ignoring more meaningful issues, and the divisive tactics used by political leaders who clung to power by exploiting racial cleavages. In 1939, Leighton's essay was republished in a mordant study of five American cities, this time with photographs reminiscent of ones that had embarrassed Birmingham in *Survey* almost three decades earlier. The pictures showed squalid dwellings engulfed by smoke from nearby furnaces, dismal street scenes, and tumbledown houses half hidden by a sign proclaiming the ironic message, "World's Highest Standard of Living: There's No Way Like the American Way." Why tourists would want to see Vulcan's Alabama home under such circumstances was anybody's guess.[10]

II

Sloss-Sheffield shared Birmingham's miseries. Conditions were particularly bad in 1932. Nationally, pig iron production sank to the lowest point since 1896, and the company incurred an unprecedented net loss of $956,312.61. In 1933, red ink flowed again, but the deficit was only $363,242.07. During the next two years the hemorrhage continued to subside as losses totaled about $41,000. Overall, Sloss-Sheffield's output of pig iron fell from 407,782 tons in 1930 to 205,843 in 1935; by-product coke production shrank from 549,315 to 322,685 tons. Much of the company's equipment lay idle, especially in the Sheffield area, where outmoded facilities continued to be abandoned and dismantled. Even in Jefferson County, the No. 4 furnace at North Birmingham was out of blast for almost eight years. Other facilities, including the Bessie and Ruffner mines, were dormant even longer.[11]

By retrenching, stopping dividend payments, withdrawing cash from surplus, and squeezing as much income as possible from severely contracted output, Morrow steered the company through the crisis. In a striking display of fiscal conservatism, he even managed by degrees to liquidate its bonded debt. By 1935, he had retired the last of the gold notes issued in 1919 to absorb earlier obligations; soon thereafter, he paid off the bonds that had financed the takeover of firms acquired by merger in the mid-1920s. Profits reemerged in 1936, when Sloss-Sheffield posted net earnings of $868,462.63 and paid dividends on preferred stock. Dividends on common were resumed in 1937 as the company earned $1,474,827.[12]

Following a recession that began in 1937, profits fell, and both North Birmingham furnaces were blown out. Still, the firm earned $589,645 and $943,178 in 1938 and 1939 respectively. By this time, because of a national preparedness drive as the country responded to the onset of World War II, the depression had finally ended. Net profits surged to $1,677,794 in 1940 and might have been even better in 1941 but for excess profits taxes that shrank earnings to $1,261,501. Accumulated surplus, which had sunk to $6,083,318 in 1935, exceeded $8 million by the time the Japanese bombed Pearl Harbor in December 1941.[13]

Morrow's austerity program had helped to carry Sloss-Sheffield through the depression, but the main reason that the company survived was the diversification and modernization program it had completed by 1931. But for this, the firm would have gone bankrupt. Income from coke by-products was crucial in getting it through the early years of the decade, when demand for pig iron was lowest. The Birmingham Gas Company continued to buy fuel from the North Birmingham plant, whose surplus electricity was also used by the Alabama Power Company. Coke derivatives went into a motor fuel, Woco-Pep, which was refined by the Wofford Oil Company and sold by a chain of southeastern gas stations. Blast furnace by-products, including slag used in making concrete, also stretched income.[14]

In 1939, as the preparedness drive intensified, Morrow continued to diversify by contracting with the Cuban-American Manganese Corporation to produce ferromanganese, a strategic alloy that eliminated harmful oxides and counteracted adverse effects of sulfur when used in foundry cupolas. Demand for this product was growing because foundries were using more scrap metal with less dependable chemical properties than those provided by pig iron. Because the contract assured Sloss-Sheffield a steady supply of Cuban manganese, the No. 3 North Birmingham furnace, which had been idle since 1937, was modified and reactivated.[15]

Marketing became more aggressive during the 1930s in response to hard times. *Pig Iron Rough Notes* continued to be a valuable promotional tool; it carried numerous articles highlighting the special suitability of Sloss pig iron for such foundry products as air compressors, boiler grates, diesel engine components, high-pressure valves, ornamental urns, piston rings, sludge pumps, sluice gates, textile machinery, and piping.[16] Calling cast iron the "Metal Eternal" and pointing to the varied roles it had played throughout history, Russell Hunt, who continued to edit the publication, assured customers that Sloss-Sheffield's sole mission was catering to foundrymen. Trying to end confusion stemming from its identity as the Sloss-Sheffield Steel and Iron Company, he emphasized that "Sloss Does Not Make Steel."[17] Stressing the special fitness of southern raw materials for foundry pig iron, he also began a drive to combat "phosphobia," which he defined in one issue as "a morbid fear of phosphorus." Some articles even contained menus showing readers how they could get enough of this chemical element in their diet.[18]

Inevitably, the depression created tension among stockholders. Trouble arose in 1936 when dividends were resumed on preferred stock. Led by Joseph Bryan's son, John Stewart Bryan, Richmond investors claimed that Sloss-Sheffield had actually earned profits in 1934 and 1935 and converted them into fictitious losses by charging excessive amounts to depreciation and using other bogus accounting practices. Before dividends were resumed on common stock, the Virginians insisted on compensation for themselves and other preferred shareholders for payments not issued between 1930 and 1935. After demanding an outside audit, Bryan and his group were rebuffed by the directors, but a Richmond accounting firm was finally given access to the books. Its report, listing various improprieties, was disputed by the regular auditors. As Bryan's group prepared to sue, Sloss-Sheffield appointed a prominent Wall Street attorney, John Foster Dulles, to find a way out of the impasse.

Dulles recommended a stock reorganization plan replacing the company's original noncumulative, noncallable preferred stock with new cumulative, callable shares. Implemented early in 1937, it resulted in a diminished annual yield but provided for a special initial dividend to mollify Richmond investors. Shortly after the plan went into effect, the Bryan family began to be represented on the board of directors for the first time in many years. In 1939, after W. H. Kettig died, Alexander Hamilton Bryan, one of Joseph Bryan's grandsons, replaced him. Even this change, however, did not prevent Richmond stockholders from continuing to bombard the firm with demands for periodic audits.[19]

III

Although the once-dominant Virginia connection was now only an annoying presence, Sloss-Sheffield remained a highly paternalistic firm, clinging to the region's historic low-wage policy and other southern customs that were as deeply entrenched in the 1930s as they had been during Joseph Bryan's lifetime. Mechanization notwithstanding, furnace crews were still solidly black except for supervisors, who were unvaryingly white. As in Maben's day, most blacks working at the City Furnaces continued to live in the squalid area, known as Sloss Quarters, bounded by Thirty-second and Thirty-fifth streets and First and Third avenues. Because few owned automobiles or could afford public transportation, this proximity made it easy for them to get to work and return home at the end of a shift. It also benefited the company, which could summon help at any time of the day or night in case of emergency. Occasionally, a burden might hang in one of the stacks before falling suddenly into the crucible, spraying slag and molten iron through the tuyéres with a deafening roar. After such a breakout, "shack rousters" were sent to the quarters to organize a cleanup. Amid the smoke and fire, the social ethos of antebellum plantations survived in a much different setting.[20]

A few white workers also lived in or near the quarters, occupying better housing that was segregated from the rest. This situation was also true at mining camps such as Flat Top, where whites lived in an area known as Silk Stocking Row. Houses for white workers were painted yellow; those occupied by blacks were colored red. Both types provided only minimal comfort, but the ones inhabited by blacks were particularly wretched; floors sagged, and wind blew through cracks in the walls. "Some of the houses was pretty good," recalled Will Prather, a black furnace keeper for forty-two years, "but [in] some of them you could stand in the house and look outdoors. I have set in my house many times and see[n] folk pass by." Until the late 1930s, when municipal officials began to enforce sanitary codes, such shacks had no running water or indoor plumbing; noisome privies lined the alleys, and water was scooped from wells. Blacks paid rents of $2.50 to $4.00 per month; whites paid up to $10.00.

Some black families patronized a "dago store," as they called it, on the corner of Thirty-fifth Street. Most, however, bought groceries and other supplies at the commissary near the stacks. Continuing time-honored practice, they paid by having earnings deducted from future wages or using company coinage known as "clacker" or "dougaloo." Local merchants hated the system because they lost trade from it.[21] Stretching their incomes as far as they could, workers living in the quarters supplemented their food supply by keeping small gardens where they raised okra, lentils, and greens. Despite their hardships, they found positive things to say when interviewed many years afterward, recalling "chitterling suppers," barbecues, Fourth of July picnics, church functions, and an absence of violence and crime. "We kept this quarter in shape," said Clarence Dean, who worked for Sloss-Sheffield for three decades. "We didn't have no rowdies. Oh, we took care of the rowdies. . . . We had a committee, we'd go there to them and talk to them. . . . You had to shape up."

Recreational activities included baseball, played by such company teams as the Raggedy Roaches, which competed with rival squads sponsored by ACIPCO, Stockham Pipe and Valve, and other local firms. Medical care was provided by company physicians. Taking their inferior status for granted, workers made the best of things. "I made good time here," stated Abraham Williams, a former sand caster who came to the City Furnaces from Sheffield after the northwest Alabama plants were shut down. Williams could have had jobs in the North, but "it would have cost too much with the children and all to move up there."

Despite clinging to traditional ways, Sloss-Sheffield was ultimately forced to yield to changing times. Dean's early years with the company coincided with bitter labor disturbances that plagued the district throughout the depression. These struggles brought unaccustomed pressures to bear upon the owners. For a time, Franklin D. Roosevelt followed policies grounded in the legacy of cooperation between government and industry that had been handed down from the World War I experience, but as hard times continued his administration swung ever more to the left as he attacked "economic royalists" and forged a new coalition between former progressives, blacks, union members, and disadvantaged urban dwellers. As this sequence of developments unfolded, New Deal policy became increasingly opposed to the low-wage strategy pursued by generations of industrialists in a region that Roosevelt saw as "The Nation's Number One Economic Problem," leading by 1938 to a fundamental assault on North-South wage differentials in the form of the Fair Labor Standards Act. As one of the main foundations of southern economic institutions began to crumble, a bitter reaction set in among Morrow and other Democrats, setting the stage for such later developments as the Dixiecrat movement and a switch to right-wing Republicanism that still lay far down the road. [22]

In 1933, Sloss-Sheffield subscribed to a Code of Fair Competition for the ferrous metals industry under the National Industrial Recovery Act (NIRA). [23] Because section 7-A of this law was aimed at promoting unionization, the Amalgamated Association of Iron, Steel, and Tin Workers (AA), the United Mine Workers (UMW), and the International Union of Mine, Mill and Smelter Workers (IUMMSW) launched drives to organize Alabama's coal, iron, and steel industries. Stubbornly opposing these campaigns, TCI, Sloss-Sheffield, Woodward, and Republic created company unions in an effort to comply with federal regulations without risking fundamental change. Union organizers, assailed by the local press as Communists and outside agitators, were threatened, beaten, and run out of town.

Badly demoralized after the collapse of its efforts to organize the steel industry in 1919, AA was far too weak to mount an effective counterattack against the operators. The UMW and IUMMSW, however, were more successful. In 1934, the district was convulsed by strikes that shut down most coal and ore mines. The Alabama Fuel and Iron Company, headed by Charles F. DeBardeleben, was particularly violent in its response, fighting back with barbed wire, dynamite, and machine guns and turning its installations into veritable fortresses. Along with other large corporations, however, Sloss-Sheffield yielded to government pressure and recognized the UMW. In the process, the operators won a revised NIRA code permitting them to pay lower wages than those prevailing in most other parts of the country, and the miners were denied a closed shop. Morrow played a key role in persuading NIRA officials in Washington that rates of compensation in the Birmingham District should remain lower than those prevailing in such places as Pittsburgh and came back home with what a Montgomery newspaper called "the laurels of a conspicuous victory." But the triumph was fleeting. After NIRA was declared unconstitutional in 1935, miners won further gains under the National Labor Relations Act, better known as the Wagner Act, passed later that same year. In 1939, after a protracted walkout, UMW won a closed shop from every Alabama operator except DeBardeleben, who left the coal business rather than submit. [24]

Meanwhile, despite company unions, intimidation, and armed force used by employers, the more radical IUMMSW managed to organize Alabama's ore miners. Sloss-Sheffield's ore miners were the first in the state to affiliate with the union, forming Local 109 on 17 July 1933, but TCI was the main target of a strike that broke out in May 1934. It was settled in late June with limited gains to the workers, who were backed by Secretary of Labor Frances Perkins. The IUMMSW failed to gain recognition from the operators, who continued to deal with a corporate-sponsored organization whose very name—the Brotherhood of Captive Mine Workers—unintentionally epitomized subservience, but after further strikes in 1935 and 1937, most Alabama operators recognized IUMMSW as sole bargaining agent for ore miners. Sloss-Sheffield was among the last to yield; a walkout by nearly 600 miners forced it to capitulate in September 1937.[25]

Because of AA's weakness, and its identification with skilled craft traditions that were foreign to most of Alabama's iron and steel workers, Sloss-Sheffield, TCI, and other local firms succeeded in delaying unionization of furnace and mill workers but could not prevail indefinitely against the trend of the times. In 1936 the newly formed Congress of Industrial Organizations (CIO) forced the older and more traditional American Federation of Labor (AFL) to cooperate in establishing a Steel Workers Organizing Committee (SWOC). Bridging racial barriers despite segregation laws that prohibited black and white workers from holding joint meetings, William Mitch and other Alabama labor leaders organized strikes that forced most local companies, including Sloss-Sheffield and Woodward, to grant union recognition in March 1937. The lone holdout was Republic, which turned Gadsden into a battleground with the ferocity of its efforts to stem the tide. In the end, however, it too gave in. By 1942, the United Steelworkers of America, as SWOC was now known, had won a closed shop throughout Alabama.[26]

Because of unionization, Sloss-Sheffield was forced to raise wages and grant fringe benefits. In 1936, the company began giving a one-week paid vacation to all employees who had been in service for five years or more. Five years later, despite opposition from Morrow and other local businessmen, the Department of Labor forced owners to pay miners from the time they came to work until they went home at the end of a shift—the portal-to-portal principle—instead of basing wages strictly on time spent in the seams. Bitterly resented by workers, this practice had prevailed in the district more than fifty years.[27]

Morrow and other Big Mules did not submit tamely to the new order. Their opposition to change became particularly vehement after the Fair Labor Standards Act went into force in 1938. As Gavin Wright has indicated, this law was aimed specifically at erasing the southern low-wage differential and marked a historic turning point in regional economic development, helping explain the vigorous tactics used by the operators in resisting it.[28] Anti-Semitism, racism, and red-baiting marked campaigns conducted by Morrow and other local industrialists, who fought relentlessly to discredit the New Deal and its working-class allies. In the process, Big Mules established such ultraconservative periodicals as *Alabama* and *Southern Outlook* and created a Constitutional Educational League to disseminate their reactionary point of view. In 1938, together with anti-Roosevelt forces in other parts of the state, they succeeded in having Frank M. Dixon, a Birmingham corporation lawyer and magnetic orator who regarded unions as "un-American," elected governor. Despite the fact that downstate landowners had benefited from the New Deal's agricultural programs, the industrialists succeeded in forging alliances with Black Belt planters to block further change. Even before America entered World War II, this counterattack had put organized labor on the defensive. Ties between industrialists and wealthy planters that dated back to the days of the Broad River group still lived on.[29]

IV

Despite their mutual hatred, workers and managers alike became increasingly involved in preparations for war. Long before the Pearl Harbor attack, *Pig Iron Rough Notes* was emphasizing the importance of cast iron to national defense and asserting Sloss-Sheffield's readiness for whatever it might be called upon to do in getting ready for the impending hostilities. When the United States finally declared war, a series of articles showed how the foundry industry was girding for victory, giving readers detailed information about the fourteen ordnance districts into which the country was divided, the locations of arsenals and military depots, the types of cast-iron products required by the armed forces, and the procedures that foundries would have to follow in securing defense contracts.[30]

Under wartime regulations, Sloss-Sheffield fell under the jurisdiction of the pig iron unit of the War Production Board's iron and steel branch, which was charged with coordinating the output of approximately 3,000 foundries scattered throughout the United States. As these installations switched to wartime operations, *Pig Iron Rough Notes* hailed their achievements, as in lauding a Virginia pipe foundry for being the first firm in the industry to win the Army-Navy "E" Award. To dramatize the contribution that cast iron was making to the war effort, Hunt devoted much of one issue to the roles women were playing as foundry managers; another number featured articles showing how cast iron helped supply food for wartime needs through its use in agricultural equipment, cookware, and containers.[31]

Nothing aroused Hunt's anger as much as statements that the global conflict was a "steel war" in which cast iron was "out of place." He used every opportunity to show that cast iron was essential to victory. Col. J. E. Getzen, chief of the Army's Birmingham Ordnance District, told in one article how vast quantities of cast iron were used in fragmentation and incendiary bombs, hand grenades, antipersonnel mines, mortar shells, marine engines, and tank components. About 60 percent of all hand grenades used by American forces in the war, Getzen pointed out, came from the Birmingham area. Another article, "Cast Iron Goes to War," included a long list of strategic and defense-related items that were made by foundries, ranging from air compressors, bed plates, brake drums, camshafts, cylinder heads, and flywheels to military stoves, soil pipe for army bases, and products that kept railroads running on schedule. Even though airplanes and ships were made largely of aluminum and steel, the trade magazine pointed out, they could not exist without cast-iron machine parts used in their manufacture.[32]

Virtually all of Sloss-Sheffield's output from 1941 to 1945 went into muni-

Workers at Sloss-Sheffield's City Furnaces, 1939. Workers and supervisors at Sloss-Sheffield's City Furnaces pose in September 1939 for a photograph after winning a trophy from the National Safety Council. The spirit of togetherness that appears to prevail in the picture masked deep racial cleavages in a social order in which African-Americans were severely disadvantaged and denied opportunities for upward mobility in the workplace. (Courtesy of Sloss Furnaces National Historic Landmark, Birmingham, Alabama.)

tions. A sign in front of the City Furnaces stated, "This is a U.S. Arsenal." Hunt took special pride in the number of company employees in the armed forces. A "Sloss Roll of Honor" that appeared in *Pig Iron Rough Notes* in 1943 included the names of 381 persons "serving on every front." By war's end, about 500 employees had seen duty. [33]

Wartime commitments did not lessen animosity between Sloss-Sheffield and the unions with which it was forced to deal. In 1943, Morrow and other Big Mules won passage of a state law, the Bradford Act, mandating open shops, banning picketing, outlawing secondary boycotts, and forbidding union contributions to political candidates. As Robert J. Norrell has said, the measure had a major impact despite its eventual judicial nullification, for it "kept the Alabama labor movement fighting for its existence in the mid-1940s, rather than devoting its energy to organizing in new fields and extending its influence in politics." [34]

Employment boomed in Birmingham during the war. TCI was running at full capacity by 1939 and subsequently expanded its coal, ore, and steel mills. A large new blast furnace was built at Fairfield, and the tinplate mill that had been finished in 1938 adopted continuous electrolytic methods to facilitate the production of containers for the massive food shipments required by the armed forces. Modification of such aircraft as the Boeing B-29 Superfortress also gave a substantial fillip to the local job market. [35]

As labor shortages resulted from wartime demand, the district was torn by strife. Despite a ban issued by the War Labor Board (WLB) in 1942, sporadic walkouts took place; in 1943 alone, Sloss-Sheffield lost more than 421,000 man hours because of wildcat strikes. Reacting to production losses, federal officials seized control of all coal mines in the Birmingham area for thirteen months after 1 May 1943. In 1945, as the end of the war approached, another wave of strikes broke out, causing Sloss-Sheffield to lose more than 317,000 man hours of production. Meanwhile, the company fared poorly in court battles to reverse federal imposition of the portal-to-portal principle. Morrow complained about this situation in annual reports, decrying wage hikes that Sloss-Sheffield had been forced to grant. [36]

Workers were not the only people who suffered from Sloss-Sheffield's hard-bitten attitudes. James P. Dovel, who throughout the war was confined to his bed by severe arthritis, received only a meager pension from the company. Periodically he was visited by his granddaughter, Dudley. Despite feeling bitter, he told her that he felt no personal animosity toward Morrow, who, as he put it, "did not have to give me a pension." As in other respects, the firm had been too concerned with cost cutting to provide old-age security even to senior officers, creating a situation in which Dovel looked upon a voluntary pittance as largesse. Indomitably, he continued to advise furnace operators by telephone about the use of his inventions, but much of his time was spent in lawsuits against corporations that infringed on his patents. His death, which came three years after the war, brought an ironic end to the life of an engineer who had once been hailed by the Birmingham press as "living proof that opportunity still exists in this country."[37]

Not only because of sharply mounting costs, which rose from $10,736,749 in 1940 to $16,160,179 in 1945, but also because of price controls and other government inflation-fighting policies, Sloss-Sheffield's earnings during World War II were modest. In 1942, after paying excess profits taxes of $57,452, it posted a net gain of $1,140,341 on sales of $16,341,774. Earnings declined to $727,831 in 1943, $668,142 in 1944, and $361,976 in 1945. Sales reached a wartime peak of $17,301,618 in 1944 before ebbing to $16,160,179 in 1945. The company was healthy at war's end, but its growth had been at best marginal.[38]

V

A far-reaching change occurred at Sloss-Sheffield on 16 December 1942, when United States Pipe and Foundry Company (USP&F) assumed operating control by acquiring, for an undisclosed sum, the 54,500 shares of common stock that had previously been held by Allied Chemical and Dye.[39] Headquartered at Burlington, New Jersey, USP&F traced its history to the early nineteenth century, when David C. Wood and others had built a foundry at Millville in the Garden State's iron-rich southern pine barrens. According to company tradition, this was the first plant in America to be set up exclusively for making cast-iron pipe. By the Civil War, the Wood family had established a network of foundries in southern New Jersey and southeastern Pennsylvania.[40]

As their operations expanded, the Woods diversified by making valves, hydrants, storage tanks, and equipment for waterworks and gas distributors. In the process, their technology changed. For many years, pipe was "cast on the side," with cores positioned horizontally. In 1854, the Woods started using a vertical casting technique developed in England. This practice spread throughout the Delaware Valley, from which new methods of pipe casting were later transferred to other parts of the country. In 1899, the steady growth of the Wood family's interests resulted in a consolidated enterprise that was initially known as the United States Cast Iron Pipe and Foundry Company. Later, the words "Cast Iron" were deleted, giving the firm its permanent name. At the time it absorbed Sloss-Sheffield, its assets included two foundries in New Jersey and plants in eight other states. Continuing to import European technology, it introduced new methods of centrifugal casting in America; these included the deLavaud process, through which pipe was cast in revolving metal molds that could be used repeatedly before being discarded. Plants using this technique included two Alabama installations, located at Bessemer and North Birmingham; in 1942, they employed almost 2,000 persons. At that time, USP&F's total assets, valued at $11,659,833, were much greater than those of Sloss-Sheffield, listed at $4,898,230.

USP&F, however, lacked access to its own raw materials. Acquiring Sloss-Sheffield was a significant step toward vertical integration.[41]

In its deal with Allied Chemical and Dye, USP&F acquired 54.87 percent of Sloss-Sheffield's 94,318 outstanding shares of common stock and 42.6 percent of its preferred and common.[42] The change of ownership produced an influx of new directors, and Catchings stepped down after a long tenure as board chairman. No basic changes, however, took place. Birmingham residents kept four seats on the board of directors, continuing past practice. Alexander Hamilton Bryan represented the Richmond investors, preserving ties going back to his grandfather's day. Meanwhile, Morrow and other veteran managers retained control of day-to-day operations. The merger thus reaffirmed Sloss-Sheffield's historic roots and intensified its foundry orientation.[43]

Shortly after the war, a new generation of executives came to power in Birmingham as the aging process took its toll. Still, continuity was preserved, and Sloss-Sheffield's deep southernness was unaffected. Claude S. Lawson, a South Carolinian who had joined the company as chemist after graduating from Clemson in 1915 and had advanced from one post to another during more than thirty years of service, was promoted to vice president, taking charge of operations. Late in 1946, after nearly fifty years of service Hunt retired and moved to Florida; two veteran employees, W. O. McMahon and M. L. Carl, took his place as coeditors of *Pig Iron Rough Notes*. Hunt's marketing duties were assumed by his son-in-law, Charles S. Northen, Jr., a Georgia Tech graduate who had previously supervised one of the Avondale textile mills in Sylacauga, Alabama. In May 1945, he was appointed sales manager to be groomed as Hunt's successor. Two years later, he became vice president for sales.[44]

In April 1948, Morrow moved up to chairman of the board. He had been president for twenty-three years, the longest term in the company's history. In naming Lawson chief executive, the board of directors again chose a man who had advanced through the ranks, reaffirming its commitment to local control of day-to-day operations. Lawson's popularity guaranteed an easy transition. Familiar with every aspect of the enterprise, he knew most employees by name and was well liked by the rank and file.[45]

As Morrow had done in 1925, Lawson began his administration at a good time. By 1948, Sloss-Sheffield was enjoying the highest profits it had posted in many years. Net earnings for 1946, $442,061 on sales of $17,248,220, were up only marginally from $361,976 the preceding year. In 1947, however, as sales escalated to $22,359,968, profits soared to $2,040,622. During each of the next two years, sales topped $25 million, and earnings rose accordingly, reaching $2,639,095 in 1948 and increasing slightly to $2,652,694 in 1949. In 1950, profits surged to an unprecedented $3,907,728.[46]

Ownership by USP&F gave Sloss-Sheffield a steady market for pig iron within the parent firm's network of pipe foundries and other installations.

Foundry iron, however, was only one of the products that sustained the company's remarkable prosperity in the immediate postwar era. Besides supplying fuel from its coke ovens to the Birmingham Gas Company, Sloss-Sheffield continued to market surplus electricity, produced benzene for motors, made railroad ballast and concrete from slag, and manufactured ferromanganese. It also created a quick-drying black aluminum-based paint used as a coating for pipe and fittings, sold slag wool for insulation, and marketed a bug killer named "DeaDinsecT" after its main active ingredient, DDT. This insecticide was a by-product of xylol from the coke plant. A slag granulator that was built in the late 1940s to process furnace waste into by-products added a new and distinctive element to the skyline of the City Furnaces. Two Ingersoll-Rand electric turbo-blowers were installed in an addition to the building housing the old steam-powered blowing engines, replacing these venerable units in supplying power for the company's various needs.[47]

VI

While Sloss-Sheffield prospered, Birmingham declined. By mid-century, it was obvious that the hopes of the city's boosters would never be fulfilled. Although Birmingham was the heart of the American foundry industry and the merchant pig iron trade, it was otherwise a pawn of U.S. Steel, which had long subordinated its interests to those of Pittsburgh and other northern cities. In 1948, after the Supreme Court ruled against basing point pricing, the steel corporation finally abandoned discriminatory freight rate differentials that had injured the Birmingham District since the early twentieth century. Even after this ruling, however, the giant firm continued pressing for changes in the Federal Trade Commission Act so that such practices could be resumed.[48] Meanwhile, Birmingham's reputation had suffered so badly from decades of poor civic leadership and corporate social irresponsibility that outside investment had almost stopped.

Employment opportunities during the war partly contributed to Birmingham's population growth from 267,583 to 326,037 in the 1940s. The city was still almost as large as its great rival, Atlanta, which had 331,314 people in 1950. But Atlanta, much wealthier and far more cosmopolitan than Birmingham, was poised for a remarkable period of sustained growth. By contrast, Birmingham's population had nearly peaked, and the city was stagnating. As an article in *Pig Iron Rough Notes* admitted, "of all the industries that had moved to the South during 1948 and 1949, not a single one had located in the District." Widely known as the "Johannesburg of America," Birmingham was a byword, even among southerners, for civic and economic malaise.[49]

Critics continued to find the city a convenient target for scorn. In 1947, Irving Beiman, a local newspaper reporter, published a scorching essay characterizing it as a "steel giant with a glass jaw." Though born in New York,

Beiman was almost a native; he had come to Alabama as a two-year-old and was an alumnus of Birmingham Southern College. After working as a reporter for a local newspaper, the *Post*, he served in the Office of War Information and returned after 1945 to write for the *News*. The mordant analysis that he presented in 1947 was all the more troubling because he knew the city so well.

Birmingham had "civic anemia," Beiman declared. "Exploited by absentee landlords and real-estate opportunists, the city has failed to develop a real community spirit. There is no civic symphony orchestra, there are no outstanding parks, there are few buildings of distinguished architectural note, and a Little Theater movement was permitted to die. Even the zoo in Avondale Park, where Miss Fancy the elephant once gave enjoyment to thousands of children, was sold, without a single protest."[50]

One by one, Beiman recounted the city's shortcomings. Its narrow-mindedness was legendary. An outstanding collection of Parisian sculpture given to the library was likely to be lost because officials had refused to display a bust of the donor, reputedly an atheist. Mostly because of the way in which unregulated loan sharks preyed upon small debtors, but also because of illegal interest rates charged by local banks, the city was known as "the bankruptcy capital of the United States." Politically, it was administered by mediocrities, typified by police commissioner Eugene ("Bull") Connor, a former sports announcer and appliance salesman whose officers were noted for the heavy-handed way in which they broke up meetings of suspected radicals and terrorized black people.[51]

Beiman was harsh in his treatment of the city's business leaders. "The financial and industrial interests that rule Birmingham keep unceasing vigilance over all phases of the city's life," he said. "No weapon is too small, no act too mean, when their opposition is aroused." As example, he told how senior businessmen, led by steel executives, had ended a Junior Chamber of Commerce drive to raise funds for diversification. What such men did not actually oppose, he charged, they killed through indifference, as by refusing to support a proposed Southern Research Institute aimed at turning the region's raw materials into new products. At virtually every level, the large corporations that dominated the city hindered its development. Because of their commissaries, retail sales in Birmingham were 40 percent less than in Atlanta, which had only 12,000 more people.

Beiman held TCI in particular contempt, asserting that "Big Steel always has treated Birmingham as a stepchild." Because of competition from Sloss-Sheffield and Woodward, pig iron cost less in the district than it did in Pittsburgh, but the reverse was true of steel. Because of basing point pricing, still being employed when Beiman's essay appeared, the entire Birmingham area, including Gadsden, produced only 4 percent of the steel made in the United States, despite its obvious cost advantages. Steel from western Pennsylvania was sold within 200 miles of Birmingham; some local industries imported it from West Virginia because of the way U.S. Steel kept TCI from competing effectively with faraway plants. Like other critics, Beiman noted how arbitrary freight differentials had prevented diversification in the district. "Perhaps Birmingham's progress would have been greater," he concluded, "had U.S. Steel ignored the financially faltering T.C.I. and left it a privately operated concern, free to grow or die as economic factors may have dictated."[52]

At the time Beiman's article appeared, Big Mules were using a familiar tactic by exploiting racial animosities to keep workers divided. Because of the better wages and working conditions won by unions under the Fair Labor Standards Act, industrial employment was suddenly more attractive to whites, and large companies in the district were hiring more whites than blacks. Nearly 9,000 new industrial openings were created in the 1940s; about five of every six were filled by whites. Mechanization, aimed at reducing costs, further shrank employment opportunities for blacks by eliminating many unskilled and semiskilled jobs to which they had formerly been relegated. As a result, black voting power in labor unions was undermined.[53]

Industrialists and recently hired white union members exploited this situation for their own respective ends. During the 1930s, desperate economic conditions and the drive to unionize previously unorganized companies had fostered class solidarity among black and white workers, but this fellow feeling began to evaporate even before the war under the red-baiting and racist attacks of the Big Mules. During the war, hearings held in Birmingham by the Fair Employment Practices Committee (FEPC) worsened the situation by arousing fears among white workers that they might lose their stranglehold on higher-paying, skilled jobs. Exploiting this anxiety in such company-owned journals as *Alabama*, industrialists aroused concern about the potential loss of white supremacy and raised the specter of black bosses supervising racially mixed shops. After the war, growing tensions between the United States and the Soviet Union increased the public appeal of the anticommunist rhetoric at which Big Mules excelled. Conflicts within organized labor itself played into their hands; determined to rid themselves of any taint of radicalism, unions conducted purges and expelled past or present communist sympathizers.

Yet another development worsened the political situation affecting black union members. Despite efforts by leaders in the Alabama State Federation of Labor to preserve unity, many white unionists decided in the late 1940s that status considerations outweighed bread-and-butter issues. Unfortunately for the blacks who worked for such companies as TCI and Sloss-Sheffield, USW locals were affected by this trend, reversing the results of efforts that Mitch and other labor leaders had made in the 1930s to foster racial cooperation, class solidarity, and the pursuit of common economic objectives. As a result, Big Mules who realized better than many workers that "the best way in the world to keep the wages of whites down was to keep blacks a little lower," as a frustrated union official put it, could manipulate the situation for their own purposes.

As had been true in the past, TCI set the pattern that other companies, including Sloss-Sheffield, followed. Soon after the war, recently hired white workers at TCI's Ishkooda and Muscoda mines, having failed to take control of black-dominated IUMMSW locals from within, began a secession movement aimed at affiliating with the USW. Echoing managers by charging that IUMMSW was controlled by Communists, they received thinly veiled support from both TCI and the USW, leading to an election preceded by a campaign in which white workers resorted to intimidation, race-baiting, and violence. Governmental and legal institutions helped insure the outcome; after a black IUMMSW member lost an eye when USW advocates forcibly kept him from making a radio address to rally support, his assailant got off with a $200 fine. The election, held in April 1949, resulted in a narrow victory for the USW, which became the sole bargaining agent for all Alabama ore miners.[54]

The determination of USW locals throughout the district to maintain white supremacy, despite repeated efforts by national and state labor leaders to convince them otherwise, blocked any chance that black workers might rise above ill-paying unskilled or semiskilled jobs, no matter how able they might be. Clarence Dean's experience at Sloss-Sheffield's City Furnaces was typical. He started in 1933, when he was eighteen years old, as a scrapper, working in front of the mixer as white-hot iron ran into the pig-molding machine. "It was a man-killer," he later recalled, requiring him to endure temperatures as high as 150 degrees Fahrenheit. Not satisfied with the few promotions for which a black could qualify, he tried repeatedly to become a pourer, which would have put him in charge of a crew. His fitness was incontestable, because he had trained whites for the position, but he was told that "we ain't going to give niggers no white folks' jobs."[55]

Birmingham, nevertheless, was beginning to feel pressures for change that would ultimately prove irresistible. The Cold War, which did much to foster hysteria and repression, also focused world attention on aspects of American society contradicting the image of freedom and democracy that the country was trying to project in its battle against Soviet totalitarianism. In this context, overt racism became a national liability.[56] Just as the Fair Labor Standards Act had undermined the distinction between northern and southern wages, which had been one of the chief props of the region's economic order, judicial decisions in the late 1940s and early 1950s began to attack segregation, and the southern social system started to crumble. The chief catalyst for change was the Supreme Court's 1954 decision in *Brown v. Board of Education of Topeka*, followed within a year by an order requiring states to integrate public schools "with all deliberate speed."

Recognizing the magnitude of the Magic City's economic problems, Sloss-Sheffield's new leaders had already joined other business executives in a development drive conducted by the Greater Birmingham Development Committee, also known as the Committee of 100. Organized in 1950 by the Birmingham Chamber of Commerce, it aimed to diversify the district's monolithic industrial structure. By 1954 it had lured into the district seventy-four new firms with yearly payrolls of approximately $50 million. Sloss-Sheffield's Lawson was on the executive board; Northen was also active. Using typical booster rhetoric, *Pig Iron Rough Notes* said that "the long term goal of the Committee is a population of one million people in metropolitan Birmingham and a balanced business and industrial economy adequate to support this population through thick and thin."[57]

As usual, however, industrialists displayed no social vision to match their economic hopes. Morrow was typical in continuing to denounce the legacy of the New Deal and defend traditional southern values. Speaking to members of the American Iron and Steel Institute at New York City in 1949, he stated that "the United States will have to look to the South to save the nation from alien thought and alien action" and upheld state's rights as "the liberty of being governed at home on things belonging to the home." Throughout the 1950s, racial tensions mounted, exacerbated by a lack of communication between black and white constituencies in the wake of the school desegregation decision. Instead of grappling creatively with demands for change, political leaders interpreted "with all deliberate speed" to mean "never." Connor's victory over a somewhat moderate opponent, Robert E. Lindbergh, in the 1957 municipal election for police commissioner was only one sign of hardening positions. Big Mules actively resisted racial accommodation; throughout the decade, no public figure in Alabama more actively opposed school integration than Morrow, whose move from the presidency of Sloss-Sheffield to the board chairmanship increased the time he could devote to White Citizens' Councils and other ultraconservative activities.

Joining hands with the segregationist elite, white union members remained steadfast in their determination to preserve white supremacy and block the upward aspirations of black workers like Clarence Dean. Connor's victory over Lindbergh was due to heavy support that he received from white workers in the city's most heavily blue-collar wards. Technological change intensified hard feelings between the races. Black employment in local industries declined precipitously as mechanization and importation of foreign ores reduced job opportunities in steel mills and mines. As Norrell has indicated, TCI virtually stopped hiring blacks after 1953.[58]

As the district expended more and more of its energy in social conflict, the development drive fizzled. By 1963, the Committee of 100 claimed to have helped bring 200 new firms into the district, adding about 20,000 jobs and resulting in capital investment estimated at $142 million. Despite this effort, during the 1950s Birmingham's population increased by less than 15,000 while that of Atlanta grew more than ten times faster, reaching 487,455 in 1960 as against Birmingham's 340,887. As usual, Birmingham ranked at or near the bottom of cities in its class in social welfare services for both blacks and whites. Efforts by a few daring citizens to start a chapter of the National Urban League died in 1949 as the result of fears that this would promote racial mixing; in 1955, a biracial committee operating under the aegis of the

Community Chest was disbanded amid the frenzy produced by the school integration crisis. As Virginia Hamilton later stated, "Birmingham entered the era of racial revolution without a single channel of communication through which whites and blacks might negotiate."[59]

VII

During the 1950s the foundry industry continued to flourish, providing low-paying but nonetheless welcome job opportunities for blacks that partially offset the loss of openings in steel plants. Buoyed by the creation of new alloys and rising demand for traditional products ranging from nuts and bolts to iron piping and machine components, Sloss-Sheffield continued to prosper, launching a major expansion of its furnace capacity in 1956.[60] Ironically, however, Sloss-Sheffield's very earning power resulted in a loss of corporate identity. Throughout the postwar era, despite its much smaller size, it consistently made better profits than its parent firm, USP&F. Frustrated by USP&F's lackluster showing, the directors decided to merge with Sloss-Sheffield. Sacking Victor C. Armstrong, a retired general who had been put in charge of USP&F after the war, they summoned Lawson to come north and take command. Lawson did not want to leave Birmingham but had no choice. To the regret of the local business community, he resigned as president of Sloss-Sheffield in April 1952 and moved to New Jersey, where he made sweeping changes in USP&F's upper-level management. Until the impending merger was consummated, Fred Osborne, a close friend of Lawson and a fellow Clemson alumnus, succeeded him in Birmingham.

Osborne had begun working for Sloss-Sheffield in 1926 and spent most of his career on its payroll. He had the distinction of being the last person ever to serve as president of Sloss-Sheffield as a corporation in its own right. On 12 September 1952, the merger was finalized, effective 1 August 1953. Norman F. S. Russell, a longtime USP&F executive, became chairman of the board, Morrow vice chairman, and Lawson president. Among other changes, Alexander Hamilton Bryan was dropped from the board, ending the Virginia connection that had begun in 1887. USP&F was the surviving corporation. After seven decades, the enterprise founded by James W. Sloss in 1881 was now merely a division within a larger entity.[61]

So far as control of day-to-day operations was concerned, however, nothing changed. As a writer for the Birmingham *News* said, it would be hard to get used to Sloss-Sheffield's loss of identity as a corporation in its own right, but local citizens could take pride in wondering which party to the merger had actually absorbed the other. Having demonstrated superior profit-making ability, Sloss-Sheffield's managerial team replaced USP&F's previous administrative hierarchy. Besides Osborne and Northen, the surviving corporation's new vice presidents included Robert E. Garrett, a University of Alabama alumnus who had been Lawson's assistant. Even more important, as a condition of taking the presidency, Lawson had demanded that USP&F move its headquarters to Birmingham. To the satisfaction of his many friends in Alabama, he now returned to the Magic City. From a local point of view, Jonah had swallowed the whale.[62]

To implement Lawson's conditions, a large office building was built near the City Furnaces, where Sloss Quarters had previously stood, to serve as USP&F's new corporate nerve center. The shantytown that had housed generations of employees suddenly disappeared as old, tumbledown dwellings were put up for sale with the understanding that successful bidders would move them to new locations or dismantle them for lumber. By the end of the decade, all company housing in North Birmingham had also been sold.[63]

VIII

While the new headquarters was being built, a major modernization program, the first that Sloss-Sheffield had undertaken since 1931, got underway in North Birmingham. A battery of 120 new Koppers coke ovens, more efficient than the earlier Semet-Solvay types, was added to the by-product plant, doubling its capacity. A new coal-crushing plant with electronic controls, the first of its type to be installed in the merchant pig iron industry, was built to improve mixing of different types of coal for conversion into coke. Coal from individual mines was stored in ten large silos prior to being crushed and blended.[64]

Most impressive, however, was a huge new blast furnace that was built in North Birmingham between 1956 and 1958. In deciding upon its location, Lawson was influenced by the fact that the City Furnace tract, hemmed in by railroad lines, and the site of the two older North Birmingham furnaces, now surrounded by residential areas, were both too cramped for a major new installation. He therefore selected a tract of land immediately adjacent to the by-product coke ovens. Among other operational economies, this would permit coke to be brought to the new furnace by conveyor belt instead of by rail.[65]

Starting with an initial budget of $15 million, Lawson spared no expense in planning the new facility, which was designed with aid from a prominent industrial consulting firm, Senior, Juengling, and Knall. The installation was intended to be state-of-the-art; everything about the furnace, when it was finally completed and put in operation, set it apart as the supreme technological achievement in the company's history. Built by about 500 workers who spent an estimated 1 million man-hours in its completion, it included 28,678 cubic yards of concrete, 5,000 tons of steel, and huge quantities of lumber, reinforcing bars, and other materials. Towering 225 feet above ground level, it sat upon a foundation containing 4,000 cubic feet of concrete and 250 tons of steel reinforcing rods.

The hearth, 25 feet in diameter, had a capacity of approximately 1,000 tons per day. It was lined with 638,000 fire bricks, weighing about 2,400 tons. The tuyéres received preheated air from three hot-blast stoves, approximately 123 feet high and 26 feet in internal diameter. Designed by a West German firm, the stoves were automatically controlled, eliminating the need for manual operation of valves and other components. Each stove contained a vast number of bricks—seventy-five carloads per unit—that were heated by recycled gases. Having a total surface area of 270,685 square feet, they formed a checkerwork with 2,500 passageways through which the heated air blast circulated en route to the furnace.

The boiler plant was equipped with three steam generators fired by blast furnace gas, coke gas, and pulverized coal. The steam was used chiefly to supply wind for the furnace by means of an electric turbo-blower with a rated capacity of 125,000 cubic feet of air per minute. The addition of a second blowing unit was planned for expanded operations envisioned in the future. Steam from the boilers provided power for a 12,500-kilowatt turbo-generator producing electricity for the entire plant. Exhaust steam from the turbo-blower and turbo-generator passed through barometric condensers, leading to cooling towers through which 31,500 gallons of water circulated each minute.

Ore, limestone, and dolomite were brought to the plant in rail cars. Upon arrival, they were analyzed by laboratory personnel before being sent to a rotary dumping apparatus that completely encircled a car and turned it upside down over one of twenty storage bins. Meanwhile, coke arrived by conveyor belt from the by-product plant. After being deposited in storage bins, it was rescreened to eliminate small particles, known as breeze, that might cause the furnace to clog. It was then measured by volume in a hopper with two electrodes.

Ready for smelting, the raw materials were transferred to a skip hoist that ran up the stack at a 60-degree incline and dumped them into the charging bell at the top of the furnace. At any given time, 1,640,000 pounds of ore, 950,000 pounds of coke, and 385,000 pounds of flux descended through the furnace in various stages of reduction. Hot air from the three stoves, reaching 2,100 degrees Fahrenheit, was blended with cold air in a mixing chamber to be reduced to operating temperatures of 1,200 to 1,700 degrees before being conducted to a circular ductwork, known as a bustle pipe, running around the furnace. From here it was distributed through pipes to the tuyéres, through which it was pumped into the area directly above the hearth at great velocity. Heated to incandescence by the action of the blast, the coke gasified and passed upward through the stack at temperatures ranging up to 3,600 degrees. After reducing the ore and calcining the limestone and dolomite, gases passed out of the furnace top as a mixture of carbon dioxide, carbon monoxide, methane, hydrogen, and residual dust particles. Descending through a downcomer, impurities were cleaned by a dust catcher and electrostatic precipitator before the gases were recycled to the boiler plant to produce more steam or to a gas washer to be reused in the stoves.

At the base of the furnace, slag and molten iron collected in the hearth.

The slag was periodically released, cooled, and taken to a nearby facility for conversion into railroad ballast or concrete blocks. The iron flowed into one of four 150-ton ladles and was mixed as the ladles were transported on rail cars to a pig-casting machine with a double strand of molds. As the molten iron was poured into individual molds, the latter moved in endless succession up an incline. Water was sprayed on them to cool the thirty-pound pigs, which fell from the top of the incline into rail cars for shipment to customers. Samples were taken from each ladle for quality testing.

IX

Designed to maximize automatic control over every possible aspect of the smelting operation, the new furnace was the largest and most modern installation ever built in the United States for making foundry pig iron. It climaxed nearly eight decades of development since the founding of the Sloss Furnace Company in 1881, but its construction was ill timed. Even as it was being built, the American iron and steel industry was beginning to face severe foreign competition. Some of it came from West German plants that had been destroyed by Allied bombing attacks during World War II and rebuilt under the Marshall Plan. In addition to being ultramodern, these had lower labor costs than older American installations. Meanwhile, new Japanese plants were also exporting low-cost iron and steel to American markets that were no longer protected by high tariffs.[66]

As USP&F faced increasingly ruinous foreign competition, markets also changed. Aluminum was beginning to replace iron in automobile engine blocks. Stoves, once a mainstay of the merchant pig iron industry, were no longer made in significant quantities. Steam radiators had gone out of use. Ornate designs for many cast-iron products were out of style, further reducing demand for the high fluidity of phosphorus-rich pig iron.

The main cause of the disaster that overtook American merchant pig iron producers, however, was what Richard Foster has called "technological discontinuity": the wrenching impact of an innovation, such as the change from vacuum tubes to semiconductors, that upsets fundamental realities so profoundly as to lead to death for once-prosperous firms.[67] Suddenly, gray iron, on which the fortunes of such companies as Sloss-Sheffield and Woodward were based, became obsolete. In 1948, three inventors, two Americans and an Englishman, developed a new type of cast iron by adding cerium or magnesium to the mix.[68] Unlike gray iron, the new alloy, known as ductile iron, contained graphite in the form of tiny spheroidal nodules instead of flakes. While retaining some of gray iron's desirable features, including a low melting point and easy machinability, ductile iron also had greater compressive strength. Combining various features of cast iron and steel, it became the material of choice used by foundries to mold products that had earlier been made from gray iron. The appearance of this remarkable new alloy created a major crisis for such merchant pig iron producers as USP&F. Because its carbon content was significantly lower than that of pig iron, foundries now used increased quantities of scrap steel in their mixes. This reduced demand for foundry pig iron at the same time that foreign pig iron became cheaper than the American product. The result was disastrous, not only for USP&F but also for such competitors as Woodward. Suddenly, the foundations on which the American merchant pig iron industry had been based for decades collapsed.

Although the new North Birmingham furnace was completed by early 1958, the market for foundry pig iron was so bad that it did not go into blast until August of that year. Meanwhile, the two older North Birmingham furnaces were banked in January 1958 until conditions improved. Because improvement never happened, they were torn down in 1960. Morrow died on 6 September 1960; Lawson took his place as USP&F's board chairman, and Garrett became president. Unlike Lawson, who had taken office under favorable circumstances, Garrett was elevated at the worst possible time. During the 1960s, as merchant pig iron producers reeled under catastrophically changed circumstances, USP&F's competitive condition worsened steadily, and the firm became a likely target for a hostile takeover.[69]

X

While USP&F fought to survive, Birmingham went through a climactic decade in which the long-range results of its fragmented social order reached tragic denouement. By 1960, nine decades of racial injustice had produced a bitter harvest in Jones Valley. Three out of every four whites were skilled, but only one of six blacks could claim such status. Forming nearly 35 percent of Birmingham's work force, blacks made up only 5 percent of the sales force and 7 percent of the managerial class. Even these proportions were possible only because of black-run businesses that catered solely to blacks. In times of economic distress, black workers were two and one-half times more likely than whites to be unemployed. Because of a 1951 amendment to the state constitution, designed to circumvent a Supreme Court ruling that had thrown out a previous amendment restricting black suffrage, less than 10 percent of the city's black people could vote. Rigidly enforced segregation laws blocked every avenue of meaningful communication between the races. As Lee E. Bains, Jr., later stated, blacks "lived within two concentric circles of segregation. One imprisoned them on the basis of color, while the other confined them within a separate culture of poverty."[70]

Because its economic and social malaise was so pronounced even by regional standards, Birmingham became a major battleground of the civil rights movement that undermined many of the South's most deeply rooted customs and institutions in the decade after *Brown v. Board of Education*. In 1956, after Gov. John Patterson won an injunction forcing the National Association for the Advancement of Colored People (NAACP) to close its Alabama offices, a black minister, Fred Shuttlesworth, began a drive for racial justice in Birmingham by founding the Alabama Christian Movement for Human Rights (ACMHR). While Morrow and other business leaders formed White Citizens' Councils to defend the status quo by legal means, other white groups resorted to violence. Homes of black leaders, including Shuttlesworth, were dynamited. Instead of trying to prevent such incidents, the municipal government sanctioned private and police brutality. In 1961, knowing that Freedom Riders would arrive in town on Mother's Day, law officers deliberately arrived late at the bus station so that Klan members could inflict a merciless beating on the outsiders. Through every possible means, demands for change met unyielding opposition.[71]

As usual, Birmingham came under a barrage of condemnation. In 1960, a first-page article in the New York *Times* by Harrison E. Salisbury compared the corrosive impact of the prevailing social environment to the "acid fog belched from the stacks of the Tennessee Coal and Iron Company's Fairfield and Ensley works." Visiting the city after sit-ins at five downtown lunch counters had unleashed a wave of cross-burnings and other forms of intimidation, Salisbury reported that "every inch of middle ground has been fragmented by the emotional dynamite of racism, reinforced by the whip, the razor, the gun, the bomb, the torch, the club, the knife, the mob, the police and many branches of the state's apparatus." Recently, hoodlums armed with "iron pipes, clubs and leather blackjacks into which razor blades were sunk" had savagely beaten the mother of a black protester. After swearing out a warrant for their arrest, she had been horrified when law officers who visited her hospital bed to receive testimony included two of her assailants. "I just don't understand the white people around here," a black citizen told Salisbury. "They seem to act so crazy. . . . Don't they know there is a limit to what people will stand?"[72]

As had been true since the late 1930s, white members of USW and other unions were active in thwarting integration despite educational campaigns conducted by national AFL-CIO leaders to convince them that the tactics used by industrialists to divide workers injured the economic interests of both races. George Wallace's successful campaign for governor in 1962, followed by his inaugural pledge to maintain "segregation now, segregation tomorrow, segregation forever," owed much to support from white union members in cities like Birmingham. In Norrell's words, "to most white unionists, Wallace was a good labor man defending the position of the white working class from incursions by blacks who intended to take away whites' superior status position."[73]

Gradually, some local commercial and industrial leaders began to recognize that Birmingham's racial intolerance and violent resistance to change were stopping outside investment. Sidney W. Smyer, a strongly segregationist businessman whose family roots went far back into the early history of the district, typified such persons. In 1961, while president of the Birmingham

Chamber of Commerce, he visited Japan with a delegation of executives from other parts of the United States and realized for the first time how badly the city's image had injured its ability to attract capital. Shaken by what he had learned, he returned convinced that things must change.[74]

Mobilizing support among progressive members of the chamber, Birmingham Bar Association, and Young Men's Business Club, Smyer mounted a campaign to replace the existing form of government, under which a weak mayor and three powerful commissioners were elected at large, with one led by a strong mayor who might lead the city into orderly accommodation with changed realities. Deliberately refraining from attempting to enlist black aid, a group calling itself Birmingham Citizens for Progress organized a petition drive that succeeded in forcing a referendum. In a closely contested vote in November 1962, the plan was adopted by a small margin. After a challenge to the constitutionality of the impending change of government was defeated, a bitterly contested election to fill newly created municipal offices failed to produce a majority for any of the three candidates running for mayor. In the ensuing runoff, on 2 April 1963, Albert Boutwell, a moderate segregationist, defeated Connor, who vowed unyielding resistance to integration. Candidates with views similar to those of Boutwell also won seats on the new city council. Although their ballots had not been courted by the moderates, eligible black voters had recognized where their interests lay and played a key role in the outcome.

Far from producing a smooth transition to a new order, these developments precipitated the worst crisis in the history of the Birmingham District. Refusing to yield authority to Boutwell and the new council members, Connor and other incumbents insisted on serving out their unexpired terms. While legal proceedings were begun to evict them from City Hall, Birmingham was left floundering with nobody in undisputed control. Within this context, events proceeded to a momentous confrontation that swiftly followed.

Unknown to either moderates or extremists in the recent electoral campaign, black civil rights leaders had been secretly planning massive demonstrations to intensify pressure on the city to end segregationist policies. Their determination resulted from frustration over events that had taken place during the previous year. Throughout much of 1962, while Smyer and his associates were fighting to change the municipal government, the ACMHR, Southern Christian Leadership Conference (SCLC), and other black organizations had led a boycott of downtown merchants in an effort to end various forms of racial discrimination in the city. By September, when the SCLC held its annual convention in Birmingham, a limited degree of desegregation had been achieved in department stores through secret negotiations between black spokesmen and a Senior Citizens' Committee composed of prominent whites who recognized the need for concessions to avoid violent demonstrations. Later in the year, however, merchants had reneged on commitments under pressure from Connor and his followers, resuming segregationist policies and causing frustrated blacks to plan resumption of the boycott. At this point,

SCLC leader Martin Luther King, Jr., accepted an invitation from ACMHR to come to Birmingham early in 1963 and spearhead renewed demonstrations. King had recently failed to attain civil rights objectives in leading demonstrations at Albany, Georgia, largely because law officials there had exercised restraint in dealing with protesters and there had been no white violence to arouse television viewers. By going to Birmingham, King was attempting to recapture momentum for the civil rights movement, which seemed to be faltering throughout the nation.[75]

Not wanting to undermine white moderates trying to unseat Connor and other extremist incumbents, King and his local allies repeatedly delayed starting the boycott until the spring balloting had ended. Immediately after Boutwell's victory, King came to Birmingham, and a series of well-planned demonstrations got underway. Still clinging to his role as police commissioner, Connor initially tried to use restraint in dealing with the protests, which got off to a slow start. In an effort to galvanize support, King got himself incarcerated by taking part in an illegal march on City Hall. This act led to one of his best-known pronouncements, "Letter from Birmingham Jail," in which he made a classic statement of his views on nonviolence and civil disobedience. The immediate impact of his action on local events, however, was marginal at best; it failed to arouse the desired sense of urgency in the black community, which was badly divided over how vigorously to pursue the demonstrations. Amid mounting anxiety, King and other black leaders decided to use schoolchildren in mass protest marches to create fresh momentum. After more than 500 young people had been jailed, waves of others took their place, and events gained the force that King and black militants wanted to generate.

"What has happened down here," Randy Newman later sang in a haunting evocation of regional torment, "is that the winds have changed."[76] Nothing better epitomized the meaning of his words than what happened on 3 May 1963, when an ugly confrontation took place between a throng of black demonstrators gathered in Kelly Ingram Park, many of them students in the first or second grades, and a crowd of hostile onlookers shouting racial epithets and obscenities. As firemen stood by and police struggled to push a group of jeering students back with the aid of leashed attack dogs, bricks and bottles began to be hurled. When some of these missiles fell on law officers, Connor ordered the dogs to be unleashed and commanded firemen to deluge protesters with high-pressure hoses under the glare of national publicity. As millions of television viewers saw children being bitten by the dogs and pinned against buildings by blasts of water sprayed with enough force to rip bark from trees, a social order that had endured for almost a century finally began to crumble and wash away.

The debacle epitomized how postindustrial technology was helping to undermine institutions typical of the industrial age. "Television instantly communicated events to the entire nation," Norrell later observed. "It was an

'action' medium that demanded movement; conflict made for good television, and violence produced virtually irresistible viewing. . . . Television
caused a mass revulsion from racial violence that aided immeasurably the
civil rights cause. Like nothing else before it, television exposed the conflict
between liberty and white supremacy." King saw the same thing, giving television much of the credit for the success of his 1963 March on Washington:
"Millions of whites, for the first time, had a clear, long look at Negroes
engaged in a serious occupation. For the first time millions listened to the
informed and thoughtful words of Negro spokesmen."[77]

Observing events from the Oval Office and becoming increasingly sickened
by what was transpiring in Birmingham, John F. Kennedy sent Assistant
Attorney General Burke Marshall to the scene to help negotiate a settlement
and prepared for federal intervention if necessary. Predictably, Connor's handling of the demonstrations had provoked outrage in such places as Ghana
and Nigeria, and Moscow was doing its best to capitalize upon what had
happened in radio broadcasts aimed at African listeners.[78] Going to Washington, Smyer urged restraint and returned home with the message that "if
we're going to have good business in Birmingham, we better change our way
of living." To all white moderates who would listen to him, he stated, "I'm a
segregationist from bottom to top, but gentlemen, you can see what's happening. I'm not a damn fool. . . . We can't win. We are going to have to stop and
talk to these folks."[79]

Behind the scenes, the Senior Citizens' Committee that had helped defuse
the crisis of the previous summer negotiated a shaky settlement with King and
moderate leaders of the black business community. Although this was condemned by Connor and his supporters and caused Shuttlesworth and other
black militants to feel betrayed, it stopped the demonstrations. Months of
strife remained, climaxed by a dynamite blast at the Sixteenth Street Baptist
Church in which four black children died. In time, however, Boutwell's new
administration took charge, and the crisis abated. Thanks to a large degree to
the way in which the horrifying scene at Kelly Ingram Park had aroused the
consciences of Americans who had witnessed it on television, the Civil Rights
Act of 1964 was passed by Congress and signed by Lyndon B. Johnson. A year
later, the Voting Rights Act of 1965 made it impossible for states to restrict
black suffrage.

Banding together under the name "Operation New Birmingham," a coalition of civic leaders began communicating across racial lines to institute a
new social order that gradually improved the city's tarnished image. Epitomizing the change was the appointment of Arthur Shores, a black attorney whose
house had been twice bombed, to the city council in 1968. A graduate of
Talladega College who had once been the only black lawyer practicing in
Birmingham, he had refused to leave the city because, as he said, "this was
my home."[80]

XI

While Birmingham emerged phoenix-like from disgrace, USP&F had increasing difficulty surviving the cataclysmic economic and technological changes overtaking the merchant pig iron industry. By the mid-1960s, Clarence Dean had finally secured the promotion to pourer that he had wanted for so long. But his career as a furnace worker was nearly over, because the furnaces themselves would soon go out of blast forever.[81]

The beginning of the end took place in 1968, when Harry E. Figgie, Jr., launched a raid on USP&F stock. Within five years, Figgie had transformed Automatic Sprinkler Corporation of America (ASCA) from a company with sales of about $23 million to a massive conglomerate with gross sales approximating $325 million. His skills in acquiring firms through heavily leveraged buyouts had made him one of the most closely watched figures in American business, noted for a "nucleus theory" of expansion that became a popular model for combining quick growth with high profits. He made it plain that he would sweep USP&F's current executives out of their jobs if his takeover bid succeeded.[82]

By December 1968, Figgie's empire, which made products ranging from fire hydrants and vacuum cleaners to baseball gloves, was collapsing. Some of the firms that it had acquired were plagued by unexpected weaknesses, and it was impossible to exercise effective control over an ill-assorted jumble of hastily acquired enterprises. Wall Street investors became leery of wildly exaggerated projections emanating from ASCA's Cleveland, Ohio, headquarters as profits, $1.43 per share in 1967, tumbled to ten cents per share. In the end, these circumstances fatally compromised Figgie's bid to take over USP&F. It crested early in 1969, by which time he had gained a 23.7 percent holding. Nevertheless, it had been a close call, and constant proxy battles had left USP&F in an unsettled state.[83]

As the struggle with Figgie reached its climax, a "white knight," Jim Walter, appeared on the scene offering a friendly takeover. As a young navy veteran in 1946, Walter had gone into partnership with a Tampa builder, O. L. Davenport, to produce unfinished shell homes that could be acquired at low cost and completed by their owners. Aided by the housing boom of the late 1940s and by federal legislation that stimulated home buying by low-income purchasers, the business thrived. Organizing their own financial arm, the Mid-State Investment Corporation, to assume mortgages, Walter and Davenport went public in the mid-1950s and raised fresh capital for expansion. Using a new name, the Jim Walter Corporation (JWC), they got backing on Wall Street from Walter E. Heller & Company, which liked the way they screened customers and the speed with which they foreclosed on homeowners who fell behind in their payments.

Such practices, combined with high interest rates on mortgage loans, not only enabled JWC to survive an industry shakeout in the early 1960s but also made it necessary for it to diversify in order to utilize surplus earnings. In

1962, JWC took control of Celotex Corporation, which made fiberboard, roofing, siding, and other building products and was involved in drilling for offshore gas and oil. In 1967, JWC absorbed another major producer of building materials, Allied Chemical and Dye's Barrett Division, which had deep ties with Sloss-Sheffield and USP&F. JWC now not only sold and financed homes but also profited from materials used in building them, thus pyramiding its earnings.[84]

By 1969, as USP&F continued to flounder while trying to fight off Figgie's takeover bid, JWC was ready to be transformed from a mere seller and builder of homes into a full-fledged conglomerate. USP&F was a logical target for acquisition. Besides manufacturing sewer pipe and other products closely related to JWC's existing operations, it also owned significant mineral reserves. By this time, Alabama iron ore deposits that had once seemed limitless had been severely depleted by decades of exploitation and were being replaced by imported ores. The discovery of Venezuela's Cerro Bolivar and La Frontiera ore deposits in 1950 was especially important, permitting high-grade ore to be shipped to Birmingport via Mobile and the Warrior River. Even the once-valuable Russellville brown ore deposits would soon be abandoned; filling with water, the massive No. 6 ore pit ultimately became a public recreational facility, appropriately named Sloss Lake. But Alabama's coal reserves were still extremely valuable in an increasingly energy-starved world. As a major holder of southern coal lands, USP&F was a tempting prize for this reason alone.[85]

Assured that its corporate executives would be permitted to retain their jobs, USP&F jumped at the chance to join Jim Walter's empire. In the spring of 1969, JWC acquired Figgie's stake in USP&F for a reported $32 million, and the company's embattled officers delivered the firm into his hands. Unfortunately, Walter, whose Tampa office was adorned by a sign proclaiming "he who got the gold makes the rules," did not wait long to introduce sweeping managerial changes. Despite his earlier assurances, Garrett and other persons who had long been associated with Sloss-Sheffield were soon fired or shunted aside. Those who remained, grumbling that their new superiors knew nothing about the foundry business, took orders from a cadre of executives who moved in from Tampa. Swiftly, the local control for which Birmingham interests had fought from the time of Rucker and Seddon to that of McQueen, Morrow, and Lawson became a thing of the past.

Buoyed by escalating demands for coal in the 1970s as turmoil in the Middle East caused petroleum prices to rise, Walter's conglomerate moved quickly to exploit the mineral lands it had acquired. Creating a new corporate division, Jim Walter Resources (JWR), and aided by Japanese firms that wanted better access to metallurgical coke, it invested $350 million over a twelve-year period to transform USP&F's mining properties into much larger operations producing 2 million tons of coal per year by 1979. Using new European technologies, JWR developed four of the deepest coal mines on the North American continent in northern Alabama's Blue Creek seam.[86]

XII

USP&F's new owners also continued for a short time to make pig iron, but the game was not worth the candle because of foreign competition and the growing use of ductile iron. Operations at the Russellville ore mines were discontinued after 1975. Using imported ore, the huge new furnace that USP&F had built during the late 1950s remained only intermittently in blast until it was blown out for the last time in 1980. Soon, it was dismantled and sold for scrap. Before passing into oblivion, it had become the last merchant blast furnace still operating in the district.

By the time the massive North Birmingham furnace was torn down, the City Furnaces had been silent for a decade. Early in 1970, the No. 2 furnace was banked pending a major overhaul to replace its heat-seared lining and restore the hearth, which had become damaged from constant use. The No. 1 furnace, however, was still in operation, yielding costs comparable to those of its much larger counterpart in North Birmingham. Nevertheless, in June 1970, it too went out of blast. Federal environmental regulations were later blamed for the shutdown, but the facility was in fact operating within existing standards governing air and water pollution. The real reason for its closing was that worldwide energy shortages had raised the price of metallurgical coke to a point where more money could be made by selling the output of the North Birmingham by-product plant directly to outside markets than by using it to make pig iron.[87]

Once operations had stopped at both City Furnaces, various considerations prevented their resumption. Because the merchant pig iron industry was virtually dead, it was hard to justify investing the capital needed to repair the No. 2 furnace, regardless of what price coke might command on the world market. Also, public pressure had mounted against relighting the stacks. Climaxing generations of complaint about the clouds of soot and smoke with which these and other industrial installations had fouled the city, concerned citizens had recently launched a protest movement, using the acronym GASP—Greater Birmingham Alliance to Stop Pollution—to dramatize their discontent.[88]

On 11 May 1971, a congeries of local organizations including the Greater Birmingham Convention and Visitors Bureau, the Birmingham Area Chamber of Commerce, the Downtown Action Committee, and Operation New Birmingham submitted "A Proposal for the Creation of a Blast Furnace Museum and Related Exhibits in Birmingham, Alabama" to B. F. Harrison, president of JWC's USP&F subsidiary. Noting that Birmingham was "world capital of the cast iron pipe business" and that the City Furnaces were "deep seated in the hearts of the people of this community and carry a great deal of historical significance," the document urged that the idle facilities be converted into "a full scale plant museum of one of the basic industrial processes on which Birmingham has been built." The site, it was said, might also accommodate a railroad museum contemplated by the Heart of Dixie Railroad Club, and visitors might be taken on rail tours of nearby industrial installations still operating in the district. Pointing out how important tourism had become to the local economy, proponents likened the potential impact of such a complex to that of Chicago's Museum of Science and Industry and stressed the educational role that it could play.[89]

After studying the tax advantages involved, Jim Walter officials consented to the plan. Without consulting state or local authorities, who learned about the decision only upon being invited to the public ceremony at which it was announced, JWC concluded that the Alabama State Fair Authority (ASFA) was the most appropriate recipient of the property. The ceremony was held at the furnace property on 1 June 1971, and the transfer became official with a deed of gift conveyed by USP&F to ASFA on September 3.[90] From then on, the historic stacks would survive, not as productive units but as monuments to an industrial past from which Birmingham was escaping.

16 Preserving the Heritage

The television cameras that riveted international attention on Birmingham in 1963 were part of an emerging era far different from the one in which the Magic City had been born. Their power to generate images that millions of viewers could see at the same time helped create a new reality best described as postindustrial consciousness. Having played an industrial role for nine decades, Sloss Furnaces now began to play a postindustrial function as one such image, a symbol of an identity toward which a greatly changed city was beginning to grope. A thirst for such symbols, underlying the historic preservation movement that began to burgeon at this time, was part of a yearning for rootedness pervading a society caught up in changes so powerful that everything seemed to be swept away at once, leaving people clutching for something permanent on which to gain a foothold.

Perhaps the very stolidity of a massive nineteenth-century blast furnace complex encouraged Birmingham's citizens to see it as worthy of preservation. Just as a mighty Corliss steam engine that was displayed at the Philadelphia Centennial Exhibition in 1876 had epitomized the aspirations of an early industrializing society that gloried in transforming the environment and measured progress in quantitative terms, a rusting hulk with a grandeur all its own now seemed a fitting icon of what such urges had cost and the lessons that might be learned from the conditions they had produced.[1] But like so much else in the history of Sloss Furnaces, the heritage would not be preserved without a struggle.

I

After USP&F's City Furnaces stopped production in 1970, Birmingham's character changed dramatically. One index of the transformation that took

450

place was that within a single decade only 10 percent of jobs in the city involved iron and steel. Related to this was the greatly shrunken size of TCI's operations despite U.S. Steel's belated adoption of such postwar innovations as the basic oxygen process. Throughout the decade the air in Birmingham, once almost unbreathable, improved enormously, and the natural beauty of the city's surroundings became apparent. Less than eight years after being disgraced on national television, Birmingham won *Look* magazine's All-American City award.

Along with TCI, important firms based on ferrous metal production did remain. ACIPCO continued making a wide variety of pipes, tubes, valves, and other products for offshore drilling rigs, television antennae for the John Hancock Building in Chicago, and carriers for Saturn rockets. McWane, Inc., and USP&F remained among the world's largest producers of pressure pipe. Connors Steel, which numbered the son of a Civil War cavalry commander among its founders and had begun by making cotton ties and barrel hoops, now made sophisticated alloys in electric furnaces and completed a major expansion program in 1980. O'Neal Steel, which had started with a $2,000 investment and a west end shack in 1921, was a leading maker of stainless steel and other special alloys. These enterprises in themselves, however, indicated the degree to which the district had become a major producer of diversified finished and semifinished goods, even in the steel industry.

Birmingham was now better known among informed people as an educational and medical center than as an industrial city. To a considerable degree this reputation was due to the University of Alabama at Birmingham, which had started in the 1930s with extension classes administered from the main campus in Tuscaloosa, been chartered by the state as a medical college in 1943, and become an independent institution in 1969. Much of its progress was due to its first president, Joseph F. Volker, who came to Birmingham from Boston in 1948 and created outstanding programs in cancer research and cardiology and the first public diabetes hospital in the United States. Under UAB's aegis, health care replaced ferrous metals as the city's biggest single employer.

Other outstanding educational institutions included Birmingham-Southern College, a Methodist-related school that led the Deep South in the proportion of its students who went on to graduate and professional training after receiving baccalaureate degrees, and Samford University, a Baptist institution that ranked first in the state in number of graduates listed in *Who's Who in America*. Samford had a well-known law school and an internationally renowned a capella choir. By 1980, it had nearly completed development of an attractive 400-acre campus in Shades Valley, with thirty-one classic red brick structures including a major fine arts complex. Another sign of change was the establishment of the Alabama School of Fine Arts, where gifted children from all over the state could receive training in such subjects as creative writing, dance, music, painting, and sculpture. Initially occupying quarters at Birmingham-

Southern, it secured its own campus, centered around a building that had once sheltered young professional women from the temptations and vices of the industrial city.

Increasingly, Birmingham's population was composed of professionals, not blue-collar workers. This was due not only to educational and medical institutions but to the city's large number of communications, electronic, engineering, and insurance firms. In 1968, Birmingham became headquarters of a newly created entity, South Central Bell, at that time the largest corporation ever chartered; it had 40,000 employees, and the network it serviced included 5.5 million telephones. In 1972, it moved its headquarters to a thirty-story building, then Alabama's biggest structure, in Birmingham's urban core. South Central Bell's work force, composed of computer operators and other technically oriented employees, was much different from the low-wage operatives who had once formed the city's economic base.

Typical of engineering companies was Blount, Inc., a giant construction firm that in 1985 completed the largest fixed-price project ever awarded, an academic campus at Riyadh, Saudi Arabia. The company ultimately moved its headquarters to Montgomery, where Winton Blount, who had been postmaster general in the Eisenhower administration, led the creation of the Alabama Shakespeare Festival, housed in an impressive theater on a tastefully manicured 100-acre tract. Still headquartered in Birmingham, however, was another huge construction company, the Harbert Corporation, whose president, John Murdoch Harbert III, the only Alabamian to make *Forbes* magazine's list of the 400 wealthiest Americans in the mid-1980s, played a leading role in revitalizing the Magic City's downtown core. In 1986, Harbert spearheaded the creation of AmSouth/Harbert Plaza at the corner of Sixth Avenue and Nineteenth Street, featuring a skyscraper office complex and fashionable specialty shops. Subsidiaries of Harbert's business empire were scattered all over the world, including Chile, Colombia, Ecuador, Egypt, and England.

Birmingham was also the home of Protective Life Insurance Company, incorporated by former governor William D. Jelks in 1907. Making an important marketing switch in 1963 by deciding to concentrate on upper-income markets, it became a nationwide business under the leadership of William J. Rushton III and moved to a new, dramatically situated home office building on a mountainside east of town in 1976. Torchmark, among the nation's fifty largest insurance firms, was another prominent local employer. Earlier known as Liberty National and led by Frank P. Samford, Jr., it acquired smaller firms in Oklahoma, Missouri, and Texas in a $500 million expansion program and had the longest continuous record of earnings and dividend increases of any company listed on the New York Stock Exchange.[2]

II

Birmingham's involvement in the growing trend toward professional and service-based occupations was typical of a postindustrial economy. The most dramatic transformation that occurred in the city during the 1970s, however, was the increasingly active political role now played by its black citizens. Nothing could have better symbolized the torrent of postindustrial change sweeping across the face of America in that decade than the election of its first black mayor in 1979, only sixteen years after Bull Connor had turned police dogs and fire hoses on black demonstrators.

Throughout its history, Birmingham had developed a group of black businessmen who catered to the needs of its African-American population. A. G. Gaston, Birmingham's first black millionaire and founder of the Booker T. Washington Burial Insurance Company, best epitomized this entrepreneurial class. Moving to Birmingham from Demopolis, where he "began his business career at age eleven selling rides on a swing . . . in exchange for buttons and pins," Gaston got a job with TCI after World War I and increased his earnings by selling popcorn and peanuts to workers during lunch hours. His rise to wealth, in the best Horatio Alger tradition, showed what determination could achieve even under the worst of circumstances. Despite a conservative orientation emphasizing the "Green Power" of money instead of the "Black Power" of organized protest, he made a key contribution to the civil rights movement by building a motel that served as a nerve center for leaders of the 1963 demonstrations.[3]

During the 1970s and 1980s, however, blacks moved away from the segregated syndrome that Gaston had been forced to accept and entered the mainstream of the city's business, cultural, and political life. Nothing better epitomized this trend than the career of Richard Arrington.

Arrington was born on a tenant farm in west Alabama's Sumter County in 1934. In 1940 his family moved to Fairfield, where his father became a steelworker for TCI. Influenced by E. J. Oliver, characterized by Jimmie Lewis Franklin as a "prototype of the progressive black principal in the American South," Arrington excelled in studies at Fairfield Industrial High School. In 1951, he matriculated at Miles College, founded in 1898 by the Colored Methodist Episcopal Church; one of its early campuses was located at TCI's Docena mining village. Majoring in biology, he went on to receive a master's degree in zoology at the University of Detroit, graduating near the head of his class. He then earned a Ph.D. in the same discipline at the University of Oklahoma, where he received a coveted honor, the Ortenburger Award, "given to a graduate student who showed unselfish interest in the welfare of other students, had a good relationship with his peers, and maintained high academic standards."

Nothing in Arrington's life to this point, including his doctoral dissertation, "Comparative Morphology of Some Dryopoid Beetles," indicated the political career that lay ahead of him when he returned to Birmingham in 1966. After he served four years as chair of the Department of Natural Sciences and dean of the college at his alma mater, his administrative skills won him appointment as executive director of the Alabama Center for Higher Education (ACHE), established under a Ford Foundation grant to foster cooperation between nine predominantly black Alabama colleges. Starting with a single secretary and a budget of $40,000, he proved adept at raising money and built ACHE within nine years to an operation with thirty staff members and a budget exceeding $1 million. His increasing prominence in Birmingham's black community resulted in his being asked to run for the city council in 1971. Supported by the *Post-Herald*, he won in a runoff, placing second among six candidates.

Arrington's election to the council was a revolution in itself, marking the first time a black had been elected to that body; as earlier noted, Arthur Shores had won membership in 1968 by appointment. Rejecting the tokenism of incumbent mayor George Seibels—there were still only 62 blacks on Birmingham's 1,200-member police force and no black department heads in the city government—Arrington championed the Affirmative Action program established by Lyndon B. Johnson in 1964. After Seibels vetoed a proposed ordinance aimed at changing municipal hiring practices, Arrington won unanimous council approval for a compromise measure that Seibels signed under pressure from the black community. Following several years of foot-dragging by Seibels in implementing the law, Arrington helped his chief ally on the council, David Vann, become mayor by a narrow margin over Seibels in 1975 by mustering support for him among black voters. Arrington himself was reelected councilman without a runoff, and another black candidate also won a seat. Having shown skill in deliberations involving television cable franchises, public housing, and community-police relations, Arrington had established himself as one of the city's ablest and most influential black leaders.

During the Seibels administration, Birmingham's police force continued its past record of brutality toward black citizens. Arrington, who received eighteen complaints about mistreatment of African-Americans in his first six months on the council, fought vigorously for public investigation and disciplinary action after beatings and killings of black suspects including Willis ("Bugs") Chambers, Jr., who was shot to death by a white policeman in 1972, and John Sullivan, whose eye was put out by a patrolman in 1975. Arrington's efforts were vehemently opposed by the Fraternal Order of Police and by Russell Yarbrough, a white councilman who took a strident law and order position and once referred to black demands for equality as "freedom crap." But Arrington also had trouble getting support from conservative members of his own race, including Gaston. He got none at all from Seibels, who, in Franklin's words, "generally left investigations to the police department and then rubber-stamped its findings."

Arrington hoped for better things from Vann, who had been a leader in bringing about a moderate city government in the early 1960s and owed his mayoral victory to support from the black community. He was keenly disappointed, therefore, by Vann's response to a tragic killing in June 1979, four months before the next mayoral election was to take place. In the incident, Bonita Carter, a young black woman, was riddled by police bullets while sitting in a car at the scene of a crime that had just taken place at a convenience store. George Sands, the officer who shot her, had been repeatedly accused of harshness toward black citizens. After appointing a blue-ribbon committee whose members delivered an unequivocal verdict that "sufficient justification did not exist" for Carter's killing, Vann dismayed previous black supporters by refusing to dismiss Sands. Protest marches were held, and resentment in the black community boiled over to a point at which a repetition of the 1963 demonstrations seemed likely. This did not happen, but the episode produced a groundswell among blacks to draft Arrington, who had acted responsibly throughout the crisis while pressing Vann for action against Sands, to run for mayor. After thinking things over, Arrington consented.

In the ensuing primary campaign, Arrington faced Vann and three other candidates, one of whom was black. Despite efforts by most candidates and their managers to keep the canvass from being dominated by racial rhetoric, it was characterized by appeals to law and order and slogans linking black people with crime. Asserting that only a small number of police officers were guilty of brutality, Arrington pledged to work for greater professionalism on the force, community involvement in public safety, and mutual respect between citizens and law officers. Aided by a determined and well-organized grassroots movement among his supporters, he piled up enormous margins in black precincts and won 31,521 votes, far more than any other candidate. Deserted by blacks, Vann came in fourth with only 11,450; Frank Parsons, an ultraconservative who had once been a lawyer for USP&F, ran second with 12,135. In the runoff, Arrington was supported by both major Birmingham newspapers, the *News* and the *Post-Herald*; the *News* called him a person of "unimpeachable integrity" and predicted that he would "be able to unify the city by establishing goals and policies that will benefit all persons." Vann also endorsed Arrington despite the estrangement that had come between them. On election day, 30 October 1979, Arrington won by a margin of 44,798 to 42,814 votes. Having drawn only 3.5 percent of the white vote in the primary, he received about 10 percent in the runoff, largely from the liberal south side of town, where most young professionals lived. What would have been unthinkable only a few years before had happened: a black mayor occupied City Hall.

Arrington's victory focused national attention on a city whose name had been synonymous with bigotry and hate. After being congratulated at the White House by Jimmy Carter, Arrington prepared for inauguration ceremonies attended by Alabama governor Fob James, U.S. senators Howell Heflin and Donald Stewart, and a host of dignitaries representing all shades of political opinion from every corner of the state. Striking a note of moderation, he defused fears that he was bent on revenge for past wrongs and indicated that he wanted only to lead the city to a better future. "My election is a clear indication of our progress in human relations," he said. "I know that the Birmingham of today is very different from the Birmingham of yesteryear which was wracked by racial strife. Although there is still work to be done to improve race relations and to bring about full racial justice, we no longer deserve the image of the Birmingham of the early 60's."[4]

Virginia Hamilton, an authority on Alabama history, was among scholars wrestling to understand what was taking place in a city that had witnessed so much racial discord. Noting that "widespread social integration" had not occurred in Birmingham, she stated that "although blacks and whites may mix socially at will, members of the two races generally tend to keep to their separate neighborhoods, clubs, churches, and lifestyles. Well-to-do whites in Mountain Brook continue to live in almost total residential segregation save for the daylight presence of maids, gardeners, and caddies." Still, she said, blacks and whites had "accepted . . . daily elbow-rubbing with a grace which neither would have dreamed possible twenty years ago. The smoothness with which the white South acceded to integration when the die was finally cast has been the envy of many a northern community."

Hamilton found several possible causes for the new ambience, including a quality of "personal friendship which has always crossed racial lines in the South, perhaps patronizing on the part of whites or subservient on the part of blacks, but nonetheless real." She also mentioned "the sunny side of southerners, the genuine warmth and kindness which has coexisted alongside fear, passion and hatred in their psyche," and "the neighborly quality of southern life, a luxury almost entirely lost to the lonely crowd on the eastern seaboard."[5]

As the South elected a wave of moderate governors including Jimmy Carter in Georgia and Terry Sanford in North Carolina, contributors to *You Can't Eat Magnolias*, a book prepared under the aegis of the progressive L. Q. C. Lamar Society, also searched in the 1970s for aspects of the southern tradition that might explain the racial accommodation now taking place in cities like Birmingham. Writers of essays for the book pointed to deeply rooted traits of civility, courtesy, and personalism that were present in the region and held them up as something special that it could offer to the country as a whole. Hamilton reflected the same spirit in referring to the "turnabout" achieved by the South "at a time when holier-than-thou Yankees were finding racial accommodation so difficult."[6]

The persistence of such incidents as the Bonita Carter case, the experience of southern liberals like Vann in trying to deal with them, and policies pursued by a wave of ultraconservative southerners who took office under the Republican label during the 1980s suggest that the air of regional self-congratulation pervading *You Can't Eat Magnolias* was premature. What had happened in Birmingham was less attributable to southern virtue than to a mixture of circumstances resulting from the industrial past and the altered realities of the postindustrial age. As the shutdown of Sloss Furnaces and the city's declining numbers of iron and steel workers indicated, Birmingham's former industrial order was dying in the 1970s. Nevertheless, much of its historical impact remained, including the demographic results of a strategy that had depended on the intensive use of low-cost black labor to compete with other parts of the country and of social forces, effectively analyzed by such historians as Paul Worthman, that had immobilized black workers and their families in the urban core while whites found it easier to flee to outlying areas. Under these circumstances, once the national government decreed that southern restrictions on black voting were illegal, blacks were bound to become sufficiently powerful to swing elections by attracting even a small percentage of white ballots, just as they did in many northern cities.

That a relatively few white voters were willing to cast their ballots for a black candidate was no doubt partly due to liberalism that had always been present in the South. But it also reflected powerful forces of historical change, epitomized by Birmingham's altered employment mix, the computer programmers now working for South Central Bell, and the public health workers at UAB. Perhaps the most important changes of all involved America's international position in a world, tightly laced by electronic media, that was no longer tolerant of overt racial discrimination. At least some of Birmingham's white residents recognized the imperatives of doing business in such a climate. In analyzing the 1979 mayoral campaign, Franklin stressed Arrington's recognition of the need to appeal for white votes from the south side of the city, where UAB was located and young professional and service workers lived, against east side blue-collar districts where attitudes formed in the old industrial order were still entrenched. Southside residents, along with such business and professional leaders as Vann, were more likely than white voters in other parts of the city to appreciate the penalties faced by Birmingham's high-tech firms if the city refused to submit to postindustrial realities. Instead of reflecting old traditions of neighborliness and civility, the new situation resulted mainly from the low-wage labor strategies of entrepreneurs who were long dead, the social forces that had concentrated black workers in the city while affluent whites fled to places like Mountain Brook, and powerful winds of change blowing in from the future. Within two decades, similar forces would be dismantling apartheid in F. W. de Klerk's South Africa, the world's last bastion of what Birmingham had once stood for.

III

While Birmingham and the world around it changed, conflict raged about what to do with the tract on which the idle City Furnaces stood. For almost five years after the Alabama State Fair Authority took control, little or nothing was done. In July 1972, the site was placed on the National Register of Historic Places, but no funds were available for its development, and the State Fair Authority had no apparent interest in its future. Derelicts slept on the grounds at night, vandals removed brass and copper parts, and the plant became a public eyesore. Civic leaders talked about converting it into a theme park, similar to Six Flags Over Georgia or Opryland USA, but nothing was done because the eighteen-acre location was too small and the projected $65 million cost too high.

In 1975, the spell of inactivity surrounding the property was momentarily broken when the pig-casting machinery that had been installed in the early 1930s was sold to ACIPCO. For the most part, however, the rusting facility suffered the same neglect that had been paid to the Vulcan statue for many years before it was finally salvaged and placed atop Red Mountain. Sloss Furnaces had become Birmingham's "iron elephant," a symbol of uselessness and futility. In March 1976, after being advised by a study commission that the property was a "health and safety hazard," ASFA decided to tear down the furnaces, sell the components for scrap, and put the tract on the market for further industrial development. Justifying such a move, Jack Beasley, ASFA's board chairman, told newspaper reporters that a sudden windstorm might tip over one of the stacks, imperiling traffic on the First Avenue Viaduct.[7]

Despite winning endorsement from the Birmingham Area Chamber of Commerce, the plan aroused a storm of opposition, much of it due to increased concern about historical preservation that was typical of a postindustrial society. As Leah Rawls Atkins has noted, "the 1970s saw the development of a greater appreciation and awareness for Birmingham and Jefferson County history." Much of the old hard-bitten industrial spirit remained that had once consigned the Vulcan statue to oblivion; such landmarks as the Morris and Tutwiler hotels, the Terminal Station, the Temple Theatre, and Vestavia were demolished in the 1960s and early 1970s. Even structures that survived were sometimes rudely treated; one of these, the 1890 Steiner Bank, became temporarily surrounded by a cagework of anodized aluminum. That even more damage was not done owed much to the state legislature's creation of the Jefferson County Historical Commission in 1971. Its first chairman, John E. Bryan, spearheaded the identification and marking of historical structures. The commission also had a statue of Thomas Jefferson placed in front of the county courthouse and helped preserve and publish historically valuable manuscripts.[8]

All over the city, the postindustrial spirit became evident. Civic organizations including the Arlington Historic Association, Birmingham Historical Society, Birmingham-Jefferson Historical Society, and West Jefferson County Historical Society were revitalized or established for the first time; Charles A. Brown led the creation of a Birmingham Branch of the Association for the Study of Afro American Life and History; historically significant houses were saved from the maul; the Birmingham Public Library appointed historian Marvin Y. Whiting to set up a department of archives and manuscripts. Supported by the National Endowment for the Humanities, a citizen group launched "Birmingfind," a community self-discovery project. Directed by Robert J. Norrell, a history professor at the University of Alabama, it published booklets about ethnic minorities and their traditions, sponsored photographic exhibits, and gathered transcripts of oral interviews with citizens whose memories of the past helped preserve vital parts of the city's industrial heritage. Atkins's book on Birmingham's history, *The Valley and the Hills*, was a product of the time, as was the collection by Wayne Flynt, then teaching at Samford, of interviews with workers that later enriched his classic study of Alabama's working-class whites, *Poor but Proud*.[9]

Such activities were part of an efflorescence of museums and other historical preservation efforts taking place not only all over the United States but throughout the postindustrial world. The citizen effort that saved Sloss Furnaces from destruction was an integral part of this new cultural ambience. After ASFA announced its decision to tear the old stacks down, voices were raised in their defense. W. Warner Floyd, executive director of the Alabama Historical Commission, argued that the furnaces were "both a visual evidence and a symbol of the iron and steel industry of Birmingham, one of America's youngest and greatest cities." Members of the Jefferson County Historical Association also protested demolition of the stacks. County commissioners Tom Gloor and Chriss Doss proposed that federal revenue-sharing funds be used for limited restoration of the furnaces so that they might become part of a "history tour" of local industrial sites, including Tannehill. Mayor David Vann urged a bond issue to rescue the City Furnaces.[10]

Despite opposition from Yarbrough, who epitomized the old order and remained a vigorous opponent of plans to save the furnaces from this time onward, the city council passed a resolution asking the State Fair Authority to delay action pending a survey by the National Park Service to determine what could be done to salvage the site. Still hoping to demolish the plant, the authority countered by organizing a public tour of the facility to demonstrate its hazardous condition. But this effort backfired when "many of the 100 people who braved mud, ants and threatening skies to make the tour," as the *Post-Herald* reported, "left with a resolve to fight to save the furnaces." "If there is any way to save it we have to," argued George Brown, a former furnace worker. "It would mean so much to the younger generation. If they do something with it, I'd even come back to work for free." An engineer in the tour group observed that much of the structural damage was only superficial. "Beautiful," said another participant. "It's just like the pyramids."

As the tour ended, a spokesman for the authority announced that plans for demolition would proceed on schedule. Seizing his bullhorn, local restaurant owner Randal Oaks denounced the move and asked members of the crowd to voice their opposition. After most of those present supported him, Oaks and other concerned citizens called a public meeting at his place of business and organized the Sloss Furnace Association (SFA) to "stop the bulldozers." Its members, including "housewives, schoolteachers, a physicist, an architect, former Sloss employees, local businessmen, and an engineer," elected Oaks chairman and petitioned the city council to delay demolition until the issue received further study. Other civic groups, including the Birmingham Jaycees and the local chapter of the American Institute of Architects, joined the cause. A survey of thirty-six city neighborhoods indicated that twenty-nine wanted to save the furnaces.[11]

Consciousness of the need to preserve historic sites was running high in the summer of 1976 because the nation was celebrating its Bicentennial. At Tannehill, the Fourth of July was observed that year by relighting the old antebellum blast furnaces, which had been carefully restored and repaired. Amid renewed concern about recapturing the past, the Historic American Engineering Record (HAER) conducted a detailed survey of Sloss Furnace property, jointly funded by the city council and the National Park Service, to assess its historical significance and prepare a permanent architectural record. Gary B. Kulik, curator of the Slater Mill Historic Site in Pawtucket, Rhode Island, and a Ph.D. candidate at Brown University, spent the summer in Birmingham writing a short history of the furnaces, based on the best sources then available.[12] Five architectural experts, including a graduate student from Auburn University who was badly stung by a nest of bees before jumping off a fifteen-foot crane, climbed over the abandoned structures and made detailed drawings for preservation by the Smithsonian Institution. Attending a special meeting of the city council in early July, at which a number of proposals for saving the installation were presented, members of the HAER team added their voices to the ongoing rescue effort.[13]

Sharp differences of opinion about how best to utilize the property, even if the furnaces escaped demolition, clouded its future. Vann, who was committed to creating a major tourist attraction with significant revenue potential, had been arguing that "if you simply make a museum out of the furnace, you'll probably have a lemon." Implementation of his proposal, which would have resulted in the creation of an amusement park similar to Six Flags Over Georgia, would have cost about $75 million, financed in part by new cigarette taxes currently being considered by the state legislature. A much less expensive alternative was proposed by councilwoman Angi Grooms Proctor, who gave a slide-illustrated presentation showing how the furnaces might be surrounded by spotlights playing on specific features at night while an audience, seated at a distance, heard tapes interpreting their significance. Pointing out that similar techniques were used in Europe, Proctor argued that her ideas, costing little to implement, could provide revenue for future site development as funds accumulated.[14]

A proposal embodying a concept similar to Vann's was presented to the State Fair Authority on behalf of the Birmingham Planning Commission. It would have created a vast theme park similar to Disneyland. After being taken through an Indian village, visitors would tour a plantation house. Taking a ride in a replica of the Confederate submarine *Hunley*, they would witness mock attacks by a Union warship and a giant squid. To appreciate industrial developments after the Civil War, they would then ride through a simulated mine in an ore car and be carried by skip hoist to the top of the No. 1 furnace for a dramatic view of Jones Valley. Another vehicle would take them on an exhilarating descent through the brilliantly illuminated interior of the stack, to the accompaniment of roaring noises simulating the action of the hot blast and the fusion of ore, coke, and flux. Boarding a ladle car, they would travel through the dustcatcher, pass around and through the boilers, descend into an area under the blowing engines, and tour the hot-blast stoves. At successive stages of the trip, hidden loudspeakers would imitate the sound of air rushing through pipes, the pounding of pistons, and the whine of turbines, and animated workers would be seen tending machinery. At the end of the tour, visitors would emerge into a display area to view textblocks, artifacts, and dioramas interpreting Birmingham's industrial past.[15]

Rejecting such visions as impractical, the State Fair Authority continued to press for demolition. Its attorney, John Foster, urged that lawsuits might result from injury to persons who ventured onto the property or incurred damage by passing it on the viaduct in cars. "We have to be realistic, because after all the verbiage we still have a rusting pile of scrap metal sitting out there," Beasley declared. Unable to resolve the dispute, the council adjourned pending further consideration.

Shortly after the council meeting, USP&F, which had conducted a study of its own, decided that at least part of the installation could be restored at a cost of $600,000. The Jim Walter Corporation offered to donate $100,000 of this if the remainder could be raised through public contributions. Submitted to the council at a special meeting on 8 July 1976, the proposal called for repairing the No. 1 furnace, along with its blowing engine building, cast house, gas washer, loading bins, stock trestle, and stoves, and razing everything else.

Reversing its previous stand, the State Fair Authority now pledged $100,000 in support of the plan. Vann and Gloor agreed to ask the city and county for matching amounts. Vann was confident that the remaining $200,000 could be secured from private sources, and the city council appointed a Sloss Furnace Study Committee to help raise donations. At last, it seemed that the installation, or at least part of it, would be saved. Despite residual enthusiasm for a theme park, things also seemed to be reverting toward the original concept of turning the site into a historical museum.[16]

With the assistance of Milo Howard, director of the Alabama Department of Archives and History, SFA mounted a fund-raising drive. Mailing lists were assembled, and newsletters were sent to prospective donors. An old car parade was organized; led by a fire engine and covered by radio and television, it started at the First Avenue viaduct, wound around the south side of the city, proceeded along the Red Mountain Expressway, and ended at Oaks's restaurant on Morris Avenue. Despite such efforts, three months passed, and the $600,000 goal had not been attained. In October, the study committee recommended that the fate of the furnaces remain in abeyance for nine months while further efforts were made to raise the money. Meanwhile, Vann, who still wanted to create a theme park, escorted a delegation from the Taft Broadcasting Corporation around the site but failed to arouse its interest in pursuing the idea. "Rusting and continuing to deteriorate," the *Post-Herald* commented in February 1977, "the old Sloss Furnaces on Birmingham's skyline still stand idle, slowly sinking in a sea of proposals, none of which have any monetary backing."[17]

By spring, things were moving again. Noting that a projected $62.5 million bond issue would be submitted to the voters in May for a number of projects involving outlays for schools, sewers, fire protection, street lights, the public library, the municipal airport, and other purposes, Vann persuaded the council to earmark $3 million of this sum for refurbishing the furnaces and turning them into an industrial museum. If this was approved in the referendum and the resulting museum succeeded in attracting a large number of visitors, Vann believed, a theme park might be created at some future time. Backing the plan, members of SFA put up posters around the city, gave slide presentations at church and civic functions, appeared on television, and circulated hand-bills stating that the furnaces should be preserved as a museum because "school children would learn from it, industries would be proud of it, artists would glory in it, visitors would acclaim it, Birmingham would profit from it." Partly because of this campaign, the proposition was approved by the voters on 10 May 1977. Celebrating the outcome, SFA held a victory party at Oaks's restaurant.[18]

At this point, an embarrassing turn of events took place. Vann had assumed that if the bond issue passed, the State Fair Authority would willingly relinquish the furnace property to the city. This was not the case. Asserting that the land and the structures on it were worth $1.5 million, State Fair officials demanded that the city buy the facility, lease it, or give the authority a quid pro quo in the form of other municipal property. When the city council met in early July to approve selling $35 million worth of the bonds authorized in the referendum, Yarbrough and another member abstained, using the impasse between Vann and the authority as an excuse. Although Vann intended to use condemnation proceedings if necessary to take possession of the tract, he took a conciliatory public stance, expressing confidence to reporters that matters would be worked out to the satisfaction of all parties concerned. The deadlock finally ended when the city offered the authority a number of concessions, including municipal funding of a long-range development plan for the state fairgrounds and assurances that the authority could retain the money it had received from ACIPCO for the pig-casting machinery. On 29 August 1977, the authority finally yielded title of the furnace tract to the city.[19]

But municipal ownership of the site did not produce immediate results. There was still no consensus about how to proceed. Even if the $3 million resulting from passage of the bond referendum would not permit creation of a theme park, some civic leaders continued to believe that such features as a miniature golf course or a swimming pool were necessary to optimize the tract's economic potential. Other enthusiasts wanted to go much further in departing from the museum concept. While debate continued, Vann secured federal funding under Title 6 of the Comprehensive Employment Training

Act for cleaning, fencing, painting, and repairing the property. After municipal workers had started these operations, work was temporarily suspended so
that film producers could use the tract as a location for a futuristic motion
picture, *The Ravagers*, starring Richard Harris and Ernest Borgnine. To create
an "illusion of antiquity," the camera crews tried to accentuate the rustiness of
the furnaces. "That was no illusion," quipped reporter Harold Jackson of the
Post-Herald. In a modest effort to improve the property, the Seaboard Air
Line Railroad agreed to modify the elevation of trackage in the area to facilitate entry to the grounds. In essence, the principal illusion surrounding the
property was that of progress itself.[20]

In June 1978, trying to generate momentum, Vann appointed David
McMullin, a forty-five-year-old native of Birmingham with "a background in
finance and entertainment" and degrees in literature from Oxford and the
University of Alabama, to "get the Sloss project into gear." Visiting Chicago,
McMullin discovered that the Museum of Science and Industry was dismantling a major exhibit of scale models pertaining to the steel industry. He
persuaded officials to give it to Birmingham, along with a Walt Disney film
showing how steel was made. After the exhibit had been moved to Alabama at
substantial cost and put in storage, Vann declared that it would make a
"magnificent beginning" for future development of the furnace site, "more
than we ever expected a year ago." Talking to reporters, McMullin stated that
he wanted the tract to have "a re-creation of an African village, with exhibits
of African art and culture, including food," as well as "a re-created coal
mine." "I want a place for people to be able to express themselves, where kids
will want to hang out, learn some things," he said. "But it's got to be interesting, and it's got to involve a lot of participation."[21]

While McMullin traveled and consulted with outside experts, Vann tried to
interest the Ford Foundation in adding to funds remaining from the May
1977 bond issue. At one point, a foundation representative visited Birmingham and called the furnaces "a sensational symbol" of industrial society. After
he returned to New York, however, nothing happened. Two professors from
the University of Alabama at Birmingham recommended that the city use
San Francisco's Exploratorium, described by the *News* as "an adventure in
mirrors, lights, prism and colors," as a model for developing the furnace tract,
pointing out that it attracted 500,000 visitors per year. Nothing came of their
idea, either.

In March 1979, after months of secrecy, McMullin unveiled a preliminary
plan for development of the site. Estimated to cost about $18 million to
implement, it called for three main attractions, including the furnaces themselves, a "Museum of Science and Industry," and a "Museum of Culture,"
with "separate Asian, European, and African pavilions." McMullin claimed
that African exhibits would be an effective way to counter arguments by Klan
members that "African culture is no culture at all." Similarly, the European
pavilion might help stimulate "a renewal of European music coming out of
the South." He also envisioned that "there would be banjo pickers, gospel

singers, jugglers, and fire-eaters available to entertain passers-by" and a German restaurant with live "oompah" music. The pavilion, he said, "would be a nice place to pass the day, sipping good coffee. We want the retired folk to come and the young chess players too, but we don't want bums so we will charge a steep price for that first cup."[22]

Local reactions to McMullin's visions were, at best, mixed. He had already alienated some taxpayers by allegedly failing to submit accurate accounts of his travel expenses, and his plan intensified opposition from critics who had thought all along that efforts to save the furnace were a boondoggle. Even newspapers taking a more moderate view were offended. Expressing skepticism in an editorial, "Sloss Dreaming," the *News* stated that McMullin's ideas sounded "more like a script from the fantasy film *Lost Horizon* than a serious and well-thought-out plan for the historical site." Characterizing the proposal as "a patchwork of themes," it argued that "the time has come to get serious about the Sloss project. . . . The site is already unique and it is silly to try to change its character into a fantasy land of entertainment. It has a stronger claim to becoming a museum of science and industry than an imitation backlot of a Hollywood studio."[23]

Undeterred by criticism, McMullin consulted with "a number of companies and corporations," which, following a model already established by the Chicago Museum of Science and Industry, would "design and build their own exhibits for the science museum, and pay the costs for maintaining them." Early in July, he presented his ideas at a council meeting at which he encountered much skepticism. Nevertheless, he persuaded a majority of the members to appropriate $190,000 for a feasibility study to be conducted by experts including Thomas Hoving, former curator of the Metropolitan Museum of Art; New York lawyer John H. Blum; James Gardner, a distinguished English museum designer; and Hardy Holzman Pfeiffer Associates of New York, which had planned a number of cultural institutions including the Brooklyn Children's Museum. As usual, Yarbrough cast a dissenting vote, but he was not alone in his negativism. A state representative from Homewood urged that plans to develop the furnaces be scrapped. "We could be dealing with the biggest white elephant in Birmingham's history," he declared. "I have found support for this project to be almost non-existent."[24]

This statement seriously underestimated public sentiment for doing something to develop the furnace site as an educational or recreational attraction. In March 1979, realizing that a long battle was in prospect, SFA reorganized itself, established six classes of membership ranging from individual ($5) to life ($500), and applied for incorporation. Receiving a charter on 20 July 1979, it secured tax-exempt status several months later. Adopting the motto "Embrace the Past—Enhance the Future," it fought to keep the museum concept alive. In October, while the mayoral campaign that led to Arrington's election was taking place, the *Post-Herald* took a poll of 350 registered voters and found that a plurality wanted to preserve the furnaces as an industrial museum: 46 percent were in favor, 41 percent were opposed, and 13 percent expressed no opinion. Black voters were highly supportive, with 60 percent in favor; whites were evenly divided.

Arrington's victory in November was a welcome development to those who favored saving the furnaces. During the campaign, he, along with all but one of the candidates for mayor, had supported development of the facility into a museum or some other form of tourist attraction. Shortly after the election, McMullin gave Arrington a progress report, and the latter reappointed him for six months. [25]

Early in 1980, a steady stream of consultants came to Birmingham to advise McMullin, SFA, and the city council about developing the furnace tract. Gardner's visit drew particular attention. Wearing corduroy slacks and a turtle-neck sweater, the white-haired, seventy-two-year-old Englishman, who had designed the Museum of the Jewish Diaspora in Tel Aviv, spoke enthusiastically about saving the installation at a public meeting held at UAB's Spain Auditorium. Calling it a "great fantastic monument," he stated that one could imagine it having been "used by giants." Further support for the drive to preserve the facility came when the city council, over dissenting votes from Yarbrough and another member, authorized spending $863,000 from the proceeds of the bond issue to pay the Jim Walter Corporation for an additional 13.2 acres of land adjoining the tract. [26]

McMullin's final proposal embarrassed even his supporters. Unveiled to the public late in June, just before his contract expired, it was even more lavish than the one that had aroused so much debate in 1979. Although its cost was specified at about $75 million, the *Post-Herald* estimated that implementation would cost at least $100 million. As McMullin saw the future of the site, the two furnaces would become focal point for a "Museum of Modern Times." The east stack, Sloss No. 1, would be restored and interpreted to the public by "expert guides," while the other would become "part of a less formal experience in which the public is free to wander among stabilized ruins." After looking at the stacks, visitors would proceed through a maze of subterranean corridors lined with exhibits marking various phases of human history. In "Corridors of Ancient Time," they would view three-dimensional mixed media presentations bringing alive such places as "Athens at the time of the trial of Socrates; Jerusalem just before the birth of Christ; or Rome under Belisarius struggling to survive the barbarian hordes." In the "Foyer of the Middle Ages," they would study displays progressing from "the murky times of the Arthurian struggles with Saxon invaders" through the advent of the Black Death. Related exhibits would be devoted to scientific and technological subjects including agriculture, chemistry, dentistry, horsemanship, and textiles.

This was merely the beginning of McMullin's dreams. The "Foyer of the Middle Ages" would lead to a "Foyer of the Renaissance," which would branch off into no fewer than twelve "Corridors of Nations" in which viewers could explore the history of such countries as France, Great Britain, and Poland. These passageways would interconnect with sub-corridors devoted to the development of exploration, discovery, colonization, and emigration. Next would be "Corridors of America," a "Birmingham Corridor," and an "Afro-American Pavilion." There would also be a separate "Science and Technology Museum for the Southeastern United States," containing the steelmaking exhibits that McMullin had brought back from Chicago. Separate pavilions, funded by private donors, would be devoted to such subjects as chemistry, mining, and transportation. Banks and other commercial enterprises would be invited to create displays telling about the history of money and insurance; law firms would sponsor exhibits dealing with the United States Constitution and the court system. Reflecting Birmingham's emergence as a major medical center, a "Pavilion of Biology and Medicine" would contain exhibits on human disorders and other related topics.

McMullin also planned to have the museum include an Omnimax Theater, similar to ones already developed in several American cities. Here visitors would see breathtaking 180-degree films with stereoscopic sound. The Omnimax would also house a branch of the Cinematheque Francaise, the largest film archive in the world. There would be an outdoor "Mixed-Media Theater," capable of combining live performances with stunning auditory and visual effects, and an indoor Roman theater for classical plays and dances. Cafes and restaurants would be scattered throughout the grounds, and wandering musicians, poets, and performers would entertain the public. Athletic facilities would include AAU and Olympic swimming pools suitable for collegiate and international competition and a gymnasium especially equipped for the handicapped. A children's museum would enable parents to enjoy the main exhibits by themselves or go shopping downtown. The entire tract would be landscaped as a vast public park. McMullin likened his concept to that of the Place Pompidou in Paris and stressed that the result would be unique to the southeastern region.[27]

In sheer scope and panache, the plan might have appealed to Atlanta, but it had no chance of realization in Birmingham, where even members of the council who were sympathetic to the idea of a public museum had decided that its budget would have to stay within the shrinking funds—about $1.8 million—remaining from the 1977 bond issue. Some citizens who heard McMullin's presentation walked out in disgust before he finished speaking. The *News* called the event "something of a public relations disaster" and pointed out that, in his "pie-in-the-sky thinking," McMullin had forgotten to provide for such mundane needs as adequate parking space. Critics of the entire museum concept pounced eagerly on the debacle. Yarbrough, who wanted to "dismantle said furnace, sell metal of said furnace for scrap and build an industrial park on the site," was supported in opposition to McMullin's plan by such groups as the Birmingham Area Chamber of Commerce and the Metropolitan Development Board.

At a council meeting in mid-July, beleaguered supporters of the museum idea bought time by voting for a forty-five-day study to determine how at least part of the plan might be realized, but bitter words were exchanged. Councilman Pete Clifford expressed a widely shared view by saying, "I think I've been had." Arrington was more philosophical. "It's a dream, but all good things begin with dreams," he said. "We need in Birmingham something of which we can be proud."[28]

IV

Although it was not clear at the time, the preservation drive had reached its nadir, and better times lay ahead. McMullin, whose contract had been renewed for thirty days, quit his post at the end of July, spurning an offer to retain his services on a day-to-day basis pending a final decision on his proposal. As everybody knew, it was dead. In a last-ditch effort to salvage something manageable from the wreckage, SFA began working on a way of implementing the first phase of McMullin's scenario, dealing with restoration of the furnaces themselves, and of staying within the financial limits imposed by the funds remaining from the 1977 bond issue. Their hopes were buoyed by news from Washington that the furnaces had been nominated for designation as a National Historic Landmark.

Emboldened by the fiasco created by McMullin's report, critics of the museum concept maneuvered to destroy its implementation once and for all. In the upcoming November elections, another bond referendum was scheduled to take place, involving $27 million worth of new municipal securities earmarked for helping build a new county jail, completing a sports arena, establishing industrial parks, redeveloping slum areas, and constructing new parking decks. Confident that the public had turned against the entire Sloss project, Yarbrough urged a referendum on its future at the same time. Otherwise, he claimed, the bond issue might be lost due to public exasperation with plans for the site. Arrington disagreed. Fearing that complicating the November balloting with the Sloss issue would have exactly the same effect, he thought it best to settle the matter separately. His view was shared by the council's finance chairman, David Herring. In the end, voters never got a chance to turn thumbs up or down on continuing the museum project.[29]

The Sloss Furnace Association benefited from Arrington's position. Because of public indignation about McMullin's report, the idea of converting the furnace site into a theme park was no longer tenable, producing a much clearer focus about what could realistically be done. The keynote of SFA's effort was now "simply to save Sloss and appreciate it for what it is." This proved to be an effective strategy. In a series of meetings at City Hall in October, Jim Waters, an architect who had become president of SFA, outlined a plan to restore the site, increase its accessibility to the public, improve illumination at night, and renovate the employee bath house for use as an administrative office and visitor center. The projected cost, $1,718,710, was within the limits imposed by remaining funds and was therefore acceptable to a majority of the council. A succession of outside experts, including Robert Vogel of the Smithsonian Institution's National Museum of History and Technology, visited Birmingham to support the plan. Relieved to see a realistic solution of the protracted conflict at last in prospect, council members who had supported the museum concept from the beginning rallied to support Waters's ideas. As usual, Yarbrough opposed them. "Anything you do," he said, "is going to fall flat on its face and I'm going to live to say I told you so." Disregarding him, council voted on 26 November 1980 to approve the plan, by a margin of six to one, with one member abstaining and another absent. The furnaces had been saved.[30]

Elated by the decision, SFA immediately began preparing the site for limited public access by 12 April 1982, the centennial of the first blowing-in by Colonel Sloss. For more than sixteen months, sandblasting and painting, grading and seeding, paving and lighting, removal or repair of hazardous structures, renovation of the bath house, and creation of textblocks and other graphic displays proceeded. In May 1981, the tract became one of only eighty-seven sites in the United States to be designated as a National Historic Landmark. At a public ceremony in November, Dwight L. Young, director of the southern regional office of the National Trust for Historic Preservation, helped dedicate the facility with a speech in which he told how he had first seen the furnaces while driving through Birmingham in 1969 on a trip to Mississippi. "I was vaguely aware of . . . a big hulking mass of smokestacks and ovens and sheds," he recalled, "and then suddenly a part of that mass erupted in flame and sparks and red-tinted steam." Jamming on his brakes to take in the spectacle, he had almost caused a pileup on the First Avenue Viaduct. Evoking such experiences, he stated, was what historic preservation was all about; it was "just having the good sense to hang on to things that are important." Monuments like Sloss, he said, helped prevent a community from becoming a "victim to amnesia," serving as "tangible reminders of who we were, how we lived, and where we were headed."[31]

Through persistent hard work, SFA succeeded in having the site ready for the 12 April 1982 celebration commemorating the first run of iron that had flowed from the original No. 1 Sloss Furnace a century earlier. Most repairs had concentrated on the east stack and adjoining cast shed, but the former employee bath house had been converted into a visitors' center. In addition, Birmingfind had arranged a gallery of photographs pertaining to coal mining, iron manufacturing, and railroading in the district. About 200 persons attended the Monday afternoon event, at which Arrington officiated. Congratulating SFA and recalling varied proposals that had been made for the tract, ranging from outright demolition to McMullin's grandiose visions, the *News* stated that "the present middle road, which will result in a more modest preservation of the furnace . . . is the result of a labor of love by Birmingham citizens committed to the idea that the heritage of the city is closely tied to the imposing skyline of Sloss."[32]

The afternoon celebration was followed after dark by a fireworks exhibition that had unintentionally dramatic consequences. After all but two members of the crowd had gone home, flames suddenly engulfed a wooden cooling tower. A motorist going to work when he saw the blaze from the Red Mountain Expressway shortly after nine o'clock that evening later reported that "it was roaring with such force that it looked like the fire was being fed by a giant gas burner." A wall of flame estimated to be fifty feet wide shot across the First Avenue Viaduct, but fortunately no one was hurt. After only half an hour, the blaze, which could be seen for miles, was extinguished. Considering the limited nature of the damage, it was perhaps only appropriate that the old ironworks, which had provided so many vivid spectacles for local residents throughout its history, had erupted once more on its one-hundredth birthday.[33]

Preparations continued throughout what remained of 1982 to open the installation to the public on a permanent basis. During the summer, a youth employment program provided work for twenty-eight young people and two supervisors who restored the stock trestle, rebuilt the elevated rail system that had once dumped raw materials into underground storage bins, and renovated an old service station where Woco-Pep had been sold to motorists in the 1930s. New metal roofing was provided for the east and west casting sheds. Former Sloss employees, some of whom had worked for the company for more than three decades, prepared to serve as tour guides. Meanwhile, engineering societies began visiting the grounds by special arrangement to study the facility.[34]

In December, SFA and the city government announced the appointment of Randall G. Lawrence, who was completing a Ph.D. in American history at Duke University, as first director of Sloss Furnaces National Historic Landmark. A native of West Virginia who had once dug slag, Lawrence had an impressive record of scholarly expertise in the history of the coal mining industry and the development of organized labor. He had been liaison officer for the Appalachian Regional Council and the President's Commission on Coal and had directed West Virginia's coal-mining exhibit at the Knoxville World's Fair. Lawrence said that his guiding principle would be "to protect the integrity of the furnace, so that it remains a working symbol of Birmingham." Talking to reporters, he stressed the uniqueness of the facility as a museum of ironmaking technology in the early twentieth century.[35]

Under Lawrence's leadership, preparations continued for the permanent opening of the new museum, fittingly scheduled to take place on Labor Day weekend, 1983. Progress was complicated by a national economic recession that produced a budgetary crisis at City Hall, and longtime critics of the project persisted in complaining about its implementation. At a council meeting in early December 1982, Yarbrough cast the only vote against an appropriation for furnishing the museum offices, stating that he had "never . . . seen more of a waste of money." Continued support came from the National Trust for Historic Preservation, which early in 1983 provided $68,000 worth of paint, but controversy continued to swirl in March, when SFA successfully

Randall G. Lawrence. Director of Sloss Furnaces National Historic Landmark from 1983 to 1991, Lawrence provided outstanding leadership that made the site a model for other industrial museums to emulate. Without his vision, this book would not have been possible. A striking photograph, taken early in his administration, evokes the badly deteriorated condition of the facility that existed at the time and suggests the artistic qualities possessed by this highly charismatic man. (Courtesy of the Birmingham *News* and Gene T. and Bernice S. Lawrence, Lithonia, Georgia. Copyright December 8, 1982, the Birmingham *News*. All rights reserved. Reprinted by permission.)

requested that the city create four new positions, including an education director and an assistant curator. In June, bids exceeding $350,000 were approved for structural repairs to the eastern stack's slag granulator and removal of asbestos from the engine rooms housing the blowing turbines.

In August, as opening day approached, Lawrence took a reporter, Elma Bell, up a catwalk to the top of the gas washer to get a panoramic view of the site. Years before, Bell's grandmother had taken her to see the furnaces in operation to give her a warning of what Hell would be like if she misbehaved. Traumatized by the experience and wanting to exorcise it, she had long wanted to see the plant razed. Now, after surveying the scene below, she wrote that "the magic of the place finally hit me. Tear down those majestic structures? That is the silliest idea I ever heard."[36]

On Friday evening, 2 September, four days of opening festivities began when the Birmingham Heritage Band gave a jazz concert in the casting shed amphitheater. About 500 people came. As vocalist Sheryl Goggans sang "Birmingham, Birmingham, This is My Home" under pink stage lights, listeners sipped wine and munched French bread and cheese. "I love trains and I love jazz," said Dick Gilchrist, chaplain to Episcopal students at Auburn University, as a locomotive and a string of freight cars passed by as if saluting the event. "What more could you ask for?"

On Saturday morning, guided tours began, SFA served barbecue, bagpipe music was played, and a blacksmith from Tannehill State Park gave a metalworking demonstration in the plaza under the water tower. In the evening, a bluegrass concert ended the day's festivities. Similar events took place the following afternoon. The four-day celebration reached its climax on Labor Day, 4 September, with three hours of activities honoring former furnace employees. Included on the program were performances by the Birmingham Community Band, the Sterling Jubilee Singers, and the Delta Aires.[37]

Not long before, Lawrence had underscored the significance of the new museum by noting that "every city needs a symbol, a place that functions not only as a symbol of the city but also serves as a center for urban life." The new museum, he hoped, would meet this need for Birmingham. His comments were echoed in a salute to the entire project that appeared in the Birmingham *News*. In an editorial, "Sloss Stirs Again," the paper stated that Lawrence and his staff had rekindled a fire in the historic facility. "It is not the coke-fed blaze of a blast furnace which animates the old mill this time" it said, "but the bright spark of community involvement." The same idea, that a particular structure or historic district could provide symbolic resonance for an entire city and epitomize its sense of identity, was expressed shortly thereafter by architectural analyst and critic Wolf von Eckardt, at the opening of "Buildings Reborn," a photographic exhibit of Birmingham's historical landmarks. Upon hearing "Paris," he said, he thought of the Eiffel Tower; "Baltimore" evoked Harbor Place. Now, he stated, whenever he heard "Birmingham," he would think of Sloss Furnaces.[38]

V

Throughout the 1980s, Lawrence and his staff of historians, curators, and volunteers used various means to fulfill their mission of preserving, developing, and interpreting the site and its structures for the increasing number of people, approaching 75,000 by 1991, who came to see it each year. Much of their effort was aimed simply at encouraging citizens to look upon Sloss Furnaces as an appropriate place for public gatherings, a focal point for community events. The Alabama Symphony Orchestra gave concerts in the east casting shed, and rock groups also performed there. In an effort to enhance the image of the property as a focus for civic pride, the Birmingham Historical Society established its heaquarters on the site in a historic house that was moved from another location.

From the outset, Lawrence viewed the museum not merely as a place for public gatherings and the dissemination of existing knowledge but as a center for research and scholarship aimed at developing new perspectives on the industrial history of Birmingham, north Alabama, and the southeastern United States. Continuing a program that SFA had already begun before he came, former employees of Sloss-Sheffield and USP&F were interviewed to preserve their recollections of the past. This resulted not only in a growing repository of information but also in publications, including a booklet, *Like It Ain't Never Passed*, evoking memories of life in Sloss Quarters. Scholarly groups, such as the Society for Industrial Archaeology, held meetings at the site. Conferences, including one dealing with Alabama's pioneer industrial historian, Ethel Armes, also took place.

Early in 1985, while visiting Auburn University, Lawrence asked the author of this book to write a short guide to the history of Sloss Furnaces to sell to museum visitors. Despite the research of such persons as Armes and Kulik, he said, much remained unknown about the succession of enterprises that had made iron on the site. Curator Robert Casey, a metallurgical engineer with a master's degree from a graduate program in industrial history maintained by the University of Delaware and the Hagley Museum and Library, had gathered a variety of source materials, including files of *Pig Iron Rough Notes* and family papers pertaining to the career of Edward A. Uehling. A local historian, Marjorie L. White, had prepared a valuable guide to the industrial history of the Birmingham District, and staff member Barbara Nunn had compiled a chronology of significant events in the tract's evolution. Interviews with former Sloss employees had begun to yield an accumulation of insights into what it had been like to work at the furnaces. Many important topics, however, remained to be explored before the history of the installation could be brought into clearer focus or seen in the context of other developments that had taken place locally, regionally, and nationally. In particular, historians had not paid adequate attention to the persistence of cast iron as a key industrial commodity and to the significance of the foundry trade as a critically important sector of the ferrous metals industry.

I accepted Lawrence's challenge. As my work proceeded, Lawrence saw that it would be impossible to write a guidebook accurately interpreting the history of the furnaces until the subject had been more thoroughly explored and secured funding for a larger study on which a guidebook could later be based. During the winter of 1985–86, Marvin Whiting learned that many business records of the Sloss-Sheffield Steel and Iron Company had survived and arranged with the Jim Walter Corporation to have them placed in his custody at the Birmingham City Archives. Study of these materials raised questions requiring further research in local newspapers, engineering journals, and trade publications. Lists of stockholders also showed that much of the history of Sloss Furnaces would have to be traced in Richmond, whence so much capital had come. The Virginia Historical Society's collection of Joseph Bryan letterbooks, supplemented by insights derived from the pioneering work of historian Maury Klein in unraveling the complex development of the Richmond and Danville railroad systems, proved indispensable in assessing how Sloss Furnaces had been an integral part of the drive to foster southern industrialization after the Civil War. Research in Birmingham newspapers yielded long-forgotten aspects of the installation's history.

Lawrence, a cancer victim, died on 15 May 1991, only thirty-six years old. Paige Wainwright, the museum's assistant director, paid tribute to his vision, saying that he had "turned a local landmark into an international model for the preservation of historic sites." Saluting his accomplishments, the *Post-Herald* stated that, more than any other person, he had given direction to the dreams that enthusiasts had entertained for the future of the "cast-off industrial relic below the First Avenue North Viaduct."[39] On 1 March 1992, James A. Burnham, formerly chief conservator of Henry Ford Museum and Greenfield Village, succeeded Lawrence as executive director of Sloss Furnaces National Historic Landmark.[40]

Although he had read articles and conference papers stemming from the research on this book and seen preliminary drafts of all but two chapters of the manuscript resulting from his proposal, Lawrence did not live to see the completed study in print. This makes it all the more appropriate that it is published as a memorial of his life and career. Because of his patience and wisdom in recognizing that such a study was necessary before a satisfactory guidebook to the museum could be written, it is now possible to see the history of the furnace site in clearer perspective.

17 In Retrospect: The Southernness of Sloss

Still standing against all odds, Sloss Furnaces National Historic Landmark remains a monument to a distinctive southern style of industrialization that was already taking shape before rebel guns opened fire on Fort Sumter. Like white-columned antebellum homes and statues of Confederate soldiers, the hulking presence along Birmingham's First Avenue Viaduct is a visible reminder that however much the South may share with the rest of the United States, it has long pursued its own separate path. Since the Great Depression, the region has entered increasingly into the national mainstream. Characteristics that it has always held in common with the country at large have become more and more pronounced. Still, it seems destined to remain indefinitely conscious of an enduring otherness. As the industrial era comes to an end, the epic of Sloss Furnaces is a significant part of the heritage that the South, still struggling with its troubled past, will bring to the postindustrial future.

I

Interplay between change and continuity creates a tension that is fundamental to the emerging pattern of historical events. Change is the more exciting of these two underlying forces, but continuity is perhaps the more powerful. Because Birmingham's origins were deeply rooted in the antebellum period, it is paradoxical that the city became one of the best-known symbols of the New South. That term, made famous by Henry W. Grady in a speech delivered at the annual dinner of the New England Society of New York City in 1886, was from the beginning intended mainly for northern consumption, a rhetorical device to convince Yankee investors that a region once dominated by slave-holding planters was now a "perfect democracy" and a hive of free institutions.[1] But the leopard had not really changed his spots. As John T. Milner's writings reveal, Birmingham would have been a center of slave-operated industries if the South had won the Civil War. Even as it was, the city perpetuated most features of antebellum industrial slavery. From the time of its founding, its promoters showed a consistent preference for low-paid, servile black workers. Convict labor, mostly black, was an important weapon in the district's economic warfare with northern manufacturers.

As previously indicated, Birmingham was an overgrown iron plantation. The squalid section known as Sloss Quarters, which endured into the 1950s, was the post–Civil War equivalent of the hovels that had housed bondsmen and their families on large agricultural estates in antebellum times. It was only natural for Joseph Bryan, who remained steadfastly true to his rural origins, to compare the condition of the Sloss Iron and Steel Company under Thomas Seddon's administration to that of an "old, run down plantation." Significantly, Bryan appointed Seddon president of Sloss at least in part because he had run a sugar plantation in Louisiana.

From a southern perspective, there was good reason for Bryan to believe Seddon's background pertinent to the managerial role that the latter assumed in Birmingham, even though Seddon knew little or nothing about coal mining and ferrous metals. Sugar plantations were agricultural and industrial establishments, which combined cane cultivation with grinding and refining operations conducted in mills that were staffed by slaves. As Carroll W. Pursell has pointed out, such mills were among the earliest large-scale users of stationary steam engines in America during the antebellum era.[2] Nor is it odd that Bryan, the son of a Virginia tobacco planter, became one of the South's leading industrialists after the Civil War. His enthusiasm for manufacturing seems strange only if one accepts Raimondo Luraghi's misleading view that the prewar South was "a backward agricultural society" and that planters were "hostile to industrialization."[3] Even if members of the Broad River group did not want to see Alabama copy too closely what was taking place in northern textile centers like the Merrimack Valley and wanted to preserve close ties between industry and agriculture, they did not oppose manufacturing per se. Some planters undoubtedly did take a negative attitude, as did Jacksonian farmers. After the Civil War, persons of this persuasion followed Jefferson Davis in wanting to keep the South a rural Arcadia. Many more, however, supported Robert E. Lee in welcoming industrial growth.[4]

Like James R. Powell, whose father had raised tobacco, Bryan was born on a Virginia plantation devoted to the same crop. This estate, Eagle Point, was part of a far-flung transatlantic network of business relationships in which agriculture and industry were tightly connected. The Stewarts, with whom Bryan joined forces in promoting industrialization after the Civil War, were involved in the same system, as were the owners of Richmond's slave-operated tobacco factories. Alabama's Broad River planters, whose activities have been skillfully traced by J. Mills Thornton, were part of this international network that linked British and American textile mills, countinghouses in New York City, Daniel Pratt's cotton gin factory and industrial complex at Prattville, and mercantile establishments in such places as Montgomery and Mobile. Broad River associates were typical of wealthy Whig slaveowners who thought in large terms. As land speculators, they were accustomed to buying stock in risky ventures with freely transferrable shares. Despite being outnumbered and sometimes outmaneuvered by the Jacksonians who controlled the Alabama state legislature, Broad River people and other affluent southern landowners produced most of the marketable surplus that, in Carl Degler's words, underlay "the culture and civilization of the region." Such persons were "adventurously, and often ruthlessly, bent on profit and reinvestment," as Degler says; he calls them "agricultural entrepreneurs in a capitalist economy." To put it most succinctly, they were agribusinessmen. Such were the forebears of the later southern industrialists who followed Lee's advice by investing in railroads, blast furnaces, coal mines, and textile mills after the Civil War.[5]

Contrary to an influential study by Jonathan Wiener, *Social Origins of the New South*, postwar Alabama planters, as a class, did not oppose industrialization. The very name of the Elyton Land Company, the enterprise that spearheaded the development of Birmingham, epitomized a deeply ingrained tendency among wealthy planters to speculate in real estate. This venture was not a departure from antebellum values, but a fulfillment of the prewar dreams of industrialization that had inspired Gilmer, Milner, and other Broad River associates. Similarly, the heavy involvement of Bryan, Dooley, Johnston, Maben, Seddon, and other Virginians in the Georgia Pacific Railway and their role in developing the network of coal mines and blast furnaces that they consolidated into the Sloss Iron and Steel Company contradict Wiener's assertion that the persons who led the drive for southern industrialization after 1865 "had their social origins outside the mainstream of planter society" and came from "families that were not enthusiastic about slavery or secession."[6]

It is not true, as Wiener claims, that John W. DuBose and other propagandists were less than candid in stating that Birmingham represented a continuation of Old South traditions. Defenders of the New South creed were not trying to gain a spurious legitimacy among interests that Wiener regards as more authentically representative of a prewar agricultural heritage. To identify Confederate veterans' organizations and southern churches with opposition to industrialization, as Wiener does, is inaccurate. Most of Alabama's leading industrialists, including Nathan Bedford Forrest, Henry F. DeBardeleben, Joseph Forney Johnston, Edmund W. Rucker, and Daniel S. Troy, had fought bravely for southern independence. So also had Bryan and most of his Virginia associates, including Algernon Buford, James Dooley, John W. Johnston, and John C. Maben. Bryan, every inch an aristocrat in the Old South tradition, became the Old Dominion's leading Episcopal layman. Nothing about his deep solicitude for the church or his ceaseless activity in veterans' groups bears out Wiener's thesis. Throughout his career, Bryan was constantly attending reunions with former comrades; raising funds for monuments honoring Confederate leaders; working on behalf of needy veterans, including his former commander, John Singleton Mosby; and doing countless other things to keep sacred the memory of the Lost Cause.

Wiener acknowledges that wealthy Alabama slaveholders invested in manufacturing in the antebellum era but mistakenly views postwar planters as being hostile toward Birmingham. In fact, no basic cleavage existed between plantation owners and the industrialists who founded the Magic City. Attacks on Birmingham that Wiener quotes from what he misleadingly calls the "planter press" need not be interpreted as displaying an underlying animus against industrialization. Such barbed comments can more plausibly be seen as expressions of hostility toward the Mineral District among boosters of rival urban centers that were jealous of its progress, continuing a tradition already strongly in evidence before the Civil War. Criticism of Birmingham in Mobile, for example, merely shows how deeply that Alabama seaport, which remained depressed for many years after the war, resented the way in which the north-central part of the state had wrested dominance from the Gulf Coast and Tombigbee Valley.

Similarly, the decline of anti-Birmingham rhetoric in south Alabama during the 1890s does not show that downstate planters were belatedly accepting industrialization, as Wiener claims. Instead, it reflected Mobile's long-delayed emergence from its postwar doldrums and the fact that its interests had finally become congruent with those of the Magic City. This compatibility stemmed from the fact that such Birmingham-based firms as Sloss and TCI, which had previously concentrated on sending iron north or east through Cincinnati, Louisville, and Savannah, had begun after 1893 to export a significant part of their output to foreign markets via the Gulf of Mexico. TCI's move into steel production and its alliance with Mobile shipbuilders at the turn of the century further solidified good relations with downstate interests, as did the subsequent canalization of the Warrior and Tombigbee rivers to Birmingport.

Much of Wiener's research involved eleven Alabama counties in particular.[7] Six of these—Colbert, Franklin, Lawrence, Marion, Lamar, and Fayette—lay west of the main Warrior coal deposits and had no reason to promote Birmingham, but they did not oppose industrial growth. When their limited coal reserves and substantial brown ore deposits were opened for exploitation in the late 1880s, they tried to aggrandize such manufacturing centers as Florence and Sheffield. The other five counties that Wiener emphasizes—Sumter, Greene, Hale, Perry, and Marengo—were tied to Mobile, Selma, Tuscaloosa, and Meridian. None of these communities stood to gain from boosting Birmingham. Even Tuscaloosa, which was located closest to the Magic City, wanted to compete with it in coke and pig iron production and was beginning to do so by the end of the nineteenth century with the emergence of industrial plants in such satellite towns as Holt and Kellerman. Urban rivalry, not planter opposition to industrial growth, provides the best explanation for whatever hostility appeared toward manufacturing in general, or toward Birmingham in particular, in the counties that Wiener emphasizes.

Nor is it strange, despite the fact that Montgomery was the mother of Birmingham, that Wiener found evidence of hostility toward the community's offspring in capital city newspapers like the *Advertiser*. It was common for any urban center in Alabama to deplore the good fortune of another, even one that it had earlier promoted. Backstabbing by Montgomery was highly predictable in view of the fact that Birmingham expected to become the state capital and had set aside a public park for the necessary buildings.[8] During the winter of 1870–71, however, when Birmingham was still only a dream and powerful interests in Montgomery were hoping to profit from its creation, things were different. At that time, the *Advertiser* strongly supported Gilmer, Milner, Powell, Tate, Wallace, and other railroad promoters who were trying to build an ironmaking center in Jones Valley. The newspaper's favorable attitude toward industry was indicated by an extremely laudatory article that it published late in 1870 about Daniel Pratt, who soon thereafter became the chief investor in an effort to restart the dormant furnace at Oxmoor.[9]

Not all planters in the Montgomery area favored the backers of the South and North Railroad in their battle against the Stantons. As Wiener indicates, some cotton growers believed that it would be better for coal and iron ore to be exported from Alabama to the outside world via Chattanooga or Mobile than to be converted into coke and pig iron within the state. But this attitude does not prove that planters as a class opposed Birmingham's creation or that the city should be regarded as a departure from Old South precedents. Birmingham was just as much a product of prewar capitalist agribusiness as was Montgomery itself.

Luraghi's heavily negative analysis of southern economic development prior to the Civil War, *The Rise and Fall of the Plantation South*, also fails to explain how antebellum aspirations underlay a postwar alternative to northern-style industrialization that made sense in its time, however regressive it may seem from a twentieth-century perspective. Typical of Luraghi's viewpoint is his statement that "industrialization requires enormous capital investment."[10] This need was not as clear at the time of Birmingham's founding as it is today. The city resulted from a distinctively southern strategy of industrial development that avoided huge capital outlays as much as possible by taking advantage of other regional assets.

To understand why such persons as Bryan or the Broad River associates became involved in city planning, railroading, coal mining, and iron manufacturing, one needs only to see the world through their eyes. To them it made sense to believe that the South could pursue a developmental strategy much different from that which was being followed above the Mason-Dixon Line but still compete on roughly equal terms with northern enterprise while preserving traditional southern values. Servile workers, labor-intensive techniques, and abundant raw materials in a state of close proximity offered a formula for success at realistic levels of funding that would not compromise southern control.

Conditions in mining and ferrous metals at the time that Birmingham was conceived were congruent with such expectations. In the 1850s, when Gilmer and Milner began to dream about a great iron manufacturing center in Jones Valley, the revolution brought about by the use of anthracite in iron smelting and refining was still in its early stages. Railroads had just begun to tap the vast coal deposits located west of the Allegheny Mountains. In many parts of the North, blast furnaces remained predominantly rural installations. Slave-operated iron plantations were common along Virginia's Blue Ridge, and rolling mills and foundries had long employed black workers successfully in Richmond, most notably at the Tredegar Iron Works. Slave labor was profitably utilized in a host of other industries. Even after the Civil War, when slave labor could no longer be used, prewar experience seemed to justify confidence on the part of Milner and other entrepreneurs that southern managers would know how to handle freedmen if the region could regain control of its political and social institutions under white supremacist leaders. This condition was fulfilled in 1874 with the election of a Redeemer government headed by Colonel Sloss's ally, George S. Houston.

During the early postwar decades, the techniques by means of which ferrous metals were produced did not change so rapidly as to make the visions of Milner, DeBardeleben, and Sloss seem unrealistic at the time. The scale of furnaces, stoves, and other up-to-date equipment for making pig iron was still sufficiently small, even in many parts of the North, for southern industrialists to believe that they could afford to build and operate competitive installations with limited help from trusted northern and foreign investors. Steel was used only for rails, and wrought and cast iron were still the main ferrous metal products. Although the tempo of change was accelerating in western Pennsylvania and Ohio's Mahoning Valley, Alabama ironmasters were not really trying to compete with those places. Instead, they possessed potentially lucrative markets much closer to home in the lower Ohio Valley. They could also undersell outmoded anthracite-burning furnaces in eastern Pennsylvania that were technologically inferior to new southern installations. If the South lacked large numbers of highly trained engineers and technicians, it had a long tradition of successfully attracting such persons from outside and could afford to pay them high wages because other costs were low. This was particularly true because the use of a labor-intensive strategy would not require a high degree of mechanization. Breakthroughs that made it possible to produce steel in large quantities had only recently been transferred from Europe to the United States. Southern promoters had no doubt that ways could be found to adapt Bessemer and open-hearth methods to regional raw materials. What later became known as the Duquesne Revolution was only beginning. For all these reasons, it was not hard to believe that the South could produce ferrous metals in its own distinctive way, at a good profit.

As Luraghi has indicated, the South lacked home markets for industrial goods.[11] But southern wealth had always been based on exports of such basic commodities as cotton and tobacco. Coal and iron ore could be shipped outside the region as raw materials, but converting them into coke, pig iron, and other ferrous metal products would command a better price. If steel could be made from southern raw materials, the rewards would become even more lucrative.

Luraghi has aptly described the type of economy that emerged in the South as "seigneurial."[12] From the outset, the rewards of industrialization were intended to accrue to aristocrats like Bryan and the wealthy Alabama planters who organized the Elyton Land Company. But this is simply the way it had always been, again underscoring the degree to which the New South was rooted in the Old. Before the war, big planters had made money by shipping agricultural staples outside the region. After Appomattox, a relatively few investors would profit by making pig iron—a semifinished commodity analogous to baled cotton and plug tobacco—for mostly outside consumption while the great mass of southerners—both black and white—gained little but a bare subsistence.

During the 1890s, upstate industrialists and downstate planters reconfirmed this approach in Alabama by crushing a populist revolt that, if it had been successful, might have redistributed income across racial lines in such a way as to create large home markets for consumer goods. Instead, when a depression caused demand to drop in midwestern and northeastern markets that had previously absorbed most southern iron, industrial leaders did exactly what Joseph Anderson had done in the antebellum era at his Tredegar works in Richmond: they developed a foreign export trade and sold much of their remaining output to the United States Navy to be used in building such warships as the *Texas* and the *Oregon*.

Japanese and German entrepreneurs were pursuing very much the same strategy at the turn of the century as they also forged alliances with military leaders and large landowners. Wiener may be at least superficially correct, therefore, in saying that Alabama followed a "Prussian Road" to industrialization. Again, however, it is not true that planters waited until the 1890s to join forces with Birmingham's Big Mules. Agriculturalists and industrialists had been marching along the same path since antebellum times, pursuing what Wiener's mentor, Barrington Moore, has aptly called "conservative modernization." To the extent that Alabama and Prussia were similar, the "marriage of iron and rye" that took place in nineteenth-century Germany paralleled a long-established union of entrepreneurial interests in the Deep South.[13]

II

By stressing the profitability of antebellum slave-operated plantations and industries, scholars including Stanley Engerman, Robert Fogel, and Robert S. Starobin have made it possible to see why Bryan and other southern industrialists were hopeful about the prospective rewards to be gained from using a servile, ill-rewarded, and predominantly black work force. Nonetheless, the evidence presented by these historians must be used with care if one is to understand why hopes that Jones Valley would become the world's dominant source of iron and steel did not materialize. It is possible that legal constraints surrounding chattel slavery permitted a degree of social control that helped make plantation agriculture and industrial slavery profitable. Even this hypothesis, however, is moot. In a searching essay, the late Herbert G. Gutman argued perceptively that slavery entailed severe disincentives that were incompatible with efficient, productive labor. If Gutman was correct, southern agriculture and manufacturing were profitable before the war not because of slavery, but in spite of it.[14]

After the war, without the legal sanctions of slavery, southern managers had an even harder time getting dependable work from disadvantaged black operatives. As Gutman indicated, it was difficult everywhere to discipline workers who were unaccustomed to an industrial regime. But the task may have been even more formidable in the South where disincentives were particularly severe because of the combination of a low-wage strategy and the suppression of black people as a matter of deliberate political, economic, and social policy. The seriousness of the labor problems facing southern managers is indicated by Colonel Sloss's testimony to senate investigators in the 1880s about the extreme difficulty of keeping his facilities adequately staffed, a situation that required him to hire twice as many operatives as he would otherwise have needed. Also pointing in the same direction are Uehling's mordant remarks, both in *Cassier's Magazine* and later in his autobiography, about the poor quality of work that was done in southern casting sheds under his direction when he was furnace superintendent for the Sloss Iron and Steel Company in the early 1890s. Thomas Culverhouse experienced similar difficulties in the same decade while managing the mining camps that Seddon had established at Brookside, Cardiff, and other Jefferson County locations. Indeed, it was partly because of the unreliability of black operatives that Seddon resorted to hiring immigrants. After TCI embraced steel production, Don Bacon and Frank Crockard encountered the same problems in unsuccessful attempts to match northern standards of efficiency. Even the expensive program of welfare capitalism that was instituted at TCI by George Gordon Crawford could not bring absenteeism down to levels prevailing in other parts of the country.

This accumulation of evidence indicates that the severe disincentives suffered by Birmingham's black work force, doubly disadvantaged by low wages and racism, were ultimately counterproductive to the welfare of the district. The resulting bitter harvest helped frustrate naive dreams that the Magic City would become the steelmaking capital of the world. Elbert Gary's failure to capitalize more effectively on the natural advantages offered by Jones Valley was at least in part a rational response to the district's incapacity to provide the levels of industrial discipline appropriate to basic steel production, in which worker regularity and dependability were essential for high-volume, standardized operations. This condition alone would have made it hard for Gary to justify the large investments that would have been necessary to fulfill Birmingham's hopes for greatness, even had U.S. Steel been less concerned about industry-wide stability, more inclined to innovate, and less afraid of federal trust-busting.

The double yoke of the low-wage system and severe racial discrimination also produced such pronounced inequities in income distribution that the Birmingham District, like the South as a whole, could not develop the consumer demand that might have permitted a thriving economy to emerge. As a careful study by H. H. Chapman and other economists at the University of Alabama concluded, this lack of broad-based purchasing power acted as a brake on the expansion of the southern steel industry by preventing the growth of mass markets for durable goods. Eugene D. Genovese, Raimondo Luraghi, Robert J. Norrell, and Gavin Wright have reached similar conclusions.[15] Here, again, one can observe the long-range adverse impact of economic strategies inherited from the Old South, where plantation owners reaped profits while slaves toiled merely to survive.

Nevertheless, one should be wary of imposing twentieth-century thinking upon nineteenth-century entrepreneurs who faced their own peculiar set of circumstances. However unfortunate a southern strategy based on low wages and labor-intensive methods may have been in the long run, it made sense at the time Birmingham was founded. The Magic City was born in a bitterly competitive age of unregulated economic activity when costs had to be cut to the bone. A dog-eat-dog attitude helped the community survive. Southern capital could not support intensive mechanization. There was a vast pool of cheap black labor available in rural districts where poverty was so abject that even life in Jefferson County's squalid mining camps and urban slums seemed preferable. Although slavery was no longer possible, antebellum experience encouraged the use of black industrial labor. Black males were thought particularly suitable for jobs requiring brute strength. The scale of casting and other operations was such that manual labor could keep up with furnace output, albeit with increasing difficulty. Such selective mechanization as was necessary could be financed without undue strain. Even in the 1890s, southern companies could ignore costly labor-saving innovations of the type patented by Uehling. Instead, they expanded by adopting relatively less expensive techniques that included washing coal and mixing red hematite with strip-mined brown ore. Primary emphasis continued to be placed on cost-cutting as the key to meeting northern competition. Meanwhile, the traditional labor system remained undisturbed without evident penalty to the district's vital interests.

Birmingham's leaders did not realize that high-volume steel production would create conditions for which the district's long-established labor strategy would be ill suited. They did not foresee how swiftly cutthroat competition would give way to trade associations, government regulation, and the drive for industrial stability that followed establishment of control over railroads and steel production by J. P. Morgan and other northern bankers. Nor did boosters envision how Pittsburgh Plus and other ways of mitigating destructive competition would undermine basic assumptions on which the district had pinned its hopes in an age of savage industrial strife, producing an increasingly stunted economy in the twentieth century. Neither civic officials nor industrial operators anticipated how the cruelly inequitable social system that resulted from a combination of racism and the low-wage strategy would cause Birmingham to lurch toward the tragic climax that took place at Kelly Ingram Park in 1963.

Because its hopes were heavily invested in the fantasy of salvation through steel, Birmingham carried its own special burden of frustration as it became increasingly obvious that Golconda would never materialize. Expectations of industrial glory were still flourishing even as late as the 1920s, but from the Great Depression onward it was clear that the district had been the victim of a grand illusion. Growing consciousness that steel had not been the panacea about which boosters had dreamed could not help but exacerbate the racial tensions that plagued the district. Over the years, particularly after population growth began to slow down by about 1950, it was especially hard to live in the shadow of a nearby metropolis that had won the wealth and power for which Birmingham had fought a losing battle. Angry charges were made in the Magic City, and continue to be heard there to this very day, that Atlanta had used unfair tactics to become the South's greatest urban success story. But Birmingham was responsible for its own problems. Atlanta had not pinned its future on the glamour alloy for which Jones Valley's human and natural resources had proved to be so ill suited, and the Magic City had no legitimate reason to complain that its great Georgia rival had followed superior strategies for long-range growth.

Had it not been for cast iron, a commodity for the production of which the raw materials available in the Birmingham District were particularly well fitted, things would have been much worse for the Magic City. Following the traditional southern approach to industrialization continued to make sense, even after World War II, if owners and managers did not try to enter markets for which the district's raw materials and human resources were inappropriate. Sloss-Sheffield had made a wise move in 1902 by deciding to specialize in foundry pig iron, a commodity made in relatively small batches for customers that continued to prefer a sand-cast product. Among its other beneficial results, this strategy spared the company from many of the labor-related problems with which TCI had been forced to reckon as a result of its involvement in high-volume steelmaking. By remaining in a traditional field of manufacturing that was free from northern freight-fixing and congruent with the peculiarities of southern ores, Sloss-Sheffield continued to be a successful enterprise despite its regressive labor practices. It was able to delay the adoption of automatic casting and other forms of mechanization, such as the use of skip hoists, until the late 1920s. Even then, it adopted advanced techniques only as part of a larger package aimed at reducing costs by better utilizing raw materials and overcoming the inefficient performance of unmotivated workers. Significantly, Woodward, a much more profitable firm than Sloss-Sheffield, waited until the same time to switch to automatic pig-casting, even though its superior earnings had enabled it to adopt by-product coke manufacture a great deal more quickly than the firm headed by McQueen and Morrow. The path followed by both companies demonstrated that mechanized casting was not necessary to their characteristically southern style of production because they remained committed to pig iron and refrained from trying to make steel, for which northern conditions were better suited.

III

By focusing on the southern low-wage system, Gavin Wright has made a valuable contribution to understanding some of the problems the region faced as a result of the strategies that its industrial leaders adopted to meet outside competition. Wright's contention that "the absence or small size of the indigenous regional technological community" retarded southern industrial growth, however, requires reexamination.

Birmingham suffered greatly in its early years from a chronic shortage of skilled workers. The "coke famine" about which local newspapers complained constantly in the 1880s could be traced in large part to the fact that too many construction projects were underway at the same time for the district's immature infrastructure to handle. Seddon voiced a typical cry when he said that he "had to start in a short time, a very large plant, literally in the woods, and it was impossible to get trained men. . . . We had to take green men and teach them the business." The "shabby workmanship" about which Uehling complained in his autobiography when he discussed the condition of the North Birmingham furnaces upon his arrival in Jones Valley was probably traceable, at least in part, to the same circumstances.

But any emerging frontier was bound to experience similar problems, regardless of where it might happen to be located. What must be recognized, in evaluating Wright's argument, is that implementation of the southern industrial strategy in the late nineteenth century did not require the type of technological infrastructure that was necessary in the North, where there was sufficient capital for intensive mechanization and where conditions were ideal for a vast scale of standardized operations for which southern conditions were unsuited. Because of the low cost of their raw materials and their cheap labor supply, southern industrialists could afford to follow a policy that was already well established in the antebellum period, that of hiring high-paid foreign and northern engineers and technicians. Except in times when mushrooming growth temporarily outstripped their needs, the supply of such persons seems to have been sufficient to meet their requirements.

Alabama's industrial development might have been accelerated, both in the antebellum and postwar periods, had more native southerners possessed Milner's yen for engineering or had the legislature provided more support to university officials in Tuscaloosa for developing an engineering curriculum. But the state's failure to build more railroads and manufacturing installations in the prewar years was due more to the vast extent of its natural waterways, the superior rewards available in agriculture, and Jacksonian opposition to large enterprises than to a lack of engineers and other technical personnel. Actually, the state had a great deal of success in attracting skilled workers from Great Britain and the northern states. After the war, the number of foreign and northern engineers and technicians in the Birmingham District must have been at least as great as that of southern-born stockbrokers on Wall Street. The railroad and the oceangoing steamship made it easy for persons with technological training and skills to come to the South. Many made the trip.

The manner in which the production of ferrous metals took place in Alabama supports Jack R. Bergstresser's observation that the state was part of a much larger technological community within which people and knowledge circulated freely. Outside technical help was always available. For this reason, and considering other circumstances that were more important in preventing the South from adopting a northern model of industrialization, it is not clear that the region suffered more than the other major underdeveloped part of the nation—the West—from a lack of outside engineering and technical assistance. As Bergstresser indicates, the methods and equipment that were utilized in Alabama during most of the nineteenth century were quite up-to-date for a part of the country that has often been stigmatized as technologically backward.

Because the concept of southern retardation is so pervasive and is applied to the region even when it is not fully appropriate, it is worthwhile to pursue Bergstresser's point in greater detail. An impressive number of northern iron-masters and engineers came to Alabama in the antebellum period. They included David Deshler, Joseph Heslip, Daniel Hillman, the Riddle brothers, Moses and Jacob Stroup, and Horace Ware. During the Civil War, a fresh cadre of northerners, including Josiah Gorgas and Wallace McElwain, contributed to the Confederate cause at such places as Selma and Irondale. In the late nineteenth and early twentieth centuries, Birmingham received a steady flow of engineering and technical assistance from northern emigrants Truman Aldrich, James P. Dovel, W. H. Hassinger, J. H. McCune, Erskine Ramsay, James Thomas, Edward Uehling, and Willard Warner. Yankee inventors, including James Henderson and Jacob Reese, often visited the Birmingham District. Northern design firms, such as Gordon, Stroebel and Laureau and J. P. Witherow and Company, also conducted a great deal of business there.

To the foregoing list must be added a large number of foreign-born engineers, technicians, and scientists who figured significantly in the development of Alabama's ferrous metals industry. These included John Adams, James Bowron, Giles Edwards, William L. Gould, Harry Hargreaves, Arthur Moxham, Samuel Noble, George Peacock, and Michael Tuomey. Foreign-based firms, such as the Whitwell Company, considered the South no more off the beaten track than any other part of the world. As Bergstresser has noted, the first Whitwell hot-blast stoves in the United States were installed at Rising Fawn, Georgia.

Wright's claims about the "absence or small size of the indigenous regional technological community" must also be reexamined in light of a significant number of engineers and technicians who came from within the region itself. Ethel Armes found that southerners had contributed heavily to Alabama's coal and iron industries. David Hanby, a machinist from Virginia who played a key role in early mining and ironmaking operations, was one of these native sons. In discussing the Selma arsenal, Armes mentioned several skilled workmen from various parts of the region, including Thomas S. Alvis, a furnaceman from Virginia who had worked at Tredegar and went to Briarfield after the war to assist Gorgas; Simon Gay, a gunmaker from the Old Dominion who had also practiced his craft at Tredegar and made ordnance in Chesterfield County; and George, John, and William Veitch, sons of a North Carolina ironmaster, who were "all brought up to the trade, and located in various sections throughout the Southern States." The "Veitch boys," as Armes called them, "were among the first furnacemen and foundrymen of the Birmingham and Sheffield districts." Their offspring were still active in Alabama's ferrous metal trades at the time Armes wrote her book. Levin S. Goodrich, who played a key role in rebuilding Oxmoor Furnace in the mid-1870s, also figured significantly in her narrative. A native of Greenup County, Kentucky, he learned ironmaking under his father and worked at furnaces in four states before coming to Birmingham after the war.[16]

John T. Milner, Sam Tate, and Campbell Wallace, all of whom were involved in building the South and North Railroad, were among other southern-born engineers and construction experts who contributed to Birmingham's development. William Henry Ruffner and Edward M. Tutwiler, both natives of Virginia, typified scientists and engineers brought to Alabama by the development of the Georgia Pacific Railway. John David Hanby, descended from a line of Virginians who had played key roles in Alabama's early history, was one of several southerners, including Thomas Culverhouse, who worked as mine superintendents for the Sloss Iron and Steel Company in the late nineteenth century. George Gordon Crawford, a southern-born engineer trained at Georgia Tech and in Germany, was perhaps the most significant figure in the entire history of TCI. Later, a chemist from South Carolina, Claude Lawson, became president of Sloss-Sheffield during its most prosperous years and climaxed his career by being chosen chief executive officer of U.S. Pipe and Foundry.

Other native southerners who formed part of Alabama's technological infrastructure might be identified through careful research in the records of such firms as the Shelby and Woodward iron companies. Additional names could be uncovered in census materials and regional trade publications. An abundance of material in one journal alone, the *Manufacturers' Record*, indicates that the South was richer in skilled human resources than accounts emphasizing regional backwardness might lead one to expect. With regard to such resources, a telling comment by Fogel and Engerman in their classic study, *Time on the Cross,* may be apt in a much broader context than the one to which they applied it. As they indicate, the South was not so much a poor part of the world prior to the Civil War as the North was a very wealthy one. "Far from being poverty-stricken, the South was quite rich by the standards of the antebellum era," they have stated. "If we treat the North and South as separate nations and rank them among the countries of the world, the South would stand as the fourth richest nation in the world in 1860."[17]

Without a wealth of skilled workers on which to draw, it is hard to see how the foundry industry could have become firmly established in the South in the late nineteenth and early twentieth centuries. Moldenke's articles and books reveal the high degree of skill that was required of persons who staffed pipe works and other installations where cast iron was molded into intricate shapes. Some of these workers, as Armes indicates, probably came to Alabama's postwar foundries fresh from a record of wartime experience gained at Selma; others may have come from such places as Richmond, home of Tredegar, or from Rome, Georgia, where the foundry trade was already well established in the antebellum era.

Once again, however, a great deal of technical help was available from outside the region. A large number of articles in *Pig Iron Rough Notes* indicate that emigrants from Great Britain—Hamilton T. Beggs, Robert Campbell, and William H. Martin, among others—played important roles in establishing the foundry trade in Alabama. Studies of Anniston also note that skilled British ironworkers were recruited to help operate the industrial plants of that city. Although Anniston's foreign-born inhabitants comprised only 2.7 percent of its population in 1900, compared to 6.3 percent in 1880, their actual numbers had increased considerably in the intervening two decades; during the 1880s, the population of the community grew by more than 26 percent per year. Skilled workmen from such places as England, which was Anniston's largest source of immigrants in 1880, may therefore have made a more significant contribution to the city's early development than mere percentages might appear to indicate.[18]

Such evidence, added to the presence of Peacock and other British technicians at Selma, indicates that a closer study of the transfer of skilled workers from Great Britain to the South could be highly worthwhile. Unfortunately, studies by Rowland Berthoff and Clifton K. Yearley of British immigration to America and the role played by Britons in American labor devote little attention to the South.[19] Because of the deep cultural ties binding that region to the British Isles, it was probably the part of the United States in which newcomers from England, Northern Ireland, Scotland, and Wales felt most at home. Various clues provide further grounds for believing that this line of research would be worth pursuing, including the strong connections that existed between Bryan and second-generation British merchants in New York, such as Goadby and Mortimer; the roles played by Bowron and other Britons in Alabama's industrial development; and the way in which British emigrants flooded into such places as the Sequatchie Valley in east Tennessee.

A close study of the development of the southern foundry trade, of the workers that it attracted, and of the sources from which they came might also test whether the southern low-wage system was imposed on all sectors of Alabama's ferrous metals industry or whether it was applied more selectively. Such a study might investigate whether or not foundry workers demanded higher wages than those earned by operatives at furnaces, mines, textile factories, and tobacco plants. Berthoff and Yearley stress the independent-mindedness of skilled British immigrants and their tendency to demand high wages. To the extent that such workers were drawn into the southern foundry trade, they may have constituted a relatively privileged class. Because blast furnace managers could draw on a vast pool of unskilled workers who flooded into the Birmingham District from impoverished rural areas, they could easily tell complaining employees, "Don't come back, there's a barefoot man at the gate waiting to be hired."[20] Foundry owners, who needed skilled personnel, probably could not take such an attitude. This situation may help explain why such foundry owners as ACIPCO's John Eagan adopted innovative employment practices.

In this connection, more research should be devoted to the social and economic development of southern communities that figured importantly in the growth of the foundry trade. Anniston, chief bastion of the soil pipe industry, has been studied in an illuminating way by two historians, Grace Hooten Gates and Samuel Noble McCaa. Their findings reveal managerial attitudes toward labor that were quite different from those of Birmingham's blast furnace and mine operators. As Gates indicates, relations between owners and workers in Anniston were generally good despite occasional strikes. Because he depended so heavily on skilled workmen, it is significant that Anniston's founder, Samuel Noble, did not disdain accepting honorary membership in a labor union, the United States Association of Charcoal Iron Workers. In 1885, he invited its secretary, John Birkinshire, to visit his furnaces. Recognition of the type of labor on which he depended may also have been part of the reason Noble believed "that an industrial community like Anniston should have the best schools possible," resulting in his establishment of educational institutions for both boys and girls. Such social responsibility on the part of a leading Alabama industrialist provides a clue that important differences existed between skilled labor towns and places where low-paid unskilled workers predominated.

The contrast between Anniston and Birmingham also indicates that different circumstances attending the birth of individual towns and cities in Alabama had a significant long-range impact on race relations and other aspects of social life in separate industrial settings. Anniston was founded by a Cornish emigrant family and was based largely on northern capital; Birmingham was created by southern landholders who came from a slaveholding background and was based partly on southern funding. Anniston was conceived as a model community; Birmingham, as an urban iron plantation. African-Americans were not employed at Anniston's early charcoal furnaces. Blacks became temporarily numerous at some of the city's foundries in the mid-1890s, but by the end of the nineteenth century more than 93 percent of the workers in the local iron industry were white. Despite lacking nearby coal deposits, Anniston had convenient access to brown ore. As they prospected in search of the latter commodity, however, Anniston's business leaders did not build a network of rough-and-tumble mining camps like those dotting the Jefferson County hills, where crews composed mainly of black convicts lived behind stockades and labored in dangerous underground occupations.

As a result of such contrasting circumstances, Anniston and Birmingham developed along different lines despite sharing undeniable similarities. African-Americans, who formed about 30 percent of Anniston's population in 1880 and 40 percent in 1900, were kept under watchful control by white citizens who regarded them as inferior and employed them mainly as unskilled laborers and domestic servants. Residential and other forms of segregation, which were not practiced early in Anniston's history, became common there by the late 1890s, as was increasingly true throughout the South as a whole. Nevertheless, a survey by Gates of social life in Anniston from 1872 to 1900 provides a stark contrast to the climate of fear and malaise pervading Birmingham during much of the same period. The sole episode of racial violence in Anniston, a riot between black and white troops stationed at an army camp that was established near the site of the Hercules Pipe Company during the Spanish-American War, did not originate locally. Throughout most of the city's early development, in Gates's words, "no serious racial incidents marred the comparatively tranquil scene," and whites often displayed "compassion" and "sympathetic concern" toward the black minority. Meanwhile, black people seemed disposed "to cooperate with the political decisions of the white leaders while participating to a minor degree in the process and to adjust to the rules of a predominantly white community while striving to build both a business and a social identity of their own." These findings by Gates, a careful historian, indicate that additional studies might fruitfully be made of significant differences between places in Alabama that pursued contrasting strategies of industrialization under varying styles of leadership.[21]

IV

A serious flaw in most discussions of the South's postwar economic development—at least as far as ferrous metals are concerned—is their failure to deal in adequate depth with problems involving raw materials. It is impossible to understand how the Birmingham District differed from other iron- and steel-producing regions of the United States without understanding how specific difficulties created by the peculiar nature of local natural resources affected the design and functioning of blast furnaces and other industrial installations. From the beginning, adaptation to Jones Valley's unique geological endowments was the key to success or failure for enterprises that were established there. The question was not so much whether considerable numbers of foreign, northern, and southern engineers and technicians could be attracted to the Birmingham District, for they obviously were. It was whether they were capable of dealing with problems that arose involving the performance of furnaces that might have functioned satisfactorily elsewhere but did not perform well in north-central Alabama because of lean raw materials with unusual chemical characteristics.

In some cases, as the troubled history of the early furnaces at North Birmingham shows, some problems were indeed traceable to the immaturity of the local infrastructure. It is also likely that even such capable outside contractors as James P. Witherow were accepting too many orders at one time, amid the frenzied boom of the mid-1880s, to do their best work. For the most part, however, even highly seasoned engineers and technicians did not appreciate the degree to which regional raw materials differed from those in other areas, or understand how such differences would affect the performance of the equipment they installed. *Iron Age* put it best by stating that "constructing Southern furnaces without regard to the difference in fuel and ores from those worked by the Northern furnaces, whose lines have been copied, has been . . . the great error of Southern iron-makers, and accounts for the irregularity in many cases of output, both in quantity and quality, as well as the short life of the furnaces in this and other Southern territory." Bryan's inability to advise his brother, St. George T. C. Bryan, how to cope with the abrasive effect of local raw materials on the inner walls of the City Furnaces or to deal with the excessively high quantities of slag that they produced typified perplexities that took many years to solve.

Bergstresser has done more than any other scholar to show how the Birmingham District's lean raw materials, specific features of their chemical makeup, and other peculiar local conditions resulted in important long-range problems. Industrialists were forced to cope with these difficulties as they gradually developed a distinctive style of southern blast furnace technology.[22] As Bergstresser has shown, it is misleading and unfair to look on southern ironmakers as being backward by comparison with their northern counterparts. Instead, they should be admired for having made a creative response to difficulties inherent in regional conditions. Southern furnace practice was a world unto itself. Long before the first runs of coke-smelted iron heralded the success of the Oxmoor experiment in 1876, the geological evolution of the Birmingham District had created a strange mixture of opportunities and headaches for its future promoters. Only by maximizing the benefits of proximity in such a way as to compensate for a multitude of problems could local entrepreneurs hope to succeed.

Scholars have long recognized how high percentages of phosphorus and silica in southern raw materials frustrated early attempts at steelmaking in the Birmingham District. But they have not realized that problems connected with a far less complex task, that of making pig iron, long prevented Sloss Furnaces and other local enterprises from achieving satisfactory output. Until the late 1890s, these problems forced such executives as Seddon to rely on selling coal until the learning curve had reached a point at which ironmaking could yield dependable profits. The scaffolding and thick, lava-like slag that baffled such furnace superintendents as St. George T. C. Bryan typified problems indigenous to the district. As Bergstresser has emphasized, the quality of red ore could vary considerably, depending upon the specific mines from which it was taken or the depths from which it was retrieved. Ultimately, it was necessary to mix red hematite with brown ore from outside the district before satisfactory results could be achieved. This necessity underscores how crucial it was for Joseph Bryan and his partners to acquire the Cahaba and Russellville ore beds, which not only contained larger percentages of metallic iron than Red Mountain deposits but could also be hydraulically strip-mined.

Raw materials had a direct bearing on the labor policies practiced in the Birmingham District. Because local natural resources made it impossible to build blast furnaces as large as those that were erected in such places as western Pennsylvania, levels of output did not reach a point at which it was imperative to use mechanized methods to clear casting floors of the iron resulting from one melt before the next was ready to be poured. Alabama furnace operators could therefore retain labor-intensive casting methods much longer than could operators in the North. The decisions made by Sloss-Sheffield and Woodward to specialize in foundry pig iron, due particularly to the high phosphorus content of local ores, intensified differences between the techniques that they employed and the methods used by northern firms that made pig iron for steel mills. The resulting contrast does not indicate that either Sloss-Sheffield or Woodward was technologically backward because they did not adopt mechanized methods that became common in the North. The Duquesne Revolution was simply irrelevant to them until the 1920s. It was part of a prodigiously expensive northern style of manufacturing that brought high-grade Lake Superior ore to chemically superior Connellsville coking coal so that extremely wealthy owners could produce standardized ferrous metals in heavily mechanized facilities for mass markets that simply were not available to southern merchant operators.

Even after the foundry trade began to move south, opening up important new outlets in the process, merchant pig iron producers had no incentive to copy northern techniques. To begin with, the mechanical strength of Warrior coal would not permit either hard driving or the large furnaces required for this practice. Even had this not been true, markets for foundry pig iron were not sufficiently standardized to make production in large batches feasible. Foundries required a wide range of chemical specifications, depending on whether the pig iron that they purchased was to be used for pipes, stove iron, enamel-coated appliances, or engine components. As previously noted, Clarence Mason, who superintended furnaces at Sloss-Sheffield for many years, compared meeting orders from specific customers to custom-filling prescriptions at a pharmacy. Huge, heavily mechanized furnaces were uneconomical in such circumstances because managers had to shift back and forth from one heat to another, depending upon the varying characteristics of individual melts. According to Mason, such realities produced serious problems after the enormous No. 5 furnace in North Birmingham was built. This situation helps explain why the old City Furnaces were not more quickly phased out.[23]

In one critical sector of technological development, the cleaning of raw materials, the Birmingham District actually played a pioneering role because it was forced into it. In this case, necessity was truly the mother of invention. Faced with coal and ore containing a large number of unwanted chemical constituents, local companies had to innovate, and did. Erskine Ramsay's cone washer, Elwood Stewart's jig washer, and James P. Dovel's gas washer were benchmarks in a chain of devices that made north-central Alabama the focal point of experiments that attracted attention throughout the industrial world; Stewart's apparatus was copied in western Pennsylvania, and Dovel's equipment won him international fame. Indeed, the reason that Stewart came from Ohio to Alabama was because he could see that the state was the best possible place to test his ideas. A distinctive location had produced an equally distinctive technological tradition.

V

Technological change occurs within a matrix of political, economic, social, and cultural forces. This fact is borne out by the contrast between the careers of the two most outstanding engineers who were involved in the development of Sloss Furnaces: Edward A. Uehling and James Pickering Dovel.

Uehling's career illustrates the contrast between the forces that affected the development of northern and southern blast furnace technique. After launching a promising career in the North, he was attracted to Birmingham by the high wages that were available in the South for northern and foreign engineers. Once he had arrived in the Magic City, he was repelled by the onerous labor demanded of low-paid, severely unmotivated black workers in its casting sheds. The heavily mechanized alternative that Uehling devised, however, was a typically northern method that had no chance of adoption in the South during the 1890s, even though Bryan admired his work and supported other steps that he took to modify the Sloss Iron and Steel Company's equipment.

In his adversarial relationship with Seddon, Uehling is reminiscent of John Hall, a northern inventor whose antebellum career has been searchingly examined by Merritt Roe Smith. Against the wishes of James Stubblefield, a southern aristocrat who directed the federal armory at Harpers Ferry, Virginia, Hall introduced techniques of interchangeable parts manufacture in the assembly of firearms. Craft-oriented gunsmiths joined Stubblefield in opposing Hall's innovations, which demanded standardized equipment that was antithetical to traditional ways of producing rifles. Hall succeeded in getting his way because he had the backing of powerful officials in the ordnance department of the United States Army who were actively committed to promoting changes that were congruent with military needs. In addition, funding for the equipment required was available from the federal treasury.[24] Uehling had neither of these advantages.

Hall was a New Englander, and his Yankee ways grated on Stubblefield's southern sensibilities. Similarly, Uehling's midwestern background may have lessened his chances of getting along with such southerners as Rucker and Seddon. But many northerners adapted to southern life without evident difficulty. Significantly, J. H. McCune, whom Seddon and his friends picked to replace Uehling as furnace superintendent, was a Union veteran who had fought at Gettysburg with a Pennsylvania regiment. Furthermore, Bryan and other Richmond investors held Uehling in great respect and continued to use him as a consultant after he had resigned under pressure from Seddon. The main reason Uehling had no future in Alabama was that his casting apparatus was much too expensive for southern pocketbooks, was not needed to achieve southern standards of output, and was incompatible with the social and economic imperatives that helped perpetuate the southern commitment to a servile black work force using labor-intensive methods. Consequently, he had to leave the South for places where high-volume production made better economic sense and social conditions were not adverse to his innovations. Both in western Pennsylvania and in central Europe, where his fluency in German was undoubtedly a major asset, Uehling and his inventions were outstandingly successful.

Uehling was well educated and had an engineering degree from Stevens Institute of Technology. Dovel had little formal education and became an engineer through on-the-job training. After Dovel was employed by Sloss-Sheffield, this background may have helped him get along with such executives as James W. McQueen, who had also worked his way upward through the ranks in the Horatio Alger tradition. But Dovel also identified with the South, which Uehling did not. Dovel had a Virginia ancestry and came from a part of Ohio where southern traditions were strongly entrenched. This heritage probably contributed to his success in Alabama, to which he had moved in a spirit of homecoming.

Dovel, however, succeeded mainly because he was in the right place at the right time and because his efforts were congruent with southern needs and conditions. By the time his career reached its climax, Sloss-Sheffield had come under the control of a new group of owners whose roots were in the chemical and electrical industries and who were supportive of technical innovations. Conditions were prosperous in the 1920s, and money was not so much a problem as it had been in Bryan's day. Falling prices for pig iron had resulted in an industry-wide shakeout that required even successful foundry-oriented producers like Sloss-Sheffield to become much more efficient in their operations. Most important of all, the lifelong thrust of Dovel's career was toward dealing creatively with southern raw materials that had long posed serious difficulties for furnace operators in the Birmingham District. Because of his ingenuity, he was able to bring a distinctively southern style of blast furnace practice to a climax and win worldwide recognition for his achievements without leaving the region, as Uehling had been forced to do.

VI

Among the most important considerations affecting the southern strategy of industrialization that emerged in the late nineteenth century was the fact that southern capital could not support the highly mechanized alternative that was developed in the North. This, plus a desire to retain control of the region's social institutions, played a major role in shaping the labor-intensive policies followed by Bryan and other southern entrepreneurs. In any case, northern capital was not readily available in the South for anything but railroads for many years. As the history of Sloss Furnaces demonstrates, some funding for southern manufacturing did come from trusted northern allies. Capital from abroad also helped finance such enterprises as the South Pittsburg complex in the Sequatchie Valley and the blast furnaces built by Sir Titus Salt and other British capitalists at Dayton, Tennessee. But most of the money for southern industry up to the end of the century came from within the region itself, as I. W. Avery noted when he protested to the *Manufacturers' Record* that Yankees were getting too much credit for their contributions to its economic development.

How pockets of southern investment capital survived the Civil War, sufficient to develop such enterprises as the ones that Bryan and his Virginia associates controlled, is a subject deserving additional study. Because Richmond was such a pocket, Virginia capital had an impact on the growth of north Alabama that has not previously been realized. Here again, the antebellum roots of postwar industrial development are important. It is likely that the links with trusted northern financiers that enabled Bryan and his associates to pursue their industrial activities in Alabama had already been forged before the Civil War by members of the Stewart family in their roles as middlemen in the international tobacco trade. The removal of funds to New York City by Daniel K. Stewart during the war and the subsequent return of the family fortune to Virginia after the conflict furnished much of the capital upon which Bryan drew as executor of the Stewart family estate. But this situation does not explain where other Virginians who joined forces with Bryan obtained the relatively large amounts of money and credit that were at the disposal of such persons as Christian, Dooley, Pace, and members of the Branch family. Presumably these investors also had northern and foreign contacts antedating the war, but they may also have had wealth of their own that they had somehow husbanded through the four-year conflict.

Nothing in the foregoing discussion is intended to belittle the role of northern capital in the developments discussed in this book. The availability of northern funding for southern railroads was in itself a crucial contribution to the region's postwar industrialization. As indicated by the railroad reservation that ran through the heart of the city, Birmingham could not have existed without the Iron Horse. The furnaces that Colonel Sloss built along the eastern edge of the city grew out of the alliance he had forged with the Louisville and Nashville, which had profited from federal largesse during the war. Similarly, Milton H. Smith of the L&N, who made a crucial contribution to the growth of the district by interlacing it with trackage, drew heavily on northern bankers like August Belmont and Jacob Schiff. Bryan and his associates depended heavily on Wall Street connections with such northern railroad promoters as Clyde and Oakman. Furthermore, railroad companies that leaned heavily on northern banking ultimately funneled money directly into industrial projects. The indispensable support given to TCI's venture into basic steel by one of J. P. Morgan's chief southern lieutenants, Samuel Spencer, is a good example of this funding.

Not all railroad leaders, northern or southern, pursued policies that were equally beneficial to industrial development. The history of Sloss Furnaces confirms the usefulness of the distinction that Maury Klein has made between speculative and developmental strategies in the growth of southern railroads.[25] The differential impact of such strategies on industrial firms was just as great as it was on transportation networks. Sloss Iron and Steel and Sloss-Sheffield benefited greatly from Bryan's strong commitment to a conservative developmental strategy, just as TCI suffered badly from Inman's penchant for speculation.

Klein has also made an important contribution to a better understanding of the way in which northern capital flowed southward into both railroads and industrial enterprises by stressing the importance of the roles played by southerners on Wall Street. This subject deserves more study than it has yet received. The way in which such persons as Maben and Seddon represented Bryan and the Stewarts in the New York financial district was of critical importance to the development of the enterprises that Virginians spearheaded in the Birmingham District. By serving as watchdogs for southern interests, these intermediaries played a major role not only in channeling funds into the region but also by trying to make sure that the South remained in control of its economic destiny, insofar as that was possible.

Because of family ties, strong cultural sympathies, and close antebellum connections stemming from the cotton and tobacco trades, southern capitalists found Wall Street investors with a British background to be particularly congenial allies. The Stewart family's Scottish roots were an important factor in the development of Sloss Furnaces. As executor of the estates amassed by John and Daniel K. Stewart, Bryan regularly distributed dividends to relatives in Scotland. Bryan's alliances with Goadby and Mortimer, whose parents had emigrated from Great Britain to New York City before the Civil War, probably stemmed from business ties between their families and the Stewarts. If so, this type of relationship provides further evidence of cultural and economic ties between southern entrepreneurs and British investors dating back to antebellum times. Bryan was an ardent anglophile. Given a choice between dealing with a Briton or a Yankee, he would certainly have chosen the former.

Did southern entrepreneurs deliberately seek British capital and technical assistance as a counterpoise to northern influence? This inference is plausible. DeBardeleben's financial association with David Roberts is only one of a number of relationships pointing in this direction; other manifestations of the same tendency include the heavy investments that Bowron's father and his associates made in the Sequatchie Valley, Thomas Whitwell's close southern ties, and the funding that Salt and his associates provided for the Dayton, Tennessee, blast furnaces. How funds flowed from Great Britain into industrial activities below the Mason-Dixon Line is a subject that needs to be studied in much more detail, not only for its economic importance but also for its cultural significance as an indication that southerners and Britons had a particular fondness for dealing with one another. [26]

Returning, however, to the theme with which this section began, the fact that southern investors had to lean on northern and foreign help in funding railroad systems, coal mines, and manufacturing facilities should not obscure the extent to which the South bore the cost of its own industrialization after the Civil War, particularly up to the end of the nineteenth century. The roles played by Bryan and other Virginians in the industrial empire that they created in north Alabama indicate that there was more money in the postwar South than traditional interpretations of regional growth might lead one to expect. One should remember once more the broader implications of Fogel and Engerman's remark that the fundamental difference between the North and the South in antebellum times was not so much that one was wealthy and the other poor, but instead that one was rich and the other not quite so rich.

Even with regard to the twentieth century, when the South came to be considered the nation's "number one economic problem," Fogel and Engerman's statement is still apt. One indication of this pertinence is the continuing importance of southern-born investors—Baruch, Catchings, McAdoo, and Sloan—in Sloss-Sheffield's development after World War I. The role played by southern capital and entrepreneurship in the tobacco and soft drink industries also supports this line of argument. So, too, does the way in which southern fortunes amassed in such enterprises as Coca-Cola and R. J. Reynolds contributed in turn to the emergence of corporate giants like Delta Air Lines. [27]

In this same context, it would be potentially instructive to study the sources of funding that built the southern foundry trade in the late nineteenth and twentieth centuries. Whether the pipe mills that proliferated in such places as Anniston and Birmingham grew largely out of southern wealth or depended upon large infusions of northern capital is beyond the scope of this book but remains an important question.

Despite the problems caused by Seddon's inexperience, the history of Sloss Furnaces and other industrial enterprises in the Birmingham District shows that the South drew upon a large supply of effective managers who came from within the region. This fact is not surprising considering the capacity for strong leadership that had been demonstrated by high-ranking Confederate officers during the Civil War. As shown by historians including William B. Hesseltine and Maury Klein, former military commanders—Edward Porter Alexander is an excellent case in point—turned to railroading and manufacturing after 1865 as a natural outlet for talents that had been honed in coordinating the movements of great masses of troops and attending to their logistical needs.[28] But the operation of antebellum plantations also required a considerable degree of managerial skill and should not be overlooked as a source of postwar capacity for guiding big enterprises through difficult times. Joseph Bryan, only a private in the war, drew in part upon his rearing at Eagle Point for the acumen he displayed at various junctures, including the crisis that confronted Sloss-Sheffield in 1902. The strong leadership given to TCI and Sloss-Sheffield by Crawford, McQueen, Morrow, and Lawson demonstrates that the capacity of the South to produce outstanding managers continued unabated after the generational transition that took place at the turn of the twentieth century.

Regional pride played an undeterminable yet unquestionably significant role in stimulating men like Bryan and John W. Johnston to develop such enterprises as the Georgia Pacific and Sloss Iron and Steel amid daunting circumstances. However much Bryan and his associates drew on the help of northern allies, they were quite consciously ex-Confederate nationalists who were passionately proud of their involvement in the Lost Cause and believed that their native region was capable of competing with northern industry on roughly equal terms. The events that gradually undermined this dream, beginning in the depression-ridden 1890s, should not obscure the depth of the group's commitment to the survival of southern power and influence in the enterprises over which they exerted control, even after the turn of the century. The persistence with which Bryan, Maben, and other Virginians clung to power in Sloss-Sheffield's boardroom up to 1917, and their refusal to expose themselves to the type of outside ownership that absorbed TCI, was partly a result of their determination to hold fast to what was left of the visions of regional competitiveness that had driven them in the 1880s. Hundreds of letters scattered throughout Bryan's correspondence leave no doubt that pecuniary motives alone are inadequate to explain the zeal with which southern industrialists tried to actualize the economic aspirations for the South that were enunciated by such persons as Henry W. Grady, Richard Hathaway Edmonds, and Robert E. Lee. Any remaining doubts about this will be dispelled by a close reading of Edmonds's journal, the *Manufacturers' Record*. The southern onslaught in the 1880s on northern markets that had previously been dominated by anthracite iron was not merely an episode in business competition, but a continuation of the Civil War by economic means.

Except for providing many indications that Bryan and his associates actively resisted the growth of what they regarded as undue northern influence, the history of Sloss Furnaces lends little support to C. Vann Woodward's argument that the South was colonized by the North during and after the 1890s. Indeed, the development of merchant blast furnaces and the foundry trade to which they catered is one of the best case studies that can be cited in opposition to this thesis. Sloss-Sheffield's decision to concentrate on foundry iron in 1902 was made by southerners in response to the southward movement of the industry, which was in turn traceable to the peculiarities of southern raw materials. Evidence in *Pig Iron Rough Notes* strongly suggests that most of the southern foundries served by Sloss-Sheffield were founded and financed by southern entrepreneurs. Even after the collapse of the Virginia dynasty, Sloss-Sheffield's destinies were controlled by yet another generation of southerners who had established themselves on Wall Street. Day-to-day administration of the enterprise was always entrusted to southern hands by the new owners, and local representation on the board of directors actually became stronger after 1917. Pittsburgh Plus and other discriminatory types of basing point pricing had no effect on the company because it stayed out of steel manufacturing. Distinctively southern labor practices were followed because the traditional orientation of the foundry trade made this possible. In the end, Birmingham even became the national headquarters of USP&F, tearing that firm away from its previous northern base. None of these facts indicates that Sloss-Sheffield was a "colonial" enterprise, even if its directors were headquartered in New York.

This is not to say, however, that the colonization theory is incorrect. What the history of Sloss Furnaces shows is that colonization could be avoided in specific cases by southern investors and managers who capitalized adroitly on the region's advantages and avoided costly technical solutions to problems that could be dealt with in less expensive and more traditional ways. A detailed study of the Woodward Iron Company, which was much more profitable than Sloss-Sheffield, might provide yet more evidence to support this argument. An enormous, little-used collection of its records at the University of Alabama would furnish the basis for a valuable, book-length monograph. By failing to heed the lessons implicit in Woodward's experience and significantly, by becoming fixated on steel, Birmingham's boosters played directly into the hands of northern interests and thereby provided C. Vann Woodward with one of the outstanding examples on which his colonization theory is based.[29]

VII

Developments recounted in this book are relevant to a number of themes that figure importantly in scholarly works on American urban development. One such theme is that of power relationships between younger and older cities. DeBardeleben's attempts to exploit divisions between backers from Cincinnati and Louisville so that he might maintain a balance of power on behalf of Birmingham's interests typified situations that later arose as the Magic City tried to manage its own internal affairs, usually with little or no success. The struggles that took place between Seddon, Rucker, and Sims on one hand,

and directors from Richmond and New York on the other, are also revealing in this regard. Further cases in point are the tensions that arose between TCI's local executives and absentee owners in the 1890s over the issue of basic steel production and the uproar that greeted Hopkins's dismissal in 1902 after Sloss-Sheffield's directors had decided that the firm should specialize in foundry iron. McQueen's later power struggle with John C. Maben, Jr., showed that battles pitting insiders against outsiders continued well into the twentieth century.

Because of a tendency among even some of the best books about postwar southern economic development to focus on northern dominance, it is important to emphasize that the outside forces against which Birmingham's business leaders struggled were often southern. Even when power was exerted from the North, it was often mediated through cities such as Richmond. This focus once more exemplifies the great extent to which southern capital and credit underlay at least the earliest phases of southern industrialization. Because TCI quickly became a football for Wall Street speculators, it probably felt the impact of northern interference more directly than did most of Birmingham's enterprises. In the case of the Sloss Iron and Steel Company, wrangling over issues involving local versus absentee control was largely a family affair between persons speaking with the same accents.

Boosterism, quintessentially important to this book, is another theme whose fruitfulness is abundantly demonstrated by the history of the Birmingham District. Throughout the late nineteenth century, boosters of the Magic City fought vigorously not only to aggrandize it at the expense of older Alabama cities, such as Mobile and Montgomery, but also to injure the interests of other centers of ferrous metal manufacture, wherever they might be. Targets included not only Hokendauqua, the main center of Pennsylvania's anthracite iron industry, but also rival southern cities like Chattanooga. Despite its positive aspects in building civic pride, boosterism was a deadly weapon in the merciless economic warfare that was fought between protagonists of various communities in the late nineteenth century. The cargo vessel *City of Birmingham*, launched in Chester, Pennsylvania, to beggar the fortunes of iron-producing cities in the same state, was almost as much a warship as the *Oregon* or the *Texas*.

As this book indicates, boosters often had distorted ideas about what was in the long-range best interests of the cities they promoted. Birmingham's fixation on steel as a panacea for the city's ills and an answer to all its hopes is a good example of such misguided thinking. Studies documenting other instances in which comparable growth strategies went awry might make valuable additions to urban history.

VIII

Whether outside pressures came from the North or from other parts of the South, the degree to which they constrained decision-making by Birmingham's business establishment must be taken into account in understanding the lack of elite involvement in local problems that was lamented by Ethel Armes and other critics. This type of pressure was particularly apparent in the early decades of the city's development, when it was only a pawn in a much larger game of entrepreneurial chess. Sheer dedication of local business leaders to self-interest, on which Armes blamed many of Birmingham's problems, is insufficient to explain why executives who lived in the community took so little apparent interest in its political, social, and cultural welfare. Birmingham was constantly engaged in a life-and-death struggle amid economic crises and fluctuations over which such men had little control. A few local business leaders, particularly Bowron, did try to make the city a better place in which to live, but most of his fellow executives seemed to believe that keeping enough revenue coming in to meet their payrolls was a sufficient contribution to civic welfare, especially given how tough an assignment such a role could be.

The problems associated with keeping Birmingham economically viable, however, do not absolve the city's business leaders of blame for its all-too-obvious deficiencies. Even after a relatively powerful managerial elite had emerged in Birmingham by the early twentieth century, most of its members took little interest in solving the municipality's glaring social problems. Business leaders imbued with Ward's progressive ideas, Eagan's commitment to the Social Gospel, or Crawford's zeal to implement welfare capitalism were atypical, as indicated by the chorus of hostile reactions to the 1911 *Survey* articles and the widespread denunciation that greeted Weatherly's drive to end smoke pollution. After the progressive impulse had waned, civic administration was abdicated to demagogues while members of the elite retreated to sanctuaries like Mountain Brook and Vestavia.

Meanwhile, most business leaders fought doggedly to put down union organizers, keep wages as low as possible, and preserve racial barriers that kept workers from making common cause against conditions that oppressed both blacks and whites. Vehement opposition to the New Deal brought these tendencies to a climax in the 1930s. As studies by such writers as Edward S. LaMonte, Horace Huntley, and Robert J. Norrell have indicated, industrialists shunned responsibility to provide relief during the Great Depression and deliberately fomented racial divisions during and after World War II.[30] Such was the long-range social legacy of the southern strategy of industrialization and the antebellum heritage in which it was rooted.

Even more than TCI, which bore the brunt of the criticisms leveled against Birmingham by such writers as Beiman and Leighton, Sloss-Sheffield epitomized the rigidly conservative attitudes toward social welfare that were deeply embedded in the southern strategy and the Old South tradition. Perhaps the most glaring example was Sloss-Sheffield's continued use of convict labor, long after TCI abandoned it. Repeated public exposure of terrible conditions at Coalburg and Flat Top produced little or no change as legislators in Montgomery made common cause with industrialists to save money for taxpayers. Only against unyielding opposition from such firms as Sloss-Sheffield was the use of convict labor abolished in the 1920s following a legalized murder that took place shortly after Morrow became president. James C. Cobb is fully justified in asserting that "there was no more damning evidence of . . . indifference to human suffering and exploitation on the part of many of the South's industrialists and public officials than the convict lease system."[31]

No less reprehensible were the methods that Sloss-Sheffield and other companies used against free workers. Owners and managers alike used heavy-handed tactics to break strikes and suppress unions. The anti-Semitism, red-baiting, beatings, and other forms of intimidation that marked the relentless struggle waged against the New Deal in the 1930s were merely later manifestations of tactics that had already been used for decades against strikers and labor organizers. In addition to the increasing poverty of the region, these forms of suppression were among the worst results of the southern low-wage strategy and the flinty determination of industrialists to perpetuate it. As Cobb has indicated, efforts of this sort continue to the present day. A leading southern university was still conducting seminars on union-busting in the late 1970s. Meanwhile, chambers of commerce remained bastions of opposition to organized labor. During World War II, a Danville, Virginia, worker quipped that he was "not worried a-tall about the Japs movin' in on us; the Chamber of Commerce has kept unions out, and I reckon as how they can keep the Japs out, too."[32] Today, the same organizations use right-to-work laws, low taxes, and other socially counterproductive inducements to lure the Japanese in.

An important goal of this book has been to help explain why events happening in Jones Valley exemplified racism at its worst and why Birmingham, rather than any other southern city, was the scene of the tragic events that took place at Kelly Ingram Park in 1963. As already noted, industrial strategies that concentrated severely disadvantaged black males in the district played a major role in this regard, heightening social and sexual fears of which W. J. Cash remains the best interpreter. Once again, however, it bears repeating that frustration of Birmingham's inflated ambitions to become the industrial center of the world, coupled with the stagnation that increasingly mocked such pretensions after World War II, also helped produce the racial debacles of the early 1960s. Technological and economic failure—the collapse of the dream of "salvation through steel"—was part of the psychological burden that Birmingham carried into the struggle against integration and part of the reason why the outcome was so damning to the city's reputation.

IX

Recent books and articles by Wayne Flynt, William W. Rogers, Richard A. Straw, and Robert D. Ward have helped focus attention on the conflict-wracked development of Alabama's coal industry.[33] The history of Sloss Furnaces shows that it is misleading to look at the succession of enterprises known by this name merely as producers of pig iron. Ethel Armes showed good judgment in entitling her classic book *The Story of Coal and Iron in Alabama*, giving both commodities equal billing. Until almost the end of the nineteenth century, Bryan's industrial empire gained most of its profits from selling coal. Similarly, during the Great Depression Sloss-Sheffield would have gone bankrupt but for the revenue it earned from by-product coke. Significantly, the firm modernized its mining operations in the 1920s before doing much to update its blast furnaces. In the end, coal, not iron, kept what was left of the enterprise in operation after it had been acquired by the Jim Walter Corporation.

The role played by coal in American industrial development has received much attention from scholars including Alfred D. Chandler, Priscilla Long, Charles Kenneth Sullivan, Anthony F. C. Wallace, and Clifton K. Yearley.[34] Little notice, however, has been paid to an extremely significant product with which this book has been much concerned: cast iron. Such inattention is mostly due to a fixation on steel. Without denying the importance of that crucial alloy, cast iron deserves much more recognition than historians have paid to it. Because of the continued usefulness of cast iron, long after the Bessemer and open-hearth methods made steel available in large quantities, the South became the heart of the American foundry trade, and Alabama became the world's largest producer of soil and pressure pipe. In the process, Sloss-Sheffield earned a starring role in the one area of ferrous metal manufacture in which the South ultimately became dominant.

Why has the importance of cast iron been largely ignored? Why is there no scholarly history of the American foundry trade? Why has the significance of the merchant blast furnace industry escaped attention? Why are such potential scholarly treasures as the collections of the Woodward Iron Company little consulted? Why the overweening attention to steel? Why is Alexander Lyman Holley's name a household word among historians of technology while few would recognize that of his analogue, Richard Moldenke?

At least part of the answer to such questions has been suggested by John M. Staudenmaier in a probing study, *Technology's Storytellers*. Staudenmaier emphasizes the need for historians of technology to escape what he calls "Whiggism." This term has many shades of meaning, but one common manifestation of the phenomenon about which Staudenmaier complains is studying technological developments without adequately exploring their social context. Another characteristic sign of Whiggism is a tendency to accept notions about industrial progress that equate it with herculean achievements in size and scale, particularly if a shift from hand labor to machinery happens to be involved. Still another mark of Whiggism is a preoccupation with success stories, in which "success" is defined as earning profits far beyond any previous levels, registering feats of productivity reaching previously undreamed-of heights, and otherwise meeting exceptional quantitative standards.[35]

To combat Whiggism, Staudenmaier has called for studies of "roads not taken," or "historical studies of technological failures." So insidious is Whiggism, however, that scholars may neglect the significance of roads that were in fact taken, ignoring them simply because they were traditionally focused and hence old-fashioned. Similarly, historians may fail to recognize successes that were in fact won, merely because they did not conform to Whiggish expectations.

Staudenmaier's argument helps explain why cast iron has played the role of a historiographical Cinderella while steel has received far more attention. Steelmaking, as practiced in the late nineteenth and early twentieth centuries, was a technology well calculated to delight the Whiggish imagination. Enormous increases of productivity, intensive mechanization, and extreme rationalization were its hallmarks. Carnegie's giant steel mill at Homestead, Pennsylvania, exemplified Whiggish values in a way that few installations could equal; only Ford's River Rouge works in Michigan comes as quickly to mind. It was because steelmaking so completely epitomized Whiggish values that it thoroughly captivated Birmingham's boosters, who were Whiggish to the core.

By contrast, the foundry trade was unexciting, even stodgy. It was steel's modern image that made Birmingham's boosters mourn to see a young and aggressive entrepreneur, Hopkins, depart for Evansville after he had failed to convince Sloss-Sheffield's directors to produce the wonder alloy. Conversely, it was cast iron's old-fashioned connotations that led reporters to give Maben a cold shoulder when he arrived from New York City to implement the board's 1902 decision to concentrate on foundries and their needs. The same symbolic resonances also help account for citizens' shunning a cast-iron statue that projected the wrong image of Birmingham's greatness because of the very material from which it was made. The glamorous associations of the word "steel" led Bryan and his successors to keep calling their enterprise the Sloss-Sheffield Steel and Iron Company despite the fact that it made only pig iron. Steel's prestige also caused reporters to make the mistake of referring to it in their efforts to puff Sloss-Sheffield and its leaders. This went so far that the writer of an obituary called Maben a "steel man" after his death in 1924, even though the fallen leader had been one of Sloss-Sheffield's strongest advocates of a foundry orientation in 1902.

I myself have witnessed the long-range impact of the gospel of salvation through steel at first hand. During the course of writing this book, I spoke to a service club about the history of Sloss Furnaces. I stressed repeatedly that the aging complex had never made steel and did my best to show that Birmingham's development was based not solely on that alloy but on its equally versatile cousin, cast iron. After the presentation, a member of the audience talked at length about growing up in Birmingham and being impressed by the fiery spectacle that occurred whenever Sloss Furnaces were tapped. "It was really dramatic to see them pouring steel," he said. So deeply had boosters impressed upon him and other residents of the city that steel was the only metal that really counted for much that this impression is apparently ineradicable, no matter how hard one tries to show otherwise.

X

Staudenmaier's strictures about Whiggism are characteristic of a post-industrial age in which the past is beginning to be seen in a new light. Sloss Furnaces National Historic Landmark stands as a monument to much that was ugly about the industrial syndrome of which the massive stacks and stoves once formed an integral part. They polluted Birmingham's atmosphere. Their owners subjected severely unmotivated black workers to labor more fit for dumb animals, as Uehling remarked; clung to a low-wage system that compounded southern problems by shrinking potential mass markets for consumer goods; resisted the abolition of convict labor long after enlightened opinion had seen it as unworthy of a civilized society; used heavy-handed tactics to thwart unionization; shirked social responsibility to the community to which they belonged; resorted to the worst sort of demagoguery and heavy-handed force in fighting the New Deal; and helped perpetuate racial divisions that caused Birmingham to become known as the Johannesburg of America. Nothing can prettify such aspects of the history of Sloss Furnaces. Persons who thought they saw images of Hell in the searing heat of the casting sheds as they passed them while driving along the First Avenue Viaduct were not mistaken.

Nevertheless, as was noted at the beginning of this book, a dynamic tension is inherent in all archetypal images, even those of Hell. Part of the legacy of Sloss Furnaces is positive, something that the postindustrial world can celebrate. Let us call it a heritage of effective entrepreneurial decision making. Paradoxically, despite the damage that they once inflicted on the local environment, Sloss Furnaces remind us that investing huge amounts of money in high-tech efforts to escape the limitations of natural resources, as TCI did, may be less wise than choosing relatively inexpensive ways of adapting to the realities of the world in which we live. Bryan and other directors who opposed Hopkins in 1902 knew nothing about "appropriate technology" but were unwittingly practicing it. Basing economic growth on the advantages of distinctive regional assets is a good idea for the postindustrial age.

XI

In the 1970s and 1980s, as Birmingham began to escape from its industrial past, it became a more hopeful place. One of Alabama's leading historians, Virginia Hamilton, commented that the city was imbued with a "new lightness of the heart."[36] The air became steadily more fit to breathe. People who would have once worked in the sweltering heat of unmechanized casting sheds were now employed by medical, educational, and professional firms. Born in an era when manufacturers preached the virtue of high tariffs shielding American enterprise from foreign competition, the Magic City was now part of a postindustrial world in which free trade was opening new international markets despite cries for a return to protectionism. Industrial concern for quantity gave ground to a postindustrial quest for excellence as Japanese automobiles began to outnumber American models on Birmingham's busy streets, and the Malcolm Baldridge Award suddenly became a coveted prize among managers who only recently had learned to wrestle with the theories of quality-control pioneer W. Edwards Deming.

Another postindustrial term, miniaturization, replaced bigness as a watchword in Birmingham office buildings where such executives as Claude Lawson had once measured progress by invoking the towering dimensions of new blast furnaces. Animated discussions about so many millions of tons of coal, iron, and steel shipped to this or that outside market were replaced by management seminars at South Central Bell about subjects involving megabytes of information on tiny computer chips. Children who at an earlier stage of the city's development might have worked ten hours a day at Comer's Avondale Mills presented performances of Humperdinck's *Hansel and Gretel* at the Alabama School of Fine Arts. White athletes who once would have played on segregated teams made a role model of Bo Jackson, who had grown up in poverty in Bessemer but was now a millionaire because of opportunities that had previously been closed to members of his race who may have been just as athletically talented. SONAT pumped money into the annual Vulcan Run, and corporations donated increasing amounts of money to cultural advancement. Some employers even went so far as to pursue high-wage growth strategies, despite criticism from Big Mules who continued to think that outside capital could best be lured into the area by right-to-work laws, low levels of social services and other survivals of the industrial spirit.

"Our philosophy rests on two overriding values, quality and growth, which by tradition and choice transcend all others," stated Protective Life's chief executive officer, William Rushton, who had obviously heard of Deming and recognized that something counted besides quantity. "We are dedicated to creating value for our customers, reward and fulfillment for our people, and profit for our owners."[37] The first, and particularly the third, of these objectives would have been quite understandable to the former managers of such firms as Sloss-Sheffield. The second would have been incomprehensible to them.[38]

In the 1970s, despite ominous developments in the Middle East that spelled the end of an era based on cheap energy, it was still easy to believe that the future was bright with promise. "The hardest thing to find in the South these days is a pessimist," said a distinguished sociologist, B. Eugene Griessman, at a conference held at Auburn University in April 1975. "Most southerners are not willing to admit that the old problems—poverty, race, demagoguery, climate, xenophobia, powerlessness, and intolerance—can keep the South from rising." No booster himself, Griessman found many reasons to doubt that such optimism was justified. "We shush our young people, who would bring some fresh ideas and honesty into government and business," he said. "We have kicked out teachers and ministers and professors who sounded a bit liberal—or, God forbid, radical." The title of Griessman's presentation—"Will the South Rise Again or Just Roll Over?"—exemplified his skepticism.[39]

At the time Griessman spoke, southerners were not alone in taking a bullish view of future prospects. In a best-selling book, *Future Shock*, Alvin Toffler anticipated the "Death of Technocracy," the "Humanization of the Planner," and "Anticipatory Democracy." Richard Brautigan published poetic visions of a "cybernetic forest" where pine trees and electronic devices co-existed harmoniously, "where deer stroll peacefully past computers as if they were flowers with spinning blossoms." Robert Blauner, a sociologist, celebrated the potential of electronic technology to liberate workers from drudgery and replace alienation with freedom at a time when carpal tunnel syndrome was yet unheard of and electronic eavesdropping was only beginning to become widespread. Junk bonds had not become the basis of "a whole new order in American finance," Michael Milken was an obscure figure, and Drexel Burnham Lambert was as respectable as it had been in the late nineteenth century when J. P. Morgan had helped found it.[40]

Despite the glow of optimism, signs of malaise were not hard to find if one took the trouble to look for them. The late 1970s brought a darkening international scene as conditions in the Middle East continued to worsen. Citizens learned a new economic term, "stagflation," and Jimmy Carter tried without much success to stimulate a sense of urgency about an "energy crisis" that many people dismissed as nonexistent. In the early 1980s, oil prices skyrocketed to $34 a barrel, and 250,000 American automobile workers were laid off. In 1985, by which time the nation's industrial heartland had become known as the Rust Belt, Honda was the fourth-largest automobile manufacturer in the United States, outstripping moribund American Motors, and Nissan began production at a big Tennessee factory located not far from Birmingham. The national debt rose to previously unthinkable levels, the trade deficit burgeoned, and a wave of hostile takeovers swept the business world. American colleges and universities seemed to be as good as they ever were, but educational experts pointed to a disturbing decline in literacy and other measures of scholastic achievement in a country that had once prided itself on the quality of its elementary and high schools. Drug traffic, once regarded complacently, had become a national scourge.[41]

It was easy not to think about such disturbing portents if one chose to ignore them. Oil was back to $15 a barrel by 1986, credit cards were as good as income, and the United States was in the midst of the longest sustained economic expansion in history under an extremely popular president, Ronald

Reagan, whose profession was acting. Even a momentary free fall in the stock market late in the decade and a scandal about arms that had been sold by government officials to Iran to raise money for anticommunist rebels in Nicaragua did not discourage most Americans, in Alabama or elsewhere. The early 1990s brought continued complacency despite the onset of the worst economic recession in many years. The Berlin Wall fell, yellow ribbons blossomed all over Birmingham in the wake of Operation Desert Storm, and the Soviet Union fell apart. Belief in the future outweighed doubt. The collapse of communism in one former bastion after another stimulated pride in the benefits of capitalism, and privatization became a watchword among social scientists.

Continuation of the recession into 1991 and 1992, however, raised fresh fears about where the country was heading, and a wave of layoffs took place as one company after another underwent restructuring. For a time, it seemed better to do nothing than to take actions that could prove unwise. Reagan's successor, George Bush, decided to pursue no fresh domestic initiatives, and a spokesman, Marlin Fitzwater, informed Congress that it might as well adjourn and go home. As a lethargic economy failed to pull out of its doldrums, however, public alarm began to mount. In 1992, Bush lost to Arkansas governor Bill Clinton in a reversal of power that had seemed inconceivable less than two years before. The postindustrial future that had once seemed so bright had suddenly become enigmatic at best.

How Birmingham would fare in that future was equally open to question. Troubling social and economic indicators raised fears that the city, and Alabama as a whole, were failing to keep up with the demands of a postindustrial society. "Between 1985 and 1987 four of Birmingham's high schools had no student who scored at or above the national average in reading, language, and science," reported Wayne Flynt, a trenchant critic of the antiquated tax laws that make it impossible for Alabama to fund education at levels comparable to other southern states. In the early 1990s, when the average southeastern expenditure per student was $5,114, Alabama spent $3,662. State funding for elementary and secondary education had risen by less than 2.5 percent in the 1980s after adjusting for inflation; only Louisiana ranked lower in this regard. Teachers' salaries in Alabama, which ranked twenty-seventh in the nation in 1987, sank to forty-third by 1992 and to tenth among the eleven southeastern states. Approximately 42,500 college graduates had left Alabama for other places during the 1980s, and only West Virginia had a smaller number of such people in its work force. Noting that the Southern Regional Education Board had predicted that 200,000 of the 300,000 new jobs that would be created in Alabama by the year 2000 would require more than a high school education, Flynt concluded that "in the world economy of the 21st century, one described by almost every observer as a global economy based on information acquisition, one can hardly conceive of a state in a less favorable competitive position."[42]

Flynt's statistics indicated that Alabamians would do well to heed David

Halberstam's warnings of a "harsh, unforgiving new world" that lay waiting in the postindustrial future. It would not be wise to take comfort in a widely heralded switch from blue-collar to service jobs, Halberstam indicated, because there were two types of these: "high service ones for privileged Americans with good educations . . . and low service . . . for those without much education." In such a society, Halberstam warned, "there was danger of the nation's becoming more sharply divided along class lines, with a diminishing middle class and a chasm between the educated few and the unlettered many."[43]

XII

Whatever hopeful or disturbing long-range trends were underway in the early 1990s, postindustrial Birmingham was a more attractive place than it had been in the days when Sloss Furnaces had lit the sky with a fiery glow. This attraction was partly because, unlike Atlanta, the Magic City had not lost so many fine old structures that merited preservation. The Steiner Bank remained from 1890 as the city's best example of Richardsonian architecture. The Brown-Marx and Woodward buildings still stood, now dwarfed by more modern highrise buildings. City National Bank saved from destruction the sixteen-story edifice that had formerly housed the Empire Improvement Company, and the American Trust and Savings Bank tower into which Sloss-Sheffield had moved its offices after 1912 continued to occupy the corner of First Avenue and Twentieth Street. It had a new name—the John Hand Building—but was now free of the industrial grime that had once encrusted its facade. St. Paul's Cathedral, an impressive example of Victorian Gothic architecture dating from 1893, survived at the corner of Third Avenue and Twenty-second Street. The Sixteenth Street Baptist Church, its bomb damage long since repaired, continued to confront passers-by with its solemn twin towers, reminding older residents of the courage and determination shown by children who had marched and suffered for racial justice.[44]

Late in 1992, Birmingham dedicated a $12 million Civil Rights Institute. Facing the Sixteenth Street Baptist Church, it formed part of a "Civil Rights District" that featured predominantly black businesses and a renovated Kelly Ingram Park. On the evening of 14 November a crowd gathered to witness the unveiling of a bronze statue of Rev. Fred Shuttlesworth, who made a speech in which he stated that "I'd like to see Birmingham become the city where brotherly love is the ideal." Nine days of formal dedication ceremonies began on 15 November, starting with an address by Andrew Young, former mayor of Atlanta and U.S. ambassador to the United Nations; other scheduled speakers included Coretta Scott King and Rev. Jesse Jackson. With 58,000 square feet of floor space, the green-domed institute contained exhibits relating to the long struggle to topple segregation and archives from which researchers could glean fresh information about the city's troubled past. Dioramas evoked conditions that had prevailed in the old days. "Banks of video screens bombard the visitor with often violent images and sounds: Freedom Riders being beaten by Klansmen and marchers being set upon by police dogs and fire hoses," stated a newspaper account. At the end of the tour, however, a "triumphant note" was reached "as plaster figures of ordinary citizens of both races march confidently toward a sunny window, through which the haunting facade of the 16th Street Baptist Church can be seen." Some citizens wondered why it was necessary to "open old wounds," but Odessa Woolfolk, the institute's board president, stated that she considered its creation "wholesome" and "healthy." An article in the New York *Times* agreed. "Instead of running away from its troubled past," it said, "Birmingham has chosen to embrace it and move on."[45]

A month after the Civil Rights Institute was dedicated, a much different ceremony took place in Montgomery. Climaxing a meticulous restoration project that had cost $28.5 million, the historic Alabama state capitol, which had been closed to the public for six years while undergoing repair, was reopened amid a firestorm of controversy. Members of the Alabama Legislative Black Caucus had demanded that the Confederate battle flag be no longer flown over the classic white dome and were threatening to go to court over the issue, but Republican governor Guy Hunt was determined to keep the banner flying. No hostile demonstrations took place, as had been widely feared, but only 2,000 persons were on hand to witness the reopening, which was supposed to have attracted five times that many spectators.

At the dedication, historical pomp was mingled with incidents revealing the tensions that remained in a state whose troubled past continued to cloud its transition into an uncertain future. A contingent of Civil War buffs fired nineteenth-century field howitzers amid the drone of bagpipes. Six people belonging to a white supremacist group wore traditional gray Confederate uniforms. Another person draped himself in the controversial battle flag after being told not to wave it because he was obstructing the view of other spectators. "My family is from the South, and I'll display the flag anywhere anytime," he stated defiantly.

In a brief ceremonial address, Hunt declared that the restored building, on the steps of which Jefferson Davis had taken his presidential oath in 1861, had once been "the symbol of a divided nation." Now, Hunt declared, it represented the unity that existed among all Alabama citizens, black and white alike. But the sparse attendance at the event belied his claims. Hunt was vigorously criticized by the Montgomery *Advertiser*, which editorialized that the controversial battle flag had become "a symbol of racial oppression and hate" and should therefore no longer be displayed. Pointing out that many business and political organizations had opposed the governor's stand, the newspaper attacked his decision as a threat to Alabama's economic progress and charged that "by defending the indefensible, he himself has become a major barrier to growth and bringing the races together." Robert Perdue, a retired black truck driver, agreed. "No doubt about it, the flag is doing damage and creating animosity," he said as he recalled hearing "nigger this and nigger that" throughout his career on the nation's highways.[46]

Perdue was only one of many black people who had endured countless indignities while making important but unheralded contributions to Alabama's growth. One of the state's greatest sources of strength lay in its vast wealth of largely anonymous persons of his race, who had worked in hot and dangerous occupations for low wages, endured water hoses and attack dogs, and yet continued to believe in its future. Such people had already suffered much, and their battle was not yet over.

If there ever was a New South, its prophet was not Henry W. Grady but W. E. B. Du Bois, the African-American educator whose brilliant career was climaxed in 1903 with the publication of his classic manifesto, *The Souls of Black Folk*. Part of the burden of being black in America, Du Bois wrote, was a peculiar type of dual consciousness, a "sense of always looking at one's self through the eyes of others, of measuring one's soul by the tape of a world that looks on in amused contempt and pity." In words that spoke poignantly about the inner war resulting from this sense of having two identities, Du Bois declared, "The history of the American Negro is the history of this strife,— the longing to attain self-conscious manhood, to merge his double self into a better and truer self. In this merging he wishes neither of the older selves to be lost. He would not Africanize America, for America has too much to teach the world and Africa. He would not bleach his Negro soul in a flood of white Americanism, for he knows that Negro blood has a message for the world. He simply wishes to make it possible for man to be both a Negro and an American, without being cursed and spit upon by his fellows, without having the doors of Opportunity closed roughly in his face."[47]

On an evening in July 1985, after just recently discovering the importance of the foundry trade to Birmingham's development, I spoke about that subject at Sloss Furnaces. My audience was mostly made up of retired employees who knew much more about foundry pig iron than I did. After the talk was over, most of those present wandered onto the furnace grounds, which were brilliantly illuminated by floodlights. Bending down and running his fingers through what had once been a slag pile, George Brown, an African-American who worked many years on the premises, picked up an apparently meaningless clod and immediately identified it as all that remained of a piece of Russellville ore. For several minutes he discoursed in detail about its chemical properties and the way in which other specimens of ore from the same area in Franklin County had been mixed with local red hematite to meet the requirements of specific markets. Knowing nothing as yet about the enormous difference that the acquisition of the Russellville deposits had made to Sloss-Sheffield in the early twentieth century, I could only listen and learn. I now realize that people like George Brown were the central figures in the drama that played itself out over the course of nine decades along the First Avenue Viaduct. Birmingham owes much to them.

XIII

Birmingham, Birmingham
The greatest city in Alabam'
You can travel 'cross this entire land
But there ain't no place like Birmingham[48]

From his perch on Red Mountain, Vulcan continues to look down on the city that the spirit of industry built in Jones Valley. However impressive it may seem from a distance, the giant statue is deteriorating. While putting it in place in the late 1930s, workers filled its hollow interior with concrete up to the shoulders. Because concrete and iron do not expand and contract at the same rate, cracks have formed, and mortar protrudes through the openings. Repairs were made to correct this flaw in the 1960s, but the problem has returned, and cracks have appeared in new places. Rust from rainwater penetrating the interior has aggravated the problem; fresh torrents are constantly pouring through a hole in Vulcan's head. "Like the smokestack industry he represented," a writer stated in *Alabama Heritage*, "the colossus of Vulcan is in decline. Untended and left to the elements, one day he may subside into the iron-rich earth from which he sprang."[49]

Perhaps so, but perhaps not. Vestavia was destroyed, but the temple that Ward erected to Sybil still stands. So too does Sloss Furnaces National Historic Landmark, spared from demolition and continuing to play a vital role in the economic, social, and cultural life of the city. It is just as possible that the same respect for the past that saved the site for posterity will also preserve the gigantic image of Vulcan so that it may continue pointing toward the installation where so much of its cast-iron body originated. There is more to the soul of Birmingham than the attitudes of a DeBardeleben, who wanted the statue to be sold for scrap, or a Yarbrough, who wished the same fate on Sloss Furnaces. Whatever happens to Armes's vision of a union between Hephaestus and Aphrodite or their Roman cousins, Vulcan and Venus, Moretti's huge creation may yet survive long into the postindustrial age to gaze across a scene increasingly different from the one produced by the spirit of the founders.

Abbreviations

AH	*Agricultural History*
AHQ	*Alabama Historical Quarterly*
AIME	American Institute of Mining Engineers; American Institute of Mining and Metallurgical Engineers (after 1918)
ALAH	*Alabama Heritage*
AR	*Alabama Review*
ASA	Alabama State Archives
BA	*Birmingham Age*
BA-H	*Birmingham Age-Herald*
BBHS	*Bulletin of the Business Historical Society*
BHR	*Business History Review*
BL	*Birmingham Ledger*
BN	*Birmingham News*
BP-H	*Birmingham Post-Herald*
BPLA	Birmingham Public Library, Department of Archives and Manuscripts
BSH	*Birmingham State Herald*
Bulletin	*Bulletin of the American Iron and Steel Association*
BWI	*Birmingham Weekly Independent*
BWIA	*Birmingham Weekly Iron Age*
CE	*Colliery Engineer*
CFC	*Commercial and Financial Chronicle*
CM	*Cassier's Magazine*
CV	*Confederate Veteran*
DAB	*Dictionary of American Biography*
EMJ	*Engineering and Mining Journal*
HAER	*Historic American Engineering Record*
HMRB	Henry Morton Record Books, S. C. Williams Library, Stevens Institute of Technology, Hoboken, N.J.
IA	*Iron Age*
ITR	*Iron Trade Review*
JAH	*Journal of American History*
JBHS	*Journal of the Birmingham Historical Society*
JBL	Joseph Bryan Letterbooks, Virginia Historical Society

JNH	*Journal of Negro History*
JSH	*Journal of Southern History*
MA	*Montgomery Advertiser*
MR	*Manufacturers' Record*
NCAB	*National Cyclopedia of American Biography*
NYT	*New York Times*
PIRN	*Pig Iron Rough Notes*
Poor's Manual	*Poor's Manual of the Railroads of the United States*
RD	*Richmond Dispatch*
RG	*Railroad Gazette*
RQ	*Richmond Quarterly*
RT	*Richmond Times*
RT-D	*Richmond Times-Dispatch*
SAQ	*South Atlantic Quarterly*
SFNHL	Sloss Furnace National Historic Landmark
SISC	Sloss Iron and Steel Company
SSSIC	Sloss-Sheffield Steel and Iron Company
SSSICR	Sloss-Sheffield Records, Birmingham Public Library, Department of Archives and Manuscripts
T&C	*Technology and Culture*
VHS	Virginia Historical Society, Richmond
VMHB	*Virginia Magazine of History and Biography*
VSLA	Virginia State Library and Archives, Richmond

Notes

1. The Inheritance

1. Raymond J. Rowell, Sr., *Vulcan in Birmingham* (Birmingham: Birmingham Park and Recreation Board, 1972); George Clinton Thompson, "Vulcan: Birmingham's Man of Iron," *ALAH* 20 (Spring 1991): 2–17.
2. On the use of Sloss iron in the statue, see *Birmingham's Vulcan* (Birmingham: Birmingham Publishing Company, 1937), 2–3, and *PIRN*, Summer 1937, 4.
3. On the mythological and symbolic associations of Hephaestus and Vulcan, see Jean S. Bolen, *Gods in Everyman* (San Francisco: Harper & Row, 1989), 127–28, 219–50; J. E. Circlot, *A Dictionary of Symbols* (New York: Philosophical Library, 1962), 361–62; *Encyclopaedia Britannica*, 1968 ed., s.v. "Hephaestus" and "Vulcan"; and Michael Stapleton, *A Dictionary of Greek and Roman Mythology* (New York: Bell Publishing Co., 1978), 93, 210.
4. Interviews of Nicholas E. Botta, Michael Burelle, J. David Hagan, and Taylor D. Littleton; Leah Rawls Atkins, "Growing Up Around Edgewood Lake," *AR* 44 (April 1991): 83–100; A. P. Holland, "Highland Avenue," *The Southsider* 3 (Summer 1985): 2.
5. Paul M. Gaston, *The New South Creed: A Study in Southern Mythmaking* (New York: Alfred A. Knopf, 1970), passim.
6. Among many other sources dealing with this transformation, see Harold G. Vatter, *The Drive to Industrial Maturity: The U.S. Economy, 1860–1914* (Westport, Conn.: Greenwood Press, 1975).
7. The interpretation of north-central Alabama's geological history that follows is based on interviews and correspondence with Thomas J. Carrington, Department of Geology, Auburn University; Charles Copeland, Alabama Geological Survey; and Lewis S. Dean, Alabama Petroleum Board.
8. *Encyclopaedia Britannica*, 1968 ed., s.v. "flux," "dolomite," "limestone."
9. William Battle Phillips, *Iron Making in Alabama*, 3d ed. (University: Geological Survey of Alabama, 1912), 27–98; Henry McCalley, "Northern Alabama . . . Its Topography, Geology and Natural Resources," in *Northern Alabama: Historical and Biographical* (Birmingham: Smith & De Land, 1888), 24–29; Virginia Van der Veer Hamilton, *Alabama: A History* (New York and London: W. W. Norton & Co., 1984), 12. On the derivation of the term "hematite," see Rodney Cotterill,

The Cambridge Guide to the Material World (Cambridge: Cambridge University Press, 1985), 110.

10. Ethel M. Armes, *The Story of Coal and Iron in Alabama* (1910; facsimile ed., Birmingham: Book-keepers Press, 1972), 257.

11. Among the many sources that discuss the characteristics and mining of Alabama's red ores and problems connected with their utilization, see H. H. Chapman et al., *The Iron and Steel Industries of the South* (University: University of Alabama Press, 1953), 35–41; W. R. Crane, "Iron Mining in the Birmingham District, Alabama," *EMJ*, 9 February 1905, 274–77; W. E. Curran, "Trend of the Southern Pig-Iron Business," AIME *Transactions* 131 (1938): 40–41; R. H. Ledbetter, "Blast Furnace Practice in Birmingham District," *IA*, 30 October 1924, 1128–30; and Phillips, *Iron Making*, 27–98. On scaffolding, see John Jermain Porter, "The Manufacture of Pig Iron," in A. O. Backert, ed., *The ABC of Iron and Steel* (Cleveland, Ohio: Penton Publishing Co., 1915), 83, 87. According to Curran, who was a blast furnace superintendent for Republic Steel Corporation, making one ton of pig iron with red ore required three tons of ore and 2,500 pounds of coke, "about one ton more ore and 600 to 700 lb. more coke than is required in Northern practice." Ledbetter, a blast furnace superintendent for the Tennessee Coal, Iron, and Railroad Company, reported that local furnaces produced "more than 2300 lb. of slag for each gross ton of iron," a volume that he called "excessively high." My discussion of Alabama raw materials and problems connected with their utilization owes much to Jack Roland Bergstresser, Sr., "Raw Material Constraints and Technological Options in the Mines and Furnaces of the Birmingham District: 1876–1930" (Ph.D. diss., Auburn University, 1993), which was written under my direction.

12. In addition to the sources cited above, see *Encyclopaedia Britannica*, 1968 ed., s.v. "blast furnace," "charcoal," "coal and coal mining," and "coke, coking and high temperature carbonization."

13. Among many discussions of Alabama coals, see particularly Chapman et al., *Iron and Steel Industries*, 54–57, and William E. Ward II and Francis E. Evans, Jr., *Coal: Its Importance to Alabama* (University: Geological Survey of Alabama, 1972), passim. On the geological features of the three coal fields, see "Coal Districts in Alabama," *MR*, 8 August 1885, 813; "The Warrior Coal Field," in ibid., 5 November 1887, 546; Joseph Squire, *Report on the Cahaba Coal Field* (Montgomery, Ala.: Brown Printing Co., 1890); A. M. Gibson, *Report on the Coosa Coal Field* (Montgomery, Ala.: Roemer Printing Co., 1895); and Henry McCalley, *Report on the Warrior Coal Fields* (Montgomery, Ala.: Barrett & Co., 1886).

14. Frederick Overman, *The Manufacture of Iron*, 2d ed. (Philadelphia, 1854), quoted in Peter Temin, *Iron and Steel in Nineteenth Century America: An Economic Inquiry* (Cambridge, Mass.: MIT Press, 1964), 56.

15. Russell S. Poor, "The Geologic Making of the Birmingham District, Alabama," *Birmingham-Southern College Bulletin* 23 (September 1930): 7–31; Thomas A. Simpson, *Structural Geology of the Birmingham Red Iron Ore District, Alabama* (University: Geological Survey of Alabama, 1963), 1–17.

16. Carter L. Hudgins, "A Natural History of Village Creek," *JBHS* 9 (December 1985): 50–59; Morris Knowles, "Water and Waste: The Sanitary Problems of a Modern Industrial District," *Survey* 27 (6 January 1912): 1492; Kathryn Tucker Windham, "Cahaba," in *Rivers of Alabama* (Huntsville, Ala.: Strode Publishers, 1968), 124–25.

17. Hamilton, *Alabama*, 12; Malcolm C. McMillan, *The Land Called Alabama* (Austin, Tex.: Steck-Vaughan Co., 1975), 14–28.

18. The account of the history of ironmaking that follows is based on Leslie Aitchison, *A History of Metals*, 2 vols. (London: MacDonald & Evans, 1960);

Douglas A. Fisher, *The Epic of Steel* (New York: Harper & Row, 1963), 5–32; W. David Lewis, *Iron and Steel in America* (Wilmington, Del.: Hagley Museum, 1976), 9–16; and James M. Swank, *History of the Manufacture of Iron in All Ages* (1892; reprint, New York: Burt Franklin, n.d.), 1–99.

19. Carlo M. Cipolla, *Guns, Sails, and Empires: Technological Innovation and the Early Phases of European Expansion 1400–1700* (New York: Pantheon Books, 1965), passim; William H. McNeill, *The Pursuit of Power: Technology, Armed Force, and Society since A.D. 1000* (reprint, Chicago: University of Chicago Press, 1982), 81–89, 99–101, 166–69.

20. On the rise of foundries, see Bruce L. Simpson, *History of the Metal-Casting Industry*, 2d ed. (Des Plaines, Ill.: American Foundrymen's Society, 1948), passim.

21. The account that follows is based on Lewis, *Iron and Steel in America*, 17–23. On ironmaking in colonial America, see also Arthur C. Bining, *Pennsylvania Iron Manufacture in the Eighteenth Century* (Harrisburg: Pennsylvania Historical Commission, 1938), and Swank, *Iron in All Ages*, 100–190.

22. Richard H. Schallenberg, "Evolution, Adaptation and Survival: The Very Slow Death of the American Charcoal Iron Industry," *Annals of Science* 32 (1975): 342–45.

23. Armes, *Coal and Iron*, 22–23.

24. Hamilton, *Alabama*, 10, 12, 103–19, 125–27; Charles S. Davis, *The Cotton Kingdom in Alabama* (Montgomery: Alabama State Department of Archives and History, 1939), 1–45; McMillan, *Land Called Alabama*, 30–139.

25. John T. Milner, *Alabama: As It Was, As It Is, and As It Will Be* (Montgomery, Ala.: Barrett & Brown, 1876), 3–40; Stanley L. Engerman, "A Reconsideration of Southern Economic Growth, 1770–1860," *AH* 49 (April 1975): 343–61; Robert W. Fogel and Stanley L. Engerman, *Time on the Cross: The Economics of American Negro Slavery* (1974; reprint, Boston: University Press of America, 1984), passim; Eugene D. Genovese, *The Political Economy of Slavery: Studies in the Economy and Society of the Slave South* (1961; 2d ed., Middletown, Conn.: Wesleyan University Press, 1989), passim; Gavin Wright, *The Political Economy of the Cotton South: Households, Markets, and Wealth in the Nineteenth Century* (reprint, New York: W. W. Norton, 1978), 10–127.

26. J. Mills Thornton, *Politics and Power in a Slave Society: Alabama, 1800–1860* (Baton Rouge: Louisiana State University Press, 1978), xix, 7–20, 47, 50–57, 66–67, 78–79, 88–89, 99–105. On the Whig affiliations of wealthy planters and urban officials in the antebellum era, see also Malcolm C. McMillan, *The Disintegration of a Confederate State: Three Governors and Alabama's Wartime Home Front, 1861–1865* (Macon, Ga.: Mercer University Press, 1986), 2.

27. Richard W. Griffin, "Cotton Manufacture in Alabama to 1865," *AHQ* 18 (Fall 1956): 290–93, 302–3; Cecelia Jean Thorn, "The Bell Factory: Early Pride of Huntsville," *AR* 32 (January 1979): 28–37.

28. My analysis of Pratt is based on Randall M. Miller's perceptive article, "Daniel Pratt's Industrial Urbanism: The Cotton Mill Town in Ante-Bellum Alabama," *AHQ* 34 (Spring 1972): 5–35, read in conjunction with the emphasis on the Jacksonian roots of secessionism in Thornton, *Politics and Power*, 267–461.

29. Griffin, "Cotton Manufacture," 293–99; Weymouth T. Jordan, *Ante-Bellum Alabama: Town and Country* (1957; reprint, Tuscaloosa: University of Alabama Press, 1987), 140–60.

30. Genovese, *Political Economy of Slavery*, 180–220; Fred Bateman, et al., "The Participation of Planters in Manufacturing in the Antebellum South," *AH* 48 (April 1974): 277–97.

31. Walter L. Fleming, "Reorganization of the Industrial System in Alabama After the Civil War," *American Journal of Sociology* 10 (January 1905): 473–99;

Ulrich B. Phillips, "The Origin and Growth of the Southern Black Belts," *American Historical Review* 11 (July 1906): 798, 803.

32. Genovese, *Political Economy of Slavery*, 221–39; Robert S. Starobin, *Industrial Slavery in the Old South* (New York: Oxford University Press, 1970), 3–34 and passim; Gavin Wright, *Old South, New South: Revolutions in the Southern Economy Since the Civil War* (New York: Basic Books, 1986), passim.

33. Wright, *Political Economy of the Cotton South*, 51, 116–17.

34. Gary B. Kulik, "The Sloss Furnace Company, 1881–1931: Technological Change and Labor Supply in the Southern Pig Iron Industry" (unpublished research report, Historic American Engineering Record, 1976), 49–56; Gary Kulik, "Black Workers and Technological Change in the Birmingham Iron Industry, 1881–1931," in Merl E. Reed, Leslie S. Hough, and Gary M. Fink, eds., *Southern Workers and Their Unions, 1880–1975* (Westport, Conn.: Greenwood Press, 1981), 23–42.

35. Michael Tuomey, *First Biennial Report on the Geology of Alabama* (Tuscaloosa, Ala.: M. D. J. Slade, 1850), 42–43, and Tuomey, *Second Biennial Report on the Geology of Alabama*, ed. by J. W. Mallet (Montgomery, Ala.: N. B. Cloud, 1858), 97–98; Jack R. Bergstresser, Sr., "A Brief History of the Birmingham District," in "The Birmingham District: An Assessment" (Birmingham: Birmingham Historical Society, 1992), 1:23; Hamilton, *Alabama*, 13.

36. The discussion here and in the following paragraphs is based on Armes, *Coal and Iron*, 27–35, 58–103; and Joseph H. Woodward II, *Alabama Blast Furnaces* (Woodward, Ala.: Woodward Iron Company, 1940), 50–55, 74, 97–99, 116–19, 120–27, 136–39, 160. On Heslip's Cedar Creek installation, see particularly Richard C. Sheridan, "Alabama's First Iron Furnace," *Journal of Muscle Shoals History* 12 (1988): 19–24.

37. Robert H. McKenzie, "Horace Ware: Alabama Iron Pioneer," *AR* 26 (July 1973): 157–63; Richard D. Wallace, "A History of the Shelby Iron Company, 1865–1872" (master's thesis, University of Alabama, 1953), 4–8; Woodham W. Cauley, "A Study of the Accounting Records of the Shelby Iron Company" (master's thesis, University of Alabama, 1949), passim; James F. Doster, "The Shelby Iron Works Collection in the University of Alabama Library," *BBHS* 26 (December 1952), 214–17; Bergstresser, Sr., "Brief History of Birmingham District," 24–25, including a photograph of the machine shop. On Cane Creek Furnace and its operations, see also J. D. B. DeBow, *The Industrial Resources . . . of the Southern and Western States* (New Orleans: DeBow's Review, 1853), 1:59.

38. James R. Bennett, *Old Tannehill: A History of the Pioneer Ironworks in Roupes Valley* (Birmingham: Jefferson County Historical Commission, 1986), passim.

39. Bergstresser, Sr., "Raw Material Constraints," 9–10; Wright, *Old South, New South*, 172.

40. John W. DuBose, *Jefferson County and Birmingham, Alabama: Historical and Biographical* (Birmingham: Teeple & Smith, 1887), 138–45; Armes, *Coal and Iron*, 46, 106, 113; *DAB*, s.v. "John Turner Milner"; "Former Senator Milner is Dead," *BA-H*, 19 August 1898; Dorothea O. Warren, *The Practical Dreamer: A Story of John T. Milner, His Family and Forebears* (Birmingham: Southern Family Press, 1959), 73–100; Marjorie Longenecker White, *The Birmingham District: An Industrial History and Guide* (Birmingham: Birmingham Historical Society, 1981), 14–31. On Gilmer's connection with the Broad River group, see Thornton, *Politics and Power*, 272.

41. Armes, *Coal and Iron*, 40–57; Howard N. Eavenson, *The First Century and a Quarter of American Coal Industry* (Pittsburgh: privately printed, 1942), 293–96; Hamilton, *Alabama*, 14.

42. Tuomey, *First Biennial Report*, 87, 95–96.

43. Virginia Knapp, "William Phineas Browne, Business Man and Pioneer Mine

Operator of Alabama," *AR* 3 (April, July 1950): 108–22, 193–94; Browne to A. Saltmarsh, 24 September 1859, in Browne Papers, Alabama Department of Archives and History, quoted in Starobin, *Industrial Slavery*, 23, 240.

44. Armes, *Coal and Iron*, 62–64; Walter B. Jones, *History and Work of Geological Surveys and Industrial Development in Alabama* (University: Geological Survey of Alabama, 1935), 12–14; James B. Sellers, *History of the University of Alabama* (University: University of Alabama Press, 1953), 1: 53, 86.

45. Armes, *Coal and Iron*, 100–103; Lewis S. Dean, "Michael Tuomey and the Pursuit of a Geological Survey of Alabama," *AR* 43 (April 1991): 101–11, with photograph of Tuomey; Jones, *History and Work*, 15–16; Robert J. Norrell, *A Promising Field: Engineering at Alabama, 1837–1987* (Tuscaloosa: University of Alabama Press, 1990), 23–24; Sellers, *University of Alabama*, 1:81–82.

46. Miller, "Daniel Pratt's Industrial Urbanism," 13–14.

47. Eugene S. Ferguson, "Steam Transportation," in Melvin Kranzberg and Carroll W. Pursell, eds., *Technology in Western Civilization* (New York: Oxford University Press, 1967), 1:291–301; George Rogers Taylor, *The Transportation Revolution, 1815–1860* (1951; reprint, New York: Harper & Row, 1968), 74–103.

48. Alfred D. Chandler, "The Organization of Manufacturing and Transportation," in David T. Gilchrist and W. David Lewis, eds., *Economic Change in the Civil War Era* (Wilmington, Del.: Eleutherian Mills–Hagley Foundation, 1965), 137–51. For a more detailed analysis, see Chandler's *The Visible Hand: The Managerial Revolution in American Business* (Cambridge: Harvard University Press, 1977), passim.

49. See table and map in Taylor, *Transportation Revolution*, 79, 87. With regard to the impact of railroads on the outcome of the Civil War, see particularly Richard N. Current, "God and the Strongest Battalions," in David Donald, ed., *Why the North Won the Civil War* (1960; reprint, New York: Collier Books, 1962), 24, 27, 30; and Richard E. Beringer et al., *Why the South Lost the Civil War* (Athens: University of Georgia Press, 1986), passim. Both sources indicate that the South benefited from relatively short interior supply lines but that the superiority of the northern rail system eventually worked to the disadvantage of the Confederacy.

50. Harriet E. Amos, *Cotton City: Urban Development in Antebellum Mobile* (Tuscaloosa: University of Alabama Press, 1985), 26, 122; Alston Fitts III, *Selma: Queen City of the Black Belt* (Selma, Ala.: Clairmont Press, 1989), 16–17; Norrell, *A Promising Field*, 9–11. On the strong preference of southern railroad builders for slave labor in the antebellum era, see Starobin, *Industrial Slavery*, 28.

51. Hamilton, *Alabama*, 14; Thornton, *Politics and Power*, 268.

52. Thornton, *Politics and Power*, 280–91. On the revival of banking during this period, see also Larry Schweikart, "Alabama's Antebellum Banks: New Interpretations, New Evidence," *AR* 38 (July 1985): 202–21.

53. Amos, *Cotton City*, 196–204; Jordan, *Ante-Bellum Alabama*, 20–21; Grace Lewis Miller, "The Mobile and Ohio Railroad in Ante Bellum Times," *AHQ* 7 (Spring 1945): 37–59; John F. Stover, "Southern Ambitions of the Illinois Central Railroad," *JSH* 20 (November 1954): 499–505.

54. For a useful compendium of legislation pertaining to the lines discussed here, see State of Alabama, *The Charter of the South & North Ala. R.R. Co. and Acts Amendatory Thereof . . . and also Laws Regulating Railroads in Alabama* (Montgomery, Ala.: Advertiser Book and Job Printing Office, 1876), 5–38, 105–18. For additional perspective, see John T. Milner, *Report of the Chief Engineer to the President and Board of Directors of the South and North Alabama Railroad Co., on the 26th of November, 1859* (Montgomery: Advertiser Steam Printing House, 1859), passim; J. Allen Tower, "The Changing Economy of Birmingham and Jefferson County," *JBHS* 1 (January 1960): 4; and Warren, *Practical Dreamer*, 96–100.

55. Armes, *Coal and Iron,* 104–20; Fitts, *Selma,* 19, 30–31; Norrell, A *Promising Field,* 19–29; Thornton, *Politics and Power,* 268–71.
56. Jordan, *Ante-Bellum Alabama,* 106–39. In addition to the analysis given in Thornton's *Politics and Power,* my discussion of broad support among various groups, including planters, for railroad building and other types of economic development in Alabama during the 1850s draws upon an unpublished paper, "Politics and Industrialism in the New South: Alabama, 1850–1900," by Robert J. Norrell, which thoughtfully challenges previous interpretations emphasizing the contrast between northern and southern industrialization, either by alleging planter opposition to industrial development in the antebellum period or, in the case of Jonathan Wiener, claiming that a still-powerful planter class harbored antiindustrial sentiments after 1865. For a short summary of Norrell's paper, see W. David Lewis, "Industrialization and Urbanization—the SENCSA Conference, University of Alabama at Birmingham, April 1987," *T&C* 30 (July 1989): 633. More recently, Norrell has advanced the same thesis in a paper, "Distant Prosperity: Modernization in Nineteenth-Century Alabama," which he has kindly shared with me. For Wiener's views, see Jonathan M. Wiener, *Social Origins of the New South: Alabama, 1860–1885* (Baton Rouge: Louisiana State University Press, 1978), passim.
57. Thornton, *Politics and Power,* 306–21.
58. Ibid., 274–80, 321–30.
59. Ibid., 292–94.
60. Carl N. Degler, *Place over Time: The Continuity of Southern Distinctiveness* (Baton Rouge: Louisiana State University Press, 1977), 17.
61. Amos, *Cotton City,* 211–12 and passim; Jordan, *Ante-Bellum Alabama,* 1–21.
62. On the Alabama cities mentioned, see Fitts, *Selma,* 1–39; Wayne Flynt, *Montgomery: An Illustrated History* (Woodland Hills, Calif.: Windsor Publications, 1980), 1–25; Hamilton, *Alabama,* 119–25; G. Ward Hubbs, *Tuscaloosa: Portrait of an Alabama County* (Northridge, Calif.: Windsor Publications, 1987), 17–37; and Jordan, *Ante-Bellum Alabama,* 1–21. On antebellum southern urbanization generally, see Don H. Doyle, *New Men, New Cities, New South: Atlanta, Nashville, Charleston, Mobile, 1860–1910* (Chapel Hill: University of North Carolina Press, 1990), 1–7; David R. Goldfield, *Cotton Fields and Skyscrapers: Southern City and Region, 1607–1980* (Baton Rouge: Louisiana State University Press, 1982), 12–79; Carville Earle and Ronald Hoffman, "The Urban South: The First Two Centuries," and David R. Goldfield, "Pursuing the American Dream: Cities in the Old South," in Blaine A. Brownell and David R. Goldfield, eds., *The City in Southern History: The Growth of Urban Civilization in the South* (Port Washington, N.Y.: Kennikat Press, 1977), 23–91; and Lawrence H. Larsen, *The Rise of the Urban South* (Lexington: University Press of Kentucky, 1985), 2–9.
63. Hamilton, *Alabama,* 104, quoting in part from Baldwin.
64. Hamilton, *Alabama,* 55–71; Starobin, *Industrial Slavery,* 109–15 and passim.
65. Milner, *Report of the Chief Engineer,* 3–10; Milner, *Alabama As It Was,* 3–63.
66. Milner, *Report of the Chief Engineer,* 15–50.
67. Ibid., 51–54.
68. Ibid., 11–13.
69. Milner, *Alabama As It Was,* 157–61.
70. Milner, *Report of the Chief Engineer,* 13–14.
71. Robert S. Henry, "Railroads of the Confederacy," AR 6 (January 1953): 3–13.
72. Martin L. Everse, "The Iron Works at Briarfield: A History of Iron Making in Bibb County, Alabama" (master's thesis, Samford University, 1984), 3–26; Grace Hooten Gates, *The Model City of the New South: Anniston, Alabama 1872–1900* (Huntsville, Ala.: Strode Publishers, 1978), 21–23; Woodward, *Ala-*

bama Blast Furnaces, 46–49, 58–61, 90–93, 97–99, 105, 120–27, 136–39, 160. For a general overview, see also Joseph H. Woodward II, "Alabama Iron Manufacturing, 1860–1865," *AR* 7 (July 1954): 199–207.

73. Hamilton, *Alabama*, 15; McKenzie, "Horace Ware," 164–165; Frank E. Vandiver, "The Shelby Iron Company in the Civil War: A Study of a Confederate Industry," *AR* 1 (January 1948): 12–26; 1 (April 1948): 111–27; and 1 (July 1948): 203–17.

74. For differing accounts, see Leah Rawls Atkins, *The Valley and the Hills: An Illustrated History of Birmingham and Jefferson County* (Woodland Hills, Calif.: Windsor Publications, 1981), 46–47, and Warren, *Practical Dreamer*, 101.

75. Armes, *Coal and Iron*, 161–67; Marilyn Davis Barefield, *A History of Mountain Brook, Alabama & Incidentally of Shades Valley* (Birmingham: Birmingham Publishing Company, 1989), 21–28; Eavenson, *American Coal Industry*, 296; Warren, *Practical Dreamer*, 100–101; Woodward, *Alabama Blast Furnaces*, 58, 82–84, 106–8. According to Eavenson, the first known case of coke manufacture in Alabama took place in Tuscaloosa County in 1854.

76. Eavenson, *American Coal Industry*, 296–98.

77. The account of the Selma complex given here and in the following paragraphs is based on Armes, *Coal and Iron*, 134–38; Barefield, *History of Mountain Brook*, 21–28; Victor S. Clark, *History of Manufactures in the United States* (New York: McGraw Hill, 1929), 2:41–47; Charles S. Davis, *Colin J. McRae: Confederate Financial Agent* (Tuscaloosa, Ala.: Confederate Publishing Company, 1961), 13–35; Fitts, *Selma*, 49–54; Walter M. Jackson, *The Story of Selma* (Birmingham: Bimingham Printing Co., 1954), 201–7, 213; Edwin T. Layton, "Colin J. McRae and the Selma Arsenal," *AR* 19 (April 1966): 125–36; Walter W. Stephen, "The Brooke Guns from Selma," *AHQ* 20 (Fall 1958): 462–75; Richard J. Stockham, "Alabama Iron for the Confederacy: The Selma Works," *AR* 21 (July 1968): 163–72; Sol H. Tepper, *"Torpedoes? Damn!" Ordeal at Selma Gun Foundry and the Battle of Mobile Bay* (Selma, Ala.: Selma Printing Co., 1979); and Frank E. Vandiver, *Ploughshares into Swords: Josiah Gorgas and Confederate Ordnance* (Austin: University of Texas Press, 1952), 170–71. On concurrent developments at Tredegar, see Charles B. Dew, *Ironmaker to the Confederacy: Joseph R. Anderson and the Tredegar Iron Works* (New Haven, Conn.: Yale University Press, 1966), 108–290.

78. W. Stanley Hoole, "The Confederate Armory at Tallassee, Alabama, 1864–1865," *AR* 25 (January 1972): 3–29.

79. James P. Jones, *Yankee Blitzkrieg: Wilson's Raid through Alabama and Georgia* (Athens: University of Georgia Press, 1976), passim. See also Armes, *Coal and Iron*, 181, 189–94; Barefield, *History of Mountain Brook*, 29–30; Bennett, *Old Tannehill*, 55–76; Fitts, *Selma*, 54–61; and Woodward, *Alabama Blast Furnaces*, 48, 84, 98, 105, 126, 139. On Croxton's Tuscaloosa raid, see Hubbs, *Tuscaloosa*, 43–45, and Norrell, *A Promising Field*, 30–32.

80. Malcolm C. McMillan, *Yesterday's Birmingham* (Miami, Fla.: E. A. Seemann Publishing Co., 1975), 17.

2. James W. Sloss and the Birth of Birmingham

1. Armes, *Coal and Iron*, 234.

2. My emphasis on railroads and competitiveness with other urban centers as a key to Birmingham's development in this and the following chapters owes much to Maury Klein and Harvey A. Kantor, *Prisoners of Progress: American Industrial Cities, 1850–1920* (New York and London: Macmillan, 1976), 1–114.

3. For overviews of economic conditions in the South during this period, see Clark, *History of Manufactures*, 2:54–61; John F. Stover, *The Railroads of the South,*

1865–1900: A Study in Finance and Control (Chapel Hill: University of North Carolina Press, 1955), 39–58; and C. Vann Woodward, *Origins of the New South, 1877–1913*, 2d ed. (Baton Rouge: Louisiana State University Press, 1971), 107–11.

4. On urban developments discussed here, see Joseph B. Mahan, *Columbus: Georgia's Fall Line "Trading Town"* (Northridge, Calif.: Windsor Publications, 1986), 65; Douglas Southall Freeman, "John Stewart Bryan: A Biography" (unpublished MS, Virginia Historical Society, c. 1947), 8–13, excellent on postwar Richmond; Gerald M. Capers, Jr., *The Biography of a River Town. Memphis: Its Heroic Age* (Chapel Hill: University of North Carolina Press, 1939), 162–86; Doyle, *New Men, New Cities, New South*, 22–65, 71–76; Goldfield, *Cotton Fields and Skyscrapers*, 126–27.

5. Justin Fuller, "Alabama Business Leaders: 1865–1900," part 1, *AR* 16 (October 1963): 279; William J. Cooper, Jr., "The Cotton Crisis in the Antebellum South: Another Look," *AH* 49 (April 1975): 388–89.

6. Fuller, "Alabama Business Leaders," 280–81; Doyle, *New Men, New Cities, New South*, 65–71; Flynt, *Montgomery*, 27–55; John H. Napier III, "Montgomery During the Civil War," *AR* 41 (April 1988): 103–31; Hubbs, *Tuscaloosa*, 54.

7. Barefield, *History of Mountain Brook*, 30–31; Clark, *History of Manufactures*, 2:68–69; Rhoda Coleman Ellison, *Bibb County Alabama: The First Hundred Years, 1818–1918* (Tuscaloosa: University of Alabama Press, 1984), 162–64; William B. Hesseltine, *Confederate Leaders in the New South* (Baton Rouge: Louisiana State University Press, 1950), 67; Frank E. Vandiver, "Josiah Gorgas and the Briarfield Iron Works," *AR* 3 (January 1950): 5–21; Woodward, *Alabama Blast Furnaces*, 48, 74, 85–86.

8. Gates, *Model City*, 13–36; Samuel Noble McCaa, "Samuel Noble: Founder of Anniston" (master's thesis, Auburn University, 1966), 1–33.

9. Robert H. McKenzie, "Reconstruction of the Alabama Iron Industry, 1865–1880," *AR* 25 (July 1972): 178–91; John B. Ryan, Jr., "Willard Warner and the Rise and Fall of the Iron Industry in Tecumseh, Alabama," *AR* 24 (October 1971): 261–79; Woodward, *Alabama Blast Furnaces*, 126–27, 140.

10. *Bulletin*, 23 February 1881.

11. On the properties of anthracite as a blast furnace fuel, see Frederick Overman, *Treatise on Metallurgy* (1852; 6th ed., New York: D. Appleton and Co., 1882), 389–93. On the transfer of hot-blast technology from Great Britain to the United States, see Darwin H. Stapleton, *The Transfer of Early Industrial Technologies to America* (Philadelphia: American Philosophical Society, 1987), 169–201.

12. Alfred D. Chandler, "Anthracite Coal and the Beginnings of the Industrial Revolution in the United States," *BHR* 46 (1972): 141–81. See also Chandler's *Visible Hand*, 76–77. For further discussions of anthracite and its importance to the iron industry, see W. David Lewis, "The Early History of the Lackawanna Iron and Coal Company: A Study in Technological Adaptation," *Pennsylvania Magazine of History and Biography* 96 (1972): 424–68, and W. Ross Yates, "Discovery of the Process for Making Anthracite Iron," ibid. 98 (1974): 206–23.

13. On the growing use of bituminous coal and coke in iron smelting and the relative profitability of using these fuels in preference to anthracite, see particularly Temin, *Iron and Steel in Nineteenth-Century America*, 51–80, 200–206, 246–56. For statistics on blast furnace fuels, see *Bulletin*, 23 February 1881. On the continued usefulness of charcoal iron, see Schallenberg, "Evolution, Adaptation and Survival," 341–58.

14. The account given here and in the following paragraphs is based on Jeanne McHugh, *Alexander Holley and the Makers of Steel* (Baltimore: Johns Hopkins University Press, 1980); Elting Morison, *Men, Machines, and Modern Times* (Cambridge, Mass.: MIT Press, 1966), 123–205; William J. Hogan, *Economic*

History of the Iron and Steel Industry in the United States (Lexington, Mass.: D. C. Heath and Co., 1971), 1:31–38; Lewis, *Iron and Steel in America*, 35–41; and Thomas J. Misa, "Science, Technology and Industrial Structure: Steelmaking in America, 1870–1925" (Ph.D. diss., University of Pennsylvania, 1987), 1–35.

15. Hogan, *Economic History*, 1:17–22; Lewis, *Iron and Steel in America*, 30–31; David A. Walker, *Iron Frontier: The Discovery and Development of Minnesota's Three Ranges* (St. Paul: Minnesota Historical Society Press, 1979), 1–15; Terry S. Reynolds, "Iron Ore Ranges" and "Iron Ore Transportation," in Paul S. Paskoff, ed., *Iron and Steel in the Nineteenth Century* (New York: Facts on File, 1989), 194–98, 199–201.

16. Justin Fuller, "From Iron to Steel: Alabama's Industrial Evolution," AR 17 (April 1964): 138.

17. For a detailed discussion of markets for iron and steel in the late nineteenth century, see Hogan, *Economic History*, 1:111–72, 303–41.

18. *Bulletin*, 23 February 1881; Justin Fuller, "History of the Tennessee Coal, Iron, and Railroad Company, 1852–1907" (master's thesis, Emory University, 1958, hereafter cited as Fuller, "TCI—Emory"), 1–13.

19. Doyle, *New Men, New Cities, New South*, 103–4; Gilbert E. Govan and James W. Livingood, *The Chattanooga Country, 1540–1976: From Tomahawks to TVA*, 3d ed. (Knoxville: University of Tennessee Press, 1977), 339–51; Richard A. Bowron, "James Bowron," *JBHS* 4 (November 1963): 17–18; Robert J. Norrell, *James Bowron: The Autobiography of a New South Industrialist* (Chapel Hill: University of North Carolina Press, 1991), 31–57. On the political, economic, and social environment in which these developments took place, see Constantine G. Belissary, "The Rise of Industry and the Industrial Spirit in Tennessee," *JSH* 19 (May 1953): 193–215.

20. *Bulletin*, 23 February 1881.

21. The account in this and the following paragraphs is based on Armes, *Coal and Iron*, 221–27; Henry M. Caldwell, *History of the Elyton Land Company and Birmingham, Ala.* (1892; reprint, with preface by John C. Henley III, Birmingham: Southern University Press, 1972), 1–6; and John W. DuBose, *The Mineral Wealth of Alabama and Birmingham Illustrated* (Birmingham: N. T. Green & Co., 1886), 57–69. On Morris and his background, see John S. Coleman, *Josiah Morris (1818–1891): Montgomery Banker Whose Faith Built Birmingham* (New York: Newcomen Society of North America, 1948), 9–15.

22. On the identification of McGehee and Scott with the Broad River group, see Thornton, *Politics and Power*, 10, 106, 272.

23. For details on the lives of these men, see Armes, *Coal and Iron*, 223–25, 247–49 (Powell and Tate); Willis Brewer, *Alabama: Her History, Resources, War Record and Public Men from 1540 to 1872* (Spartanburg, S.C.: Reprint Co., 1975), 243, 292, 293, 306–7 (Hall, Mudd, Powell); Mary Powell Crane, *The Life of James R. Powell and Early History of Alabama and Birmingham* (Brooklyn, N.Y.: Braunworth & Company, 1930); DuBose, *Jefferson County and Birmingham*, 311–14, 539 (Mudd, Nabers, Worthington); Mary G. Duffee, *Sketches of Alabama* (reprint, University: University of Alabama Press, 1970), 71–77 (Powell); *Memorial Record of Alabama: A Concise Account of the State's Political, Military, Professional and Industrial Progress, Together with the Personal Memoirs of Many of Its People* (Spartanburg, S.C.: Reprint Co., 1976), 2:250–51, 311, 317–19 (Caldwell, Nabers, Worthington, Powell); *Mineral Wealth of Alabama*, 95–100, 146–48 (Caldwell, Mudd, Powell); Thomas M. Owen, *History of Alabama and Dictionary of Alabama Biography* (Spartanburg, S.C.: Reprint Co., 1978), 3: 659, 726 (Gilmer, Hall); "Maj. Campbell Wallace" and "Major Campbell Wallace: A Just Tribute to a Worthy Man," MA, 17 November 1870 and 5 January 1871. On the number of shares held by the various members of the Elyton Land

Company, see Caldwell, *Elyton Land Company*, 5. I am also indebted to Hines H. Hall III, a member of the History Department at Auburn University, for information about Bolling Hall.

24. Marjorie Longenecker White, "The Grid and the Garden," in Philip A. Morris and Marjorie Longenecker White, *Designs on Birmingham: A Landscape History of a Southern City and Its Suburbs* (Birmingham: Birmingham Historical Society, 1989), 8, including the earliest survey of the Jones Valley site.

25. Marvin Y. Whiting, "James R. Powell and 'This Magic Little City of Ours': A Perspective on Local History," *JBHS* 8 (1983): 39.

26. The discussion of Alabama railroads here and in the following paragraphs is based chiefly on Horace Mann Bond, *Negro Education in Alabama: A Study in Cotton and Steel* (1939; reprint, New York: Octagon Books, 1969), 35–62; A. B. Moore, "Railroad Building in Alabama During the Reconstruction Period," *JSH* 1 (November, 1935): 421–41; John R. Scudder, Jr., "The Alabama and Chattanooga Railroad Company, 1868–1871" (master's thesis, University of Alabama, 1951); and Stover, *Railroads of the South*, 61, 88–92. For an overview of aid to railroads in the South in this period, see Mark W. Summers, *Railroads, Reconstruction, and the Gospel of Prosperity: Aid under the Radical Republicans, 1865–1877* (Princeton: Princeton University Press, 1984).

27. On the use of Chinese labor in Alabama, see particularly Lucy M. Cohen, *Chinese in the Post-Civil War South: A People Without a History* (Baton Rouge: Louisiana State University Press, 1984), 90–95, and Brian S. Wills, *A Battle from the Start: The Life of Nathan Bedford Forrest* (New York: Harper Collins, 1992), 354, 360–61.

28. On political and economic developments in Alabama leading up to and including Houston's election, see Bond, *Negro Education*, 40–57; Alan Johnston Going, *Bourbon Democracy in Alabama, 1874–1890* (University: University of Alabama Press, 1951), 1–19; John Craig Stewart, *The Governors of Alabama* (Gretna, La.: Pelican Publishing Co., 1975), 120–27; and Sarah Woolfolk Wiggins, *The Scalawag in Alabama Politics, 1865–1881* (University: University of Alabama Press, 1977), 72–107. I have also drawn on a paper by Faye A. Axford, "Alabama's 'Bald Eagle': Governor George Smith Houston," delivered 24 April 1993 at the annual meeting of the Alabama Historical Association, Florence, Alabama, stressing Houston's background as a plantation owner and slaveholder. On the centrality of issues involving control over black labor in southern politics during the Reconstruction period, see Eric Foner, *Reconstruction: America's Unfinished Revolution, 1863–1877* (New York: Harper & Row, 1988), passim.

29. The account that follows is based largely upon Armes, *Coal and Iron*, 215–21; DuBose, *Jefferson County and Birmingham*, 142–44; Maury Klein, *History of the Louisville and Nashville Railroad* (New York: Macmillan, 1972), 116–20; and Warren, *Practical Dreamer*, 151–58. Of these discussions, Klein's is particularly lucid. See also Atkins, *Valley and Hills*, 47–53, raising perceptive questions about the need for reexamination of the account presented by Armes; and Bond, *Negro Education*, 40–47. Bond's account is a valuable corrective to the "heroic" view generally taken of the episode, in that he sees both sides in the conflict as playing the same venal game.

30. *Charter of the South & North Ala. R.R. Co.*, 26–27.

31. I am indebted to Leah Rawls Atkins and Marvin Y. Whiting for helping me review available evidence pertaining to the siting of Birmingham and reaffirming my conclusion that Stanton had chosen the plot of land on which the city became located, as earlier accounts by Armes, DuBose, and Warren indicate. For a misleading analysis of Stanton's role in the episode, see Martha C. Mitchell, "Birmingham: Biography of a City of the New South" (Ph.D. diss., University of Chicago, 1946), 17. In denying the validity of earlier accounts, Mitchell inac-

curately states that Stanton did not convey his claims to Morris until 20 January 1871, citing an entry on p. 37 of William P. Barker's Book of Survey for the Elyton Land Company as evidence. As preceding entries in Barker's survey indicate, Morris had acquired all of Stanton's options prior to 20 December 1870; the document signed by Stanton on 20 January 1871 was not a conveyance but a formal renunciation of any previous claim to the site chosen for the city. Nothing in the survey book contradicts earlier interpretations of the struggle between Milner and Stanton, but the number of sections held by Mudd, Nabers, and Worthington shows clearly why they were included among the founding members of the Elyton Land Company while Jefferson County landowners who held smaller portions of the 4,700-acre tract—William W. Brown, George W. Clift, Robert N. Green, Daniel Hillman, Alburto Martin, John T. Milner, Thomas Peters, William A. Walker, Sr., and James M. Ware—were not. See entries and plot of holdings in Elyton Land Company Records, Book of Survey, BPLA, 32–38.

32. *DAB*, s.v. "James Withers Sloss"; DuBose, *Jefferson County and Birmingham*, 185–86; Owen, *History of Alabama*, 4:1572–73; obituary of Sloss in *IA*, 5 June 1890.

33. *Charter of the South & North Ala. R.R. Co.*, 16.

34. Bond, *Negro Education*, 40. On Sloss's role during this period, see also Bond's article, "Social and Economic Forces in Alabama Reconstruction," *JNH* 23 (July 1938): 316–19, 321, 324.

35. Klein, *Louisville & Nashville*, 1–101.

36. On Fink's significance as a pioneer in railroad management and strategy, see not only Klein's discussion but also David T. Gilchrist, "Albert Fink and the Pooling System," *BHR* 34 (Spring 1960): 24–49, and Chandler, *Visible Hand*, 116–17, 138–48.

37. On the developments discussed here and in the following paragraph, see Armes, *Coal and Iron*, 245–49, and Klein, *Louisville & Nashville*, 120–22.

38. On the difficulties of the Alabama and Chattanooga and its takeover by the Erlanger syndicate, see Moore, "Railroad Building in Alabama," 421–41, and Stover, *Railroads of the South*, 88–92.

39. Jean E. Keith, "The Role of the Louisville and Nashville Railroad in the Early Development of Alabama Coal and Iron," *BBHS* 26 (September 1952), 165–74; Klein, *Louisville & Nashville*, 223–313.

40. The account of Jefferson County's early history and the founding of Birmingham in this and the following paragraphs is based on Atkins, *Valley and Hills*, 9–56; Floelle Y. Bonner, "Arlington: Mudd-Munger Mansion," in Birmingham Branch, National League of American Pen Women, *Historic Homes of Alabama and Their Traditions* (Birmingham: Birmingham Publishing Co., 1935), 221–30; Caldwell, *Elyton Land Company*, 5–12; Paul Johnson, *The Birth of the Modern: World Society, 1815–1830* (New York: Harper Collins, 1991), 31–32; McMillan, *Yesterday's Birmingham*, 9–19; William S. Rutledge, "An Economic and Social History of Ante-Bellum Jefferson County, Alabama" (master's thesis, University of Alabama, 1939), passim; and White, *Birmingham District*, 15–25.

41. W. Stanley Hoole, ed., "Elyton, Alabama, and the Connecticut Asylum: The Letters of William H. Ely, 1820–1821," *AR* 3 (January 1950): 36–69.

42. Among descriptions of Elyton, see particularly Duffee, *Sketches of Alabama*, 31–38.

43. Caldwell, *Elyton Land Company*, 8–9.

44. Atkins, *Valley and Hills*, 72; McMillan, *Yesterday's Birmingham*, 16–19.

45. Milner, *Alabama As It Was*, 191–99.

46. Woodward, *Alabama Blast Furnaces*, 107–8; "Stirring Sketch of Col. H. F. De-Bardeleben's Life Shows Picturesqueness," *BA-H*, 7 December 1910.

47. Armes, *Coal and Iron*, 238–42, 343; *DAB* s.v. "Henry Fairchild DeBardeleben"; Justin Fuller, "Henry F. DeBardeleben, Industrialist of the New South," *AR* 39 (January 1986), 3–18; "Col. H. F. DeBardeleben's Life Ends After Many Years of Stirring Events" and "Stirring Sketch," *BA-H*, 7 December 1910. As indicated, I am indebted to Prof. Wayne Flynt of the History Department at Auburn University, for suggesting the symbolism inherent in the marriage between Henry F. DeBardeleben and Ellen Pratt.

48. McMillan, *Yesterday's Birmingham*, 19–20; Klein, *Louisville & Nashville*, 130; Caldwell, *Elyton Land Company*, 13; *Memorial Record of Alabama*, 2: 250–51; Crane, *Life of Powell*, 246–307. On Jordan's work and background, see Bonner, "Arlington," 225.

49. On these developments and the events that followed, see Armes, *Coal and Iron*, 255–61, and Klein, *Louisville & Nashville*, 133–34. On Troy, see DuBose, *Jefferson County and Birmingham*, 179–82.

50. DuBose, *Jefferson County and Birmingham*, 556–58.

51. Bergstresser, Sr., "Raw Material Constraints," 104–5, 161–63.

52. Ethel Miller Gorman, *Red Acres* (Birmingham: Vulcan Press, 1956), 302.

53. Compare Armes, *Coal and Iron*, 261, and *BWIA*, 16 March 1876. The account in *BWIA* indicates that James Thomas, not Goodrich, superintended the Oxmoor experiment. Armes placed Thomas at the Irondale furnace at this time.

54. *IA*, 30 March 1876.

55. On the fund-raising activities discussed here and in the following paragraph, see Armes, *Coal and Iron*, 261–62, and Klein, *Louisville & Nashville*, 134–35.

56. Keith, "Role of the Louisville and Nashville," 169.

57. In addition to the discussion in Armes, *Coal and Iron*, 262–63, see also Woodward, *Alabama Blast Furnaces*, 108.

58. Milner, *Alabama As It Was*, passim.

59. Armes, *Coal and Iron*, 266–82; Fuller, "Henry F. DeBardeleben," 5–6; Owen, *History of Alabama*, 3:16.

60. Armes, *Coal and Iron*, 283–87; Woodward, *Alabama Blast Furnaces*, 37–38.

61. Robert H. McKenzie, "The Great Birmingham Iron Boom, 1880–1892," *A Journal of History* [West Jefferson County Historical Society, Bessemer, Ala.] 3 (April 1975): 201–11; Woodward, *Alabama Blast Furnaces*, 160–62.

62. James R. Alexander, *Jaybird: A. J. Moxham and the Manufacture of the Johnson Rail* (Johnstown, Pa.: Johnstown Area Heritage Association, 1991), 13–16; "Birmingham Rolling Mill and Surroundings from 1880 to 1900" and other related materials in Hill Ferguson Collection, BPLA; Armes, *Coal and Iron*, 355.

63. Alice M. Bowsher, et al., *Cinderella Stories: Transformations of Historic Birmingham Buildings* (Birmingham: Birmingham Historical Society, n.d.), 26, 29.

64. DuBose, *Mineral Wealth of Alabama*, passim; Atkins, *Valley and Hills*, 77–89; McMillan, *Yesterday's Birmingham*, 35–38.

65. Armes, *Coal and Iron*, 287–88.

3. The Sloss Furnace Company

1. Armes, *Coal and Iron*, 287–88. On growing demand for iron starting in 1879, related in part to a four-year cycle of expanded railroad building that began that year, see particularly "Irregularity in Railroad Building and Recent Business Depressions," *IA*, 18 September 1884.

2. Armes, *Coal and Iron*, 287–88; Articles of Incorporation, Sloss Furnace Company, Jefferson County Corporation Record A (1876–86), Office of the Probate Judge, Jefferson County Court House, Birmingham, Ala. (hereafter cited as Articles of Incorporation), 37–38.

3. Articles of Incorporation, 38–40; *BWIA*, 30 March and 11 May 1881. On Guthrie, see Klein, *Louisville & Nashville*, 134, 177–78, 197. On the amount directly invested by the L&N, see Keith, "Role of the Louisville and Nashville Railroad," 170.

4. Woodward, *Origins of the New South*, 1–74 passim; Going, *Bourbon Democracy in Alabama*, 13–18; Clark, *History of Manufactures*, 2:159–61.

5. See particularly Wright, *Political Economy of the Cotton South*, 158–84. For other discussions of the problems involved, see Julius Rubin, "The Limits of Agricultural Progress in the Nineteenth-Century South," *AH* 49 (April 1975): 362–73; Harold D. Woodman, "New Perspectives on Southern Economic Development: A Comment," ibid., 374–80; William J. Cooper, Jr., "The Cotton Crisis," ibid., 381–91; Roger Ransom and Richard Sutch, "The 'Lock-in' Mechanism and Overproduction of Cotton in the Postbellum South," ibid., 405–25; and Joseph D. Reid, Jr., "Sharecropping in History and Theory," ibid., 426–40.

6. Cooper, "The Cotton Crisis," 388–89; Milner, *Alabama As It Was*, 37–63.

7. Wiener, *Social Origins of the New South*, 196–209; Going, *Bourbon Democracy*, 47.

8. Gaston, *New South Creed*, 43–116; Woodward, *Origins of the New South*, 144–47.

9. Justin Fuller, "Boom Towns and Blast Furnaces: Town Promotion in Alabama, 1885–1893," *AR* 29 (January 1976): 37–48; Stuart Seely Sprague, "Alabama and the Appalachian Iron and Coal Town Boom, 1889–1893," *AHQ* 37 (Summer 1975): 85–91; James C. Cobb, *Industrialization and Southern Society, 1877–1984* (Lexington: University Press of Kentucky, 1984), 19; Wayne Flynt, *Mine, Mill & Microchip: A Chronicle of Alabama Enterprise* (Northridge, Calif.: Windsor Publications, 1987), 102–3.

10. For representative articles, all published in 1881, see *BWI*, 4 June; 3 September; 26 November; and *BWIA*, 6, 13, 20 April; 11, 18 May; 1, 8, 15 June; 3, 11, 25 August; 8, 15 September; 13 October; 1, 17 November; and 1 December.

11. *BWI*, 12 November 1881.

12. Except where otherwise noted, the details on the furnace plant and its construction that follow are taken from *BWIA*, 6 July 1881, 13 October 1881, 30 March 1882, and *BWI*, 21 January 1882, 11 March 1882. On the donation of the tract, see Elyton Land Company Minute Book 1871–92, 199–200, in BPLA.

13. J. E. Johnson, Jr., *Blast Furnace Construction in America* (New York: McGraw-Hill, 1917), 190–92.

14. Armes, *Coal and Iron*, 288–89. The country from which Hargreaves originated is uncertain. Armes states that he was born in Switzerland, but a contemporary newspaper account stated that he was Swedish: see *BWIA*, 13 October 1881. On British preeminence in furnace design up to about 1880 and the gaining of American leadership in the next decade, see particularly John Jermain Porter, "Manufacture of Pig Iron," in Backert, ed., *ABC of Iron and Steel*, 79–80.

15. "A Sketch of the History of Hot-Blast Brick Stoves," *IA*, 31 July 1879. See also Victor O. Stroebel, "Gordon's Improved Whitwell-Cowper Fire Brick Hot-Blast Stove," *AIME Transactions* 14 (June 1885–May 1886): 159–72.

16. Bergstresser, Sr., "Raw Material Constraints," 14–16.

17. *BWIA*, 18 August 1881; Atkins, *Valley and Hills*, 61; Barefield, *History of Mountain Brook*, 31. On blowing equipment at this time generally, see Overman, *Treatise on Metallurgy*, 397–411.

18. Clark, *History of Manufactures*, 2:254; James H. Bridge, *The Inside History of the Carnegie Steel Company: A Romance of Millions* (New York: Aldine Book Co., 1903), 54–70; Joseph Frazier Wall, *Andrew Carnegie* (1970; reprint, Pittsburgh: University of Pittsburgh Press, 1989), 307–60; Overman, *Treatise on Metallurgy*,

523. For some representative dimensions of Pennsylvania blast furnaces between 1880 and 1885, see John Birkinbine, "Comparisons of Blast Furnace Records," AIME *Transactions* 15 (May 1886–February 1887): 147–70.

19. In the absence of original drawings, this description of the furnace is purposely vague regarding a few details. For discussions of furnace design that would apply to virtually any blast furnace built in America at this time, see Overman, *Treatise on Metallurgy*, 505–25, and Johnson, *Blast Furnace Construction in America*, 233–84. On breakouts, see also Porter, "Manufacture of Pig Iron," 85.

20. W. Galloway, "Genesis and Development of the Coking Oven," *EMJ* 88 (1909): 11–13; Mable D. Mills, *Coke Industry in Alabama* (University, Ala.: Bureau of Business Research, 1947), passim; Richard Moldenke, *The Coke Industry of the United States, As Related to the Foundry* (Washington, D.C.: U.S. Bureau of Mines, 1910), passim; Overman, *Treatise on Metallurgy*, 383–89. See also *Encyclopaedia Britannica*, 1968 ed., s.v. "coke, coking, and high temperature carbonization." "Annotated Bibliography on Beehive Coke Ovens" (unpublished research report, Auburn University, 1986), by Jack Roland Bergstresser, Sr., has contributed significantly to my analysis of this subject.

21. *BWI*, 21 January 1882; testimony of James W. Sloss in United States Senate, *Report of the Committee of the Senate Upon the Relations between Labor and Capital*, 5 vols. (hereafter cited as *Labor and Capital*) (Washington, D.C.: Government Printing Office, 1885), 4:281.

22. *BWIA*, 20 April 1882.

23. *BWIA*, 26 August 1882, 23 November 1882, and 31 May 1883; *PIRN*, Spring 1935, 2, with photograph of medal won at Louisville.

24. *BWIA*, 26 September 1882 and 31 May 1883.

25. "Dixie's Hub," *BWIA*, 1 March 1883; Armes, *Coal and Iron*, 254; McMillan, *Yesterday's Birmingham*, 22–25.

26. "Another Furnace Horror. Awful Death of Two Negroes Last Night at Sloss Furnace," *BWIA*, 23 November 1882.

27. "A Horrible Suicide," ibid.

28. Kathryn Tucker Windham, *The Ghost in the Sloss Furnaces* (Birmingham: Birmingham Historical Society, 1978), passim. Windham fictionalizes Jowers's name, calling him "Theophilus Calvin Jowers."

29. DuBose, *Mineral Wealth of Alabama*, 109; Robert J. Norrell and Otis Dismuke, *The Other Side: The Story of Birmingham's Black Community* (Birmingham: Birmingfind, n.d.), 1–2; Carl V. Harris, *Political Power in Birmingham, 1871–1921* (Knoxville: University of Tennessee Press, 1977), 34.

30. Klein and Kantor, *Prisoners of Progress*, 14–15; Wright, *Old South, New South*, 7–8 and passim.

31. Similarities between Birmingham and Johannesburg were suggested to me by hearing Dalvan Coger's presentation, "Johannesburg on the Veldt: The Birth of the City of Gold," at a conference in Birmingham in 1987 and by South African analogies in an article by Robert J. Norrell that appeared about the same time. See Lewis, "Industrialization and Urbanization," 628–34, and Norrell's "Caste in Steel: Jim Crow Careers in Birmingham, Alabama," *JAH* 73 (December 1986): 669–94. For perspectives on the history of Johannesburg, see Nigel Mandy, *A City Divided: Johannesburg and Soweto* (New York: St. Martin's Press, 1984). For pictures of Johannesburg in the late nineteenth century that are eerily similar to ones taken in Birmingham during the same era, see Peter Kallaway and Patrick Pearson, *Johannesburg: Images and Continuities* (Braamfontein, South Africa: Raven Press, 1986).

32. Bruce Levine et al., *Who Built America? Working People and the Nation's Economy, Politics, Culture, and Society* (New York: Pantheon Books, 1989), 1:216–17;

Charles H. Wesley, *Negro Labor in the United States, 1850–1925* (1927; reprint, New York: Russell & Russell, 1967), 142; Starobin, *Industrial Slavery,* 107–8.

33. Kathleen Bruce, *Virginia Iron Manufacture in the Slave Era* (1930; reprint, New York: Augustus M. Kelley, 1968), 231–58; Richard C. Wade, *Slavery in the Cities: The South, 1820–1860* (New York: Oxford University Press, 1964), 37.
34. Wade, *Slavery in the Cities,* 34–35.
35. Dew, *Ironmaker to the Confederacy,* 22–32; Charles B. Dew, "Southern Industry in the Civil War Era: Joseph Reid Anderson and the Tredegar Iron Works, 1859–1867" (Ph.D. diss., Johns Hopkins University, 1964), 346–85.
36. *Labor and Capital,* 4:45–48; Miller, "Daniel Pratt's Industrial Urbanism," 26–32.
37. Woodward, *Alabama Blast Furnaces,* 17, 46, 51, 74, 84, 87, 105, 120, 122, 126, 136; Bennett, *Old Tannehill,* 39.
38. John S. Ezell, *The South Since 1865,* 3d ed. (Norman: University of Oklahoma Press, 1975), 141–42. On the moral overtones involved, see particularly W. J. Cash, *The Mind of the South* (New York: Alfred A. Knopf, 1941), 113–17, and Woodward, *Origins of the New South,* 133–34.
39. Cash, *Mind of the South,* 115–20.
40. Wade, *Slavery in the Cities,* 243–81.
41. Claudia Dale Goldin, *Urban Slavery in the American South, 1820–1860: A Quantitative History* (Chicago: University of Chicago Press, 1976), passim.
42. Ibid., 3–5.
43. Starobin, *Industrial Slavery,* passim; Goldin, *Urban Slavery,* 46, 60, 64.
44. Goldin, *Urban Slavery,* 32, 47–48; Milner, *Alabama As It Was,* 155–61.
45. Kulik, "Black Workers and Technological Change," 25.
46. Ibid., 26–27; Kulik, "The Sloss Furnace Company," 7–8; Edward A. Uehling, "Pig-Iron Casting and Conveying Machinery: Its Development in the United States," *CM* 24 (June 1903): 115–20, with photographs of the processes involved.
47. The discussion of the use of black labor by the Sloss Furnace Company in this and the following paragraphs is based on Sloss's testimony in *Labor and Capital,* 4:282–300.
48. Herbert Gutman, *Work, Culture, and Society in Industrializing America* (New York: Alfred A. Knopf, 1976), 15, quoted in Burns, *Workshop of Democracy,* 143.
49. Starobin, *Industrial Slavery,* 77–84.
50. On Sloss's use of contract workers at the Potter ore mine, see *Labor and Capital,* 4:285.
51. DuBose, *Mineral Wealth of Alabama,* 111.
52. Sterling D. Spero and Abram L. Harris, *The Black Worker: The Negro and the Labor Movement* (1931; reprint, Port Washington, N.Y.: Kennikat Press, 1966), 7–14.
53. DuBose, *Mineral Wealth of Alabama,* 109.
54. Wright, *Old South, New South,* passim.
55. On the discussion in this and the following paragraph, see particularly Lewis, *Iron and Steel in America,* 43–50.
56. Starobin, *Industrial Slavery,* 173–74.
57. Wall, *Andrew Carnegie,* 324–60; Chandler, *Visible Hand,* 260–69. On techniques used at the Edgar Thomson works, see also Misa, "Science, Technology and Industrial Structure," 36–44.
58. *IA,* 15 March 1883, 21 August 1884, and 21 April 1887; *Bulletin,* 22 January 1890.
59. On the progress and effects of the strike, see *IA,* 25 May, 29 June, 6 and 27 July, 10 and 24 August, 14 and 21 September, and 9 November 1882.
60. "Louisville, Ky.: The Central Metropolis of the South," *MR,* 3 September 1887.
61. See Sloss testimony in *Labor and Capital,* 4:283. The key importance of Cincin-

nati and Louisville as markets for pig iron from both Birmingham and Chattanooga can be followed in the market reports from trade correspondents in these cities in virtually every issue of *Iron Age* during the 1880s. See also the brief historical retrospective by Russell Hunt, "Sloss Iron Is Uniform Iron," *PIRN*, Mid-Winter 1934, 1, 15. On the historical background of this method of marketing pig iron, see particularly Glenn Porter and Harold C. Livesay, *Merchants and Manufacturers: Studies in the Changing Structure of Nineteenth-Century Marketing* (Baltimore and London: Johns Hopkins Press, 1971), 37–55.

62. Hunt, "Sloss Iron Is Uniform Iron," 1, 15.

63. On the problems of the iron industry in both New England and the Middle Atlantic states as discussed in this and the following paragraph, see Clark, *History of Manufactures*, 2:218–19; "American Anthracite Furnaces," *Bulletin*, 15 and 25 October 1882; "The Eastern Pig Iron Trade," *IA*, 27 November 1884; "Grievances of Eastern Pig Iron Producers," ibid., 11 December 1884; "Relief to the Eastern Iron Trade," ibid., 18 December 1884; "Eastern Manufacturers and the Anthracite Coal Companies," ibid., 29 July 1886; "An Attack on the Coal Combination," ibid., 7 October 1886; "Is Pennsylvania Losing Her Leadership in the Manufacture of Iron and Steel?" *Bulletin*, 31 July 1889; "The Cost of Producing Pig Iron," ibid., 20 and 27 February 1884, reporting cost of ironmaking in Lehigh and Schuylkill valleys; "Production of Pig Iron in the United States in 1883 and 1884," ibid., 4 and 11 February 1885; and statistical materials, ibid., 20 and 27 January 1886, showing decline in number of anthracite furnaces in blast between the beginning of 1883 and mid-1885.

64. Clark, *History of Manufactures*, 2:219–20; *IA*, 20 November 1884 and 7 October 1886; Temin, *Iron and Steel in Nineteenth Century America*, 51–80. On mining techniques in the anthracite fields and mounting problems affecting the anthracite fields in the late nineteenth century, see Donald L. Miller and Richard E. Sharpless, *The Kingdom of Coal: Work, Enterprise, and Ethnic Communities in the Mine Fields* (Philadelphia: University of Pennsylvania Press, 1985), 84–134, and Anthony F. C. Wallace, *St. Clair: A Nineteenth-Century Coal Town's Experience with a Disaster-Prone Industry* (New York: Alfred A. Knopf, 1987), 7–53, 403–45.

65. For an account of the campaign against northeastern anthracite producers up to and after 1886, see W. David Lewis, "The Invasion of Northern Markets by Southern Iron in a Decade of Boom and Bust: Sectional Competition in the 1880s," *Essays in Economic and Business History* 8 (1990): 257–69.

66. See particularly *Bulletin*, 29 March 1882, 8 November 1882, 20 and 27 January 1886; and *IA*, 21 February 1884. On the causes of the downturn, see also Clark, *History of Manufactures*, 2:294–95.

67. Woodward, *Alabama Blast Furnaces*, 162.

68. The discussion of the Woodward Iron Company in this and the following two paragraphs is based upon ibid., 155–56; Armes, *Coal and Iron*, 299–301; "Dixie's Hub," *BWIA*, 1 March 1883; *IA*, 15 April 1886; and "Iron Ore: Making and Remaking the Map of Steel," *Fortune*, May 1931, 134.

69. In addition to the sources just cited, discussions with Jack R. Bergstresser, Sr., have helped me assess the comparative performances of the Woodward and Sloss companies.

70. Testimony of Sloss in *Labor and Capital*, 4:297; "Model Furnaces," *BWIA*, 31 May 1883; Woodward, *Alabama Blast Furnaces*, 128, 160. Exactly when Sloss's second furnace went into full-scale production is not clear from available sources. Woodward states that it occurred at some point in October 1883.

71. Testimony of Sloss in *Labor and Capital*, 4:281–82, 285, 300.

72. *IA*, 12 June 1884.

73. "Pig Iron Making in the South," *MR*, 26 July 1884.

74. See particularly "The Scheme to Restrict Production," *IA*, 25 September 1884, with list of firms agreeing and refusing to participate in Hull's project.
75. "Southern Pig Iron for $12.50," *IA*, 24 April 1884.
76. "Industrial Items," ibid., 26 February, 19 and 26 November 1885; *MR*, 14 March 1885.
77. "Industrial Items," *IA*, 13 and 27 May, 24 June, 14 October 1886.
78. "Industrial Items," *IA*, 7 October 1886. The steady improvement of demand for pig iron and the consequent increase in prices for this commodity can be followed in weekly market reports from various cities published in *Iron Age* throughout the summer and fall of 1886. See for example the issues of 17 and 24 June; 29 July; 26 August; 7, 14, and 21 October; and 25 November.
79. See particularly Birmingham trade report in *IA*, 9 December 1886, stressing "almost feverish activity in real estate speculation, superinduced largely by the recent assurances of the strength of the town's position in the industrial world."
80. Armes, *Coal and Iron*, 347.

4. Joseph Bryan and the Virginia Connection

1. Virginius Dabney, *Virginia: The New Dominion* (Garden City, N.Y.: Doubleday, 1971), 353–54.
2. Armes, *Coal and Iron*, 347–48; handwritten draft of Johnston's option agreement, 6 November 1886, in SSSICR.
3. For various perspectives on the role played by outside capital in southern economic development, see W. H. Baughn, "Capital Formation and Entrepreneurship in the South," *Southern Economic Journal* 16 (October 1949), 161–69; Cobb, *Industrialization and Southern Society*, 11–26; Clarence H. Danhof, "Four Decades of Thought on the South's Economic Problems," in Melvin L. Greenhut and W. Tate Whitman, *Essays in Southern Economic Development* (Chapel Hill: University of North Carolina Press, 1964), 7–68; Ezell, *South Since 1865*, 136–53; Gaston, *New South Creed*, 71–75; and Woodward, *Origins of the New South*, 107–41, 264–320.
4. I. W. Avery, "Southern Men and Southern Money Developing the South," *MR*, 2 August 1884.
5. Wright, *Old South, New South*, 13.
6. Michael B. Chesson, *Richmond After the War, 1865–1890* (Richmond: Virginia State Library, 1981), 117–43; Virginius Dabney, *Richmond: The Story of a City* (Garden City, N.Y.: Doubleday, 1976), 119–225; Freeman, "John Stewart Bryan," 8–13.
7. Hesseltine, *Confederate Leaders in the New South*, 33–35, 95–97, 116–34.
8. For varied perspectives on the developments discussed here, see Robert G. Albion, *The Rise of New York Port* (1939; reprint, New York: Charles Scribner's Sons, 1970); Norman S. Buck, *The Development of the Organisation of Anglo-American Trade, 1800–1850* (1925; reprint, New York: Greenwood Press, 1968); Philip S. Foner, *Business & Slavery: The New York Merchants & the Irrepressible Conflict* (Chapel Hill: University of North Carolina Press, 1941); Basil Leo Lee, *Discontent in New York City, 1861–1865* (Washington, D.C.: Catholic University of America Press, 1943), ix, 1–40, 87–124; Ernest A. McKay, *The Civil War and New York City* (Syracuse, N.Y.: Syracuse University Press, 1990); and William R. Taylor, *Cavalier and Yankee: The Old South and American National Character* (1961; reprint, Garden City, N.Y.: Doubleday Anchor Books, 1963). On analogous ties between New England textile firms and southern planters in the antebellum period, see Thomas H. O'Connor, *Lords of the Loom: The Cotton Whigs and the Coming of the Civil War* (New York: Charles Scribner's Sons, 1968).
9. Hesseltine, *Confederate Leaders in the New South*, 94.

10. Paul H. Buck, *The Road to Reunion, 1865–1900* (Boston: Little, Brown and Co., 1937), 152–55.
11. Maury Klein, *Edward Porter Alexander* (Athens: University of Georgia Press, 1971), 149–200; John C. Jay, "General N. B. Forrest as a Railroad Builder in Alabama," *AHQ* 24 (Spring 1964): 16–31.
12. My discussion of the development of the Richmond and Danville and Terminal companies is based largely upon Maury Klein, *The Great Richmond Terminal: A Study in Businessmen and Business Strategy* (Charlottesville: University Press of Virginia, 1970), 1–114. See also Stover, *Railroads of the South*, 233–53.
13. For a list of the members of the "ironclad pool" and the shares they held, see John Stewart Bryan, *Joseph Bryan: His Times, His Family, His Friends* (Richmond, Va.: Whittet and Shepperson, 1935), 246.
14. Klein, *Great Richmond Terminal*, 12–13, 16–29, 288–94.
15. In addition to the accounts of the early history of the Georgia Pacific in Klein, *Great Richmond Terminal*, 91–92, and Stover, *Railroads of the South*, 240, see Burke Davis, *The Southern Railway: Road of the Innovators* (Chapel Hill: University of North Carolina Press, 1985), 191–92; Fairfax Harrison, *A History of the Legal Development of the Railroad System of Southern Railway Company* (Washington, D.C.: n.p., 1901), 402–25; and Allen P. Tankersley, *John B. Gordon: A Study in Gallantry* (Atlanta: Whitehall Press, 1955), 322–25. On Gordon's visions regarding the new line, see *BWIA*, 13 April 1881.
16. Robert R. Russel, "The Pacific Railway Issue in Politics prior to the Civil War," *Mississippi Valley Historical Review* 12 (September 1925): 187–201; Paul N. Garber, *The Gadsden Treaty* (Philadelphia: University of Pennsylvania Press, 1921), passim; Harrison, *Legal History*, 437; Klein, *Great Richmond Terminal*, 92. On the early progress of construction on the line, see *BWIA*, 10 April; 11, 18 May; and 3, 11 August, all in 1881. On Gordon's further involvement and eventual withdrawal, see Tankersley, *John B. Gordon*, 325.
17. "The Georgia Western Road," *BWIA*, 13 April 1881; "The Georgia Western and Birmingham," ibid., 20 April 1881.
18. "Major John W. Johnston" and "Brave Soldier Answers Call," *RT-D*, 23 May 1905; "Maj. John William Johnston," *CV* 19 (March 1911): 116–17.
19. Richard Wilmer, *In Memoriam: John Stewart of Rothesay* (New York: n.p., 1887), passim; obituary of John Stewart in *RD*, 12 March 1881; obituary of Daniel K. Stewart, ibid., 13 August 1889; Bryan, *Joseph Bryan*, 176–89; Daniel K. Stewart Letterbook, 1863–64, Stewart Family Papers, VHS; Chesson, *Richmond after the War*, 59; Freeman, "John Stewart Bryan," 25.
20. The account of Bryan's life in this and the following paragraphs is taken largely from Bryan, *Joseph Bryan*. The passage quoted appears on p. 151. Among other accounts of Bryan's life, see particularly W. Gordon McCabe, "Joseph Bryan: A Brief Memoir," *VMHB* 17 (April 1909): iii–xxix, and Virginius Dabney, *Pistols and Pointed Pens: The Dueling Editors of Old Virginia* (Chapel Hill, N.C.: Algonquin Books, 1987), 160–82.
21. See particularly Daniel J. Boorstin, *The Americans: The Democratic Experience* (New York: Random House, 1973), 53–64.
22. Wiener, *Social Origins of the New South*, 209.
23. The interpretation of Bryan's values and attitudes in this and the paragraph that follows is based partly upon James M. Lindgren, "'First and Foremost a Virginian': Joseph Bryan and the New South Economy," *VMHB* 96 (April 1988): 157–80, and partly on my own research in Bryan's letterbooks at VHS.
24. J. Bryan III, *The Sword over the Mantel: The Civil War and I* (New York: McGraw-Hill, 1960), 26.
25. Dabney, *Pistols and Pointed Pens*, 173–74.

26. Bryan, *Joseph Bryan*, 248.
27. On Bryan's presence in Anniston for the inauguration of the Georgia Pacific's service to that city, see handwritten notes on his trip, 16–21 April 1883, in Joseph Bryan Commonplace Book, Bryan Family Papers, VHS. On the progress of the Georgia Pacific toward Anniston, see *CFC*, 17 June 1882, 15 July 1882, 9 September 1882, 7 October 1882, and 17 February 1883; and *RG*, 19 January 1883, 16 February 1883, 9 March 1883. On the opening of Anniston to public investment, which began in June 1883, see Gates, *Model City*, 56–81.
28. On the lives and careers of these men, see R. A. Brock, *Virginia and Virginians* (Richmond, Va., and Toledo, Ohio: H. H. Hardesty, 1888), 2:779–80; Charles M. Carvati, *Major Dooley* (Richmond, Va.: n.p., 1978); "Edmund Dunscomb Christian," in William Couper, *The V.M.I. New Market Cadets: Biographical Sketches of All Members of the Virginia Military Institute Corps of Cadets Who Fought in the Battle of New Market, May 15, 1864* (Charlottesville, Va.: Michie Co., 1933), 37; Dabney, *Richmond*, 204, 248–49; William H. Gaines, Jr., *Biographical Register of Members, Virginia State Convention of 1861* (Richmond: Virginia State Library, 1969), 19; Edward Alvey, Jr., "James B. Pace: A Richmonder of Many Talents," *RQ* 4 (Fall 1981): 29–35; Lyon G. Tyler, *Men of Mark in Virginia* (Washington, D.C.: Men of Mark Publishing Co., 1906), 1:167–69; "William R. Trigg's Useful Life Ended," *RT-D*, 17 February 1903; and "William Robertson Trigg," in Philip Alexander Bruce, ed., *Virginia: Rebirth of the Old Dominion* (Chicago: Lewis Publishing Co., 1929), 5:393–94.
29. Material on Maben in this and the following paragraphs is based on "John Campbell Maben," *Encyclopedia of American Biography*, new ser., 16:291–92; letters in Maben Family Papers, VSLA, particularly letter from C. Martin to David Maben, 10 January 1850; Armes, *Coal and Iron*, 453–455; obituaries of Maben in *BN*, 1 September 1924, and *BA-H*, 2 September 1924; credit reports on Maben in R. G. Dun Collection, Baker Library, Harvard University, Boston, Mass., vol. 341, pp. 200, A/B, A/140, used by permission of Dun & Bradstreet and the Baker Library. Bryan's letterbooks at VHS contain hundreds of letters to Maben.
30. Maury Klein has analyzed the important roles played by a number of southerners on Wall Street in the late nineteenth century. In addition to his *Great Richmond Terminal*, previously cited, see his article, "Southern Railroad Leaders, 1865–1893: Identities and Ideologies," *BHR* 42 (Autumn 1968): 288–310.
31. Obituary of Goadby, *NYT*, 5 July 1925; Bryan, *Joseph Bryan*, 281; credit reports on Goadby in R. G. Dun Collection, vol. 422, pp. 626, 700, A4, A43, used by permission of Dun & Bradstreet and the Baker Library.
32. Obituary of Richard Mortimer, *NYT*, 31 May 1882. I have been unable to determine whether this Richard Mortimer was the father or grandfather of Richard Y. Mortimer. On Mortimer's being a bondholder in the Georgia Pacific see Bryan to John C. Maben, 3 March 1888, JBL.
33. Elie Weeks, "Sabot Hill," *Goochland County Historical Magazine* 10 (Autumn 1978): 6–7; "Death of Thomas Seddon Comes Suddenly and Shocks the Entire Community," *BSH*, 12 May 1896; "Death of Thomas Seddon, Esq.: A Distinguished and Successful Career Suddenly Ended," *RT*, 12 May 1896; Armes, *Coal and Iron*, 350–52.
34. Joseph Bryan to John Randolph Bryan, Jr., 2 July 1882, in Bryan Family Papers, Personal Papers Collection (Accession 24882), Archives and Records Division, VSLA.
35. Bryan to "My Dear Boys," 12 July 1882, ibid.
36. See biographical sketch of Clyde in Klein, *Great Richmond Terminal*, 35–36.
37. *NCAB*, s.v. "Walter George Oakman."

38. On the role played by the Central Trust Company in Danville and Terminal Company affairs, see Klein, *Great Richmond Terminal*, 95, 136, 140, 141, 240.

39. Among many other sources on Ruffner, see particularly Anne H. R. Barclay, "William Henry Ruffner, LL.D.," *West Virginia Historical Magazine Quarterly* 2 (January 1902): 33–43; E. L. Fox, "William Henry Ruffner and the Rise of the Public Free School System of Virginia," *The John P. Branch Papers of Randolph-Macon College* 2 (June 1910): 124–44; Walter J. Fraser, Jr., "William Henry Ruffner and the Establishment of Virginia's Public School System, 1870–1874," *VMHB* 79 (July 1971): 259–79; Walter J. Fraser, Jr., "William Henry Ruffner: A Southern Liberal in the North," *The Citadel Monograph Series* 9 (February 1972): 1–29; and C. Chilton Pearson, "William Henry Ruffner: Reconstruction Statesman of Virginia," *SAQ* 20 (January, April 1921): 25–32, 137–51.

40. Biographical sketch of Logan in Klein, *Great Richmond Terminal*, 42–43; Hesseltine, *Confederate Leaders*, 125–27.

41. Joseph K. Roberts, *Annotated Geological Bibliography of Virginia* (Richmond, Va.: Dietz Press, 1942), 39–42; Joseph K. Roberts, "Contributions of Virginians to the Geology of the State," *Virginia Journal of Science* 1 (February–March 1940): 71–73; Ruffner diary and handwritten notes in William Henry Ruffner Papers, VSLA; John L. Campbell and William Henry Ruffner, *A Physical Survey Extending from Atlanta, Ga., across Alabama and Mississippi to the Mississippi River, along the Line of the Georgia Pacific Railway, Embracing the Geology, Topography, Minerals, Soils, Climate, Forests, and Agricultural and Manufacturing Resources of the Country* (New York: E. F. Weeks, 1883). I am grateful to Robert Yuill of Springville, Alabama, for sharing with me the results of his own research in the Ruffner papers at VSLA.

42. *RG*, 6 April 1883; *BWIA*, 5 April 1883.

43. "Birmingham's Biggest Boom," *BWIA*, 17 May 1883; handwritten notes, box 2, Ruffner Papers, shared with me by Yuill; Going, *Bourbon Democracy in Alabama*, 179. On the Coalburg purchase, see also White, *Birmingham District*, 282–83.

44. "Birmingham's Biggest Boom," *BWIA*, 17 May 1883; Joseph Bryan to John W. Johnston, 18 May 1883, JBL.

45. Articles of Incorporation of Coalburg Coal and Coke Company, Office of the Probate Judge, Jefferson County Courthouse, Birmingham, Ala., Corporate Records Section, vol. A, pp. 153–56.

46. Armes, *Coal and Iron*, 439–40; Couper, *V.M.I. New Market Cadets*, 210–12; George M. Cruikshank, *A History of Birmingham and Its Environs* (Chicago and New York: Lewis Publishing Co., 1920), 1:21–22; Thomas C. McCorvey, "Henry Tutwiler and the Influence of the University of Virginia on Education in Alabama," in *Alabama Historical Sketches* (Charlottesville: University of Virginia Press, 1960), 3–32; Sellers, *University of Alabama*, 43–45.

47. Elyton Land Company Minute Book, BPLA, 217–24; *RG*, 13 July 1883, 31 August 1883, 2 November 1883; *CFC*, 11 August 1883, 29 September 1883; *BWIA*, 3 May, 6 September 1883.

48. *BWIA*, 22 November 1883; *RG*, 23 November 1883; *CFC*, 24 November 1883; *Poor's Manual of Railroads*, 1883, 468, and 1884, 452.

49. Joseph Bryan to George S. Scott, 22 November 1883, JBL; Klein, *Great Richmond Terminal*, 99–100; *BWIA*, 29 November 1883; *RG*, 7 December 1883. For a complete list of the officers and directors chosen at the November 29 meeting, see *Poor's Manual of Railroads*, 1884, 452.

50. *RG*, 16 November 1883, 11 July 1884, 25 July 1884, 18 December 1885. On the recession, which began in March 1882 and became particularly pronounced in 1884, see Rendig Fels, *American Business Cycles, 1865–1897* (Chapel Hill: University of North Carolina Press, 1959), 124–36.

51. Joseph Bryan to John W. Johnston, 24 January 1884, 23 February 1884, 26 March 1884, 4 April 1884, 15 April 1884, and 19 April 1884, JBL.

52. *Poor's Manual of Railroads*, 1885, 441; obituary of John Stewart, *RD*, 12 March 1885; *RG*, 18, 25 December 1885.

53. *Poor's Manual of Railroads*, 1885, 440–41, and 1886, 535–36.

54. Fels, *American Business Cycles*, 137.

55. Armes, *Coal and Iron*, 298–99; Woodward, *Alabama Blast Furnaces*, 149–50.

56. *RG*, 29 May 1885, 28 August 1885, 18 December 1885; *Poor's Manual of Railroads*, 1888, 588. On the deepening investments of Bryan and Daniel K. Stewart in the Extension Company and Bryan's efforts to secure new equipment, see Joseph Bryan to Richard Irvin, Jr., 12 February 1886; to John W. Johnston, 27 February 1886; to W. G. Oakman, 20 March 1886; and to Norman S. Walker, 17 April 1886, JBL.

57. Klein, *Great Richmond Terminal*, 105–14.

58. *Poor's Manual of Railroads*, 1887, 590.

59. Ibid.; 1887, 590; *RG*, 18 September 1886.

60. *IA*, 26 August, 30 September, 7, 21 October, 9 December 1886 and 31 March 1887. For a general overview of the 1886–87 boom in Birmingham, see Armes, *Coal and Iron*, 330–59; on the spasm of urban growth that occurred in mid-America generally in the 1880s, but particularly during this brief economic upsurge, see Carl L. Degler, *Out of Our Past: The Forces That Shaped Modern America* (1959, 3d ed., New York: Harper & Row, 1984), 335.

61. "North Birmingham: Soon to Become a Great Suburb of the Magic City," *BA*, 30 December 1886; Articles of Incorporation, North Birmingham Land Company, North Birmingham Building Association, and South Highlands Company, Probate Office, Jefferson County Courthouse, vol. A, pp. 381–82, 509–10, 512–14. The articles of incorporation of the North Birmingham Street Railway Company do not appear in these records.

62. Joseph Bryan to John W. Johnston, 25 September 1886; to Thomas Seddon, 4 November 1886; to J. Y. Sage, 13 November 1886; and to Daniel K. Stewart, 6 December 1886, JBL.

63. "Another Furnace" and "Coalburg Doings," *BA*, 3 November and 7 December 1886.

64. *DAB*, s.v. "Reese, Jacob." On puddling, see particularly Paul F. Paskoff, "Puddling," in Paskoff, ed., *Iron and Steel in the Nineteenth Century*, 286–87; on the Cambria firm and the work of John Fritz, see particularly Stephen H. Cutcliffe, "Cambria Iron Company," ibid., 38–42.

65. *DAB*, s.v. "Reese, Jacob"; McHugh, *Alexander Holley and the Makers of Steel*, 342, 347, 349, 359–60; "The Basic Steel Process," *BA*, 2 November 1886; Fuller, "From Iron to Steel," 139.

66. "The Basic Steel Process," *BA*, 2 November 1886; Charter of Jefferson Iron Company, Alabama State Legislature, Act No. 297, H.B. 290, approved 14 February 1885, copy in SSSICR.

67. Charter of Sloss Iron and Steel Company, Alabama State Legislature, Act No. 297, H.B. 260, 14 February 1885, as amended by Act No. 149, H.B. 143, 29 November 1886, copy in SSSICR.

68. Armes, *Coal and Iron*, 349; "Johnston, Joseph Forney," in Owen, *History of Alabama*, 3:918; Jessie Mae Roberts, "Joseph Forney Johnston" (unpublished bachelor's thesis, Howard College, 1931), 1–12.

69. McMillan, *Yesterday's Birmingham*, 50; Owen, *History of Alabama*, 4:1472; Wills, *Battle from the Start*, 207–8, 227, 366–69.

70. *BA*, 20 November 1886.

5. Takeover, Expansion, and Recession

1. Armes, *Coal and Iron*, 347–48.
2. Ibid., 453, identifying Bryan as first vice president of the Sloss-Sheffield Steel and Iron Company in 1902.
3. *BA-H*, 21 November 1908; *BN*, 21 November 1908.
4. See pp. 300–301.
5. See pp. 260–66.
6. Handwritten letter from John W. Johnston to James W. Sloss, 6 November 1886, SSSICR.
7. Telegram and letter from Joseph Bryan to John W. Johnston, both dated 4 November 1886; letters from Bryan to Thomas Seddon and T. M. Logan, 4 November 1886, JBL.
8. Joseph Bryan to John W. Johnston, 15 and 16 November 1886, and to Wilson, Colston and Company, 16 November 1886, ibid.
9. Joseph Bryan to J. A. Upshur, 5 January 1887, ibid.
10. Klein, *Great Richmond Terminal*, 38–40.
11. Joseph Bryan to C. E. Whitlock, 1 December 1886; to J. B. Pace, 1 December 1886; to Daniel K. Stewart, 6 and 9 December 1886; to Joseph Forney Johnston, 6 December 1886; to Walter G. Oakman, 21 and 23 December 1886, 4 and 6 January 1887; and to John C. Maben, 30 December 1886, JBL.
12. Joseph Bryan to John W. Johnston, 5 and 6 January 1887; to Thomas Pinckney, 7 January 1887; to Walter G. Oakman, 17 January 1887; to Wilson, Colston & Co., 18 January 1887; to Charles H. Cocke, 31 January 1887; and to C. C. Page, 31 January 1887, JBL; General Mortgage of the Sloss Iron and Steel Company, SSSICR.
13. Original stock registers of Sloss Iron and Steel Company, dated 31 January and 16 March 1887, SSSICR. In cases where persons resident in Birmingham, New York, or other places were actually Virginians, I have included them in the Virginia category.
14. *IA*, 13 January, 24 February, 28 April 1887.
15. For the best accounts of the developments that follow, see Fuller, "TCI—Emory," 14–28, and Justin Fuller, "TCI—History of the Tennessee Coal, Iron and Railroad Company, 1852–1907" (Ph.D. diss., University of North Carolina at Chapel Hill, hereafter cited as "TCI—Chapel Hill"), 35–79. See also Armes, *Coal and Iron*, 360–93.
16. William H. Moore, "Preoccupied Paternalism: The Roane Iron Company in Her Company Town—Rockwood, Tennessee," *East Tennessee Historical Society Publications* 39 (1967): 56–70.
17. On the events just recounted, see also Fuller, "Alabama Business Leaders," Part 1, 286, and Part 2, 63, and Fuller, "Henry F. DeBardeleben," 7–8.
18. The account of Smith's career in this and the following paragraphs is taken from Armes, *Coal and Iron*, 278–82 and passim; Klein, *Louisville & Nashville Railroad*, 223–70; and H. C. Nixon, *Lower Piedmont Country: The Uplands of the Deep South* (1946; reprint, Tuscaloosa: University of Alabama Press, 1984), 44–54.
19. Nixon, *Lower Piedmont Country*, 45; Armes, *Coal and Iron*, 281–82.
20. Armes, *Coal and Iron*, 295–96, 306–8; Fuller, "TCI—Chapel Hill," 91; Woodward, *Alabama Blast Furnaces*, 100.
21. "A Retrospect . . . Bessemer, Ala.," *MR*, 30 June 1888. On the origins of the venture, see also Fuller, "Henry F. DeBardeleben," 9–12.
22. Armes, *Coal and Iron*, 330–33; Fuller, "TCI—Chapel Hill," 91–98; Ellen Slaughter, "DeBardeleben, Bessemer and the Montezuma Hotel," *JBHS* 9 (Spring 1988): 49–61; White, *Birmingham District*, 106–10; Woodward, *Alabama Blast Fur-*

naces, 43; "Bessemer's Growth," *MR,* 30 June 1888; "The Twentieth Century Exposition of the City of Bessemer, Alabama," *Bessemer Weekly,* 18 May 1901 (reprint, Bessemer, Ala.: Bessemer Hall of History, 1977), 1.

23. Armes, *Coal and Iron,* 173–77, 212. For an extended discussion of David Thomas and his work, see Stapleton, *Transfer of Early Industrial Technologies,* 169–201.

24. White, *Birmingham District,* 131–33; Robert Casey and Marjorie Longenecker White, "A Look at Thomas, An Alabama Iron Town," *Canal History and Technology Proceedings* 9 (17 March 1990): 121–41; "Two More Furnaces for Alabama," *MR,* 25 February 1888; Woodward, *Alabama Blast Furnaces,* 142.

25. Caldwell, *Elyton Land Company,* 15–27; Lyn Johns, "Early Highland and the Magic City, 1884–1893, Including Willis J. Milner's 'History of Highland Avenue,'" *JBHS* 6 (July 1979): 33–43; "The Magic City of the South: What a Special Correspondent Says of Birmingham and Its Industries," *MR,* 2 October 1886.

26. Armes, *Coal and Iron,* 300–301; Caldwell, *Elyton Land Company,* 25–26; "Great Dividends," *MR,* 14 April 1888; Woodward, *Alabama Blast Furnaces,* 156–58.

27. *IA,* 28 October, 4 November, and 2 December 1886; "To Build Two Additional Furnaces and Steel Works," *MR,* 1 January 1887.

28. *BWA,* 30 December 1886; White, *Birmingham District,* 150–51.

29. "St. George T. C. Bryan," *CV* 24 (June 1916): 275; W. Gordon McCabe, "St. George Tucker Coalter Bryan of Richmond, 1843–1916," paper read at meeting of VHS, 17 March 1917, at VHS; Bryan to John W. Johnston, 16 August 1886, JBL.

30. Bryan to John W. Johnston, 5 January 1887; to Charles H. Cocke, 31 January 1887; to W. C. Seddon, 7 February 1887, and to Thomas Seddon, 20 April 1887; copy of letter from Joseph F. Johnston to Thomas Seddon, 27 April 1887; Bryan to St. George T. C. Bryan, 30 May 1888, JBL. On the varying composition of the ore and coal upon which Sloss depended at this time, see "American Institute of Mining Engineers: The Scranton Meeting," *IA,* 24 February 1887.

31. "Capt. John Randolph Bryan," *CV* 26 (April 1918): 168; "Balloon Used for Scout Duty. Terrible Experiences of a Confederate Officer who saw the Enemy from Dizzy Heights," *Southern Historical Society Papers* 33 (1905): 32–42; Joseph Jenkins Cornish III, *The Air Arm of the Confederacy* (Richmond, Va.: Richmond Civil War Centennial Committee, 1963), 19, 30–33. I am indebted to Capt. Roy F. Houchin II, United States Air Force, for advice in studying the details of Bryan's mission.

32. Bryan to J. R. Bryan, 29 July 1886, 12 March 1887, 19 April 1888, JBL.

33. C. B. Bryan, "Dr. John Randolph Page," in Homer Richey, ed., *Memorial History of the John Bowie Strange Camp, United Confederate Veterans* (Charlottesville, Va.: Michie Co., 1920), 83–84; Joseph Bryan to Dr. J. R. Page, 5 June 1888, JBL.

34. Joseph Bryan to Thomas Seddon, 3 May, 9 May, 3 June 1889; to John W. Johnston, 4 January 1887; to A. B. Johnston, 15 January 1887; and to J. R. Bryan, Jr., 12 March 1887, JBL.

35. Joseph Bryan to John W. Johnston, 3 December 1886, JBL; Walter C. Whitaker, *History of the Protestant Episcopal Church in Alabama, 1763–1891* (Birmingham: Roberts & Son, 1898), 153–63; Wilmer, *In Memoriam,* passim.

36. *IA,* 14 April and 19 May 1887.

37. *IA,* 1 December 1887; Woodward, *Alabama Blast Furnaces,* 130.

38. Stock Ledger, Sloss Iron and Steel Company, Issue of 18 July 1887, SSSICR.

39. *IA,* 24 March 1887.

40. Fels, *American Business Cycles,* 137–58; *IA,* 3 November 1887.

41. *IA,* 1 September, 20 October, and 3 November 1887.

42. Bryan to W. C. Seddon, 12 March 1887, JBL; *IA*, 3, 17 February, 3 March, 14, 28 April, 9 June 1887.
43. *IA*, 3 March, 15 September, and 3 November 1887.
44. Philip Taft, *Organizing Dixie: Alabama Workers in the Industrial Era*, rev. and ed. Gary M. Fink (Westport, Conn.: Greenwood Press, 1981), 8; Cobb, *Industrialization and Southern Society*, 87–88; Wayne Flynt, *Poor but Proud: Alabama's Poor Whites* (Tuscaloosa: University of Alabama Press, 1989), 138–39.
45. *IA*, 6 January, 3, 10, 17 February, 26 May, 30 June, 14, 21, 28 July, 4 August, 1 September 1887.
46. Cobb, *Industrialization in Southern Society*, 68–69.
47. The historical account of Alabama's convict labor system in this and the following paragraphs is based on Jack Leonard Lerner, "A Monument to Shame: The Convict Lease System in Alabama" (master's thesis, Samford University, 1969); Elizabeth Boner Clark, "The Abolition of the Convict Lease System in Alabama, 1913–1928" (master's thesis, University of Alabama, 1949), 8–17; Mary Ann Neeley, "Painful Circumstances: Glimpses of the Alabama Penitentiary, 1846–1852," *AR* 44 (January 1991): 3–16; and Fannie Ella Sapp, "The Convict Lease System in Alabama (1846–1895)" (master's thesis, George Peabody College for Teachers, 1931).
48. Blake McKelvey, "Penal Slavery and Southern Reconstruction," *JNH* 20 (April 1935): 159–61.
49. Quoted in Lerner, "A Monument to Shame," 101, and George W. Cable, "The Convict Lease System in the Southern States," *Century Magazine* 27 (February 1884): 595.
50. Clark, "Abolition," 16–17; Kulik, "Sloss Furnace Company," 35.
51. Bryan to Seddon, 4 May 1887, JBL; *IA* 12 May, 25 October 1887.
52. *IA*, 1, 8 December 1887; *MR*, 9 June 1888.
53. Fuller, "TCI—Chapel Hill," 79–82.
54. "To Build 30 Furnaces," *MR*, 29 January 1887; Armes, *Coal and Iron*, 350; Bryan to Thomas Seddon, 22 April 1887, JBL.

6. A Sea of Troubles

1. Bryan to Seddon, 22 April 1887, JBL.
2. Annual Report of the President, SISC, year ending 1 February 1888, SSSICR; Armes, *Coal and Iron*, 350.
3. Louis Galambos and Joseph Pratt, *The Rise of the Corporate Commonwealth: U.S. Business and Public Policy in the Twentieth Century* (New York: Basic Books, 1988), 23, 29–32.
4. Annual Report of the President, SISC, year ending 1 February 1888, SSSICR, passim; "The Blast Furnaces of the United States on February 1," *MR*, 18 February 1888; Bryan to Seddon, 31 May 1888, JBL.
5. Annual Report of the President, SISC, year ending 1 February 1888, SSSICR, passim.
6. Bryan to Seddon, 13, 20, 23, 30 January; 4, 16 February; and 5, 9, 10 March 1888, JBL.
7. Consolidated Mortgage Bond Orders, 1888–1890, SSSICR; "Sloss Iron and Steel Company Consolidated Second Mortgage Loan," JBL.
8. Report to Directors, SISC, year ending 1 February 1889, in Bound Volume of Financial Statements of North Alabama Furnace Company, 1901–1904, and of Sloss Sheffield Steel and Iron Company, 1892–1902, SSSICR; Comparison of Cost Sheets, SISC, fiscal years 1887–1891, ibid.
9. Report to Directors, SISC, year ending 1 February 1889, SSSICR.
10. Edward A. Uehling, "Autobiography of Edward A. Uehling, M.E., Dr. of Engi-

neering, 1849–1939," unpublished ms. supplied by Edward R. Uehling (hereafter cited Uehling Autobiography), 39.

11. Ibid.
12. Ms. autobiography of James Bowron, Jr., in W. S. Hoole Special Collections Library, University of Alabama (hereafter cited Bowron Autobiography), 1:274–75.
13. Bryan to Seddon, 31 May 1888; Bryan to Dr. John R. Page, 4 June 1888; Bryan to Seddon, 13 June 1888, JBL; "Coke-Making by the Sloss Iron and Steel Company," *CE* 9 (July 1889): 271.
14. Bryan to Maben, 8 May 1888, and to Seddon, 30 May and 13 June 1888, JBL.
15. Bryan to Seddon, 4 June 1888, ibid.
16. Bryan to Seddon, 19 April 1888, and to John W. Johnston, 14 May 1888, JBL; Register of Stock, Sloss Iron and Steel Company, SSSICR.
17. Klein, *Great Richmond Terminal*, 173–204, 257; Stover, *Railroads of the South*, 245–50.
18. *Poor's Manual of Railroads*, 1888, 588, and 1889, 575; Joseph Bryan to Central Trust Company, 15 June 1888; statement from law office of John J. Dennis, Starkville, Miss., 6 July 1888, JBL; *CFC*, 29 December 1888.
19. Bryan to Maben, 28 April 1888, and to John W. Johnston, 28 April 1888, JBL.
20. "New Steamers for the South," *MR*, 9 June 1888; "Birmingham s Growtn," ibid., 23 June 1888.
21. *IA*, 12, 19 April 1888; 10, 24 May 1888; 7, 14 June 1888.
22. On general features of the upswing, which continued in America until mid-1890, see Fels, *American Business Cycles*, 159–71. On conditions in the iron industry, see *IA*, 16, 23, 30 August; 6, 13 September; and 27 December 1888; 3, 10, 24 January; 28 February; 16, 23, 30 May; 5 September; 10 October; and 19 December 1889, as well as the general overview in "Rapid Growth of the American Iron Trade in 1888 and 1889" in *Bulletin*, 22 January 1890. On the rising British demand that was checking exports to the United States and contributing to the upswing there, see particularly "The English Iron Market," *IA*, 21 November 1889.
23. Comparison of Cost Sheets, SISC, fiscal years 1887–1891, SSSICR.
24. Ibid.; Report to Directors, SISC, year ending 1 February 1889, SSSICR; Armes, *Coal and Iron*, 349; Bryan to Seddon, 30 May and 4 June 1888, JBL.
25. Bergstresser, Sr., "Raw Material Constraints," 64–67.
26. Report to Directors, SISC, years ending 1 February 1889 and 1 February 1890, SSSICR; "Coke-Making by the Sloss Iron and Steel Company," 271; "Coalburg, Ala.: Coal Mines of the Sloss Iron and Steel Company," *CE* 8 (February 1888): 153. On increasing coal production in Alabama during this period, see Flynt, *Mine, Mill & Microchip*, 80.
27. *CFC*, 11 May and 22 June 1889.
28. Joseph Bryan to Mrs. Jane Stewart McIntosh, 16 August 1889, and to Bryce Stewart, 17 and 10 August 1889, JBL; obituary of Daniel K. Stewart, *RD*, 13 August 1889.
29. *Poor's Manual of Railroads*, 1890, 408–9; Klein, *Great Richmond Terminal*, 227. On Bryan's reluctance to see Scott relinquish the presidency of the Danville, see particularly Bryan to Scott, 11 November 1889, JBL; on his opposition to Inman's expansion plans, see Bryan to Maben, 31 January 1890, ibid.
30. Bryan to Johnston, 14 December 1889 and 2 February 1890; to Maben, 14 December 1889; and to J. R. Bryan, 25 February 1890, JBL; *BN*, 4 January and 26 June 1890.
31. Bryan to Oakman, 26 February 1890, and to Sage, 8 March 1890, JBL; "Captain Sage Resigns," and "A Genuine Surprise: The Georgia Pacific Boys Show Their Love for Captain Sage," *BA-H*, 25 July and 1 August 1890.

32. *Poor's Manual of Railroads*, 1890, 409, and 1891, 466, showing removal of Johnston and Scott from the Georgia Pacific board in 1890; Bryan to Johnston, 15 February 1890, JBL. For biographical details on Brice and Seney, see Klein, *Great Richmond Terminal*, 30–31, 48–49.

33. Bryan to Alexander Brown & Sons, 30 January and 5, 11 February 1890; to Wilson, Colston & Co., 30 January and 12 February 1890; to Johnston, 12 February 1890; to Inman, 3 March and 23 April 1890; to C. W. Watson, 28 March and 27 June 1890; to Maben, 7, 14, and 17 April 1890; and to Central Trust Company, 24 April 1890, JBL; "Work to Begin on the New Georgia Pacific Shops Within the Next Thirty Days," *BN*, 25 June 1890.

34. "Coke-Making By the Sloss Iron and Steel Company," *CE* 9 (June 1889): 271.

35. Comparison of Cost Sheets, SISC, fiscal years 1887–1891, SSSICR.

36. Obituaries of Morton in New York *Herald* and *NYT*, 10 May 1902.

37. Biographical sketch of Dashiell in *NCAB*, 34: 96–97; Uehling Autobiography, 37; Morton to Uehling, 14 February 1890, in collection of letters from Morton to Uehling supplied by Edward R. Uehling (hereafter cited as Morton-Uehling Correspondence).

38. Bryan to St. George T. C. Bryan, 24 July 1889; to Kenneth Robertson, 24 and 29 July 1889; to John A. Rutherfurd, 27 July 1889; to Edward Engle, 30 October 1889; to John C. Maben, 1 and 25 November 1889; to Thomas Seddon, 1 November and 23 December 1889, JBL.

39. Bryan to Morton, 7, 10, 11 March 1889, JBL; Uehling Autobiography, 36–37; entries of 16–23 March 1890, HMRB; letters from Morton to Uehling, 28 February and 8, 25 March 1890, Morton-Uehling Correspondence.

40. Letter of recommendation for Uehling by Morton, 14 August 1878, and letter from Morton to Uehling, 25 March 1890, Morton-Uehling Correspondence; Uehling Autobiography, 1–28 and passim.

41. Uehling Autobiography, 29–36. The second of the two outstanding engineers who figured significantly in the development of the firm was James Pickering Dovel, whose life and achievements will be traced in chapter 14.

42. Wright, *Old South, New South*, 172.

43. "27,000 is the Approximate Official City Census," *BN*, 26 June 1890; McMillan, *Yesterday's Birmingham*, 50.

44. For statistics on American pig iron production, I have relied especially upon tables in *Bulletin*, 23 February 1881 and 22 January 1890 based upon net tons of 2,000 pounds. For statistical perspectives on pig iron production in Alabama as compared with other states, see also William W. Sweet, "Iron and Steel Manufacture," in Eleventh Census of the United States, *Report on Manufacturing Industries in the United States at the Eleventh Census, 1890*, Pt. 3 (Washington, D.C.: Government Printing Office, 1895), 383–483.

45. Atkins, *Valley and Hills*, 71; Caldwell, *Elyton Land Company*, 29–30; McMillan, *Yesterday's Birmingham*, 35, 38, 40–41, 51; "Most Auspiciously Does 1890 Open Up for Birmingham" and "Heavy Traffic: Enormous Amount Over the Mineral Road," *BN*, 10 and 13 January 1890.

46. "Production of Pig Iron in the United States in 1888 and 1889" in *Bulletin*, 22 January 1890; Fuller, "TCI—Chapel Hill," 80; Atkins, *Valley and Hills*, 70–71.

47. Fuller, "TCI—Chapel Hill," 81–89, and "TCI—Emory," Appendix 4.

48. *BN*, 5 and 6 May 1890.

49. "J. W. Sloss," *IA*, 5 June 1890.

50. Alexander, *Jaybird*, 17–19, 45–50; McMillan, *Yesterday's Birmingham*, 52.

51. *MR*, 3 December 1887; "The Henderson Steel," ibid., 17 March 1888; "Our Alabama Letter," ibid., 24 March 1888; "Henderson Steel," ibid., 31 March 1888; Bernice Shield Hassinger, *Henderson Steel: Birmingham's First Steel* (Bir-

mingham: Gray Printing, 1978), 1–11. Brady Banta, "Henderson Steel & Manufacturing Company," in Paskoff, ed., *Iron and Steel in the Nineteenth Century*, 154–55.

52. Hassinger, *Henderson Steel*, 11–41; "Basic Steel" and "Fires Lighted," *BN*, 4 and 30 January 1890; "The Steel Test: The Question Will Soon Be Settled," "The Steel Test: It is All Right and Results Very Satisfactory," "Making Basic Steel," and "Steel in Chattanooga," *BA-H*, 20 August and 18 September 1890; Fuller, "From Iron to Steel," 140–41.

53. Uehling Autobiography, 37; Bryan to Maben, 7, 14, and 17 April 1890, JBL.

7. Turmoil and Tenacity

1. Bryan to Morton, 2 June, 5 July 1890, JBL.

2. Bergstresser, Sr., "Raw Material Constraints," 90–92; G. R. Delamater, "Bituminous Coal Washing," *Mines and Minerals* 27 (August, September 1907): 7–10, 62–65. On the previous use of coal washing at Coalburg, see above, p. 125.

3. Uehling Autobiography, 37. On Seddon's being in White Sulphur Springs during part of the summer, see Bryan to Seddon, 2 August 1890, JBL.

4. Uehling Autobiography, 37; entries of 27–30 September 1890, HMRB; biographical sketch of Dashiell in *NCAB*, 34:96–97.

5. Uehling Autobiography, 37; Bryan to Seddon, 30 December 1890, JBL. For other indications of Bryan's fund-raising activities during this period, see Bryan to Seddon, 6, 12, 14, 15 November 1890; 12 December 1890; and 1, 8, 16, 22, 29, and 31 January 1891, ibid.

6. Armes, *Coal and Iron*, 351–52.

7. Statement of Business, SISC, six months ending 31 July 1891, and Annual Report of Secretary and Treasurer, SISC, fiscal year ending 31 January 1892, including comparison of cost sheets from 1887 through 1891, SSSICR. For examples of statements by Bryan praising Uehling's work, see Bryan to Morton, 3 January, 6 February 1891, JBL. On the sluggish nature of demand throughout much of 1891, see Fels, *American Business Cycles*, 171–76.

8. Statement of Business, SISC, six months ending 31 July 1891, SSSICR; Bergstresser, Sr., "Brief History of Birmingham District," 64–68, with accompanying photographs.

9. For an overview of conditions in the mining camps maintained by Sloss, TCI, and other Alabama companies during the late nineteenth century, see Flynt, *Poor but Proud*, 115–24. In addition to material gleaned from ledgers, day books, and journals in SSSICR, I have also benefited from insights in Jingsheng Dong, "The Social and Working Lives of Alabama Coal Miners and Their Families, 1893–1925" (master's thesis, Auburn University, 1988), 39–40 and passim.

10. Alfred F. Brainerd, "Colored Mining Labor," AIME *Transactions* 14 (June 1885–May 1886): 78–79.

11. "Coalburg, Ala.: Coal Mines of the Sloss Iron and Steel Company," *CE* 8 (February 1888): 153; "A Pennsylvania Miner's Impressions of Alabama," ibid. 9 (July 1889), 281; "Mining Towns: Dots About Coalburg, Blossburg, Brookside and Other Places," *BN*, 17 June 1890, 1; Statement of Business, SISC, six months ending 31 July 1891, SSSICR.

12. "Mining Towns," *BN*, 17 June 1890, 1; "Order Restored," ibid., 6; White, *Birmingham District*, 278–79 and passim; Balance Sheet, SISC, year ending 31 January 1897, SSSICR; Bergstresser, Sr., "Brief History of Birmingham District," 64.

13. "Mining Towns," *BN*, 17 June 1890; "Sloss Coal Mines," ibid., 9 June 1890; "Bloody Tragedy," ibid., 6 January 1890; White, *Birmingham District*, 272–73, 276–77 and passim; Balance Sheet, SISC, year ending 31 January 1897, SSSICR.
14. White, *Birmingham District*, 146, 234, 276–78.
15. Ibid., 223–31; Balance Sheet, SISC, year ending 31 January 1897, SSSICR.
16. White, *Birmingham District*, 227–28; *Bessemer Weekly*, 18 May 1901 (reprint, July 1977), 57; Balance Sheet, SISC, year ending 31 January 1897, SSSICR.
17. Report of J. W. Castleman to Seddon, March 1896, SSSICR.
18. Brainerd, "Colored Mining Labor," 79.
19. Balance Sheet, SISC, year ending 31 July 1891, SSSICR; on the lack of compulsion upon workers to trade at the company stores, see Report of Superintendent for Brookside, Brazil, and Cardiff mines, SISC, year ending 31 January 1896, ibid.
20. Armes, *Coal and Iron*, 21–23, 48–50, 457–58; *Bessemer Weekly*, 18 May 1901 (reprint, 1977), 57. For a specimen of Culverhouse's views, see Report of Superintendent for Brookside, Brazil, and Cardiff mines, year ending 31 January 1896, SSSICR.
21. On antebellum precedents of the system described here, see Starobin, *Industrial Slavery*, 105–9.
22. "No Discrimination," *Bessemer Weekly*, 18 May 1901 (reprint, July 1977), 56.
23. Paul Worthman, "Working Class Mobility in Birmingham, Alabama, 1880–1914," in Tamara K. Hareven, ed., *Anonymous Americans: Explorations in Nineteenth-Century Social History* (Englewood Cliffs, N.J.: Prentice-Hall, 1971), 196–97.
24. Ibid., 180–85; Barbara J. Mitchell, "Steel Workers in a Boom Town: Birmingham, 1900," *Southern Exposure* 12 (November–December, 1984): 56–60. The title of Mitchell's otherwise excellent demographic study of Sloss workers in this period is a misnomer, in that Sloss did not make steel.
25. See particularly Balance Sheet, SISC, six months ending 31 July 1890 and 31 July 1891, and accounts in Ledger C, SISC, 1890–92, SSSICR.
26. Balance Sheet, SISC, six months ending 31 July 1891, SSSICR.
27. Report of William Kent to John C. Maben, 10 March 1892, SSSICR.
28. Bryan to J. J. Newman, 31 December 1890; 7, 14, 24 January; 6 February 1891, JBL. Newman was general manager of the North Carolina Steel & Iron Company in Greensboro, N.C.
29. Undated proposal, "To the stockholders of the Tennessee Coal, Iron & R.R. Co. and the stockholders of the Sloss Iron & Steel Co.," late 1891–early 1892, Alfred Montgomery Shook Papers, BPLA.
30. Leslie Hannah, "Mergers," in Porter, ed., *Encyclopedia of American Economic History*, 2:641.
31. Manuscript diary of James Bowron, Jr., W. S. Hoole Special Collections Library, University of Alabama (hereafter cited Bowron Diary), entries for 26 February, 1 and 2 March 1892; Bowron Autobiography, 1:317–19.
32. *BA-H*, 5 March 1892.
33. Klein, *Great Richmond Terminal*, 235–51; *NYT*, 15 March 1892.
34. Bryan's letterbook for the period 1 November 1891 through 26 July 1892 shows that he was moving almost constantly between New York, Richmond, and Baltimore in the first three weeks of March 1892. He was in New York from 29 February through 7 March; returned to Richmond to take part in a "very important" meeting of the Richmond and Kanawha Lumber Company, of which he was president, on 8 March; went to Baltimore on 10 March to meet with the trustees of the Chesapeake and Ohio Canal; and was back in New York on 16 March. That

he kept in touch with what was going on in New York while he was away from the city is evident; see for example copy of telegram from Bryan to Maben, 9 March 1892, JBL, asking if his presence there was necessary on that extremely important date in the merger discussions.

35. Bowron Diary, entries of 1, 2, 4, 8 March 1892; *NYT*, 8 March 1892. For a biographical sketch of Thomas, see Klein, *Great Richmond Terminal*, 50–51.

36. "Coal and Iron Properties To Be United in One Great Consolidation," *BA-H*, 5 March 1892; "Alabama Iron: Three Great Producing Companies Come Together," ibid., 8 March 1892; "The Deal to be Made," ibid., 9 March 1892; "The Big Deal Will be Consummated Today," ibid., 10 March 1892; "Signed: The Consolidation Scheme Now a Fixed Fact," ibid., 12 March 1892. For national reportage, see particularly "The Proposed Southern Consolidation," *IA*, 10 March 1892, and "Three to Make One," *NYT*, 12 March 1892.

37. Bowron Autobiography, 1:318–19; Bowron Diary, entries of 4, 5, 7, 8, 9, 10, 11, 12 March 1892; Bryan to Maben, 10 March 1892, JBL. Among previous secondary sources, the best account of the negotiations, told from the point of view of TCI, is in Fuller, "TCI—Chapel Hill," 102–3. Fuller, however, refers mistakenly to the president of Sloss as "William Seddon" and is also unaware of Bryan's guiding presence behind the scenes.

38. "Closed is the Great Deal in Iron Property" and "The Great Deal," *BA-H*, 14 March 1892; "Coal Companies Consolidated," *NYT*, 14 March 1892; "The Southern Consolidation," *IA*, 17 March 1892; Bowron Autobiography, 1:319; Bowron Diary, entries of 14, 15, 16, 17, 18, 26 March 1892; Fuller, "TCI—Chapel Hill," 103–8; Armes, *Coal and Iron*, 423–24.

39. Bowron Diary, entries of 14, 15, 18 March 1892; "Mr. Seddon Insisted on Conditions for the Sloss Company Which Were Not Agreed To," *BA-H*, 15 March 1892; Armes, *Coal and Iron*, 423; Fuller, "TCI—Chapel Hill," 109–12, 384–88; Fuller, "Henry F. DeBardeleben," 12–16.

40. "An Elaborate Scheme: The Richmond Terminal Organization Plan Made Public by the Olcott Committee . . . President Inman Retires," *NYT*, 17 March 1892; "Richmond Terminal: The New President Takes Hold," ibid., 18 March 1892; Klein, *Great Richmond Terminal*, 254–55.

41. Klein, *Great Richmond Terminal*, 255–58.

42. Bryan's correspondence in JBL between July and December 1892 is rich in letters relating to this project.

43. Klein, *Great Richmond Terminal*, 259–84.

44. Freeman, "John Stewart Bryan," 483; Davis, *Southern Railway*, 192. For a perceptive analysis of Bryan's conviction that the Morgan-backed reorganization represented a victory for stability, security, and developmental goals, see Anne Brockenbrough, "Joseph Bryan from Planter to Industrialist: Portrait of a Conservative Southern Democrat" (unpublished research paper written at the University of Virginia, April 1986, copy at VHS, 12–15).

45. Bryan to Mrs. M. S. Peterkin, 20 September 1894, JBL.

46. Bryan to Maben, 12 November 1894, JBL.

47. Davis, *Southern Railway*, 19–44, 116, 197. On Andrews, see biographical sketch by Bennett L. Steelman in William S. Powell, ed., *Dictionary of North Carolina Biography* (Chapel Hill: University of North Carolina Press, 1979), 1:34–36.

48. Bryan to Maben, 8 September 1894, JBL.

49. Bryan, *Joseph Bryan*, 265–70; Lindgren, "'First and Foremost a Virginian,'" 171–73.

50. Dabney, *Virginia: The New Dominion*, 415–16; Allen W. Moger, *Virginia: Bourbonism to Byrd, 1870–1925* (Charlottesville: University Press of Virginia, 1968), 84. On Laburnum, see Bryan, *Joseph Bryan*, 230–45. I am grateful for assistance

from Merle L. Colglazier, medical librarian, Richmond Memorial Hospital, for sharing with me notes resulting from research that he has conducted on the history of the mansion.

51. Bryan to Messrs. Hardy & Co., Birmingham, 4 October 1894, JBL. The impetus for the public listing seems to have come from Thomas Seddon and his Birmingham allies through a Baltimore firm run by relatives of Seddon. See Bryan to W. C. Seddon, 6 October 1894, ibid.

52. Fels, *American Business Cycles*, 184–92; Robert Sobel, *Panic on Wall Street: A History of America's Financial Disasters* (New York: Macmillan, 1968), 245–72.

53. Among numerous accounts of these developments, see particularly Page Smith, *The Rise of Industrial America: A People's History of the Post-Reconstruction Era* (New York: McGraw-Hill, 1984), 508–53.

54. Clark, *History of Manufactures*, 2:229–30, 302–3, and 3:28–29. On the impact of the depression upon the anthracite iron industry, see particularly Temin, *Iron and Steel in Nineteenth Century America*, Appendix C. By 1897, the percentage of pig iron smelted with anthracite alone was zero, but some furnaces were still using a mixture of anthracite and coke.

55. In addition to the installations named, the roster of plants that went out of blast during the decade included both the coke-fired Woodstock furnaces at Anniston, Bibb Furnace at Briarfield, Clara Furnace at North Birmingham, Decatur Furnace in Morgan County, Eagle Furnace at Attalla, Edwards Furnace in Bibb County, the Etowah and Gadsden furnaces in Etowah County, Philadelphia Furnace at Florence, Stonewall and Tecumseh furnaces in Cherokee County, Talladega Furnace at Talladega, and Trussville Furnace in Jefferson County. In addition, Piedmont Furnace in Calhoun County, begun in 1890, was never completed. Woodward, *Alabama Blast Furnaces*, 39, 48–49, 62–63, 68–73, 75–76, 96, 111–13, 132–35, 140–41, 147, 150–54.

56. Fuller, "TCI—Chapel Hill," 111–12; Bowron Autobiography, 1:386.

57. Balance Sheets, SISC, fiscal years ending 31 January 1894, 1895, 1896, and 1897, SSSICR.

58. Ibid.

59. Quoted in Stuart Bruchey, *The Wealth of the Nation: An Economic History of the United States* (reprint, New York: Harper & Row, 1988), 123.

60. Bryan to J. C. Brooks, 28 March 1895; Bryan to Sol Haas, 20 January 1897, JBL.

61. Report of Superintendent for Brookside, Brazil, and Cardiff mines, SISC, year ending 31 January 1896, SSSICR.

62. Balance Sheet, SISC, fiscal years ending 31 January 1894 and 1895, SSSICR.

63. Porter and Livesay, *Merchants and Manufacturers*, 131–47.

64. Armes, *Coal and Iron*, 351–52; "A Most Powerful Developing Agency," BA-H, 5 August 1900; Annual Report of Secretary and Treasurer, SISC, fiscal year ending 31 January 1897, SSSICR; Bowron Autobiography, 1:387, 389. On the marketing efforts of the Richmond Locomotive Works in Finland and Sweden, see Bryan to Charles W. Dear, 10 November 1898, JBL, and "Locomotives for Work North of the Arctic Circle," *IA*, 17 August 1899.

65. Armes, *Coal and Iron*, 352. On the voyage itself, see A. B. Feuer, "The Odyssey of the Battleship *Oregon*," *Battleships at War* 1 (1991): 76–83.

66. Uehling Autobiography, 38, 39a.

67. Ibid., 39–40.

68. Bryan to Seddon, 4 February 1895, JBL.

69. Bryan had praised McCune in a letter to Seddon in 1891 after McCune had agreed to perform services for the company that Bryan did not specify. Bryan to Seddon, 31 January 1891, JBL.

70. Morton to Uehling, 5 February 1894, Morton-Uehling Correspondence.

71. Uehling Autobiography, 40; Armes, *Coal and Iron,* 300; Morton to Uehling, 17 March 1894, Morton-Uehling Correspondence; Bryan to Maben, 8 February 1895, JBL.

72. The foregoing discussion of problems involved in manual casting and the description of Uehling's machine that follows are based on Uehling, "Pig-Iron Casting and Conveying Machinery," 115–18, and specification and drawings in Edward A. Uehling, "Apparatus for and Method of Casting and Conveying Metals," U.S. Patent no. 548,146, 15 October 1895. I am indebted to Melvin C. Smith, a graduate student at Auburn University, for assistance in studying the processes involved in both manual and mechanical casting.

73. Uehling Autobiography, 79–133; Uehling, "Pig-Iron Casting and Conveying Machinery," 118.

74. Edward A. Uehling and Alfred Steinbart, "Pyrometer," U.S. Patent no. 554,323, 11 February 1896; Uehling Autobiography, 44–45.

75. Uehling Autobiography, 45–46, 67; Morton to Uehling, 28 February 1896, Morton-Uehling Correspondence.

76. Uehling Autobiography, 48–50.

77. Balance sheets for fiscal years ending 31 January 1893, 1894, and 1895, SSSICR.

78. A journal of the Sloss Iron and Steel Company for 1891–94 in SSSICR is rich in quarterly entries relating to expenses incurred in connection with Coalburg and other convict operations conducted by the company. See for example entries of 30 November 1891, folio 53; 30 April 1892, folio 159; 30 June 1892, folios 206 and 215; and 31 December 1892, folio 347. For additional perspectives on the system, see Flynt, *Mine, Mill & Microchip,* 106, 149; Flynt, *Poor but Proud,* passim; Mitchell, "Steel Workers in a Boom Town," 57; Taft, *Organizing Dixie,* passim; and Mitch Mentzer and Mike Williams, "Images of Work: Birmingham, 1894–1937," *JBHS* 7 (June 1981): 10–23, with photographs.

79. Robert David Ward and William Warren Rogers, *Labor Revolt in Alabama: The Great Strike of 1894* (University: University of Alabama Press, 1965), 30–74.

80. For the best accounts of the rise of radical agrarianism in Alabama, see Sheldon Hackney, *Populism to Progressivism in Alabama* (Princeton, N.J.: Princeton University Press, 1969), and William Warren Rogers, *The One-Gallused Rebellion: Agrarianism in Alabama, 1865–1896* (Baton Rouge: Louisiana State University Press, 1970). On Bourbon politics, see Going, *Bourbon Democracy in Alabama.* On the supposed connection between radical agrarianism and the beginnings of the Big Mule–Black Belt Alliance, see Wiener, *Social Origins of the New South,* 222–27.

81. Ward and Rogers, *Labor Revolt in Alabama,* 75–102; Birmingham *Item,* 21 May 1894, 2 and 4 June 1894.

82. Ward and Rogers, *Labor Revolt in Alabama,* 103–38.

83. Thomas D. Parke, M.D., *Report on Coalburg Prison . . . Rendered to the Committee of Health of Jefferson County Medical Society* (Birmingham: Roberts & Son, Printers, 1895), passim; "Facts About Coalburg Prison: Doctor Parke's Report Reviewed," pamphlet by J. W. Castleman and F. P. Lewis, M.D. (Birmingham: Patch Stationery, c. 1895), copy in SSSICR. On the use of similar hospital facilities by industrial slaveholders prior to the Civil War, see Starobin, *Industrial Slavery,* 69–71.

84. Report of J. W. Castleman, Assistant to President, SISC, fiscal year ending 31 January 1896, SSSICR; letter from F. P. Lewis, M.D., to S. B. Trapp, Alabama Board of Convict Inspectors, 15 October 1898, Alabama State Archives (hereafter cited ASA). For an account of the entire episode, see Robert David Ward and William Warren Rogers, "Racial Inferiority, Convict Labor, and Modern Medicine: A Note on the Coalburg Affair," *AHQ* 44 (Fall–Winter 1982): 203–10.

Ward and Rogers, however, are mistaken in stating that the dispute "failed to produce any results that were immediately beneficial," not realizing that Sloss was forced to undertake at least some limited remedial action.

85. "Thomas Seddon" and "Death of Thomas Seddon," *BSH,* 12 May 1896, 4, 8; obituary of Seddon, *RT,* 12 May 1896; "Tribute to Mr. Seddon," newspaper clipping from *BSH,* 31 May 1896, copy in SSSICR.

8. Brown Ore, Basic Steel, and the Emergence of Sloss-Sheffield

1. Woodward, *Origins of the New South,* 291–320.
2. Obituary of Haas in *RT-D,* 25 November 1909.
3. Bryan's correspondence prior to 1896 contains many letters to Haas. See, for example, Bryan to Haas, 19 July, 5 August, 7 October 1890, 29 July 1892, and 22 August 1895, JBL.
4. Bryan to Haas, 6 July 1896, ibid.
5. Bryan to Haas, 20 and 26 January 1897, ibid. On the significance of marketing as a theme in the history of Sloss Furnaces, see Claire-Louise Datnow, "The Sloss Company: Symbol of the New South" (master's thesis, University of Alabama at Birmingham, 1987), 102–26.
6. Bryan to Maben, 13 January 1897, ibid. On the loss about which Bryan complained, see Balance Sheet, SISC, year ending 31 January 1897, SSSICR.
7. *ITR,* 3 June 1897.
8. Ibid., 6 May 1897. On Strong's activities during this period, see Klein, *Great Richmond Terminal,* 245, 246, 253, 255, 259–60n, 261, 262, 265, 269.
9. William C. Oates, "Industrial Development of the South," *North American Review* 161 (November 1895): 569; Balance Sheet, SISC, year ending 31 January, 1897, SSSICR.
10. Dabney, *Virginia: The New Dominion,* 425; Bryan to Haas, 22 December 1896 and 26 January 1897, JBL.
11. Geoffrey H. Moore, "Business Cycles, Panics, and Depressions," in Porter, ed., *Encyclopedia of American Economic History,* table on p. 152; Sobel, *Panic on Wall Street,* 269–71; Sidney Ratner, et al., *The Evolution of the American Economy: Growth, Welfare, and Decision Making* (New York: Basic Books, 1979), 390.
12. Balance Sheet, SISC, year ending 31 January 1898, and report of Robertson to Haas, 5 April 1899, in SSSICR.
13. Approximate Estimate of expenses, SISC, twelve months ending 31 January 1899, SSSICR; Bryan to Haas, 10 February 1898; Bryan to Alexander B. Andrews, 10 February 1898; Bryan to J. W. Johnston, 14 February 1898; Bryan to Maben, 12, 21 April 1898; Bryan to Haas, 3 or 4 June 1898, JBL.
14. Estimate of expenses, SISC, twelve months ending 31 January 1899, SSSICR; Report of J. W. Castleman to Alabama Industrial and Scientific Society, Birmingham, 10 February 1899, copy in ibid.; Jack Roland Bergstresser, Sr., "The Primary Iron and Steel Industry of Jefferson County" (unpublished research report, SFNHL, 1989), 13; Chapman et al., *Iron and Steel Industries,* 42–45; P. B. McDonald, "Iron Ore Mining in the South," *ITR,* 22 October 1914, 759–62; Dwight E. Woodbridge, "Iron Mining Methods and Costs in Alabama," *IA,* 11 September 1913, 588–89.
15. Bryan to Haas, 29 July 1897, JBL.
16. Resolution, Adjourned Meeting of Directors, SISC, 10 November 1897, SSSICR; Minutes, Board of Directors, SISC, November–December 1897, ibid.; Bryan to Thomas Pinckney, 3 December 1897; Bryan to Central Trust Company of New York, 4, 8, 11 December 1897; Bryan to Haas, 12 and 14 January 1898, JBL.

17. Bryan to Haas, 10 February, 10 March, 12, 13 April, 9 June 1898; Bryan to Maben, 4 and 5 March, 12 April, 14 July 1898, JBL.
18. Bryan to Haas, 14, 18 July 1898; Bryan to Maben, 14, 25 July 1898, ibid.
19. Bryan to Haas, 3 October 1898, ibid.; Castleman, Report to Alabama Industrial and Scientific Society, passim.
20. Bryan to Maben, 29 August 1898, JBL; Haas, Estimate of Expenses, SISC, twelve months ending 31 January 1899, SSSICR.
21. Bryan to W. H. Goadby and Co., 3 November 1898, JBL.
22. Nathan Rosenberg and L. E. Birdzell, Jr., *How the West Grew Rich: The Economic Transformation of the Industrial World* (New York: Basic Books, Inc., 1986), 220–28. For statistical data on the merger movement, see Naomi R. Lamoreaux, *The Great Merger Movement in American Business, 1895–1904* (Cambridge: Cambridge University Press, 1985), 1–5. Among many other sources dealing with the escalation of industrial mergers during this period, see particularly Chandler, *Visible Hand*, 315–44, and Leslie Hannah, "Mergers," in Porter, ed., *Encyclopedia of American Economic History*, 2: 640–44. On mergers in the iron and steel industry, see Lewis, *Iron and Steel in America*, 50.
23. Chandler, *Visible Hand*, 287, 294–95, 312–36, 355–59, 409, 472–73. For a shorter exposition of the same tendencies in this period, see Chandler, "Rise and Evolution of Big Business," in Porter, ed., *Encyclopedia of American Economic History*, 2: 625–30.
24. Armes, *Coal and Iron*, 336–37, 356–57; White, *Birmingham District*, 128–29.
25. Lewis, *Iron and Steel in America*, 43–50.
26. See particularly Wright, *Old South, New South*, 172–73.
27. Charles H. Frye, comp., *Birmingham Illustrated: Showing Photographic Views of Many of the Manufactories, Mining and Iron Industries, Public Men and Places of Interest* (Birmingham: Commercial Club, 1896), 6–7.
28. For a brief but lucid discussion of the problems involved in basic steel manufacture, see *Encyclopedia Britannica*, 1968 ed., s.v. "Iron and Steel Industry." On the legal situation involving the Thomas and Reese patents, see Fuller, "TCI—Chapel Hill," 252; McHugh, *Alexander Holley and the Makers of Steel*, 358–60; and Wall, *Andrew Carnegie*, 502–4. On European methods that required even more phosphorus than southern ores contained, see Wright, *Old South, New South*, 173.
29. McHugh, *Alexander Holley and the Makers of Steel*, 360; Wright, *Old South, New South*, 165.
30. Bowron Autobiography, 1:372–73; Clark, *History of Manufactures*, 2:243.
31. Clark, *History of Manufactures*, 2:243; Fuller, "TCI—Chapel Hill," 256. For the analysis that follows, I have drawn largely upon Fuller's handling of the subject. See also Armes, *Coal and Iron*, 431–35, 461–67.
32. Bowron Autobiography, 1:372–73; Fuller, "TCI—Chapel Hill," 256–57.
33. For historical perspective on these developments, see Bergstresser, Sr., "Raw Material Constraints," 92–94; H. S. Geismer, "Coal Washing Practice in Alabama," AIME *Transactions* 71 (1925): 1088–1105; and David Hancock, "Coal Washing in Alabama," in Phillips, *Iron Making in Alabama*, 233–45. My description of the Robinson-Ramsay machine is based on Erskine Ramsay, "Coal and Mineral Washer," U.S. Patent no. 528,803, 6 November 1894. See also the account of its background and functioning in James Saxon Childers, *Erskine Ramsay: His Life and Achievements* (New York: Cartwright & Ewing, 1942), 199–207, which is consistent with the details in Ramsay's patent. On improvements made to the same machine by Ramsay and an assistant after 1894, see Erskine Ramsay and Ernest Dreyspring, "Coal and Mineral Washer," U.S. Patent no. 579,840, 30 March 1897. For a description of a Robinson-Ramsay machine

that included some of the features shown in Ramsay's patent drawings but was different from it in many respects, see J. J. Ormsbee, "Coal-Washing: Notes on a Southern Coal-Washing Plant," in *Colliery Engineer and Metal Miner* 16 (September 1895): 27–29. The installation shown in Ormsbee's article, located at TCI's No. 2 Pratt mine, may be an earlier version of the design patented by Ramsay in 1894, which is how I have interpreted it in the account given here. For additional discussions showing Robinson-type washers with other features than those shown in Ramsay's patent drawings, see L. A. Harding and G. R. Delamater, "Bituminous Coal Washing," *Mines and Minerals* 25 (July 1905): 580, and Ernest Prochaska, *Coal Washing* (New York: McGraw-Hill Book Company, 1921), 53–57. Both of these sources contain drawings of coal-washing equipment made by the Jeffrey Manufacturing Company, an American firm that first introduced Robinson's apparatus to the United States. Evidently the original British prototype spawned a number of models that bore a family resemblance to one another but differed in certain details, suggesting the need for further exploration. I am indebted to Bergstresser for sharing with me the results of his research and for his help in assessing problems raised by the sources just cited.

34. Bowron Autobiography, 1:373–75.
35. Ibid., 1:379–86; Fuller, "TCI—Chapel Hill," 258–59.
36. Bowron Autobiography, 1:396–408; *BA-H*, 7 July 1899; Fuller, "TCI—Chapel Hill," 259–60. For a photograph of workers and other people posing with the first batch of steel ingots produced at the new mill, see Flynt, *Mine, Mill & Microchip*, 117.
37. Bowron Autobiography, 1:409–10.
38. For another account of the episode that follows, based largely on sometimes misleading information in Armes, see Klein, *Louisville & Nashville Railroad*, 273–76. In cases of discrepancy, I have relied on Bowron's reminiscences and Fuller, "TCI—Chapel Hill."
39. Klein, *Louisville & Nashville*, 270–72. On the earlier history of the Anniston & Atlantic and the Anniston & Cincinnati, see Gates, *Model City*, 88–90, 100–101.
40. Bowron Autobiography, 1:414.
41. Ibid., 1:414–16. For copies of the executive committee's resolution and subsequent letters on the matter, copied from its minutes, see Armes, *Coal and Iron*, 463–64.
42. Bowron Autobiography, 1:417–19.
43. Ibid., 1:419–36; Fuller, "TCI—Chapel Hill," 117.
44. For a detailed description of the operation of the plant, see "Mr. Bowron on Rapid Steel Manufacture," *BA-H*, 4 December 1899, reprinted from New York *Times*. In addition, see Fuller, "TCI—Chapel Hill," 263–64.
45. "The Steel Mill in Operation," *BA-H*, 1 December 1899; "New Steel Works in Alabama," ibid., 3 December 1899, reprinted from New York *Times*. For further details, see Bowron, *Autobiography*, 1:480; Clark, *History of Manufactures*, 3:49–50; and Fuller, "TCI—Chapel Hill," 265.
46. Armes, *Coal and Iron*, 413–14; "The Other Birmingham: Sheffield and Its Advantages as a Manufacturing Town," *BA-H*, 9 December 1886; *Northern Alabama: Historical and Biographical*, 409–10, 417–18.
47. "The Founding of a Town," *MR*, 17 May 1884.
48. "Points of Information in Reference to the Proposed Great Iron Manufacturing City of Sheffield, Alabama," ibid., 26 April 1884; *Northern Alabama: Historical and Biographical*, 411.
49. "Sheffield's Progress," "Sheffield's Big Consolidation," and "A Big Enterprise," in *MR*, 11 December 1886, 10, 24 September 1887; Armes, *Coal and Iron*, 415;

Northern Alabama: Historical and Biographical, 410–11; Woodward, *Alabama Blast Furnaces*, 56, 75; dedication program of observances marking erection of a historical marker at Furnace Hill, 4 April 1976, supplied to author by Richard C. Sheridan.

50. Armes, *Coal and Iron*, 412; Woodward, *Alabama Blast Furnaces*, 75, 96.
51. "To Build 100-Ton Furnace," "Florence, Alabama," and broadside, "Florence, Ala.: County Seat of Lauderdale County," MR, 16, 23 April 1887 and 13 August 1887; *Northern Alabama: Historical and Biographical*, 288–92; Woodward, *Alabama Blast Furnaces*, 103–4, 111–12.
52. "Sheffield, Colbert County, Ala.: The Iron Manufacturing Center of the South," MR, 24 August 1887; Armes, *Coal and Iron*, 415–16.
53. "Sheffield, Ala.," MR, 7 July 1888.
54. Woodward, *Alabama Blast Furnaces*, 56–57, 76, 96, 104, 112.
55. DuBose, *Mineral Wealth of Alabama*, 32; *Northern Alabama: Historical and Biographical*, 102. On the nature and characteristics of the Russellville deposits, see E. F. Burchard, "The Brown Iron Ores of the Russellville District, Alabama," in U.S. Geological Survey, *Contributions to Economic Geology*, Bulletin 315-D (Washington, D.C., 1907).
56. Balance Sheet, SISC, fiscal year ending 31 January 1900, SSSICR.
57. Minutes of the Board of Directors, Sloss Iron and Steel Company, 1899, SSSICR; Willie Jo Wood, "Sloss Lake, A Plantation Mine," in *The Source* [Franklin Printing Co., Russellville, Ala.] 3 (Winter 1993): 9–12; Bowron Autobiography, 1:470–71.
58. Woodward, *Alabama Blast Furnaces*, 112; Bowron Autobiography, 1:473; Minutes, Board of Directors, SISC, 1899, SSSICR; "Sloss Company to Get Ore Mines," BA-H, 31 May 1903; Annual Report of the President, SSSIC, March 1904, SSSICR.
59. BA-H, 13 July 1899; 18 July 1899, 5; 30 July 1899; Minutes, Meetings of Stockholders and Directors, SSSIC, 30 August and 10 November 1899, SSSICR; list of original stockholders in Sloss-Sheffield, SSSICR. For a complete list of the property holdings of the new company, see President's Report, Annual Meeting of Sloss-Sheffield Steel and Iron Company, 19 March 1901, SSSICR. I am indebted to the State Library of Louisiana for providing manuscript census materials identifying Seixas as a resident of New Orleans in 1900.
60. List of original stockholders in Sloss-Sheffield, SSSICR. On the involvement of Hopkins in northwestern Alabama properties acquired by Sloss-Sheffield, see Minutes, Board of Directors, SSSIC, 12 June 1906, SSSICR. The possibility of a family connection between Parsons and the Virginia entrepreneur mentioned here is logical because Parsons was ultimately succeeded on Sloss-Sheffield's board of directors by a son named Henry. I have not been able, however, to trace such a relationship. On the life and career of Henry C. Parsons, a native of Vermont who settled in West Virginia after the Civil War and became active in southern railroading, see NCAB, 14:497.
61. For a complete list of the new board, see *Moody's Manual of Industrial and Miscellaneous Securities*, First Annual Number, 1900 (New York: O. C. Lewis Co., 1900), 461.

9. The Turning Point

1. My use of the mirror-image concept to characterize the differing roles played by Sloss-Sheffield and TCI in this and other places is borrowed from Edwin T. Layton, "Mirror-image Twins: The Communities of Science and Technology in 19th Century America," T&C 12 (October 1971): 562–80, with permission.

2. "Former Senator Milner is Dead" and "Funeral of J. T. Milner," *BA-H*, 19 and 20 August 1898.
3. Martha Mitchell Bigelow, "Birmingham's Carnival of Crime, 1871–1910," *AR* 3 (April 1950): 123–33.
4. Atkins, *Valley and Hills*, 48–49, 96; Frye, *Birmingham Illustrated*, 26–27; McMillan, *Yesterday's Birmingham*, 35, 48, 52–55, 72; Marjorie L. White, *Downtown Birmingham: Architectural and Historical Walking Tour Guide* (Birmingham: Birmingham Historical Society and First National Bank of Birmingham, 1980), 39, 80. On the significance of the marketing shift mentioned here, see Klein and Kantor, *Prisoners of Progress*, 77–84.
5. Atkins, *Valley and Hills*, 79, 88; Bowron Autobiography, 1:376, 388, 394, 402, 453; Frye, *Birmingham Illustrated*, 14–15, 25; McMillan, *Yesterday's Birmingham*, 68, 71; Anne G. Pannell and Dorothea E. Wyatt, *Julia S. Tutwiler and Social Progress in Alabama* (Tuscaloosa: University of Alabama Press, 1961), 1–16 and passim.
6. Annie Woodbridge, "Mary Johnston: A Universal Virginian," *RQ* 4 (Summer 1981): 23–33. Among other assessments, see Edward Wagenknecht's laudatory "The World and Mary Johnston," in *Sewanee Review* 44 (April–June 1936): 188–206, and the dismissive treatment in Robert A. Lively, *Fiction Fights the Civil War* (Chapel Hill: University of North Carolina Press, 1957). For a catalog of Johnston's works and critical estimates of her writings, see George C. Longest, *Three Virginia Writers; Mary Johnston, Thomas Nelson Page and Amelia Rives Troubetzkoy: A Reference Guide* (Boston: G. K. Hall & Co., 1978).
7. Harris, *Political Power in Birmingham*, 59–63.
8. Ibid., 186–89.
9. Ibid., 105–07; McMillan, *Yesterday's Birmingham*, 50.
10. Frye, *Birmingham Illustrated*, 5–6; Flynt, *Mine, Mill & Microchip*, 94–104; McMillan, *Yesterday's Birmingham*, 50; "Phenomenal Growth of Birmingham District During Past Twelve Months," *BA-H*, 5 August 1900.
11. "The Great Tennessee Coal, Iron and Railroad Co.," *BA-H*, 2 August 1900.
12. "A Most Powerful Developing Agency," *BA-H*, 5 August 1900; "Sloss-Sheffield Deal has been Consummated . . . A New Steel Plant in View," ibid., 2 August 1899; "Will Build a Steel Plant," ibid., 8 August 1899; "North Birmingham Land Company . . . Rumored that Steel Plant of Sloss-Sheffield Company Will be Located On their Lands," ibid., 19 August 1899. On the attempted land acquisition from The University of Alabama, see "Exposure Created a Big Sensation," "Great Indignation Felt in Tuscaloosa," "A Hot Reply from Mr. Fitts," "Sale of the University Lands Severely Condemned by Alumni," "Trustees Declare the Sale of the University Lands Illegal," "Great Victory for the Alumni," and other articles in *BA-H*, 18, 20, 21, 27, 29 November and 6, 8, 20, 21, 22 December 1899. For Stewart's assessment of the episode, see his *Governors of Alabama*, 146.
13. Balance Sheet, SISC, year ending 31 January 1900, with attached clipping from unidentified newspaper of 5 April 1900, SSSICR.
14. See table of cyclical fluctuations in the American economy between 1834 and 1975 in Moore, "Business Cycles, Panics, and Depressions," in Porter, ed., *Encyclopedia of American Economic History*, 1:152.
15. Bryan to P. H. Earle, 18 December 1899; to Maben, 18 December 1899; and to George Arents, 3 February 1900, JBL.
16. Bryan to Oakman, 27 November and 7 December 1899; to George S. Bertine, assistant secretary, Central Trust Company, 27 November 1899; to Maben, 29 November and 15 December 1899; to Haas, 29 November 1899; to Alabama National Bank, 4 December 1899; to Birmingham Trust & Savings Company, 4 December 1899; to W. H. Goadby and Company, 11 December 1899; to Earle, 18 December 1899, ibid.

17. Bryan to W. H. Goadby and Company, 19, 20, 22, 26, 27, 28, 30 December 1899, ibid.
18. Bryan to Maben, 4 January 1900; to Bertine, 4 January 1900; to Goadby and Company, 4 and 5 January 1900; to N. S. Walker, 20 January 1900, ibid.
19. Bryan to Goadby and Company, 29 January 1900; Bryan to Alabama National Bank, 2 March 1900; to Birmingham Trust and Savings Company, 2 March 1900; to Central Trust Company, 30 March, 4 April 1900; to Maben, 30 March, 3 April 1900; to Frederic P. Olcott, 31 March 1900; to Bertine, 3 April 1900, ibid.
20. Bryan to Guaranty Trust Company of New York, 30 March 1900; to Ladenburg, Thalman and Company, 4, 5 April 1900; to Central Trust Company, 4 April 1900; to Bertine, 5 April 1900; and to Maben, 11 April 1900, ibid. Goadby was also willing to loan Bryan $50,000 and take further shares of Sloss-Sheffield as collateral. Bryan, however, was able to decline this aid, probably in part because of Maben's action in settling his debt. See Bryan to Goadby and Company, 3, 5 April 1900, ibid.
21. Bryan to W. C. LeGendre, 17 March 1900, ibid.; Dabney, *Richmond*, 268; Lindgren, "'First and Foremost A Virginian,'" 172–73.
22. Bryan to Maben, 17, 21 February 1900, JBL; "Sloss Company Stock Listed on the New York Stock Exchange," *BA-H*, 17 June 1900. In the final outcome, all but 30 of 50,000 shares of the Sloss Iron and Steel Company's capital stock were acquired by Sloss-Sheffield. See *Moody's Manual*, 1900, 457.
23. Bryan to Haas, 7 February 1900, and to Maben, 7 February 1900, JBL. On Hopkins's background, see *A History of Vanderburgh County, Indiana* (n.p.: Brant & Fuller, 1889), 212–13, and *BA-H*, 10 August 1900. I am indebted to Marvin Guilfoyle, librarian, University of Evansville, and Eric L. Mundell, head of reference services, Indiana Historical Society, for assistance in largely unsuccessful efforts to learn more about the involvement of Hopkins in northwest Alabama enterprises prior to the beginning of his association with Sloss-Sheffield.
24. Bryan to Haas, 7, 11 December 1899; Bryan to Oakman, 11 December 1899; Bryan to Maben, 26 February 1900, JBL; "T. H. Aldrich Goes with the Sloss Co.," *BA-H*, 18 December 1899. On the departure of Aldrich from TCI, see Fuller, "TCI—Chapel Hill," 128–29.
25. Bryan to Haas, 5, 17, 21 April 1900, JBL.
26. Bryan to Haas, 28 April 1900; to Maben, 28 April and 3 May 1900; to Rucker, 28 April 1900; and to Oakman, 4 May 1900, ibid.
27. "$75,000 Fire at Sloss Furnace," *BA-H*, 6 May 1900.
28. "No. 2 Furnace Again in Blast," *BA-H*, 11 May 1900; "Again in Blast," ibid., 17 May 1900; "Enterprise of the Management: Damage to Sloss Company's Furnaces Quickly Overcome, Only Few Days were Lost," ibid., 1 June 1900; "Other Railroads Assisted Sloss Company in Repairing Damage to Furnaces," ibid., 2 June 1900.
29. Bryan to Aldrich, 7 May 1900; to Maben, 7 May 1900; to Rucker, 7, 15 May 1900; and to Oakman, 15 May 1900, JBL.
30. Bryan to Haas, 15, 21, 24, 28, 29 May and 11, 21 June 1900, ibid.
31. "Mr. Haas Resigns the Presidency," *BA-H*, 3 August 1900.
32. "E. O. Haskins Elected President," *BA-H*, 10 August 1900; *Moody's Manual*, 1900, 461.
33. "People and Things," *BL*, 12 March 1902.
34. "President Hopkins Here," *BA-H*, 5 September 1900; "Resignation of T. H. Aldrich," ibid., 26 September 1900; "Officials Named by E. O. Hopkins," ibid., 1 October 1900.
35. Report of the President, SSSIC, 1900 Fiscal Year, SSSICR; Minutes, Meetings of Stockholders and Directors, SSSIC, 1899–1918, ibid.
36. Report of President, SSSIC, 1901 Fiscal Year, SSSICR.

37. Fuller, "TCI—Chapel Hill," 132–34 and Table A, 384.
38. Minutes, Executive Committee, SSSIC, 14 November 1901 and 9 January 1902, SSSICR.
39. Bowron Autobiography, 1:442, 446, 451, 495, 523.
40. *BL*, 18 March 1902; *BN*, 28 March 1902; *BA-H*, 29 March 1902.
41. "More Resignations to Follow in Sloss Company," *BA-H*, 30 March 1902; "Hopkins in Demand and He May Accept," ibid., 1 April 1902; "People and Things," *BL*, 1 April 1902; "Mr. Hopkins to Sail June 25," *BA-H*, 11 April 1902.
42. Minutes, Annual Meeting, SSSIC, 20 March 1902, SSSICR. On Brown's attempts to have Bowron elected president of Sloss-Sheffield and his subsequent attempt to enlist Bowron in a takeover of TCI, see Bowron Autobiography, 1:471, 598.
43. "Maben Arrives in Birmingham," *BA-H*, 28 April 1902.
44. "President Maben Now in Charge," *BA-H*, 1 May 1902.
45. Bryan to Seixas, 3 February 1903, JBL.
46. Fuller, "TCI—Chapel Hill," 137–38.
47. Woodward, *Alabama Blast Furnaces*, 157–58; "Woodward Plant Improvements," *ITR*, 12 November 1914.
48. James Bowron, "The Southern Iron and Steel Industry: Beginnings of Concentration of Interests—Introduction of Steel Making—Low Pig-Iron Prices," *IA*, 19 November 1914, 1184–85; Second Biennial Report, Alabama State Inspector of Mines, ASA. On declining southern production of rolled iron, see also James Harvey Dodd, *A History of Production in the Iron and Steel Industry in the Southern Appalachian States (1901–1926)* (Nashville: George Peabody College for Teachers, 1928), 113–14.
49. Lamoreaux, *Great Merger Movement*, 16–20.

10. Progress and Paradox

1. *BA-H*, 11, 28 April 1902; Armes, *Coal and Iron*, 453–55; "John Campbell Maben," *Encyclopedia of American Biography*, new ser., 16:291–92; picture of Maben in *The Book of Birmingham and Alabama* (Birmingham: Birmingham Ledger, 1914), 165. The article about Maben in the *Encyclopedia* is actually a composite of information about him and his son, John Campbell Maben, Jr.
2. Carolyn Green Satterfield, *Historic Sites of Jefferson County, Alabama* (Birmingham: Gray Printing Company, 1976), 23; White, *Downtown Birmingham*, 34. On the genesis of the style itself, see Wayne Andrews, *Architecture, Ambition and Americans* (1947; reprint, New York: Free Press, 1964), 211–13, and Carl W. Condit, *American Building* (Chicago: University of Chicago Press, 1968), 121–30.
3. *Birmingham City Directory, 1903* (Atlanta, Ga.: Mutual Publishing Company, 1903), 565; Satterfield, *Historic Sites*, 28; White, *Downtown Birmingham*, 34; "A $600,000 Office Building for Birmingham, Ala.," *MR*, 26 March 1908; Armes, *Coal and Iron*, 228.
4. "A Splendid Terminal: Birmingham's New Railroad Station Is An Impressive Edifice," *MR*, 11 July 1907; "Terminal Dedicated with Impressive Ceremonies," *BL*, 6 April 1909; "True Tales of Birmingham: The Terminal Station," *BN*, *BP-H*, 25 May 1991. On the older depot, see McMillan, *Yesterday's Birmingham*, 33, and White, *Downtown Birmingham*, 11. On the increase of automobiles in Birmingham in the opening decade of the century, leading to a Good Roads Tour sponsored by the Birmingham *Ledger* in 1910, see Atkins, *Valley and Hills*, 124.
5. Unless otherwise noted, the account of the Vulcan episode in this and the following paragraphs is based on Rowell, *Vulcan in Birmingham*; Carolyn Green Satterfield, "J. R. McWane: Pipe and Progress," *AR* 35 (January, 1982): 14–29; George

Clinton Thompson, "Vulcan: Birmingham's Man of Iron," *ALAH* 20 (Spring 1991): 2–17; and Marvin Y. Whiting, "Landmarks: Giuseppe Moretti," *JBHS* 9 (January 1985): 5–13, 60.

6. In addition to material on Moretti's career in the sources just cited, see Jennifer Willard, "Giuseppe Moretti," *ALAH* 20 (Spring 1991): 18–33.

7. On the use of Sloss pig iron in the casting, see Russell Hunt, "Sloss No. 2 Soft Pig Iron Was Used in Casting Vulcan," *PIRN*, Summer 1937, 4, based upon company records that have since disappeared. The wording used by Hunt is sufficiently ambiguous to raise the possibility that the statue was cast from a mixture of pig iron produced at a number of local furnaces. This mixing might well have been done in a project involving so much civic pride.

8. The field was actually known as the "Slag Pile Ball Park." On its location, see "Birmingham Rolling Mill and Surroundings from 1880 to 1900," Hill Ferguson Collection, BPLA.

9. Noel Hynd, *The Giants of the Polo Grounds: The Glorious Times of Baseball's New York Giants* (New York: Doubleday, 1988), 123–29.

10. *BN*, 17, 18, 19, 21 March 1904. McGinnity was at the peak of his career, having won thirty-one games in 1903 and adding thirty-five more victories in 1904. See Joseph L. Reichler, ed., *The Baseball Encyclopedia*, 5th ed. (New York: Macmillan, 1982), 1860.

11. *BA-H*, 24 March 1904; *BN*, 24 March 1904. For information about the players named, see Reichler, *Baseball Encyclopedia*, 690, 701, 1131, 1148, 2090.

12. *Encyclopaedia Britannica*, 1968 ed., s.v. "Cast Iron" and "Cupola Furnace." In addition to the various sources cited for the paragraphs that follow, see W. David Lewis, "Sloss Furnaces and the Southern Foundry Trade: A Case Study of Industrialization in the New South," in Howard L. Hartman, ed., *Proceedings of the 150th Anniversary Symposium on Technology and Society: Southern Technology, Past, Present, and Future* (Tuscaloosa: University of Alabama College of Engineering, 1988), 58–80.

13. Richard Moldenke, "Manufacture of Gray Iron Castings," in Backert, ed., *ABC of Iron and Steel*, 207–8, 210.

14. In addition to Jack R. Bergstresser, Sr., and Gary B. Kulik, whose contributions have already been cited, Louis C. Hunter has written briefly but incisively about the importance of cast iron and the foundry trade in *A History of Industrial Power in the United States, 1780–1930*, vol. 2: *Steam Power* (Charlottesville: University Press of Virginia, 1985), 183–93. By contrast, an enormous amount of scholarly attention has been paid to the steel industry. For a variety of published scholarly perspectives on the progress of steel manufacture in the late nineteenth and twentieth centuries, see Duncan L. Burn, *The Economic History of Steelmaking: 1867–1939* (Cambridge: Cambridge University Press, 1940); David S. Landes, *The Unbound Prometheus: Technological Change and Industrial Development in Western Europe from 1750 to the Present* (Cambridge: Cambridge University Press, 1969), 249–69; Lewis, *Iron and Steel in America*, 35–60; McHugh, *Alexander Holley and the Makers of Steel*, passim; Morison, "Almost the Greatest Invention," in *Men, Machines, and Modern Times*, 123–205, on Bessemer; Allan Nevins, *Abram S. Hewitt, with Some Account of Peter Cooper* (New York: Harper & Brothers, 1935), a full-scale biography of the person who introduced open-hearth technology to the United States; H. R. Schubert, "The Steel Industry," in Charles Singer et al., *A History of Technology, vol. 5 of The Late Nineteenth Century, c. 1850 to 1900*, 53–71; Cyril Stanley Smith, "Mining and Metallurgical Production, 1800–1880" and "Metallurgy: Science and Practice before 1900," in Kranzberg and Pursell, *Technology in Western Civilization*, 1:349–67, 592–602; Temin, *Iron and Steel in Nineteenth Century America*, 125–230; and Wall, *Andrew Carnegie*, passim. None of these books, my own included, contain

more than an occasional hint about the continuing importance of cast iron; most do not mention the subject at all. The same criticism can be made with regard to Clark, *History of Manufactures*, 2:192–366 and 3:15–135, and to Hogan, *Economic History of the Iron and Steel Industry in the United States*, vol. 1; Clark, for example, devotes part of one page (3:85) to foundries in the volume covering 1893 to 1928, and even that is chiefly devoted to steel castings. Temin devotes some attention to cast iron (see pp. 27–29, 31–34, 35–49, 61–62, 214–16) but concentrates largely on other ferrous metals, particularly steel. This emphasis is more understandable in his case because his book is devoted to the nineteenth century. Temin does, however, miss the upswing in the American foundry trade that occurred very late in that period as the advantages of southern pig iron over the Scotch variety began to become apparent. Paskoff's *Iron and Steel Industry* epitomizes continuing neglect; his encyclopedic work contains no articles on either cast iron or foundries, on such firms as Sloss or Woodward, or on such major figures as Seth Boyden and David C. Wood. Predictably, TCI, because of its role in the steel industry, is the only Alabama firm to which an article is devoted (see Brady Banta, "Tennessee Coal, Iron & Railroad Company," pp. 330–33.) For good perspectives on the importance of cast iron and the foundry trade, see two nonscholarly but nonetheless useful books commissioned by the trade itself: Clyde A. Sanders and Dudley C. Gould, *History Cast in Metal: The Founders of North America* (n.p.: American Foundrymen's Society, Cast Metals Institute, 1976), and Bruce L. Simpson, *History of the Metal-Casting Industry* (Chicago: American Foundrymen's Association, 1948). In a more specialized vein, see John Gloag and Derek Bridgwater, *A History of Cast Iron in Architecture* (London: G. Allen and Unwin, 1978).

15. Moldenke, "Manufacture of Gray Iron Castings," 212–16.
16. Richard Moldenke, "Manufacture of Malleable Castings," in Backert, ed., *ABC of Iron and Steel*, 217–28.
17. Moldenke, "Manufacture of Gray Iron Castings," 207.
18. For a lucid discussion of the nature and properties of both cast iron and ductile iron, see *Encyclopaedia Britannica*, 1968 ed., s.v. "cast iron." For a good nineteenth-century analysis, see Overman, *Treatise on Metallurgy*, 499–505. On properties that made cast iron useful in the early twentieth century, see Moldenke, "Manufacture of Gray Iron Castings," 209–11.
19. Albert F. Hill, "Influence of Phosphorus," *IA*, 13 April 1882.
20. The role played by phosphorus in cast iron is exhaustively discussed in successive issues of *PIRN*. See particularly Rebecca Hall, "The Effect of Phosphorus on the Growth of Cast Iron," Summer 1934, 2–4, and "The Effect of Phosphorus on the Wear Resistance of Cast Iron," Autumn 1934, 4–8; M. L. Karl, "Kish—Excess Carbon," Spring–Summer 1945, 23–25; E. Piwowarsky, "Phosphorus in Cast Iron," Winter 1948, 2–9, and continuation of same article in Spring–Summer 1948, 18–23. Along with J. E. Stead of Great Britain, Piwowarsky, a metallurgist from Aachen, Germany, was one of the two key European metallurgists whose work explored the role of phosphorus in conferring beneficial properties upon cast iron. On the same subject, see also James W. McQueen, "Southern Foundry Pig Iron," *Year Book of the American Iron and Steel Institute* (1914), 450–51. For reasons of space and because of the frequency with which the virtues of phosphorus are discussed in *PIRN*, I have deliberately singled out this chemical constituent, although three others—silica, sulphur, and manganese—were also significant. For analyses of their respective roles, see J. T. MacKenzie, "Silicon in Cast Iron," Winter 1941, 5–9; "Sulphur in Gray Cast Iron," Summer 1941, 13–16; and "Manganese in Gray Cast Iron," Autumn 1941, 5–8.
21. Lewis, *Iron and Steel in America*, 56; Moldenke, "Manufacture of Malleable Castings," 218.

22. *Annual Statistical Report of the American Iron and Steel Institute for 1915* (New York: American Iron and Steel Institute, 1916), 7; James R. Boyle and Thomas J. Joiner, *Minerals in the Economy of Alabama* (Washington, D.C.: U.S. Dept. of the Interior, Bureau of Mines, 1978), 11–12 and passim.

23. For various perspectives, see Bradford C. Colcord, *The History of Pig Iron Manufacture in Alabama* (Woodward, Ala.: Woodward Iron Company, 1950); William Davis Moore, *Development of Cast Iron Pressure Pipe Industry in the Southern States, 1800–1938: An American Pilgrimage* (Birmingham: Birmingham Publishing Company, c. 1939); and Henry Jeffers Noble, *Development of Cast Iron Soil Pipe in Alabama* (Birmingham: Sloss-Sheffield Steel and Iron Company, 1941), passim. For comparative statistics on American production of cast iron and steel immediately prior to World War I, see Backert, *ABC of Iron and Steel*, 210, 270.

24. See particularly Chapman et al., *Iron and Steel Industries*, 103–5, and Noble, *Cast Iron Soil Pipe*, 7–13. Among other relevant articles in *PIRN* on the nature, properties, and history of soil pipe, see H. Y. Carson, "Permanence of Cast Iron Pipe," Autumn 1935, 7–9, and Arthur R. McWane, "Silent Transportation: The Story of Cast Iron Pipe at Work," Autumn 1950, 28–34.

25. Noble, *Cast Iron Soil Pipe*, 8, 11, 14–17; Gates, *Model City*, 111–16; Russell Hunt, "Henry Bascom Rudisill: Pioneer Cast Iron Soil Pipe Manufacturer," *PIRN*, June 1928, n.p.; "C. A. (Tobe) Hamilton . . . from Water-Boy to Anniston's Greatest Manufacturer," September 1928, n.p.; "Charles Anglin Hamilton," Summer 1942, 5.

26. Rebecca L. Thomas, "John J. Eagan and Industrial Democracy at Acipco," *AR* 43 (October 1990): 270–88.

27. Satterfield, "J. R. McWane," 32–37.

28. Russell Hunt, "A Romance of Birmingham," *PIRN*, August 1938, 4, and "Our 'Foundry Family' Page . . . Dan B. Dimmick," November 1927, n.p.; W. D. Moore, "Birmingham's Place in the Cast Iron Pipe Industry," June 1929, n.p.; Atkins, *Valley and Hills*, 96; Flynt, *Mine, Mill & Microchip*, 126–27, 278–79; Norrell, *Bowron*, 189–92.

29. Hunter, *Steam Power*, 190–92.

30. Draft report, "Hardie-Tynes Manufacturing Company," prepared in 1992 by Tanya English for the Historic American Engineering Record. I appreciate her kindness in supplying me with a preliminary copy of her paper prior to its submission to *HAER*.

31. Russell Hunt, "Our 'Foundry Family' Page: Benjamin Edward McCarthy . . . President, Phillips & Buttorff Mfg. Co.," *PIRN*, May 1927, n.p.; *IA*, 13 March 1884, 24, 28 November 1887; list of Sloss customers prior to 1886 in *PIRN*, 50th Anniversary Edition (1932), 5.

32. Russell Hunt, "Rome, Georgia: 'The Stove Center of the South'" and other articles in *PIRN*, July 1937, passim; Hunt, "The St. Louis District" and other articles in ibid., September 1929, passim.

33. Russell Hunt, "Berry Harrison Hartsfield and Sam Jones Price: The Men Who Put Birmingham on the Map as a Stove Producing Center," *PIRN*, December 1928, n.p.; Hunt, "While Sheffield has Muscle Shoals, She also has Will Martin," May 1928, n.p.; Hunt, "Our 'Foundry Family' Page: Otto Agricola, Northeast Alabama's Foremost Citizen," April 1927, n.p.

34. Russell Hunt, "Gasteam Radiators: A Notable Development by Clow," *PIRN*, July 1930, n.p.; Hunt, "In Memoriam," February 1929, n.p.

35. Russell Hunt, "The Country has Gone Bath Tub! And Nobody Waits for Saturday Night," *PIRN*, April 1930, n.p.; "The Magician of the Bathtub!" October 1930, n.p.

36. Russell Hunt, "Packard: Ask the Man Who Owns One," *PIRN*, Spring 1937, 3, and accompanying photographs and captions on pages 2, 4; J. H. Cheetham, "Pig

Iron to Piston Rings," Winter 1935, 22–25; C. M. Dale, "Piston Rings: How They are Made," Spring 1942, 17–19, 22.

37. Erskine Ramsay and Charles E. Bowron, "Coal Washing in Alabama," *Mines and Minerals* 25 (December 1904): 227–31; Delamater, "Bituminous Coal Washing," 64–65; Elwood A. Stewart, "Coal Washer," U.S. Patent no. 657,184, 4 September 1900; Bergstresser, Sr., "Raw Material Constraints," 94–96; White, *Birmingham District*, 274.

38. Kulik, "Sloss Furnace Company," 39–40. Most of the other six blowing machines now standing in the engine house also date from the turn of the twentieth century but were acquired second-hand by Sloss-Sheffield in the 1920s. As Kulik indicates, the eight specimens, which were about to be restored in 1993 as this book was being completed, "are the oldest and most important pieces of surviving technology at the site" and "represent the first type of modern blowing engine used on a large scale by coke-fueled, metal plate blast furnaces." On the Rust boiler, see Johnson, *Blast Furnace Construction in America*, 112–13.

39. Sobel, *Panic on Wall Street*, 301.

40. Annual Report of the President, SSSIC, March 1903, SSSICR.

41. Annual Report of the President, SSSIC, March 1904, SSSICR; Bryan to W. E. Peters, 23 October 1903, JBL.

42. Annual Report of the President, SSSIC, March 1905, SSSICR. For an overview of the ups and downs of the business cycle during these years, see Moore, "Business Cycles, Panics, and Depressions," 152.

43. Annual Report of the President, SSSIC, March 1906, SSSICR.

44. Ibid.; on the Bessie mine and the location of the Maben post office, see also White, *Birmingham District*, 85, 144, 146, 238, 273, 274, 289.

45. Annual Report of the President, SSSIC, March 1907, SSSICR.

46. Annual Report of the President, SSSIC, March 1908, SSSICR.

47. Annual Report of the President, SSSIC, March 1906, SSSICR. On policies that created shortages of cars in places like Birmingham, see William G. McAdoo, *Crowded Years* (Boston and New York: Houghton-Mifflin Co., 1931), 449, reflecting McAdoo's experience as federal railroad administrator in World War I.

48. Kenneth Warren, *The American Steel Industry, (1850–1970: A Geographical Interpretation* (Oxford: Clarendon Press, 1970), 185.

49. Ibid., 186; Fuller, "TCI—Chapel Hill," 133–41, 267–68, 384; M. P. Gentry Hillman, "The Modern Development of the Iron and Steel Industry in the South," *Year Book of the American Iron and Steel Institute* (1914), 445–46. For detailed analysis of the duplex process, see also J. Allen Tower, "The Industrial Development of the Birmingham Region," *Bulletin of Birmingham-Southern College* 46 (December 1953): 9–10.

50. Fuller, "TCI—Chapel Hill," 145–48; John Stewart Bryan, *Joseph Bryan*, 273. The Bryan letterbook for this period has been lost, making it impossible to trace Sloss-Sheffield's role in these negotiations in further detail. On the acquisition of the Pioneer Company by Republic, see Casey and White, "A Look at Thomas," 130–31.

51. Fuller, "TCI—Chapel Hill," 148–52; Warren, *American Steel Industry*, 186–87.

52. Lewis, *Iron and Steel in America*, 51. Among various accounts of the formation of U.S. Steel, see particularly Vincent P. Carosso and Rose C. Carosso, *The Morgans: Private International Bankers, 1854–1913* (Cambridge: Harvard University Press, 1987), 466–74, and Wall, *Andrew Carnegie*, 765–93.

53. Fuller, "TCI—Chapel Hill," 152; Carosso and Carosso, *The Morgans*, 502–3; Klein, *History of the L&N Railroad*, 311–12; Larry Schweikart, "John Warne Gates," in Paskoff, ed., *Iron and Steel in the Nineteenth Century*, 147.

54. Sobel, *Panic on Wall Street*, 299–305; Fuller, "TCI—Chapel Hill," 153–54.

55. Carosso and Carosso, *The Morgans*, 535–44; Sobel, *Panic on Wall Street*, 305–21.
56. James MacGregor Burns, *The Workshop of Democracy* (New York: Alfred A. Knopf, 1985), 348–52.
57. For detailed discussions of these events, and the takeover in general, see William Glenn Moore, "The Acquisition of the Tennessee Coal and Iron Company by the United States Steel Corporation in 1907" (master's thesis, University of Alabama, 1951) and Betty Hamaker Mullins, "The Steel Corporation's Purchase of the Tennessee Coal, Iron & Railroad Company in 1907" (master's thesis, Samford University, 1970).
58. Carosso and Carosso, *The Morgans*, 544–45; Ron Chernow, *The House of Morgan: An American Banking Dynasty and the Rise of Modern Finance* (New York: Atlantic Monthly Press, 1990), 126–28; Fuller, "TCI—Chapel Hill," 156–63; Sobel, *Panic on Wall Street*, 317–20; Moore, "Acquisition of the Tennessee Coal and Iron Company," 53, 74; Warren, *American Steel Industry*, 186.
59. Fuller, "TCI—Chapel Hill," 163–65; Mullins, "Steel Corporation's Purchase," 33–34.
60. Marlene H. Rikard, "An Experiment in Welfare Capitalism: The Health Care Services of the Tennessee Coal, Iron and Railroad Company" (Ph.D. diss., University of Alabama, 1983), passim; Flynt, *Mine, Mill & Microchip*, 135; Mullins, "Steel Corporation's Purchase," 77–99.
61. Hugh W. Sanford, "The South and the Steel Corporation," *MR*, 3 September 1908.
62. On the antitrust proceedings, see particularly Mullins, "Steel Corporation's Purchase," 61–76.
63. Adam Smith, *The Wealth of Nations* (London and New York: Everyman's Library, 1910), 1:402. I am indebted to Prof. Edwin T. Layton for suggesting the pertinence of this observation to the diverging experiences of TCI and Sloss-Sheffield.
64. Warren, *American Steel Industry*, 188–94, 278; Wright, *Old South, New South*, 172–97.

11. Divergent Paths

1. For a recent analysis of the period as a whole, see John M. Cooper, Jr., *Pivotal Decades: The United States, 1900–1920* (New York: W. W. Norton, 1990).
2. For perspectives on southern progressivism, see Woodward, *Origins of the New South*, 369–95, and George B. Tindall, *The Emergence of the New South, 1913–1945* (Baton Rouge: Louisiana State University Press, 1967), 1–32. On progressive reform in Alabama, see Atkins, *Valley and Hills*, 95–103; Hackney, *Populism to Progressivism in Alabama*, passim; and McMillan, *Yesterday's Birmingham*, 73–79. "Brother" Bryan was not related to Joseph Bryan.
3. Rogers, *One-Gallused Rebellion*, passim; C. Vann Woodward, *The Strange Career of Jim Crow* (1955; 3d ed., New York: Oxford University Press, 1974), 67–109.
4. Hackney, *Populism to Progressivism*, 147–229 and passim; Hamilton, *Alabama*, 92–96; Norrell and Dismuke, *The Other Side*, 9.
5. Hackney, *Populism to Progressivism*, 72, 74–75, 245–48, 276–77, 306–7.
6. Flynt, *Mine, Mill & Microchip*, 128–30; Kulik, "Sloss Furnace Company," 43–44, 47.
7. John A. Fitch, *The Steel Workers* (1911; reprint, with introduction by Roy Lubove, Pittsburgh: University of Pittsburgh Press, 1989), 30.
8. The shifting membership of the board can be traced in Minutes, Board of Directors, 1899–1918, bound volume in SSSICR. For the composition of the ex-

ecutive committee during the same period, see ibid. Among other business activities, Oakman had become heavily involved in the Hudson Tunnel systems being built by William G. McAdoo. See the latter's *Crowded Years*, 75, 77, 86, 89, 91–94.

9. Bryan to Goadby, 31 December 1902, JBL.

10. Bryan to Davis & Bryan, Philadelphia, Pa., 2 January 1903; to Goadby, 3, 5, 6 January 1903; to C. H. Ackert, general manager, Southern Railway, Washington, D.C., 3 January 1903, ibid.

11. Bryan to Ackert, 14, 31 January 1903; to Maben, 14, 16, 23 January 1903; to Goadby, 17, 22 January 1903; to Davis & Bryan, 17 January 1903, ibid.

12. Bryan to Ackert, 9 February 1903; to Davis & Bryan, 10 February 1903; to Goadby, 13, 16, 17 February 1903; to Maben, 17, 18 February 1903, ibid.

13. Bryan to Goadby, 6, 21 April, 10 November 1903; to Haas, 13 April 1903; to Ackert, 21 April, 7 July 1903; to James W. McQueen, 24 April 1903; to W. H. Owens, 1 June 1903; to Maben, 2, 7 July, 15 October 1903, ibid.

14. Bryan to McQueen, 15 April, 11, 19 May, 1, 6, 10 June, 20 October, 17 December 1903; to J. W. Worthington, 8 May, 19 June, 11 July 1903; to Oakman, 25 May 1903; to Warren Y. Soper, 4, 17 June 1903; to Maben, 11 July, 16, 31 December 1903; to Goadby, 12, 16, 27 October 1903; to E. R. Stettinius, 5 December 1903, ibid.

15. Kevin H. Siepel, *Rebel: The Life and Times of John Singleton Mosby* (New York: St. Martin's Press, 1983), 271–76; Bryan to St. George T. C. Bryan, 28 October 1903, JBL.

16. "William R. Trigg's Useful Life Ended" and "William Roberson Trigg," *RT-D*, 17 February 1903; "Major John W. Johnston" and "Brave Soldier Answers Call," ibid., 23 May 1905.

17. Bryan to W. G. Elliott, 7 October 1903, JBL; Bryan, *Joseph Bryan*, 344.

18. Bryan, *Joseph Bryan*, 345–68.

19. "Laburnum, Home of Jos. Bryan, Burns to Ground," *RT-D*, 8 January 1906; "Transcription of Telephone Conversation with D. Tennant Bryan . . . about the Laburnum Mansion," by Merle Colglazier, copy in possession of author; Bryan, *Joseph Bryan*, 354.

20. Bryan, *Joseph Bryan*, 347–48, 350, 354, 358, 371–73.

21. "Pay Tributes to Joseph Bryan," *RT-D*, 22 November 1908, and "Thousands Pay Tribute to Dead," ibid., 23 November 1908; "Memorial Service to Joseph Bryan and Dedication of the Window given by his Fellow-Laborers on The Times-Dispatch," 27 June 1909, copy of program at VHS; "Bryans Present Rosewood to City: Give 262-Acre Tract as Memorial to Joseph Bryan," *RT-D*, 4 December 1909; "The Character of the Citizen Is the Strength of the State" and "Loyal Tribute Lovingly Paid to Joseph Bryan," ibid., 11 June 1911; "Joseph Bryan in Bronze," ibid., 10 June 1911.

22. "Times-Dispatch Owner is Dead" and "Death of Joseph Bryan," *BA-H*, 21 November 1908; "South Loses a Friend" and "Death of Joseph Bryan," *BN*, 21 November 1908.

23. "Henry F. DeBardeleben," "Col. H. F. DeBardeleben's Life Ends After Many Years Of Stirring Events," "Stirring Sketch of H. F. DeBardeleben's Life Shows Picturesqueness," "One of Alabama's Greatest Men," "Hundreds Call to Pay Respects to the Departed," and "Final Tribute to Col. DeBardeleben," *BA-H*, 7, 8, 9 December 1910; "Henry F. DeBardeleben Home," Hill Ferguson Collection, BPLA. On the extent of DeBardeleben's estate, see Fuller, "Henry F. DeBardeleben," 16.

24. "Final Tribute to DeBardeleben," *BA-H*, 9 December 1910.

25. Edward S. LaMonte, *George B. Ward: Birmingham's Urban Statesman* (Birmingham: Birmingham Public Library, 1974), 11–12.

26. Morris Knowles, "Water and Waste: The Sanitary Problems of a Modern Industrial District," *Survey* 27 (6 January 1912): 1485–92; Rikard, "Experiment in Welfare Capitalism," 48–49, 131–33.

27. Harris, *Political Power in Birmingham*, 159–60; Mitchell, "Birmingham: Biography of a City," 158–61, 172, 174.

28. Bryan to Haas, 13 April 1903, JBL; McMillan, *Yesterday's Birmingham*, 75; Harris, *abo30litical Power in Birmingham*, 229.

29. Satterfield, *Historic Sites of Jefferson County*, 129, 134; McMillan, *Yesterday's Birmingham*, 75.

30. W. M. McGrath, "Conservation of Health," *Survey* 27 (6 January 1912), 1510–12; Harris, *Political Power in Birmingham*, 160.

31. LaMonte, *George B. Ward*, 28–30.

32. John A. Fitch, "Birmingham District: Labor Conservation," *Survey* 27 (6 January 1912): 1532. On Fitch and his career, see Roy Lubove, "John A. Fitch, *The Steel Workers*, and the Crisis of Democracy," in Fitch, *Steel Workers*, vii–xiv.

33. "Birmingham on the Rebound," *Survey* 27 (3 February 1912): 1661–62; "Many Industrial Men Resent Work of Muckrakers," *BA-H*, 14 January 1912.

34. LaMonte, *George B. Ward*, 9–31; Glenn T. Eskew, "Demagoguery in Birmingham and the Building of Vestavia," *AR* 42 (July 1989): 192–98.

35. "Walter Moore is Wrong Says Ward," *BA-H*, 15 January 1912; Harris, *Political Power in Birmingham*, 189–96.

36. Harris, *Political Power in Birmingham*, 198–202.

37. Ibid., 202–7; Lerner, "A Monument to Shame," passim.

38. "No More Convicts Will be Kept at Coalburg Mines," *BA-H*, 23 March 1902; "Convicts May Be Taken From Coalburg Mines" and "To Move Convicts From Coalburg Rapidly as Possible," ibid., 11 April 1902.

39. Annual Report of the President, SSSIC, March 1903, SSSICR.

40. Ibid.; Lerner, "Monument to Shame," 148; Samuel H. Lea, "Flat Top Mine," *Mines and Minerals* 25 (March 1905): 395.

41. Miller and Sharpless, *Kingdom of Coal*, 83–134.

42. Lea, "Flat Top Mine," 395.

43. For discussions of mining procedures used in Alabama and other places at this time, see Flynt, *Poor but Proud*, 127–29, and Priscilla Long, *Where the Sun Never Shines: A History of America's Bloody Coal Industry* (New York: Paragon House, 1989), 26, 36–37.

44. Flynt, *Poor but Proud*, 127, 129; Lea, "Flat Top Mine," 395–96.

45. For perspective on such organizations, see Clarence E. Bonnett, *History of Employers' Associations in the United States* (New York: Vantage Press, 1956), which is unfortunately weak in coverage of southern associations and contains no information at all about Alabama.

46. Annual Reports of the President, SSSIC, 1905, 1906, 1907, SSSICR; Robert David Ward and William Warren Rogers, *Convicts, Coal, and the Banner Mine Tragedy* (Tuscaloosa: University of Alabama Press, 1987), 47–50; Taft, *Organizing Dixie*, 26–29; Richard A. Straw, "Soldiers and Miners in a Strike Zone: Birmingham, 1908," *AR* 38 (October 1985): 289–308; George R. Leighton, "Birmingham, Alabama: City of Perpetual Promise," *Harper's Magazine* 175 (August 1937): 235.

47. Richard A. Straw, "The Collapse of Biracial Unionism: The Alabama Coal Strike of 1908," *AHQ* 37 (Summer 1975): 92–114.

48. Ethel M. Armes, "The Spirit of the Founders," *Survey* 27 (6 January 1912): 1453–63. For an account of Armes's career, including the details I have given here, see Hugh G. Bailey, "Ethel Armes and *The Story of Coal and Iron in Alabama*," *AR* 22 (July 1969): 188–99.

49. Armes, "The Spirit of the Founders," 1457–58.

50. Fitch, *Steel Workers*, 45–53; David Brody, *Steelworkers in America: The Non-union Era* (1960; reprint, New York: Harper Torchbooks, 1969), 37, 87–91. For additional perspective, see "Machinery and the Displacement of Skill," in George E. Barnett, *Chapters on Machinery and Labor* (1926; reprint, Carbondale, Ill.: Southern Illinois University Press, 1969), 116–38.

51. Stuart D. Brandes, *American Welfare Capitalism 1880–1940* (Chicago: University of Chicago Press, 1976), 10–19; Stanley Buder, *Pullman: An Experiment in Industrial Order and Community Planning, 1880–1930* (New York: Oxford University Press, 1967), 39–45. See also David Brody, "The Rise and Decline of Welfare Capitalism," in Brody's anthology, *Workers in Industrial America: Essays on the 20th Century Struggle* (New York: Oxford University Press, 1980), 48–81.

52. Flynt, *Mine, Mill & Microchip*, 116–17.

53. Rikard, "An Experiment in Welfare Capitalism," 54–55.

54. Ibid., 54–56, 133–34, 138–39.

55. Ibid., 53–59, 64–66; White, *Birmingham District*, 99–101, 252–53. For a more negative assessment of the condition of Ensley in 1912, see Knowles, "Water and Waste," 1489.

56. Rikard, "An Experiment in Welfare Capitalism," 30–34, 57; Ida M. Tarbell, *The Life of Elbert H. Gary: The Story of Steel* (New York and London: D. Appleton and Co., 1925), 152–77; Brandes, *American Welfare Capitalism*, 30, 32, 84, 86.

57. Rikard, "An Experiment in Welfare Capitalism," 34–35; Marlene H. Rikard, "George Gordon Crawford: Man of the New South," *AR* 31 (July 1978): 163–64; Edwin Mims, *The Advancing South: Stories of Progress and Reaction* (Garden City, N.Y.: Doubleday, Page & Co., 1926), 92–94.

58. Rikard, "George Gordon Crawford," 20.

59. Rikard, "An Experiment in Welfare Capitalism," 99–113; Marlene H. Rikard, "Wenonah: The Magic Word," *JBHS* 7 (January 1981): 2–9; Marlene H. Rikard, "'Take Everything You Are . . . and Give It Away': Pioneer Industrial Social Workers at TCI," *JBHS* 7 (November 1981): 25–41.

60. Buder, *Pullman*, passim. On Anniston's experience as a model community, see Gates, *Model City*, 32–55.

61. Tarbell, *Life of Elbert H. Gary*, 152.

62. Ibid., 91–122; White, *Birmingham District*, 116–24; Graham Romeyn Taylor, "Birmingham's Civic Front," *Survey* 27 (6 January 1912): 1467–68.

63. Rikard, "An Experiment in Welfare Capitalism," 239–67; White, *Birmingham District*, 256–58.

64. Knowles, "Water and Waste," 1489; White, *Birmingham District*, 256–60.

65. Rikard, "An Experiment in Welfare Capitalism," 205–38; Clyde W. Ennis, *The Industrial Center of the South: Ensley, Fairfield, Wylam, Pratt City, Alabama* (Birmingham: n.p., 1927), 8; Tarbell, *Life of Elbert H. Gary*, 313.

66. Brandes, *American Welfare Capitalism*, 37. For an informing overview of connections between business and reform in this period, see Robert H. Wiebe, *Businessmen and Reform: A Study of the Progressive Movement* (Cambridge: Harvard University Press, 1962), passim.

67. Flynt, *Mine, Mill & Microchip*, 117; Rikard, "An Experiment in Welfare Capitalism," 105.

68. For apt assessments of the incongruities in Comer's character and activities, see Flynt, *Mine, Mill & Microchip*, 94, 104–5, 128–30. For an unflattering description of conditions at Comer's Avondale mill community, see Taylor, "Birmingham's Civic Front," 1467.

69. Harris, *Political Power in Birmingham*, 107–11.

70. Ward and Rogers, *Convicts, Coal, and the Banner Mine Tragedy*, 6–93.

71. Ibid., 113–16; "Tennessee Co. to Stop Using Convict Labor," *BA-H*, 2 December 1911. TCI continued for a short time to use local convicts until existing contracts

expired; see "An Analysis of Method Used by Alabama in Handling Convicts," *BA-H*, 11 February 1912.

72. "Maben Says Mines are Best Place to Put the Convicts," *BA-H*, 19 January 1912.
73. "Sloss Contract is to be Investigated by the Officials," *BA-H*, 24 January 1912; "Defer Probe of Sloss Contract," ibid., 25 January 1912; "Sloss Contract May be Modified," ibid., 17 February 1912.
74. "Col. Maben Talks of Recent Effort to Move Slagpile," *BA-H*, 24 November 1912; Harris, *Political Power in Birmingham*, 229; Minutes, Executive Committee, SSSIC, 13 January 1913, SSSICR.
75. "Impossible to Prevent Smoke, Say Local Manufacturers," *BA-H*, 28 January 1913; "New Smoke Ordinance is Last Straw; Small Plants Prepare to Leave City," ibid., 30 January 1913; "Smoke Not as Bad as Dust, Say Well Known Citizens," ibid., 4 February 1913.
76. Harris, *Political Power in Birmingham*, 230.
77. Minutes, Executive Committee, SSSIC, 10 January 1913 through 17 September 1914, SSSICR.

12. The End of an Era

1. Rikard, "George Gordon Crawford," 56–60; "Big Ensley Improvements," *MR*, 17 September 1908.
2. Knowles, "Water and Waste," 1492–94, with accompanying pictures of the dam; Rikard, "George Gordon Crawford," 56–58; Arthur V. Wiebel, *Biography of a Business* (Birmingham: Tennessee Coal and Iron Division, United States Steel Corporation, 1960), 39–40. In cases of discrepancies between these sources on the size and capacity of this system, I have relied on Knowles. On Knowles's background, see "Contributors of Birmingham Articles," *Survey* 27 (6 January 1912): 1450.
3. Rikard, "George Gordon Crawford," 58–63.
4. Ibid., 63–64; Wiebel, *Biography of a Business*, 41.
5. Annual Reports of the President, SSSIC, July 1908, March 1909, 1911, 1912, 1913, SSSICR; *BA-H*, 31 January 1912.
6. "Woodward Plant Improvements," 905–10.
7. Annual Report of the President, SSSIC, March 1909, SSSICR. On the duration of the downturn, which began in late spring 1907 and did not end until June 1908, see Moore, "Business Cycles, Panics, and Depressions," in Porter, ed., *Encyclopedia of American Economic History*, 1:152.
8. Annual Report of the President, SSSIC, March 1911, SSSICR. The annual report for fiscal 1909 is missing, and developments taking place at that time must be inferred from references to it in the report for fiscal 1910.
9. Annual Report of the President, SSSIC, March 1912, SSSICR.
10. Armes, *Coal and Iron*, 455.
11. Ibid., 455, 458; Pascal G. Shook, "Southern Foundry Pig Iron," *Year Book of the American Iron and Steel Institute* (1914), 457.
12. *BN*, 15 November 1911.
13. "John Campbell Maben," in *Encyclopedia of American Biography*, New Ser., 16:291–92; "New Coal Co. at Davis Creek," *BL*, 22 February 1902; Minutes, Board of Directors, SSSIC, 23 August 1911, SSSICR; *BA-H*, 16 December 1911; *BN*, 15 November 1911. On the death of Seixas, which took place 19 May 1911, see New Orleans *Daily Picayune*, 21 May 1911.
14. Minutes, Executive Committee, SSSIC, 19 December 1905, SSSICR.
15. "M. M. Richey Chosen First Vice President," "Mr. Richey's Successor," and "Richey Returns from New York," *BA-H*, 20 and 22 December 1905.
16. Minutes, Executive Committee, SSSIC, 21 September and 17 October 1906,

SSSICR; Minutes, Annual Meeting of the Board of Directors, 14 March 1907, ibid.

17. Minutes, Executive Committee, SSSIC, 17 November 1911, SSSICR; Minutes, Board of Directors, SSSIC, 12 December 1911, ibid.

18. Minutes, Annual Meeting of the Board of Directors, SSSIC, 13 March 1912, ibid.; "Maben Says Mines are Best Place to Put the Convicts," *BA-H*, 19 January 1912. According to the *Birmingham City Directory for the Year Commencing February 1, 1912* (Birmingham: R. L. Polk & Co., 1912), 838, Maben was still living at his 701 South Twenty-first Street residence at this time. The *Age-Herald* article, however, indicated that, as of January 1912, he had already been gone for "a few weeks" in New York, "where he now permanently resides."

19. Minutes, Executive Committee, SSSIC, 12 April, 16 July, 1 and 19 August, 23 September, 8 and 21 October, and 20 December, 1912, SSSICR; *BA-H*, 2 February 1913.

20. Annual Report of the President, SSSIC, March 1913, SSSICR.

21. Ibid., March 1914, SSSICR.

22. Chernow, *House of Morgan*, 472; Moore, "Business Cycles, Panics, and Depressions," 152; Sobel, *Panic on Wall Street*, 322–49.

23. Clark, *History of Manufactures*, 3:104–6; Annual Report of the President, March 1915, SSSICR.

24. Minutes, Executive Committee, SSSIC, 29 December 1913 and 29 January 1914, SSSICR.

25. Minutes, Annual Meeting of Stockholders and Directors, SSSIC, 11 March 1914, 10 March 1915, and 15 March 1916, SSSICR.

26. Minutes, Executive Committee, SSSIC, 6 January, 24 June, and 17 August 1915, SSSICR.

27. Minutes, Annual Meeting of Stockholders and Directors, 13 March 1912 and 15 March 1916, SSSICR; biographical sketch of Andrews by Steelman in Powell, ed., *Dictionary of North Carolina Biography*, 1:34–36; obituary of Camp in *NYT*, 17 July 1924.

28. Minutes, Executive Committee, SSSIC, 24 June 1915 and 2 February 1916, SSSICR.

29. Sobel, *Panic on Wall Street*, 338; Clark, *History of Manufactures*, 3:307; "Record Is Broken by Steel and Iron Sent Over Seas," *BA-H*, 19 November 1916.

30. Annual Report of the President, SSSIC, March 1916, and Balance Sheet, fiscal year ending 30 November 1915, SSSICR.

31. Balance Sheet, fiscal year ending 30 November 1916, SSSICR.

32. The discussion of coal-tar derivatives and the history of by-product technology in this and succeeding paragraphs is based largely upon Williams Haynes, *American Chemical Industry: A History* (New York: D. Van Nostrand Co., 1945–49), 1:270–73 and 2:21–22, 124–39, and H. Cole Estep, "The Manufacture of By-Product Coke," in Backert, ed., *ABC of Iron and Steel*, 69–76. See also *Encyclopaedia Britannica*, 1968 ed., s.v. "Coal Tar" and "Coke, Coking and High-Temperature Carbonization."

33. In addition to discussion in Haynes, see "The Recovery of By-Products in Coke Manufacture," *IA*, 31 March 1892.

34. "The Industries of Birmingham," *BA-H*, 31 July 1898; "The Great Work at Ensley City," ibid., 20 October 1898; Bergstresser, Sr., "Raw Material Constraints," 111; Erskine Ramsay, "The By-Products Coke Industry and Its Future," in Childers, *Erskine Ramsay*, 388–90. Ramsay misdates the opening of the plant by stating that it went on line in 1895 but supplies valuable information about the nature of the agreement between Semet-Solvay and TCI.

35. Estep, "Manufacture of By-Product Coke," 69–70; Rikard, "George Gordon

Crawford," 68–69; "$15,000,000 Developments Planned in Irondale and Shades Valley Sections by Big Interests," *BA-H*, 4 November 1916.

36. Bergstresser, Sr., "Raw Material Constraints," 101, 107–14.

37. Haynes, *American Chemical Industry*, 2:129; Estep, "Manufacture of By-Product Coke," 69.

38. Chapman et al., *Iron and Steel Industries*, 144; "Woodward Plant Improvements," 910; Woodward, *Alabama Blast Furnaces*, 147–48.

39. Minutes, Special Meeting, Board of Directors, SSSIC, 2 February 1916, SSSICR; Minutes, Regular Meeting, Executive Committee, SSSIC, 28 February 1916, SSSICR.

40. Minutes, Executive Committee, SSSIC, 25 May, 1 June, 16 August, and 14 November 1916, SSSICR.

41. "Many Lives Lost At Bessie Mines," *BA-H*, 5 November 1916; "Thirty Were Killed in the Bessie Mine Disaster Saturday" and "Miner Relates Harrowing Story of Escape of Party from Bessie Mine," ibid., 6 November 1916; "Fire at Sloss Mines Causes Shutdown," ibid., 18 November 1916.

42. "$15,000,000 Developments Planned in Irondale and Shades Valley Sections," "Germans Say Anglo-French Somme Loss Reaches 600,000," "Armored Submarine U-57 Will Convoy Deutschland on Return," and other war-related stories, *BA-H*, 4 November 1916.

43. "Sloss Company Plans Battery of Coke Ovens Costing About $2,500,000," *BA-H*, 29 November 1916.

44. "Col. Maben Denies Report that Sloss Is to Change Hands," *BA-H*, 30 November 1916; "Sloss Officials Arrive Today to Make Inspection," ibid., 4 December 1916; "No Big Dividend Till Debt Is Paid," ibid., 7 December 1916; "Sloss Company to Set New Record in Iron Production," ibid., 14 December 1916.

45. David M. Kennedy, *Over Here: The First World War and American Society* (1980; reprint, New York: Oxford University Press, 1982), 3–15.

46. Minutes, Executive Committee, SSSIC, 15 and 23 March 1917, SSSICR; "Waddill Catchings Chosen President of Sloss Company," *BA-H*, 24 March 1917; sketch of Catchings's career by Broadus Mitchell, *DAB*, Supp. 8 (1988), 78–79. On the Central Iron and Coal Company's operations, see Hubbs, *Tuscaloosa*, 56–57.

47. "Changes in Sloss Company Rumored," *BA-H*, 20 March 1917; "Waddill Catchings Chosen President of Sloss Company," ibid., 24 March 1917.

48. "Waddill Catchings Chosen President of Sloss Company," *BA-H*, 24 March 1917; "Catchings Announces Sewell on Sloss Board of Directors," ibid., 27 March 1917.

49. "Parsons Makes No Announcements of a Definite Nature," *BA-H*, 1 April 1917; "Waddill Catchings Returns to Duties," ibid., 8 April 1917; "Sloss Company Elects its Vice Presidents," ibid., 10 April 1917.

50. "Sloss Ovens on First Avenue are to be Relighted by City Consent," ibid., 18 April 1917.

51. For accounts of the prevailing atmosphere, see Kennedy, *Over Here*, 93–143, and Robert D. Cuff, *The War Industries Board: Business-Government Relations during World War I* (Baltimore and London: Johns Hopkins University Press, 1973), 13–67.

52. "Sloss-Sheffield to Declare Dividend," *BA-H*, 4 January 1917; "Catchings Wins in His Dividend Fight," ibid., 18 April 1917. On the disappearance of Ames from the company, see Minutes, Annual Meetings of Board of Directors, SSSIC, 14 March 1917 and 13 March 1918, SSSICR.

53. Wiebel, *Autobiography of a Business*, 49; map and article on new Fairfield works in *BA-H*, 24 July 1917.

54. Minutes, Special Meeting, Board of Directors, SSSIC, 4 April 1917; Minutes, Regular Meetings, Board of Directors, of 7 December 1917 and 18 January 1918, SSSICR.
55. Minutes, Regular Meeting, Board of Directors, SSSIC, 18 January 1918, SSSICR; Minutes, Special Meeting, Board of Directors, 8 March 1918, ibid.; Minutes, Annual Meetings of Stockholders and Directors, 13 March 1918, ibid.
56. "Rumor Prevails Catchings to Leave Sloss-Sheffield Co.," *BA-H*, 11 December 1917; biographical sketch of Catchings by Mitchell in *DAB*, 78; Kennedy, *Over Here*, 129–33; Minutes, Special Meeting, Board of Directors, SSSIC, 15 February 1918; Minutes, Annual Meeting of Stockholders, 13 March 1918, ibid.
57. "Sloss-Sheffield Co.'s New Executive Head" and "McQueen Is Elected to the Presidency of Sloss-Sheffield Co.," *BA-H*, 16 February 1918.

13. McQueen in Command

1. "Birmingham's Population Passes the 200,000 Mark," *BA-H*, 24 December 1916; "Exodus of Negroes Has Not Stopped," ibid., 20 November 1916; "Exodus of Negro May be Beneficial," ibid., 2 December 1916; "Chicago a Mecca for Negroes from South," ibid., 11 May 1917; "Cash Pay Rolls Here Now Reach Enormous Figure," ibid., 21 May 1917; "Birmingham Expansion on Very Great Scale," ibid., 13 August 1917. For general perspective on the black exodus and on strikes reflecting increased militancy on the part of African-American workers in the South, see Edward E. Lewis, *The Mobility of the Negro: A Study in the American Labor Supply* (New York: Columbia University Press, 1931), passim; Carole Marks, *Farewell—We're Good and Gone: The Great Black Migration* (Bloomington: Indiana University Press, 1989), passim; David Montgomery, *The Fall of the House of Labor: The Workplace, the State, and American Labor Activism, 1865–1925* (Cambridge: Cambridge University Press, 1987), 378–83; and Tipton Ray Snavely, "The Exodus of Negroes from the Southern States: Alabama and North Carolina," in U.S. Department of Labor, Division of Negro Economics, *Negro Migration in 1916–17* (1919; reprint, New York: Negro Universities Press, 1969), 51–74.
2. Atkins, *Valley and Hills*, 106–7; McMillan, *Yesterday's Birmingham*, 78–79. For further information on the project, see "Barge Line for Warrior River" and "Warrior To Be Taken Over By McAdoo," *BA-H*, 11 July 1918.
3. Atkins, *Valley and Hills*, 93; White, *Downtown Birmingham*, 97; "Detailed Plans Given Out for Magnificent New Tutwiler Hotel," *BA-H*, 2 December 1912.
4. McMillan, *Yesterday's Birmingham*, 110–11; Satterfield, *Historic Sites of Jefferson County, Alabama*, rev. ed. (1985), 25–35; White, *Downtown Birmingham*, 33, 36–37.
5. Chandler, *Visible Hand*, Appendix A, 509; American Iron and Steel Institute, *Directory of Iron and Steel Works of the United States and Canada*, 18th ed. (New York: American Iron and Steel Institute, 1916), 297–98; *Poor's Manual of Industrials*, Ninth Annual Number, 1918 (New York: Poor's Manual Company, 1918), 2575–77.
6. John N. Ingham, *The Iron Barons: A Social Analysis of an American Urban Elite, 1874–1965* (Westport, Conn.: Greenwood Press, 1978), passim.
7. Fuller, "Alabama Business Leaders," part 2, 66–71.
8. For details on McQueen drawn upon in this and the following paragraphs, see Armes, *Coal and Iron*, 359, 458–60; *BA-H*, 21 and 23 April 1925; *IA*, 21 February 1918.
9. Coal Operators Association Minutes, 7 June 1902, SSSICR; *IA*, 21 February 1918; Bryan to McQueen, 30 May, 10 June 1898; 30 December 1899; 6, 13 January, 21, 22 February, 1, 5, 8 March, 18 June 1900, JBL.

10. Armes, *Coal and Iron*, facing page 458; interview of Marvin Y. Whiting.
11. Obituary of Sevier in *BN*, 27 November 1942, clipping in BPLA; "L. Sevier, Vice-President of Sloss-Sheffield Co.," *BA-H*, 14 March 1918; "Sloss-Sheffield Meeting," *IA*, 21 March 1918.
12. "Funeral Services of Mrs. J. W. M'Queen," *BA-H*, 31 July 1918.
13. "McQueen Returns from Conference With Directors," *BA-H*, 15 May 1918; Annual Report of the President, SSSIC, March 1919, SSSICR.
14. C. R. M. F. Cruttwell, *A History of the Great War, 1914–1918* (1934; 2d ed., London: Granada Publishing, 1982), 486–535; Kennedy, *Over Here*, 190.
15. Haynes, *American Chemical Industry*, 2:90–92, 104–6; "T.C.I. Company to Build By-Product Coke Ovens To Secure War Materials," *BA-H*, 4 April 1918. On the Muscle Shoals project, see Daniel Schaffer, "War Mobilization in Muscle Shoals, Alabama, 1917–1918," *AR* 39 (April 1986): 110–46.
16. "Sloss-Sheffield Company to Build $5,000,000 By-Product Coke Oven Plant in District," "Feeling of Optimism Regarding Conditions Accompanying Peace," and "Industrial Conditions Now Being Readjusted Along Lines of Peace," in *BA-H*, 13 July 1918; 4 November 1918; and 9 December 1918. For a synopsis of the contract, see addendum in Minutes, Executive Committee, SSSIC, 17 January 1919, SSSICR. On difficulties in labor recruitment and other problems in northwest Alabama at the time, see Schaffer, "War Mobilization in Muscle Shoals," 127–37.
17. Cruttwell, *History of the Great War*, 543–97.
18. Annual Report of the President, SSSIC, March 1919, SSSICR; addendum in Minutes, Executive Committee, 17 January 1919, ibid.
19. Minutes, Executive Committee, SSSIC, 13 March 1919, SSSICR; "New Ovens Start Work Tomorrow," *BA-H*, 14 April 1920; "Sloss-Sheffield Battery of Semet-Solvay Ovens to Start This Week," ibid., 27 May 1920; Annual Report of the President, March 1920, SSSICR.
20. See foldout, "Bee-hive and By-product Coke Made in Alabama and the Number of Ovens in Existence, 1892–1940," in Childers, *Erskine Ramsay*, between pp. 390–91.
21. "Sloss-Sheffield Battery of Semet-Solvay Ovens to Start This Week," *BA-H*, 27 May 1920; W. B. Bridge and J. M. Hastings, Jr., "New By-Product Coke Ovens at Birmingham," *IA*, 12 August 1920; Richard A. Johnston, "A Visit to the New By-Product Plant of the Sloss-Sheffield Steel & Iron Company," ibid., 30 May 1920. On the role of the Barrett Company in the distillation and marketing of coal tar products, see Haynes, *American Chemical Industry*, 4:210. On specific features of the design of Semet-Solvay ovens, see H. Cole Estep, "The Manufacture of By-Product Coke," in Backert, ed., *ABC of Iron and Steel*, 72–73.
22. "Enables Birmingham Iron to Travel Far Afield," *BA-H*, 30 May 1920; *Moody's Manual of Railroads and Corporation Securities*, Nineteenth Annual Number, Industrial Section (1918), 1480, and Twenty-first Annual Number, Industrial Section (1920), 1616; Minutes, Executive Committee, SSSIC, 8 July 1919, SSSICR.
23. Entry on Jarvie in *Who Was Who in America* (Chicago: Marquis, 1943), 1:630; obituaries of Wallace and Davison, *NYT*, 12 October 1919 and 17 June 1953.
24. Obituary of Sloss in *BA-H*, 17 January 1919; obituary of Oakman in *NYT*, 19 March 1922; Carvati, *Major Dooley*, 58.
25. George Soule, *Prosperity Decade: From War to Depression, 1917–1929* (New York: Rinehart & Co., 1947), 81–95.
26. Ibid., 91, 190–96; Clark, *History of Manufactures*, 3:333–35. On business fluctuations generally during the 1920s, see table in Moore, "Business Cycles, Panics, and Depressions," in Porter, ed., *Encyclopedia of American Economic History*, 1:152.

27. Richard A. Straw, "The United Mine Workers of America and the 1920 Coal Strike in Alabama," *AR* 28 (April 1975), 104–28; Rikard, "George Gordon Crawford," 130–48. For national perspective on the 1919 disturbances, see particularly David Brody, *Labor in Crisis: The Steel Strike of 1919* (Philadelphia and New York: J. B. Lippincott Co., 1965).
28. Annual Report of the President, SSSIC, March 1920, SSSICR; Certified Balance Sheet, 31 December 1919, ibid.
29. "Iron Market is Throbbing with Record Business," *BA-H*, 30 May 1920; Soule, *Prosperity Decade*, 96–106; Annual Report of the President and Certified Balance Sheet, SSSIC, April 1921, SSSICR.
30. Annual Reports of the President, SSSIC, March 1920 and April 1922, 1923, 1924, and 1925, SSSICR.
31. Ibid., March 1920 and April 1921.
32. Ibid., April 1921, 1922, and 1923.
33. Soule, *Prosperity Decade*, 138–43; Leslie Hannah, "Mergers," in Porter, ed., *Encyclopedia of American Economic History*, 2:644–45.
34. Minutes, Annual Meeting, Board of Directors, SSSIC, 11 April 1923, and Annual Report of the President, April 1924, SSSICR; Woodward, *Alabama Blast Furnaces*, 56–57. On Gayley, see also Wall, *Andrew Carnegie*, 610, 632, 667, 746–47, 752, 771–72.
35. Minutes, Board of Directors, SSSIC, 5 and 28 November 1924, SSSICR; Annual Report of the President, April 1925, ibid.
36. "An Important Factor in Alabama History," *BA-H*, 5 August 1900; "Alabama Company Keeps Properties at High Standard," ibid., 26 November 1916; "Alabama Co. Has Doubled Output," ibid., 23 December 1916; *IA*, 23 April 1925; Kulik, "Sloss Furnace Company," 51; Woodward, *Alabama Blast Furnaces*, 35, 68–69, 79, 81, 100.
37. "Sloss-Sheffield Steel & Iron Co. to Acquire Alabama Co. Properties," *IA*, 16 October 1924; "Alabama Co.'s Improvements," ibid., 12 May 1921; "Foundry Pig Iron in Birmingham District," ibid., 7 April 1921; Kulik, "Sloss Furnace Company," 51.
38. Minutes, Board of Directors, SSSIC, 4 September 1924, SSSICR; obituaries of Camp and Evans in *NYT*, 17 July 1924 and 30 August 1924.
39. "Gen. Edmund W. Rucker . . . is Dead in his 88th Year," *BA-H*, 14 April 1924.
40. Minutes, Board of Directors, SSSIC, 22 May 1924, SSSICR.
41. Dan Cragg, *Guide to Military Installations*, 2d ed. (Harrisburg, Pa.: Stackpole Books, 1988), 5–6.
42. "Death of Maben in East Throws Pall over City: Pioneer Steel Man Succumbs At Atlantic City," *BA-H*, 2 September 1924. See also obituary of Maben in *BN*, 1 September 1924.
43. On the formation of Allied, see Haynes, *American Chemical Industry*, 3:426 and 6:9–11.
44. "James W. McQueen, Steel Head, Dead," *NYT*, 21 April 1925; "McQueen Funeral Is Set," "James William McQueen," and "Sloss Workers in Last Tribute to J. W. McQueen," *BA-H*, 21 and 23 April 1924; obituary of McQueen in *IA*, 23 April 1924; commemorative booklet honoring McQueen, SFNHL.

14. Morrow and Modernization

1. Atkins, *Valley and Hills*, 129; John R. Hornady, *The Book of Birmingham* (New York: Dodd, Mead and Co., 1921), 375; Neil M. Clark, "Birmingham—The Next Capital of the Steel Age: A Smoking Metropolis on a Mountain of Iron," *World's Work*, 53 (March 1927): 534–45.

2. Blaine A. Brownell, "Birmingham, Alabama: New South City in the 1920s," *JSH*, 38 (February 1972): 22, 47–48; James F. Sulzby, Jr., *Birmingham Sketches from 1871 through 1921* (Birmingham: Birmingham Printing Co., 1945), 181–88.

3. Harris, *Political Power in Birmingham*, 33–34.

4. Carl Carmer, *Stars Fell on Alabama* (1934; reprint, Tuscaloosa: University of Alabama Press, 1985), 79–81. On Carmer's background, see the accompanying introduction by Wayne Flynt, xi–xviii.

5. Eskew, "Demagoguery in Birmingham," passim.

6. Marjorie Longenecker White, "The Grid and the Garden," Marvin Y. Whiting, "Robert Jemison, Jr.: A Tribute," and C. Chappell Jarrell, "From Farmscape to Suburban Yard: Robert Jemison, Jr.'s Spring Lake Farms," in Morris and White, eds., *Designs on Birmingham*, 25–27, 38–41, 42–45; Barefield, *History of Mountain Brook*, 70–73, 202.

7. LaMonte, *George B. Ward*, 31–45; Bailey, "Ethel Armes and *The Story of Coal and Iron in Alabama*," 188–99.

8. LaMonte, *George B. Ward*, 7; Carmer, *Stars Fell on Alabama*, 79; Eskew, "Demagoguery in Birmingham," 206–11; Mittie Owen McDavid, "Vestavia: Classic Home of George Battey Ward," in Birmingham Branch, National League of American Pen Women, *Historic Homes of Alabama and Their Traditions*, 308–14; George R. Leighton, "Birmingham, Alabama," 226.

9. Blaine A. Brownell, "The Notorious Jitney and the Urban Transportation Crisis in Birmingham in the 1920s," *AR* 25 (April 1972): 105–18.

10. McMillan, *Yesterday's Birmingham*, 126; Brownell, "Birmingham," 22–25, 33–36; article in Birmingham *Post*, 19 December 1928, quoted by Edward S. LaMonte, "Politics and Welfare in Birmingham, Alabama: 1900–75" (Ph.D. diss., University of Chicago, 1976), 152.

11. Kenneth T. Jackson, *The Ku Klux Klan in the City, 1915–1930* (New York: Oxford University Press, 1967), 82–83; Brownell, "Birmingham," 39–42; LaMonte, "Politics and Welfare," 135–37; William R. Snell, "Fiery Crosses in the Roaring Twenties: Activities of the Revised Klan in Alabama, 1915–1930," *AR* 23 (October 1970): 256–76.

12. Atkins, *Valley and Hills*, 132–34; Wayne Flynt, "Religion in the Urban South: The Divided Mind of Birmingham, 1900–1930," *AR* 30 (April 1977): 125–29; Snell, "Fiery Crosses," 275.

13. LaMonte, "Politics and Welfare," 127–33.

14. Hornady, *Book of Birmingham*, 56–67, 130–34, 213–27, 283–98, 326–27.

15. Atkins, *Valley and Hills*, 135–37; Sulzby, *Birmingham Sketches*, 169–75; Wesley Phillips Newton, "Lindbergh Comes to Birmingham," *AR* 26 (April 1973): 105–21.

16. Edwin C. Eckel, "The Iron and Steel Industry of the South," *Annals of the American Academy of Political and Social Science* 153 (January 1931): 61–62; "Iron Ore—Making and Remaking the Map of Steel," *Fortune* 3 (May 1931): 84.

17. Thomas M. Doerflinger and Jack L. Rivkin, *Risk and Reward: Venture Capital and the Making of America's Great Industries* (New York: Random House, 1987), 107–113.

18. The analysis that follows is based principally on George W. Stocking, *Basing Point Pricing and Regional Development: A Case Study of the Iron and Steel Industry* (Chapel Hill: University of North Carolina Press, 1954), passim. It also draws, however, on other sources presenting varying perspectives on the Pittsburgh Plus controversy, including Chapman et al., *Iron and Steel Industries*, 378–85; Cobb, *Industrialization and Southern Society*, 20; Flynt, *Mine, Mill & Microchip*, 130–31; Norrell, *Bowron*, xxx–xxxii; Woodward, *Origins of the New South*, 302,

315–17; and Wright, *Old South, New South,* 168–70. For a thoughtful discussion that attempts to minimize damage done to the region by Pittsburgh Plus and other freight differentials, see Rikard, "George Gordon Crawford," 157–90.

19. Stocking, *Basing Point Pricing,* 68–69, 102–3, 110. For lists of average annual prices of foundry pig iron in Ohio and the Birmingham District from 1920 to 1950 that bear out Stocking's analysis, see Chapman et al., *Iron and Steel Industries,* 388.

20. Hornady, *Book of Birmingham,* 130–44; Warren, *American Steel Industry,* 189; Wiebel, *Biography of a Business,* 49–56.

21. Kulik, "Sloss Furnace Company," 47–48; Lewis, "Invasion of Northern Markets," 261, 263; Norrell, *Bowron,* xxx–xxxi. For general perspective, see Roland B. Eustler, "Transportation Developments and Economic and Industrial Changes," in William J. Carson, ed., *The Coming of Industry to the South* (Philadelphia: American Academy of Political and Social Science, 1931), 202–9, and David M. Potter, "The Historical Development of Eastern-Southern Freight Rate Relationships," *Law and Contemporary Problems* 12 (Summer 1947), 416–48.

22. Chapman et al., *Iron and Steel Industries,* 236.

23. Ibid., 192–93; Wright, *Old South, New South,* 176.

24. Minutes, Board of Directors, SSSIC, 14 May 1925, SSSICR; "Hugh Morrow is Elected Head of Sloss Company" and "Native Son Promoted," *BA-H,* 15 May 1925; "Choice of Morrow No Surprise," *BN,* 15 May 1925; Articles of Incorporation, Sloss Furnace Company, Jefferson County Corporation Record A (1876–86), Office of the Probate Judge, Jefferson County Courthouse, Birmingham, Ala.

25. Details on Morrow's life and career in this and the following paragraphs are based chiefly upon the newspaper articles previously cited and a collection of clippings, photographs, and other materials pertaining to him in the Hill Ferguson Collection, BPLA. Among the latter, see particularly obituaries in *BN* and *BP-H,* 7 September 1960, and "Late Hugh Morrow's Many Friends Recall His Humor, Wisdom, Charity," in *BP-H,* 8 September 1960. For additional biographical data, see also *IA,* 14 May 1925; Harris, *Political Power in Birmingham,* 93–94; and *Who's Who in America,* 16 (Chicago: A. N. Marquis Company, 1930), 1611.

26. *Alabama Blue Book and Social Register* (Birmingham: Blue Book Publishing Co., 1929), 148; White, *Birmingham District,* 26.

27. Obituary of Morrow, *BN,* 7 September 1960, in Hill Ferguson Collection.

28. Harris, *Political Power in Birmingham,* 107, 254; Ward and Rogers, *Convicts, Coal, and the Banner Mine Tragedy,* 67–69, 92.

29. On Morrow's involvement in negotiations with municipal officials concerning the smoke emission and other issues at the time, see Minutes, Executive Committee Meeting, SSSIC, 10 January 1913, SSSICR.

30. Obituary of Sevier in *BN,* 27 November 1942, clipping in BPLA. On the Associated Industries of Alabama, see particularly Flynt, *Mine, Mill & Microchip,* 163, 231.

31. "Hugh Morrow Is Elected Head Of Sloss Company," *BA-H,* 15 May 1925.

32. "Sloss-Sheffield Adds 2 Officials," *BN,* 18 May 1925; "Sloss Officials Named," *BA-H,* 19 May 1925; *IA,* 21 May 1925; list of officials in *Twenty-seventh Annual Report of the Sloss-Sheffield Steel and Iron Company for the Calendar Year ended December 31, 1926,* 2. For the years 1926 and later I have relied on published annual reports in the Angelo Bruno Business Library, University of Alabama, Tuscaloosa.

33. Robert F. Himmelberg, *The Origins of the National Recovery Administration: Business, Government, and the Trade Association Issue, 1921–1933* (New York: Fordham University Press, 1976), 54–74. For other perspectives on the same theme, see Otis L. Graham, Jr., *An Encore for Reform: The Old Progressives and*

the New Deal (New York: Oxford University Press, 1967); Ellis W. Hawley, *The Great War and the Search for a Modern Order: A History of the American People and Their Institutions, 1917–1933* (New York: St. Martin's Press, 1979); and Ellis Hawley et al., *Herbert Hoover and the Crisis of American Capitalism* (Cambridge, Mass.: Schenkman Publishing Company, 1973). For a complete list of Sloss-Sheffield's directors at about the time Morrow became president, see *Twenty-seventh Annual Report*, 3.

34. *Who's Who in America*, 16:602, 1242, 1270; obituaries of Crawford and Kettig in *NYT*, 20 February 1941 and 4 August 1939; "W. W. Crawford is Sloss Director," *BA-H*, 23 January 1918. On Kettig's views concerning the smoke emission issue, see *BA-H*, 4 February 1912.

35. For a short account of Baruch's life and career, see sketch by Robert D. Cuff in *DAB*, Supp. 7 (1961–65), 34–37.

36. McAdoo, *Crowded Years*, passim; obituary of Sloan in *NYT*, 15 June 1945; *Who's Who in America*, 16:2033; C. M. Stanley, "Famous Auburn Boy and Girl," *MA*, 3 June 1951, and obituary of Lottie Lane Sloan from New York *World-Telegram and Sun*, 24 May 1951, clippings in Auburn University Archives.

37. "Edward Magruder Tutwiler," *BA-H*, 23 April 1925.

38. "William H. Goadby . . . Dies at 75," *NYT*, 5 July 1925.

39. Clark, *History of Manufactures*, 3:335.

40. Annual Report of the President, SSSIC, March 1926, SSSICR.

41. Annual Report of the President, SSSIC, March 1927, SSSICR; List of Iron and Steel Prices, 1913 through 1929 (hereafter referred to as "List"), *IA*, 2 January 1930.

42. Annual Report of the President, SSSIC, March 1928 and 1929, SSSICR; George Soule, *Prosperity Decade*, 275; Moore, "Business Cycles, Panics, and Depressions," in Porter, ed., *Encyclopedia of American Economic History*, 1:152; *IA*, 2 January 1930.

43. For Kulik's hypothesis, see his essay, "Black Workers and Technological Change," in Reed et al., *Southern Workers and Their Unions*, 33–36, and Kulik, "Sloss Furnace Company," 55–56. For evidence that the mechanization carried out at North Birmingham immediately after World War I was partly related to unsettled labor conditions, see Annual Reports of the President, SSSIC, 1919, 1920, 1921, SSSICR. On the substantial growth of black population in Birmingham during the 1920s, which was consistent with the experience of such other cities as Atlanta and Memphis, see table in United States Department of Commerce, Bureau of the Census, *Negroes in the United States 1920–1932* (Washington, D.C.: Government Printing Office, 1935), 55. In assessing the degree to which outmigration of black workers may have influenced Sloss-Sheffield's modernization drive in the 1920s, I have drawn upon James L. Sledge III, "The Great Migration, Alabama Blacks, and the Birmingham District: An Overview" (unpublished seminar paper, Auburn University, 1993). Sledge's research indicates that the Great Migration was not simply a movement from South to North but also an exodus from rural to urban areas within the South itself, continuing a trend already well established since the end of the Civil War. In addition, his findings show that a large proportion of the African-Americans who came to Birmingham in the 1920s were males in age categories suitable for unskilled industrial labor. His paper, along with an absence of complaints about labor problems in annual presidential reports written after 1921, has strengthened my belief that such executives as McQueen and Morrow were mainly concerned about increasing efficiency and reducing costs, and that whatever labor-related difficulties they were trying to mitigate through mechanization were of so familiar a nature, being related to the same lack of discipline that had always prevailed among severely disadvantaged black workers, that they saw no reason to mention them.

44. Annual Reports of the President, SSSIC, 1919–25, SSSICR.
45. Annual Reports of the President, SSSIC, 1926–31, SSSICR.
46. Meeting, Board of Directors, 22 October 1925, SSSICR; "Rebuilding Alabama Blast Furnaces," *IA*, 24 June 1926; "Birmingham's Most Modern Iron Maker" and "Sloss New No. 2 Furnace," *PIRN*, August–September 1927; "Another Iron Maker to Serve the Foundryman," ibid., January 1929. See also Datnow, "The Sloss Company," 145–47; Kulik, "Sloss Furnace Company," 53–54; and Woodward, *Alabama Blast Furnaces*, 131. Estimates of the capacities of the four Birmingham and North Birmingham furnaces vary; I have used the best figures available.
47. Material on Dovel's earlier life in this and the following paragraphs is taken largely from a collection of newspaper clippings, some from unidentified sources, at SFNHL. These were used in preparation of a permanent installation, with accompanying textblock, honoring his memory in 1989. See particularly Bem Price, "Indomitable Inventor Helps Steel Industry With Ideas," from unidentified Birmingham newspaper, 19 February 1940.
48. Monte A. Calvert, *The Mechanical Engineer in America, 1830–1910: Professional Cultures in Conflict* (Baltimore: Johns Hopkins University Press, 1967).
49. Interview of Dudley Dovel Shearburn, granddaughter of James P. Dovel, by author.
50. James P. Dovel and John J. Shannon, "Gas-Washer," U.S. Patent no. 1,001,739, 29 August 1911; James P. Dovel, "Gas-Cleaning Apparatus," U.S. Patent no. 1,001,740, 29 August 1911; James P. Dovel and Charles C. Glidden, "Amalgamator," U.S. Patent no. 1,123,116, 29 December 1914.
51. John J. Shannon and James P. Dovel, "Air-Cooled Blast Furnace Stack," U.S. Patent no. 1,090,574, 17 March 1914.
52. James P. Dovel, "Blast-Furnace Stack Reinforcement," U.S. Patent no. 1,316,085, 16 September 1919; "Twyer and Cooling Box," U.S. Patent no. 1,354,032, 28 September 1920.
53. James P. Dovel, "Ore-Washing Plant," U.S. Patent no. 1,252,414, 8 January 1918; "Power-Shovel," U.S. Patent no. 1,444,670, 6 February 1923; "Settling Apparatus," U.S. Patent no. 1,485,452, 4 March 1924; "Pig-Breaking Machine," U.S. Patent no. 1,622,029, 22 March 1927; James P. Dovel, "Reasons Why Foundry Iron Should Be Sand Cast," *IA*, 21 April 1921.
54. The foregoing discussion is based partly on a series of patents for which Dovel applied for protection in the 1920s and partly on information in visual displays scattered about Sloss Furnaces National Historic Landmark. The patents, cited in chronological and numerical order, are "Gas Cleaner," U.S. Patent no. 1,609,611, 7 December 1926; "Blast Furnace," U.S. Patent no. 1,703,517, 26 February 1929; "Blast Furnace," U.S. Patent no. 1,703,518, 26 February 1929; "Method for the Protection of Blast-Furnace Jackets," U.S. Patent no. 1,703,519, 26 February 1929; "Apparatus for the Protection of Blast-Furnace Jackets," U.S. Patent no. 1,703,520, 26 February 1929; "Blast Furnace," U.S. Patent no. 1,783,416, 2 December 1930; and "Gas Cleaner," U.S. Patent no. 1,797,906, 24 March 1931. Other relevant materials are a number of clippings and reprints at SFNHL. These include Dovel's article "Improved Furnace on Southern Ores," *IA*, 22 September 1927; "Dovel Type Blast Furnace Put on Test" and "Improvement of Existing Blast Furnaces," from the December 1928 and August 1930 issues of *Blast Furnace and Steel Plant*, and "Marked Operating Economies Obtained with Improved Type Blast Furnace," from an unidentified journal. On the cooling pond, which was undergoing restoration at the time this book was nearing completion, see also "Facelift steams ahead at Sloss Furnaces," *BN*, 23 May 1993.

55. Database of statistical information on blast furnaces and coal mines in the Birmingham District compiled by Jack R. Bergstresser, Sr.; interview of Bergstresser.

56. Field Inventory, Fixed Capital Ledgers, SSSIC, SSSICR; "Birmingham's Most Modern Iron Maker," *PIRN*, August–September 1927, with accompanying illustrations.

57. Louis C. Hunter, "The Influence of the Market upon Technique in the Iron Industry in Western Pennsylvania up to 1860," *Journal of Economic and Business History* 1 (February 1929): 241–81.

58. Interview of Clarence E. Mason.

59. "Iron Ore—Making and Remaking the Map of Steel," 134.

60. "Sloss Uni-Pigs: A Better Pig Iron," *PIRN*, January 1929.

61. Field Inventory, Fixed Capital Ledgers, SSSIC, SSSICR; interviews of George Brown and Robert Casey. For a photograph of the ladle, see *PIRN*, Autumn 1934, 16–17; a drawing appears on the cover of the issue for Mid-Winter 1934. For a photograph of the automatic casting apparatus, see ibid., Mid-Winter 1934, 16. The term "Bull Ladle" seems first to have been used, at least in print, in 1935 when the phrase "Caught Hot From The Bull Ladle" began to be featured in the company magazine. See ibid., Winter 1935, 1. On the process itself, see "Sloss Ladle-Mixed Pig Iron," *PIRN*, Autumn 1934, 15; on the size of the pigs, see *PIRN*, Winter 1937, 4. On the Jones Mixer, see particularly Wall, *Andrew Carnegie*, 532, 642–43. On the characteristics of Heyl-Patterson machines, see Uehling, "Pig-Iron Casting and Conveying Machinery," 129–31.

62. Uehling, "Pig-Iron Casting and Conveying Machinery," 113–33; Bergstresser, Sr., "Raw Material Constraints," 185–86; interview of Clarence E. Mason regarding skimming process, which took place in the trough leading from the mixing ladle to the pig-casting machine.

63. Among numerous scholarly treatments of Taylorism, see particularly Gail Cooper, "Frederick Winslow Taylor and Scientific Management," in Carroll W. Pursell, Jr., ed., *Technology in America: A History of Individuals and Ideas*, 2d ed. (Cambridge, Mass.: MIT Press, 1990), 163–76.

64. Kulik, "Sloss Furnace Company," 56; "Sloss Ladle-Mixed Pig Iron," *PIRN*, Fiftieth Anniversary Issue, 1932, 24–26.

65. On the early history of the magazine, see Russell Hunt, "Rough Notes Begins Ninth Year," *PIRN*, Autumn 1934, 1.

66. See advertisement, "Pig Iron with A Wide Range of Analyses," in *PIRN*, November 1926 and all succeeding issues through November 1929. On "Clifton" pig iron, see the article on this particular brand in *PIRN*, November 1926.

67. *PIRN*, July 1927.

68. *PIRN*, January, February, July 1927. Early issues of the periodical, which contained eight pages at most, were unpaginated.

69. "Special Soil Pipe Edition," *PIRN*, March 1927; announcement of NAM meeting, October 1927; "Can the Gray Iron Foundryman meet the Test," December 1928; statements of standards adopted by Gray Iron Institute, June, July, August 1929.

70. Y. A. Dyer, "Cupola Metallurgy and Operation," *PIRN*, November 1926 through September 1927; "Iron Oxide Chiefly Due to Poor Cupola Practice," in ibid., October 1927.

71. "American Foundrymen's Association Bestows the First Seaman and McFadden Medals," *IA*, 17 September 1925, including biographical sketch of Moldenke; "Biographical Sketches of Contributors," in Backert, ed., *ABC of Iron and Steel*, xv; "Foundry Practice and Cupola Operation: Special Announcement," *PIRN*, December 1927. On Holley's analogous work, see McHugh, *Alexander Holley and the Makers of Steel*, passim.

72. "The Lesson of Defective Castings," *PIRN*, January 1928; "The Purchase of Foundry Raw Materials," February 1928; "Gating Molds," March 1928; "Molding Sand," April 1928; "Cores," May 1928; "Foundry Coke," June 1928; "The Cupola Melting Process—I," July 1928; "The Cupola Melting Process—II," August 1928; "The Cupola Melting Process—III," September 1928; "The Cupola Melting Process—IV," October 1928; "High Test Cast Iron," November 1928; "Pin Holes in Castings," December 1928; "Burnt Iron," September 1929; "The Gating of Molds," September 1929; "Cleaning Castings," May 1930; and "Foundry Blackings," June 1930, all by Richard Moldenke. Among articles by other contributors, see Eugene W. Smith, Jr., "Foundry Sands," February, March 1929; G. S. Evans, "What Happens in the Cupola," April through August 1929; and Harry W. Dietert, "Pouring Time of Castings," April 1930.

73. Minutes, Board of Directors, SSSIC, 30 August 1928, SSSICR.

74. In the summer of 1987 the author and Robert Casey, then curator at SFNHL, explored many of the sites of former mining camps maintained by the company in northern Jefferson County, including Bessie, Brookside, Blossburg, Cardiff, Coalburg, and Flat Top. This characterization is based upon our observations.

75. Lerner, "A Monument to Shame," 165–66; Stewart, *Governors of Alabama*, 173–74.

76. Lerner, "A Monument to Shame," 168–70; Clark, "Abolition," 99–112; Stewart, *Governors of Alabama*, 174–75.

77. Kulik, "Sloss Furnace Company," 51–54; Woodward, *Alabama Blast Furnaces*, 69, 81, 112.

78. The analysis of the background of the Great Depression in this and the following paragraphs draws upon Lester V. Chandler, *America's Greatest Depression, 1929–1941* (New York: Harper & Row, 1970), 1–29 and passim; Peter Fearon, *The Origins and Nature of the Great Slump, 1929–1932* (Atlantic Highlands, N.J.: Humanities Press, 1979), passim; John Kenneth Galbraith, *The Great Crash: 1929* (Boston: Houghton Mifflin, 1979); John A. Garraty, *The Great Depression* (San Diego: Harcourt Brace Jovanovich, 1986), 2–49; William K. Klingaman, *1929: The Year of the Great Crash* (New York: Harper & Row, 1989), passim; Broadus Mitchell, *Depression Decade: From New Era Through New Deal, 1929–1941* (New York: Rinehart & Co., 1947), 25–38; and Sobel, *Panic on Wall Street*, 350–91.

79. Galbraith, *Great Crash*, 141–42; Degler, *Out of Our Past*, 412–13.

80. Minutes, Board of Directors, SSSIC, 18 December 1929, SSSICR; Annual Report of the President, SSSIC, 7 March 1930, ibid.

81. *Thirty-first Annual Report of the Sloss-Sheffield Steel and Iron Company for the Calendar Year ended December 31, 1930*, 5–6.

82. *Thirty-second Annual Report of the Sloss-Sheffield Steel and Iron Company for the Calendar Year ended December 31, 1931*, 5–6.

15. From Hugh Morrow to Jim Walter

1. Irving Beiman, "Birmingham: Steel Giant with a Glass Jaw," in Robert S. Allen, ed., *Our Fair City* (New York: Vanguard Press, 1947), 105; LaMonte, "Politics and Welfare," 127.

2. LaMonte, "Politics and Welfare," 160–65; Barefield, *History of Mountain Brook*, 78; Eskew, "Demagoguery in Birmingham," 214–15.

3. John Williams, "Struggles of the Thirties in the South," in Bernard Sternsher, ed., *The Negro in Depression and War: Prelude to Revolution 1930–1945* (Chicago: Quadrangle Books, 1969), 167–78. For an overview of communism in Alabama during this period, see Robin D. G. Kelley, *Hammer and Hoe: Alabama*

Communists During the Great Depression (Chapel Hill: University of North Carolina Press, 1990), passim.

4. McMillan, *Yesterday's Birmingham*, 147–48; LaMonte, "Politics and Welfare," 127–37; Atkins, *Valley and Hills*, 142–46.
5. Rikard, "George Gordon Crawford," 191–223, 244–60.
6. Wiebel, *Biography of a Business*, 56–61.
7. Doerflinger and Rivkin, *Risk and Reward*, 112–13.
8. "Iron Ore—Making and Remaking the Map of Steel," 92, 134, 137.
9. Rowell, *Vulcan in Birmingham*, 43–46.
10. Leighton, "Birmingham, Alabama," 225–42; George R. Leighton, *Five Cities: The Story of Their Youth and Old Age* (New York: Harper & Brothers, 1939), 100–139.
11. *Thirty-third Annual Report of Sloss-Sheffield Steel and Iron Company for the Calendar Year ended December 31, 1932*, 5–6; *Thirty-fifth Annual Report . . . 1934*, 5–6; *Thirty-sixth Annual Report . . . 1935*, 5–6. On the protracted shutdowns of the No. 4 furnace at North Birmingham and the Bessie and Ruffner mines, see Minutes, Executive Committee Meetings, 27 February 1940 and 18 December 1941, SSSICR, and *Fortieth Annual Report . . . 1939*, 6.
12. *Thirty-sixth Annual Report . . . 1935*, 5; *Thirty-seventh Annual Report . . . 1936*, 5, 10; *Thirty-eighth Annual Report . . . 1937*, 5, 10; Executive Committee Minutes, 27 April 1938, SSSICR. On the way in which Morrow reduced bonded indebtedness by degrees in the mid-1930s, see Executive Committee Minutes, 26 January 1934, 26 April 1934, 6 August 1934, 22 October 1934, 22 April 1935, and 17 July 1935, SSSICR.
13. *Thirty-Sixth Annual Report . . . 1935*, 10; *Thirty-ninth Annual Report . . . 1938*, 5; *Fortieth Annual Report . . . 1939*, 5; *Forty-first Annual Report . . . 1940*, 5; *Forty-second Annual Report . . . 1941*, 5, 14.
14. Minutes, Board of Directors, SSSIC, 10 February 1938 and 26 July 1939, SSSICR; Minutes, Executive Committee, SSSIC, 7 May, 28 August, and 4 December 1940, ibid.; *BA-H*, 17 and 24 May 1925.
15. Minutes, Executive Committee, SSSIC, 25 October 1939, SSSICR; *Fortieth Annual Report . . . 1939*, 5. On the function of ferromanganese, see "Sloss Ferromanganese," *PIRN*, Autumn–Winter 1949–50.
16. See for example *PIRN*, Spring 1934, 7, 9, 10, 15, 19; Summer 1934, 7, 11, 15, 17, 24; Autumn 1934, 2–3, 6, 8, 25, 30; Mid-Winter 1934, 4, 5, 7; Spring 1935, 11, 14, 21, 22, 25; Summer 1935, 4, 16, 17, 20, 30, 31; Autumn 1935, 16, 30, 34, 45.
17. "Cast Iron—The Metal Eternal," *PIRN*, Summer 1935, 29–30; "Sloss Does Not Make Steel," Spring 1935, 4; "Specializing in Foundry Pig Iron," Spring 1938, 3. For a specimen of Hunt's many historical articles, see "Cast Iron Pipe Is as Enduring as the Eternal Hills," *PIRN*, Autumn 1935, 3–5.
18. F. E. Fisher, "Phosphobia," *PIRN*, Summer 1938, 9–12; Rebecca Hunt, "Phosphorus Again Proves Its Merit," Winter–Spring 1936, 25; W. O. McMahon, "Had Your PHOSPHORUS Today?," Spring 1937, 33–35.
19. Minutes, Board of Directors, SSSIC, 14 May, 10 June, 12 August, 1 September, 14 October, 18 November, 1 December 1936 and 15 January, 20 January, and 10 February 1937, SSSICR, with numerous supporting documents including "Certificate of Decrease, Readjustment and Change of Capital Stock and Amendment" and amended certificate of incorporation, both dated 1 March 1937. On the continuation of the conflict and the election of Alexander Hamilton Bryan as a director, see Minutes, Board of Directors, 14 April 1938, 6 May 1938, and 30 August 1939, ibid.
20. Material on Sloss Quarters in this and the following paragraphs is taken from *Like*

It Ain't Never Passed: Remembering Life in Sloss Quarters (Birmingham: Sloss Furnaces National Historic Landmark, 1985), based on oral interviews with eleven former Sloss Furnace employees.

21. On opposition among merchants to the use of scrip and other forms of company currency, see particularly Beiman, "Birmingham," 111–12. Ultimately, as Beiman indicates, the state legislature abolished the practice.

22. Anthony S. Campagna, *U.S. National Economic Policy, 1917–1985* (New York: Praeger, 1987), 107–54; Cobb, *Industrialization and Southern Society*, 63, 149–56. Interpretations of the period stressing differences between the First New Deal and Second New Deal are numerous; for a thoughtful assessment and historiographical overview, see Degler, *Out of Our Past*, 412–50, 605–8.

23. Minutes, Executive Committee, SSSIC, 12 December 1933, SSSICR.

24. Taft, *Organizing Dixie*, 82–95; clipping from unidentified Montgomery newspaper in Hill Ferguson Collection, BPLA.

25. Taft, *Organizing Dixie*, 110–16; Horace Huntley, "Iron Ore Miners and Mine Mill in Alabama, 1933–1952" (Ph.D. diss., University of Pittsburgh, 1977), 42–122.

26. Taft, *Organizing Dixie*, 96–110.

27. *Thirty-seventh Annual Report . . . 1936*, 5–6; *Thirty-eighth Annual Report . . . 1937*, 5; *Fortieth Annual Report . . . 1939*, 5; *Forty-first Annual Report . . . 1940*, 6; *Forty-second Annual Report . . . 1941*, 6–7.

28. Wright, *Old South, New South*, 216–25 and passim.

29. Robert J. Norrell, "Labor at the Ballot Box: Alabama Politics from the New Deal to the Dixiecrat Movement," *JSH* 57 (May 1991): 216–20; Stewart, *Governors of Alabama*, 187.

30. "Preparedness," *PIRN*, Summer 1940, 3; "Cast Iron for the Duration—And Thereafter," Spring 1941, 3; "The Southeast is Fast Becoming Industrialized: Its Defense Effort One of Immense Import," Summer 1941, 3; "War Procurement," Spring 1942, 5–13.

31. "Excellence Is the Word," *PIRN*, Winter 1943, 5–6; "Kansas Foundrywoman," "Alabama Foundrywoman," "Illinois Foundrywoman," "Ohio Foundrywoman," and "Georgia Foundrywoman," Autumn 1942, 3, 5–6, 13, 23, 29–31, 33; "Sloss on the Food Front," and other related articles, Autumn 1943, 3–4, 5–8, 14–16, 21–29, 48–59.

32. J. E. Getzen, "Birmingham District Foundries War Effort," *PIRN*, Autumn 1945, 5–8; W. O. McMahon, "Good Cast Iron Marches On," *PIRN*, Spring 1943, 9; "Cast Iron Goes to War," *PIRN*, Winter 1943, 9–12.

33. "Roll of Honor of Sloss," *PIRN*, Winter 1942, 20–21; *Forty-sixth Annual Report . . . 1945*, 8.

34. Norrell, "Labor at the Ballot Box," 220–21.

35. Wiebel, *Biography of a Business*, 62–64; Flynt, *Mine, Mill & Microchip*, 177–78.

36. Norrell, "Caste in Steel," 679; *Forty-fourth Annual Report . . . 1943*, 6; *Forty-fifth Annual Report . . . 1944*, 6; *Forty-sixth Annual Report . . . 1945*, 6.

37. Price, "Indomitable Inventor Helps Steel Industry with Ideas"; copy of news release, Sloss Furnaces National Historic Landmark, 1990, on opening of exhibit relating to Dovel; interviews of Dudley Dovel Shearburn.

38. For various details on Sloss-Sheffield's assets, earnings, dividends, and surplus accumulation during World War II, see *Forty-third Annual Report . . . 1942*, 5–9, 12–14; *Forty-fourth Annual Report . . . 1943*, 5–9, 12–15; and *Forty-fifth Annual Report . . . 1944*, 5–9, 12–14; *Forty-sixth Annual Report . . . 1945*, 5–9, 12–14. For other information utilized here, see *Moody's Manual of Investments*, 1941, 2244–46; 1943, 1706–7; and 1944, 2178–79.

39. "Sloss-Sheffield Stock Majority Acquired by U.S. Pipe & Foundry," *BN*, 26 and 27 December 1942; *Moody's Manual of Investments*, 1943, 1706.

40. On the discussion of USP&F in this and the following paragraphs, see Norman F. S. Russell, *"U.S.P.&F."*: *Cast Iron Pressure Pipe in New Jersey* (New York: Newcomen Society of North America, 1951), 8–17. For additional material on the Woods, see Sanders and Gould, *History Cast in Metal*, 340, 425.

41. *Moody's Manual of Investments*, 1943, 1706, 2203.

42. Ibid., 1706.

43. For the new composition of the board of directors and the executive committee, see *Forty-fourth Annual Report . . . 1943*, 3.

44. *Forty-seventh Annual Report . . . 1946*, 4; "Claude S. Lawson," *PIRN*, Spring–Summer 1948, 17; "Our New Sales Manager," ibid., Spring–Summer 1945, 5; "RH," ibid., Winter 1946, 3; "Florida Success Story," ibid., Spring–Summer 1947, 16–17.

45. "Two Are Elevated at Annual Session of Sloss-Sheffield," *BN*, 14 April 1948; *Forty-ninth Annual Report . . . 1948*, 8; "Hugh Morrow" and "Claude S. Lawson," *PIRN*, Spring–Summer 1948, 16–17; interview of Clarence E. Mason.

46. *Forty-sixth Annual Report . . . 1945*, 5; *Forty-seventh Annual Report . . . 1946*, 5; *Forty-eighth Annual Report . . . 1947*, 5; *Forty-ninth Annual Report . . . 1948*, 5; *Fiftieth Annual Report . . . 1949*, 5; *Fifty-first Annual Report . . . 1950*, 5, 14.

47. In addition to references to such products in Morrow's and Lawson's annual reports, cited above, see advertisements in *PIRN*, Autumn 1948, 32; Winter 1948, 32; Autumn–Winter 1948, 28; and Autumn 1950, 46. On "Sloss Dea-DinsecT," introduced in 1945, see *Forty-sixth Annual Report . . . 1945*, 7. On the physical changes made to the City Furnace plant, see Kulik, "Sloss Furnace Company," 56–57. As Kulik points out, the two turbo-blowers, along with a gas-washer installed in 1949, constitute the newest machinery on the site.

48. Stocking, *Basing Point Pricing*, 156–89.

49. Edward F. Haas, "The Southern Metropolis, 1940–1976," in Brownell and Goldfield, eds., *The City in Southern History*, 173, table 6-1; "The Vision of a Few," *PIRN*, Fall–Winter 1954–55, 27; Huntley, "Iron Ore Miners," 16.

50. Beiman, "Birmingham," 99.

51. On Connor's background and career through 1952, see William A. Nunnelley, *Bull Connor* (Tuscaloosa: University of Alabama Press, 1991), 1–47.

52. Beiman, "Birmingham," 99–122.

53. The analysis presented in this and the following paragraphs is based on Norrell, "Caste in Steel," 669–94.

54. Huntley, "Iron Ore Miners," 123–74.

55. Norrell, "Caste in Steel," 669.

56. On this point, see particularly Robert J. Norrell, "One Thing We Did Right: Protest, History, and the Civil Rights Movement," in Armistead L. Robinson and Patricia Sullivan, *New Directions in Civil Rights Studies* (Charlottesville: University Press of Virginia, 1991), 72.

57. "The Vision of a Few," 27–31; membership list of the Committee of 100, BPLA.

58. "Hugh Morrow Sees Nation's Hope in the South," "Great Citizen," and other clippings, Hugh Morrow File, Hill-Ferguson Collection; Nunnelley, *Bull Connor*, 48–67; Norrell, "Caste in Steel," 686–89, and "Labor at the Ballot Box," 208–9.

59. "Committee of 100," informational summary, BPLA; Haas, "The Southern Metropolis," 173, table 6-1; LaMonte, "Politics and Welfare," 238–78; Hamilton, *Alabama*, 140.

60. Norrell, "Caste in Steel," 686; Theron D. Parker, "Molybdenum—Versatile Alloy

for Cast Iron," *PIRN*, Autumn 1956, 2–9; J. Dodd, "Unalloyed and Low Alloy White Irons," ibid., 10–19; "U.S. Pipe Building New Blast Furnace," ibid., 17; C. K. Donoho, "Cast Iron Bolts and Nuts for Pipe Joints," ibid., 34–39.

61. "Agreement of Merger by and between United States Pipe and Foundry Company and Sloss-Sheffield Steel and Iron Company Dated September 12, 1952," SSSICR. Interviews of Clarence E. Mason have contributed to my efforts to understand better the dynamics of these changes.

62. Robert W. Kincey, "Business and Industry," *BN*, 1 November 1952; "'Sloss' Merged with U.S. Pipe," *PIRN*, Winter–Spring 1954, 2–3; Mason interview.

63. *Like It Ain't Never Passed*, 10–11.

64. "Sloss Coke Developments," *PIRN*, Autumn 1960, 23–24.

65. The description of the new North Birmingham blast furnace that follows is based on "U.S. Pipe's Number 5 Blast Furnace," *PIRN*, Autumn 1960, 2–22. I am grateful to Clarence E. Mason for additional insight pertaining to the operation of the furnace.

66. The analysis in this and the following paragraphs is based on interviews with Clarence E. Mason, the last superintendent of Sloss-Sheffield's City Furnaces.

67. Richard Foster, *Innovation: The Attacker's Advantage* (New York: Summit Books, 1986), 45–86, 115–35.

68. The discussion of ductile iron that follows is based upon *Encyclopaedia Britannica*, 1968 ed., s.v. "ductile iron," and interviews by author of Clarence E. Mason.

69. "Hugh Morrow Sr. Dies, Rites to be Tomorrow," *BN*, 7 September 1960; Mason interview.

70. Lee E. Bains, Jr., "Birmingham, 1963: Confrontation over Civil Rights," in David J. Garrow, ed., *Birmingham, Alabama, 1956–1963: The Black Struggle for Civil Rights* (Brooklyn, N.Y.: Carlson Publishing, 1989), 165–66. On restrictions limiting black voting, see Hamilton, *labama*, 96–97.

71. In addition to the analysis of these developments in Bains, cited above, see Nunnelley, *Bull Connor*, 68–128.

72. Harrison E. Salisbury, "Fear and Hatred Grip Birmingham," *NYT*, 12 April 1960.

73. Robert J. Norrell, "Labor Trouble: George Wallace and Union Politics in Alabama," in Robert Zieger, ed., *Organized Labor in the Twentieth-Century South* (Knoxville: University of Tennessee Press, 1991), 250–72.

74. The account of events in the early 1960s in this and the following paragraphs is taken chiefly from Glenn T. Eskew, "The Alabama Christian Movement for Human Rights and the Birmingham Struggle for Civil Rights, 1956–1963," in Garrow, ed., *Birmingham, Alabama, 1956–1963*, 3–114; Lewis W. Jones, "Fred L. Shuttlesworth, Indigenous Leader," ibid., 115–50; and Bains, "Birmingham 1963," ibid., 151–289. See also Atkins, *Valley and Hills*, 157–59, and Nunnelley, *Bull Connor*, 129–64.

75. On the developments discussed in this paragraph and the ones that follow, see also David J. Garrow, *Bearing the Cross: Martin Luther King, Jr., and the Southern Christian Leadership Conference* (New York: William Morrow and Co., 1986), 173–286.

76. "Louisiana 1927" (Randy Newman) © 1974 Warner-Tamerlane Publishing Corp. All rights reserved. Used by permission.

77. Norrell, "One Thing We Did Right," 72–73.

78. John Walton Cotman, *Birmingham, JFK and the Civil Rights Act of 1963: Implications for Elite Theory* (New York, Bern, Frankfurt am Main, and Paris: Peter Lang, 1989), 100–102.

79. Quoted in Atkins, *Valley and Hills*, 158.

80. Ibid., 159.

81. Norrell, "Caste in Steel," 690.

82. See particularly "Chief of 'Automatic' Sprinkler Sets Sights High," *NYT*, 23 December 1967. For a brief synopsis of Figgie's career, see *Who's Who in America*, 1972–73, 1:998.
83. "Give Me Time," *Forbes Magazine*, 1 April 1969; William Simon Rukeyser, "Why Rain Fell on 'Automatic' Sprinkler," *Fortune* 79 (1 May 1979): 88–91, 126–29; Mason interview.
84. Randall Williams and Hilda Dent, "Billion Dollar Shell Game," *Southern Exposure* 8 (Spring 1980): 86–91; "Jim Walter Corporation," in Milton Moskowitz, Michael Katz, and Robert Levering, eds., *Everybody's Business, An Almanac: The Irreverent Guide to Corporate America* (New York: Harper & Row, 1980), 172.
85. On the impact of Venezuelan ore on the Birmingham District, see particularly Atkins, *Valley and Hills*, 159. On the abandonment and subsequent development of the Russellville tract, see Wood, "Sloss Lake," 9–12.
86. Moskowitz, Katz, and Levering, *Everybody's Business*, 172; Mason interview.
87. Mason interview.
88. Interview by author of J. David Hagan, M.D., a participant in the movement.
89. James W. Sledge III, "Birmingham's Iron Elephant: The Creation of Sloss Furnaces Museum" (term paper, Department of History, Auburn University, 1989), 2; copy of the proposal, shared with me by Sledge.
90. Sledge, "Birmingham's Iron Elephant," 2–3; "Summary of Events between May, 1971 and the Present"; clipping, "U.S. Pipe Donates Historic Landmark to City," from *BN*, 1 June 1971; and copy of deed of gift, 3 September 1971, shared with me by Sledge.

16. Preserving the Heritage

1. On the Corliss engine as an icon of industrialism, see particularly John F. Kasson, *Civilizing the Machine: Technology and Republican Values in America* (1976; reprint, New York: Penguin Books, 1977), 164.
2. Atkins, *Valley and Hills*, 152–53, 159, 164, 186, 190–91, 206, 211, 214–15; Flynt, *Mine, Mill & Microchip*, 217, 232, 270–71, 278–81, 286–87, 298–99, 306–7, 334–35.
3. Atkins, *Valley and Hills*, 157–58, 168; Flynt, *Mine, Mill & Microchip*, 204, 218; Arthur George Gaston, *Green Power: The Successful Way of A. G. Gaston* (1968; reprint, Troy, Ala.: Troy State University Press, 1977), passim.
4. Jimmie Lewis Franklin, *Back to Birmingham: Richard Arrington, Jr., and His Times* (Tuscaloosa: University of Alabama Press, 1989), 3–180.
5. Hamilton, *Alabama*, 147–48.
6. Ibid., 147; H. Brandt Ayers and Thomas H. Naylor, eds., *You Can't Eat Magnolias* (New York: McGraw Hill, 1972), passim.
7. Sledge, "Birmingham's Iron Elephant," 3–4; "Jim Walter Park in U.S. Register," *BN*, 26 July 1972; "Sloss Furnace Now 17-acre Liability?" ibid., 15 August 1972; "Furnace, Museum Plans Both Rusting," ibid., 24 November 1975. I am grateful to Sledge for sharing with me his collection of newspaper clippings and other source materials, upon which I have drawn extensively in this and the paragraphs that follow.
8. Atkins, *Valley and Hills*, 159–60; Bowsher, Morris, and White, eds., *Cinderella Stories*, 10. The building presently known as the Tutwiler Hotel is the former Ridgely Apartments, renovated in 1986–88.
9. Atkins, *Valley and Hills*, 160; publications of Birmingfind including *Birmingham's Lebanese: "The Earth Turned to Gold"*; *Elyton-West End: Birmingham's First Neighborhood*; *The Italians: From Bisacquino to Birmingham*; and *The Other Side: The Story of Birmingham's Black Community*; Flynt, *Poor but Proud*, previously cited.

10. Sledge, "Birmingham's Iron Elephant," 4; Chris Conway, "Fair Authority Decides to Raze Old Sloss Furnaces," *BN*, 17 March 1976; "Chamber's Conclusion: Furnace Land As Park Ruled Out," ibid., 18 March 1976; "Keep Furnaces, Historical Panel Says," ibid., 28 April 1976; "Several Groups Trying to Halt Plan to Level Sloss Furnaces; Many Voice Ideas for Using It," ibid., 24 May 1976; "Vann Would Keep Furnaces," *BP-H*, 28 May 1976.

11. Sledge, "Birmingham's Iron Elephant," 4–6; "City Wants to Have a Say on Furnaces," *BN*, 25 May 1976; "Delay Asked on Furnaces," ibid., 26 May 1976; Bill Conway, "Many Hope to Save Sloss Furnaces," *BP-H*, 28 May 1976; Frances Spotswood, "Save-Sloss Group Takes 'Stop Bulldozers' as Theme," *BN*, 31 May 1976; "Neighborhood Groups Favor Saving Furnace," ibid., 22 June 1976.

12. See Kulik, "Sloss Furnace Company."

13. Sledge, "Birmingham's Iron Elephant," 7; "Furnaces Study Long Awaited," *BP-H*, 13 July 1976; Cheryl Blackerby, "Furnaces No Picnic for Engineering Team," ibid., 31 August 1976; Bill Crowe, "Study Group Holds Future of Historic Sloss," *BN*, 5 September 1976.

14. Bill Crowe, "Vann Sees Sloss Furnace as Lemon if Only Museum; Urges Theme Park," *BN*, 24 June 1976; Bob Johnson, "Multitude of Proposals on Sloss Furnaces Heard," *BP-H*, 9 July 1976.

15. "Sloss Furnace Tract: A Preliminary Study of Site Development Prepared by the Birmingham Planning Commission for the Alabama State Fair Authority," SFNHL.

16. Sledge, "Birmingham's Iron Elephant," 7–8; Ron Casey, "Fair Authority: Would Let about Half of Sloss Stand," *BN*, 13 July 1976, and Dennis Washburn, "C of C Exec: Don't Use Tax Monies to Restore Sloss Furnace Facility," ibid., 16 July 1976; Bill Steverson, "Fair Board Reverses Itself: To Restore Sloss Furnaces," *BP-H*, 13 July 1976; Barbara Crane, "Historic Sloss Furnaces Will Be Saved, Mayor Vann Says," ibid., 15 July 1976; Bill Crowe, "Sloss Site Group Leans to Museum," *BN*, 12 August 1976.

17. Sledge, "Birmingham's Iron Elephant," 7–8; Helen Mabry, "Sloss Furnace Association, 1976–1986," manuscript shared with author by Sledge; Chris Conway, "Future of Sloss Furnaces Is Unclear," *BP-H*, 4 February 1977.

18. Sledge, "Birmingham's Iron Elephant," 8–9; Bill Crowe, "Sloss 'Hot Potato' in Public's Lap after Juggling by City Officials," *BN*, 8 May 1977; handbill circulated by SFA in referendum, copy provided by Sledge.

19. Sledge, "Birmingham's Iron Elephant," 9–10; Richard Friedman, "Plan for Sloss Furnace Museum Hits Snag over Ownership of Site" and "Mayor 'Confident' City Will Get Title to Furnace," *BN*, 7 and 14 July 1977; Mabry, "Sloss Furnace Association," 7.

20. Sledge, "Birmingham's Iron Elephant," 10; Mabry, "Sloss Furnace Association," 7; Harold Jackson, "Sloss Furnace Plans Get Boost," *BP-H*, 7 September 1978.

21. Jackson, "Sloss Furnace Plans Get Boost," *BP-H*, 7 September 1978.

22. Richard Friedman, "Little Money, High Hopes for Sloss," *BN*, 18 February 1979; "Sloss Secrets," ibid., 22 February 1979; Friedman, "Oompah and Dominoes: McMullin Sloss Plan Envisions Park to Rival Fantasy Island," ibid., 11 March 1979.

23. "Sloss Dreaming," *BN*, 13 March 1979.

24. Tommy Black, "Group Rekindles Plan for Sloss Museum," *BN*, 27 June 1979; Ingrid Kindred, "City Officials: Sloss Museum Must Be Top-Notch or Dump Idea," ibid., 8 July 1979; "City OKs $190,000 Sloss Feasibility Study," ibid., 12 July 1979; "Scrapping Sloss Plans Urged," ibid., 26 July 1979.

25. Mabry, "Sloss Furnace Association," 9–10; Thomas Hargrove, "Voters Tend to Favor Saving Sloss Furnace," *BP-H*, 5 October 1979.

26. Sledge, "Birmingham's Iron Elephant," 11–12; Mabry, "Sloss Furnace Associa-

tion," 10; Kitty Frieden, "Gardner: Sloss is 'Fantastic Monument'" and "Sloss Furnace Future Project?" *BN*, 20 May 1980.

27. David McMullin and Hardy Holzman Pfeiffer Associates, "Sloss Furnace: A Museum of Modern Times" (Birmingham: privately published, 1980) copy at SFNHL.

28. Sledge, "Birmingham's Iron Elephant," 14–15; Mabry, "Sloss Furnace Association," 11; Tommy Black, "Sloss Furnace: Project Director to Unveil Development Plan," *BN*, 23 May 1980; Mitch Mendelson, "Grandiose Development Scheme Unlikely to Get City Council OK," *BP-H*, 20 June 1980; "Cost Questions Raised over Plans for Sloss," *BN*, 25 June 1980; "Chamber, MDB Oppose Sloss Museum Proposal," ibid., 27 June 1980; "Sloss Questions," ibid., 7 July 1980; Kitty Frieden, "Arrington for Jail as County Cuts Share of City Bill," ibid., 17 July 1980.

29. Mabry, "Sloss Furnace Association," 12–13; "Sloss Project Director's Contract Expires," *BN*, 1 August 1980; Kitty Frieden, "Bond Approval May Hinge on Voter Say-so on Sloss," ibid., 1 October 1980.

30. Sledge, "Birmingham's Iron Elephant," 15–16; Mabry, "Sloss Furnace Association," 13; Kitty Frieden, "Simple 'Save Sloss' Philosophy Scales Down Museum Plans," *BN*, 12 October 1980; Mitch Mendelson, "Scaled-Down Plan Raises Hopes," *BP-H*, 18 November 1980; Frieden, "Council OKs Plan to Save Sloss Furnace," *BN*, 26 November 1980.

31. Gayle McCracken, "Sloss Furnaces Opens for Tours in the Spring," *BP-H*, 27 November 1980; "Major Restoration of Sloss to Begin," ibid., 25 June 1981; "Firms Can Enter Sloss Bids Shortly," *BN*, 25 July 1981; Tommy Black, "Group Trying to Spruce Up Historic Sloss," ibid., 2 August 1981; "Furnaces to Get Landmark Tag," ibid., 31 October 1981; Nancy Campbell, "Museum Pictured as a Way to Bank Memory of Sloss," ibid., 9 November 1981.

32. Walter Bryant, "Sloss Furnaces to Roar Back to Life . . . in a Way," *BN*, 4 April 1982; "Bid Day Near for New-Look Sloss Furnace," ibid., 11 April 1982; "Sloss' New Life," ibid., 12 April 1982.

33. "Unlicensed Fireworks Display Believed Cause of Sloss Fire," *BN*, 13 April 1982.

34. *BN*, 22 August 1982.

35. Kitty Frieden, "Man Who Knows History, Coal Mining Hired to Direct Sloss Museum Project," *BN*, 8 December 1982; interview of Lawrence by author.

36. Ibid.; "Sloss Furnace to Get Free Paint," *BN*, 2 January 1983; "'Budget Crisis' Puts Bugs in Grand Plans at Sloss," ibid., 9 March 1983; "Sloss, Museum Bids Accepted," ibid., 1 June 1983; Elma Bell, "Sloss: Fires of Hell or One Big Sculpture," ibid., 29 August 1983.

37. Walter Bryant, "Sloss Finally Stirs as Old Memories Fire Up on Friday for a Public Hello," *BN*, 28 August 1983; Monique Van Landingham, "Sloss' Ancient Furnaces Rekindled by Jazz," ibid., 3 September 1983; interview of Paige Wainwright by author.

38. "Preserving a Symbol," Anniston *Star*, 20 August 1983; "Sloss Stirs Again," *BN*, 2 September 1983; "Sloss Furnace Gets International Praise," *The Preservation Report* 11 (November–December 1983): 1.

39. "Lawrence, Sloss Furnaces Director, Dies," *BN*, 17 May 1991; "Randy Lawrence," *BP-H*, 18 May 1991.

40. Frederick Kaimann, "Taking on a New Life: New Director Guides Sloss Furnaces on Its New Mission," *BN*, 7 June 1992.

17. In Retrospect: The Southernness of Sloss

1. For a text of the speech, see Joel Chandler Harris, *Life of Henry W. Grady, Including his Writings and Speeches* (New York: Cassell Publishing Co.,

1890), 83–93. On the background of the speech and public reaction to it, see Raymond B. Nixon, *Henry W. Grady: Spokesman for the New South* (New York: Alfred A. Knopf, 1943), 237–60.

2. Carroll W. Pursell, *Early Stationary Steam Engines in America: A Study in the Migration of a Technology* (Washington, D.C.: Smithsonian Institution Press, 1969), 75.

3. Raimondo Luraghi, *The Rise and Fall of the Plantation South* (New York: New Viewpoints, 1978), 108–9.

4. Hesseltine, *Confederate Leaders*, 27–147.

5. Thornton, *Politics and Power*, passim; Degler, *Out of Our Past*, 178–79.

6. The extended discussion that follows is a response to Wiener, *Social Origins of the New South*, 137–227.

7. See map, ibid., facing p. 1.

8. Atkins, *Valley and Hills*, 72.

9. For examples of sympathetic coverage by the *Advertiser* of activities involving Gilmer, Hall, Milner, Powell, Tate, Wallace, and other persons connected with the South and North Railroad, see particularly "Letter from Col. J. R. Powell," *MA*, 16 November 1870; "Major Campbell Wallace: A Just Tribute to a Worthy Man," ibid., 17 November 1870; "Stockholders Meeting of the South and North Railroad," ibid., 23 December 1870; and "Maj. Campbell Wallace," ibid., 5 January 1871. For the *Advertiser's* opinions on Pratt and his activities, see "Daniel Pratt, Esq." ibid., 18 September 1870. Calling Pratt an "excellent gentleman" and "respected citizen," the writer of this article declared, "Perhaps there is no man in Alabama who is more thoroughly esteemed in all the relations of life than the benevolent and energetic founder of the flourishing village of Prattville."

10. Luraghi, *Rise and Fall*, 112.

11. Ibid., 107.

12. Ibid., 7 and passim.

13. Barrington Moore, Jr., *Social Origins of Dictatorship and Democracy: Lord and Peasant in the Making of the Modern World* (Boston: Beacon Press, 1966), 289–92, 301–5, 435–42, and passim.

14. Herbert G. Gutman, "Enslaved Afro-Americans and the 'Protestant' Work Ethic," in Ira Berlin, ed., *Power & Culture: Essays on the American Working Class* (New York: Pantheon Books, 1987), 298–325. Comparison of Gutman's essay with Fogel and Engerman, *Time on the Cross*, 103–7 is illuminating in this respect.

15. Chapman, *Iron and Steel Industries*, 231–35; Genovese, *Political Economy of Slavery*, 158; Luraghi, *Rise and Fall*, 107; Norrell, *Bowron*, xx–xxi; Wright, *Old South, New South*, passim.

16. Armes, *Coal and Iron*, 178–94, 256–57.

17. Fogel and Engerman, *Time on the Cross*, 249.

18. Gates, *Model City*, 33, 148–51, 197; McCaa, "Samuel Noble: Founder of Anniston," 27.

19. Rowland T. Berthoff, *British Immigrants in Industrial America, 1790–1950* (Cambridge: Harvard University Press, 1953), 65, 113–14; Clifton K. Yearley, Jr., *Britons in American Labor: A History of the Influence of the United Kingdom Immigrants on American Labor, 1820–1914* (Baltimore: Johns Hopkins Press, 1957), 142–47.

20. Flynt, *Poor but Proud*, 157–58.

21. Gates, *Model City*, 90–91, 122–25, 146–265. See also McCaa, "Samuel Noble: Founder of Anniston," passim.

22. The ensuing discussion is based partly on Bergstresser, Sr., "Raw Material Constraints," passim, and partly on my own research in the sources that Bergstresser has consulted.

23. Mason interview.

24. Merritt Roe Smith, *Harpers Ferry Armory and the New Technology: The Challenge of Change* (Ithaca, N.Y.: Cornell University Press, 1977), 184–218.
25. Klein, *Great Richmond Terminal,* passim.
26. For already-published insights into the role of British investment in southern industry, see particularly Woodward, *Origins of the New South,* 118–20, 126, and Dorothy R. Adler, *British Investment in American Railways, 1834–1898,* ed. Muriel E. Hidy (Charlottesville: University Press of Virginia for Eleutherian Mills–Hagley Foundation, 1970).
27. For representative sources on the developments mentioned here, see Charles H. Candler, *Asa Griggs Candler* (Emory University, Ga.: Emory University Press, 1950); W. David Lewis and Wesley Phillips Newton, *Delta: The History of an Airline* (Athens: University of Georgia Press, 1979); Patrick Reynolds and Tom Shachtman, *The Gilded Leaf: Triumph, Tragedy, and Tobacco. Three Generations of the R. J. Reynolds Family and Fortune* (Boston: Little, Brown and Co., 1989); Nannie M. Tilley, *The R. J. Reynolds Tobacco Company* (Chapel Hill: University of North Carolina Press, 1985); Pat Watters, *Coca-Cola: An Illustrated History* (Garden City, N.Y.: Doubleday & Co., 1978); and John K. Winkler, *Tobacco Tycoon: The Story of James Buchanan Duke* (New York: Random House, 1942).
28. Hesseltine, *Confederate Leaders,* passim; Klein, *Edward Porter Alexander,* 141–200; Klein, *Great Richmond Terminal,* 45–46, 161–84.
29. Woodward, *Origins of the New South,* 300–301, 315–17.
30. Hundley, "Iron Ore Miners," LaMonte, "Politics and Welfare," and Norrell, "Caste in Steel," all previously cited.
31. Cobb, *Industrialization and Southern Society,* 68.
32. Ibid., 91.
33. See Flynt, *Poor but Proud;* Straw, "Collapse of Biracial Unionism," "Soldiers and Miners," and "The United Mine Workers"; and Ward and Rogers, *Convicts, Coal, and the Banner Mine Tragedy,* all previously cited.
34. In addition to Chandler, "Anthracite Coal," Long, *Where the Sun Never Shines,* and Wallace, *St. Clair,* all previously cited, see Charles Kenneth Sullivan, *Coal Men and Coal Towns: Development of the Smokeless Coalfields of Southern West Virginia, 1873–1923* (New York and London: Garland Publishing, 1989), and Clifton K. Yearley, Jr., *Enterprise and Anthracite: Economics and Democracy in Schuylkill County, 1820–1875* (Baltimore: Johns Hopkins Press, 1961).
35. John M. Staudenmaier, *Technology's Storytellers: Reweaving the Human Fabric* (Cambridge, Mass.: MIT Press, 1985), 162–201. See also Staudenmaier's article, "What SHOT Hath Wrought and What SHOT Hath Not: Reflections on Twenty-five Years of the History of Technology," *T&C* 25 (October, 1984): 707–30.
36. Hamilton, *Alabama,* 148.
37. Quoted in Flynt, *Mine, Mill & Microchip,* 287.
38. For discussion of the postindustrial syndrome into which the newer terminology fits, see John Naisbitt and Patricia Aburdene, *Re-inventing the Corporation* (New York: Warner Books, 1985), 79–119.
39. B. Eugene Griessman, "Will the South Rise Again or Just Roll Over?" in Lewis and Griessman, *Southern Mystique,* 125–31.
40. Alvin Toffler, *Future Shock* (New York: Random House, 1970), 3, 430; Richard Brautigan, *The Pill Versus the Springhill Mine Disaster* (New York: Delacorte Press, 1968); Robert Blauner, *Alienation and Freedom: The Factory Worker and His Industry* (Chicago: University of Chicago Press, 1964), both cited in W. David Lewis, "Technology, Community, and Humanity: The Big Picture," in Lewis and Griessman, *Southern Mystique,* 26–29. On junk bonds, Milken, and Drexel Burnham Lambert, see especially the analysis in Toffler's *Power Shift: Knowledge, Wealth, and Violence at the Edge of the 21st Century* (New York: Bantam Books, 1990), 45–59.

41. On most of these trends, see David Halberstam, *The Reckoning* (New York: William Morrow and Co., 1986), 673–728.
42. Wayne Flynt, "Alabama Fares Very Poorly in Marketplace Competition," *MA*, 22 July 1992.
43. Halberstam, *Reckoning*, 726–28.
44. White, *Downtown Birmingham*, passim.
45. "A Monumental Achievement," "Dedication Today for Civil Rights Institute," and other articles in *BN*, 15 November 1992, including special section, "Civil Rights Institute: A Guide to Its Galleries, History and Mission"; "Facing Up to Racial Pains of Past, Birmingham Moves On," *NYT*, 15 November 1992.
46. "March Through Time: Capital Opens to Public" and "He's the Barrier: Hunt Stand Inconsistent with Post," *MA*, 13 December 1992; Tom Lindley, "Capitol Reopens Amid Talk of Unity" and "Capitol Really Is Important Place in State History," *BN*, 13 December 1992.
47. W. E. B. Du Bois, *The Souls of Black Folk* (1903; paperback ed., with introduction by Henry Lewis Gates, Jr., New York: Bantam Books, 1989), xxiii, 3.
48. "Birmingham" (Randy Newman) © 1974 Warner-Tamerlane Publishing Corp. All rights reserved. Used by permission.
49. Thompson, "Vulcan: Birmingham's Man of Iron," 17.

Bibliography

Archives and Manuscript Collections

Alfred Montgomery Shook Papers, Birmingham Public Library, Department of Archives and Manuscripts, Birmingham, Alabama.

Biennial Reports, Alabama State Inspector of Mines, Alabama State Archives, Montgomery, Alabama.

Bryan Family Papers, Virginia Historical Society, Richmond, Virginia.

Bryan Family Papers, Virginia State Library and Archives, Richmond, Virginia.

Corporate Records, Probate Office, Jefferson County Courthouse, Birmingham, Alabama.

Daniel K. Stewart Letterbooks, Virginia Historical Society.

Edward A. Uehling, unpublished autobiography and letters supplied by Edward R. Uehling, Tuscaloosa, Alabama.

Elyton Land Company Minute Books, Birmingham Public Library, Department of Archives and Manuscripts.

Henry Morton Record Books, Samuel C. Williams Library, Stevens Institute of Technology, Hoboken, New Jersey.

Hill Ferguson Collection, Birmingham Public Library, Department of Archives and Manuscripts.

James Bowron Autobiography and Diaries, W. S. Hoole Special Collections, University of Alabama Library.

James Pickering Dovel Collection, Sloss Furnaces National Historic Landmark, Birmingham, Alabama.

Joseph Bryan Letterbooks, Virginia Historical Society.

Maben Family Papers, Virginia State Library and Archives.

Matthew Scott Sloan Materials, Auburn University Archives, Auburn, Alabama.

R. G. Dun Collection, Baker Library, Harvard University Graduate School of Business Administration, Boston, Massachusetts.

Sloss-Sheffield Steel and Iron Company Records, Birmingham Public Library, Department of Archives and Manuscripts.

Stewart Family Papers, Virginia Historical Society.

William Henry Ruffner Papers, Virginia State Library and Archives.

Government Publications

Burchard, E. F. "The Brown Iron Ores of the Russellville District, Alabama." In U.S. Geological Survey, *Contributions to Economic Geology*, Bulletin 315-D. Washington, D.C.: U.S. Geological Survey, 1907.

Snavely, Tipton Ray. "The Exodus of Negroes from the Southern States: Alabama and North Carolina." In United States Department of Labor, Division of Negro Economics, *Negro Migration in 1916–1917*. 1919. Reprint. New York: Negro Universities Press, 1969.

State of Alabama. *The Charter of the South & North Ala. R.R. Co. and Acts Amendatory Thereof . . . and also Laws Regulating Railroads in Alabama*. Montgomery: Advertiser Book and Job Printing Office, 1876.

United States Bureau of the Census. *Report on Manufacturing Industries in the United States at the Eleventh Census* (by William W. Sweet) in *Eleventh Census of the United States*, Part 3. Washington, D.C.: Government Printing Office, 1895.

United States Department of Commerce, Bureau of the Census. *Negroes in the United States, 1920–32*. Washington, D.C., Government Printing Office, 1935.

United States Senate. *Report of the Committee of the Senate Upon the Relations between Labor and Capital*, vol. 4. Washington, D.C.: Government Printing Office, 1885.

United States Patents

Dovel, James P. "Apparatus for the Protection of Blast-Furnace Jackets." U.S. Patent no. 1,703,250, 26 February 1929.
———. "Blast Furnace." U.S. Patent no. 1,703,517, 26 February 1929.
———. "Blast Furnace." U.S. Patent no. 1,703,518, 28 July 1927.
———. "Blast Furnace." U.S. Patent no. 1,783,416, 2 December 1930.
———. "Blast-Furnace Stack Reinforcement." U.S. Patent no. 1,316,085, 16 September 1919.
———. "Gas Cleaner." U.S. Patent no. 1,609,611, 7 December 1926.
———. "Gas Cleaner." U.S. Patent no. 1,797,906, 24 March 1931.
———. "Gas-Cleaning Apparatus." U.S. Patent no. 1,001,740, 29 August 1911.
———. "Method for the Protection of Blast-Furnace Jackets." U.S. Patent no. 1,703,519, 26 February 1929.
———. "Ore-Washing Plant." U.S. Patent no. 1,252,414, 8 January 1918.
———. "Pig-Breaking Machine." U.S. Patent no. 1,622,029, 22 March 1927.
———. "Power-Shovel." U.S. Patent no. 1,444,670, 6 February 1923.
———. "Settling Apparatus." U.S. Patent no. 1,485,452, 4 March 1924.
———. "Twyer and Cooling-Box." U.S. Patent no. 1,354,032, 28 September 1920.
Dovel, James P., and Charles C. Glidden. "Amalgamator." U.S. Patent no. 1,123,116, 29 December 1914.
Dovel, James P., and John J. Shannon. "Gas-Washer." U.S. Patent no. 1,001,739, 29 August 1911.
Ramsay, Erskine. "Coal and Mineral Washer." U.S. Patent no. 528,803, 6 November 1894.
Ramsay, Erskine, and Ernest Dreyspring. "Coal and Mineral Washer." U.S. Patent no. 579,840, 30 March 1897.
Shannon, John J., and James P. Dovel. "Air-Cooled Blast-Furnace Stack." U.S. Patent no. 1,090,574, 17 March 1914.
Stewart, Elwood A. "Coal Washer." U.S. Patent no. 657,184, 4 September 1900.
Uehling, Edward A. "Apparatus for and Method of Casting and Conveying Metals." U.S. Patent no. 548,146, 15 October 1895.

Uehling, Edward A., and Alfred Steinbart. "Pyrometer." U.S. Patent no. 554,323, 11 February 1896.

Directories

American Iron and Steel Institute. *Directory of Iron and Steel Works of the United States and Canada.* 18th ed. New York: American Iron and Steel Institute, 1916.
Birmingham City Directory, 1903. Atlanta, Ga.: Mutual Publishing Company, 1903.
Birmingham City Directory for the Year Commencing February 1, 1912. Birmingham: R. L. Polk & Co., 1912.

Interviews

Atkins, Leah Rawls. Interviews with author. Auburn, Alabama.
Bergstresser, Jack Roland, Sr. Interviews with author. Birmingham, Alabama.
Botta, Nicholas E. Interview with author. Auburn, Alabama.
Bryan, D. Tennant. Interview with author. Richmond, Virginia.
Burelle, Michael. Interview with author. Auburn, Alabama.
Carrington, Thomas J. Interviews with author. Auburn, Alabama.
Casey, Robert. Interviews with author. Detroit, Michigan.
Colglazier, Merle E. Interview with author. Richmond, Virginia.
Copeland, Charles. Interview with author. Tuscaloosa, Alabama.
Dean, Lewis S. Interview with author. Tuscaloosa, Alabama.
Flynt, Wayne. Interview with author. Auburn, Alabama.
Hagan, J. David. Interview with author. Auburn, Alabama.
Hall, Hines H., III. Interview with author. Auburn, Alabama.
Klein, Maury. Interview with author. Kingston, Rhode Island.
Lawrence, Randall G. Interviews with author. Birmingham, Alabama.
Littleton, Taylor D. Interview with author. Auburn, Alabama.
Mason, Clarence E. Interviews with author. Auburn, Alabama.
Shearburn, Dudley Dovel. Interviews with author. Winston-Salem, North Carolina.
Wainwright, Paige. Interview with author. Birmingham, Alabama.
White, Marjorie Longenecker. Interviews with author. Birmingham, Alabama.
Whiting, Marvin Y. Interviews with author. Birmingham, Alabama.

Newspapers

Anniston *Star,* 1983.
Bessemer Weekly, 1901 (reprint, 1977).
Birmingham *Age,* 1886.
Birmingham *Age-Herald,* 1886–1925.
Birmingham *Item,* 1894.
Birmingham *Ledger,* 1902–9.
Birmingham *News,* 1890–1993.
Birmingham *Post-Herald,* 1960–81.
Birmingham *State Herald,* 1896.
Birmingham *Weekly Age,* 1886.
Birmingham *Weekly Independent,* 1881–82.
Birmingham *Weekly Iron Age,* 1876–83.
Montgomery *Advertiser,* 1870–1992.
New Orleans *Daily Picayune,* 1911.
New York *Herald,* 1902.
New York *Times,* 1882–1992.

Richmond *Dispatch*, 1881–89.
Richmond *Times*, 1896.
Richmond *Times-Dispatch*, 1903–11.

Commercial and Trade Publications

Annual Statistical Report of the American Iron and Steel Institute, 1915.
Bulletin of the American Iron and Steel Association, 1881–90.
Colliery Engineer, 1888–89.
Commercial and Financial Chronicle, 1882–83, 1888–89.
Iron Age, 1876–1930.
Iron Trade Review, 1897–1914.
Manufacturers' Record, 1884–1908.
Moody's Manual of Industrial and Miscellaneous Securities, 1900.
Moody's Manual of Investments, 1941–43.
Moody's Manual of Railroads and Corporation Securities, 1918, 1920.
Pig Iron Rough Notes, 1926–60.
Poor's Manual of Industrials, 1918.
Poor's Manual of the Railroads of the United States, 1883–90.
Railroad Gazette, 1883–86.
Year Book of the American Iron and Steel Institute, 1914.

Miscellaneous Reference Works

Dictionary of American Biography
Encyclopaedia Britannica, 200th Anniversary Edition, 1968
Encyclopedia of American Biography
National Cyclopedia of American Biography
Who's Who in America
Who Was Who in America

Books

Adler, Dorothy R. *British Investment in American Railways, 1834–1898.* Edited by Muriel E. Hidy. Charlottesville: University Press of Virginia, 1970.
Aitchison, Leslie. *A History of Metals.* 2 vols. London: MacDonald & Evans, 1960.
Alabama Blue Book and Social Register. Birmingham: Blue Book Publishing Co., 1929.
Albion, Robert G. *The Rise of New York Port.* 1939. Reprint. New York: Charles Scribner's Sons, 1970.
Alexander, James R. *Jaybird: A. J. Moxham and the Manufacture of the Johnson Rail.* Johnstown, Pa.: Johnstown Area Heritage Association, 1991.
Amos, Harriet E. *Cotton City: Urban Development in Antebellum Mobile.* Tuscaloosa: University of Alabama Press, 1985.
Andrews, Wayne. *Architecture, Ambition and Americans.* 1947. Reprint. New York: Free Press, 1964.
Armes, Ethel M. *The Story of Coal and Iron in Alabama.* 1910. Facsimile edition. Birmingham: Book-keepers Press, 1972.
Atkins, Leah Rawls. *The Valley and the Hills: An Illustrated History of Birmingham and Jefferson County.* Woodland Hills, Calif.: Windsor Publications, 1981.
Ayers, H. Brandt, and Thomas H. Naylor, eds. *You Can't Eat Magnolias.* New York: McGraw-Hill, 1972.

Barefield, Marilyn Davis. *A History of Mountain Brook, Alabama & Incidentally of Shades Valley.* Birmingham: Birmingham Publishing Co., 1989.

Barnett, George E. *Chapters on Machinery and Labor.* 1926. Reprint. Carbondale: Southern Illinois University Press, 1969.

Bennett, James R. *Old Tannehill: A History of the Pioneer Ironworks in Roupes Valley.* Birmingham: Jefferson County Historical Commission, 1986.

Beringer, Richard E., Herman Hattaway, Archer Jones, and William N. Still, Jr. *Why the South Lost the Civil War.* Athens: University of Georgia Press, 1986.

Berthoff, Rowland T. *British Immigrants in Industrial America, 1790–1950.* Cambridge: Harvard University Press, 1953.

Bining, Arthur C. *Pennsylvania Iron Manufacture in the Eighteenth Century.* Harrisburg: Pennsylvania Historical Commission, 1938.

Birmingham's Lebanese: "The Earth Turned to Gold." Birmingham: Birmingfind, n.d.

Blauner, Robert. *Alienation and Freedom: The Factory Worker and His Industry.* Chicago: University of Chicago Press, 1964.

Bolen, Jean S. *Gods in Everyman.* San Francisco: Harper & Row, 1989.

Bond, Horace Mann. *Negro Education in Alabama: A Study in Cotton and Steel.* 1939. Reprint. New York: Octagon Books, 1969.

Bonnett, Clarence E. *History of Employers' Associations in the United States.* New York: Vantage Press, 1956.

The Book of Birmingham and Alabama. Birmingham: Birmingham *Ledger,* 1914.

Boorstin, Daniel J. *The Americans: The Democratic Experience.* New York: Random House, 1973.

Bowsher, Alice M., Philip A. Morris, and Marjorie L. White. *Cinderella Stories: Transformations of Historic Birmingham Buildings.* Birmingham: Birmingham Historical Society, n.d.

Boyle, James R., and Thomas J. Joiner. *Minerals in the Economy of Alabama.* Washington, D.C.: U.S. Dept. of the Interior, Bureau of Mines, 1978.

Brandes, Stuart D. *American Welfare Capitalism 1880–1940.* Chicago: University of Chicago Press, 1976.

Brautigan, Richard. *The Pill Versus the Springhill Mine Disaster.* New York: Delacorte Press, 1968.

Brewer, Willis. *Alabama: Her History, Resources, War Record and Public Men from 1540 to 1872.* Spartanburg, S.C.: Reprint Co., 1975.

Bridge, James H. *The Inside History of the Carnegie Steel Company: A Romance of Millions.* New York: Aldine Book Co., 1903.

Brock, R. A. *Virginia and Virginians.* Richmond, Va., and Toledo, Ohio: H. H. Hardesty, 1888.

Brody, David. *Labor in Crisis: The Steel Strike of 1919.* Philadelphia and New York: J. B. Lippincott Co., 1965.

———. *Steelworkers in America: The Nonunion Era.* 1960. Reprint. New York: Harper Torchbooks, 1969.

Bruce, Kathleen. *Virginia Iron Manufacture in the Slave Era.* 1930. Reprint. New York: Augustus M. Kelley, 1968.

Bruce, Philip Alexander, ed. *Virginia: Rebirth of the Old Dominion.* 5 vols. Chicago: Lewis Publishing Co., 1929.

Bruchey, Stuart. *The Wealth of the Nation: An Economic History of the United States.* Reprint. New York: Harper & Row, 1988.

Bryan, J., III. *The Sword over the Mantel: The Civil War and I.* New York: McGraw-Hill, 1960.

Bryan, John Stewart. *Joseph Bryan: His Times, His Family, His Friends.* Richmond, Va.: Whittet and Shepperson, 1935.

Buck, Norman S. *The Development of the Organisation of Anglo-American Trade, 1800–1850.* 1925. Reprint. New York: Greenwood Press, 1968.

Buck, Paul H. *The Road to Reunion, 1865–1900.* Boston: Little, Brown and Co., 1937.

Buder, Stanley. *Pullman: An Experiment in Industrial Order and Community Planning, 1880–1930.* New York: Oxford University Press, 1967.

Burn, Duncan L. *The Economic History of Steelmaking: 1867–1939.* Cambridge: Cambridge University Press, 1940.

Burns, James MacGregor. *The Workshop of Democracy.* New York: Alfred A. Knopf, 1985.

Caldwell, Henry M. *History of the Elyton Land Company and Birmingham, Ala. 1892.* Reprint, with preface by John C. Henley III. Birmingham: Southern University Press, 1972.

Calvert, Monte A. *The Mechanical Engineer in America, 1830–1910: Professional Cultures in Conflict.* Baltimore: Johns Hopkins University Press, 1967.

Campagna, Anthony S. *U.S. National Economic Policy, 1917–1985.* New York: Praeger, 1987.

Campbell, John L., and William Henry Ruffner. *A Physical Survey Extending from Atlanta, Ga., across Alabama and Mississippi to the Mississippi River, along the Line of the Georgia Pacific Railway, Embracing the Geology, Topography, Minerals, Soils, Climate, Forests, and Agricultural and Manufacturing Resources of the Country.* New York: E. F. Weeks, 1883.

Candler, Charles H. *Asa Griggs Candler.* Emory University Ga.: Emory University Press, 1950.

Capers, Gerald M., Jr. *The Biography of a River Town. Memphis: Its Heroic Age.* Chapel Hill: University of North Carolina Press, 1939.

Carlton, David L. *Mill and Town in South Carolina 1800–1920.* Baton Rouge and London: Louisiana State University Press, 1982.

Carmer, Carl. *Stars Fell on Alabama.* 1934. Reprint, with preface by Wayne Flynt. Tuscaloosa: University of Alabama Press, 1985.

Carosso, Vincent P., with Rose C. Carosso. *The Morgans: Private International Bankers, 1854–1913.* Cambridge: Harvard University Press, 1987.

Carvati, Charles M. *Major Dooley.* Richmond, Va.: n.p., 1978.

Cash, W. J. *The Mind of the South.* New York: Alfred A. Knopf, 1941.

Chandler, Alfred D. *The Visible Hand: The Managerial Revolution in American Business.* Cambridge: Harvard University Press, 1977.

Chandler, Lester V. *America's Greatest Depression, 1929–1941.* New York: Harper & Row, 1970.

Chapman, H. H., W. M. Adamson, H. D. Bonham, H. D. Pallister, and E. C. Wright. *The Iron and Steel Industries of the South.* University: University of Alabama Press, 1953.

Chernow, Ron. *The House of Morgan: An American Banking Dynasty and the Rise of Modern Finance.* New York: Atlantic Monthly Press, 1990.

Chesson, Michael. *Richmond after the War, 1865–1890.* Richmond: Virginia State Library, 1981.

Childers, James Saxon. *Erskine Ramsay: His Life and Achievements.* New York: Cartwright & Ewing, 1942.

Cipolla, Carlo M. *Guns, Sails, and Empires: Technological Innovation and the Early Phases of European Expansion 1400–1700.* New York: Pantheon Books, 1965.

Circlot, J. E. *A Dictionary of Symbols.* New York: Philosophical Library, 1962.

Clark, Victor S. *History of Manufactures in the United States.* 3 vols. New York: McGraw-Hill, 1929.

Cobb, James C. *Industrialization and Southern Society, 1877–1984.* Lexington: University Press of Kentucky, 1984.

Cohen, Lucy M. *Chinese in the Post-Civil War South: A People Without a History.* Baton Rouge: Louisiana State University Press, 1984.

Colcord, Bradford C. *The History of Pig Iron Manufacture in Alabama*. Woodward, Ala.: Woodward Iron Company, 1950.

Coleman, John S. *Josiah Morris (1818–1891): Montgomery Banker Whose Faith Built Birmingham*. New York: Newcomen Society of North America, 1948.

Condit, Carl. *American Building*. Chicago: University of Chicago Press, 1968.

Cooper, John M. *Pivotal Decades: The United States, 1900–1920*. New York: W. W. Norton, 1990.

Cornish, Joseph Jenkins, III. *The Air Arm of the Confederacy*. Richmond, Va.: Richmond Civil War Centennial Committee, 1963.

Cotman, John Walton. *Birmingham, JFK and the Civil Rights Act of 1963: Implications for Elite Theory*. New York, Bern, Frankfurt am Main, and Paris: Peter Lang, 1989.

Cotterill, Rodney. *The Cambridge Guide to the Material World*. Cambridge: Cambridge University Press, 1985.

Couper, William. *The V.M.I. New Market Cadets: Biographical Sketches of All Members of the Virginia Military Institute Corps of Cadets Who Fought in the Battle of New Market, May 15, 1864*. Charlottesville, Va.: Michie Company, 1933.

Cragg, Dan. *Guide to Military Installations*. 2d ed. Harrisburg, Pa.: Stackpole Books, 1988.

Crane, Mary Powell. *The Life of James R. Powell and Early History of Alabama and Birmingham*. Brooklyn, N.Y.: Braunworth & Company, 1930.

Cruikshank, George M. *A History of Birmingham and Its Environs*. Chicago and New York: Lewis Publishing Co., 1920.

Cruttwell, C. R. M. F. *A History of the Great War, 1914–1918*. 1934. 2d ed. London: Granada Publishing, 1982.

Cuff, Robert D. *The War Industries Board: Business-Government Relations during World War I*. Baltimore and London: Johns Hopkins University Press, 1973.

Dabney, Virginius. *Pistols and Pointed Pens: The Dueling Editors of Old Virginia*. Chapel Hill, N.C.: Algonquin Books, 1987.

———. *Richmond: The Story of a City*. Garden City, N.Y.: Doubleday, 1976.

———. *Virginia: The New Dominion*. Garden City, N.Y.: Doubleday, 1971.

Davis, Burke. *The Southern Railway: Road of the Innovators*. Chapel Hill: University of North Carolina Press, 1985.

Davis, Charles S. *Colin J. McRae: Confederate Financial Agent*. Tuscaloosa, Ala.: Confederate Publishing Company, 1961.

———. *The Cotton Kingdom in Alabama*. Montgomery: Alabama State Department of Archives and History, 1939.

DeBow, J. D. B. *The Industrial Resources . . . of the Southern and Western States*. New Orleans: DeBow's Review, 1853.

Degler, Carl. *Out of Our Past: The Forces That Shaped Modern America*. 1959. 3d ed. New York: Harper & Row, 1984.

———. *Place over Time: The Continuity of Southern Distinctiveness*. Baton Rouge: Louisiana State University Press, 1977.

Dew, Charles B. *Ironmaker to the Confederacy: Joseph R. Anderson and the Tredegar Iron Works*. New Haven, Conn.: Yale University Press, 1966.

Dodd, James Harvey. *A History of Production in the Iron and Steel Industry in the Southern Appalachian States (1901–1926)*. Nashville: George Peabody College for Teachers, 1928.

Doerflinger, Thomas M., and Jack L. Rivkin. *Risk and Reward: Venture Capital and the Making of America's Great Industries*. New York: Random House, 1987.

Doyle, Don H. *New Men, New Cities, New South: Atlanta, Nashville, Charleston, Mobile, 1860–1910*. Chapel Hill: University of North Carolina Press, 1990.

Du Bois, W. E. B. *The Souls of Black Folk*. 1903. Reprint, with introduction by Henry Lewis Gates, Jr. New York: Bantam Books, 1989.

DuBose, John W. *Jefferson County and Birmingham, Alabama: Historical and Biographical*. Birmingham: Teeple & Smith, 1887.

———. *The Mineral Wealth of Alabama and Birmingham Illustrated*. Birmingham: N. T. Green and Co., 1886.

Duffee, Mary G. *Sketches of Alabama*. Reprint. University: University of Alabama Press, 1970.

Eavenson, Howard N. *The First Century and a Quarter of American Coal Industry*. Pittsburgh: privately printed, 1942.

Eller, Ronald D. *Miners, Millhands, and Mountaineers: Industrialization of the Appalachian South, 1880–1930*. Knoxville: University of Tennessee Press, 1982.

Ellison, Rhoda Coleman. *Bibb County, Alabama: The First Hundred Years, 1818–1918*. Tuscaloosa: University of Alabama Press, 1984.

Ennis, Clyde W. *The Industrial Center of the South: Ensley, Fairfield, Wylam, Pratt City, Alabama*. Birmingham: n.p., 1927.

Ezell, John S. *The South Since 1865*. 3d ed. Norman: University of Oklahoma Press, 1975.

Fearon, Peter. *The Origins and Nature of the Great Slump, 1929–1932*. Atlantic Highlands, N.J.: Humanities Press, 1979.

Fels, Rendig. *American Business Cycles, 1865–1897*. Chapel Hill: University of North Carolina Press, 1959.

Fisher, Douglas A. *The Epic of Steel*. New York: Harper & Row, 1963.

Fitch, John A. *The Steel Workers*. 1911. Reprint, with introduction by Roy Lubove. Pittsburgh: University of Pittsburgh Press, 1989.

Fitts, Alston, III. *Selma: Queen City of the Black Belt*. Selma, Ala.: Clairmont Press, 1989.

Flynt, Wayne. *Mine, Mill & Microchip: A Chronicle of Alabama Enterprise*. Northridge, Calif.: Windsor Publications, 1987.

———. *Montgomery: An Illustrated History*. Woodland Hills, Calif.: Windsor Publications, 1980.

———. *Poor but Proud: Alabama's Poor Whites*. Tuscaloosa: University of Alabama Press, 1989.

Fogel, Robert W., and Stanley L. Engerman. *Time on the Cross: The Economics of American Negro Slavery*. 1974. Reprint. Boston: University Press of America, 1984.

Foner, Eric. *Reconstruction: America's Unfinished Revolution, 1863–1877*. New York: Harper & Row, 1988.

Foner, Philip S. *Business & Slavery: The New York Merchants & the Irrepressible Conflict*. Chapel Hill: University of North Carolina Press, 1941.

Ford, Lacy K., Jr. *Origins of Southern Radicalism: The South Carolina Upcountry, 1800–1860*. New York: Oxford University Press, 1988.

Foster, Richard. *Innovation: The Attacker's Advantage*. New York: Summit Books, 1986.

Franklin, Jimmie Lewis. *Back to Birmingham: Richard Arrington, Jr., and His Times*. Tuscaloosa: University of Alabama Press, 1989.

Frye, Charles S., compiler. *Birmingham Illustrated: Showing Photographic Views of Many of the Manufactories, Mining and Iron Industries, Public Men and Places of Interest*. Birmingham: Commercial Club, 1986.

Gaines, William H., Jr. *Biographical Register of Members, Virginia State Convention of 1861*. Richmond: Virginia State Library, 1969.

Galambos, Louis, and Joseph Pratt. *The Rise of the Corporate Commonwealth: U.S. Business and Public Policy in the Twentieth Century*. New York: Basic Books, 1988.

Galbraith, John Kenneth. *The Great Crash: 1929*. Boston: Houghton Mifflin, 1979.

Garber, Paul N. *The Gadsden Treaty.* Philadelphia: University of Pennsylvania Press, 1921.

Garraty, John A. *The Great Depression.* San Diego: Harcourt Brace Jovanovich, 1986.

Garrow, David J. *Bearing the Cross: Martin Luther King, Jr., and the Southern Christian Leadership Conference.* New York: William Morrow and Co., 1986.

Gaston, Arthur George. *Green Power: The Successful Way of A. G. Gaston.* 1968. Reprint. Troy, Ala.: Troy State University Press, 1977.

Gaston, Paul M. *The New South Creed: A Study in Southern Mythmaking.* New York: Alfred A. Knopf, 1970.

Gates, Grace Hooten. *The Model City of the New South: Anniston, Alabama 1872–1900.* Huntsville, Ala.: Strode Publishers, 1978.

Genovese, Eugene D. *The Political Economy of Slavery: Studies in the Economy and Society of the Slave South.* 1961. 2d ed. Middletown, Conn.: Wesleyan University Press, 1989.

Gibson, A. M. *Report on the Coosa Coal Field.* Montgomery, Ala.: Roemer Printing Company, 1985.

Gloag, John, and Derek Bridgwater. *A History of Cast Iron in Architecture.* London: G. Allen and Unwin, 1978.

Going, Alan Johnston. *Bourbon Democracy in Alabama, 1874–1890.* University: University of Alabama Press, 1951.

Goldfield, David R. *Cotton Fields and Skyscrapers: Southern City and Region, 1607–1980.* Baton Rouge: Louisiana State University Press, 1982.

Goldin, Claudia Dale. *Urban Slavery in the American South, 1820–1860: A Quantitative History.* Chicago: University of Chicago Press, 1976.

Gorman, Ethel Miller. *Red Acres.* Birmingham: Vulcan Press, 1956.

Govan, Gilbert E., and James W. Livingood. *The Chattanooga Country, 1540–1976: From Tomahawks to TVA.* 3d ed. Knoxville: University of Tennessee Press, 1977.

Graham, Otis L., Jr. *An Encore for Reform: The Old Progressives and the New Deal.* New York: Oxford University Press, 1967.

Gutman, Herbert G. *Work, Culture, and Society in Industrializing America.* New York: Alfred A. Knopf, 1976.

Hackney, Sheldon. *Populism to Progressivism in Alabama.* Princeton, N.J.: Princeton University Press, 1969.

Halberstam, David. *The Reckoning.* New York: William Morrow and Co., 1986.

Hamilton, Virginia Van der Veer. *Alabama: A History.* New York and London: W. W. Norton and Co., 1984.

Hamrick, Peggy, and Robert J. Norrell. *Elyton-West End: Birmingham's First Neighborhood.* Birmingham: Birmingfind, n.d.

Harris, Carl V. *Political Power in Birmingham, 1871–1921.* Knoxville: University of Tennessee Press, 1977.

Harris, Joel Chandler. *Life of Henry W. Grady, Including His Writings and Speeches.* New York: Cassell Publishing Co., 1890.

Harrison, Fairfax. *A History of the Legal Development of the Railroad System of Southern Railway Company.* Washington, D.C.: n.p., 1901.

Hassinger, Bernice Shield. *Henderson Steel: Birmingham's First Steel.* Birmingham: Gray Printing, 1978.

Hawley, Ellis W. *The Great War and the Search for a Modern Order: A History of the American People and Their Institutions, 1917–1933.* New York: St. Martin's Press, 1979.

Hawley, Ellis W., Murray N. Rothbard, Robert F. Himmelberg, and Gerald D. Nash. *Herbert Hoover and the Crisis of American Capitalism.* Cambridge, Mass.: Schenkman Publishing Company, 1973.

Haynes, Williams. *American Chemical Industry: A History.* 6 vols. New York: D. Van Nostrand Co., 1945–49.

———. *American Chemical Industry: The World War I Period, 1912–1922.* New York: D. Van Nostrand Co., 1945.

Hesseltine, William B. *Confederate Leaders in the New South.* Baton Rouge: Louisiana State University Press, 1950.

Himmelberg, Robert F. *The Origins of the National Recovery Administration: Business, Government, and the Trade Association Issue, 1921–1933.* New York: Fordham University Press, 1976.

A History of Vanderburgh County, Indiana. N.p.: Brant & Fuller, 1889.

Hogan, William J. *Economic History of the Iron and Steel Industry in the United States.* 5 vols. Lexington, Mass.: D. C. Heath and Co., 1971.

Hornady, John R. *The Book of Birmingham.* New York: Dodd, Mead and Co., 1921.

Hubbs, G. Ward. *Tuscaloosa: Portrait of an Alabama County.* Northridge, Calif.: Windsor Publications, 1987.

Hunter, Louis C. *A History of Industrial Power in the United States, 1780–1930.* Vol. 2, *Steam Power.* Charlottesville: University Press of Virginia, 1985.

Hynd, Noel. *The Giants of the Polo Grounds: The Glorious Times of Baseball's New York Giants.* New York: Doubleday, 1988.

Ingham, John N. *The Iron Barons: A Social Analysis of an American Urban Elite, 1874–1965.* Westport, Conn.: Greenwood Press, 1978.

Jackson, Kenneth T. *The Ku Klux Klan in the City, 1915–1930.* New York: Oxford University Press, 1967.

Jackson, Walter M. *The Story of Selma.* Birmingham: Birmingham Printing Co., 1954.

Johnson, J. E., Jr. *Blast Furnace Construction in America.* New York: McGraw-Hill, 1917.

Johnson, Paul. *The Birth of the Modern: World Society, 1815–1830.* New York: Harper Collins, 1991.

Jones, James P. *Yankee Blitzkrieg: Wilson's Raid Through Alabama and Georgia.* Athens: University of Georgia Press, 1976.

Jones, Walter B. *History and Work of Geological Surveys and Industrial Development in Alabama.* University: Geological Survey of Alabama, 1935.

Jordan, Weymouth T. *Ante-Bellum Alabama: Town and Country.* 1957. Reprint. Tuscaloosa: University of Alabama Press, 1987.

Kallaway, Peter, and Patrick Pearson. *Johannesburg: Images and Continuities.* Braamfontein, South Africa: Raven Press, 1986.

Kasson, John F. *Civilizing the Machine: Technology and Republican Values in America.* 1976. Reprint. New York: Penguin Books, 1977.

Kelley, Robin D. G. *Hammer and Hoe: Alabama Communists During the Great Depression.* Chapel Hill: University of North Carolina Press, 1990.

Kennedy, David M. *Over Here: The First World War and American Society.* 1980. Reprint. New York: Oxford University Press, 1982.

Klein, Maury. *Edward Porter Alexander.* Athens: University of Georgia Press, 1971.

———. *The Great Richmond Terminal: A Study in Businessmen and Business Strategy.* Charlottesville: University Press of Virginia, 1970.

———. *History of the Louisville and Nashville Railroad.* New York: Macmillan. 1972.

Klein, Maury, and Harvey A. Kantor. *Prisoners of Progress: American Industrial Cities, 1850–1920.* New York and London: Macmillan, 1976.

Klingaman, William K. *1929: The Year of the Great Crash.* New York: Harper & Row, 1989.

LaMonte, Edward S. *George B. Ward: Birmingham's Urban Statesman.* Birmingham: Birmingham Public Library, 1974.

Lamoreaux, Naomi R. *The Great Merger Movement in American Business, 1895–1904.* Cambridge: Cambridge University Press, 1985.

Landes, David S. *The Unbound Prometheus: Technological Change and Industrial Development in Western Europe from 1750 to the Present.* Cambridge: Cambridge University Press, 1969.

Larsen, Lawrence H. *The Rise of the Urban South.* Lexington: University Press of Kentucky, 1985.

Lee, Basil Leo. *Discontent in New York City, 1861–1865.* Washington, D.C.: Catholic University of America Press, 1943.

Leighton, George R. *Five Cities: The Story of Their Youth and Old Age.* New York: Harper & Brothers, 1939.

Levine, Bruce, Stephen Brier, David Brundage, Edward Countryman, Dorothy Fennell, Marcus Rediker, and Joshua Brown. *Who Built America? Working People and the Nation's Economy, Politics, Culture, and Society.* New York: Pantheon Books, 1989.

Lewis, Edward E. *The Mobility of the Negro: A Study in the American Labor Supply.* New York: Columbia University Press, 1931.

Lewis, W. David. *Iron and Steel in America.* Wilmington, Del.: Hagley Museum, 1976.

Lewis, W. David, and Wesley Phillips Newton. *Delta: The History of an Airline.* Athens: University of Georgia Press, 1979.

Like It Ain't Never Passed: Remembering Life in Sloss Quarters. Birmingham: Sloss Furnaces National Historic Landmark, 1985.

Lively, Robert A. *Fiction Fights the Civil War.* Chapel Hill: University of North Carolina Press, 1957.

Long, Priscilla. *Where the Sun Never Shines: A History of America's Bloody Coal Industry.* New York: Paragon House, 1989.

Longest, George C. *Three Virginia Writers. Mary Johnston, Thomas Nelson Page and Amelia Rives Troubetzkoy: A Reference Guide.* Boston: G. K. Hall & Co., 1978.

Luraghi, Raimondo. *The Rise and Fall of the Plantation South.* New York: New Viewpoints, 1978.

McAdoo, William G. *Crowded Years.* Boston and New York: Houghton-Mifflin Company, 1931.

McCalley, Henry. *Report on the Warrior Coal Fields.* Montgomery, Ala.: Barrett & Co., 1886.

McHugh, Jeanne. *Alexander Holley and the Makers of Steel.* Baltimore: Johns Hopkins University Press, 1980.

McKay, Ernest A. *The Civil War and New York City.* Syracuse, N.Y.: Syracuse University Press, 1990.

McMillan, Malcolm C. *The Disintegration of a Confederate State: Three Governors and Alabama's Wartime Home Front, 1861–1865.* Macon, Ga.: Mercer University Press, 1986.

————. *The Land Called Alabama.* Austin, Tex.: Steck-Vaughan Company, 1975.

————. *Yesterday's Birmingham.* Miami, Fla.: E. A. Seeman Publishing Co., 1975.

McNeill, William H. *The Pursuit of Power: Technology, Armed Force, and Society since A.D. 1000.* Paperback reprint, Chicago: University of Chicago Press, 1982.

Mahan, Joseph B. *Columbus: Georgia's Fall Line "Trading Town."* Northridge, Calif.: Windsor Publications, 1986.

Mandy, Nigel. *A City Divided: Johannesburg and Soweto.* New York: St. Martin's Press, 1984.

Marks, Carole. *Farewell—We're Good and Gone: The Great Black Migration.* Bloomington: Indiana University Press, 1989.

Memorial Record of Alabama: A Concise Account of the State's Political, Military, Professional and Industrial Progress, Together with the Personal Memoirs of Many of Its People. Spartanburg, S.C.: Reprint Co., 1976.

Miller, Donald L., and Richard E. Sharpless. *The Kingdom of Coal: Work, Enterprise, and Ethnic Communities in the Mine Fields.* Philadelphia: University of Pennsylvania Press, 1985.

Mills, Mable D. *Coke Industry in Alabama.* University, Ala.: Bureau of Business Research, 1947.

Milner, John T. *Alabama: As It Was, As It Is, and As It Will Be.* Montgomery, Ala.: Barrett & Brown, 1876.

————. *Report of the Chief Engineer to the President and Board of Directors of the South and North Alabama Railroad Co., on the 26th of November, 1859.* Montgomery, Ala.: Advertiser Steam Printing House, 1859.

Mims, Edwin. *The Advancing South: Stories of Progress and Reaction.* Garden City, N.Y.: Doubleday, Page & Co., 1926.

Mitchell, Broadus. *Depression Decade: From New Era through New Deal, 1929–1941.* New York: Rinehart & Co., 1947.

Moger, Allen W. *Virginia: Bourbonism to Byrd, 1870–1925.* Charlottesville: University Press of Virginia, 1968.

Moldenke, Richard. *The Coke Industry of the United States, As Related to the Foundry.* Washington, D.C.: U.S. Bureau of Mines, 1910.

Montgomery, David. *The Fall of the House of Labor: The Workplace, the State, and American Labor Activism, 1865–1925.* Cambridge: Cambridge University Press, 1987.

Moore, Barrington, Jr. *Social Origins of Dictatorship and Democracy: Lord and Peasant in the Making of the Modern World.* Boston: Beacon Press, 1966.

Moore, William Davis. *Development of Cast Iron Pressure Pipe Industry in the Southern States, 1800–1938: An American Pilgrimage.* Birmingham: Birmingham Publishing Company, c. 1939.

Morison, Elting. *Men, Machines, and Modern Times.* Cambridge, Mass.: MIT Press, 1966.

Moskowitz, Milton, Michael Katz, and Robert Levering, eds. *Everybody's Business, An Almanac: The Irreverent Guide to Corporate America.* New York: Harper & Row, 1980.

Naisbitt, John, and Patricia Aburdene. *Re-inventing the Corporation.* New York: Warner Books, 1985.

Nevins, Allan. *Abram S. Hewitt, with Some Account of Peter Cooper.* New York: Harper & Brothers, 1935.

Nixon, H. C. *Lower Piedmont Country: The Uplands of the Deep South.* 1946. Reprint. Tuscaloosa: University of Alabama Press, 1984.

Nixon, Raymond B. *Henry W. Grady: Spokesman of the New South.* New York: Alfred A. Knopf, 1943.

Noble, Henry Jeffers. *Development of Cast Iron Soil Pipe in Alabama.* Birmingham: Sloss-Sheffield Steel and Iron Co., 1941.

Norrell, Robert J. *James Bowron: The Autobiography of a New South Industrialist.* Chapel Hill: University of North Carolina Press, 1991.

————. *A Promising Field: Engineering at Alabama, 1837–1987.* Tuscaloosa: University of Alabama Press, 1990.

Norrell, Robert J., and Karen Rolen. *The Italians: From Bisacquino to Birmingham.* Birmingham: Birmingfind, n.d.

Norrell, Robert J., and Otis Dismuke. *The Other Side: The Story of Birmingham's Black Community.* Birmingham: Birmingfind, n.d.

Northern Alabama: Historical and Biographical. Birmingham: Smith & De Land, 1888.

Nunnelley, William A. *Bull Connor.* Tuscaloosa: University of Alabama Press, 1991.

Oakes, James. *The Ruling Race: A History of American Slaveholders.* New York: Alfred A. Knopf, 1982.

O'Connor, Thomas H. *Lords of the Loom: The Cotton Whigs and the Coming of the Civil War.* New York: Charles Scribner's Sons, 1968.

Overman, Frederick. *Treatise on Metallurgy.* 1852. 6th ed. New York: D. Appleton and Co., 1882.

Owen, Thomas M. *History of Alabama and Dictionary of Alabama Biography.* 4 vols. Spartanburg, S.C.: Reprint Co., 1978.

Pannell, Anne G., and Dorothea E. Wyatt. *Julia S. Tutwiler and Social Progress in Alabama.* Tuscaloosa: University of Alabama Press, 1961.

Parke, Thomas D., M.D. *Report on Coalburg Prison* ... *Rendered to the Committee of Health of Jefferson County Medical Society.* Birmingham: Roberts & Son, Printers, 1895.

Phillips, William Battle. *Iron Making in Alabama.* 3d ed. University: Geological Survey of Alabama, 1912.

Porter, Glenn, and Harold C. Livesay. *Merchants and Manufacturers: Studies in the Changing Structure of Nineteenth-Century Marketing.* Baltimore and London: Johns Hopkins University Press, 1971.

Pursell, Carroll W. *Early Stationary Steam Engines in America: A Study in the Migration of a Technology.* Washington, D.C.: Smithsonian Institution Press, 1969.

Ratner, Sidney, James H. Soltow, and Richard Sylla. *The Evolution of the American Economy: Growth, Welfare, and Decision Making.* New York: Basic Books, 1979.

Reichler, Joseph L., ed. *The Baseball Encyclopedia.* 5th ed. New York: Macmillan, 1982.

Reynolds, Patrick, and Tom Shachtman. *The Gilded Leaf: Triumph, Tragedy, and Tobacco: Three Generations of the R. J. Reynolds Family and Fortune.* Boston: Little, Brown and Co., 1989.

Roberts, Joseph K. *Annotated Geological Bibliography of Virginia.* Richmond, Va.: Dietz Press, 1942.

Rogers, William Warren. *The One-Gallused Rebellion: Agrarianism in Alabama, 1865–1896.* Baton Rouge: Louisiana State University Press, 1970.

Rosenberg, Nathan, and L. E. Birdzell, Jr. *How the West Grew Rich: The Economic Transformation of the Industrial World.* New York: Basic Books, 1986.

Rowell, Raymond J., Sr. *Vulcan in Birmingham.* Birmingham: Birmingham Park and Recreation Board, 1972.

Russell, Norman F. S. *"U.S.P.&F.": Cast Iron Pressure Pipe in New Jersey.* New York: Newcomen Society of North America, 1951.

Sanders, Clyde A., and Dudley C. Gould. *History Cast in Metal: The Founders of North America.* N.p.: American Foundrymen's Society, Cast Metals Institute, 1976.

Satterfield, Carolyn Green. *Historic Sites of Jefferson County, Alabama.* Birmingham: Gray Printing Company, 1976.

Sellers, James B. *History of the University of Alabama.* University: University of Alabama Press, 1953.

Shore, Laurence. *Southern Capitalists: The Ideological Leadership of an Elite, 1832–1885.* Chapel Hill and London: University of North Carolina Press, 1986.

Siepel, Kevin H. *Rebel: The Life and Times of John Singleton Mosby.* New York: St. Martin's Press, 1983.

Simpson, Bruce L. *History of the Metal-Casting Industry.* 2d ed. Des Plaines, Ill.: American Foundrymen's Society, 1948.

Simpson, Thomas A. *Structural Geology of the Birmingham Red Iron Ore District, Alabama.* University: Geological Survey of Alabama, 1963.

Smith, Adam. *The Wealth of Nations.* Reprint. 2 vols. London and New York: Everyman's Library, 1910.

Smith, Merritt Roe. *Harpers Ferry Armory and the New Technology: The Challenge of Change*. Ithaca, N.Y.: Cornell University Press, 1977.

Smith, Page. *The Rise of Industrial America: A People's History of the Post-Reconstruction Era*. New York: McGraw-Hill, 1984.

Sobel, Robert. *Panic on Wall Street: A History of America's Financial Disasters*. New York: Macmillan, 1968.

Soule, George. *Prosperity Decade: From War to Depression, 1917–1929*. New York: Rinehart & Co., 1947.

Spero, Sterling D., and Abram L. Harris. *The Black Worker: The Negro and the Labor Movement*. 1931. Reprint. Port Washington, N.Y.: Kennikat Press, 1966.

Squire, Joseph. *Report on the Cahaba Coal Field*. Montgomery, Ala.: Brown Printing Co., 1890.

Stapleton, Darwin H. *The Transfer of Early Industrial Technologies to America*. Philadelphia: American Philosophical Society, 1987.

Stapleton, Michael. *A Dictionary of Greek and Roman Mythology*. New York: Bell Publishing Co., 1978.

Starobin, Robert S. *Industrial Slavery in the Old South*. New York: Oxford University Press, 1970.

Staudenmaier, John M. *Technology's Storytellers: Reweaving the Human Fabric*. Cambridge, Mass.: MIT Press, 1985.

Stocking, George. *Basing Point Pricing and Regional Development: A Case Study of the Iron and Steel Industry*. Chapel Hill: University of North Carolina Press, 1954.

Stover, John F. *The Railroads of the South, 1865–1900: A Study in Finance and Control*. Chapel Hill: University of North Carolina Press, 1955.

Sullivan, Charles Kenneth. *Coal Men and Coal Towns: Development of the Smokeless Coalfields of Southern West Virginia, 1873–1923*. New York & London: Garland Publishing, 1989.

Sulzby, James F., Jr. *Birmingham Sketches from 1871 through 1921*. Birmingham: Birmingham Printing Co., 1945.

Summers, Mark W. *Railroads, Reconstruction, and the Gospel of Prosperity: Aid under the Radical Republicans, 1865–1877*. Princeton, N.J.: Princeton University Press, 1984.

Swank, James M. *History of the Manufacture of Iron in All Ages*. 1892. Reprint. New York: Burt Franklin, n.d.

Taft, Philip. *Organizing Dixie: Alabama Workers in the Industrial Era*. Revised and edited by Gary M. Fink. Westport, Conn.: Greenwood Press, 1981.

Tankersley, Allen P. *John B. Gordon: A Study in Gallantry*. Atlanta: Whitehall Press, 1955.

Tarbell, Ida. *The Life of Elbert H. Gary: The Story of Steel*. New York and London: D. Appleton and Co., 1925.

Taylor, George Rogers. *The Transportation Revolution, 1815–1860*. 1951. Reprint. New York: Harper & Row, 1968.

Taylor, William R. *Cavalier and Yankee: The Old South and American National Character*. 1961. Reprint. Garden City, N.Y.: Doubleday Anchor Books, 1963.

Temin, Peter. *Iron and Steel in Nineteenth Century America: An Economic Inquiry*. Cambridge, Mass.: MIT Press, 1964.

Tepper, Sol H. *"Torpedoes? Damn!" Ordeal at Selma Gun Foundry and the Battle of Mobile Bay*. Selma, Ala.: Selma Printing Co., 1979.

Thornton, J. Mills. *Politics and Power in a Slave Society: Alabama, 1800–1860*. Baton Rouge: Louisiana State University Press, 1978.

Tilley, Nannie M. *The R. J. Reynolds Tobacco Company*. Chapel Hill: University of North Carolina Press, 1985.

Tindall, George B. *The Emergence of the New South, 1913–1945.* Baton Rouge: Louisiana State University Press, 1967.

Toffler, Alvin. *Future Shock.* New York: Random House, 1970.

———. *Power Shift: Knowledge, Wealth, and Violence at the Edge of the 21st Century.* New York: Bantam Books, 1990.

Tuomey, Michael. *First Biennial Report on the Geology of Alabama.* Tuscaloosa, Ala.: M. D. J. Slade, 1850.

———. *Second Biennial Report on the Geology of Alabama.* Edited by J. W. Mallet. Montgomery, Alabama: N. B. Cloud, 1858.

Tyler, Lyon G. *Men of Mark in Virginia.* Vol. I. Washington, D.C.: Men of Mark Publishing Company, 1906.

Vandiver, Frank E. *Ploughshares into Swords: Josiah Gorgas and Confederate Ordnance.* Austin: University of Texas Press, 1952.

Vatter, Harold G. *The Drive to Industrial Maturity: The U.S. Economy, 1860–1914.* Westport, Conn.: Greenwood Press, 1975.

Wade, Richard C. *Slavery in the Cities: The South, 1820–1860.* New York: Oxford University Press, 1964.

Walker, David A. *Iron Frontier: The Discovery and Development of Minnesota's Three Ranges.* St. Paul: Minnesota Historical Society Press, 1979.

Wall, Joseph Frazier. *Andrew Carnegie.* 1970. Reprint. Pittsburgh: University of Pittsburgh Press, 1989.

Wallace, Anthony F. C. *St. Clair: A Nineteenth-Century Coal Town's Experience with a Disaster-Prone Industry.* New York: Alfred A. Knopf, 1987.

Wallenstein, Peter. *From Slave South to New South: Public Policy in Nineteenth-Century Georgia.* Chapel Hill and London: University of North Carolina Press, 1987.

Ward, Robert David, and William Warren Rogers. *Convicts, Coal, and the Banner Mine Tragedy.* Tuscaloosa: University of Alabama Press, 1987.

———. *Labor Revolt in Alabama: The Great Strike of 1894.* University: University of Alabama Press, 1965.

Ward, William E., II, and Francis E. Evans, Jr. *Coal: Its Importance to Alabama.* University: Geological Survey of Alabama, 1972.

Warren, Dorothea O. *The Practical Dreamer: A Story of John T. Milner, His Family and Forebears.* Birmingham: Southern Family Press, 1959.

Warren, Kenneth. *The American Steel Industry, 1850–1970: A Geographical Interpretation.* Oxford: Clarendon Press, 1970.

Watters, Pat. *Coca-Cola: An Illustrated History.* Garden City, N.Y.: Doubleday & Co., 1978.

Wesley, Charles H. *Negro Labor in the United States, 1850–1925.* 1927. Reprint. New York: Russell & Russell, 1967.

Whitaker, Walter C. *History of the Protestant Episcopal Church in Alabama, 1763–1891.* Birmingham: Roberts & Son, 1898.

White, Marjorie Longenecker. *The Birmingham District: An Industrial History and Guide.* Birmingham: Birmingham Historical Society, 1981.

———. *Downtown Birmingham: Architectural and Historical Walking Tour Guide.* Birmingham: Birmingham Historical Society and First National Bank of Birmingham, 1980.

Wiebe, Robert H. *Businessmen and Reform: A Study of the Progressive Movement.* Cambridge: Harvard University Press, 1962.

Wiebel, Arthur V. *Biography of a Business.* Birmingham: Tennessee Coal and Iron Division, United States Steel Corporation, 1960.

Wiener, Jonathan M. *Social Origins of the New South: Alabama, 1860–1885.* Baton Rouge: Louisiana State University Press, 1978.

Wiggins, Sarah Woolfolk. *The Scalawag in Alabama Politics, 1865–1881*. University: University of Alabama Press, 1977.

Wills, Brian S. *A Battle from the Start: The Life of Nathan Bedford Forrest*. New York: Harper Collins, 1992.

Wilmer, Richard. *In Memoriam: John Stewart of Rothesay*. New York: n.p., 1887.

Windham, Kathryn Tucker. *The Ghost in the Sloss Furnaces*. Birmingham: Birmingham Historical Society, 1978.

Winkler, John K. *Tobacco Tycoon: The Story of James Buchanan Duke*. New York: Random House, 1942.

Woodward, C. Vann. *Origins of the New South, 1877–1913*. 2d ed. Baton Rouge: Louisiana State University Press, 1971.

———. *The Strange Career of Jim Crow*. 3d ed. New York: Oxford University Press, 1974.

Woodward, Joseph H., II. *Alabama Blast Furnaces*. Woodward, Ala.: Woodward Iron Company, 1940.

Wright, Gavin. *Old South, New South: Revolutions in the Southern Economy Since the Civil War*. New York: Basic Books, 1986.

———. *The Political Economy of the Cotton South: Households, Markets, and Wealth in the Nineteenth Century*. Reprint. New York: W. W. Norton, 1978.

Yearley, Clifton K., Jr. *Britons in American Labor: A History of the Influence of the United Kingdom Immigrants on American Labor, 1820–1914*. Baltimore: Johns Hopkins Press, 1957.

———. *Enterprise and Anthracite: Economics and Democracy in Schuylkill County, 1820–1875*. Baltimore: Johns Hopkins Press, 1961.

Chapters and Sections of Books

Bains, Lee E., Jr. "Birmingham, 1963: Confrontation over Civil Rights." In David J. Garrow, ed., *Birmingham, Alabama, 1956–1963: The Black Struggle for Civil Rights*, 151–289. Brooklyn, N.Y.: Carlson Publishing, 1989.

Banta, Brady. "Henderson Steel & Manufacturing Company." In Paul S. Paskoff, ed., *Iron and Steel in the Nineteenth Century*, 154–55. New York: Facts on File, 1989.

———. "Tennessee Coal, Iron, & Railroad Company." In Paskoff, ed., *Iron and Steel in the Nineteenth Century*, 330–33.

Beiman, Irving. "Birmingham: Steel Giant with a Glass Jaw." In Robert S. Allen, ed., *Our Fair City*, 99–122. New York: Vanguard Press, 1947.

Bonner, Floelle Y. "Arlington: Mudd-Munger Mansion." In Birmingham Branch, National League of American Pen Women, *Historic Homes of Alabama and Their Traditions*, 221–30. Birmingham: Birmingham Publishing Co., 1935.

Brody, David. "The Rise and Decline of Welfare Capitalism." In David Brody, ed., *Workers in Industrial America: Essays on the 20th Century Struggle*, 48–81. New York: Oxford University Press, 1980.

Bryan, C. B. "Dr. John Randolph Page." In Homer Richey, ed., *Memorial History of the John Bowie Strange Camp, United Confederate Veterans*, 83–84. Charlottesville, Va.: Michie Co., 1920.

Chandler, Alfred D. "The Organization of Manufacturing and Transportation." In David T. Gilchrist, and W. David Lewis, eds., *Economic Change in the Civil War Era*, 137–65. Wilmington, Del.: Eleutherian Mills–Hagley Foundation, 1965.

———. "Rise and Evolution of Big Business." In Glenn Porter, ed., *Encyclopedia of American Economic History: Studies of the Principal Movements and Ideas*, vol. 2, 619–38. New York: Charles Scribner's Sons, 1980.

Cooper, Gail. "Frederick Winslow Taylor and Scientific Management." In Carroll W.

Pursell, Jr., ed., *Technology in America: A History of Individuals and Ideas*, 163–76. Second ed. Cambridge, Mass.: MIT Press, 1990.

Current, Richard N. "God and the Strongest Battalions." In David Donald, ed. *Why the North Won the Civil War*, 15–32. 1960. Reprint. New York: Collier Books, 1962.

Cutcliffe, Stephen H. "Cambria Iron Company." In Paskoff, ed., *Iron and Steel in the Nineteenth Century* (see above), 38–42.

Danhof, Clarence H. "Four Decades of Thought on the South's Economic Problems." In Melvin L. Greenhut and W. Tate Whitman, *Essays in Southern Economic Development*, 7–68. Chapel Hill: University of North Carolina Press, 1964.

Earle, Carville, and Ronald Hoffman, "The Urban South: The First Two Centuries." In Blaine A. Brownell, and David R. Goldfield, eds., *The City in Southern History: The Growth of Urban Civilization in the South*, 23–51. Port Washington, N.Y.: Kennikat Press, 1977.

Eskew, Glenn T. "The Alabama Christian Movement for Human Rights and the Birmingham Struggle for Civil Rights, 1956–1963." In Garrow, ed., *Birmingham, Alabama, 1956–1963* (see above), 3–114.

Estep, H. Cole. "The Manufacture of By-Product Coke." In A. O. Backert, ed., *The ABC of Iron and Steel*, 69–76. Cleveland, Ohio: Penton Publishing Company, 1915.

Eustler, Roland B. "Transportation Developments and Economic and Industrial Changes." In William J. Carson, ed., *The Coming of Industry to the South*, 202–9. Philadelphia: American Academy of Political and Social Science, 1931.

Ferguson, Eugene S. "Steam Transportation." In Melvin Kranzberg and Carroll W. Pursell, eds., *Technology in Western Civilization*. Vol. 1, 291–301. New York: Oxford University Press, 1967.

Goldfield, David R. "Pursuing the American Dream: Cities in the Old South." In Brownell and Goldfield, eds., *The City in Southern History* (see above), 52–91.

Griessman, B. Eugene. "Will the South Rise Again or Just Roll Over?" In W. David Lewis and B. Eugene Griessman, eds., *The Southern Mystique: Technology and Human Values in a Changing Region*, 125–31. University: University of Alabama Press, 1977.

Gutman, Herbert. "Enslaved Afro-Americans and the 'Protestant' Work Ethic." In Ira Berlin, ed. *Power & Culture: Essays on the American Working Class*, 298–325. New York: Pantheon Books, 1987.

Haas, Edward F. "The Southern Metropolis, 1940–1976." In Brownell and Goldfield, eds., *The City in Southern History* (see above), 159–91.

Hancock, David. "Coal Washing in Alabama." In Phillips, *Iron Making in Alabama* (see above), 233–45.

Hannah, Leslie. "Mergers." In Porter, ed. *Encyclopedia of American Economic History*. Vol. 2 (see above), 639–51.

Jarrell, C. Chappell. "From Farmscape to Suburban Yard: Robert Jemison, Jr.'s Spring Lake Farms." In Philip A. Morris and Marjorie Longenecker White, eds., *Designs on Birmingham: A Landscape History of a Southern City and Its Suburbs*, 42–45. Birmingham: Birmingham Historical Society, 1989.

Jones, Lewis W. "Fred L. Shuttlesworth, Indigenous Leader." In Garrow, ed., *Birmingham, Alabama, 1956–1963* (see above), 115–50.

Kulik, Gary. "Black Workers and Technological Change in the Birmingham Iron Industry, 1881–1931." In Merl E. Reed, Leslie S. Hough, and Gary M. Fink, eds., *Southern Workers and Their Unions, 1880–1975*, 22–42. Westport, Conn.: Greenwood Press, 1981.

Lewis, W. David. "Sloss Furnaces and the Southern Foundry Trade: A Case Study of Industrialization in the New South." In Howard L. Hartman, ed. *Proceedings of* the 150th Anniversary Symposium on Technology and Society: Southern Technology, Past, Present, and Future, 58–80. Tuscaloosa: University of Alabama College of Engineering, 1988.

———. "Technology, Community, and Humanity: The Big Picture." In Lewis and Griessman, *The Southern Mystique*, 15–32.

McCalley, Henry. "Northern Alabama . . . Its Topography, Geology and Natural Resources." In *Northern Alabama: Historical and Biographical*, 24–29. Birmingham: Smith & De Land, 1888.

McCorvey, Thomas C. "Henry Tutwiler and the Influence of the University of Virginia on Education in Alabama." In *Alabama Historical Sketches*, 3–32. Charlottesville: University of Virginia Press, 1960.

McDavid, Mittie Owen. "Vestavia: Classic Home of George Battey Ward." In Birmingham Branch, National League of American Pen Women, *Historic Homes of Alabama and Their Traditions* (see above), 308–14.

Moldenke, Richard. "Manufacture of Gray Iron Castings" and "Manufacture of Malleable Castings." In Backert, ed., *ABC of Iron and Steel* (see above), 207–16, 217–28.

Moore, Geoffrey H. "Business Cycles, Panics, and Depressions." In Porter, ed., *Encyclopedia of American Economic History*. Vol. 1 (see above), 151–56.

Norrell, Robert J. "Labor Trouble: George Wallace and Union Politics in Alabama." In Robert Zieger, ed., *Organized Labor in the Twentieth Century South*, 250–72. Knoxville: University of Tennessee Press, c. 1991.

———. "One Thing We Did Right: Reflections on the Movement." In Armistead L. Robinson and Patricia Sullivan, eds., *New Directions in Civil Rights Studies*, 65–80. Charlottesville and London: University Press of Virginia, 1991.

Paskoff, Paul. "Puddling." In Paskoff, ed., *Iron and Steel in the Nineteenth Century*, 286–87.

Porter, John Jermain. "Manufacture of Pig Iron." In Backert, ed., *ABC of Iron and Steel* (see above), 77–92.

Reynolds, Terry S. "Iron Ore Ranges." In Paskoff, ed., *Iron and Steel in the Nineteenth Century*, 194–98.

———. "Iron Ore Transportation." In Paskoff, ed., *Iron and Steel in the Nineteenth Century*, 199–201.

Schubert, H. R. "The Steel Industry." In Charles Singer, et al., eds., *A History of Technology*. Vol. 5, *The Late Nineteenth Century, c. 1850 to 1900*, 53–71. Oxford: Clarendon Press, 1954.

Schweikart, Larry. "John Warne Gates." In Paskoff, ed., *Iron and Steel in the Nineteenth Century* (see above), 146–47.

Smith, Cyril Stanley. "Mining and Metallurgical Production, 1800–1880" and "Metallurgy: Science and Practice before 1900." In Kranzberg and Pursell, eds., *Technology in Western Civilization*, Vol. 1 (see above), 349–67, 592–602.

Steelman, Bennett L. "Andrews, Alexander Boyd." In William S. Powell, ed., *Dictionary of North Carolina Biography*. Vol. 1, 34–36. Chapel Hill: University of North Carolina Press, 1979.

White, Marjorie Longenecker. "The Grid and the Garden." In Morris and White, eds., *Designs on Birmingham* (see above), 6–20.

Whiting, Marvin Y. "Robert Jemison, Jr.: A Tribute." In Morris and White, eds., *Designs on Birmingham* (see above), 38–40.

Williams, John. "Struggles of the Thirties in the South." In Bernard Sternsher, ed., *The Negro in Depression and War: Prelude to Revolution, 1930–1945*, 167–78. Chicago: Quadrangle Books, 1969.

Windham, Kathryn Tucker. "Cahaba." In *Rivers of Alabama*, 123–34. Huntsville: Strode Publishers, 1968.

Worthman, Paul. "Working Class Mobility in Birmingham, Alabama, 1880–1914." In Tamara K. Hareven, ed., *Anonymous Americans: Explorations in Nineteenth-Century Social History*, 172–213. Englewood Cliffs, N.J.: Prentice-Hall, 1971.

Articles in Journals

Alvey, Edward, Jr. "James B. Pace: A Richmonder of Many Talents." *Richmond Quarterly* 4 (Fall 1981): 29–35.

Armes, Ethel M. "The Spirit of the Founders." *Survey* 17 (6 January 1912): 1453–63.

Atkins, Leah Rawls. "Growing Up Around Edgewood Lake." *Alabama Review* 44 (April 1991): 83–100.

Bailey, Hugh G. "Ethel Armes and *The Story of Coal and Iron in Alabama*." *Alabama Review* 22 (July 1969): 188–99.

"Balloon Used for Scout Duty. Terrible Experiences of a Confederate Officer Who Saw the Enemy from Dizzy Heights." *Southern Historical Society Papers* 33 (1905): 32–42.

Barclay, Anne H. R. "William Henry Ruffner, LL.D." *West Virginia Historical Magazine Quarterly* 2 (January 1902): 33–43.

Bateman, Fred, James Foust, and Thomas Weiss. "The Participation of Planters in Manufacturing in the Antebellum South." *Agricultural History* 48 (April 1974): 277–97.

Baughn, W. H. "Capital Formation and Entrepreneurship in the South." *Southern Economic Journal* 16 (October 1949): 161–69.

Belissary, Constantine G. "The Rise of Industry and the Industrial Spirit in Tennessee." *Journal of Southern History* 19 (May 1953): 193–215.

Bigelow, Martha Mitchell. "Birmingham's Carnival of Crime, 1871–1910." *Alabama Review* 3 (April 1950): 123–33.

Birkinbine, John. "Comparisons of Blast Furnace Records." American Institute of Mining Engineers *Transactions* 15 (May 1886–February 1887): 147–70.

Bond, Horace Mann. "Social and Economic Forces in Alabama Reconstruction." *Journal of Negro History* 23 (July 1938): 290–348.

Bowron, James. "Iron, Steel and Coal in Dixie." Parts 1, 2. *Iron Trade Review* (29 October, 5 November 1914), 813–16, 825, 862–65.

———. "The Southern Iron and Steel Industry." *Iron Age*, 12, 19, 26 November 1914, 1126–28, 1184–86, 1228–30.

Bowron, Richard A. "James Bowron." *Journal of the Birmingham Historical Society* 4 (November 1963): 16–31.

Brainerd, Alfred F. "Colored Mining Labor." American Institute of Mining Engineers *Transactions* 14 (June 1885–May 1886): 78–79.

Brownell, Blaine A. "Birmingham, Alabama: New South City in the 1920s." *Journal of Southern History* 38 (February 1972): 21–48.

———. "The Notorious Jitney and the Urban Transportation Crisis in Birmingham in the 1920s." *Alabama Review* 25 (April 1972): 105–18.

Cable, George W. "The Convict Lease System in the Southern States." *Century Magazine* 27 (February 1884): 582–99.

"Capt. John Randolph Bryan." *Confederate Veteran* 26 (April 1918): 168.

Carson, H. Y. "Permanence of Cast Iron Pipe." *Pig Iron Rough Notes* 63 (Autumn 1935): 7–9.

Casey, Robert, and Marjorie Longenecker White. "A Look at Thomas, an Alabama Iron Town." *Canal History and Technology Proceedings* 9 (17 March 1990): 121–41.

Chandler, Alfred D. "Anthracite Coal and the Beginnings of the Industrial Revolution in the United States." *Business History Review* 46 (1972): 141–81.

Clark, Neil M. "Birmingham—The Next Capital of the Steel Age: A Smoking Metropolis on a Mountain of Iron." *World's Work* 53 (March 1927): 534–45.

Colby, Albert Ladd. "The Advantages of Machine-Cast Pig Iron." *Cassier's Magazine* 24 (June 1903): 134–38.

Collins, Herbert. "The Southern Industrial Gospel Before 1860." *Journal of Southern History* 12 (August 1946): 386–401.

Connors, George W. "Modern Development of the Iron and Steel Industry in the South." *Year Book of the American Iron and Steel Institute* (1914), 436–40.

Cooper, William J., Jr. "The Cotton Crisis in the Antebellum South: Another Look." *Agricultural History* 49 (April 1975): 381–91.

Crane, W. R. "Iron Mining in the Birmingham District, Alabama." *Engineering and Mining Journal*, 9 January 1905, 274–77.

Cunningham, S. A., ed. "The Last Roll: Maj. John William Johnston." *Confederate Veteran* 19 (March 1911): 116–17.

Curran, W. E. "Trend of the Southern Pig-Iron Business." American Institute of Mining and Metallurgical Engineers *Transactions* 131 (1938): 40–41.

Davidson, Philip G. "Industrialism in the Ante-Bellum South." *South Atlantic Quarterly* 27 (October 1928): 405–25.

Dean, Lewis S. "Michael Tuomey and the Pursuit of a Geological Survey of Alabama." *Alabama Review* 43 (April 1991): 101–11.

Delamater, G. R. "Bituminous Coal Washing." *Mines and Minerals* 27 (August, September 1907): 7–10, 62–65.

Doster, James F. "The Shelby Iron Works Collection in the University of Alabama Library." *Bulletin of the Business Historical Society* 26 (December 1952): 214–17.

Dovel, J. P. "Improved Furnace on Southern Ores." *Iron Age*, 22 September 1927, 20.

———. "Reasons Why Foundry Iron Should be Sand Cast." *Iron Age*, 21 April 1921, 1035–37.

Dyer, Y. A. "Cast-Iron Pipe Manufacture in the South." *Iron Age*, 23 November 1916, 1159–62.

Eckel, Edwin C. "The Iron and Steel Industry of the South." *Annals of the American Academy of Political and Social Science* 153 (January 1931): 54–62.

Engerman, Stanley L. "A Reconsideration of Southern Economic Growth, 1770–1860." *Agricultural History* 49 (April 1975): 343–61.

Eskew, Glenn T. "Demagoguery in Birmingham and the Building of Vestavia." *Alabama Review* 42 (July 1989): 192–217.

Feuer, A. B. "The Odyssey of the Battleship *Oregon*." *Battleships at War* 1 (1991): 76–83.

Fishback, Price V. "Did Coal Miners 'Owe their Souls to the Company Store'? Theory and Evidence from the Early 1900s." *Journal of Economic History* 46 (December 1986): 1011–29.

Fitch, John A. "Birmingham District: Labor Conservation." *Survey* 27 (6 January 1912): 1527–40.

Fleisig, Heywood. "Slavery, the Supply of Agricultural Labor, and the Industrialization of the South." *Journal of Economic History* 36 (September 1976): 572–92.

Fleming, Walter L. "Industrial Development of Alabama During the Civil War." *South Atlantic Quarterly* 3 (July 1904): 260–72.

———. "Reorganization of the Industrial System in Alabama After the Civil War." *American Journal of Sociology* 10 (January 1905): 473–99.

Flynt, Wayne. "Religion in the Urban South: The Divided Mind of Birmingham, 1900–1930." *Alabama Review* 30 (April 1977): 108–34.

Fox, E. L. "William Henry Ruffner and the Rise of the Public Free School System of Virginia." *The John P. Branch Papers of Randolph-Macon College* 2 (June 1910): 124–44.

Fraser, Walter J., Jr. "William Henry Ruffner and the Establishment of Virginia's Public School System, 1870–1874." *Virginia Magazine of History and Biography* 79 (July 1971): 259–79.

———. "William Henry Ruffner: A Southern Liberal in the North." *Citadel Monograph Series* 9 (February 1972): 1–29.

Fuller, Justin. "Alabama Business Leaders: 1865–1900." Parts 1, 2. *Alabama Review* 16 (October 1963): 279–86; 17 (January 1964): 63–75.

———. "Boom Towns and Blast Furnaces: Town Promotion in Alabama, 1885–1893." *Alabama Review* 29 (January 1976): 37–48.

———. "From Iron to Steel: Alabama's Industrial Evolution." *Alabama Review* 17 (April 1964): 137–48.

———. "Henry F. DeBardeleben, Industrialist of the New South." *Alabama Review* 39 (January 1986): 3–18.

Galloway, W. "Genesis and Development of the Coking Oven." *Engineering and Mining Journal* 88 (1909): 11–13.

Gates, Grace Hooten. "Anniston: Model City and Rival City." *Alabama Review* 31 (January 1978): 33–47.

Geismer, H. S. "Coal Washing Practice in Alabama." American Institute of Mining and Metallurgical Engineers *Transactions* 71 (1925): 1088–1105.

Genovese, Eugene D. "Yeomen Farmers in a Slaveholders' Democracy." *Agricultural History* 49 (April 1975): 331–42.

Gilchrist, David T. "Albert Fink and the Pooling System." *Business History Review* 34 (Spring 1960): 24–49.

"Give Me Time." *Forbes Magazine* 103 (1 April 1969): 18–19.

Glenn, Thomas K. "Modern Development of the Iron and Steel Industry in the South." *Year Book of the American Iron and Steel Institute* (1914), 426–35.

———. "The Southern Iron and Steel Industry." *Iron Trade Review*, 29 October 1914, 992–94.

Graves, Charles Marshall. "Tribute to a Virginia Gentleman." *Virginia Magazine of History and Biography* 59 (July 1951): 309–14.

Griffin, Richard W. "Cotton Manufacture in Alabama to 1865." *Alabama Historical Quarterly* 18 (Fall 1956): 289–307.

Henry, Robert S. "Railroads of the Confederacy." *Alabama Review* 6 (January 1953): 3–13.

Hillman, M. P. Gentry. "Birmingham Furnace Practice Development." *Iron Trade Review*, 29 October 1914, 994–96.

———. "The Modern Development of the Iron and Steel Industry in the South." *Year Book of the American Iron and Steel Institute* (1914), 441–47.

Holland, A. P. "Highland Avenue." *The Southsider* 3 (Summer 1985): 2.

Hoole, W. Stanley. "The Confederate Armory at Tallassee, Alabama, 1864–1865." *Alabama Review* 25 (January 1972): 3–29.

———, ed. "Elyton, Alabama, and the Connecticut Asylum: The Letters of William H. Ely, 1820–1821." *Alabama Review* 3 (January 1950): 36–69.

Hudgins, Carter L. "A Natural History of Village Creek." *Journal of the Birmingham Historical Society* 9 (December 1985): 50–59.

Hunter, Louis C. "The Influence of the Market upon Technique in the Iron Industry in Western Pennsylvania up to 1860." *Journal of Economic and Business History* 1 (February 1929): 241–81.

Huntley, Horace. "The Status Quo against Interracial Unionism, 1933–1949." *Journal of the Birmingham Historical Society* 6 (January 1979): 5–13.

"Iron Ore—Making and Remaking the Map of Steel." *Fortune* 3 (May 1931): 84–92.

Jay, John C. "General N. B. Forrest as a Railroad Builder in Alabama." *Alabama Historical Quarterly* 24 (Spring 1964): 16–31.

Johns, Lyn. "Early Highland and the Magic City, 1884–1893, Including Willis J. Milner's 'History of Highland Avenue.'" *Journal of the Birmingham Historical Society* 6 (July 1979): 33–43.

Keith, Jean E. "The Role of the Louisville and Nashville Railroad in the Early Development of Alabama Coal and Iron." *Bulletin of the Business Historical Society* 26 (September 1952): 165–74.

Klein, Maury. "Southern Railroad Leaders, 1865–1893: Identities and Ideologies." *Business History Review* 42 (Autumn 1968): 288–310.

Knapp, Virginia. "William Phineas Browne, Business Man and Pioneer Mine Operator of Alabama." *Alabama Review* 3 (April, July 1950): 108–22, 193–94.

Knowles, Morris. "Water and Waste: The Sanitary Problems of a Modern Industrial District." *Survey* 27 (6 January 1912): 1485–92.

Layton, Edwin T. "Colin J. McRae and the Selma Arsenal." *Alabama Review* 19 (April 1966): 125–36.

———. "Mirror-image Twins: The Communities of Science and Technology in 19th Century America." *Technology and Culture* 12 (October 1971): 562–80.

Lea, Samuel H. "Flat Top Mine." *Mines and Minerals* 25 (March 1905): 395.

Ledbetter, R. H. "Blast Furnace Practice in Birmingham District." *Iron Age*, 30 October 1924, 1128–30.

Leighton, George. "Birmingham, Alabama: City of Perpetual Promise." *Harper's Magazine* 175 (August 1937): 225–42.

Lewis, W. David. "The Early History of the Lackawanna Iron and Coal Company: A Study in Technological Adaptation." *Pennsylvania Magazine of History and Biography* 96 (October 1972): 424–68.

———. "Industrialization and Urbanization—The SENCSA Conference, University of Alabama at Birmingham, April 1987." *Technology and Culture* 30 (July 1989): 628–34.

———. "The Invasion of Northern Markets by Southern Iron in a Decade of Boom and Bust: Sectional Competition in the 1880s." *Essays in Economic and Business History* 8 (1990): 257–69.

———. "Joseph Bryan and the Virginia Connection in the Industrial Development of Northern Alabama." *Virginia Magazine of History and Biography* 98 (October 1990): 613–40.

Lindgren, James M. "'First and Foremost a Virginian': Joseph Bryan and the New South Economy." *Virginia Magazine of History and Biography* 96 (April 1988): 157–80.

McCabe, W. Gordon. "Joseph Bryan: A Brief Memoir." *Virginia Magazine of History and Biography* 17 (April 1909): iii–xxix.

McDonald, P. B. "Iron Ore Mining in the South." *Iron Trade Review*, 22 October 1914, 759–62.

McGrath, W. M. "Conservation of Health." *Survey* 17 (6 January 1912): 1510–12.

McKelvey, Blake. "Penal Slavery and Southern Reconstruction." *Journal of Negro History* 20 (January 1935), 153–79.

McKenzie, Robert H. "The Great Birmingham Iron Boom, 1800–1892." *A Journal of History* [West Jefferson County Historical Society, Bessemer, Ala.] 3 (April 1975): 201–11.

———. "Horace Ware: Alabama Iron Pioneer." *Alabama Review* 26 (July 1973): 157–63.

———. "Reconstruction of the Alabama Iron Industry, 1865–1880." *Alabama Review* 25 (July 1972): 178–91.

McQueen, James W. "Southern Foundry Pig Iron." *Year Book of the American Iron and Steel Institute* (1914), 448–53.

McWane, Arthur R. "Silent Transportation: The Story of Cast Iron Pipe at Work." *Pig Iron Rough Notes* 112 (Autumn 1950): 28–34.

McWhiney, Grady. "The Revolution in Nineteenth-Century Alabama Agriculture." *Alabama Review* 31 (January 1978): 3–32.

Mentzer, Mitch, and Mike Williams. "Images of Work: Birmingham, 1894–1937." *Journal of the Birmingham Historical Society* 7 (June 1981): 10–23.

Miller, Grace Lewis. "The Mobile and Ohio Railroad in Ante Bellum Times." *Alabama Historical Quarterly* 7 (Spring 1945): 37–59.

Miller, Randall M. "Daniel Pratt's Industrial Urbanism: The Cotton Mill Town in Ante-Bellum Alabama." *Alabama Historical Quarterly* 34 (Spring 1972): 5–35.

Mitchell, Barbara J. "Steel Workers in a Boom Town: Birmingham, 1900." *Southern Exposure* 12 (November–December 1984): 56–60.

Moldenke, Richard. "Cast Iron Pipe Manufacture in the South." *Iron Age* (18 September 1924): 697–98.

Moore, A. B. "Railroad Building in Alabama During the Reconstruction Period." *Journal of Southern History* 1 (November 1935): 421–41.

Moore, W. D. "Birmingham's Place in the Cast Iron Pipe Industry." *Pig Iron Rough Notes* 32 (June 1929): n.p.

Moore, William H. "Preoccupied Paternalism: The Roane Iron Company in Her Company Town—Rockwood, Tennessee." *East Tennessee Historical Society Publications* 39 (1967): 56–70.

Napier, John H., III. "Montgomery During the Civil War." *Alabama Review* 41 (April 1988): 103–31.

Neeley, Mary Ann. "Painful Circumstances: Glimpses of the Alabama Penitentiary, 1846–1852." *Alabama Review* 44 (January 1991): 3–16.

Newton, Wesley Phillips. "Lindbergh Comes to Birmingham." *Alabama Review* 26 (April 1973): 105–21.

Norrell, Robert J. "Caste in Steel: Jim Crow Careers in Birmingham, Alabama." *Journal of American History* 73 (December 1986): 669–94.

———. "Labor at the Ballot Box: Alabama Politics from the New Deal to the Dixiecrat Movement." *Journal of Southern History* 57 (May 1991): 201–34.

Oates, William C. "Industrial Development of the South." *North American Review* 161 (November 1895): 566–74.

Pearson, C. Chilton. "William Henry Ruffner: Reconstruction Statesman of Virginia." Parts 1, 2. *South Atlantic Quarterly* 20 (January, April 1921): 25–32, 137–51.

Phillips, Ulrich B. "Conservatism and Progress in the Cotton Belt." *South Atlantic Quarterly* 3 (January 1904): 1–10.

———. "The Origin and Growth of the Southern Black Belts." *American Historical Review* 11 (July 1906): 798–816.

Poor, Russell S. "The Geologic Making of the Birmingham District, Alabama." *Birmingham-Southern College Bulletin* 23 (September 1930): 7–31.

Potter, David M. "The Historical Development of Eastern-Southern Freight Rate Relationships." *Law and Contemporary Problems* 12 (Summer 1947): 416–48.

Ramsay, Erskine, and Charles E. Bowron. "Coal Washing in Alabama." *Mines and Minerals* 25 (December 1904): 227–31.

Ransom, Roger, and Richard Sutch. "The 'Lock-in' Mechanism and Overproduction of Cotton in the Postbellum South." *Agricultural History* 49 (April 1975): 405–25.

Reid, Joseph D., Jr. "Sharecropping in History and Theory." *Agricultural History* 49 (April 1975): 426–40.

Rikard, Marlene H. "George Gordon Crawford: Man of the New South." *Alabama Review* 31 (July 1978): 163–81.

———. "'Take Everything You Are . . . and Give It Away': Pioneer Industrial Social Workers at TCI." *Journal of the Birmingham Historical Society* 7 (November 1981): 25–41.

————. "Wenonah: The Magic Word." *Journal of the Birmingham Historical Society* 7 (January 1981): 2–9.

Roberts, Joseph K. "Contributions of Virginians to the Geology of the State." *Virginia Journal of Science* 1 (February–March 1940): 71–73.

Rubin, Julius. "The Limits of Agricultural Progress in the Nineteenth-Century South." *Agricultural History* 49 (April 1975): 362–73.

Rukeyser, William Simon. "Why Rain Fell on 'Automatic' Sprinkler." *Fortune* 79 (1 May 1979): 88–91, 126–29.

Russel, Robert R. "The Pacific Railway Issue in Politics prior to the Civil War." *Mississippi Valley Historical Review* 12 (September 1925): 187–201.

Ryan, John B., Jr. "Willard Warner and the Rise and Fall of the Iron Industry in Tecumseh, Alabama." *Alabama Review* 24 (October 1971): 261–79.

"St. George T. C. Bryan." *Confederate Veteran* 24 (June 1916): 275.

Satterfield, Carolyn Green. "J. R. McWane: Pipe and Progress." *Alabama Review* 35 (January 1982): 14–29.

Schaffer, Daniel. "War Mobilization in Muscle Shoals, Alabama, 1917–1918." *Alabama Review* 39 (April 1986): 110–46.

Schallenberg, Richard H. "Evolution, Adaptation and Survival: The Very Slow Death of the American Charcoal Iron Industry." *Annals of Science* 32 (1975): 341–58.

Schweikart, Larry. "Alabama's Antebellum Banks: New Interpretations, New Evidence." *Alabama Review* 38 (July 1985): 202–21.

Sheridan, Richard C. "Alabama Chemists in the Civil War." *Alabama Historical Quarterly* 37 (Winter 1975): 265–74.

————. "Alabama's First Iron Furnace." *Journal of Muscle Shoals History* 12 (1988): 19–24.

Shook, Pascal G. "Southern Foundry Pig Iron." *Year Book of the American Iron and Steel Institute* (1914), 457–62.

Slaughter, Ellen. "DeBardeleben, Bessemer, and the Montezuma Hotel." *Journal of the Birmingham Historical Society* 9 (Spring 1988): 49–61.

"Sloss Furnace Gets International Praise." *The Preservation Report* 11 (November–December 1983): 1.

Snell, William R. "Fiery Crosses in the Roaring Twenties: Activities of the Revised Klan in Alabama, 1915–1930." *Alabama Review* 23 (October 1970): 256–76.

Sprague, Stuart Seely. "Alabama and the Appalachian Iron and Coal Town Boom, 1889–1893." *Alabama Historical Quarterly* 37 (Summer 1975): 85–91.

Staudenmaier, John M. "What SHOT Hath Wrought and What SHOT Hath Not: Reflections on Twenty-five Years of the History of Technology." *Technology and Culture* 25 (October 1984): 707–30.

Stephen, Walter W. "The Brooke Guns from Selma." *Alabama Historical Quarterly* 20 (Fall 1958): 462–75.

Stockham, Richard J. "Alabama Iron for the Confederacy: The Selma Works." *Alabama Review* 21 (July 1968): 163–72.

Stover, John F. "Southern Ambitions of the Illinois Central Railroad." *Journal of Southern History* 20 (November 1954): 499–505.

Straw, Richard A. "The Collapse of Biracial Unionism: The Alabama Coal Strike of 1908." *Alabama Historical Quarterly* 37 (Summer 1975): 92–114.

————. "Soldiers and Miners in a Strike Zone: Birmingham, 1908." *Alabama Review* 38 (October 1985): 289–308.

————. "The United Mine Workers of America and the 1920 Coal Strike in Alabama." *Alabama Review* 28 (April 1975): 104–28.

Stroebel, Victor O. "Gordon's Improved Whitwell-Cowper Fire Brick Hot-Blast Stove." *American Institute of Mining Engineers Transactions* 14 (June 1885–May 1886): 159–172.

Stuyvesant, H. R. "Alabama Co.'s Improvements." *Iron Age*, 12 May 1921, 1240–41.

Taylor, Graham Romeyn. "Birmingham's Civic Front." *Survey* 17 (6 January 1912): 1467–68.

Temin, Peter. "The Post-Bellum Recovery of the South and the Cost of the Civil War." *Journal of Economic History* 36 (December 1976): 898–907.

Thomas, Rebecca L. "John J. Eagan and Industrial Democracy at Acipco." *Alabama Review* 43 (October 1990): 270–88.

Thompson, George Clinton. "Vulcan: Birmingham's Man of Iron." *Alabama Heritage* 20 (Spring 1991): 2–17.

Thorn, Cecelia Jean. "The Bell Factory: Early Pride of Huntsville." *Alabama Review* 32 (January 1979): 28–37.

Tower, J. Allen. "The Changing Economy of Birmingham and Jefferson County." *Journal of the Birmingham Historical Society* 1 (January 1960): 2–6.

―――. "The Industrial Development of the Birmingham Region." *Bulletin of Birmingham-Southern College* 46 (December 1953): 9–10.

Uehling, Edward A. "Pig-Iron Casting and Conveying Machinery: Its Development in the United States." *Cassier's Magazine* 24 (June 1903): 113–33.

Vandiver, Frank E. "Josiah Gorgas and the Briarfield Iron Works." *Alabama Review* 3 (January 1950): 5–21.

―――. "The Shelby Iron Company in the Civil War: A Study of a Confederate Industry." Parts 1–3. *Alabama Review* 1 (January, April, July 1948): 12–26, 111–27, 203–17.

Vedder, Richard K., and David C. Stockdale. "The Profitability of Slavery Revisited: A Different Approach." *Agricultural History* 49 (April 1975): 392–404.

Wagenknecht, Edward. "The World and Mary Johnston." *Sewanee Review* 44 (April–June 1936): 188–206.

Ward, Robert David, and William Warren Rogers. "Racial Inferiority, Convict Labor, and Modern Medicine: A Note on the Coalburg Affair." *Alabama Historical Quarterly* 44 (Fall–Winter 1982): 203–10.

Weeks, Elie. "Sabot Hill." *Goochland County Historical Magazine* 10 (Autumn 1978): 6–7.

Weiher, Kenneth. "The Cotton Industry and Southern Urbanization." *Explorations in Economic History* 14 (1977): 120–40.

Whiting, Marvin Y. "James R. Powell and 'This Magic Little City of Ours': A Perspective on Local History." *Journal of the Birmingham Historical Society* 8 (1983): 38–41.

―――. "Landmarks: Giuseppe Moretti." *Journal of the Birmingham Historical Society* 9 (January 1985): 5–13, 60.

Willard, Jennifer. "Giuseppe Moretti." *Alabama Heritage* 20 (Spring 1991): 18–33.

Williams, Randall, and Hilda Dent. "Billion Dollar Shell Game." *Southern Exposure* 8 (Spring 1980): 86–91.

Wood, Willie Jo. "Sloss Lake, a Plantation Mine." *The Source* [Franklin Printing Co., Russellville, Ala.] 3 (Winter 1993): 9–12.

Woodbridge, Annie. "Mary Johnston: A Universal Virginian." *Richmond Quarterly* 4 (Summer 1981): 23–33.

Woodbridge, Dwight E. "Iron Mining Methods and Costs in Alabama." *Iron Age*, 11 September 1913, 588–89.

Woodman, Harold D. "New Perspectives on Southern Economic Development: A Comment." *Agricultural History* 49 (April 1975): 374–80.

Woodward, Joseph H., II. "Alabama Iron Manufacturing, 1860–1865." *Alabama Review* 7 (July 1954): 199–207.

"Woodward Plant Improvements." *Iron Trade Review*, 12 November 1914, 905–10.

Wright, E. C. "The Economics of Raw Material Supplies in the Birmingham District." *Mining Engineering*, December 1950, 1214–20.

Yates, W. Ross. "Discovery of the Process for Making Anthracite Iron." *Pennsylvania Magazine of History and Biography* 98 (1974): 206–23.

Recorded Music

Newman, Randy. *Good Old Boys*. Warner Brothers Records, Record Album MS 2193, 1974.

Theses and Dissertations

Bergstresser, Jack Roland, Sr. "Raw Material Constraints and Technological Options in the Mines and Furnaces of the Birmingham District: 1876–1930." Ph.D. diss., Auburn University, 1993.

Cauley, Woodham W. "A Study of the Accounting Records of the Shelby Iron Company." Master's thesis, University of Alabama, 1949.

Clark, Elizabeth Boner. "The Abolition of the Convict Lease System in Alabama, 1913–1928." Master's thesis, University of Alabama, 1949.

Datnow, Claire-Louise. "The Sloss Company: Symbol of the New South." Master's thesis, University of Alabama at Birmingham, 1987.

Dew, Charles B. "Southern Industry in the Civil War Era: Joseph Reid Anderson and the Tredegar Iron Works, 1859–1867." Ph.D. diss., Johns Hopkins University, 1964.

Dong, Jingsheng. "The Social and Working Lives of Alabama Coal Miners and Their Families, 1893–1925." Master's thesis, Auburn University, 1988.

Everse, Martin L. "The Iron Works at Briarfield: A History of Iron Making in Bibb County, Alabama." Master's thesis, Samford University, 1984.

Fuller, Justin. "History of the Tennessee Coal, Iron, and Railroad Company, 1852–1907." Master's thesis, Emory University, 1958.

———. "TCI—History of the Tennessee Coal, Iron and Railroad Company, 1852–1907." Ph.D. diss., University of North Carolina at Chapel Hill, 1967.

Huntley, Horace. "Iron Ore Miners and Mine Mill in Alabama, 1933–1952." Ph.D. diss., University of Pittsburgh, 1977.

LaMonte, Edward S. "Politics and Welfare in Birmingham, Alabama: 1900–75." Ph.D. diss., University of Chicago, 1976.

Lerner, Jack Leonard. "A Monument to Shame: The Convict Lease System in Alabama." Master's thesis, Samford University, 1969.

McCaa, Samuel Noble. "Samuel Noble: Founder of Anniston." Master's thesis, Auburn University, 1966.

Misa, Thomas Jay. "Science, Technology and Industrial Structure: Steelmaking in America, 1870–1925." Ph.D. diss., University of Pennsylvania, 1987.

Mitchell, Martha Carolyn. "Birmingham: Biography of a City of the New South." Ph.D. diss., University of Chicago, 1946.

Moore, William Glenn. "The Acquisition of the Tennessee Coal and Iron Company by the United States Steel Corporation in 1907." Master's thesis, University of Alabama, 1951.

Mullins, Betty Hamaker. "The Steel Corporation's Purchase of the Tennessee Coal, Iron & Railroad Company in 1907." Master's thesis, Samford University, 1970.

Rikard, Marlene H. "An Experiment in Welfare Capitalism: The Health Care Services of the Tennessee Coal, Iron and Railroad Company." Ph.D. diss., University of Alabama, 1983.

———. "George Gordon Crawford: Man of the New South." Master's thesis, Samford University, 1971.

Roberts, Jessie Mae. "Joseph Forney Johnston." Bachelor's thesis, Howard College, 1931.

Rutledge, William S. "An Economic and Social History of Ante-Bellum Jefferson County, Alabama." Master's thesis, University of Alabama, 1939.

Sapp, Fannie Ella. "The Convict Lease System in Alabama (1846–1895)." Master's thesis, George Peabody College for Teachers, 1931.

Scudder, John R., Jr. "The Alabama and Chattanooga Railroad Company, 1868–1871." Master's thesis, University of Alabama, 1951.

Wallace, Richard D. "A History of the Shelby Iron Company, 1865–1872." Master's thesis, University of Alabama, 1953.

Unpublished Research Manuscripts, Papers, and Reports

Axford, Faye A. "Alabama's 'Bald Eagle': Governor George Smith Houston." Paper delivered at annual meeting of Alabama Historical Association, Florence, Alabama, 24 April 1993.

Bergstresser, Jack Roland, Sr. "Annotated Bibliography on Beehive Coke Ovens." Unpublished research report, Auburn University, 1986.

———. "A Brief History of the Birmingham District." In Birmingham Historical Society, "The Birmingham District: An Assessment: A Study Prepared for the National Park Service, Department of the Interior." Birmingham, Alabama: Birmingham Historical Society, 1992.

———. "The Primary Iron and Steel Industry of Jefferson County." Unpublished research report, Sloss Furnaces National Historic Landmark, Birmingham, Alabama, 1989.

———, compiler. Database of statistical information on blast furnaces and coal mines in the Birmingham District.

Brockenbrough, Anne. "Joseph Bryan from Planter to Industrialist: Portrait of a Conservative Southern Democrat." Unpublished research paper written at the University of Virginia, 1986, copy at Virginia Historical Society, Richmond, Virginia.

Coger, Dalvan. "Johannesburg on the Veldt: The Birth of the City of Gold." Paper presented at annual meeting of Southeastern Nineteenth-Century Studies Association, Birmingham, Alabama, April 1987.

Colglazier, Merle E. Research materials and transcripts of interviews pertaining to history of Laburnum House, Richmond, Virginia, supplied to the author.

English, Tanya. "Hardie-Tynes Manufacturing Company." Unpublished research report for Historic American Engineering Record, 1992.

Freeman, Douglas Southall. "John Stewart Bryan: A Biography." Unpublished manuscript, Virginia Historical Society, Richmond, Virginia, c. 1947.

Kulik, Gary B. "The Sloss Furnace Company, 1881–1931: Technological Change and Labor Supply in the Southern Pig Iron Industry." Unpublished research report for Historic American Engineering Record, 1976.

Lee, J. L. Research notes on Georgia Pacific Railway and Richmond Locomotive Works, supplied to author.

Mabry, Helen. "Sloss Furnace Association 1976–1986." Unpublished manuscript, Sloss Furnaces National Historic Landmark.

McCabe, W. Gordon. "St. George Tucker Coalter Bryan of Richmond, 1843–1916." Paper read at meeting 17 March 1917, copy at Virginia Historical Society.

Norrell, Robert J. "Distant Prosperity: Modernization in Nineteenth-Century Alabama." Unpublished manuscript supplied to author.

———. "Politics and Industrialism in the New South: Alabama, 1850–1900." Unpublished manuscript supplied to author.

Sledge, James W., III. "Birmingham's Iron Elephant: The Creation of Sloss Furnaces Museum." Unpublished term paper, Department of History. Auburn University. 1989.

————. "The Great Migration, Alabama Blacks, and the Birmingham District." Unpublished term paper, Department of History, Auburn University, 1993.

————. Research notes for paper, "Birmingham's Iron Elephant: The Creation of Sloss Furnaces Museum," supplied to author.

Yuill, Robert. Notes derived from research in Ruffner Papers, Virginia State Library and Archives, supplied to author.

Index

Particularly significant entries are printed in **boldface** type. Page numbers for illustrations are in *italics*. Abbreviations used throughout the endnotes are also used in the index (see abbreviations list on pp. 513–14).

About the Author

W. David Lewis is Distinguished University Professor at Auburn University, where he was Hudson Professor of History and Engineering from 1971 to 1994. A native of Pennsylvania, he holds bachelor's and master's degrees from Pennsylvania State University and a doctorate from Cornell University. Among his many honors are the Leonardo da Vinci Medal, the highest honor given by the Society for the History of Technology; appointment as Charles A. Lindbergh Professor of Aerospace History, National Air and Space Museum, 1993–94; the Faculty Achievement Award in the Humanities, Auburn University, 1990; appointment as a National Humanities Fellow, University of Chicago, 1978–79; Mortar Board Favorite Teacher Award, Auburn University, 1976; and Senior Fellow in American Civilization, Cornell University, 1958–59. He has played the leading role in developing Auburn University's undergraduate and graduate programs in Technology and Civilization, and he has directed the world's first international conference on the history of civil and commercial aviation at Lucerne, Switzerland, in 1992. His previous works include *Iron and Steel in America*, coauthored histories of Delta Air Lines and All American Aviation (a precursor of USAir), several other books, and many chapters and journal articles. Currently he is writing a biography of Edward V. Rickenbacker, famed World War I combat pilot and longtime chief executive of Eastern Air Lines.